THE SEA

Ideas and Observations on Progress in the Study of the Seas

Volume 1 PHYSICAL OCEANOGRAPHY

Volume 2 COMPOSITION OF SEA WATER

COMPARATIVE AND DESCRIPTIVE OCEANOGRAPHY

Volume 3 THE EARTH BENEATH THE SEA

HISTORY

Volume 4 NEW CONCEPTS OF SEA FLOOR EVOLUTION (IN THREE PARTS)

Volume 5 MARINE CHEMISTRY

Volume 6 MARINE MODELING

Volume 7 THE OCEANIC LITHOSPHERE

Volume 8 DEEP-SEA BIOLOGY

DEEP-SEA BIOLOGY

Edited by

GILBERT T. ROWE

Brookhaven National Laboratory

THE SEA

Volume 8

A Wiley-Interscience Publication

JOHN WILEY & SONS New York • Chichester • Brisbane • Toronto • Singapore

Library of Congress Cataloging in Publication Data:

(Revised for vol. 8)
Main entry under title:

The Sea, ideas and observations on progress in the study of the seas.

Vols. 1–3 edited by M. N. Hill; vol. 4 by A. E. Maxwell; vols. 5–6 by Edward D. Goldberg [et al.]; vol. 7 by Cesare Emiliani; vol. 8 by Gilbert T. Rowe.
Vol. 4, pts. 1–2 have imprint: New York: Wiley-Interscience; vol. 5–8: New York: Wiley.
Vols. 5–8: "A Wiley-Interscience publication."
Includes bibliographies and indexes.
Contents: v. 1. Physical oceanography—v. 2. The composition of sea-water. Comparative and descriptive oceanography—[etc.]—v. 8. Deep-sea biology.
1. Oceanography. 2. Submarine geology. I. Hill, M. N. (Maurice Neville), 1919– .
GC11.S4 551.46 62-18366

ISBN 0-471-57910-6 (v. 4, pt. 1)
ISBN 0-471-04402-4 (v. 8)

Printed in the United States of America

10 9 8 7 6 5 4 3 2 1

CONTRIBUTORS

RONALD A. CAMPBELL, Department of Biology, Southeastern Massachusetts University, North Dartmouth, Massachusetts

ROBERT S. CARNEY, Moss Landing Marine Laboratory, Moss Landing, California

JAMES E. ECKMAN, Marine Sciences Research Center, State University of New York, Stony Brook, New York

RICHARD L. HAEDRICH, Department of Biology, Memorial University of Newfoundland, St. John's, Newfoundland, Canada

KENNETH R. HINGA, Graduate School of Oceanography, University of Rhode Island, Narragansett, Rhode Island

PETER W. HOCHACHKA, Department of Biology, University of British Columbia, Vancouver, British Columbia, Canada

HOLGER W. JANNASCH, Woods Hole Oceanographic Institution, Woods Hole, Massachusetts

PETER A. JUMARS, Department of Oceanography, University of Washington, Seattle, Washington

ERIC MILLS, Department of Oceanography, Dalhousie University, Halifax, Nova Scotia, Canada

MICHAEL A. REX, Department of Biology, University of Massachusetts, Harbor Campus, Boston, Massachusetts

GILBERT T. ROWE, Division of Oceanographic Sciences, Department of Energy and Environment, Brookhaven National Laboratory, Upton, New York

MYRIAM SIBUET, Centre Océanologique de Bretagne, Brest, Cedex, France

JOSEPH F. SIEBENALLER, School of Oceanography, Oregon State University, Marine Science Center, Newport, Oregon

KENNETH L. SMITH, JR., Scripps Institution of Oceanography, La Jolla, California

George N. Somero, Scripps Institution of Oceanography, La Jolla, California

Hjalmar Thiel, Institut für Fischereiwissenschaft und Hydrobiologie, Universität Hamburg, Hamburg, West Germany

V. B. Tseitlin, Shirshov Institute of Oceanology, Akademy of Sciences, Moscow, Soviet Union

Michael E. Vinogradov, Shirshov Institute of Oceanology, Akademy of Sciences, Moscow, Soviet Union

Carl O. Wirsen, Woods Hole Oceanographic Institution, Woods Hole, Massachusetts

PREFACE

The suggestion that a volume of *The Sea* be dedicated to deep-sea biology came from Dr. Edward Goldberg, an editor in the series, while we both were attending a *Marcus Wallenberg Foundation* symposium to foster international cooperation in the sciences. I believed such a volume was appropriate at the time because even the most recent syntheses on aspects of deep-sea biota failed to incorporate the new and rapidly accumulating information on rates of metabolism, behavior, and community structure at great depths in the ocean.

The year was 1977, notable because our earth sciences colleagues had just discovered the remarkable fauna around the deep Galápagos hydrothermal vents. Many of us at the meeting were still skeptical about what we now recognize as one of the sea's most startling phenomena.

In asking colleagues to contribute to *Deep-Sea Biology* an attempt was made to reflect the progress in broad general subjects made in the last few years. Complete coverage of topical issues was not possible, of course. The list of contributors includes many close associates with whom I have had the good fortune to work and identifies personal acquaintances and close colleagues with whom I have shared ships and trips to exotic places.

Life for the deep-sea biologist may have been simpler a few years ago. We no longer have the excuse that the environment is inaccessible to us, as many of the contributors to this volume have demonstrated. We are just beginning to parameterize biological phenomena in the deep ocean, including the sulfide-based systems around hydrothermal vents.

Working with the authors on this book has been a true delight for me. In the face of the intensive research activity by biologists, this set of reviews is a necessary link between the important natural history traditionally practiced with simple gear tethered by a wire to a ship and complex interdisciplinary measurements of rates of processes in studies of whole ecosystems. I trust that readers' disappointment at the omission of a favorite subject will be assuaged by the thought that deep-sea studies are accumulating information so rapidly that not too many years will go by before another compendium on new and different deep-sea biology will be required.

GILBERT T. ROWE

The Village of Port Jefferson, New York
April 1983

CONTENTS

DEEP-SEA BIOLOGY

DEEP-SEA BIOLOGY

1. PROBLEMS OF DEEP-SEA BIOLOGY: AN HISTORICAL PERSPECTIVE

ERIC L. MILLS

1. Introduction: Historical Perspectives in Oceanography

At midday on Sunday, August 31, 1898, the converted Hamburg–Amerika liner *Valdivia* cast off from the quay at Hamburg for a round-the-world cruise of exploration and deep-sea work (Fig. 1), leaving behind a crowd of dignitaries, including Sir John Murray and German Secretary of State for the Interior Count Posadowsky. The scientific leader of the expedition, Carl Chun (1900), in describing the ship's departure for a popular audience, speculated that Bismarck's thinking was foremost in the addresses given then by Murray and the secretary of state. He described how the secretary of state likened the prosperous and powerful nation, which saw the support of pure science as an ideal of sacrifice, to a well-to-do private citizen wanting to transcend the practical and monetary concerns of everyday life. The support of pure science, according to Posadowsky, was not a functionless ideal, but served to support the whole concept of the German empire.

Chun's interpretation of this incident appears to place the origins of one episode in the history of oceanography neatly in a social and political framework, a product of the newly established German empire's "authoritarian, militaristic, efficient, monarchical regime" (Brinton et al., 1969). Whether this is truly the case is a matter for historical analysis and the historian's insight into events and social forces, an approach that has been used relatively little in describing the history of the marine sciences. Without being unduly critical of earlier approaches, I would like to describe some problems of theory and interpretation that implicitly or explicitly confront the historian who attempts to evaluate the development of oceanography as a relatively coherent discipline.

One of the first attempts to provide a systematic framework for the history of oceanography, that is, in modern terms, to construct a historiographic model of the science, was by Herdman (1923), who divided the history of oceanography into three stages: the period of Edward Forbes; the period of Wyville Thomson and H.M.S. *Challenger*; and post-*Challenger* times, when modern oceanography unfolded through international cooperation in European seas. A later quite different attempt to systematize the history of oceanography, concentrating mainly on physical science, was made by Wüst (1964), who divided the history of the field into four eras—those of exploration (1873 to 1913), national surveys (1925 to 1940), new methods (1947 to 1956), and international cooperation (1957 onward). Wüst emphasized that the progress of oceanography depended on expeditions, instruments and methods, and the development of scientific theory.

Fig. 1. Cover illustration from one of the fascicles of Carl Chun's book *Aus den Tiefen des Weltmeeres* (1900) describing the deep-sea explorations of the German ship "Valdivia" (1898 to 1899). Romantic, scientific, and political attitudes toward deep-sea research are summarized in this elegantly composed colored book jacket, because Chun modeled the expedition and its published results on the voyage of H.M.S. *Challenger* (1872 to 1876) and its scientific results, which had been completed only 5 years before.

These attempts at categorizing events (building historiographic models) are probably better than a simple-minded biographical approach, provided the formulations are recognized as hypotheses about historical change and not as concrete representations of reality, even though they provide very little information about the dynamics of change. But the very fact that at least two unrelated approaches are possible with use of the same material, the facts and issues in the development of marine science, indicates that something is missing or unrecognized in these early attempts to establish a framework for the history of oceanography and that the attempts were premature in imposing "eras" or "stages" that may actually conceal the dynamics of historical change.

Although it is impossible to divorce theory from analysis, of course, detailed historical study and analysis of the development of oceanography are needed *before* a well-articulated synthesis according to some theory of historical reconstruction is attempted. Related branches of science can provide us with excellent examples of this approach, some of the most notable recent ones of which are in natural history (Allen, 1976), geology (Rudwick, 1972; Porter, 1977), physiology (Geison, 1978), and the history of scientific institutions (Berman, 1978). There are at least two stimulating recent works that apply this technique to marine science, Margaret Deacon's (1971) *Scientists and the Sea* and the portions of Schlee's (1973) *The Edge of an Unfamiliar World* dealing with oceanography in the United States during the nineteenth century. In this essay I attempt to follow the example of these earlier studies to look in detail at a rather broad spectrum of events and ideas pertaining to the knowledge of deep-sea animals between the early nineteenth century and the 1970s. Thus my essay examines the history of oceanography as a series of largely unsolved historical problems set in a reasonably complete matrix of fact.

Many previous attempts to describe the history of oceanography, such as those by Herdman and Wüst, implicitly assume that succession in time involves causation. The best known example from biological oceanography, which is discussed in detail later, is that there is a truly causal succession from Edward Forbes and his azoic theory to the Sars and then on, through Wyville Thomson and Carpenter's *Lightning* and *Porcupine* expeditions, to the great *Challenger* expedition of 1872 to 1876, from which modern oceanography unfolded in a similar succession of genetically related developments. This simplistic approach is hardly consistent with Deacon's (1971) penetrating observation that marine science has evolved not in a linear way, but in a series of phases or episodes. Deacon describes the outcome of her research:

> It became plain that there was in fact no continuous development of marine science between the seventeenth century and the present day. What we were looking at were large-scale fluctuations of interest in which activity in the marine sciences built up, flourished for a while, and then declined. Though in each succeeding period of activity observations and ideas tended to become more sophisticated than in the one before, this change was allied more to the scientific climate of the time than to one generation "improving" the work of another. Areas of particular interest shifted between one period of expansion and the next. Generally the people involved in such a cycle had only a very partial appreciation of the work which had gone on in the previous ones. Their interest in it, if any, was largely antiquarian.

As shown later in this chapter, the development of deep-sea biology has been similarly episodic, circumscribed, ahistorical, and conditioned by a variety of contingent forces, ranging from technological to economic. Even where a strong intellectual impetus or continuity exists, scientific change, or the lack of it, may be governed by bread and butter. As Deacon has said, referring to the absence of continuous change in marine science, "people could not continue their work on problems which seemed relevant without financial and institutional resources and these resources tended to be monopolized by sciences which were already established."

The last quotation leads us into related but different areas; namely, the factors that make a science what it is—and those that cause changes of scientific direction, make one discipline fashionable or imperative, another out of fashion or stagnant. If, as Barnes (1974, p. 57) suggests, "Science may be regarded as loosely associated sets of communities, each using characteristic procedures and techniques to further the metaphorical redescription of a puzzling area of experience in terms of a characteristic, accepted set of cultural resources," we would expect oceanography to be a scientific discipline with a well-established set of practices, theoretical presuppositions, and modes of thought, or to be in the process of becoming this way. I am not at all sure that present-day oceanography may be described in these terms. It frequently appears to be a patchwork fabric of relatively unrelated disciples, each with a different scientific ethos and mode of procedure. If this is the case, the dynamics of historical change may well have been quite different in the "subdisciplines" that come under the umbrella of modern oceanography. "Oceanography," regardless of whether it exists as a unified discipline, has been conditioned by changing knowledge of fluid mechanics, sedimentary geology and paleontology, physiological ecology and limiting-factor chemistry, evolutionary biology, agricultural practice, animal nutrition, the technology of machinery, and computers. The balance of forces involved in the development of models of general oceanic circulation (in physical oceanography) and the theory of animal diversity in the sea (in biological oceanography), for example, has been quite different, although unstudied in careful, analytic fashion. It is far from clear how (and how separately) these disparate disciplines have developed, and under what influences from inside or outside their boundaries.

Historians and philosophers of science have expended thousands of words in the last decade or so exploring "internal" as opposed to "external" determination of the kind and direction of scientific change. The subject is reviewed by Kuhn (1968), in the prefaces to many books (e.g., Geison, 1978; Berman, 1978), in a chapter by Cannon (1978, Chapter 5), and in a sociological context by Barnes (1974) and Ravetz (1971). I agree with Barnes, on the basis of my experience in marine science, that there is virtually a continuum from scientific change occurring for purely intellectual reasons through the intrinsic dynamics of the discipline (internal influences) and change occurring through powerful social, psychological, or economic forces generated outside the discipline ("external" influences) and that pure examples of the extremes cannot exist. In addition, Kuhn's (1968) suggestion, more than a decade old [also supported by Cannon (1978, Chapter 5)], that as a new scientific discipline develops it is at its most susceptible to external influence, absorbing ideas from outside and possibly gaining impetus through social and economic forces, and that it gains internal coherence and becomes more and more self-contained, is highly plausible for the branches of oceanography. Oceanography may, because of its nature, provide some excellent material to test these ideas.

Modern oceanography, a hybrid science, should fall—for the reasons I have outlined (including its callow youth)—closer to the externally influenced end of the spectrum than, let us say, pure mathematics. Because instruments and ships, which are expensive, have been integral to the conduct of marine science, oceanography has been dependent on economics and on the successes or failures of technology and commerce, in addition to ideas from other branches of science. Burstyn (1968, 1972), for example, gives an excellent example of how a major expedition (the *Challenger* expedition, 1872 to 1876) came to receive government support in nineteenth century Britain through a combination of political and scientific influences.

It is a reasonable working principle that the branches of oceanography have been molded by the technical abilities of each period in history as well as by the social norms of what science was conceived of as setting out to achieve at those times. But putting the principle into historical practice is far from easy at our present stage of knowledge. How did fluid mechanics enter physical oceanography? Why did biological oceanography remain virtually atheoretical for nearly the first century of its development? How can one account for the fascinating and historically unexplored tension between the motivations of the dominant leaders in oceanography (especially the leaders of the expeditions), who, to quote Barry Barnes (1974), had "confidence in science as a mode of action" and the societies in which they operated? Where does individual imagination, or the scientific ethos per se, stop and social or economic influence begin in the history of marine science, specifically in the history of biological oceanography? The questions are very difficult to answer, but I ask them, not to indicate that I will answer them, but to suggest a useful intellectual framework for the work that should follow descriptive essays like my own.

Leaving aside these questions of historiography and the sociology of scientific change, I have used a practical distinction to divide my account of the development of deep-sea biology. A sketchy dividing line between "earlier" and "later" biological oceanography is provided by Murray and Hjort's book *The Depths of the Ocean* (1912) based on the cruise of *Michael Sars* in 1910, because it is the first book that is recognizably modern to my oceanographer's eye. My account concentrates mostly on the years before 1910, but I also give an account (briefer and probably less objective) of the historically less digestible decades after 1910 because there is evidence that biological oceanography took new directions then and that the change is of unusual interest to the historian of science.

For well over 100 years the scientific concerns of deep-sea biology have been constant, pervasive and continuous through periods of rapid change like the early twentieth century. Charles Wyville Thomson stated a number of these fascinating problems in 1878, just after the *Challenger* expedition, in summarizing the results of that great exploration. He recognized, for example, after a period of controversy, that animals occur at the greatest depths and that the majority of invertebrate groups was found there, although perhaps in different proportions than in shallow water. The deepwater fauna appeared to be largely a cosmopolitan one, and its distribution was related to deep currents and to the rich faunas of the Arctic and Antarctic. In some way the fauna of the deep ocean was related to geologically older faunas, although Thomson in 1878 was beating a slow, strategic retreat from his earlier position that the deep ocean held an archaic fauna.

If we add to this summary, or extend it, by adding the ideas that some food must reach the deep sea by falling from the surface waters and that bacteria must somehow be involved in its conversion, nearly all the topics that recur and interplay in a hundred or more years of oceanographic work on the deep ocean have been outlined or implied for more detailed examination. The real questions of significance are why these topics were considered important in the first place, what was regarded as evidence, and how attitudes toward them changed under external and internal influences since the formative year, 1839, when Edward Forbes received a grant of £60 from the British Association for the Advancement of Science to prosecute "researches with the dredge" around the British Isles.

2. Light on the North Atlantic. British Marine Biology before the *Challenger* Expedition

Undoubtedly the most influential figure to turn the attention of naturalists to deep water during the early decades of the nineteenth century was the Manxman Edward Forbes (1815 to 1854), who, despite his short life, had an impressive effect on marine biology, geology, and paleontology (Rehbock, 1975b; Mills, 1978). During his time jobs for professional biologists were few or nonexistent, but there was a strong current of amateur activity. This began first in naturalists' societies and field clubs and did not take a fully professional turn, leaving behind the country clergyman and artisans, until nearly the end of the century (Allen, 1976; Turner, 1978). Forbes himself was responsible for drawing amateurs together in a common cause, carefully recorded marine dredging, and after his time there were several more or less stillborn attempts to do collaborative research in marine biology until they reached full expression in major expeditions after 1867.

Forbes was a naturalist from an early age. He collected shells and had begun to dredge before he left the Isle of Man in 1831 to study art in London. There he was refused admission by the Royal Academy, and to suit his mother's wishes, soon settled in Edinburgh to study medicine. During his years as a student the lure of natural history was great and he failed to sit his examinations in 1836. For another 5 years Forbes remained in Edinburgh or traveled on the continent, living on a small income from his father and looking sporadically and unsuccessfully for suitable jobs (e.g., the Chairs of Natural History at St. Andrews and Aberdeen). In 1841, through friends, he was offered a naturalist's position on H.M.S. *Beacon*, surveying the coast of Asia Minor and exploring the Aegean Sea. There he undertook an extensive series of dredgings and studied the shore fauna of the islands and mainland. Late in 1842 Forbes returned suddenly to England, having been offered the Chair of Botany at King's College, London, something of a windfall because his family's finances had failed.

At King's, Forbes was both overworked and seriously underpaid. To meet his expenses, he took on the additional job of curator of the Geological Society and thus became well known to the major geologists of the day. He made a natural transition in 1844 to the new Geological Survey of Great Britain as a paleontologist under Henry De La Beche, where for 10 years he worked on the relation between fossils and animal distribution in the British Isles. In 1854, just after achieving his lifelong ambition, the Regius Professorship of Natural History at Edinburgh, Forbes died, many plans unfulfilled.

Forbes' influence was largely through personality and friendships. He was the center of a learned and active group of students at Edinburgh, many of whom became distinguished professional men. In London, he was equally influential through his erudition and because he helped his younger colleagues, including

T. H. Huxley. A naturally sociable man, he centered much of his vivacity around the British Association, which in 1839 had given him a grant of £60 to undertake a program of dredging. At that early date he had already established a scheme of zonation from the shore into deeper water (Forbes, 1840) that remained virtually unchanged for many years and was used to interpret the fossil record. Forbes' year in the Aegean (1841 to 1842) on H.M.S. *Beacon* enabled him to extend this system into even deeper water and test its utility in reconstructing the fossil record (Forbes, 1844a,b; Mills, 1978). In brief, he showed that detailed knowledge of the distribution and habits of living animals could be used to establish the nature of ancient habitats and their climatic conditions. He predicted what would happen if the deep floor of the Aegean were suddenly raised and then tested his prediction on an islet near Santorini that had been raised during volcanic activity in 1707.

When he dredged in the Aegean to 130 fathoms (238 m), Forbes noted that animals became few and small below 100 fathoms. "In the deepest parts explored of this abyss, very few species were found, and it seemed as if we were approaching a region that was barren and desert, where there was no more life, unless of minute forms of low organization" (Spratt and Forbes, 1847, Vol. 2, p. 107; Forbes, 1844b). Later, in a posthumously published book (Forbes and Godwin-Austen, 1859) that has been much abused by later explorers of the deep sea, he stated the tentative hypothesis [with which the Swede Sven Lovén (1809 to 1895) agreed (Lovén, 1845)] that life is absent or much attenuated below 300 fathoms. This suggestion, Forbes' azoic theory, has been a red herring for years, focusing attention on a minor part of his work and distracting it from his conviction that "it is in the exploration of this vast deep-sea region that the finest field for submarine discovery yet remains" (Forbes and Godwin-Austen, 1859, p. 27).

The early dredging under the general guidance of Forbes was carried on in impressive volume with careful documentation (Fig. 2). Throughout the 1840s and 1850s, dredging, with token financial support from the British Association, was used to explore the faunas of Cornwall, Ireland, Wales, and Scotland. By 1850 there had been 144 investigations all around the British Isles, many by Forbes himself or due to his direct influence (Forbes, 1851), and the British faunal list had been greatly enlarged. After Forbes' death, many local societies and field clubs established dredging committees that carried on the work from Belfast, Aberdeen, Dublin, Newcastle-upon-Tyne, Manchester, and Plymouth. By 1875 the grants were few or nonexistent and the impetus had shifted from these genial collaborations of well-meaning amateurs and gifted semiprofessionals. Although the dredging committees finally flickered out as late as 1881 (Bate and Rowe, 1882), only one, led by John Gwyn Jeffreys (1809 to 1885) and his colleagues, who mounted an assault on the deepwater marine fauna of the Shetland Islands in six summers between 1861 and 1868, stands out in scale of organization and execution.

Forbes and Jeffreys had met at the British Association in 1836 while Forbes was a student in Edinburgh and Jeffreys a young, well-to-do solicitor in Swansea (Mills, 1978). Jeffreys had begun a shell collection when he was a boy, and later, during his early professional career, he accumulated a notable collection (by field work, trading, and purchase) of European shells and published notes on them. Perhaps it was Forbes' influence that sent Jeffreys to Shetland in 1841 and 1848, for the two shared an interest not just in invertebrates, but also in the history of

No. ______________

GEOLOGICAL SURVEY OF THE UNITED KINGDOM
RECENT SEABED

Date ....
Locality ....
Depth ....
Distance from shore ....
Ground ....
Region ....

Species Obtained	Number of Living Specimens	Number of Dead Specimens	Observations

Fig. 2. Example of the dredging sheets filled out by Edward Forbes and other early British dredgers inspired by his example. By summarizing the information from dozens of such sheets, Forbes compiled information that was the basis of his famous paper in 1851 on the distribution of the British marine fauna. (Copy courtesy of John Thackray, the Geological Museum, London, England.)

faunas. This lay dormant for some years, until, well after he had become a barrister and man of affairs in London in 1856, Jeffreys felt free and able to begin an ambitious series of dredging expeditions around the British Isles, using his brother-in-law's yacht, *Osprey* (Fig. 3), which he had outfitted with ropes, dredges, and screens.

Jeffreys' work had a twofold origin. He was, first and foremost, a collector and systematist. Throughout the time he was dredging in the 1860s his growing collection was the basis of his masterwork, *British Conchology* (1862–1869), still an important reference for the taxonomy of European mollusks. His letters also reveal a preoccupation with the fine points of taxonomy (Alder and Norman, 1826–1911). In addition, he believed that by dredging in deep water, below the influence of the glaciations, he could find elements of the ancient fauna that had been the rootstock of the recent British marine biota. Present-day distributions,

Fig. 3. The yacht *Osprey*, used for dredging around the Shetland Islands by John Gwyn Jeffreys in 1861, 1864, 1867, and 1868. The caption in French (with corrections) reads "Departure of the yacht Osprey (Captain Phillips) from Tenwick for Balta Sound with the scientific explorers Jeffreys and Waller (June 1868)." The yacht, built in Wales in 1852, was originally cutter rigged and was then altered to a schooner in 1868, when she was purchased by Jeffreys from his brother-in-law, W. H. Nevill of Llanelly. (Photograph of the original watercolor and information courtesy of Robin Craig, Department of History, University College, London.)

he believed, were merely modifications of the ancestral situation that had existed in warmer seas at and before the time of the Coralline Crag of East Anglia (marine deposits now placed in the Upper Pliocene) (Jeffreys, 1856). Deep water was not easily accessible from the mainland of Britain, but from Shetland it could be reached in a few hours of sailing, given favorable weather.

This was a difficult condition to realize, and Jeffreys' expeditions make a fascinating story of partially thwarted scientific ambition as they proceeded between 1861 and 1868 (Mills, 1978). Nonetheless, Jeffreys and his companions, who included Canon A. M. Norman (Mills, 1980b), made impressive progress. They dredged to 170 fathoms (311 m) in 1867, the deepest dredging to date in British waters. This was no mean feat in a 30-ton, 51-ft sailing vessel with no auxiliary power and only hemp lines for the dredges. Their strenuous labors added 156 species of nonmolluscan invertebrates and 48 mollusks to the British list during the six expeditions. Not only was there a strong northern component in the Shetland fauna, reflecting the effects of the glaciations, but also a small Mediterranean component. Most important in Jeffreys' eyes, he found a few living species in the cold waters of Shetland that he considered conspecific with fossils in the Coralline Crag deposits. Year by year their triumphs as well as their disappointments were passed on to the scientific community at the meetings of the British Association, often in lengthy reports that still have some value 120 years later (e.g., Jeffreys, 1864, 1869b; Norman, 1868, 1869b).

Jeffreys' expeditions seem to be excellent candidates for a link between Forbes and the great expeditions of the 1870s, but on close examination they fail to fill this role (Mills, 1978). Jeffreys, who certainly was influenced by Forbes, never developed Forbes' ideas, perhaps because he was preoccupied with systematics rather than theory. Moreover, the expeditions themselves, which were well known and given year-by-year publicity by their participants, came to an end without leading directly into something greater. Both Jeffreys and Norman took new directions and became caught up in the scientific schemes of Charles Wyville Thomson and W. B. Carpenter, who had found new ways to explore deep water. Jeffreys' expeditions, which he had financed almost entirely himself, were superseded both scientifically and in organization because of work in another branch of biology.

3. *Lightning, Porcupine,* and the Azoic Deep Sea

In 1823 the Irish Army Surgeon John Vaughn Thompson of Cork discovered the tiny stalked pentacrinoid stage of the crinoid *Antedon* (then called *Comatula*). Thompson's discovery, which promised to link the history of the stalked crinoids (abundant only in the fossil record) with the well-known unstalked comatulids, was taken up enthusiastically in the middle of the century when W. B. Carpenter (1813 to 1885) and Charles Wyville Thomson (1830 to 1882) (Fig. 4) began to collaborate on studies of echinoderm embryology and structure.

Carpenter, the son of a Unitarian minister, took a medical degree at Edinburgh but spent a large part of his career in research on physiology and as the registrar of the University of London (Thomas, 1971; J. E. Carpenter, 1888). During a holiday at Arran in 1855 he dredged *Antedon* and its immature stages (W. B. Carpenter, 1856), which were attached to algae. This gave him the opportunity to work out its development, and he rented a house on Holy Isle, in Lamlash Bay, Arran, to be near the source of supply. Nearly simultaneously, Thomson, who was then professor of geology and mineralogy at Queen's College, Belfast (Merriman, 1972), dredged pentacrinoids in Belfast Lough (Yonge, 1972) and began studies of the embryology of comatulids. Thomson and Carpenter were collaborating on studies of crinoids by 1863, if not much earlier, and 2 years later they published some of the results of their work (Thomson, 1865; W. B. Carpenter, 1865).

Thomson, who was to succeed G. J. Allman (1812 to 1898) as Regius Professor of Natural History at Edinburgh in 1870 and thus come directly into Edward Forbes' academic lineage (he had dredged with Forbes as a student in the 1840s), was startled in 1864 by the discovery of the stalked crinoid *Rhizocrinus lofotensis* (Fig. 5) by G. O. Sars (1837 to 1937) (Sars, 1868), whose father Michael Sars (1805 to 1869) was professor of zoology at Christiania (Sivertsen, 1968). The younger Sars, a fisheries inspector for the Norwegian government, had been dredging in the fiords for years, as had his father. By 1864 they had taken at least 92 species at 200 to 300 fathoms (Sars, 1864), far deeper than Jeffreys had been able to dredge off Shetland. Their list by 1868 was 427 species; it included *Rhizocrinus* and the sea star *Brisinga*, which was believed to be a modern relative of the Mesozoic *Protaster*, then regarded as a link between the asteroids and the echinoids (Thomson, 1873). Only two stalked crinoids, both in the genus *Pentacrinus*, were known before this from dredgings in the West Indies. Thomson, who had a profound

Fig. 4. Charles Wyville Thomson (1830 to 1882), the scientific leader of the *Challenger* expedition, 1872 to 1876, and Regius Professor of Natural History in the University of Edinburgh. It was Thomson's collaboration with W. B. Carpenter (1813 to 1885) to study the development and morphology of echinoderms that led eventually to the expedition. [Photograph from Herdman (1923).]

interest in the radiate animals (especially sponges and echinoderms), believed that the stalked crinoids were virtually extinct after their peak in the Mesozoic and that the opportunity to study *Rhizocrinus* would illuminate the structure of the fossil forms. He enthusiastically visited the Sars in 1866 and, with Carpenter at Belfast in 1868, began to consider how they might collect more material and look for other rare and zoologically important animals in deep water.

In his book *The Depths of the Sea* (1873, p. 50 ff.), Thomson has given us an account of how he wrote to Carpenter in May, 1868 outlining the Sars' discoveries, which made it very unlikely that Forbes' dictum about the absence of life in deep water was correct. He also pointed out that if rare animals representative of the fossil record could be taken in deep water, it would be possible to study their soft parts and thus "give us an opportunity of testing our determination of the zoological position of some fossil types by an examination . . . of their recent relatives."

At that time Carpenter was vice president of the Royal Society. In this capacity, as well as through his role at the University of London, he was well known to the scientific community and to the ministers of Gladstone's Liberal government.

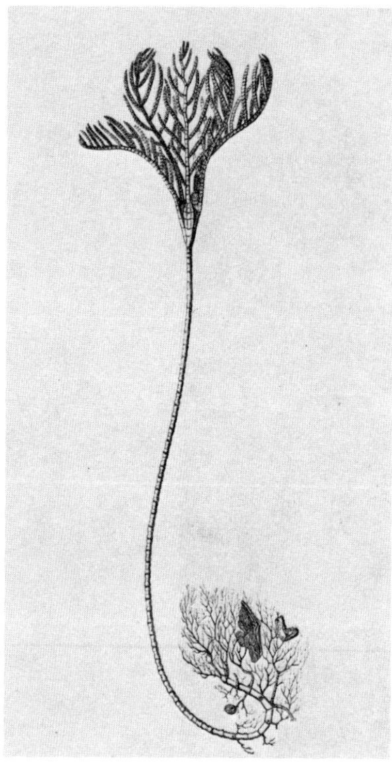

Fig. 5. *Rhizocrinus lofotensis* M. Sars, the stalked crinoid discovered in Norwegian waters in 1864 by G. O. Sars. As a representative of Mesozoic stalked crinoids, *Rhizocrinus* inspired Charles Wyville Thomson to look for archaic organisms in the deep sea. [From Thomson (1873).]

Because of his representations, the Council of the Royal Society requested a ship for deepwater dredging from the *Admiralty*. In mid-July this was granted.

The letters documenting these events, as quoted, are rather matter-of-fact; no doubt they reveal very little of the actual contacts that led to the first major British deep-sea expedition. Thomson, in his initial letter, listed purely scientific aims—discovering the structure of fossils, determining (in the constant conditions of the deep sea) the factors that changed faunal distribution through "oscillations of the earth's crust," the effects of low temperature and high pressure on animals—although his investigations were to take place in a practical age of cable laying and naval activity. The first northern Atlantic telegraph cable laid from Ireland to Newfoundland in 1858 failed almost immediately and in the next 8 years (until the success of Brunel's *Great Eastern* in 1866) there was a flurry of activity, in which the Royal Navy was involved, to find a shorter northern route for the cable (Rice et al., 1976). Ships and technically trained naval officers were available, perhaps even embarrassingly so, when the surveying was finished, so Thomson's request for a ship to carry out purely scientific pursuits was well timed, whether by luck or shrewd judgment. The private negotiations and the letters that might record them deserve some long-overdue consideration by historians of the first British oceanographic expeditions. Until then, the printed record with its biases must suffice.

Another problem from Thomson and Carpenter's time, a curiously intractable one, is the staying power of the azoic theory. Thomson, in beginning his book *The Depths of the Sea* (1873), considered it sufficiently important that he began the first chapter with "the question of a bathymetrical limit to life." Since then, this kind of approach to the history of deep-sea biology has had a remarkable persistence. Daniel Merriman (1968) has called the prevalence of the azoic theory in the 1860s, quite correctly, a "history of scientific obduracy."

When Thomson and Carpenter went to sea on their first deep-water cruise in 1868, evidence had accumulated for at least 50 years that animals could live in deep water. In 1818, Sir John Ross (1819), accompanied by Edward Sabine and James Clark Ross, had made a number of deep soundings in Baffin Bay while searching for the eastern end of a northwest passage. On September 1, 1818, just off Bylot Island, Ross sounded at an apparent depth of 1000 fathoms by using a tiny bottom grab and found on the line a living *Gorgonocephalus*. Later he recorded benthic animals from lesser depths along the west coast of Baffin Island. Rice (1975) has shown that Ross's depths were seriously in error because he had failed to detect the weight hitting bottom; nonetheless, he had found life at 500 to 600 fathoms, far below the zero that Forbes had suggested. Animals were also found at 300 to 400 fathoms in 1841 by James Clark Ross, by then in command of the Antarctic Expedition of 1839 to 1843, when he sounded and dredged in the Tasman Sea (J. C. Ross, 1847). Only 4 years later, Harry Goodsir, the naturalist on Sir John Franklin's last and ill-fated arctic expedition in H.M.S. *Erebus*, dredged abundant animals at 300 fathoms in Davis Strait. His letter mentioning the discovery, mailed from Disko before the ship disappeared forever into Lancaster Sound, reached Edward Forbes, who quoted it without comment (Goodsir, 1845; Forbes and Godwin-Austen, 1859, pp. 50–52).

In 1846, Forbes' companion Spratt (1849) dredged shells in the Mediterranean at 310 fathoms. Ten years later he found fragments of shells at 1610 fathoms between Malta and Crete (Spratt, 1865), although there was some doubt that they had been living. In 1861 a Swedish expedition to Spitzbergen under Otto Törell found animals in its bottom samples from 1400 fathoms (Keferstein, 1864; Lovén, 1865). During this time Michael Sars began a series of dredgings (later continued by his son) in the Norwegian fiords that yielded 19 species from below 300 fathoms by 1850 (Sars, 1850) and more than 400 by 1868 (Sars, 1868).

During the late 1840s and 1850s surveys began to turn up the shells of foraminifera at great depths. J. W. Bailey (1811 to 1857), who analyzed lead line samples collected by the U.S. Coast Survey, was surprised to find foraminifera in the deeper samples (at and beyond 100 m) (Bailey, 1851). This observation was repeated in later years by microscopists in Europe, as well as by Bailey himself, who examined samples from as deep as 4000 m. Bailey was not certain whether the forams (which were largely *Globigerina*) lived in the water column or on the bottom, but the great German microscopist, C. G. Ehrenberg (1795 to 1876) claimed that they were benthic, and T. H. Huxley (1825 to 1895), who examined the sediment samples taken by Dayman with H.M.S. *Cyclops* in the northern Atlantic in 1857 (Dayman, 1858; Huxley, 1858), cautiously agreed with Ehrenberg. The controversy was a long-lasting one, and its outcome is discussed in a different context later. Its significance for this discussion is that by the 1860s there was at least a strong possibility that simple forms of life lived in very deep water even if more complex ones did not.

One person who agreed with this view and carried it farther into a long-standing controversy was G. C. Wallich (1815 to 1899), one-time Indian Army surgeon, who sailed with Captain Sir Leopold M'Clintock (the discoverer of Franklin's fate) in H.M.S. *Bulldog* in 1860 during one of the northern telegraph surveys (Rice et al., 1976). In the northern Atlantic between Greenland and Rockall they sounded at 1260 fathoms and brought up on the line 13 brittle stars attached to a portion that had been lying on the bottom. Wallich (1860) made this discovery the basis of a polemic, in which he believed he established on *a priori* grounds that life could exist in the deep ocean and that neither the absence of plants nor the cold, high pressure, and darkness need prevent animals from living in deep water. He also stated his view that there could be gradual movement of animals between the great depths and the shallows and that the deep sea had a well-established and ancient fauna, some of which in the geologic past had been preserved in strata formed there.

Wallich was far from alone in his views. For example, Jeffreys (1861) was sufficiently convinced of the presence of active animals in the deep sea to suggest that telegraph cables should be protected against boring molluscs. Final proof of the presence of animals in deep water was established in 1861, when G. J. Allman (1812 to 1898) and Alphonse Milne-Edwards (1835 to 1900) identified molluscs and corals from a telegraph cable raised from 2000 to 2800 m (1093 to 1577 fathoms) between Sardinia and northern Africa (Milne-Edwards, 1861; Thomson, 1873). Not only were there abundant animals in the deep sea, but rare ones too, for at nearly the same time that *Rhizocrinus* appeared on the scene in Norwegian waters, Bocage (1867) and Wright (1868) showed that the rare sponge *Hyalonema*, previously known only from the Far East, could be dredged on the Atlantic continental slope off Portugal where shark fishers had worked for centuries. Two themes—the elucidation of fossil species and the search for life in deep water—had converged.

Because of the abundant evidence for deepwater life before the mid-1860s, only part of which I have reviewed, it is surprising that the controversies were so bitter and that there was such reluctance to admit its existence. Most of the work since the time of Sir John Ross was known to the marine scientists of the 1860s, but they refused to accept evidence on a variety of grounds that in retrospect sound strained. Wallich's observations, for example, were considered suspect because the brittle stars might have been floating above the bottom and grasped the line as it went past them. Even though each piece of evidence, such as Wallich's, might not have been definitive, by 1860 there was strong circumstantial evidence for both protozoan and metazoan life in deep water. The persistence of the azoic theory and the resistance of the scientific establishment (including Wyville Thomson himself) to evidence not gathered by the dredge suggest that the azoic theory had deep psychological roots. It was given longer life by the absence of an evolutionary viewpoint, particularly the realization that animal adaptation can have very wide limits. In fact, to many biologists of the early Darwinian era, the suggestion that adaptation had wide bounds and that variation was omnipresent was tantamount to accepting the progressionist "Lamarckian" view that species boundaries did not exist. The power of the azoic theory requires explanation, perhaps in these terms, with the use of documents from the nineteenth century. There was indeed scientific obduracy. But what provided its dynamics? The answers are far from clear.

To Wyville Thomson, at least, the controversy was nearly fully resolved during the cruise of H.M.S. *Lightning* in 1868, while Jeffreys was finishing his work at Shetland (W. B. Carpenter, 1868; Thomson, 1873). *Lightning*, a creaky old paddle-wheel surveying vessel, was ill equipped for deep-sea work, and the weather was poor throughout. The majority of Carpenter and Thomson's work was done between Stornaway, which they left on August 12; the Faeroes; and Oban, reached on September 17. During 6 weeks they had only 10 days of successful dredging and made only four collections below 500 fathoms (the deepest at 680 fathoms), in addition to determining bottom temperatures with unprotected maximum–minimum thermometers. Their aim, as Carpenter wrote to Norman, was "to make out *everything* we can as to the distribution of life at great depths in these northern seas" (Alder-Norman, 1826–1911, Letters No. 325). They found that life was abundant everywhere, even down to 680 fathoms (1189 m), the greatest depth ever dredged. Masses of sponges, echinoderms, even *Brisinga* and *Rhizocrinus*, emerged from the dredge bags. Most of the species were new, and in Thomson's estimation many were related to Tertiary fossils, perhaps even to mesozoic ones. Although they found that the depths were far from azoic, the abyss was also far from uniform, for the assemblages of species seemed to vary with bottom temperature, and at equivalent depths animals could vary greatly in abundance depending on the local environment. Most important for the events to follow, they discovered that the deepwater temperatures were very low (0 to 0.5°C) northeast of a line between the Faeroes and Shetland, but much warmer (4.5 to 8.5°C) to the southwest in the open Atlantic.

Carpenter's imagination was piqued by the deepwater temperatures, and it was this as much as the biological results that gave impetus to their organization of a new expedition. Once again the Royal Society was involved in negotiations with the secretary of the Admiralty, and once again a ship was granted—the 300-ton surveying ship *Porcupine*, under Staff-Commander E. K. Calver. Its cruises during 1869, because of the care that went into equipment, organization, and investigations, qualify as the first great oceanographic expedition (Carpenter et al., 1870; Jeffreys, 1869a, W. L. Carpenter, 1870). The overall planning was done by a committee appointed by the Royal Society, consisting of its president, General Edward Sabine (1788 to 1883); W. B. Carpenter; John Gwyn Jeffreys; and Captain George Richards (1819 to 1896), the hydrographer of the Admiralty. This was hardly a disinterested committee, but certainly one with a vested interest in the best equipment and good results. The aims of *Porcupine*'s work during the summer of 1869 were to take deepwater temperatures (especially in the area between Scotland and Faeroe), measure the gases, salinity and organic matter of seawater, determine the extinction of light with depth (not actually accomplished), and dredge as deeply as possible. The equipment of the ship (Carpenter et al., 1870; W. L. Carpenter, 1870) was ordered or designed by a committee on scientific apparatus comprised of the president of the Royal Society and its officers, Carpenter, Richards, the instrument maker C. W. Siemens (1823 to 1883), and the physicists John Tyndall (1820 to 1893) and Charles Wheatstone (1802 to 1875). Whereas *Lightning* had unprotected thermometers that required correction, *Porcupine* took the first protected Miller–Casella maximum–minimum thermometers as well as unprotected ones for comparison. Temperature could also be measured by use of a rather cumbersome electric thermometer, consisting of a resistance coil that was lowered over the side and a Wheatstone bridge on the ship balanced by adding cool or warm water to a bath surrounding a second coil. When the bridge was balanced, sea temperature was the same as that indicated by a thermometer in the bath surrounding the coil (Matthaüs, 1968). In practice, this awkward but potentially useful device never worked properly on the *Porcupine* cruises. Water samples for physical and chemical studies could be taken with a brass tube fitted with simple release valves. The water was analyzed for organic matter by using permanganate oxidation, for specific gravity by using hydrometers, and for dissolved gases by boiling them off in a sealed system in which they could be absorbed on suitable chemicals. For these complex operations, a combined physicist–chemist was provided (two of Carpenter's sons were among the three holders of this post). Even dredging was modified. The *Porcupine* added rubber band accumulators to take the ship's motion on the dredging rope, and a double-cylinder donkey engine did the hauling. Hundreds of fathoms of rope were coiled on special pins ("aunt sallies") attached to a bulwark on the dredging deck (Fig. 6). Extra large and heavy dredges were built, and at sea Calver attached to them long cotton swabs, probably from deck mops, to entangle the sponges, hydroids, and echinoderms that the dredge often brushed over. Jeffreys, who led the first cruise, had provided nested sieves, the smallest of which had a mesh size of about 0.8 mm, and brought two colleagues from the Shetland expeditions, William Laughrin of Polperro as dredger and screener and B. S. Dodd to help with sorting and preserving.

The ship was at sea from May 18 to September 15. Jeffreys directed the first

Fig. 6. The deck and dredging equipment of H.M.S. *Porcupine*. This ship was used by Wyville Thomson to dredge to 4456 m southwest of Ireland in July 1868. Note the coils of dredging rope, the accumulator, which took up undue strain on the rope, and the naturalists' dredge being put over the stern. [From W. L. Carpenter (1870).]

cruise from the western coast of Ireland to Porcupine Bank, Rockall Bank, and Belfast, during which time they dredged to 1476 fathoms (2700 m). Thomson then took over, changed plans (which had been to go north) and headed for deep water off Ireland. On July 22 to 23 they dredged at the unprecedented depth of 2435 fathoms (4456 m) about 240 nautical miles southwest of Cape Clear. Thomson (1870b) gave a graphic description of the event when he wrote to A. M. Norman from Belfast on August 7:

My dear Norman,—You are already aware that, during the first cruise of this year, Mr. Jeffreys and his party dredged and took most important thermometrical and other observations to a depth of 1476 fathoms. When I took Mr. Jeffreys' place for the second cruise, it was the intention to proceed northwards, and to work up a part of the north-west passage, north of Rockall. I found, however, on joining the vessel, the gear in such perfect order, all the arrangements so excellent, the weather so promising, and the confidence of our excellent commander so high, that, after consulting with Captain Calver, I suggested to the hydrographer that we should turn southwards, and explore the very deep water off the Bay of Biscay. I was anxious that, if possible, the great questions of the distribution of temperature &c., and of the conditions suitable to the existence of animal life, should be finally settled; and the circumstances seemed singularly favourable. No thoroughly reliable soundings have been taken beyond 2800 fathoms, and I felt that if we could approach 2500, all the grand problems would be virtually solved, and the investigation of any greater depths would be a mere matter of detail and curiosity. The Hydrographer at once consented to this change of plan; and on the 17th of July we left Belfast and steered round to Cork, where we coaled, and then stood out towards some soundings, about a couple of hundred miles south-west of Ushant, marked on the Admiralty charts 2000 fathoms and upwards. On the 20th and 21st we took a few hauls of the dredge on the slope of the great plateau, in the mouth of the Channel, in depths from 75 to 725 fathoms, and on the 22nd we sounded with the "Hydra" sounding-apparatus, the depth 2435 fathoms, with a bottom of fine Atlantic chalk-mud, and a temperature registered by two standard Miller-Six's thermometers of 36.5 Fahrenheit. A heavy dredge was put over in the afternoon, and slowly the great coils of rope melted from the "Aunt-Sallies"—as we call a long line of iron bars, with round wooden heads, on which the coils are hung. In about an hour the dredge reached the bottom, upwards of three miles off. The dredge remained down about three hours, the Captain moving the ship slowly up to it from time to time, and anxiously watching the pulsations of the accumulator, ready to meet and ease any undue strain. At nine o'clock p.m. the drums of the donkey-engine began to turn, and gradually and steadily the "Aunt-Sallies" filled up again, at the average rate of about 2 feet of rope per second. A few minutes before one o'clock in the morning 2 cwt. of iron (the weights fixed 500 fathoms from the dredge) came up, and at one o'clock precisely a cheer from a breathless little band of watchers intimated that the dredge had returned in safety from its wonderful and perilous journey of more than six statute miles. A slight accident had occurred. In going down the rope had taken a loop round the dredge-bag, so that the bag was not full. It contained, however, enough for our purpose—1½ cwt. of "Atlantic ooze"; and so the feat was accomplished. Some of us tossed ourselves down on the sofas, without taking off our clothes, to wait till daylight to see what was in the dredge.

Life extends to the greatest depths, and is represented by all the marine invertebrate groups. At 2435 fathoms we got a handsome Dentalium, one or two crustaceans, several Annelids and Gephyrea, a very remarkable new Crinoid with a stem 4 inches long (I am not prepared to say whether a mature form or a Pentacrinoid), several starfishes, two hydroid zoophytes, and many Foraminifera. Still the fauna has a dwarfed and arctic look. This is, doubtless, from the cold.

When Carpenter took over in mid-August, he headed north for the boundary between cold and warm bottom water that lay somewhere between Faeroe, the Scottish mainland, and Shetland. A long section of the report (Carpenter et al., 1870) is devoted to the results, which supported Carpenter's developing obsession with deep oceanic circulation, a study that led to a further season with *Porcupine* in the Bay of Biscay and the Mediterranean in 1870 (Carpenter and Jeffreys, 1870, 1871).

During 1870 Jeffreys took the ship from England to Gibraltar, tracing cold-water species as far as the Strait of Gibraltar and into the Mediterranean, where some of them had been found in the Tertiary deposits (now dated as Pleistocene) in Sicily. These species, he believed, could only have reached the Mediterranean by a seaway across central France (the Languedoc Canal) because the Mediterranean inflow was too shallow and warm to be a feasible route of invasion. In addition, with good fortune that recalled the origin of these pioneering explorations, he dredged a magnificent *Pentacrinus*, the first discovered in the northeastern Atlantic (Jeffreys, 1871). At Gibraltar the major biological work came to an end and Carpenter began a study of temperatures and the currents of the Straits of Gibraltar by use of thermometers and a primitive drogue that could be set anywhere between the surface and 250 fathoms (458 m). He was convinced by these means that the Atlantic inflow to the Mediterranean was underlain by a compensating outflow, an outflow that, like the cold deep water northwest of Scotland, was likely to be driven by density differences rather than the winds. This "magnificent generalization" about oceanic circulation, as Thomson called Carpenter's hypothesis, soon involved him in bitter controversy (Deacon, 1971; Mills, 1975b); most important, for the purposes of this essay, it convinced him that a grand expedition was needed to establish the causes of general oceanic circulation.

The early explorations of *Lightning* and *Porcupine* began because of zoological discoveries linking recent invertebrates with their fossil precursors. As Wyville Thomson (1873) said, summarizing the work:

> A grand new field of inquiry has been opened up . . . Every haul of the dredge brings to light new and unfamiliar forms—forms which link themselves strangely with the inhabitants of past periods in the earth's history; but as yet we have not the data for generalizing the deep-sea fauna, and speculating on its geological and biological relations; for notwithstanding all our strength and will, the area of the bottom of the deep sea which has been fairly dredged may still be reckoned by the square yard.

The next step became possible because of the very different concerns, biological and physical, of Thomson and Carpenter, and because Carpenter, with his "magnificent generalization" in mind, was again able to manipulate political and economic forces in mid-Victorian England.

4. The *Challenger* Expedition and Its Results

The voyage of H.M.S. *Challenger* (1872 to 1876) is part of the folklore of oceanography; and, like many legends, its impact often has been more symbolic than concrete. The voyage is popularly credited with beginning the science of oceanography, although as I have shown earlier, the first voyage of H.M.S. *Porcupine* in 1869 deserves the credit of being the first fully organized oceanographic expedition. I have no doubt that too great an emphasis on events such as the *Porcupine* and *Challenger* voyages has diverted attention from the long history of work in physics, chemistry, and biology that has made a science of the sea.

However, despite my skeptical remarks about the significance of "great landmarks" such as the early expeditions, it is clear that the *Challenger* voyage was significant in at least three ways: (1) as Schlee (1973) has noted, it was the report of the *Challenger* expedition (not the voyage itself) that provided an historical substratum for oceanography; (2) both the magnitude of the voyage and its success provided a model for future expeditions and a sense of community, an esprit de corps, that still has a significant influence on oceanography; and (3) the expedition involved government in a major scientific project in a way that had few parallels in nineteenth century Europe.

During the *Porcupine* expeditions W. B. Carpenter became convinced that only a major expedition working over much of the globe and taking deepwater temperatures in many ocean basins could provide enough evidence for his hypothesis that ocean currents were driven by differences in temperature (and thus in density) of the waters. A great deal of information is available on how Carpenter went about organizing support for a major expedition. Burstyn (1968, 1972) has shown how Carpenter, in his combined role as scientist and administrator, was able to enlist support from the Royal Society, the hydrographer of the Admiralty, and the politicians who were most influential, especially the First Lord of the Admiralty and the Chancellor of the Exchequer. The Royal Navy was willing to supply a ship and the funds to run it, but the Treasury had to supply the money for the scientific supplies and the salaries of the civilian scientists who would be aboard. Nearly all the resources of shakers and movers were used to produce these results. Carpenter lectured to the Royal Institution in early June 1871, warning that Britain could easily lose its lead in oceanic exploration to Germany, Sweden, and the United States, which were organizing expeditions (Anonymous, 1871b). At almost the same time, perhaps coincidentally, the Chancellor of the Exchequer, Robert Lowe, was elected a Fellow of the Royal Society. In October, at Carpenter's urging, the Royal Society appointed a committee to make recommendations for a voyage of circumnavigation. Its report was incorporated in a letter from the Royal Society to the Admiralty requesting a ship; by March 1872 the request had been granted and funds were provided by the Treasury.

During the next few months the scientific staff was appointed. It consisted of Charles Wyville Thomson as leader, J. J. Wild, secretary to Thomson and artist; J. Y. Buchanan (1844 to 1925), chemist and physicist; H. N. Moseley (1844 to 1891), naturalist; William Stirling (replaced by Rudolf von Willemoes-Suhm), naturalist; and John Murray (1841 to 1914) (Fig. 7), naturalist (Merriman, 1972). For their enlightenment, the Circumnavigation Committee provided a detailed account of the scientific objectives, which were to (1) "determine the physical conditions of the deep sea" (i.e., measure depth, temperature, specific gravity, light penetration, and circulation), (2) establish the chemical composition of seawater (salt, gases, organic matter in solution, and suspended particles), (3) determine and chart the marine sediments and establish their sources, and (4) study the distribution and abundance of organisms and determine where they had originated. In addition, there were several pages of instructions on botanical collecting in remote areas (written no doubt by J. D. Hooker). This lengthy in-

Fig. 7. Sir John Murray (1841 to 1914), who as a young naturalist accompanied H.M.S. *Challenger* during its circumnavigation. After Charles Wyville Thomson's death in 1882 Murray took over responsibility for editing the *Challenger Reports*. Later in his life Murray was an important international statesman of marine science. Under his influence Edinburgh became a leading center for oceanographic research until the beginning of the First World War. [Photograph from Herdman (1923).]

struction is still worth reading as a concise outline of marine science circa 1872 (Thomson, 1878, Vol. 1, pp. 81–98). The ship itself, the 226-ft spar-deck corvette *Challenger*, 2306 tons, was refitted for its new role by building in chemical and zoological laboratories, a darkroom, spirit stores, and accommodation for scientists and junior officers. A raised deck for dredging was added (Fig. 8) and near it a preparation room to receive the samples as they came from dredge and trawl.

The voyage itself is too well known to require more than a brief summary (Deacon, 1971; Linklater, 1972; Mills, 1975a; Schlee, 1973); it is best told in the accounts of the participants, beginning with the narrative opening the reports (Tizard et al., 1885; Thomson, 1878; Campbell, 1876; Moseley, 1880; Spry, 1877; Swire, 1938; Wild, 1878; Willemoes-Suhm, 1877). The ship headed south from Portsmouth on December 21, 1872 but, because of bad weather its first deep-sea station, was made early in 1873. During 1873 the ship sailed from Lisbon to Madeira, the Canaries, St. Thomas, Bermuda, Halifax, the Azores, the Cape

Fig. 8. The dredging equipment on H.M.S. *Challenger*. Note the steam winch on the main deck, the raised dredging platform, the accumulators, and the beam trawl, which was the gear most commonly used at all depths. [From Thomson (1880), p. 9, Fig. 7.]

Verdes, St. Paul's Rocks, Fernando de Noronha, Bahia, Tristan da Cunha, Cape Town, the Prince Edward Islands, and Crozet. In 1874 it visited Kerguelen, touched the edge of the Antarctic ice pack at 81°E, and then went northeastward to Melbourne, Sydney, Wellington, the Kermadecs, Tonga, Fiji, the New Hebrides, Cape York, Aru, Amboina, the Philippines, and Hong Kong. In the following year it returned to the Philippines and then went on to New Guinea, Yokohama, the Hawaiian Islands, Tahiti, Valparaiso, and the Straits of Magellan. Early in 1876 it reentered the Atlantic to visit the Falklands and Montevideo before arriving at Portsmouth on May 24, 1876. The *Challenger* was away for $3\frac{1}{4}$ years, steamed or sailed 68,890 nautical miles, and made 362 stations spaced at intervals of approximately 200 mi. Almost 40% of the time away was spent in port for repairs and provisioning, giving the naturalists a chance to travel or send their scientific papers home for publication.

The major results of the expedition were considerable. Animals were found to 5500 m, finally removing any lingering doubts in Thomson's mind that the great depths were inhabited. The constancy of composition of seawater was well established by using the largest number of samples that had ever been available; and the salinity, temperature, and oxygen content of the waters were used to characterize water masses and to speculate on their origins. Oceanic sediments were mapped, including the newly characterized "red clays" and a variety of other pelagic deposits. Perhaps even more important, the samples were sent to specialists in Europe and North America, a process that spread the work of the expedition to many others and encouraged the development of science in other countries. Doing so was politically difficult and also expensive. The expedition itself had cost nearly £92,000 and the production of the report between 1876 and 1895 roughly an additional £80,000 (Burstyn, 1975), in addition to much time and energy from Thomson (who was worn out by the struggle and died in 1882) and Murray, who saw the reports through to completion (Deacon, 1971, Chapter 16).

The *Challenger* years, which extend from the beginning of the voyage in 1872 to the publication of the last volume of the reports in 1895, were marked by great changes in scientific ideas. This may be seen particularly clearly in the contrast between Thomson's early reports (1878, 1880) and the summary written by Murray (1895b) near the end of the century. By comparing their ideas and examining the controversies resolved by the *Challenger* scientists, we have some insight into changes in scientific fashions between the generations represented by Thomson and Murray.

Thomson had always viewed the study of the deep sea as a way to provide "connection of the present with the past condition of the globe," in the words of the Circumnavigation Committee. His concern with the history of life dominates his first summary of the expedition's results, the book *The Voyage of the Challenger* (1878), which was written hurriedly for popular consumption before anything but shipboard analysis of the results was possible. Thomson had been convinced during the first crossing of the Atlantic between the Tenerife and Leeward Islands in 1873 that there could be no depth limit to life; however, it was now clear that the number of species and the abundance of animals and their size decreased with depth. As they worked farther and farther south, Thomson also came to believe that the abundance of fossil species—like hexactinellid sponges, primitive echinoids, and crinoids—increased, and that the Southern Ocean south of 40°S was the source of the fossils found farther north, just as it provided the cold deep water to lower latitudes.

Because temperature and food were relatively constant in the deep sea, the fauna was essentially continuous everywhere. Hence genera, at least—often species—were widely distributed. But the "peculiar conditions" of deep water imposed their effect by governing the groups that could exist there. For example, bivalves and gastropods were rare, according to Thomson; the characteristic abyssal groups were hexactinellid sponges, stalked crinoids, primitive sea urchins, and some holothurians. The midwater, in particular, formed an azoic barrier between the fauna of the surface waters and that of the near-bottom waters and the deep-sea sediments. He was willing to concede that those sediments, too, were far more complex than he had thought earlier, for the majority of deep-sea sediment was not *Globigerina* chalk but abyssal "red clay," formed (as he believed for a time) from the remains of organisms after the carbonates had gone into solution (Thomson, 1874b; 1878). Even his ideas on the origin of the *Globigerina* sediments changed once Murray's tow nettings showed, conclusively at last, that the foraminifera (also the majority of radiolarians) were found in the surface waters (Thomson, 1874a).

Two years later Thomson (1880) amplified his conclusions, partly using the same text as in the first account, but with additions that changed its effect in small but significant ways. He emphasized the extreme age and stability of the deep sea, where only below the effect of "seasonal changes"—that is, below the depth of the thermocline—the true abyssal fauna begins. There it inhabits an environment that has retained the same depth and low temperatures for very long periods. The deep-sea fauna "has a relation to the deep water fauna of the Oolite, the Chalk, and the Tertiary formations, so close that it is difficult to suppose it in the main other than the same fauna which has been subjected to a slow and continuous change under slightly varying circumstances" (Thomson, 1880, p. 49). Emphasizing the same point, he said that the deep-sea fauna is one "which has arrived at its present condition by a slow process of evolution from which all causes of rapid change have been eliminated" (Thomson, 1880, p. 50). In these words Thomson grudgingly accommodated the evolutionism of the 1880s to his earlier ideas that the deep-sea fauna represented the fauna of Cretaceous chalks, a belief that there had been, in fact, a "Continuity of the Chalk" since the Mesozoic in a changing world.

With these words, Thomson was strategically backing away from a controversy that had arisen before the *Challenger* voyage. It appears to have originated in a misunderstanding by W. B. Carpenter (1868) of Thomson's rather naively expressed views on the antiquity of the deep-sea fauna. In his report on the cruise of H.M.S. *Lightning* in 1868, Carpenter stated that, judging by that summer's results, even deeper dredging would yield animals "belonging to a Geological epoch supposed to have terminated" (Carpenter, 1868, p. 185). A little later he referred the Atlantic *Globigerina* sediments to Cretaceous chalks, citing the authority of Bailey, Huxley, and others, and stated that this sediment was "not merely *a* Chalk-formation, but a continuation of *the* Chalk-formation; so that *we may be said to be still living in the Cretaceous Epoch*," a near-literal quotation from one of Thomson's (1869) talks to laymen. If such a deposit were raised and juxtaposed to a warm-temperate fauna from the shallows, Carpenter concluded, the geologists would be confused by the apposition of Cretaceous and Recent fossils. The geologists, notably the eminent Murchison (1871) and Lyell (1871), were offended by the suggestion that their carefully worked out stratigraphic principles could be so blithely transcended. Thomson, backed into a difficult corner by his colleague's forthrightness, responded with a powerful account of his views on the antiquity of the deep-sea fauna, its descent from an ancient deepwater fauna, and the continuity of abyssal sedimentation from the Cretaceous to the present (Thomson, 1871, 1873; Chapter 10). He was able to cite a number of fossils common to Cretaceous chalks and the deep sea, although related only at the generic or familial level. Thomson's concern was never to prove the stratigraphic continuity of the Cretaceous chalks with deep-sea sediments, but rather to show the continuity of the deep-sea environment and its general types of animal over long periods, a principle that was actually compatible with Lyell's uniformitarianism as it is now understood (Rudwick, 1972, Chapter 4). As Thomson stated, "our dredgings only show that these abysses of the ocean . . . are inhabited by a special deep-sea fauna, possibly as persistent in its general features as the abysses themselves" (Thomson, 1873, p. 495). The argument dragged on for a time and caused some differences between Thomson and Jeffreys (Jeffreys, 1869a, p. 167; 1878), who found that the mollusks of the Cretaceous chalks were shallow-water ones, not deep-sea species, and that the *Globigerina* sediments contained far less calcium carbonate than the Cretaceous ones. W. B. Carpenter (1870), too, qualified his remarks (although staying on the offensive by claiming that glacial climates had been continuous in the deep sea since at least the Cretaceous). Thomson was able to blunt the effect of their early statements over a period of years when the results of *Challenger* showed that there were really relatively few living fossils or their representatives in deep water among the thousands of less exciting species (Thomson, 1876, 1878, 1880) and that the recent "Chalk" was really not very widespread. Interestingly enough, it was a form of Darwinism that allowed Thomson to keep the most important elements of his ideas, the stability of the great depths, while admitting the "descent with modification" of its inhabitants. This controversy, quieted by time, died with Thomson.

The complicated problem of the relationship between food supply to the deep sea and the nature of the earliest units of life was also affected by the results of the *Challenger* expedition. Three subjects are nearly inextricably linked by the events before *Challenger*: the source of deep calcareous sediments (the *Globigerina* oozes); the nature of life early in its evolution; and the source of food for deep-sea organisms.

During the early bathymetric surveys of the North Atlantic, as I discussed in brief earlier, it soon became clear that many deep-sea sediments were calcareous and that they were made up of the remains of foraminifera, largely *Globigerina*. Bailey (1851) of West Point, who examined the samples taken between 50 and 100 fathoms off New Jersey by the U.S. Coast Survey in 1848, was surprised to find abundant foraminifera in the deeper samples. As the soundings went deeper, he discovered that the sediments of the deep northern Atlantic over a very wide area were "literally nothing but a mass of microscopic shells" (Bailey, 1854, 1855), made up of foraminifera, radiolarian, and sponge spicules, and that they were very much like the Cretaceous chalk of England. He was uncertain whether the forams, in particular, lived at those depths or had been carried there by currents, for samples of the surface water were devoid of them and the species he found were not known from shallow-water sediments. By contrast, the German microscopist C. G. Ehrenberg, who also examined the forams, believed he could see protoplasm inside the tests [as he had with the diatoms in James Clark Ross's sediment samples taken during 1842 (Ehrenberg, 1844)] and concluded that they had been alive on the bottom.

Only two years later, in 1857, more sediment samples were collected in the North Atlantic between 1700 and 2400 fathoms (3111 to 4392 m) in the course of Lieutenant-Commander Joseph Dayman's survey for the North Atlantic telegraph (Dayman, 1858). The samples were rather briefly examined by Huxley (1858), who found them to be made up predominantly of calcareous forams. The seemingly amorphous background material proved to contain minute plates that Huxley called *coccoliths*. All—and particularly the forams—appeared to have originated on the bottom; because of their geologic continuity from at least the Cretaceous, they were likely to be able to live under the difficult conditions of the deep sea (Huxley, 1868a). In this, Huxley made an association (common since at least the time of Edward Forbes) between great geologic age and eurytopy.

The view that forams were benthic was fiercely supported by G. C. Wallich, who had gone to sea with Sir Leopold M'Clintock on a northern telegraphy survey in 1860. In his book *The North-Atlantic Sea-Bed* Wallich (1862), not only made a case for the ability of animals to survive at great depths, but claimed that simple organisms like the protozoa could manufacture their own protoplasm there by special physiological processes, which, as he described them, were chemosynthetic. Wallich (1860, 1861) also observed coccoliths attached to membranous spheres called *coccospheres*, which might have been the larvae of some foram. His conclusion that coccoliths were produced by some unicellular organism was arrived at almost simultaneously by at least one other (Sorby, 1861).

Later in the decade Carpenter and Wyville Thomson, during the voyages of *Lightning* and *Porcupine* in 1868 and 1869, found that the foraminiferal sediments northwest of the British Isles were associated with the warmer Atlantic water, not the colder northern water, and that forams were sometimes abundant just above the bottom. They were apparently unable to find them in the surface waters above the calcareous sediments. But not everyone agreed that calcareous foraminifera were benthic, notably their colleague John Gwyn Jeffreys (1869b), who had not only observed living forams in surface net samples taken at Shetland, but also accepted earlier observations that forams were widely distributed in the surface waters of the Atlantic and Indian Oceans (Owen, 1865, 1868).

The latter view was vindicated when Thomson, writing from Melbourne on March 17, 1874, finally admitted, on the basis of Murray's many surface tows during the first 15 months of the *Challenger* expedition (Thomson, 1874a,b), that living forams were abundant in the surface waters and that their abundance coincided with the areas of calcareous sediments. In addition, coccoliths and coccospheres were undoubtedly related, as the origin of the plates was probably minute algae or their spores from the surface waters, a view he had developed during the earlier expeditions (Thomson, 1873, pp. 45–48). In response, Carpenter (1875) tried to save appearances by claiming that although *Globigerina* lived at the surface as young stages, they sank to the bottom to reproduce, a view bitterly attacked by Wallich (1876) [see Rice et al. (1976)].

During the peak of the *Globigerina* debate in 1868, Huxley returned to the sediment samples from H.M.S. *Cyclops* that he had examined 10 years before. His motives are not entirely clear, but it is possible that he wanted to check Wallich's contention that coccoliths were the inorganic remains of cells, and not—as Huxley himself had suggested originally—inorganic parts that had coalesced to form the coccospheres in the sediments (Rehbock, 1975a). Whatever his motivation, he found in the preserved samples some "transparent gelatinous matter," occasionally full of coccoliths and other inclusions (Huxley, 1868b). This he identified as belonging to the Monera, the simplest of living organisms, described by Haeckel the same year. In Haeckel's honor he named the deep-sea moneran *Bathybius Haeckelii* (Fig. 9) and announced its discovery at the Norwich meeting of the British Association in August (Huxley, 1869). *Bathybius*, it appeared, formed a primitive, living sheet over much of the ocean floor.

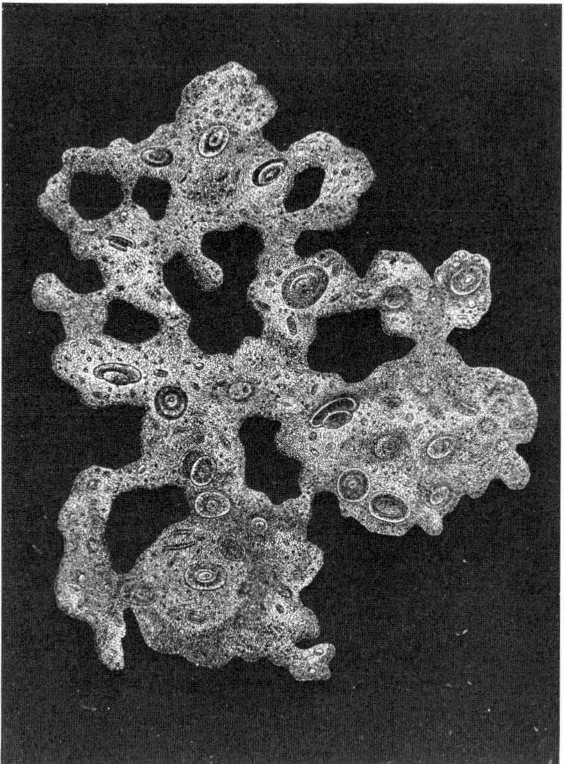

Fig. 9. *Bathybius haeckelii*, the supposedly primitive moneran discovered by T. H. Huxley in 1868 when he was reexamining cores from the North Atlantic collected more than a decade previously. Note the coccoliths embedded in the gelatinous ground material. During the cruise of H.M.S. *Challenger*, J. Y. Buchanan demonstrated that *Bathybius* was a precipitate of ethanol in sea water. *Bathybius* was not an egregious error, but a natural mistake at a time when the nature of life and its localization in protoplasm was being actively debated. [Illustration from Thomson (1873), p. 412.]

Bathybius had a short but significant life. After accepting its existence (with some reservations), Thomson wrote to Huxley from the *Challenger* on June 9, 1875, explaining that Buchanan had discovered *Bathybius* to be merely an inorganic precipitate caused by adding alcohol to seawater (Huxley, 1875). Huxley gracefully accepted his error. Despite this, the story has lived on as an example of the uncritical errors that may be made even by great scientists (e.g., Argyll, 1887). The truth, of course, is far more complex and more interesting, as Rehbock (1975a) has shown in an elegant analysis. Huxley's "discovery" of *Bathybius* came during two decades of intensive work by microscopists on protoplasm, which was replacing the cell membrane and organelles as the most important ground substance of life. Huxley himself, sometime between 1853 and 1868, had switched his allegiance to the new ideas. During the same decades, only partly stimulated by Darwinism, speculation about the abiogenetic origin of life became widespread, and it was widely regarded that any protoorganism must have been made up of undifferentiated protoplasm. Into this background of developing concepts, *Bathybius* fitted very neatly as a primitive organism, a living fossil, an evolutionary necessity, and a means of providing food to higher organisms (Rehbock, 1975a). Huxley's mistake was hardly a gratuitous error, rather a very natural discovery given the scientific emphases and concepts of the time, and as Huxley himself said (1887), only the man who does nothing at all makes no mistakes.

In narrower terms also, *Bathybius* came at an appropriate time, for it gave a new view of the problem of how deep-sea animals obtained food. Wallich (1862), as I have described, envisioned a process akin to chemosynthesis occurring in the deep sea, by which animals absorbed inorganic matter in solution and converted it to protoplasm, just as they could absorb carbonic acid to form calcareous shells. No one agreed entirely with Wallich that this process was possible, but it was very difficult to find conventional alternatives. Nearly all the attempts were made just after Huxley's announcement of *Bathybius*. For example, W. B. Carpenter (1868), reporting on the results from the *Lightning* expedition, agreed with the view that animals were dependent on green plants and ultimately the sun for nutrition, but suggested that *Bathybius* might have "so far the attributes of a Vegetable, that it is able to elaborate Organic Compounds out of the materials supplied by the medium in which it lives, and thus to provide sustenance from the Animals imbedded in its midst," a statement that may have been deliberately ambiguous about whether "the materials supplied by the medium" were organic or inorganic. Carpenter's involvement was complex because he had committed himself to the organic nature of J. W. Dawson's Precambrian *Eozoon Canadense* (O'Brien, 1970, p. 122); he believed that *Bathybius*, if it were able to secrete a test, would be just like *Eozoon*. Thus *Bathybius* was living proof of primitive acellular life through almost inconceivably long periods of time.

Thomson was not committed to *Eozoon*, nor particularly strongly to *Bathybius*, either. But during the first cruise of *Porcupine* in 1869, W. L. Carpenter's chemical work had shown that dissolved organic matter was abundant in seawater. Although there were not enough diatoms and other algae in the surface waters to provide food for deepwater animals, the deepwater fauna could base its feeding on protozoa (in which he included the Porifera) if the latter were able to absorb dissolved organic matter. Foraminifera and sponges, the early results showed, were so abundant that they could easily provide the food base for the other animals. As Thomson said, "There is no difficulty in accounting for the alimentation of the higher Animal types, with such an unlimited supply of food as is afforded by the *Globigerinae* and the *Sponges*, in the midst of which they live, and on which many of them are known to feed. Given the Protozoa, everything else is explicable" (Carpenter et al., 1870, p. 477).

As Thomson had first stated in a lecture in 1869, protozoa can absorb through their whole bodies; sponges in particular could take up such minute quantities of silica that it was surely possible for them to absorb organic matter from solution, especially if their metabolism were very low and thus suited to a sparse supply of nutrients. *Bathybius* could well fill this role. In an interesting instance of teleological reasoning, Thomson (1870a) stated that a "principal function of this vast sheet [Porifera and foraminifera] of the lowest type of animal life, which probably extends over the whole of the warmer regions of the sea, may possibly be to diminish the loss of organic matter by gradual decomposition, and to aid in maintaining in the ocean the 'balance of organic matter'." Thomson never quite gave *Bathybius* this role (although Huxley's creation had been detected in the *Por-*

cupine samples), perhaps in part because he was so deeply convinced of the importance of his pet group, the sponges, and because he believed that the blanket of *Globigerina* was benthic. Perhaps also he was more sympathetic for some reason to the dissenting voices.

Wallich was convinced from the first that *Bathybius* was decomposing organic matter (Wallich, 1869). Unlike Wallich, Jeffreys was remarkably silent on Huxley's discovery, except to note (1869a), in his account of the *Porcupine*'s work west of Ireland, that the bottom down to 1400 fathoms was covered with a "flocculent mass" that could well have been derived from salps or decaying medusae. His brand of silence was almost more telling than a direct attack on *Bathybius*. Whatever the outside influences may have been, Thomson's most detailed account of *Bathybius* in *The Depths of the Sea* (1873) contains his statement that "I feel by no means satisfied that *Bathybius* is the permanent form of any distinct living being" despite its potential convenience in providing food to the deep-water benthos. By 1878, outlining the reasons for the *Challenger* voyage, he made no reference to it whatsoever, not even as an interesting relic of the early exploration of the deep sea.

Curiously enough, although speculation about *Bathybius* and the supply of food to the deep sea were at their height in 1869, A. M. Norman, who had just received the animals taken in deep benthic hauls by the *Porcupine* off the western coast of Ireland, found a number of unknown copepods that had apparently been taken in the dredge near, but not on, the bottom. In a letter to Charles Darwin on August 2, 1869 (Norman, 1869a), he speculated that this might solve the problem of how food reached abyssal fishes. Norman's momentary glimpse of a new facet of deep-sea biology led nowhere. It was nearly 20 years before the whole question of life in the water column and its significance to the deepwater benthos became a major controversy again, as discussed in Section 6.

The final element in the story of *Challenger*'s influence on oceanography was Murray's summary of the scientific results (Murray, 1895a,b), published nearly 20 years after the return of the expedition, 10 years after Carpenter's death, and 13 years after Thomson's. Because marine science had changed so greatly in the intervening years (Deacon 1971, Chapter 16), it is almost anachronistic to make detailed comparisons between the scientific views held by Thomson and Murray. However, the relevant comparisons show the extent of the change. It was only Murray who saw the *oeuvre complète*—the complete data of the expedition in his own papers or bound in 50 volumes of *Scientific Reports*. Thus his summary [in particular his concluding discussion on the distribution of marine organisms (Murray, 1895a)] deserves study in its own right. It is interesting both as summary and as an expression of a complete view of life in the oceans.

To Thomson, the overriding reason for the great expedition was to prove the existence of life at all depths and in all places in the oceans. There was no doubt that it had succeeded. Thomson had seen it first, but Murray was able to show in copious detail that life was omnipresent in all kinds of sediments, even the abyssal clays. Animals were particularly abundant in the Southern Ocean, probably because they had an abundant food supply from the plankton. But there were peculiar changes with increasing depth that Thomson, who had never been able to look at the complete faunal lists, could not have suspected. Although the abundance of animals and the number of species declined with depth, just as Thomson had noted, Murray's compilation showed that the number of genera and species compared to the number of specimens actually increased at the great depths. Below the littoral zone there were also small but significant numbers of species common to high northern and southern latitudes but absent in the tropics (Murray listed 91 and inferred that the list must be longer). Contrary to Thomson's view, cosmopolitan species were few in the deep sea; in fact, Murray was able to show by comparing stations at comparable depths in different oceans (and even in the same ocean) that the number of species in common was very few, indeed. He was sure that there was "no striking evidence of a universal deep-sea fauna spread all over the floor of the ocean." He could show, though, that as depth increased there was a general increase in the number of species common to adjacent depth intervals (Murray used intervals of 500 fathoms for comparison), that is, as depth and the stability of the environment increased, so did the bathymetric range of the benthic organisms. Zonation was sharp only in shallow water, where the environmental fluctuations were greatest.

These seemingly disparate facts and interpretations were united by Murray's belief that the oceans had had a far shorter and more varied faunal history than Thomson would have admitted. His study of the sediments convinced Murray that the present state of the deep sea dated only from the Mesozoic. Up to that time the deep sea had been warm and anoxic, thus virtually azoic, although life had thrived in the warm shallows and at the surface of the open sea. As the seas cooled the deep ocean was colonized over and over again (accounting for the large number of species and genera) by animals from the rich feeding areas at about 100 fathoms, a depth below the effect of climatic changes where food was particularly abundant. The species from this "mud line" colonized the high latitudes as their shallow-water faunas perished and gave continuity to the faunas of high southern and northern latitudes, although mainly at the generic rather than the species level. It was clear too that "living fossils" could not occur in the deep sea because of its climatic history (they were found instead in the tropical shallow waters). Thus the failure of Thomson's expectations became explicable as part of a general theory of faunal history.

Murray had at his command more solid information on the distribution and abundance of animals than any combination of his predecessors might have assembled. He was, in general, rather cautious about generalizing (on the same grounds as Thomson had been) because it was often difficult to establish with certainty that animals had come from the bottom at a stated depth rather than from somewhere else. I have shown before (Mills, 1972) that his caution was more than warranted; for example, of the 63 species of presumably benthic amphipods described in Stebbing's (1888) *Challenger* monograph, only 44 have proved to be benthic. The sampling efficiency of the trawls used on the *Challenger* was very low in deep water, so that only 7 of the 19 species of amphipods taken below 2000 meters were actually benthic. Most of the rest of the small, mobile epifauna was poorly sampled also, partly because the small, light organisms were washed out as the nets emerged open through kilometers of water. These biases are a major deficiency of the *Challenger* data. Murray was well aware of the effect such biases might have.

Despite his problems in evaluating the data, Murray's summary deserves more attention than it has received. He was the first, as I have shown, to detect the high species diversity in the deep sea. His tabulations of data were a great advance over the qualitative presentations of his predecessors (Murray, 1895b, p. 1430), and he appears to have been the first marine scientist to use simple statistical comparisons to arrive at zoogeographic conclusions.

If Murray had taken the extra step of extending his calculations to the distribution of animals with depth, he would have discovered a pattern of faunal change that was first stated quantitatively nearly 70 years later (Vinogradova, 1962). Murray seems to have outstripped his times, and certainly the statistical resources of his era. The scientific results of his work on deep-sea faunas were practically without influence. Late nineteenth century biological oceanography repeated the pattern of the *Challenger* years without adding significantly to theoretical knowledge of life in deep water. Murray's fashionable nineteenth century inductivist belief that "whenever science is enriched by a large addition of new facts, a change in theoretical views invariably follows" was a statement of faith rather than a description of the basis of scientific change.

5. Deep-Sea Fauna in North American Oceanography During the Nineteenth Century

Nineteenth century North America did not have a *Challenger* expedition, but there were striking parallels in the development of marine science in the United States and in Britain during 1850 to 1870. This was in large part due to the commerce of ideas across the North Atlantic, although Schlee (1973) has described striking differences in popular and governmental attitudes toward science. Although government support of science was always grudging and sporadic in Britain, there was a doubly strong resistance to nonpractical science in the United States, partly on the basis of the belief that government support of science was unconstitutional. There was also a powerful sentiment that only the aspects of science that aided commerce, such as charting and hydrography, were worth doing at all (Schlee, 1973, Chapter 1).

In this unpromising context the United States Coast Survey, which had been established in 1807, was for many decades the main agent for the study of deep-water animals. During the early years of the Coast Survey its second superintendent, Alexander Dallas Bache (1806 to 1867), was in charge of an extensive survey of the Gulf Stream and its environs from the Straits of Florida to the New England coast. The sediment samples from this work were received first by J. W. Bailey in 1848, with results discussed in an earlier section. In 1857 the accumulating collections became the responsibility of Louis François de Pourtalès (1823 to 1880), a Swiss student of Louis Agassiz and nominally the director of the Coast Survey's tidal division, who eventually published a definitive map of the sediments off the east coast (Pourtalès, 1870) based on the Coast Survey's collections (Scheltema and Scheltema, 1972).

Louis Agassiz (1807 to 1873), already an eminent naturalist, came to North America in 1846 for a short visit and stayed for the rest of his life as professor of zoology and geology at Harvard, where his scientific and public influence was very great (Lurie, 1960, 1962). In summer 1847 Agassiz spent some time with a Coast Survey party in Massachusetts Bay, then again off Nantucket in 1849, and in the Florida Keys in 1851. During the later trips he was accompanied by his young son Alexander (1835 to 1910), who had joined his father in America in 1849 (Murray, 1911). The links between Pourtalès' work and Louis Agassiz's ideas are not fully documented, but the outcome of the Coast Survey's work up to 1872 shows the effect of both personalities. Alexander Agassiz later transcended both in the rigor and the extent of his work.

Pourtalès' early dredging showed that there were significant discoveries to be made in deep water. During a Gulf Stream survey about 1853 he discovered foraminifera, coral skeletons, and dead shells in sediments from 1920 m, and, to his surprise, living mollusks at 915 m, somewhat beyond the limit that Forbes (1844b) had suggested for life in the depths (Pourtalès, 1854). But it was not until 1867 that this work was continued in a significant way. In that year, using the U.S. Coast Survey steamer *Corwin* in the Straits of Florida, Pourtalès dredged to 270 fathoms (494 m), and, as he said "the highly interesting fact was disclosed, that *animal life exists at great depths, in as great diversity and as great an abundance as in shallow water*" (Pourtalès, 1867). The fauna at the greatest depths appeared to be mainly forams, but in shallower water a fragment of a *Pentacrinus* came up with the bryozoans and corals. It appeared to Pourtalès that zones of depth could easily be established by using the sessile or slow-moving animals as indicators. He also observed that because there were no plants in the depths, the animals there must be carnivores. Although the sounding line had shown only the presence of forams, Pourtalès was convinced that a rich fauna of higher invertebrates was available to the dredge (Pourtalès, 1867, 1869).

In the next two summers Pourtalès, accompanied part of the time by Louis Agassiz, extended this work, using the steamer *Bibb* in the same region. During 1868, using a winch for the first time, they dredged along six transects in the region between the Florida Keys and Grand Bahama Bank. Animals were abundant down to 517 fathoms (946 m), although they tended to be small at the greatest depth, and a stalked crinoid identified by Agassiz as Sars' *Rhizocrinus lofotensis* emerged from 237 to 306 fathoms (433 to 560 m) (Pourtalès, 1868a–c). The striking sea urchin *Pourtalesia miranda*, another of the discoveries, was closely related to a group of Cretaceous urchins according to the younger Agassiz (A. Agassiz, 1869), who took the opportunity of publication to repeat his father's beliefs that marine distributions should reflect the course of ancient currents and that Cretaceous species succeeded those of the Tertiary as depth increased in the sea.

The last season's work was summarized by Louis Agassiz (1869), who used it as a vehicle for his fertile ideas. Although *Bibb* had dredged to 800 fathoms (1555 m), there were no striking zoological discoveries like *Rhizocrinus*. Nonetheless, Agassiz noted that there was an abundance of animals in deep water, contrary to expectations, and that the fauna was markedly zoned in and around the reefs. Once below the reefs the sediments were like nothing he had seen on land; thus he inferred the permanence of the continents from the 200-fathom line upward. He could also see in outline a way of tracing the history of the western Atlantic marine fauna by extending lines of dredging north, then across the Gulf Stream off New England, and by dredging on each side of the Isthmus of Panama. The first would show the extent of the Gulf Stream in early times, for there had to be a correlation between the types of animal in the relatively constant deep water and the physical conditions adapted to their existence (note the order of adaptation, characteristic of Louis Agassiz's philosophy). By working off Panama the role of the North Equatorial current could be established, for if the same fauna existed on either side of the isthmus, it would be evident that the current had once swept into the Pacific.

The dredgings from *Bibb* had been done with the approval of Agassiz's friend Benjamin Peirce (1809 to 1880), superintendent of the Coast Survey after Bache's death in 1867. Peirce had the problem of keeping the Survey going despite the completion of much of its near shore charting on the east coast. His solution was to turn to terrestrial surveying and to begin marine charting on the west coast. To this end, the survey's ship *Hassler* was transferred to California in 1871 and Louis Agassiz was invited to join the voyage around the Horn for an expedition of deep-sea research. For a variety of reasons the ship was unsuitable and the dredging equipment nonfunctional, so Agassiz's hopes of startling discoveries were not realized. The most significant outcome was actually a letter to Peirce, written by Agassiz before the voyage, outlining his predictions about the deep-sea fauna (L. Agassiz, 1872), for it was among the last powerful expressions of Agassiz's zoological philosophy during a time that he was seriously troubled by the power of Darwinism (Lurie, 1960). The powerful links that Agassiz discovered between animal morphology, development, succession in time, geographic distribution, and the nature of the physical environment made it possible for him to predict that the deep sea would contain archaic fishes, nautiloids, belemnites, eryonid crustacea, and many other representatives of earlier creations. This letter is a striking landmark and was made even more noteworthy because deep-sea work in the United States was never much concerned with living fossils after it. Interests and events had shifted behind Agassiz's back; when he died in 1873, deep-water research (indeed all of biology) took quite different directions under different personalities.

The main significance of the *Hassler* voyage was not what Agassiz expected. Although, as I discussed earlier, it filled W. B. Carpenter with foreboding that the great British expedition he envisaged would be beaten away from the quay side, significant deep-sea work had come to a halt in the United States in 1869 and would not be resumed until 1877. Jeffreys' deepest dredging, and then the *Lightning* and *Porcupine* cruises, coincided with Pourtalès's work on *Corwin* and *Bibb*, but no American *Challenger* expedition followed. Instead, the most significant event during the 1870s was the formation of the U.S. Fish Commission in 1871 under Spencer Fullerton Baird (1823 to 1887), for it encouraged two decades of shallow-water marine biological work [see summaries in Rathbun (1892) and Schopf (1968)] that was only improved upon in the 1960s. Farther north, in the newly established Dominion of Canada (still *terra incognita* to the historian of science), a few hardy dredgers like Robert Bell (1841 to 1917), J. W. Dawson (1820 to 1899), and especially J. F. Whiteaves (1835 to 1910), encouraged by correspondence from Britain or by the occasional visit (Anonymous, 1871a), dredged in shallow water for the Geological Survey or the Department of Marine and Fisheries (Whiteaves, 1872a,b, 1873, 1874a–c, 1901). Deep-sea biology in North America resumed only when Alexander Agassiz turned his interests from business and echinoderm systematics to the practicalities of sampling in deep water.

The younger Agassiz (Fig. 10), who took degrees in engineering and natural science at Harvard University in 1857, for a time helped his father in a number of teaching enterprises and in systematic work on marine invertebrates. In 1863 he

Fig. 10. Alexander Agassiz (1835 to 1910) on the deck of the U.S. Fish Commission steamer *Albatross* in the tropical Pacific about the turn of the century. Agassiz, the dominant North American figure who was involved in deep-sea exploration from 1877 until his death, introduced wire rope into dredging and made many other technical advances, including the design of closing plankton nets. [Photograph by C. A. Kofoid, from Herdman (1923).]

worked as a mining engineer in the Pennsylvania coalfields and then from 1867 to 1868 as the superintendent (later president) of the Calumet and Hecla copper mines on Lake Superior. It was his experience in mining engineering, his wealth, his early experiences in natural history, and considerable personal unhappiness that made Alexander Agassiz a notable influence on American deep-sea biology during the last three decades of the nineteenth century. Agassiz first met Wyville Thomson just after the *Porcupine* expedition of 1869 when ill health had driven him with his family to Europe for recuperation. During a year in Europe he visited museums, advanced his research on the taxonomy of sea urchins, and kept in touch with Thomson, a kindred spirit, on deep-sea exploration and echinoderm taxonomy. When *Challenger* visited Halifax during 1873, Agassiz went up from Boston by steamer to renew acquaintances. Later the same year both his father and his wife died. He seems to have been driven by grief and loss into extensive travels for research that continued for nearly 40 years (Murray, 1911).

The most significant of Alexander Agassiz's early work was brought about by an invitation from Peirce to use the U.S. Coast Survey steamer *Blake* for dredging, beginning in the winter of 1877. Agassiz had the ship refitted at his own expense, drawing on his experience with mining technology. His most important step was to replace the hemp dredging ropes with wire rope (ca. 0.95-cm diameter), which took up only one-ninth of the space and was much easier to handle. With this change dredging and trawling were greatly speeded. Whereas *Challenger,* at best, could do one very deep station in a day, Agassiz found he could do several. Collaborating with *Blake*'s captain, he also equipped the ship with a double-beam trawl (the Sigsbee or Blake trawl) and a strengthened, modified naturalist's dredge that came to be called the *Blake dredge*. A specially designed Kelvin-type sounding machine, Miller–Casella maximum–minimum thermometers, and a variety of other pieces of gear became part of the ship's standard equipment (Sigsbee, 1878; A. Agassiz, 1888b) (see also Table I).

During the winters of 1877/1878 and 1878/1879, Agassiz worked with *Blake* in the Straits of Florida, Gulf of Mexico, and Caribbean Sea. In 1878 they trawled to 1920 fathoms (3567 m) and collected abundant animals everywhere, including two genera of crinoids, although little that had not been familiar from the *Challenger* collections (A. Agassiz, 1878a,b; A. Agassiz and Pourtalès, 1878). This impression was reinforced in Agassiz the next year, when he wrote (A. Agassiz, 1879):

> I was greatly struck with the large number of species which, if not identical, are at least closely allied to those brought home by the "Challenger"; and I was specially disappointed at the absence of types not already collected by the great English expedition. I think it can be fairly stated that the great outlines of deep sea fauna are now known, and that although many interesting forms will undoubtedly be dredged in the shallower waters, between 100 and 300 fathoms, we can hardly expect to add materially to the types discovered by the dredging expeditions of the last ten years. As has been well said by Mr. Moseley of the "Challenger", it becomes somewhat monotonous to find constantly the same associations of invertebrates in the deeper hauls, and it is only in the shallower waters that it is possible to keep up one's enthusiasm after a few months' work.

Despite the monotony (from which he soon turned enthusiastically to work on plankton and coral reefs) there were many less notable discoveries. For example, pteropod remains contributed to the calcareous oozes of the Caribbean, and land vegetation was often found in deep water (A. Agassiz, 1888b, Vol. 1, p. 291):

> It was not an uncommon thing to find at a depth of over one thousand fathoms, ten or fifteen miles from land, masses of leaves, pieces of bamboo and of sugarcane, dead land shells, and other land *debris* . . . . We frequently found floating on the surface masses of vegetation, more or less waterlogged and ready to sink. The contents of some of our trawls would certainly have puzzled a palaeontologist; between the deep-water forms of crustacea, annelids, fishes, echinoderms, sponges, etc., and the mango and orange leaves mingled with branches of bamboo, nutmegs and land shells . . . he would have found it difficult to decide whether he had to deal with a marine or a land fauna. Such a haul from some fossil deposit would naturally be explained as representing a shallow estuary surrounded by forests, and yet the depth might have been fifteen hundred fathoms. This large amount of vegetable matter, thus carried out to sea, seems to have a material effect in increasing in certain localities, the number of marine forms.

In addition, he felt justified in stating that the deep-sea fauna (with local variations) began between 300 and 350 fathoms (549 to 640 m) and that it was quite uniformly distributed, a conclusion that his collaborator W. H. Dall (1879) did not hesitate to dispute, at least as far as the distribution of the mollusks was concerned.

The third and last cruise of *Blake* in June and July 1880 was a northern one from Newport, Rhode Island and Georges Bank south to Charleston, sounding and dredging along transects from about 100 fathoms (183 m) out to a maximum of 1632 fathoms (2985 m). During that cruise Agassiz tried out a new plankton collecting device designed by Sigsbee to solve the problem of whether an "intermediate fauna" of planktonic organisms lived between the top 100 fathoms and the abyssal depths (A. Agassiz, 1880; Sigsbee, 1880). He concluded that the surface plankton "only sinks out of reach of the disturbance of the top, and does not extend downward to any depth" and that the food of deep-water benthic animals must of necessity be the remains of surface plankton that sank, supplemented, of course, in some areas by terrestrial or littoral vegetation. These concerns and many others were given emphatic form in Agassiz's volumes *Three Cruises of the "Blake"* published in 1888. This work, like Thomson's *Depths of the Sea* (1873), was based on a short series of expeditions and, like the earlier work, was an attempt to summarize its generation's state of knowledge. It was perceptibly influenced by Moseley's *Notes by a Naturalist* (1880), the most scientifically interesting of the popular accounts spawned by the *Challenger* expedition.

Three Cruises of the "Blake" is written in a choppy, brusque, ungracious style, with repetitions and contradictions. Agassiz was obviously preoccupied while writing his awkward masterpiece; nonetheless, he left a document that deserves more study in its nineteenth century context than it has received. It not only summarized the results of the cruises but also focused attention on refinements in equipment, the problem of food supply to the deep sea, and the enigma of the missing "intermediate fauna." In these there is an interesting mixture of the new and the old. For example, Agassiz gave qualified support to the continuity of the chalk but saw clearly that living fossils (certainly those of Paleozoic age) were absent in the deep sea. The nature of the sediments and the proximity of food from shallow water or the shore was critical in governing the types of animal in local habitats, but on a broader scale he believed that it was the great ocean currents (as his father had said) that governed the dispersal of animals under the constant conditions of the deep ocean.

Many of Agassiz's observations on food supply to deep water were not original, but they focused attention in a way that is still recognizable in twentieth century literature (e.g., Menzies, 1962). Agassiz knew that Jeffreys had observed terrestrial debris on the bottom during his dredgings at Shetland but had proposed that it was phytoplankton, probably after being eaten by animals, that was the food base of the deep sea (Jeffreys, 1869/1870, 1881). He found it possible to agree with Thomson that a "sort of broth" of decomposing organic matter might be fit nutrient for protozoa and sponges but that the larger animals, for lack of better sources, must depend on the rain of slowly decomposing plankton settling into deep water or on the sinking of terrestrial debris, where it was present in the enclosed seas or along coasts. It was this concern that made the search for the "intermediate fauna" so important in Agassiz's later work, for if it did not exist the other sources were the only ones possible.

6. The Problem of Deepwater Plankton

Near the end of the *Challenger* expedition John Murray had attempted to collect deepwater plankton by making vertical hauls with an open surface townet

Table I
Equipment on Major Oceanographic Vessels, 1872 to 1910[a]

	Challenger (1872 to 1876)	*Blake* (1877 to 1880)	*Albatross* (ca. 1897)	*Valdivia* (1889 to 1899)	*Princess Alice II* (1898 to 1910)
Ship	69 m (226 ft), 2306 tons; converted RN corvette. Auxiliary steam; 6 scientists, 243 crew	U.S. Coast Survey steamer, 43 m (140 ft), 350 tons; 3 to 4 scientists; crew?	71 m (234 ft), 1074 tons specially built; 2 engines—twin screw; iron hull; 2+ scientists, 80 officers and crew	98 m (320 ft), 2176 tons; converted Hamburg–Amerika liner; 13 scientists + assistants, 48 crew	73 m (240 ft), 1420 tons; specially built for oceanography; 7 to 8 scientists (maximum), crew 60
Ropes, cables	Hemp rope, 2.4-, 2.0-, + 1.6-cm diameter, 7300 m total; 10,980 m, 0.8-cm-diameter sounding rope	10,980-m, 0.9-cm-diameter steel cable with hemp core (in 2 coils); piano wire for sounding	0.95-cm ($\frac{3}{8}$-in)-diameter steel cable, total of ca. 14,640 m in shorter lengths (longest ca. 7000 m)	10,000-m steel cable 10- and 12-m-diameter sections	12,000-m tapered steel cable, maximum diameter probably ca. 1 cm; fine steel sounding lines
Sounding gear	Rope leadline + steam sounding winch—modified *Hydra* sounding machine; also cup lead	Modified Kelvin sounding machine, steam driven, with accumulators (Sigsbee sounding machine)	Sigsbee sounding machine, Tanner sounding machine (hand) for shallow depths	LeBlanc (steam driven) and Sigsbee (electrical) machines; bottom corers on line	Various designs, some original based on Kelvin and Sigsbee
Thermometers	Miller–Casella maximum–minimum protected; Siemen's electrical briefly used	Miller–Casella maximum–minimum; tried Siemen's electrical 1881	Negretti–Zambra reversing thermometers with propeller release; Miller–Casella maximum–minimum	Maximum–minimum and Negretti–Zambra reversing thermometers, Siemens electrical thermometer with 750-m cable	Negretti–Zambra reversing with messenger release (often on bottles)

Water bottles	Slip water bottle with bottom release or intermediate-depth release; Buchanan stopcock bottle for gases	Sigsbee water cup (propeller-closed water bottle)	Sigsbee water bottle, Kidder–Flint bottle (propeller-closed, tight-fitting valves)	Sigsbee, Meyer, and Pettersson bottles (last for gas analysis)	Richard reversing bottle; Buchanan bottle (for gases); microbiology sampler; Richard mercury bottle (for gases)
Current meters	Drogue available, not used	Moored current meters used in other investigations of Gulf Stream	—	—	Current drifters; moored current meters
Plankton nets	Surface and vertical nets nonclosing, 30 to 46-cm-diameter; dip nets	Surface tow nets and dip nets; Sigsbee plankton trap (cylinder)	Standard tow nets; 2 types of closing net (see text)	Open vertical nets with glass buckets; Hensen net; Chun net with propeller closing	Many, including surface tow nets (one quantitative); 9 × 9 m vertical, Bourée bathypelagic, Giesbrecht, etc.; midwater trawls, high-speed surface trawls; depth gauge on net, 1913
Benthic equipment	Naturalist's dredge, 1.4 × 0.38 m with swabs; mostly used single-beam trawl	"Blake dredge"–modified naturalist's dredge with tangles; Blake (Sigsbee) double-beam trawl	Blake dredge with tangles; oyster dredge, and others; double-beam trawl (Sigsbee); Tanner trawl (single beam)	Blake dredge; tangle dredge; double-beam trawl	Blake or naturalist's dredge; Blake trawl extensively used; otter trawls and trammel nets; longlining
Meteorology	Standard observations	—	—	Standard observations—T, winds, barometer, humidity, etc.	Standard + kites + balloons to 25,000 m

Table I (*Continued*)

	Challenger (1872 to 1876)	*Blake* (1877 to 1880)	*Albatross* (ca. 1897)	*Valdivia* (1889 to 1899)	*Princess Alice II* (1898 to 1910)
Special equipment	Hydrometers for salinity; donkey engines for winches; natural history and chemical labs; photo studio; Siemens photometer with sensitive paper	"Hilgard salinometer"–hydrometer with cup; steam winding engine; no chemical lab	Hilgard salinometer; steam winches; various fishing gear during U.S. Fish Commission work; 2 main labs; electrical lighting, sperm oil emergency lights	Hydrometer & refractometer; fishing gear, hooks, lines, etc., harpoons for whales; main lab + photographic, chemical, and bacteriological; electrical interior, deck and submersible lights	Steam winches; fishing gear—various, harpoons, etc.; 2 labs; freezer and still; electrical deck + interior lighting; various photometers used (film)

[a] For details, see Thomson (1878), A. Agassiz (1888b), Tanner (1897), Chun (1900), and Richard (1934).

from 1000 and 500 fathoms (1830 and 915 m). Because the deeper hauls contained species not seen in the shallower one, he inferred the existence of a deepwater plankton. His was the only systematic attempt until 1880 to determine the nature of deepwater plankton, although accidental collections suggested now and again that a rich and sometimes bizarre fauna might exist below the top 100 or 200 meters. For example, during a circumnavigation by the German ship *Gazelle* in 1874 to 1876 (S.M.S. *Gazelle* 1888/1890), Studer (1878) had taken siphonophores from the sounding wire at a position corresponding to 800 to 1500 fathoms (1463 to 2744 m) but had not found them in his surface collections to 200 fathoms (366 m). Alexander Agassiz (1879) had a similar experience in 1879 when the *Blake*'s sounding wire was festooned with siphonophores from some uncertain depth in the Caribbean Sea.

Determined to throw some light on the problem of this "intermediate fauna," Agassiz and Sigsbee collaborated to design a device that could be lowered to any depth, opened, and then closed and recovered (A. Agassiz, 1880; Sigsbee, 1880). This was Sigsbee's gravitating plankton trap (Fig. 11*a*), a brass cylinder containing a net that could be opened by a messenger at depth and then slid down the wire for any desired distance to capture plankton. During the *Blake* cruise of summer 1880 between Georges Bank and the Carolinas, Agassiz tested this device by assuring that it did catch plankton near the surface. Then at two stations, one south of Cape Cod and the other about 80 nautical miles southeast of Cape Lookout, he conducted more rigorous tests by using the trap at 5 to 50, 50 to 100, and 100 to 150 fathoms (9 to 91, 91 to 183, and 183 to 274 m). In both cases, although there was catchable surface plankton, the trap caught nothing below 100 fathoms and only a sparse fauna (a few radiolarians or crustacean larvae) between 50 and 100 fathoms. Agassiz wrote, "The above experiments appear to prove conclusively that the surface fauna of the sea is really limited to a comparatively narrow belt in depth, and that there is no intermediate belt, so to speak, of animal life between those living on the bottom, or close to it, and the surface pelagic fauna." He maintained these views for over 20 years in the face of increasing evidence for a varied and widespread midwater fauna.

The main opposition to Agassiz came from Europe, based first on indirect evidence and then on more and more sophisticated investigations of the water column. Although Studer's observations during the *Gazelle* voyage were widely known, it was the collections made during the Italian circumnavigation in the corvette *Vettor Pisani* (1882 to 1885) that provided stimulus for nearly three decades of work on midwater animals. During the Italian cruise (Chierchia, 1884, 1885) the ship's captain, Palumbo, designed a closing device that used the propeller-driven reversing mechanism of a Negretti–Zambra thermometer to collapse the mouth of the net. Using this original but imperfect device (which apparently left the net partially open at least some of the time) on the sounding line, Chierchia and Palumbo recovered animals from as deep as 2300 m. Curiously, it was not these collections but a number of siphonophores taken on the sounding line (nominally at 900 m) during January 1884 that excited the interest of the zoologist Carl Chun (1852 to 1914) of Königsberg when the specimens reached him after the voyage.

In August 1886 Chun went to Naples to collect deep-water plankton and continue taxonomic studies of siphonophores. At the Stazione Zoologica his friend,

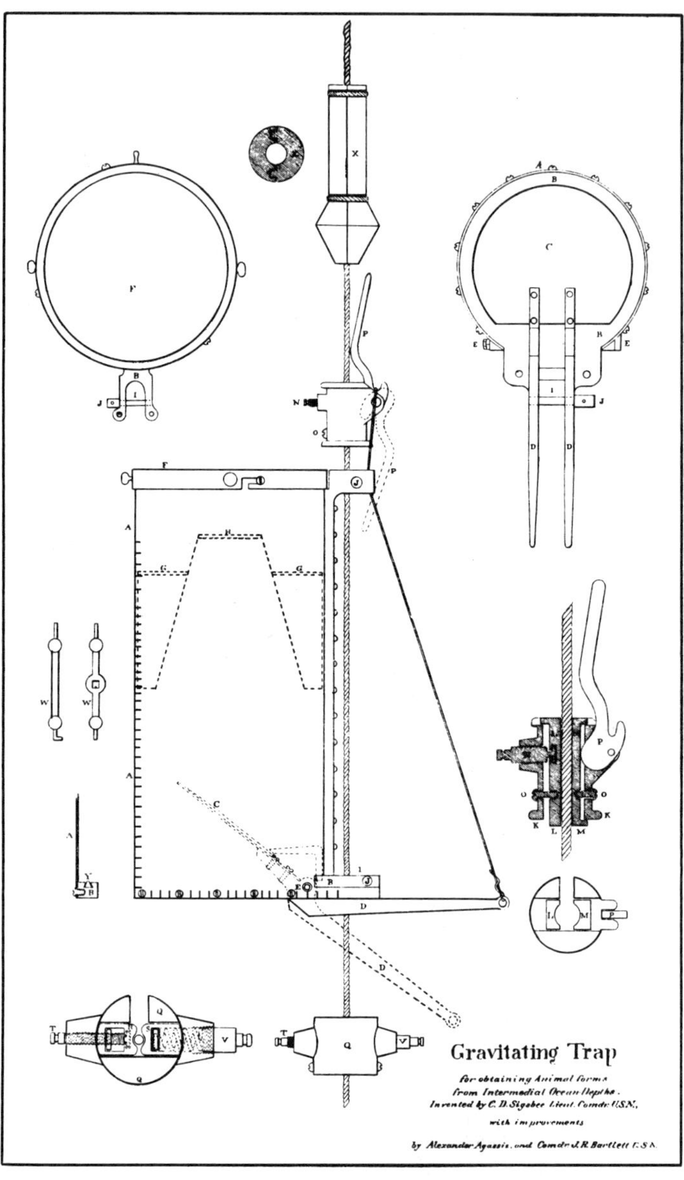

(a)

Fig. 11. Two kinds of plankton collecting device used by Alexander Agassiz. (*a*) Sigsbee's plankton trap, a device designed by Lieutenant-Commander C. D. Sigsbee and first used during the *Blake* cruise in 1880. The trap, a metal cylinder (*A*) of diameter approximately 22 cm, filled with filtered water, had a gauze mesh (*H*) inside and was lowered closed to any depth attached to a stop (*N*, *O*, *P*) on the wire. Then a messenger (*X*) opened the door (*C*, *D*) and released the trap, which slid down the wire (usually about 90 m) until it hit a stop (*Q*) that closed the door again. [Figure from Sigsbee (1880).] (*b*) The first Tanner closing net, as used by Alexander Agassiz in the eastern tropical Pacific in 1891. It was made of 0.5-in.-mesh lined with mosquito netting for half its length. The cod end was closed when a messenger released the weights (*w*), closing the lines (*l′*) around the net; *r′* is a frame to which the lines and a large weight (*b*) are attached. [Illustration from Tanner (1893).]

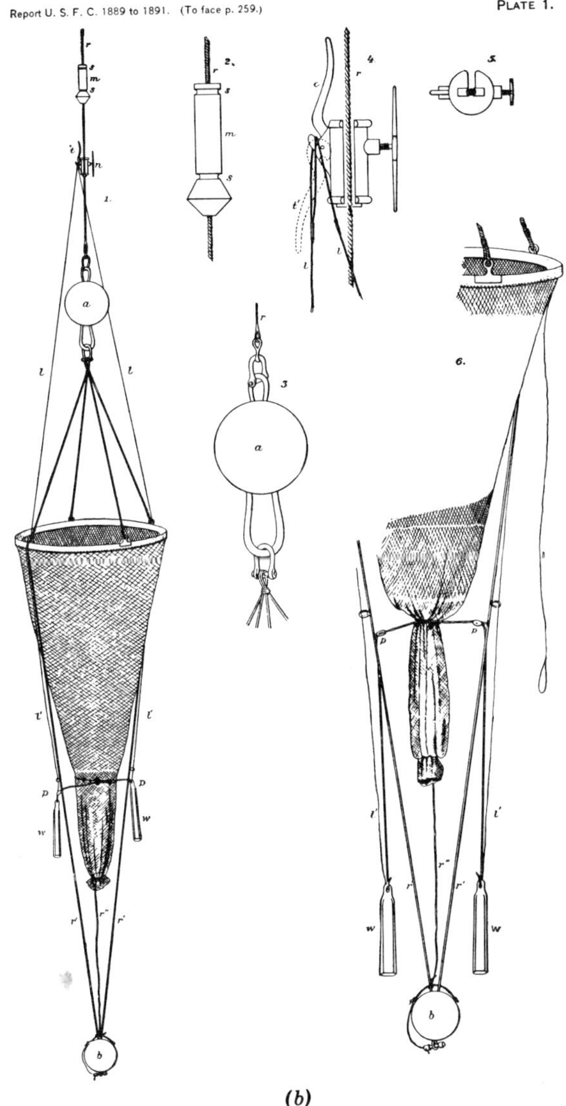

(b)

Fig. 11. (*Continued*)

the engineer Eugen von Petersen, constructed a net based on Palumbo's original design but improved sufficiently that it came to be used extensively until the turn of the century (Fig. 12). Using it, Chun made deep-water hauls to 1400 m around the Ponza Islands, Ischia, and Ventotene during his first season, towing the 1 to 1.5 m net for 15 to 20 minutes at a time. His results were the basis of a monograph, *Die pelagische Thierwelt* (Chun, 1887) in which Chun summarized his work and established a broader framework for studies of deep-water plankton. On the basis of plankton hauls off Naples, he concluded that there was a rich pelagic fauna of medusae, ctenophores, siphonophores, chaetognaths, polychaetes, copepods, euphausiids, decapods, and invertebrate larvae down to at least 1400 m. This

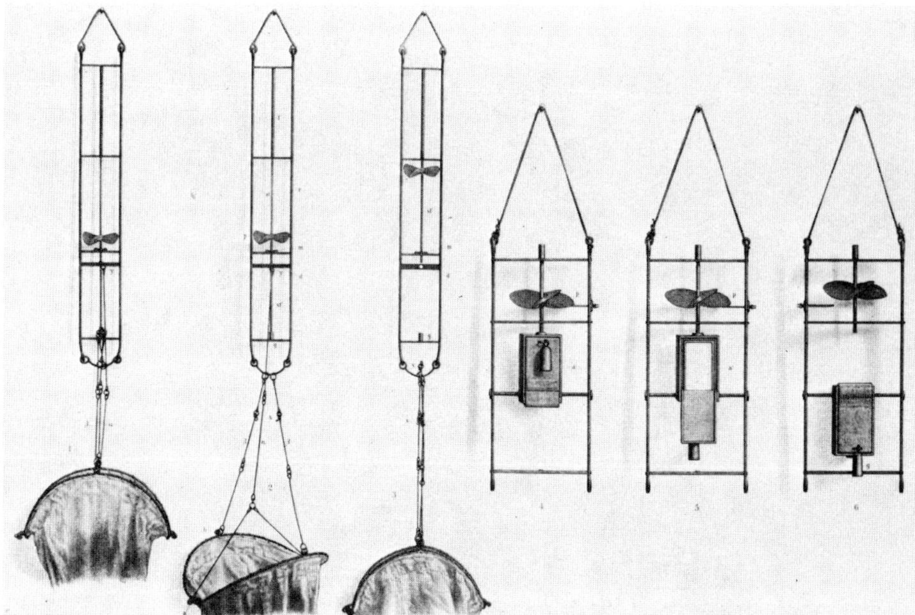

Fig. 12. On the left is the closing net (originally designed by Palumbo) that was first used by Carl Chun in 1886 to study the deep-water zooplankton of the Mediterranean. The net was lowered closed; as it descended, a propeller-driven mechanism opened the jaws; then, as the propeller continued to turn during a horizontal or vertical tow, it eventually released the central cables, closing the net for the return to the surface. The device on the right contained a photographic plate and was opened and closed by a similar mechanism to determine the depth to which light penetrated in the sea. [Illustration from Chun (1887).]

raised at least four significant issues that he believed further study could help to resolve: (1) how the deep sea was colonized; (2) how the animals are adapted (in morphology and physiology) to the uniform conditions of the deep sea; (3) how deep water animals obtain food; and (4) how these animals reproduce. Chun speculated that, given the existence of a rich intermediate fauna, it was quite possible that that fauna, rather than the surface plankton (Moseley's suggestion) had colonized both the surface waters and the deep-sea floor and that the existence of the intermediate fauna provided an additional food source, supplementing the rain of dead plankton reaching the bottom. Migrating plankton that fed at the surface on algae would provide food for the intermediate fauna; the latter, in turn, could add to the sinking carcasses that eventually reached the deepwater benthos.

Chun's idea that a "ladder of vertical migrations" (Menzies, 1962; Vinogradov, 1970) carried food to the deepwater benthos must have offended the practically minded Agassiz, for he made a point of reviewing the monograph in detail (1888a). Agassiz's criticisms were based on the belief that only a surface plankton existed [to a depth of 100—ca. 300 fathoms (183 to 549 m) depending on the local conditions] and that above the benthos in deep water was a narrow layer containing mobile pelagic animals that might occasionally venture into the virtually azoic intermediate depths. The surface plankton might vary in depth in accordance with disturbances at the surface (strong light, summer heat, wind), but it seldom, if ever, met the deep-water near-bottom fauna except in the rapidly shoaling water around oceanic islands, along the continental slopes, and in seas like the Mediterranean, where deepwater temperatures were abnormally high. These, of course, were just the sorts of place Chun had sampled. Thus Agassiz did not seriously criticize Chun's sampling methods (although he was convinced of their inadequacy), but rather the generality of his conclusion that a midwater fauna would occur in the open oceans as well as near shore and in enclosed seas. Until these criticisms could be overcome, and until the surface species became so well known that the differences between surface faunas could not be mistaken for an intermediate fauna, Agassiz believed that there must be healthy skepticism about the existence of deep-water plankton and fish.

In 1887 and 1888 Chun (1889) improved the closing device (its effectiveness had been questionable in 1886), sampled between the Bay of Biscay and the Canary Islands, and found a midwater fauna between 500 and 1600 m. Victor Hensen's (1835 to 1924) Plankton Expedition on the *National* in the northern Atlantic during 1889 (Hensen, 1895) used Petersen–Chun closing nets in vertical hauls from 3500 meters and found a sparse fauna of copepods and radiolarians, whereas more abundant animals of several groups were taken at lesser depths.

Agassiz's (1892) response to these discoveries was cool. Some of the samples had been taken near land. He also complained that the closing net was not adequately described, so its performance could not be evaluated. In addition, he believed that the deepwater plankton might be merely sinking carcasses from the surface waters. Clearly, a more definitive test was required, and this Agassiz soon undertook during a cruise in the eastern tropical Pacific using the U.S. Fish Commission steamer *Albatross* (Tanner, 1897) (Table I).

Early in 1891 Agassiz joined *Albatross*, which he was operating at his own expense, in Panama for oceanographic studies off the Pacific coast of Central America, around the Galápagos Islands, south of Mexico, and in the Gulf of California. His program was a general one of soundings, serial temperatures, water samples, deep-sea trawling and dredging, and surface plankton tows—all the operations he had begun on *Blake* several years before. But Agassiz included in the ship's equipment a Chun closing net, specially built for the cruise. To his disappointment, it failed to work properly, so in its place Captain Tanner of *Albatross* devised a new closing net that could be lowered open and closed by a messenger after a horizontal tow at any depth (Tanner, 1893, Fig. 9*b*). Tanner's first closing net worked well at the surface and when towed cod end forward, collapsed sufficiently that Agassiz believed it would not collect plankton during a vertical lowering. After a series of test hauls in the surface waters, he began work in much deeper water midway between northern Ecuador and the Galápagos Islands (A. Agassiz, 1891, 1892). Here, in depths between 200 and 1773 fathoms (366 to 3243 m) he found no animals in the closed cod end of the net after 15- to 20-minute hauls at depth. In a more rigorous test on April 8, 1891 in the same area, the closing net captured animals at 100 and 200 fathoms (183 and 366 m) but was entirely empty at 300 fathoms (549 m). A few days later off Acapulco a 300-fathom tow was entirely empty, whereas one at 175 fathoms (320 m) some miles to the northwest contained surface animals. Occasionally a very deep haul trapped an amphipod or a shrimp; if so, the net must have dropped into the near-bottom layer. He concluded that "in the open sea, even when close to the land, the surface pelagic fauna does not descend far beyond a depth of 200 fathoms, and that there is not an intermediate pelagic fauna living between that depth and the bottom, and that even the free-swimming bottom species do not rise to any great distance, as we found no trace of anything within 60 fathoms from the bottom where it had been fairly populated" (A. Agassiz, 1892, p. 55). This evidence was so compelling that in his subsequent cruises in 1899/1900 and 1904/1905 (when admittedly Agassiz's thoughts were shifting to the origin of coral reefs and islands), he seldom bothered to sample below 300 fathoms and, indeed, came to call the plankton at and around that depth the "intermediate fauna" to distinguish it from the shallowest-living plankton (A. Agassiz, 1899, 1900, 1902, 1905).

Agassiz's *idée fixe* did not much affect his American colleagues (cf. Good and Beane, 1895), nor did it impress the energetic Chun, who was involved in organizing his own deep-sea expedition on the converted passenger liner *Valdivia* (Chun, 1899, 1900; Deutsche Tiefsee Expedition, 1898/1899; Sachse, 1899, 1925; Schott, 1899) (Table I). The *Valdivia* cruise was a nationalistic expression by a prosperous, confident, and increasingly militaristic nation (see Section 1.1 of this chapter), but to Chun it was also a means of returning to his basic questions about the nature of life in the deep sea and of examining deep-sea distributions through vast distances under varied hydrologic regimes. The ship's route, from Hamburg to the African coasts, south to Antarctica, through the Indian Ocean, into the Red Sea, and finally home by way of the Suez Canal and the Mediterranean, made it possible to sample under greatly different conditions and to look for any relations that might exist between the faunas of high northern and southern latitudes where those faunas met somewhere in the equatorial abysses.

The results of the expedition appeared first in an elegantly produced semipopular book, *Aus den Tiefen des Weltmeeres* (Chun, 1900), then in 25 folio volumes (Deutsche Tiefsee Expedition, 1898/1899) that defy a brief summary. For the purposes of this chapter, it is sufficient to say that, among many other things, Chun noted that there was abundant benthos in the deep sea near coasts where terrestrial and littoral plant material could sink. He noted also the very high standing crop of Antarctic benthos and its association with intense diatom blooms (thus confusing biomass and production rates—concepts, of course, that were just being separated at that time). The deep-sea fauna, whether benthic or pelagic, began where light fails and temperature drops to winter levels (although only light was operative in the cold waters of high latitudes).

During the *Valdivia* cruise, Chun and the other scientists made more than 100 closing net samples (using nets based on his original design) to maximum depths of 5000 m. Their surface- and closing-net tows showed that there were no plants below 300 m—in fact, most were in the top 80 m. In the same depths, but perhaps extending to 600 m, was a surface fauna, replaced at greater depths by a distinctive, quite abundant, and very widespread deep-water pelagic fauna that became sparse below about 2000 m. Chun stated categorically, *"Auf Grund der Ergebnisse konnen wir positiv behaupten, dass azoische Wasserschichten zwischen Oberfläche und Meeresgrund nicht existieren"* (Chun, 1899, p. 115). He was more than ever convinced that the intermediate fauna and sinking plankton were the key elements in food supply to all the deep-water areas except those very close to land.

When Agassiz abandoned the problem of the intermediate fauna just after the turn of the century, his views must have seemed idiosyncratic to those such as Chun and the prince of Monaco (see Section 7), whose collections showed beyond any doubt that mesopelagic and bathypelagic animals were a ubiquitous component of the oceanic water column. Why did Agassiz fail to find this deep-water fauna? Not every case can be answered with full certainty, but three factors appear to be involved in Agassiz's peculiar findings: (1) the locations he chose; (2) the design of the gear; and (3) the depths sampled, especially during his expeditions in the Pacific. These factors in combination account for most of Agassiz's results (Mills, 1980a).

Agassiz's first decisive collections were made from the *Blake* using the "gravitating trap" or plankton cylinder designed by Sigsbee (Fig. 11*a*). This was a metal cylinder about 37 cm long and 22 cm in diameter, weighing about 23 kg. Inside the cylinder was a screen of approximately 900-μm mesh, and the upper end was covered with a 416-μm mesh. In use, the trap was filled with filtered, plankton-free seawater and a pair of valves closed. Then it was placed on the wire and lowered to some desired depth. A messenger then opened the valves and released the trap, which slid down the wire [usually 50 fathoms (91 m)] to a stop, where the valves closed. The sample, according to Agassiz's accounts, was carefully examined after being filtered through very fine cloth. Agassiz realized that this device sampled a very small volume. Judging by the sizes I can estimate from Sigsbee's (1880) plan, the mouth area of the trap was 0.038 m^2, and in falling through a column of 91 m it would sample 3.5 m^3. By contrast, a 70-cm net (a common size in zooplankton work) towed vertically the same distance would sample 35 m^3, exactly 10 times as much. In addition, the netting used (the finest about 416-μm mesh) was rather coarse for retaining small zooplankters, which are usually the most abundant at most depths. However, this in itself is probably not enough to give completely negative results at about 200 m without the effect of other factors.

Both Agassiz's thorough tests of Sigsbee's trap were carried out in plankton poor areas. The first, south of Nantucket, just northwest of Atlantis Canyon, was in slope water, a water mass notoriously poor in plankton. The second, in the Gulf Stream southeast of Cape Lookout, must have been in even more sparsely inhabited water. The only reasonable conclusion about these first attempts is that the gear did not sample an adequate volume of water in the plankton-poor waters that Agassiz, by chance, chose for his experiments.

Some different factors apply in the tests that Agassiz conducted in the eastern tropical Pacific early in 1891. Here he used the first of Tanner's (1893, 1897) two closing nets, a standard plankton net of approximately 91-cm diameter, 2.1 m long, of coarse mesh lined with fine mosquito net or bolting silk in the rearmost 60 to 76 cm (Fig. 11*b*). This net was lowered vertically to depth (the pressure of water kept it pressed flat according to Agassiz's observations), where it could be towed horizontally (with careful attention to the wire angle) for 15 to 20 min, and then a messenger released weights to choke off the cod end. Later, finding this design a bit too complex for reliable use, Tanner (1897) redesigned the net on a smaller scale (diameter 60 cm, length 1.5 m), having a metal frame with traveling weights that acted through ropes and pulleys to throttle the cod end. In terms of aperture and mesh size, both designs should have been good plankton catchers, and this was Agassiz's experience in the topmost 90 m or so. Even more than with the gravitating trap, his failure lies in the sampling locations.

The most thorough critical tests of the Tanner nets were made between Ecuador and the Galápagos Islands and off the Pacific coast of Mexico, where the oxygen minimum layer of the eastern tropical Pacific is well developed (Wyrtki, 1966). Between Ecuador and the Galápagos the core of this layer, with oxygen levels of about 0.25 ml/liter, lies at 400 to 500 m, just the depth where Agassiz could capture no plankton. Farther north, south of Mexico, the oxygen minimum layer is 1200 m thick, extending in places to within 50 m of the surface, and although Agassiz did only single-depth tows there at 320 and 549 m, it is not surprising that he found the closed portion of the net empty. Although the oxygen minimum layer is not entirely azoic, plankton abundance is very low and Agassiz's nets were not suited to capture the rapidly swimming, vertically migrating large zooplankton that are tolerant of the conditions in that part of the tropical Pacific. There is little doubt that Agassiz's ingenious gear was employed in two of the most plankton-poor areas he could have sampled. As a consequence, his beliefs about the nature of the deep-water fauna, nearly correct in their own context, were as particular and restricted as the ones he accused Chun of holding.

7. Deep-Sea Biology at the End of the Century

In the decades following the *Challenger* expedition, new expeditions inspired by *Challenger* or based on its results were common. Wyville Thomson himself, as Deacon (1977) describes, saw *Challenger* not as the end of an era, but as the beginning of new directions in British oceanography. After the great expedition he returned to the old problem of contiguous warm and cold deep water between Scotland and Faeroe Bank, each with its distinctive fauna. The physical basis of the separation had to be a submarine ridge, and to explore it Thomson used the well-tried mechanism of representations to the Admiralty from the Royal Society for use of a ship. During summer 1880 the 180-ton paddle steamer *Knight Errant* was employed in sounding the Faeroe–Shetland Channel (Tizard and Murray, 1882). The ridge was discovered and eventually named after Wyville Thomson.

Thomson, who was both ill and deeply involved in difficulties with the *Challenger* collections, hoped to make this the first step in detailed regional studies of the deep sea (Deacon, 1977). He envisioned research on the contiguous cold and warm areas of bottom water, which would involve currents, the bottom sediments in relation to the surface biota, and the nature of the benthic fauna. A small area of abyssal ocean with large physical differences could be studied in detail to provide organisms for the collections of the British Museum and to allow his earlier generalizations about animal distributions to be supported or modified.

The next and last step in this program, another voyage to the Faeroe–Shetland Channel, took place in 1882 in H.M.S. *Triton* under John Murray's scientific direction (Deacon, 1977). Much new hydrographic and zoological material was collected, but Thomson, the guiding spirit of the venture, had died early in the year, leaving the *Challenger* collections to Murray. Thus ended British deep-sea exploration for over 40 years. Its demise was due to a combination of circumstances, the elements of which were, in part only, Thomson's death and Murray's deep involvement with the completion of the *Challenger* reports. Government funding for deep-sea research (or any other kind of pure science) was becoming increasingly difficult to obtain, especially if any continuing commitment was wanted. At the same time the first evidence of overfishing became apparent in northern European seas. Fishery boards or commissions were established to survey the fisheries, and in this climate of interest the Marine Biological Association of the United Kingdom was formed in 1884; its laboratory at Plymouth opened in 1888. Marine science in Britain then took a whole new direction governed by near-shore marine science and the difficulties of the fisheries, not by the curiosity of scientists about problems of deep-sea biology.

It was in France in the 1880s that the British voyages had a reverberatory effect extending well into the next century. In 1878 Jeffreys visited his fellow malacologist Leopold Alexandre Guillaume, the Marquis de Folin, commandant of the port of Bayonne, who had been dredging for several years. With Jeffreys' urging, Folin applied to the Minister of Public Instruction for funds and a ship to explore the deep water of the Bay of Biscay. A commission to oversee a deep-sea expedition was established by the Academy of Sciences under its president, Henri-Milne Edwards, but the actual direction of work devolved on his son Alphonse Milne-Edwards (1835 to 1900) (Milne-Edwards, 1880). The original stimulus was certainly scientific: Jeffreys and Folin wanted to extend the work begun by *Porcupine* in the Bay of Biscay in 1870 by making a close comparison of the deep-water faunas there and in the Mediterranean. There were also good practical reasons for wanting to improve the charts of the area and to provide new information on the abundance of animals for fishermen. However, Folin at least, and probably also the majority of the other French participants, were anxious to improve French prestige through deep-sea exploration, modeling their work on *Porcupine* and *Challenger* (Folin, 1882).

There were four voyages in all, the first during July and August 1880, when Jeffreys and A. M. Norman were invited to join the company on *Travailleur*, a 900-ton paddle-wheel dispatch boat. They trawled or dredged to 2708 m (mostly much shallower) and made a series of soundings in the Bay of Biscay. The biological results mirrored those of *Porcupine*. Jeffreys and Norman were the experts on the fauna, and only their papers on the cruise of 1880 have much solid zoological information (Jeffreys, 1880; Norman, 1880). They were aware of this, for in a letter dated August 28, 1880, Jeffreys wrote to Norman "I thought you would be amused by the way the French Savants have appropriated all the knowledge they gleaned during the cruise of the Travailleur. Never mind. We can afford it!" (unpublished letter, Balfour and Newton Libraries, Cambridge University). Later, however, the confidence of the French Savants increased as the voyages proceeded (on *Travailleur* in 1881/1882 and on *Talisman* in 1883) to more distant parts of the Mediterranean, eastern Atlantic, and West African coast.

The biological results of the *Travailleur* and *Talisman* expeditions (*Travailleur* and *Talisman*, 1888–1927) were issued over a number of years as the taxonomic work was finished by specialists. The majority appeared between 1888 and 1906 as systematic monographs typical of the period. They are unusual, however, in giving a glimpse of French interests in the evolution (no doubt Lamarckian) and zoogeography of the deep-water fauna. Because these volumes have been neglected, it is worthwhile to summarize a few of the interesting (occasionally contradictory) ideas immured in them for many decades:

1. There is a distinctive deep-water ichthyofauna, and some of its members appear to migrate vertically.

2. The deep-water faunas of various parts of the eastern Atlantic are far more similar than are the shallow-water faunas on opposite sides of the Atlantic. The distribution of this fauna is governed by temperature (a suggestion made by Thomson years before).

3. The fauna of the Mediterranean was derived from that of the Atlantic by temperature changes, and some species known as fossils in the Mediterranean area are still living in the Atlantic (noted by Jeffreys during the Shetland dredgings and on the *Porcupine* in 1870).

4. Abyssal mollusks are widespread and originated in the cold waters of high northern latitudes. Very few ancient species are found in deep water, but the species there have distinctive features of size, shell structure, and color.

5. The most ancient species of annelids are widespread, but there are many rarer, localized ones that are in the process of evolution under *"l'influence des circonstances extérieures"* by what is described (in modern terms) as allopatric speciation. The result is "representative species" (a term dating back to Forbes) that have appeared through evolution.

6. The number of species of Bryozoa decreases with depth because the proper habitat, suitable sediment and sites for attachment, become less abundant with depth.

These selections from the scientific results indicate the French biologists' strong interest in zoogeography and evolution, although it is possible to find some flashes of ecological insight as well. In a more detailed study it would be fascinating to see if there are links between the evolutionary ideas of the French biologists and those of Thomson or Jeffreys. Both of the latter used the nature of the deep-sea fauna as a test of evolution. If natural selection worked, Thomson expected to find transitional forms in the deep sea—but he did not and thus concluded that although evolutionary change occurred, it could not be by Darwin's mechanism. Jeffreys (1881) searched for evidence of a struggle for survival in the deep ocean and saw none, perhaps because he thought resources were superabundant and because there was so little sexual dimorphism in marine invertebrates that sexual selection could not operate. The French biologists, of course, were working in quite a different tradition and did not see any particular problem in the application of natural selection—it was simply not relevant to their evolutionism. The context of these ideas deserves study if only because their sophistication and variety is anomalous in the biological oceanographic literature of the late nineteenth century.

Although the effect of the French biologists' theorizing was practically negligible, the effect of the expeditions themselves was not. They were an inspiration to the young prince of Monaco, who eventually combined forces with Jules Richard (1863 to 1945), one of Alphonse Milne-Edwards' young colleagues, to carry out the most technically varied deep-water oceanography of the late nineteenth and early twentieth centuries.

Albert Honoré Charles Grimaldi (1848 to 1922), better known as Albert I, Prince of Monaco, was a striking figure in oceanography whose work extended from the *Challenger* era to World War I. He summarized his approach to the ocean saying that *"on exploite tous les systèmes imaginables pour étudier la faune des mers"* (Albert I, 1896). In addition to wide-ranging studies of deep-sea animals, he made measurements of currents, undertook major charting projects, and conducted meteorological studies with balloons from oceanic islands or ships (Petit, 1970). If any person in post-*Challenger* oceanography surpassed Alexander Agassiz in ingenuity and ambition, it was certainly Albert I, whose major direct contributions were the development of new equipment, the use of older sampling gear in original ways, and the promotion of international science.

Most of the summaries of Albert's life have been purely laudatory rather than analytical (e.g., Richard, 1910, 1934; prefaces to Albert I, 1914; Petit, 1970). It is easy to be distracted by these panegyrics or by the sheer volume of the scientific contributions made by his co-workers (110 reports published between 1889 and 1950). However, Albert I should be evaluated in the light of European science, history, and politics during his time, using the vast material in Monaco rather than seductive secondary sources. In many respects he seems to personify late nineteenth century society, which, in Barbara Tuchman's (1966) vivid words, "was not so much decaying as bursting with new tensions and accumulated energies."

Prince Albert of Monaco began a sea-going career as an ensign in the Spanish navy. After the exile of the Spanish royal family in 1868, he joined the French navy and fought against Germany in the Franco-Prussian war of 1870. During the years of peace following that brief war he bought a yacht, *Pleiad*, renamed *Hirondelle*, a 200-ton schooner, in which he cruised the European seas during 1873 to 1885. In 1884 Albert collected a few surface plankton hauls in the Baltic during a cruise. In the following year he began a series of oceanographic investigations that occupied him for the rest of his life.

Hirondelle was pressed into service as a research vessel in four seasons between 1885 and 1888. In the beginning, the prince's main concern was the circulation of the northern Atlantic, using current drifters. In his second season he turned attention to the sardine fishery of northern Spain and also began trawling at great depths (to 510 m) as well as putting down baited traps. By the end of the season of 1888, the prince and his scientific collaborator, Baron Jules de Guerne, had released 1675 current floats between Europe and Newfoundland, trawled to 2870 m, used closing nets to 1850 m and a pelagic trawl to 2200 m, and experimented with submarine light. *Hirondelle* had been equipped with steel wire in 1887, but it had only a manual capstan, and the deep station at 2870 m in 1888 took the crew 20 hr of continuous work.

Late in 1891 Albert took delivery of a new ship, *Princess Alice*, a 650-ton auxiliary steam yacht, equipped with electrical lighting (pioneered by the U.S. Fish Commission steamer *Albatross* in 1882); special laboratories; a freezer and a still; and most important, steam winches and sounding machine (Albert I, 1891). Its oceanographic equipment was complete only in 1895, but before and after that, until 1897, the ship was in use each summer with various personnel, including the new scientific director, Dr. Jules Richard (who succeeded de Guerne in 1888), an artist (who was a woman on one occasion), the *Challenger*'s chemist J. Y. Buchanan, and various zoologists. They worked from Madeira to Europe in deeper and deeper water, trawling to more than 500 m, setting moored fish traps to 5285 m, making a variety of plankton collections from the surface to 1000 m, and also setting current meters in the Straits of Gibraltar. At the Azores in 1895 the prince found that he could find deep-water cephalopods in the stomachs of whales; thereafter he regularly collected at least the smaller whales and porpoises.

The next series of cruises, the longest (1898 to 1910), was in a new auxiliary steam yacht, *Princess Alice II*, of 1420 tons and 73 m (Table I). In its first season this ship took the prince and a party including Buchanan and the Scottish explorer W. S. Bruce (1867 to 1921) to Spitsbergen for charting and terrestrial exploration. A similar voyage took place the following year; then again in 1906 and 1907 the ship returned to Spitsbergen accompanied by a Norwegian expedition under Captain Isachsen. In the other years, Albert I employed the new ships in increasingly detailed studies of deep-water faunas ranging from the Norwegian coast south to the Mediterranean, the Cape Verdes and the equatorial Atlantic westward toward the Brazilian coast.

Princess Alice II was roomy and well equipped, with a crew of 60 and room for 7 or 8 scientific personnel doing specialized studies, such as Richard's research on surface plankton, or Portier's studies of oceanic bacteria (in 1901 and 1904) (Fig. 13*b*). Its deep trawling at 6035 m southwest of the Cape Verdes on August 6, 1901 was the deepest successful trawl in the Atlantic until 1947, when the Swedish *Albatross* otter trawled in the Puerto Rican Trench. By the turn of the century, Albert and Richard were quite routinely trawling, setting cages and longlining well below 4000 m, although these were depths that they could less successfully sample with horizontally towed closing nets. For deepwater pelagic sampling, they settled on the use of giant, square-framed vertically hauled nets to work out the distribution of plankton and fish (see discussion later). Year by year miscellaneous projects accumulated, depending on personnel and opportunity; these included digestive enzymes of seals, body fluids of arctic animals, glaciology, charting, sediments and soundings, the toxin of *Physalia*, arsenic content of seawater, gases, alkalinity and specific gravity of water (Buchanan's specialty), the radium content of seawater, and upper-atmosphere meteorology. By shortly after the turn of the century, the prince was concentrating on a few major projects, such as plotting bathypelagic vertical ranges of the plankton, and in 1909 he worked for 5 days and nights at one station at 5940 m west of Portugal using all the available equipment at all the appropriate depths. This, no doubt, was the longest and most complete oceanographic station made to that date.

By 1910 Albert considered that *Princess Alice II* was worn out. He replaced it with a fine steam yacht, *Hirondelle II*, 1600 tons, which began work in July, 1911 south of the Canaries and the Azores, then in the Mediterranean. The prince then concentrated almost entirely on the deep-water plankton, using giant nets vertically towed and a new high-speed net to capture fish and rapidly swimming invertebrates. He was able to confirm the vertical migrations of bathypelagic animals that he had inferred much earlier and that Murray and Hjort (1912) had described from the cruise of *Michael Sars* in 1910. These studies came to an abrupt end when, in 1914, just as *Hirondelle II* reached the Azores to begin its summer work, war was declared. The ship returned quickly to Monaco to discharge men of conscription age. The war virtually ended the prince of Monaco's work (except for a few stations in the Mediterranean in 1915), for when peace returned, Albert was too ill to continue. After Albert's death, the completion of the reports was

(a)

Fig. 13. Contrasting styles in sea-going biological laboratories: (*a*) the "natural history work room" of H.M.S. *Challenger*, 1872 to 1876 [Thomson (1878), Fig. 1]; (*b*) laboratory of *Princess Alice II* during cruise in summer 1904. The personnel are, from left to right, L. Tinayre (artist), P. Portier (bacteriologist), and Jules Richard (scientific director). During this cruise in the Mediterranean and tropical eastern Atlantic, Portier, using sterilized reversing bottles, showed that the open sea had relatively few bacteria and that none could be grown from sediment samples collected at abyssal depths (a result at variance with earlier studies) although there were always bacteria associated with deep water animals (Richard, 1943). (Photograph courtesy of Institut Océanographique, Monaco.)

directed by Richard, who had been the director of the Musée Océanographique de Monaco since 1900.

Albert's cruise results are characterized by detailed station lists and careful accounts of the expeditions (Albert I, 1932; Richard, 1934). When he began, as he said himself (Albert I, 1904), he had only the example of *Challenger* and the French expeditions on *Travailleur* and *Talisman* as models, so it is probably no accident that Alphonse Milne-Edwards' protégé Richard became involved so quickly in the scientific work. The prince himself seems to have been well-acquainted with the scientific work, although the scientific publications were always by others. In particular, he assembled a group of technically able subordinates, including Richard himself, crew members, and well-trained young naval officers, who combined the old and the new in a way that made even the innovator Agassiz take second place.

The most conservative of their equipment was that used for deep-sea trawling, based on Agassiz's beam trawl (the Blake trawl), equipped with long cotton tangles on the sides (Fig. 14). This was first used in *Hirondelle* in 1886 and again in 1888 at 2870 m during a heroic deep-sea station accomplished entirely by manual labor and a conventional capstan. With the steam winch of *Princess Alice* and *Princess Alice II* deep-water trawling became less time consuming and easier. It was Albert's custom to trawl in enlarging circles, working near the breaking strain of the steel cable while monitoring bottom contact with the strain dynamometer he used routinely from 1904 onward. It became normal to trawl at 4000 m, and on August 5, 1891 at 12°07′N, 35°53′W, the beam trawl returned successfully from 6035 m bearing a fish, a sea anemone, a worm, three brittle stars, and a sea star. This deep haul, exceeded only by Alexander Agassiz's trawl at 7631 m in the Kermadec–Tonga Trench in 1899, was not repeated. Subsequent attempts were successful to 5413 m, but no deeper. On occasion the prince tried to increase the size of the hauls by otter trawling on the bottom at abyssal depth (a technique that he used successfully in midwater), but he failed at 3465 m in 1905 and probably never successfully used the technique below 2000 m.

Trawling was supplemented very successfully by the use of baited traps or cages moored at great depths. Albert was convinced that a technique was needed to capture large, mobile predatory species that could not be taken by the beam trawl, so he experimented with metal cages (like large minnow traps) to 620 m in 1886 and 1887. The best design, a large polyhedral mesh trap of more than 1 m in each dimension, bearing smaller traps inside and baited with salt fish, offal, chicken parts, assafetida or bright scraps, he employed first in 1888 down to 2000 m (Fig. 15). The results were outstanding—107 fish from various depths and a host of amphipods. An early experiment in 1894, when both a beam trawl and a trap were employed at 1674 m, showed that whereas the trawl took slower-moving animals such as shrimp, pycnogonids, sea stars, and sea urchins, the trap took fish. The traps worked well down to 5310 m, where amphipods were abundant, and during a series of lowerings in 1897 a cage at 5285 m captured a giant lysianassid amphipod 14 cm long, the largest gammaridean known until the photographic work of the 1960s (Hessler et al., 1972) revealed that very large epibenthic amphipods were quite common in deep water. Albert's observation in 1894 that a fish captured at 4898 m had been completely eaten by amphipods was the first indication (long ignored or regarded as unimportant) that there is a mobile predatory or scavenging epifauna or hyperbenthos in deep water. Another species of large

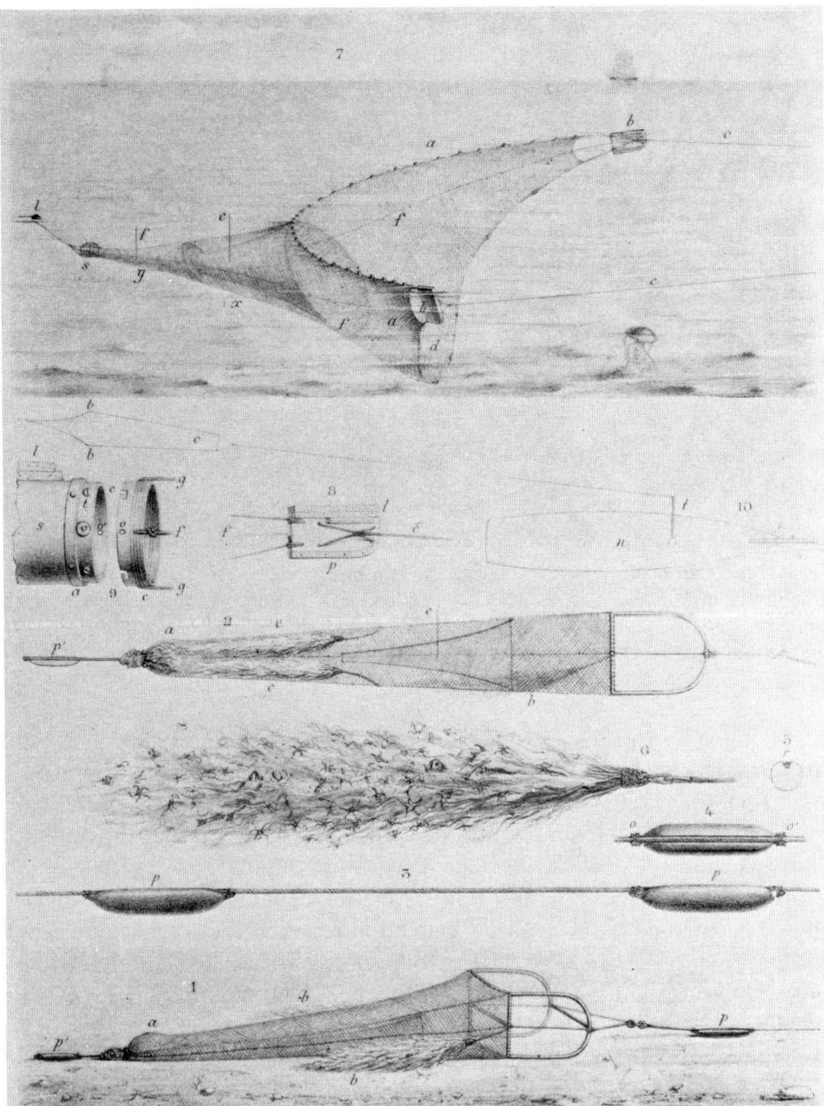

Fig. 14. Two major pieces of equipment used by Albert I, Prince of Monaco (1848 to 1922), on his ships between 1891 and 1913. Above is a large pelagic trawl, below the beam trawl used extensively for benthic work to depths as great as 6035 m (in 1891). Note the cotton swabs alongside the trawl. This sophistication, dating from the first cruise of H.M.S. *Porcupine* in 1869, was used to collect animals such as brittle stars that were not easily gathered by the trawl itself. [From Albert I (1932), plate 1; reproduced courtesy of Institut Océanographique, Monaco.]

lysianassid amphipod, *Eurythenes grillus*, which he took at 4780 m in 1903, was also discovered in a Fulmar stomach, a clue that some bathypelagic or abyssopelagic animals may range to the surface (Hessler et al., 1972).

The trap lowerings were laborious in deep water, where Albert used several 500-m lengths of steel cable joined by swivels and a lighted surface buoy as a marker. Despite the labor, the results were so spectacular that he used cages on every expedition, often supplemented by hook and line fishing (to 3970 m in 1901)

Fig. 15. A large trap being recovered during the cruise of *Princess Alice II* in 1909. During that season the Prince of Monaco used similar traps and long lines as deep as 5940 m (where amphipods were captured), took plankton hauls by using large nets to 5000 m, and beam trawled at 4600 m. [Photograph from Richard (1934); reproduced courtesy of Institut Océanographique, Monaco.]

and increasingly by the use of longlines as deep as 5940 m (1908). The Setubal shark fishermen—whose results had so interested Bocage, Wyville Thomson, and Wright during the 1860s—provided the design for these lines, which had 150 hooks on a 300-m line. In use, the line was stretched between the ship and a small boat and was then released, to be recovered hours later.

The prince of Monaco's plankton investigations (under the direct supervision of Richard) diversified as the years passed and reached their peak between 1900 and 1912 on *Princess Alice II* and *Hirondelle II*. At first Richard took only simple tows with a fine-meshed net, to which in 1895 he added quantitative tows (using a device called the *Buchet net*). Then in 1887 Richard and the prince began experimenting with Chun's propeller-driven closing net, which they were able to haul at 1850 m in the year 1888. Its problem was the jawlike closing device, which almost certainly leaked. As a replacement, Albert experimented with a small boxlike net (*filet à rideau*) with a gate that could be opened and closed by messengers at depth (Fig. 16*a*). Mechanically it worked well, and in tests off Madeira it caught a few

Fig. 16. Two of the nets used during plankton investigations by the Prince of Monaco and Jules Richard. (*a*) The curtain closing net ("filet à rideau"), which could be sent down closed, opened on striking a stop, and was closed by a messenger. Because of its small size, this net was never particularly successful; it was used for only a few cruises between 1888 and about 1896. [From Albert I (1932), plate 3.] (*b*) Because a variety of small closing nets could not adequately sample the sparse zooplankton of midwater, the Prince of Monaco and Jules Richard used overlapping vertical hauls of large open nets to establish vertical ranges of the animals, a technique used by John Murray on the *Challenger* expedition. On *Princess Alice II*, beginning in 1903, 3 × 3-, 5 × 5-, and 9 × 9-m nets were used regularly to deep abyssal depths. The net shown here, "le filet Richard à grande ouverture," appears to be the smallest of the three used in 1905. [Richard (1934), plate 3.] (Both illustrations reproduced courtesy of Institut Océanographique, Monaco.)

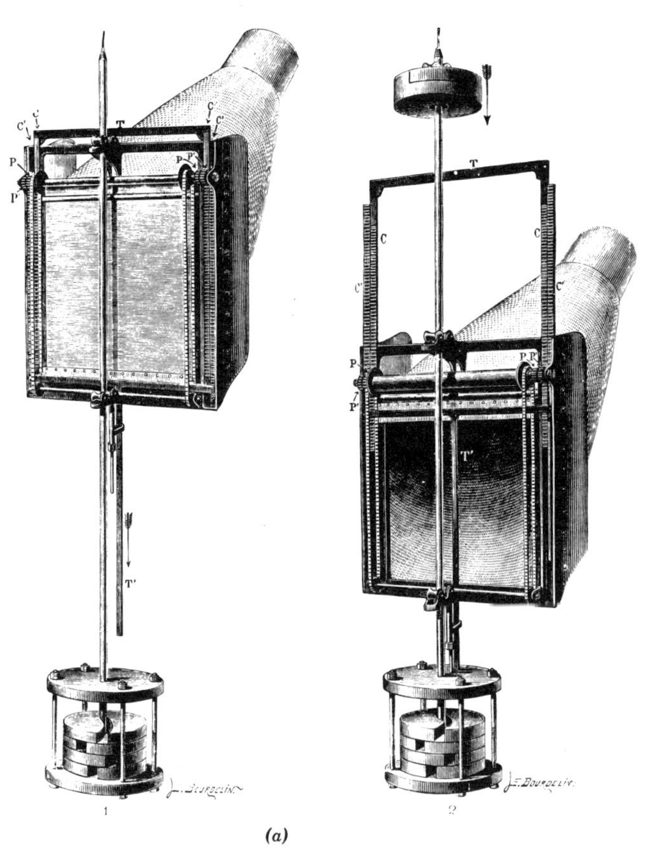

(a)

(b)

animals, but it filtered very little water and was eventually abandoned. After 1896 the main closing net used by Richard and the prince was the Giesbrecht net (which had a fine-meshed closing net in a square frame), although it was never used successfully below 1700 m.

The main problem with all horizontally towed nets was that to be towable, they had to be small. As the work on *Princess Alice II* shifted increasingly to the study of deepwater plankton and its vertical distribution, it became necessary to devise large-volume samplers. These would have to remain open and be towed vertically (so that ranges came to be worked out by doing tows in a series of overlapping depth ranges). In 1903 Richard tried a large net in a 3 × 3-m frame, which successfully sampled the water column from 1500 m to the surface. Encouraged, they went on to the use of 5 × 5- and 9 × 9-m nets (Fig. 16*b*); the latter became the mainstay of the work through which Albert and Richard were convinced by 1912 that a uniform bathypelagic fauna was widespread throughout the northern Atlantic and that animals from as deep as 4500 m might make diurnal migrations to 200 m. Their deepest hauls with these huge, unwieldly nets were to 5500 m (1909). In 1910 and thereafter the use of large nets was complemented by the use of the *Bourée net*, which used a 1-cm mesh on a square frame of 15 m^2, provided with fin stabilizers. It was designed to be towed at high speeds vertically from great depths (in fact, routinely from 2500 m; from 5100 m in 1911) to catch the fish and cephalopods that the large nets missed. As the cruises drew to a close in 1913, the depth at which these nets had been working was established directly by use of a recording manometer in a pressure case. Near the end also, in 1912, they tried out a plankton pump from the surface to 100 m, employing it to collect microscopic plankton for the phycologist on board.

Portier's work on marine microbiology, part of the water-sampling program in 1901/1902 and 1904, also used special equipment and provided results that, although not fully original, have been neglected [even by ZoBell (1946)]. The sampling device (Richard, 1907) was an evacuated glass tube, sealed and autoclaved, and then set in a metal frame that was attached to the wire (Fig. 17). Once at depth, a messenger released the frame, which reversed, shearing off the end of the glass tube, allowing the sampler to fill with 25 ml of water. When it had been retrieved, the sample was inoculated onto a medium of bouillon, gelatin, seawater, and fish or crustacean extracts. Normal bacteriologic media were apparently not satisfactory; Portier and Richard were convinced that the bacteria were not the usual ones from land or freshwater. Whatever their nature, bacteria were most abundant in ports and near the coasts, very scarce in the open sea. Abundance declined with depth as well as distance from land: in the open sea it was not rare to find no bacteria in 25 to 30 ml of seawater from 1000 m. There were always large numbers of bacteria associated with plankton, especially with decomposing organisms and with excreta, and animals from all depths had a large bacterial flora in their guts. The sediments also showed great differences; for example, whereas there might be 24,000 cells/ml at 1100 m just off Naples, none could be detected in the sediments from 300 m in the open sea. With the end of Portier's work on open ocean bacteria in about 1904, deep-sea bacteria were ignored for nearly three decades (ZoBell and Morita, 1959) and only fully studied in the 1960s as the result of an accident involving the research submersible *Alvin* (Jannasch and Wirsen, 1977).

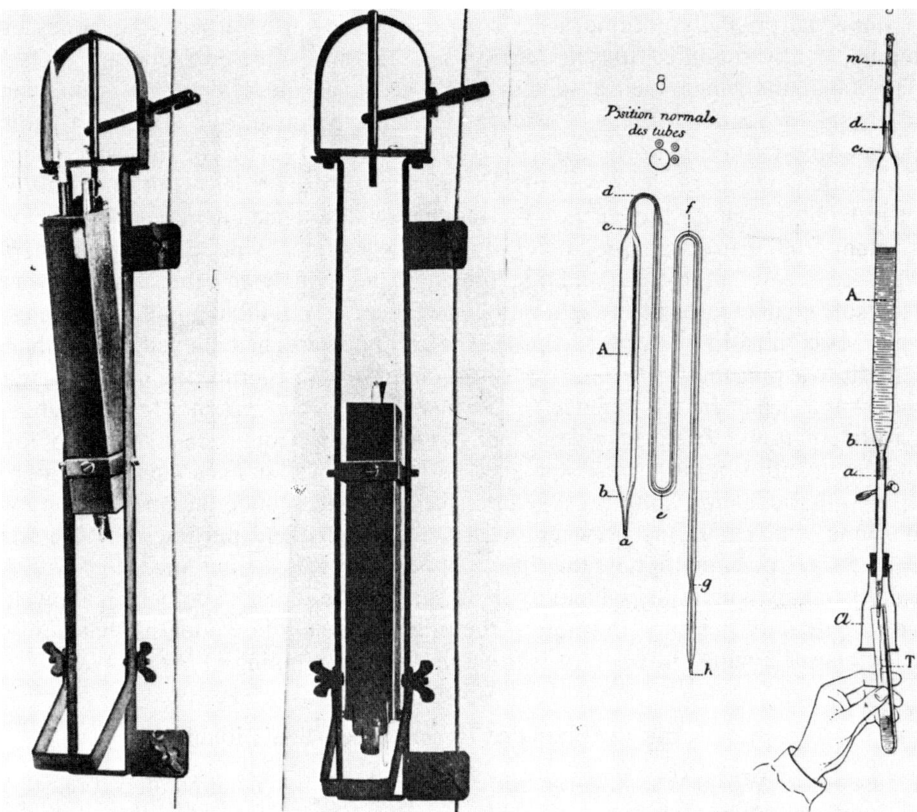

Fig. 17. The bacteriological water sampler used by Portier on *Princess Alice II* in 1904. A sealed, autoclaved, evacuated glass tube with a reservoir (right center) could be sent to any depth in a protective metal box (left), where a messenger released the box, shearing off the end of the glass tube; this allowed the reservoir to fill. At right the reservoir is being emptied into a culture tube. Note the precautions to prevent contamination. [Richard (1934), plate 3; reproduced courtesy of Institut Océanographique, Monaco.]

By describing some of the equipment and research on the prince of Monaco's ships, I have attempted to show a considerable increase in the technical complexity of his oceanographic operations from their simple beginning in 1885. This, of course, was only part of a general change in marine equipment as the age of sail and wooden ships gave way to steam and iron. Changes over a longer period, from *Challenger* to *Princess Alice II*, the most heavily used of the Prince of Monaco's ships, are shown in Table I. The prince's innovations, startling when viewed in isolation, occur in a gradient of change when they are considered in this setting. There is still much to be learned about the relationship between European engineering skills and the way they affected oceanography between the time of the *Challenger* and the last cruises of the *Hirondelle II*.

The Prince of Monaco himself is difficult to place in context, partly because he was a patrician figure in a time and a science that was becoming increasingly bourgeois. In his scientific knowledge the prince could be extremely up-to-date, such as when he discussed the decomposition of organic matter (Albert I, 1907), or rather surprisingly antique, as when he used Edward Forbes' four zones of depth, slightly modified to suit Murray's work on sediments (Albert I, 1898). Without doubt Albert was a skilled promoter of marine sciences. His popular addresses to Scottish, English, French, German, Spanish, and American audiences (Albert I, 1891, 1896, 1898, 1904, 1907, 1912, 1921) have the same basic content (changing, of course, with his advancing knowledge), but each starts cleverly by extolling the virtues of that nation's science, national virtues, or freedom from European prejudices (in the case of the United States). Albert regarded oceanography as a unifying influence, like science in general, which was "the sole emancipator of thought and conscience, an infallible guide to a strong, generous civilization" (Albert I, 1904). The great powers, he believed, rather than stupidly wasting their efforts on warfare, should be furthering the cause of science—and what better science to further than oceanography for the great maritime nations! The prince's mixture of motivations (this is also shown by his interest in anthropology) was expressed particularly well in his statement (Albert I, 1898) that oceanography is "a science well fitted to seduce the imagination by the bond which it forms between poetry, philosophy and pure science." Like many a later oceanographer, the Prince of Monaco was a romantic. He differed from the many who followed mainly in his financial independence, which allowed him to pay only passing attention to the practical scientific concerns that dominated marine science in western Europe up to the beginning of World War I.

8. Some Problems of Recent Deep-Sea Biology

Just before World War I a large body of knowledge had accumulated about the distribution and general biology of deep-sea animals. As I showed in the preceding sections, it had been gained by a varied group of investigators working with a variety of motives in quite different scientific, political, and social settings. Collecting zeal ranked very high among the motives shown by Forbes, Jeffreys, Thomson, Alexander Agassiz, Carl Chun, and Albert I of Monaco. Their personal ambitions were directed first to the fauna of some local area, such as the Isle of Man, the Shetlands, Belfast Lough, or the Mediterranean just off Monte Carlo. As it became clear that general conclusions could not be achieved by studying small areas, each investigator became involved in extensive expeditions. These could best be mounted using the financial and technical resources of the major, wealthy, confident, and often militaristic nations, especially if they had underutilized ships from peacetime navies, telegraph surveys, or coastal surveying. Except for Forbes, the early entrepreneurs (including Carpenter) either purchased the means to study the deep sea or used national prestige, international welfare, or the advancement of knowledge as levers to free money from governments for a concerted study of the sea.

Biological oceanography had taken new directions away from the study of the great depths by 1910, the year that Sir John Murray and Johan Hjort (1869 to 1948) investigated the northern Atlantic by use of the research trawler *Michael Sars* (Fig. 18). Their depleted fisheries had become a concern to nearly every nation around the North Sea (Schlee, 1973). Both a newly emerging fishery science and less practical aspects of marine science were being carried on in the marine stations founded in western Europe and North America from the 1880s onward.

Fig. 18. The Norwegian trawler "Michael Sars" anchored off Plymouth in 1910. By 1910 European marine scientists had turned their attention from the deep sea to problems of the fisheries. However, the cruise of *Michael Sars* under John Murray and Johan Hjort in the northern Atlantic during summer 1910 was a general research cruise, not an applied one. Most important, it provided new information for a textbook, Murray and Hjort's *The Depths of the Ocean*, which summarized knowledge of biological oceanography up to that time. The conjunction of *Michael Sars* and Plymouth is apt, for it was there that modern biological oceanography originated (Murray and Hjort, 1912).

Deacon (1971) and Schlee (1973) have given an outline of the events from the time of the *Challenger* to World Wars I and II in the United Kingdom and the United States. In what follows I give an impressionistic account of the period after 1910, not by following the outlines that they have provided nor by giving a detailed narrative, but rather by examining one or two threads that have had greater influence than others. What follows are my ideas about the influences that split deep-sea biology from the development of biological oceanography late in the nineteenth century and the more recent ones that are reuniting it with the rest of biology, rather than fully analyzed history. In this, the work of Murray and Hjort on the *Michael Sars* in 1910 forms an excellent historical landmark because it resulted in an influential textbook of oceanography (Murray and Hjort, 1912) summarizing the state of marine science between the time of the great expeditions and World War I.

The *Michael Sars* cruise occurred because of international cooperation to study the environment of the northern European fisheries. King Oscar II of Sweden, influenced by Gustaf Ekman (1825 to 1930) and Otto Pettersson (1848 to 1941), proposed during an International Hydrographic Congress in 1899 that a commission be set up by the major fishing nations of northern Europe to coordinate research on the sea. In 1902, as a result, the International Commission for the Exploration of the Sea was established, with Germany, Denmark, Sweden, Norway, Russia, Finland, Great Britain, and the Netherlands committed to physical, biological, and chemical studies of the northeastern Atlantic (Schlee, 1973). In Britain, the Treasury asked the Marine Biological Association to be responsible for the British contribution, whereas Norway, among its other contributions, put the steam trawler *Michael Sars* into service in 1900. During the years prior to 1909 this ship was kept busy in routine hydrologic and biological surveys, but in that year Murray attended the ICES meeting in Copenhagen and suggested a more ambitious Atlantic-wide expedition to broaden European research. Murray agreed to pay the bills, and the Norwegian government provided the ship (Fig. 18), which left Bergen on April 1, 1910, returning there in mid-August after steaming 11,500 miles between Europe, the Canaries, and Newfoundland. The work on the cruise was not particularly unusual, except perhaps in its emphasis on small phytoplankton (which H. H. Gran concentrated by using a combination of centrifuge and filters) and on details of the vertical distribution of midwater fish and invertebrates (which Hjort worked out from long horizontal tows of large open nets). In addition, Bjorn Helland-Hansen (1877 to 1957) devised a complex apparatus with filters and panchromatic film to determine the spectral composition of light in the sea and the extent of the lighted zone. The multiauthored book *The Depths of the Ocean* resulting from the cruise summarized current knowledge and thus became an important source book for the historian.

In 1910 there was little doubt that marine production depended ultimately on the nutrient elements phosphorus, nitrogen, and silicon, which had to be derived from land or by vertical advection from deep water. The chemistry of this process was derived from Karl Brandt's (1854 to 1931) application of Liebig's principles (Brandt, 1905), the biology from ICES investigations in the preceding decade. H. H. Gran, who worked on the phytoplankton from the *Michael Sars* cruise, was convinced that plant production was governed primarily by the breakdown of vertical stratification, allowing nutrients to reach the plant cells near the surface. The plant cells, in their turn, although they seemed too few for the abundance of zooplankton, must provide food to deep-water animals, supplemented by the remains or feces of surface animals and deep-water pelagic animals whose larvae might once have lived at the surface. By these means "a large proportion of the produce of the surface layers must be continually descending into the deep-sea" [Gran, cited in Murray and Hjort (1912)]. The "dust-fine mass" of plant detritus that C. G. J. Petersen (1860 to 1928) (Petersen and Jensen, 1911) believed was so important as food for the shallow-water benthos was a possible food source in deep water, but Gran could not find it by centrifuging open-ocean seawater, and thus the importance of that material, like the dissolved organic matter that Pütter believed necessary to allow zooplankton to survive (Jorgensen, 1976), was problematic.

The invertebrate benthos was summarized by A. Appellöf in a lengthy chapter that concentrated mainly on systematics and zoogeography, using evidence dating back at least to the *Lightning* and the *Porcupine* that "physical conditions" (mainly temperature) governed distributions. Animal distributions expanded in the deep sea, but Appellöf could find little evidence of cosmopolitan distributions. There was, however, a "deep-sea morphology" in groups such as crinoids, macrourid fish, elasipode holothurians, and so on that could be detected almost everywhere in deep water. It is worth noting at this point that Appellöf's conclusions, like most of the ideas in the preceding decades, were based on the study of large, trawlable animals. Only H. J. Hansen (1855 to 1936) on the Danish ship *Ingolf*, which worked in the northern North Atlantic in 1895 and 1896 (Wolff, 1967), had troubled to use very fine screens; he thereby discovered a host of new species among the small Crustacea. But the *Ingolf* techniques were not applied again for the study of deep-sea benthos until the early 1960s. It was Appellöf's [cited in Murray and Hjort (1912)] considered judgment that

> we may briefly characterise the deep-sea fauna as follows: It is largely composed of groups of forms, which morphologically differ in many essentials from the types of the littoral fauna. These groups are distributed over very extensive tracts of the deep sea, but the different species (genera, families) within the groups may be limited to more circumscribed areas. It is evident, therefore, that we can distinguish between the various faunal areas of the deep sea though we may not yet be able to fix their boundaries.

Curiously, despite the hint in Gran's chapter that food supply to the benthos might be a problem, Appellöf does not mention benthic nutrition.

By far the most impressive results were those that Hjort achieved using his huge horizontally towed mesopelagic nets. By towing for long distances to reduce the contamination by the surface fauna, he obtained huge collections that allowed him to show, for example, that two widespread fish in the genus *Cyclothone* had different vertical ranges, that the smaller fish were found nearer the surface, and that there was geographic variation in their size. He concluded that there was a zonation not just of species, but also of colors suited to the prevailing light regime. For example, blue or transparent fish appeared in the top 150 m, silvery or grayish Sternoptychidae and Stomiatidae lived between 180 and 500 m, and black fish (especially *Cyclothone*) and red prawns were found below about 500 m. These generalizations, although too simple (Marshall, 1954), should have been observed earlier by Chun and Prince Albert. Hjort sent the deep-water Copepoda to G. O. Sars for identification and listed the species at various depths; his table clearly shows an increase in species diversity with depth. This was never pursued because it fit no particular theoretical framework.

Considering the nature of the *Michael Sars* collections, it was natural that Hjort summarized the general biology by concentrating on the pelagic fauna. This had also been the focus of *Valdivia*'s work as well as that of the Prince of Monaco since the turn of the century. Pelagic tow nets and trawls were easier to handle than benthic gear, and the samples were more obviously interesting and easier to process. Although Murray and Hjort trawled at 5160 m, their frequent experience was that the trawl came up blocked with mud, yielding only a few worms or holothurians for hours of labor and wear and tear on equipment. Thus, despite years of work up to Hjort's time, he had to acknowledge that "in the open ocean conditions are still practically unexplored." He then set out without pause to speculate about how food might reach deep water. In doing so, he based his discussion on Murray's distinction between terrigenous and pelagic sediments, for Hjort, like Gran, believed that the original source of food and dissolved nutrients was land or the shallow coastal waters. If this were the case, the richest fauna should be in the northwestern Atlantic, as he believed could be observed, for there the prevailing currents carried terrestrial material and plant debris farthest from shore. The second source of food to the depths was phytoplankton, which was also most abundant near the coasts. Although detritus should be a prime source of nutrition, it was difficult to find—Gran had been unable to centrifuge it from seawater—and only the amorphous gut contents of deep-water copepods might be interpreted as plant detritus or animal remains that had sunk from the surface. Whatever the importance of the other sources, the main food source in the open ocean appeared to be the phytoplankton of the surface, which Hjort usually found confined almost entirely to the top 100 m, where it was grazed by tiny Crustacea that would eventually die and fall into deeper water. There, in the depths, lived a significant fauna of macrocrustaceans, centered in distribution near the top of the main thermocline and probably subsisting on sinking remains. The uncertainties were very great, as Hjort knew, for it was almost impossible to relate production to the standing crop of organisms at any depth, despite Hans Lohmann's (1908) attempt to show that there was surplus plant production in the surface waters during most of the year. Hjort's philosophical disposition was to do qualitative work first, next to work out relative abundances, and then if possible to apply quantitative methods, a technique that Victor Hensen and a succession of colleagues, especially Lohmann, had applied in reverse order (wrongly in Hjort's opinion) to plankton biology in the decades around 1900.

Hjort, with his command of marine science, might have gone much farther with the problems of deep-sea biology, but as he concluded the writing of *The Depths of the Sea*, he was preoccupied by research on fisheries year-class strength (Hjort, 1914), a study that revolutionized European fisheries biology but led him away from deep-sea research. He, like many other biologists, was following the problems that had led to the establishment of ICES 10 years before, not chasing the many unsolved problems that had accumulated since the *Lightning* and the *Porcupine*. This was also true in Great Britain, where a series of particularly significant events sharpened the division between deep-sea biology and what was to become biological oceanography.

The Marine Biological Association of the United Kingdom was established in 1884, partly as an outcome of the great International Fisheries Exhibition held in London the year before "for the purpose of establishing and maintaining laboratories on the Coast of the United Kingdom, where accurate researches may be carried on leading to the improvement of Zoological and Botanical Science, and to an increase of our knowledge as regards the food, life conditions, and habits of British Food-fishes and Molluscs." Its laboratory at Plymouth was completed in 1888 (the same year that the Marine Biological Laboratory was established at Woods Hole, Massachusetts), supported financially by, among other sources, the Company of Fishmongers and a small government grant, sources that for years kept the laboratory solvent, although impoverished and understaffed (Bidder, 1943; Allen and Harvey, 1928). From the earliest days the Plymouth Laboratory was a center for research on fish [e.g., E. W. L. Holt's classic study on the Grimsby trawl fishery (1895)], the pelagic and benthic biota of the English Channel, water chemistry, and hydrology. Walter Garstang's (1900) paper showing that the North Sea trawl fisheries were declining (despite powerful opinion to the contrary) was prescient, for in 1902 the Plymouth Laboratory took over the British component of the ICES surveys and immediately sent Garstang to a small new branch laboratory devoted to fisheries work at Lowestoft on the Suffolk coast. There he conducted trawl surveys, experimented with fish tagging, and attempted to transplant plaice from their inshore nursery areas to the offshore fishing grounds.

Plymouth's connection with Lowestoft was severed in 1910 when the government's Board of Fisheries and Agriculture took over the running of the fisheries laboratory on the North Sea. Almost immediately there was a striking increase in the amount of pure research being done in Plymouth, beginning with faunistic work in the English Channel. By 1915 E. J. Allen's (1866 to 1942) work on diatom culture and the substances needed for phytoplankton growth was being published, and Marie Lebour (1876 to 1971) began her studies of fish and invertebrate larvae in the plankton of the English Channel. By 1922 W. R. G. Atkins (1884 to 1959) began to work on the chemistry of seawater; this study was taken over by L. H. N. Cooper (born 1905) in 1930, freeing Atkins to work full time on underwater illumination. H. W. Harvey (1887 to 1970), who was first appointed hydrographer to the Plymouth Laboratory, soon began to follow seasonal changes in nutrients and then worked on the responses of phytoplankton to the availability of nitrate and ammonia. Between 1923 and 1925 F. S. Russell (born 1897) began studies of vertical migration, showing eventually that the diurnal migrations of many planktiers were regular and closely followed specific levels of illumination. During the 1930s the Plymouth group developed colorimetric standards for nutrient analysis and for phytoplankton standing stock. They also began to follow seasonal changes in nutrients and phytoplankton, just as Brandt and Lohmann had done around the turn of the century at Kiel. In 1935 Harvey and his co-workers published their first definitive paper on the control of phytoplankton production by nutrient abundance and grazing (Harvey et al., 1935).

The brushstrokes in this picture of scientific change are broad, but they are adequate to show that in the 30 years after 1910, when their laboratory had been released from its concentration on applied biology, the Plymouth scientists (along with a number of Europeans doing background work for ICES) established the framework of modern biological oceanography, a discipline that has, since those years, always been strongly oriented toward plankton dynamics. Deep-sea biology, which flourished before the turn of the century, became a side issue (if it were carried on at all), a specialized, difficult, and even scientifically uninteresting residuum of nineteenth century thought. Deep-sea research did not begin to develop again until after World War II. When it was resumed, it was with strong echoes of the past.

The Swedish deep-sea expedition of 1947/1948 on the *Albatross* was chronologically first in the resurgence. It had its immediate origins in Hans Pettersson's (1888 to 1966) interest in the radiochemistry of sediments (Pettersson, 1953, 1966), although its broad aim was the general study of oceanic sediment layers by the use of newly developed seismic techniques and piston coring. The ship was owned by a shipping firm, the Broström Combine; its outfitting as well as the planning of the cruise timing and route were undertaken with business-style efficiency (Christiansson, 1966; Pettersson, 1966). During their circumnavigation the Swedish scientific staff concentrated mainly on coring, seismic reflection, light in the water column, and water sampling near the bottom. It was only when the ship was returning home, during the last 3 months in the Atlantic, that deep-sea benthic work was undertaken, directed by Orvar Nybelin. Using an 11-m shrimp otter trawl, Nybelin did a series of 14 deepwater trawls of fish and invertebrates from below 4000 m. Trawling in the Puerto Rican Trench at 7625 to 7900 m on August 17/18, 1948, he took a few worm tubes and holothurians and with this effort made the deepest station in the northern Atlantic. (In 1891 the Prince of Monaco had beam trawled at 6035 m.) The depth was significant, for during the next decade a surprising amount of effort was expended in collecting in the greatest depths (a preoccupation with a long history), especially in the trenches.

The next attempts to discover life in the trenches were made soon after by scientists on the Danish research vessel *Galathea* during its round-the-world cruise of 1950 to 1952. Its results radiate a vibrancy and enthusiasm that is rare in recent accounts of oceanographic work. The history and results of the *Galathea* expedition call for attention because they show to what degree deep-sea biology had diverged from the mainstream of biological oceanography; they also give the first indications of how changes in science and society outside oceanography were to reorient the work of deep-sea biologists.

Denmark has had a long tradition of marine exploration, much of it in the late nineteenth and early twentieth centuries, either centered around its colony Greenland or carried out for ICES [for a detailed account, see Wolff (1967)]. This must have had some effect on the thinking of Anton Bruun (1901 to 1961), the leader of the *Galathea* expedition, whose first extended exposure to open-sea oceanographic work was on the round-the-world cruise of *Dana II* under Johannes Schmidt (1877 to 1933) in 1928 to 1930. In the 1930s Schmidt and August Krogh (1874 to 1949) had discussed organizing a Danish expedition to extend the work of the *Dana*, which had concentrated mainly on fish, surface plankton, and properties of the water column. The biology of the sea floor, especially at great depths, had not been one of its concerns. Bruun, who probably knew of Schmidt and Krogh's ideas, was appointed scientific leader of the Danish expedition on the yacht *Atlantide*, which worked mainly on the continental shelf of West Africa in 1945/1946 (Atlantide Reports, 1950, pp. 1966 ff; Wolff, 1967), but the events leading to the *Galathea*'s voyage began earlier, according to published accounts, when Bruun lectured on sea serpents (a topic of speculation even more venerable than the nature of the deep sea) and aroused the interest of the journalist Hakon Mielche (Spärck, 1956; Mielche, 1956). They began to plan an expedition to investigate the greatest depths of the oceans, an expedition delayed, of course, by the war and its consequences. After the war an expedition fund was established, the Swedes sold the Danes their large trawl winch from *Albatross* (also cable, water bottles, thermometers, and other equipment), and in 1949 the organizers purchased and began to outfit the former Royal Navy sloop H. M. S. *Leith* for deep-sea benthic work. The cost was considerable, about 3,000,000 kr in capital expenditure and 2,500,000 kr in operating costs (roughly U.S. $1,100,000 and $900,000 in 1980), of which the expedition fund paid nearly one-third and the Danish government the rest, not without some political embarrassment (Wolff, 1967). Both the funding and execution of the expedition had their unique aspects. The Danish Expedition Fund was allowed to buy goods overseas and sell them at increased prices in Denmark, accumulating the profit toward the costs of the expedition. In addition, Mielche's press corps on the *Galathea* ensured that the ship received the maximum of press and radio coverage at each port of call as well as in Denmark (Mielche, 1956). Truly in the latter respect the *Galathea* expedition was a twentieth century creation.

The scientific program of the *Galathea* was to be both traditional and modestly original. Its first aim was to collect the fauna of deep-sea trenches and then to capture the large, active animals of the abyss by using the largest trawls that were feasible. There would also be a program of quantitative grab sampling, working outward from the coasts under different levels of surface production into very deep water. A much longer list of scientific aims prepared for (or by?) Bruun also included plankton collections, physiology of deep-sea animals, hydrography, magnetic studies, shore collecting, terrestrial biology in remote areas, and cultural anthropology. Foreign scientists were to be invited to join the ship when their work could be fitted into its complex routines (Bruun, 1956).

The two readable, accurate accounts of the cruise by Bruun (1956) and Wolff (1967) give many details of the voyage and its early results, so I concentrate only on four subjects that have significant connections with the history of deep-sea investigation or with twentieth century biological oceanography: the study of trench faunas; quantitative benthic sampling; the use of ^{14}C to estimate primary production; and the study of deep-sea bacteria.

On July 22, 1951 Bruun and his colleagues successfully trawled at 10,190 m in the Philippine Trench; there they recovered sea anemones and amphipods, isopods, bivalves, and holothurians. By the end of the cruise they had trawled in five trenches and collected more than 115 species at depths greater than 6000 meters, decisively showing that all the remarkably persistent questions about the existence of life in the deep sea could finally be given a decent burial after nearly 150 years. On the basis of the *Galathea*'s work, Bruun postulated that there was a taxonomically distinct hadal fauna resident in the trenches below 6000 m; thereby he offset an argument about the nature of deep-sea faunas (rather than about their existence) that has spluttered fitfully on in the literature (Menzies and George, 1967; Wolff, 1970).

At the same time, the Soviets also began to work in the trenches when their research vessel *Vitiaz* was transferred to their Far Eastern Seas in 1949 (Zenkevitch, 1963; Belyaev, 1972). Thereafter that Soviet vessel, later the vessels *Ob* and *Mikhail Lomonosov*, made many grab stations in the deep water of all the major oceans, greatly extending the 65 grab samples taken by *Galathea* [see Vinogradova (1962) for a summary of Soviet work during that period]. For the first time, information about the abundance of animals at all depths rather than qualitative data began to accumulate. It is curious that both the *Galathea* and *Vitiaz* were used for quantitative sampling as well as for sampling the trenches at just that time, for scientific contact was rare between the west and the Soviet Union during the cold war in the early 1950s. The influence of personalities may have had some role, for Bruun had a wide circle of scientific friends throughout the world, and his influence, strong during his time, continued on his acquaintances for years after his death.

Open-ocean bacteria received a little flurry of attention in the late nineteenth century during the French *Travailleur* and *Talisman* expeditions (Certes, 1884a,b), Hensen's Plankton Expedition (Fischer, 1894), and on one or two of the prince of Monaco's summer cruises about 1901, when Richet and Poirier found bacteria in deep water, some deep sediments, and fish guts [see ZoBell (1946) for a brief account of early work on marine bacteria]. Although Certes had found that some marine bacteria could grow at high pressures, there was no evidence on the abundance, nutrient requirements, or pressure tolerances of deep-sea bacteria up to the time of the *Galathea* expedition. Then during the expedition, Claude ZoBell and Richard Morita (1959), working under very difficult conditions, isolated and estimated the numbers of barophilic (or at least barotolerant) bacteria from great depths that would grow on their culture media. The question they asked in 1959, "What is the ecological and oceanographic significance of the bacterial populations found in the oceanic deeps? Are they primarily passive inhabitants on the deep-sea floor or is their rate of metabolism and reproduction rapid enough for them to affect substantially chemical or biological conditions?" had also been asked by Certes (1884a,b) years before. This was very difficult to answer because the ability to recover undecompressed samples and the technology to perform experiments in deep water were both required. Technology possessed these abilities by 1960, but it was as the curious result of an accident in 1968 (Jannasch and Wirsen, 1977), rather than by a series of sequentially linked scientific investigations, that the results from the *Galathea* and earlier expeditions were extended and superseded.

ZoBell and Morita's work on microbiology falls well outside the very largely systematic and zoogeographic framework that characterizes the majority of the *Galathea* reports. So, too, does the paper by E. Steemann Nielsen and Aabye Jensen (1959) reporting at length Nielsen's recently developed method of using ^{14}C to estimate the primary production of phytoplankton. The ^{14}C technique was apparently developed very rapidly. Because it was first widely applied on a major expedition, information about primary production under widely varying hydrographic conditions became available almost immediately. It is difficult to imagine a scientific advance that has occurred much more rapidly and one that could have had a more profound effect on the whole basis of biological oceanography. That this innovation occurred on a deep-sea expedition is partly historical accident and partly the result of Steemann Nielsen's shrewdness in applying nuclear technology in marine biology.

I can hardly argue with Torben Wolff (1967) that "there can be no doubt that the timing of the Galathea Expedition was absolutely right." Although the idea for the expedition probably had a complex history and was founded in good solid Danish systematic biology, the organization, content, and execution were contingent on external forces. The postwar financial boom was favorable, especially before inflation inevitably began. (In fact, inflation rapidly increased the cost and seriously embarrassed both the organizers and their political friends.) Nuclear technology had just made ^{14}C available. Surplus ships could be purchased for relatively low prices as the world's navies retrenched. Scientific talent, dammed up and frustrated throughout occupied Europe, was ready to burst into action. It could do so in a world that was relatively stable politically and even downright friendly to a converted gunboat entering a multitude of territorial seas. National rivalry—as well as cooperation—also had a role, for Sweden had recently successfully seen its elegant motor–sailer *Albatross* home, and its scientific gear was available to an interested taker. Add to this an element of imagination, almost of juvenile play, independent of practical concerns, that is common to most of the leaders of the major deep-sea expeditions, and one sees a complex of factors that would be difficult to duplicate during the 1970s, or probably in the foreseeable future.

After the return of *Galathea*, remarkable quantitative and qualitative changes occurred in oceanography. In deep-sea biology, the 1950s were the decade of the Danish *Galathea* expedition and Soviet surveys of biomass in all the world's oceans. In the 1960s the scene shifted to the United States, to well-funded oceanographic research of a nonexpeditionary kind, and to the uneasy reunion of deep-sea biology with the rest of biological oceanography. Coming in from the wings, theoretical ecology put its stamp on nearly all the investigations of deep-sea biology that began after 1960.

In 1949 there were fewer than 100 oceanographers of all kinds in the United States, according to a committee of the National Academy of Sciences. The Korean War (1950 to 1953) further diverted money from science into warfare, but after the war ended, more or less coinciding with the International Geophysical Year(s) 1957 to 1959, U.S. oceanography began to receive unprecedented attention by the National Academy of Sciences and the government. Congress, for example, showed its awakened interest in marine science when the House of Representatives' Committee on Merchant Marine and Fisheries formed a Special Subcommittee on Oceanography. This was the immediate impulse, according to Cochrane (1978), for massive financial support to U.S. oceanography through the Office of Naval Research, Atomic Energy Commission, Bureau of Commercial Fisheries, and National Science Foundation. The federal budget for oceanography rose from $21 million in 1959 to $221 million in 1969. At least 20 oceanographic vessels and eight new laboratories were built during this boom. Oceanography became an academic subject all over North America rather than being centered at the Woods Hole Oceanographic Institution and Scripps Institution of Oceanography.

The First International Oceanographic Congress in New York in 1959 also inspired new scientific work on the oceans, notably deep-sea biology, for it brought Europeans like Bruun and Zenkevitch to a new world setting ripe for big science. In my judgment as an observer of these events, it was the Congress, along with Woods Hole Oceanographic Institution's decision to proceed with open-ocean and deep-sea research, that led to the most significant changes in deep-sea biology in recent decades. Within 5 years, because of the quantitative sampling carried out between Massachusetts and Bermuda by Howard Sanders and his colleagues at WHOI, it became clear that the deep sea, even beneath the relatively unproductive Sargasso Sea, had unexpectedly high species richness (Sanders et al., 1965), as John Murray had hinted in 1895. By using a simple new collecting device and very fine screens, the WHOI group also showed that the small infauna was numerically and biologically important in deep-sea ecosystems (Hessler and Sanders, 1967). This single step took most of the force out of the lingering arguments over the antiquity of the deep sea (Menzies and Imbrie, 1958; Menzies et al., 1961; Zenkevitch and Birstein, 1960), which had continued past the time of the Oceanographic Congress and that derived force largely from the collection of large, relatively scarce animals by trawl. From the new results a theoretical structure explaining deep-sea species diversity could be built, in terms of either environmental stability (e.g. Sanders, 1968, 1969) or in terms of environmental disturbance (Dayton and Hessler, 1972). By another rather simple change in sampling technique (Hessler and Jumars, 1974), it became possible a few years later to study in detail the patchiness of the small infauna and to relate its spatial patterns to the ecological forces that might govern high species diversity and the division of resources in the deep sea (Jumars, 1975, 1976; Thistle, 1978).

These changes, although they are far from the only ones in deep-sea biology in the last three decades, represent the most powerful North American influence on deep-sea biology since the time of Alexander Agassiz. Their proximate cause was a rather simple refinement of sampling techniques. Their effect was so influential because they could be related to a well-developed theoretical framework. This was provided in part by biological oceanography because of its emphasis on the water column as a production system. But the influence of theoretical population ecology, developed in the United States between 1940 and the 1960s [see Hutchinson (1978) for a well-documented account], was a far more important influence because it provided for the first time a series of closely linked evolutionary hypotheses about the factors governing the abundance, diversity, and interrelations of deep-sea organisms. The other chapters of this volume provide evidence of the power and problems of this approach applied to the population biology, distribution, feeding ecology, and physiology of deep-sea organisms. Those chapters, like my analysis of the history of deep-sea biology, will in their own time provide new material for analysis by the historian of oceanography.

Acknowledgments

The Social Sciences and Humanities Research Council of Canada provided financial support for much of the research leading to this chapter. I am grateful to the Master and Fellows of Corpus Christi College, Cambridge for their hospitality during critical stages of research and writing, and to Ron Hughes and Jean Sanderson of the Balfour Library, Department of Zoology, Cambridge University, for their help. Margaret Deacon and Daniel Merriman kindly read a near-final version of the manuscript. Figures 13*b* and 14 through 17 have been reproduced with permission of the Service des Publications, Institut Océanographique de Monaco. Most of all I am grateful to the Reverend Canon A. M. Norman, who, by leaving his library to the Department of Zoology at Cambridge, gave historians of science a remarkably complete collection of nineteenth century zoological publications.

I hope that Evelyn Hutchinson, Gordon Riley, and Sydney Smith will enjoy the chapter, for each in his own way has contributed to my enjoyment in writing it.

References

Agassiz, A., 1869. Preliminary report on the Echini and star-fishes dredged in deep water between Cuba and the Florida Reef, by L. F. de Pourtalès, Asst. U.S. Coast Survey. *Bull. Mus. Compar. Zool.*, **1**, 253–308.

Agassiz, A., 1878a. (Letter No. 1) to C. P. Patterson, Superintendent Coast Survey, Washington, D.C., from Alexander Agassiz, on the dredging operations of the United States Coast Survey Steamer "Blake" during parts of January and February, 1878. *Bull. Mus. Compar. Zool.*, **5**, 1–9.

Agassiz, A., 1878b. (Letter No. 2) to C. P. Patterson, Superintendent Coast Survey, Washington, D.C., from Alexander Agassiz, on the dredging operations of the United States Coast Survey Steamer "Blake" during parts of March and April, 1878, with the preliminary report on the Mollusca of the expedition, by Wm. H. Dall. *Bull. Mus. Compar. Zool.*, **5**, 55–64, figs.

Agassiz, A., 1879. (Letter No. 3) to C. P. Patterson, Superintendent United States Coast Survey, Washington, D.C., from Alexander Agassiz, on the dredging operations carried on from December

1878 to March 10, 1879, by the United States Coast Survey Steamer "Blake," Commander J. R. Bartlett, U.S.N. *Bull. Mus. Compar. Zool.*, **5**, 289–302.

Agassiz, A., 1880. (Letter No. 4) to C. P. Patterson, Superintendent United States Coast and Geodetic Survey, Washington, D.C., from Alexander Agassiz, on the dredging operations carried on during part of June and July, 1880, by the United States Coast Survey Steamer "Blake," Commander J. R. Bartlett, U.S.N. *Bull. Mus. Compar. Zool.*, **6**, 147–154.

Agassiz, A., 1888a. Bibliotheca zoologica (review of C. Chun's *Die pelagische Thierwelt). Am. J. Sci.*, Ser. 3, **35**, 420–424.

Agassiz, A., 1888b. *Three cruises of the United States Coast and Geodetic Survey Steamer* "Blake" *in the Gulf of Mexico, in the Caribbean Sea, and along the Atlantic Coast of the United States, from 1877 to 1880*. Sampson Low, London, 2 vols.

Agassiz, A., 1891. Three letters from Alexander Agassiz to the Honorable Marshall McDonald, United States Commissioner of Fish and Fisheries, on the dredging operations off the west coast of Central America to the Galapagos, to the west coast of Mexico, and in the Gulf of California, in charge of Alexander Agassiz, carried on by the U.S. Fish Commission Steamer "Albatross," Lieutenant Commander Z. L. Tanner, U.S.N., commanding. *Bull. Mus. Compar. Zool.*, **21**, 185–200.

Agassiz, A., 1892. Reports on the dredging operations off the west coast of Central America to the Galapagos, to the west coast of Mexico, and in the Gulf of California, in charge of Alexander Agassiz, carried on by the U.S. Fish Commission Steamer "Albatross," Lieutenant Commander Z. L. Tanner, U.S.N., commanding. II. General sketch of the expedition of the "Albatross," from February to May, 1891. *Bull. Mus. Compar. Zool.*, **23**, 1–89, 22 plates.

Agassiz, A., 1899. Cruise of the *Albatross*. (Explorations of the *Albatross* in the Pacific Ocean. Letters to U.S. Commissioner of Fisheries.) *Science*, N. S., **10**, 833–841.

Agassiz, A., 1900. Cruise of the *Albatross*. (Explorations of the *Albatross* in the Pacific Ocean. Letters to U.S. Commissioner of Fisheries.) *Science*, N. S., **11**, 92–98, 288–292, 574–578.

Agassiz, A., 1902. Preliminary report and list of stations. Report of the scientific results of the expedition to the tropical Pacific. *Memoirs Mus. Compar. Zool.*, **26** (1).

Agassiz, A., 1905. Three letters from Alexander Agassiz to the Hon. George M. Bowers, United States Fish Commissioner, on the cruise, in the eastern Pacific, of the U.S. Fish Commission Steamer "Albatross," Lieut.-Commander L. M. Garrett, U.S.N., commanding. *Bull. Mus. Compar. Zool.*, **46** (4), 65–84.

Agassiz, A. and L. F. Pourtalès, 1878. Reports on the results of dredging, under the supervision of Alexander Agassiz, in the Gulf of Mexico, by the United States Coast Survey Steamer "Blake," Lieutenant-Commander C. D. Sigsbee, U.S.N., commanding. II. Report on the Echini, by Alexander Agassiz, crinoids and corals, by L. F. de Pourtalès, and ophiurans, by Theodore Lyman. Preceded by a bibliographical notice of the publications relating to the deep-sea investigations carried on by the United States Coast Survey. *Bull. Mus. Compar. Zool.*, **5**, 181–238, plates.

Agassiz, L., 1869. Report upon deep-sea dredgings in the Gulf Stream during the third cruise of the U.S.S. "Bibb." *Bull. Mus. Compar. Zool.*, **1**, 363–386.

Agassiz, L., 1872. A letter concerning deep-sea dredgings, addressed to Professor Benjamin Peirce, Superintendent, United States Coast Survey, by Louis Agassiz. *Bull. Mus. Compar. Zool.*, **3**, 49–53.

Albert I, Prince de Monaco, 1891. A new ship for the study of the sea. *Proc. Roy. Soc. Edinburgh*, **18**, 295–302.

Albert I, Prince de Monaco, 1896. Voyages scientifiques du yacht *Princesse Alice*. *Report of the 6th International Geographical Congress, Paris*, pp. 437–441.

Albert I, Prince de Monaco, 1898. Oceanography of the North Atlantic. *Geogr. J.*, **12**, 445–469.

Albert I, Prince de Monaco, 1904. Les progrès de l'Océanographie. *Bull. Musée Oceanogr. Monaco*, No. 6, 9 pp.

Albert I, Prince de Monaco, 1907. *Der Fortschritt der Meereskunde*. Munich, 11 pp.

Albert I, Prince de Monaco, 1912. Les progrès de l'Océanographie. Address to Societé royale de Geographie de Madrid, 1932. *Rés. Scientif.*, No. 84, pp. 323–336.

Albert I, Prince de Monaco, 1914. *La Carrière d'Un Navigateur*, Vol. 7, 350 pp., Paris.

Albert I, Prince de Monaco, 1921. Discours sur l'océan. *Bull. Inst. Oceanogr. Monaco,* No. 392, 14 pp.

Albert I, Prince de Monaco, 1932. Recueil des travaux publiés sur les campagnes scientifiques. In *Résultats des Campagnes Scientifiques Accompliés sur Son Yacht par Albert I,* etc., No. 84, 366 pp. + 11 plates, charts.

Alder, J. and A. M. Norman, 1826 to 1911. *Scientific Correspondence of Joshua Alder and Alfred Merle Norman 1826–1911.* General Library, British Museum (Natural History), London, 7 vols.

Allen, D. E., 1976. *The Naturalist in Britain.* Allen Lane, London, 292 pp.

Allen, E. J. and H. W. Harvey, 1928. The laboratory of the Marine Biological Association at Plymouth. *J. Mar. Biol. Assoc. U.K.,* **15,** 735–751.

Anonymous, 1871a. (John Gwyn Jeffreys' visit to North America.) *Nature,* **4,** 512–513.

Anonymous, 1871b. (W. B. Carpenter's lecture to the Royal Institution on deep-sea exploration.) *Nature,* **4,** 107.

Argyll, Duke of, 1887. A great lesson. *Nineteenth Century,* **22,** 293–309.

Atlantide Reports, 1950. *Atlantide Report; Scientific Results of the Danish Expedition to the Coasts of Tropical West Africa, 1945–1946.* Danish Science Press, Copenhagen, 10 vols.

Bailey, J. W., 1851. Microscopical examination of soundings made by the U.S. Coast Survey off the Atlantic Coast of the United States. *Smithsonian Contrib. Knowledge,* **2** (3), 1–15.

Bailey, J. W., 1854. Examination of some deep soundings from the Atlantic Ocean. *Am. J. Sci.,* **17,** 176–178.

Bailey, J. W., 1855. Microscopical examination of deep soundings from the Atlantic Ocean. *Quart. J. Microsc. Sci.,* **3,** 89–91.

Barnes, S. B., 1974. *Scientific Knowledge and Sociological Theory,* Routledge and Kegan Paul, London, x + 192 pp.

Bate, C. S. and J. B. Rowe, 1882. Report on the marine fauna of the southern coast of Devon and Cornwall. *Rep. Br. Assoc. Adv. Sci., York, 1881,* pp. 198–200.

Belyaev, G. M., 1972. *Hadal Bottom Fauna of the World Ocean.* Israel Program for Scientific Translations, Jerusalem, 199 pp.

Berman, M., 1978. *Social Change and Scientific Organization. The Royal Institution, 1799–1844.* Cornell University Press, Ithaca, xxv + 224 pp.

Bidder, G. P., 1943. Edgar Johnson Allen 1866–1942. *J. Mar. Biol. Ass. U.K.,* **25,** 671–684.

Bocage, J. V. B. du, 1867. On *Hyalonema lusitanicum. Ann. Mag. Nat. Hist.,* Ser. 3, **20,** 123–127.

Brandt, K., 1905. On the production and conditions of production in the sea. *Cons. Perm. Int. Expl. Mer, Rapp. Proc.-Verb.,* Appendix D, **3,** 1–12.

Brinton, C., J. B. Christopher, and R. L. Wolff, 1969. *Civilization in the West,* 2nd ed. Englewood Cliffs, NJ, Prentice-Hall, xiv + 753 pp.

Bruun, A. F., 1956. Objects of the expedition. In *The Galathea Deep Sea Expedition 1950–1952.* A. F. Bruun et al., eds., George Allen and Unwin, London, pp. 26–28.

Burstyn, H. L., 1968. Science and government in the nineteenth century: The Challenger expedition and its report. *Bull. Inst. Océanogr. Monaco,* Special No. 2 (*Congr. Int. Hist. Oceanogr.* **1**), pp. 603–613.

Burstyn, H. L., 1972. Pioneering in large scale scientific organisation: The *Challenger* Expedition and its report. I. Launching the expedition. *Proc. Roy. Soc. Edinburgh, B,* **72,** 47–61.

Burstyn, H. L., 1975. Science pays off: Sir John Murray and the Christmas Island phosphate industry 1886–1914. *Soc. Stud. Sci.,* **5,** 5–34.

Campbell, Lord George, 1876. *Log-Letters from the Challenger.* Macmillan, London, 448 pp.

Cannon, S. F., 1978. *Science in Culture: The Early Victorian Period,* Science History Press, New York, xii + 296 pp.

Carpenter, J. E., 1888. William Benjamin Carpenter. A memorial sketch, pp. 1–152. In W. B. Carpenter, ed., *Nature and Man. Essays Scientific and Philosophical.* Kegan Paul, London, vi + 483 pp.

Carpenter, W. B., 1856. On the occurrence of the pentacrinoid larva of *Comatula rosacea*, in Lamlash Bay, Isle of Arran. *Rep. Br. Assoc. Adv. Sci., Glasgow,* 1855, trans., p. 107.

Carpenter, W. B., 1865. Researches on the structure, physiology, and development of *Antedon* (*Comatula,* Lamk.) *rosaceus*. *Proc. Roy. Soc. Lond.*, **14,** 376–378 (abstract).

Carpenter, W. B., 1868. Preliminary report of dredging operations in the seas to the north of the British islands, carried on in H.M.S. *Lightning,* by Dr. Carpenter and Dr. Wyville Thomson, Professor of Natural History in Queen's College, Belfast. *Proc. Roy. Soc. Lond.*, **17,** 168–200.

Carpenter, W. B., 1870. The geological bearings of recent deep-sea exploration. *Nature,* **2,** 513–515.

Carpenter, W. B., 1875. Remarks on Professor Wyville Thomson's preliminary notes on the nature of the sea-bottom. Procured by the soundings of H.M.S. Challenger. *Proc. Roy. Soc. Lond.*, **23,** 234–245.

Carpenter, W. B. and J. G. Jeffreys, 1870. Report on deep-sea researches carried on during the months of July, August and September 1870, in H.M. surveying-ship 'Porcupine.' *Proc. Roy. Soc. Lond.*, **19,** 146–221, charts.

Carpenter, W. B. and J. G. Jeffreys, 1871. Report on deep-sea researches carried on during the months of July, August and September, 1870, in H.M. surveying ship 'Porcupine.' *Nature,* **3,** 334–339, 415–417.

Carpenter, W. B., J. G. Jeffreys, and C. W. Thomson, 1870. Preliminary report of the scientific exploration of the deep-sea in H.M. surveying vessel 'Porcupine,' during the summer of 1869. *Proc. Roy. Soc. Lond.*, **18,** 397–492.

Carpenter, W. L., 1870. On the apparatus employed in deep-sea explorations, on board H.M.S. *Porcupine,* in the summer of 1869. *Pop. Sci. Rev.*, **9,** 281–290.

Certes, A., 1884a. Sur la culture, à l'abri des germes atmosphériques, des eaux et des sédiments rapportés par les expéditions du Travailleur et du Talisman; 1882–1883. *Compt. Rend. Acad. Sci.*, **98,** 690–693.

Certes, A., 1884b. De l'action du hautes pressions sur les phénomènes de la putrefaction et sur la vitalité des micro-organismes d'eau douce et d'eau de mer. *Compt. Rend. Acad. Sci.*, **99,** 385–388.

Chierchia, G., 1884. The voyage of the "Vettor Pisani." *Nature,* **30,** 365.

Chierchia, G., 1885. Collezione per studi di scienze naturali nel viaggio intorno al mondo dalla R. Corvett "Vettor Pisani," commandante G. Palumbo, Anni 1882–83–84–85. *Revista Marittima,* September–November 1885, 174 pp.

Christiansson, E. T., 1966. How the *Albatross* became a research vessel. *Reports Swedish Deep-Sea Expedition 1947–1948,* **I,** 125–142.

Chun, C., 1887. *Die pelagische Thierwelt in grosseren Meerestiefen und ihre Beziehungen zu der Oberflachenfauna.* Bibliotheca Zoologica, No. 1. Theodor Fischer, Cassel, 72 pp., 5 plates.

Chun, C., 1889. Bericht über eine nach den Canarischen Inseln im Winter 1887/88 ausgeführte Reise. *Sitzunsb. Königl. Preuss. Akad. Wiss. Berlin,* 1888, No. 30, 519–553.

Chun, C., 1899. Die Deutsche Tiefsee Expedition. A. Berichte des Leiters der Expedition Professor Dr. Chun an das Reichs-Amt des Innern. *Zeitsch. der Gesellsch. f. Erdkunde zu Berlin,* **34,** 75–134, Table 3.

Chun, C., 1900. *Aus den Tiefen des Weltmeeres. Schilderungen von der Deutschen Tiefsee-Expedition.* Gustav Fischer, Jena, 549 pp.

Cochrane, R. C., 1978. *The National Academy of Sciences. The First Hundred Years 1863–1963.* National Academy of Sciences, Washington, D.C., xv + 694 pp.

Dall, W. H., 1879. Report on the results of dredging, under the supervision of Alexander Agassiz, in the Gulf of Mexico, 1877-78, by the United States Coast Survey Steamer "Blake," Lieutenant-Commander C. D. Sigsbee, U.S.N., commanding. V. General conclusions from a preliminary examination of the Mollusca. *Bull. Mus. Compar. Zool.*, **6,** 85–93.

Dayman, J., 1858. *Deep-sea soundings in the North Atlantic Ocean, between Ireland and Newfoundland, made in H.M.S.* "Cyclops," Lieut.-Commander Joseph Dayman, in June and July 1857. Published by order of the Lords Commissioners of the Admiralty, London.

Dayton, P. K. and R. R. Hessler, 1972. Role of biological disturbance in maintaining diversity in the deep-sea. *Deep-Sea Res.*, **19,** 199–208.

Deacon, M., 1971. *Scientists and the Sea 1650–1900. A Study of Marine Science*, Academic Press, London, 445 pp.

Deacon, M., 1977. Staff-commander Tizard's journal and the voyages of H.M. ships *Knight Errant* and *Triton* to the Wyville Thomson Ridge in 1880 and 1882. In *A voyage of discovery. George Deacon 70th Anniversary Volume*, M. Angel, Ed., Pergamon, New York, pp. 1–14.

Deutsche Tiefsee-Expedition. 1898/1899. *Wissenschaftliche Ergebnisse der Deutschen Tiefsee-Expedition auf den Dampfer "Valdivia" 1898–1899*. Herausgegeben von Carl Chun. Vols. 1–25, 1902 ff.

Ehrenberg, C. G., 1844. On microscopic life in the ocean at the South Pole, and at considerable depths. *Ann. Mag. Nat. Hist.*, **14,** 169–178.

Fischer, B., 1894. Die Bakterien des Meeres nach den Untersuchungen der Plankton-Expedition unter gleichzeitiger Berücksichtigung einiger alterer und neuere Untersuchungen. *Ergebnisse der Plankton-Expedition der Humboldt-Stiftung*, Vol. 4, 1–83.

Folin, M. le Marquis de (L. A. Guillaume), 1882. Les explorations sous-marines de l'Aviso à vapeur Le Travailleur en 1880 et 1881. *Bulletin de la Societé des Sciences, Lettres et Arts de Pau*, **11** (2), 60 pp., 7 plates.

Forbes, E., 1840. On the associations of Mollusca on the British coasts, considered with reference to Pleistocene geology. *Edinburgh Academic Annual*, 1840, **1,** 177–183.

Forbes, E., 1844a. On the light thrown on geology by submarine researches; being the substance of a communication made to the Royal Institution of Great Britain, Friday evening, the 23rd February 1844. *Edinburgh New Philos. J.*, **36,** 318–327.

Forbes, E., 1844b. Report on the Mollusca and Radiata of the Aegean Sea, and on their distribution, considered as bearing on geology. *Rep. Br. Assoc. Adv. Sci., Cork*, 1843, pp. 130–193.

Forbes, E., 1851. Report on the investigation of British marine zoology by means of the dredge. Part I. The infra-littoral distribution of marine invertebrata on the southern, western, and northern coasts of Great Britain. *Rep. Br. Assoc. Adv. Sci. Edinburgh*, 1850, pp. 192–263.

Forbes, E. and R. Godwin-Austen, 1859. *The Natural History of the European Seas*. John Van Voorst, London, viii + 306 pp.

Garstang, W., 1900. The impoverishment of the sea. A critical summary of the experimental and statistical evidence bearing upon the alleged depletion of the trawling grounds. *J. Mar. Biol. Assoc. U.K.*, **6,** 1–69.

S.M.S. *Gazelle*, 1888–1890. *Die Forschungsreise S.M.S. "Gazelle" in den Jahren 1874 bis 1876 unter Kommando des Kapitan zur See Treiherrn von Schleinitz herausgegeben von dem Hydrographischen Amt des Reichs–Marine-Amts*. Mittler und Sohn, Berlin.

Geison, G. L., 1978. *Michael Foster and the Cambridge School of Physiology. The Scientific Enterprise in Late Victorian Society*. Princeton University Press, Princeton, NJ, 401 pp.

Goode, G. B. and T. H. Bean, 1895. Oceanic ichthyology, a treastise on the deep-sea and pelagic fishes of the world, based chiefly upon the collections made by the steamers *Blake, Albatross,* and *Fish Hawk,* in the northwestern Atlantic, with an atlas containing 417 figures. *U.S. Nat. Mus. Spec. Bull.*, xxxv + 553 pp. Atlas xxiii + 26 pp. 123 plates.

Goodsir, H., 1845. The arctic expedition under the command of Sir John Franklin. *Ann. Mag. Nat. Hist.*, **16,** 163–166.

Harvey, H. W., L. H. N. Cooper, M. V. Lebour, and F. S. Russell, 1935. Plankton production and its control. *J. Mar. Biol. Assoc. U.K.*, **20,** 407–441.

Hensen, V., 1895. Methodik der Untersuchungen. *Ergebnisse der Plankton-Expedition der Humboldt-Stiftung*, 1B Kiel.

Herdman, W. G., 1923. *Founders of Oceanography and Their Work. An Introduction to the Science of the Sea,* Edward Arnold, London, 340 pp.

Hessler, R. R., J. D. Isaacs, and E. L. Mills, 1971. Giant amphipod from the abyssal Pacific Ocean. *Science,* **175,** 636–637.

Hessler, R. R. and P. A. Jumars, 1974. Abyssal community analysis from replicate box cores in the central North Pacific. *Deep-Sea Res.*, **21,** 185–209.

Hessler, R. R. and H. L. Sanders, 1967. Faunal diversity in the deep sea. *Deep-Sea Res.*, **14,** 65–78.

Hjort, J., 1914. Fluctuations in the great fisheries of Europe. *Cons. Perm. Int. Expl. Mer, Rapp. Proc.-Verb.*, **20,** 228 pp.

Holt, E. W. L., 1895. An examination of the present state of the Grimsby trawl fishery, with especial reference to the destruction of immature fish. *J. Mar. Biol. Assoc. U.K.*, **3,** 339–448.

Hutchinson, G. E., 1978. *An Introduction to Population Ecology*. Yale University Press, New Haven, CT. xi + 260 pp.

Huxley, T. H., 1858. Deep-sea soundings in the North Atlantic Ocean. Appendix A in J. Dayman, *Deep-Sea Soundings in the North Atlantic Ocean,* London.

Huxley, T. H., 1868a. On a piece of chalk. *Macmillan's Magazine,* **18,** 396–408.

Huxley, T. H., 1868b. On some organisms living at great depths in the North Atlantic Ocean. *Quart. J. Microsc. Sci.,* N. S., **8,** 203–212.

Huxley, T. H., 1869. On some organisms which live at the bottom of the North Atlantic, in depths of 6000 to 15,000 feet. *Rep. Br. Assoc. Adv. Sci., Norwich,* 1868, p. 102 (title only).

Huxley, T. H., 1875. Notes from the *Challenger. Nature,* **12,** 315–316.

Huxley, T. H., 1887. Science and the bishops. *Nineteenth Century,* **22,** 625–641.

Jannasch, H. W. and C. O. Wirsen, 1977. Microbial life in the deep sea. *Sci. Am.,* **236** (6), 42–52.

Jeffreys, J. G., 1856. On the marine Testacea of the Piedmontese coast. *Ann. Mag. Nat. Hist.,* Ser. 2, **17,** 155–188.

Jeffreys, J. G., 1861. On a presumed case of failure in oceanic telegraphy; and on the existence of animal life at great depths in the sea. *Ann. Mag. Nat. Hist.,* Ser. 3, **7,** 254–255.

Jeffreys, J. G., 1862 to 1869. *British Conchology, or an Account of the Mollusca Which Now Inhabit the British Isles and the Surrounding Seas,* London, John Van Voorst, 5 vols.

Jeffreys, J. G., 1864. Report of the committee appointed for exploring the coasts of Shetland by means of the dredge. *Rep. Br. Assoc. Adv. Sci., Newcastle-upon-Tyne,* 1863, pp. 70–81.

Jeffreys, J. G., 1869a. The deep-sea dredging expedition in H.M.S. "Porcupine." *Nature,* **1,** 135–137, 166–168.

Jeffreys, J. G., 1869b. Last report on dredging among the Shetland Isles. *Rep. Br. Assoc. Adv. Sci., Norwich,* 1868, pp. 232–247.

Jeffreys, J. G., 1869/1870. Food of oceanic animals. *Nature,* **1,** 192, 315.

Jeffreys, J. G., 1871. On a *Pentacrinus* (*P. Wyville-Thomsoni*) from the coasts of Spain and Portugal. *Rep. Br. Assoc. Adv. Sci., Liverpool,* 1870, trans. p. 119.

Jeffreys, J. G., 1878. Address by J. Gwyn Jeffreys, LL.D., F.R.S., Treas. G. & LSS., President of the Section. *Rep. Br. Assoc. Adv. Sci., Plymouth,* 1877, trans., pp. 79–87.

Jeffreys, J. G., 1880. The French deep-sea exploration in the Bay of Biscay. *Rep. Br. Assoc. Adv. Sci., Swansea,* 1880, pp. 378–387.

Jeffreys, J. G., 1881. Deep-sea exploration. *Nature,* **23,** 300–302, 324–326.

Jorgensen, C. B., 1976. August Pütter, August Krogh, and modern ideas on the use of dissolved organic matter in aquatic environments. *Biol. Rev.,* **51,** 291–328.

Jumars, P., 1975. Environmental grain and polychaete species diversity in a bathyal benthic community. *Mar. Biol.,* **30,** 253–266.

Jumars, P., 1976. Deep-sea species diversity: Does it have a characteristic scale? *J. Mar. Res.,* **34,** 217–246.

Keferstein, W., 1864. Ueber die geographischen Verbreitung der Prosobranchien. *Nachr. königl. Gesellsch. Wiss. Gottingen,* **1864** (6), 103–110.

Kuhn, T. S., 1968. The history of science. In *International Encyclopedia of the Social Sciences,* Vol. 14, David L. Sills, ed., New York, pp. 74–83.

Linklater, Eric, 1972. *The voyage of the Challenger.* John Murray, London, 288 pp.

Lohmann, H., 1908. Untersuchungen zur Feststellung des vollständigen Gehaltes des Meeres an Plankton. *Wiss. Meeresunters., Abt. Kiel,* N. F., **10,** 129–370.

Lovén, S., 1845. On the bathymetrical distribution of submarine life on the northern shores of Scandinavia. *Rep. Br. Assoc. Adv. Sci., York,* 1884, trans., pp. 50–51.

Lovén, S., 1865. Om resultaten af de af den svenska Spetsbergs-Expeditionen 1861. *Förhandlingar vid de Skandinaviska Naturforskarnes Nionde Möte i Stockholm,* 1863, pp. 384–386.

Lurie, E., 1960. *Louis Agassiz. A life in science.* University of Chicago Press, xiv + 449 p.

Lurie, E., 1962. Editor's introduction to *Essay on classification by Louis Agassiz.* Harvard University Press, Cambridge, MA, pp. ix–xxxiii.

Lyell, Charles, 1871. *The student's Elements of Geology,* London.

Marshall, N. B., 1954. *Aspects of Deep Sea Biology.* Philosophical Library, New York.

Matthäus, W., 1968. The historical development of methods and instruments for the determination of depth-temperatures in the sea *in situ. Bull. Inst. Océanogr. Monaco,* Special No. 2 (*Congr. Int. Hist. Oceanogr.* **1**), pp. 35–47.

Menzies, R. J., 1962. On the food and feeding habits of abyssal organisms as exemplified by the Isopoda. *Int. Rev. Ges. Hydrobiol.,* **47,** 339–358.

Menzies, R. J. and R. Y. George, 1967. A re-evaluation of the concept of hadal or ultra-abyssal fauna. *Deep-Sea Res.,* **14,** 703–723.

Menzies, R. J. and J. Imbrie, 1958. On the antiquity of the deep-sea bottom fauna. *Oikos,* **9,** 192–201.

Menzies, R. J., J. Imbrie, and B. C. Heezen, 1961. Further consideration regarding the antiquity of the abyssal fauna with evidence for a changing abyssal environment. *Deep-Sea Res.,* **8,** 79–94.

Merriman, D., 1968. Speculations on life at the depths: A xixth-century prelude. *Bull. Inst. Oceanogr. Monaco,* Special No. 2 (*Congr. Int. Hist. Oceanogr.* **1**), pp. 377–385.

Merriman, D., 1972. Challengers of Neptune: The "Philosophers." *Proc. Roy. Soc. Edinburgh, B,* **72,** 15–45.

Mielche, H., 1956. Films, press, and radio on the expedition. In A. F. Bruun *et al., The Galathea Deep-Sea Expedition 1950–52,* A. F. Bruun et al., eds., George Allen & Unwin, London, pp. 282–293.

Mills, E. L., 1972. T. R. R. Stebbing, The *Challenger* and knowledge of deep-sea Amphipoda. *Proc. Roy. Soc. Edinburgh, B,* **72,** 69–87.

Mills, E. L., 1975a. The *Challenger* expedition: How it started and what it did. In *One Hundred Years of Oceanography,* E. L. Mills, ed., Dalhousie University and Nova Scotia Museum, Halifax, Canada, 89 pp., pp. 1–23.

Mills, E. L., 1975b. H.M.S. *Challenger* and the controversy about how the oceans circulate. In *One Hundred Years of Oceanography,* E. L. Mills, ed., Dalhousie University and Nova Scotia Museum, Halifax, Canada, 89 pp., pp. 54–73.

Mills, E. L., 1978. Edward Forbes, John Gwyn Jeffreys, and British dredging before the *Challenger* expedition. *J. Soc. Bibl. Nat. Hist.,* **8,** 507–536.

Mills, E. L., 1980a. Alexander Agassiz, Carl Chun and the problem of the intermediate fauna. In *Oceanography: The Past.* M. Sears and D. Merriman, eds., Springer-Verlag, New York, pp. 360–372.

Mills, E. L., 1980b. One "different kind of gentleman": Alfred Merle Norman, invertebrate zoologist. *Zool. J. Linn. Soc.,* **68,** 69–98.

Milne-Edwards, A., 1861. Observations sur l'existence de divers mollusques et zoophytes à de très grandes profundeurs dans la Mer Méditerranée. *Ann. Sci. Naturelles,* Ser. 4, *Zoologie,* **15,** 149.

Milne-Edwards, A., 1880. *Rapport sur les Travaux de la Commission Chargée par M. le Ministre de l'Instruction Publique d'Étudier la Faune Sous-marine dans les Grandes Profundeurs de Golfe du Gascogne.* Gauthier Villars, Paris, 12 pp.

Moseley, H. N., 1879. *Notes by a naturalist on the Challenger,* Macmillan. London, xiv + 821 pp.

Murchison, R. I., 1871. Geography. Address by Sir Roderick Impey Murchison, Bart., K.C.B., D.C.L., LL.D., F.R.S., F.G.S., President of the Section. *Rep. Br. Assoc. Adv. Sci., Liverpool,* 1870, pp. 158–166.

Murray, J., 1895a. General observations on the distribution of marine organisms. *Rep. Sci. Res. Voy. H.M.S. Challenger, Summ. Scient. Res.,* **2,** 1431–1462.

Murray, J., 1895b. A summary of the scientific results. *Rep. Sci. Res. Voy. H.M.S. Challenger, Summ. Sci. Res.,* **1,** i–xxv, 1–1608.

Murray, J., 1911. Alexander Agassiz: His life and scientific work. *Bull. Mus. Comp. Zool.*, **54**, 139–158.

Murray, J. and J. Hjort, 1912. *The Depths of the Ocean. A General Account of the Modern Science of Oceanography Based Largely on the Scientific Researches of the Norwegian Steamer Michael Sars in the North Atlantic.* Macmillan, London, 821 pp.

Norman, A. M., 1868. Preliminary report on the Crustacea, Molluscoida, Echinodermata, and Coelenterata, procured by the Shetland Dredging Committee in 1867. *Rep. Br. Assoc. Adv. Sci., Dundee,* 1867, pp. 437–441.

Norman, A. M., 1869a. Letter to Charles Darwin, August 3, 1869. Darwin MSS, Cambridge University Library.

Norman, A. M., 1869b. Shetland final dredging report—Part II. On the Crustacea, Tunicata, Polyzoa, Echinodermata, Actinozoa, Hydrozoa, and Porifera. *Rep. Br. Assoc. Adv. Sci., Norwich,* 1868, pp. 247–336, 341–342.

Norman, A. M., 1880. Supplementary paper (on the French deep-sea exploration in the Bay of Biscay). *Rep. Br. Assoc. Adv. Sci., Swansea,* 1880, pp. 387–390.

O'Brien, C. F., 1970. *Eozoön canadense,* the dawn animal of Canada. *Isis,* **61,** 206–223.

Owen, Samuel, R. J., 1865. On the surface fauna of mid-ocean. No. 1. Polycystina and allied rhizopods. *J. Linn. Soc. (Zool.),* **8,** 202–205.

Owen, Samuel, R. J., 1868. On the surface fauna of mid-ocean. No. 2. Foraminifera. *J. Linn. Soc. (Zool.),* **9,** 147–157.

Petersen, C. G. J. and P. B. Jensen, 1911. Valuation of the sea. I. Animal life of the sea bottom, its food and quantity. (Quantitative studies.) *Rept. Dan. Biol. Sta.,* **20,** 81 pp.

Petit, G., 1970. Albert I of Monaco (Honoré Charles Grimaldi). *Dict. Sci. Biogr.,* **I,** 92–93.

Pettersson, H., 1953. *Westward Ho with the Albatross.* Macmillan, London, xii + 205 pp.

Pettersson, H., 1966. The voyage. *Rpts. Swedish Deep-Sea Exped. 1947–1948,* **1,** 1–123.

Porter, Roy, 1977. *The Making of Geology. Earth Science in Britain 1660–1815.* Cambridge University Press, xii + 288 pp.

Pourtalès, L. F. de, 1854. Extracts from the letters of L. F. Pourtalès, Esq., Assistant in the Coast Survey to the Superintendent, upon examination of specimens of bottom obtained in the exploration of the Gulf Stream by Lieut. Cmg. T. A. Craven and Lieut. Cmg. J. N. Moffit, U.S. Navy, assistants in the Coast Survey. *Rep. Supt. U.S. Coast Survey, 1853,* Appendix 30, pp. 82–83.

Pourtalès, L. F. de, 1867. Contribution to the fauna of the Gulf Stream at great depths. *Bull. Mus. Compar. Zool.,* **1,** 103–120.

Pourtalès, L. F. de, 1868a. Contributions to the fauna of the Gulf Stream at great depths (2d series). *Bull. Mus. Compar. Zool.,* **1,** 121–142.

Pourtalès, L. F. de, 1868b. Deep-sea dredgings in the region of the Gulf Stream. *Am. J. Sci.,* Ser. 2, **46,** 413–415.

Pourtalès, L. F. de, 1868c. Report on dredging near the Florida reefs. Appendix 12, *Rep. Supt. U.S. Coast Survey for 1868,* pp. 168–170.

Pourtalès, L. F. de, 1869. Report on the fauna of the Gulf Stream in the Strait of Florida. Appendix 16, *Rep. Supt. U.S. Coast Survey for 1867,* pp. 180–182.

Pourtalès, L. F. de, 1870. Der Boden des Golfströmes und der Atlantischen Küste Nord-Amerikas. *Petermann's Mitt.,* **16,** 393–398, 20 plates.

Rathbun, R., 1892. The U.S. Fish Commission, some of its work. *Century Mag.,* **43** (New Series, **21**), 679–697.

Ravetz, J. R., 1971. *Scientific Knowledge and its Social Problems,* Oxford University Press, New York, 449 pp.

Rehbock, P. F., 1975a. Huxley, Haeckel and the oceanographers: The case of *Bathybius haeckelii. Isis,* **66,** 504–533.

Rehbock, P. F., 1975b. *Organisms in Space and Time: Edward Forbes (1815–1854) and New Directions for Early Victorian Natural History.* Johns Hopkins University, Ph.D. thesis.

Rice, A. L., 1975. The oceanography of John Ross's arctic expedition of 1818: A reappraisal. *J. Soc. Bibl. Nat. Hist.*, **7,** 291–319.

Rice, A. L., H. L. Burstyn, and A. G. E. Jones, 1976. G. C. Wallich, M.D.—megalomaniac or misused oceanographic genius? *J. Soc. Bibl. Nat. Hist.*, **7,** 423–450.

Richard, J., 1907. *L'Océanographie*. Vuibert et Nony Editeurs, Paris, 398 pp.

Richard, J., 1910. *Campagnes Scientifiques de SAS le Prince Albert I de Monaco*, Monaco, xxix + 159 pp.

Richard, J., 1934. Liste générale des stations des campagnes scientifiques du Prince Albert du Monaco. Avec neuf cartes des itinéraires et des notes et observations. *Résultats des Campagnes Scientifiques Accompliés sur son Yacht par Albert I, Prince Souverain de Monaco*. No. 79, 471 pp. + 8 plates.

Ross, J. C., 1847. *A Voyage of Discovery and Research in the Southern and Antarctic Regions, During the Years 1839–1843*. John Murray, London, 2 vols.

Ross, John, 1819. *A Voyage of Discovery Made Under the Order of the Admiralty in His Majesty's Ships "Isabella" and "Alexander" for the Purpose of Exploring Baffin Bay, and Inquiring into the Possibility of a Northwest Passage*. John Murray, London, xl + 252 pp. + cxliv.

Rudwick, M. J. S., 1972. *The meaning of fossils. Episodes in the History of Palaeontology*. Macdonald, London, 288 pp.

Sachse, W., 1899. Die Deutschen Tiefsee Expedition. C. Bericht des Navigations-Offiziers der Expedition Walther Sachse. *Zeitschr. Gesellsch. Erdkunde Berlin*, **34,** 183–192, Tables 6, 7.

Sachse, W., 1925. Ausrüstung der *Valdivia. Wissenschaftliche Ergebnisse der Deutschen Tiefsee Expedition auf den Dampfer "Valdivia" 1898/99*, Vol. 10, No. 5, pp. 207–339 + plates.

Sanders, H. L., 1968. Marine benthic diversity: A comparative study. *Am. Nat.*, **102,** 243–282.

Sanders, H. L., 1969. Benthic marine diversity and the stability-time hypothesis. In *Diversity and Stability in Ecological Systems. Brookhaven Symposia Biol.*, No. 22, pp. 71–81.

Sanders, H. L., R. R. Hessler, and G. R. Hampson, 1965. An introduction to the study of deep-sea benthic faunal assemblages along the Gay Head–Bermuda transect. *Deep-Sea Res.*, **12,** 845–867.

Sars, M., 1850. *Beretning om en i Sommeren, 1849, Foretagen Zoologisk Reise i Lofoten og Finmarken*, Christiania.

Sars, M., 1864. Bemaerkninger Overdet Dyriske Livs Udbredning i Havets Dybder. Christiania, *Videnskabs-Selskabs Forhandlinger for 1864*.

Sars, M., 1868. Fortsatte Bemaerkninger over det dyriske Livs Udbredning i Havets Dybder. Christiania, *Videnskabs-Selskabs Forhandlinger for 1868*.

Scheltema, R. S. and A. H. Scheltema, 1972. Deep-sea biological studies in America, 1846–1872—their contribution to the *Challenger* expedition. *Proc. Roy. Soc. Edinburgh, B*, **72,** 133–144.

Schlee, S., 1973. *The Edge of an Unfamiliar World. A History of Oceanography*. Dutton, New York, 398 pp.

Schopf, T. J. M., 1968. Atlantic continental shelf and slope of the United States—nineteenth century exploration. *U.S. Geol. Survey, Prof. Paper* 592-F, pp. 1–12.

Schott, G., 1899. Die Deutsche Tiefsee Expedition. B. Berichte des Oceanographen der Expedition Dr. Gerhard Schott an das Reichs–Marine-Amt. *Zeitschr. der Gesellsch. f. Erdk. Berlin*, **34,** 135–183, Tables 4, 5.

Sigsbee, C. D., 1878. Reports on the results of dredging under the supervision of Alexander Agassiz, in the Gulf of Mexico, by the United States Coast Survey steamer "Blake," Lieutenant-Commander C. D. Sigsbee, U.S.N., Commanding. I. Description of the sounding machine, water-bottle, and detacher used on board the "Blake." *Bull. Mus. Compar. Zool.*, **5,** 169–179.

Sigsbee, C. D., 1880. Reports of the results of dredging, under the supervision of Alexander Agassiz, on the east coast of the United States, by the U.S. Coast Survey steamer "Blake," Commander J. R. Bartlet, U.S.N. VII. Description of a gravitating trap for obtaining specimens of animal life from intermedial ocean-depths. *Bull. Mus. Compar. Zool.*, **6,** 155–158, plates.

Sivertsen, E., 1968. Michael Sars, a pioneer in marine biology, with some aspects from the early history of biological oceanography in Norway. *Bull. Inst. Oceanogr. Monaco*, Special No. 2 (*Congr. Int. Hist. Oceanogr.* **1**), pp. 439–452.

Sorby, H. C., 1861. On the organic origin of the so-called "Crystalloids" of the Chalk. *Ann. Mag. Nat. Hist.*, Ser. 3, **8,** 193–200.

Spärck, R., 1956. Background and origin of the expedition. In *The Galathea Deep Sea Expedition 1950–1952,* A. F. Bruun et al., George Allen & Unwin, London, pp. 11–17.

Spratt, T. A. B., 1849. On the influence of temperature upon the distribution of the fauna in the Aegean Sea. *Rep. Br. Assoc. Adv. Sci., Swansea,* 1848, trans., pp. 81–82.

Spratt, T. A. B. and E. Forbes, 1847. *Travels in Lycia, Milyas, and the Cybyratis, in Company with the Late Rev. E. T. Daniell,* John Van Voorst, London, 2 vols.

Spratt, T. A. B., 1865. *Travels and Researches in Crete.* John Van Voorst, London, 2 vols.

Spry, W. J. J., 1877. *The Cruise of HMS Challenger,* London.

Stebbing, T. R. R., 1888. Report on the Amphipoda collected by HMS *Challenger* during the years 1873–76. *Rep. Sci. Res. Voy. HMS Challenger, Zoology,* **29,** 1–1737.

Steemann Nielsen, E. and E. A. Jensen, 1959. Primary oceanic production. The autotrophic production of organic matter in the oceans. *Galathea Rep.*, **1,** 49–138.

Studer, T., 1878. Ueber Siphonophoren des tiefen Wassers. *Zeitschr. Wiss. Zool.*, **31** (1), 1–3.

Swire, H., 1938. *The Voyage of the "Challenger": A Personal Narrative of the Historic Circumnavigation of the Globe in the Years 1872–6, by H. Swire.* Illustrated with reproductions from paintings and drawings in his journals; foreword by R. Swire, introduction by G. H. Fowler. Golden Cockerel Press, London, 2 vols.

Tanner, Z. L., 1893. Report upon the investigations of the U.S. Fish Commission Steamer Albatross from July 1, 1889, to June 30, 1891. *U.S. Comm. Fish & Fisheries, Report of the Commissioner, 1889–1891.* Appendix 1, pp. 207–342, chart.

Tanner, Z. L., 1897. Deep-sea exploration, a general description of the steamer *Albatross,* her appliances and methods. *Bull. U.S. Fish. Comm. for 1896,* **16,** 257–424, 40 plates + 76 figs.

Thistle, D., 1978. Harpacticoid dispersion patterns: implications for deep-sea diversity maintenance. *J. Mar. Res.*, **36,** 377–397.

Thomas, K. B., 1971. Carpenter, William Benjamin. *Dict. Sci. Biogr.*, **3,** 87–89.

Thomson, C. W., 1865. On the embryogeny of *Antedon rosaceus* Link (*Comatula rosacea,* Lamarck). *Phil. Trans. Roy. Soc.*, **155,** 513–544.

Thomson, C. W., 1869. *The Depths of the Sea. The Substance of a Lecture Delivered by Professor Wyville Thomson, LL.D., F.R.S.E., F.G.S., on the 10th of April, 1869 in the Theatre of the Royal Dublin Society,* 13 pp., privately printed.

Thomson, C. W., 1870a. Food of oceanic animals. *Nature,* **1,** 315–316.

Thomson, C. W., 1870b. Letter from Prof. Wyville Thomson to the Rev. A. M. Norman on the successful dredging of H.M.S. "Porcupine." *Rep. Br. Assoc. Adv. Sci., Exeter,* 1869, pp. 115–116.

Thomson, C. W., 1871. The continuity of the Chalk. *Nature,* **3,** 225–227.

Thomson, C. W., 1873. *The Depths of the Sea. An Account of the General Results of the Dredging Cruises of H.M.SS. "Porcupine" and "Lightning" During the Summers of 1868, 1869, and 1870, Under the Scientific Direction of Dr. Carpenter, F.R.S., J. Gwyn Jeffreys, F.R.S., and Dr. Wyville Thomson, F.R.S.,* Macmillan, London, xx + 527 pp.

Thomson, C. W., 1874a. On dredgings and deep-sea soundings in the South Atlantic, in a letter to Admiral Richards, C.B., F.R.S., *Proc. Roy. Soc.*, **22,** 423–428.

Thomson, C. W., 1874b. Preliminary notes on the nature of the sea-bottom procured by the soundings of H.M.S. "Challenger" during her cruise in the "Southern Ocean" in the early part of the year 1874. *Proc. Roy. Soc.*, **23,** 32–49, 4 plates.

Thomson, C. W., 1876. The "Challenger" expedition. *Nature,* **14,** 492–495.

Thomson, C. W., 1878. *The Voyage of the "Challenger." The Atlantic. A Preliminary Account of the General Results of the Exploring Expedition of H.M.S. "Challenger" During the Year 1873 and the Early Part of the Year 1876.* Harper, New York, 2 vols., 391 and 340 pp.

Thomson, C. W., 1880. General introduction to the zoological series of reports. *Rep. Sci. Res. Voyage H.M.S. Challenger. Zoology,* **1,** 1–50.

Tizard, T. H., H. N. Moseley, J. Y. Buchanan, and J. Murray, 1885. Narrative of the cruise of H.M.S.

"Challenger," with a general account of the scientific results of the expedition. *Rep. Sci. Res. Voy. H.M.S. Challenger, 1873–1876, Narrative,* **1** (1, 2), 1172 pp.

Tizard, T. H., J. Murray, et al., 1882. Exploration of the Faroe Channel, during the summer of 1880 in H.M.'s hired ship "Knight Errant," with subsidiary reports. *Proc. Roy. Soc. Edinburgh,* **11,** 638–720, plates.

Travailleur and *Talisman,* 1888–1927. *Expéditions Scientifiques du "Travailleur" et du "Talisman" Pendant les Années 1880, 1881, 1882, 1883.* Ouvrage publié sous les auspices du Ministère de l'instruction publique, sous la direction de A. Milne-Edwards (de 1888 à 1900), et continué par Edmond Perrier. Paris, Masson et Cie, Paris, 9 vols.

Tuchman, B. W., 1966. *The Proud Tower. A Portrait of the World Before the War 1890–1914.* Macmillan, New York.

Turner, F. M., 1978. The Victorian conflict between science and religion: A professional dimension. *Isis,* **69,** 356–376.

Vinogradov, M. E., 1970. Vertical distribution of the oceanic zooplankton. Program for Scientific Translations. Jerusalem, Israel, 339 pp.

Vinogradova, N. G., 1962. Some problems of the study of the deep-sea bottom fauna. *J. Oceanogr. Soc. Japan, 20th Anniv. Vol.,* pp. 724–741.

Wallich, G. C., 1860. *Notes on the Presence of Animal Life at Vast Depths in the Sea; with Observations on the Nature of the Sea Bed, as Bearing on Submarine Telegraphy.* Taylor and Francis, London, 38 pp.

Wallich, G. C., 1861. Remarks on some novel phases of organic life at great depths in the sea. *Ann. Mag. Nat. Hist.,* Ser. 3, **8,** 52–58.

Wallich, G. C., 1862. *The North-Atlantic Sea-Bed. Comprising a journey on H.M.S. Bulldog in 1860.* Van Voorst, London (unfinished), 160 pp., + 6 plates.

Wallich, G. C., 1869. On the vital functions of the deep-sea Protozoa. *Month. Microsc. J.,* **1,** 32–41, 231–233.

Wallich, G. C., 1876. *Deep-Sea Researches on the Biology of Globigerina. Where do the Globigerinae Live, Multiply, and Die?* John Van Voorst, London, 75 pp. + plate.

Whiteaves, J. F., 1872a. Notes on a deep-sea dredging-expedition round the island of Anticosti in the Gulf of St. Lawrence. *Ann. Mag. Nat. Hist.,* Ser. 4, **10,** 341–354.

Whiteaves, J. F., 1872b. Report on a deep-sea dredging expedition to the Gulf of St. Lawrence, Canada. *Canada, Rep. Dept. Marine & Fisheries for 1871,* pp. 1–12.

Whiteaves, J. F., 1873. Report on a second deep-sea dredging expedition to the Gulf of St. Lawrence, with some remarks on the marine fisheries of the Province of Quebec. *Canada, Rep. Dept. Marine & Fisheries for 1872,* pp. 1–22.

Whiteaves, J. F., 1874a. On recent deep-sea dredging operations in the Gulf of St. Lawrence. *Am. J. Sci.,* **7,** 1–9.

Whiteaves, J. F., 1874b. On recent deep-sea dredging operations in the Gulf of St. Lawrence. *Canadian Naturalist,* N. S., **7,** 257–267.

Whiteaves, J. F., 1874c. Report on deep-sea dredging operations in the Gulf of St. Lawrence, with notes on the present condition of the marine fisheries and oyster beds of part of that region. *Canada, Rep. Dept. Marine & Fisheries for 1873,* pp. 1–29.

Whiteaves, J. F., 1901. *Catalogue of the Marine Invertebrata of Eastern Canada.* Geological Survey of Canada, Ottawa, 271 pp.

Wild, J. J., 1878. *At Anchor,* London, 198 pp.

Willemoes-Suhm, R. von, 1877. *Challenger Briefe,* Engelmann, Leipzig.

Wolff, T., 1967. *Danish Expeditions on the Seven Seas.* Rhodos, Copenhagen, 336 pp.

Wolff, T., 1970. The concept of the hadal or ultra-abyssal fauna. *Deep-Sea Res.,* **17,** 983–1003.

Wright, E. P., 1868. Notes on deep-sea dredging. *Ann. Mag. Nat. Hist.,* Ser. 4, **2,** 423–427.

Wüst, G., 1964. The major deep-sea expeditions and research vessels. A contribution to the history of oceanography. *Progr. Oceanogr.,* **2,** 1–52.

Wyrtki, K., 1966. Oceanography of the eastern tropical Pacific Ocean. *Oceanogr. Mar. Biol. Ann. Rev.*, **4**, 33–68.

Yonge, C. M., 1972. The inception and significance of the *Challenger* expedition. *Proc. Roy. Soc. Edinburgh, B*, **72**, 1–13.

Zenkevitch, L. A. and J. A. Birstein, 1960. On the problem of the antiquity of the deep-sea fauna. *Deep-Sea Res.*, **7**, 10–23.

Zenkevitch, L. A., 1963. *Biology of the Seas of the USSR*. Wiley, New York, 955 pp.

ZoBell, C., 1946. *Marine Microbiology*. Chronica Botanica, Waltham, MA.

ZoBell, C. and R. Y. Morita, 1959. Deep-sea bacteria. *Galathea Rep.*, **1**, 139–154.

2. RECENT ADVANCES IN INSTRUMENTATION IN DEEP-SEA BIOLOGICAL RESEARCH

GILBERT T. ROWE AND MYRIAM SIBUET

1. Introduction

A radical change in deep-sea biological research in the 1970s was precipitated by new instrumentation. Manned deep submersibles [or deep submergence research vessels (DSRVs)] as well as tethered and free unmanned devices now allow biologists to consider aspects of physiology and dynamic ecology that had long been denied them because of the remote, alien character of the deep sea.

Sample taking in the deep sea has been reviewed extensively by Menzies et al. (1973). Until recently biologists could take samples from the deep sea only with simple mechanical devices at the remote end of a wire (Spärck, 1951). Such "samples" were qualitative, as with a trawl, or quantitative, as with a grab or box core. Whatever the device used, the recovered sample allowed enumeration and identification of a standing stock of organisms, whether they were fish or microbes. Organisms brought to the surface in such devices were invariably killed by decompression or temperature change. Even when animals surfaced alive in cold water, as was occasionally the case at high latitude or in the winter (Paul, 1973), one could not trust physiological studies because the milieu of high pressure was not duplicated. Such samples did not allow one to measure rates of biological processes as they occur in the deep sea. Great quantities of taxonomic, zoogeographic, and standing-stock data accumulated, but we knew nothing from such work about metabolism, growth rates, behavior, and small spatial scales of dispersion. We could not see, touch, or feel what we were doing.

Four capabilities that have implicit importance to field and laboratory biology but that could not be accomplished remotely on the end of a wire are (1) knowing where you are (precise navigation), (2) looking at what you are doing (observing the milieu), (3) touching and turning (manipulation) of the milieu, the biota, or experimental devices, and (4) taking living things home (recovery from the sea floor). New ability in these categories has revolutionized deep-sea biology. The remainder of this chapter deals with those instruments that allow us to do one or several of these four categories. We confine our discussion to gear that is used universally rather than that being developed and used by a single specialty within the field. This will eliminate the need for various authors in ensuing chapters to discuss this general gear and allow them to address their subjects more fully.

Contribution No. 767 from the Centre Océanologique de Bretagne, Brest, France.

2. Acoustics

Sound now allows us to communicate in the ocean. Ships on the surface send and receive different frequencies to precisely locate themselves and their gear, control and retrieve simple mechanical devices, penetrate and characterize deep deposits or resuspended particles, and follow large migrations of the fauna. Echo sounders with strip-chart recorders developed prior to World War II have had a number of uses of great importance. Bottom topographic charts are as necessary to biologists as they are to geologists. The most stunning biological discovery made with sound was the deep-scattering layer (DSL). A phenomenon in all bodies of water, first visualized conclusively with acoustic records, the DSL proved to be the diurnal migration of midwater fishes and zooplankton (Vinogradov and Tseitlin, this volume, Chapter 4), surfacing at night and descending out of the light in the daytime.

Acoustics is also used to precisely control gear depth relative to the bottom. By putting a sound source on an instrument, the depth of the instrument or its proximity to the bottom can be read off the echo sounder chart record (Bandy, 1965; Rowe and Menzies, 1967; Backus, 1966; Laubier et al., 1972; Menzies et al., 1973; Hessler and Jumars, 1974). Bottom contact switches can be placed within instruments that alter "ping" frequency or repetition rate or that turn off when an instrument touches bottom or loses its vertical orientation or when some mechanism is activated. Previously this was attempted by measuring variations in wire tension with a hydraulic–electric tensiometer or spring accumulator, but it was always attempted with great difficulty and a certain amount of guesswork.

The most sophisticated application for mapping seafloor topography has been a 16-beam echo sounder called *SEABEAM* (Renard and Allenou, 1979) to cover a swath of the bottom that has a width that is three-fourths of the water depth, that is, 4.5 km when the water depth is 6 km. The echo sounder is coupled to a computer and high-speed plotter to allow immediate data reduction and real-time visualization of topography. The aerial coverage and the contour interval are controlled by the operator. Navigational accuracy is related to ship's surface navigational precision, which is much better with a satellite or a bottom-mounted acoustic navigation net, as discussed below. SEABEAM is utilized on only a few of the larger research vessels, including the *Jean Charcot*.

3. Navigation Systems

Small-scale spatial studies require precise navigation, and long-term experiments require returning to precisely the same spot after predetermined periods of time. By the "same spot," we mean down to distances of a few meters. Conventional surface-ship navigation, although accurate to 50 m or so on the surface, cannot determine precise locations on the bottom. This can now be done by setting several acoustic transponders near the bottom, each with a different frequency (Anonymous, 1977). From the surface, a ship can interrogate each one acoustically, determine their range (distance) and bearing (direction) relative to the ship, and thereby determine the ship's location relative to the transponders within a few meters. When any instrument with a transponder on it is placed within the "net" of bottom transponders, its location relative to them and the surface ship can also be monitored. In this way a submersible with its own transponder can be directed by the surface ship to any location within or near the net. This navigation is especially useful for finding free vehicles referred to as "elevators" that carry experimental gear that a DSRV or a remotely operated vehicle (ROV) will operate or set in place with its manipulator.

Ship-tethered instruments can also be located and monitored precisely when near or on the bottom (Khripounoff et al., 1982). Typical studies by biologists at the Centre Oceanologique de Bretagne (COB) illustrate maximum use of the navigation system used aboard R/V *Jean Charcot* (Fig. 1). Each transponder is interrogated automatically and continuously, as is the transponder on a trawl, camera, and so on. The data are immediately reduced by shipboard computer to give distances to each mooring, and this information is then plotted (Fig. 2).

In a single region of about 30 km^2 between the four navigation transponders, all the biological sampling can be accurately plotted as it is being carried out. For the COB group, this would include numerous box cores, several *poisson RAIE* (photographic) lowerings (Fig. 3), several beam trawls, baited traps, sediment traps, current meters, and azoic mud boxes (Fig. 1). Differences in sediment types, topography, currents, and so on, can then be related to variations in the biota on scales of meters, rather than on scales of tens of kilometers as is the usual practice without the navigation net.

In hydrothermal vents studies, as well as in typical seafloor experiments, acoustic navigational aids are left in place for extended periods of time so that long-term experiments can be relocated and retrieved after months on the bottom. The growth experiments on the fast-growing organisms at the Galápagos hydrothermal vent region require such an approach, as do recolonization studies (Desbruyérès et al., 1980; Grassle, 1977).

4. Photography

Photography and its applications in the deep sea have been discussed in a number of reviews (Holme and MacIntyre, 1972; Hersey, 1967; Heezen and Hollister, 1971; Menzies et al., 1973), but of these, biologists have used photography in four applications: (1) broad surveys of sediment surface-living organisms (and their tracks and tails) where cameras can quantify sparsely distributed animals and lebenspüren more accurately than can grabs or trawls; (2) monitoring the performance of other instruments; (3) long-term time-lapse systems to assess animal behavior or erosion and deposition; and (4) short-term baited time-lapse deployments for studying the feeding behavior of organisms that are difficult to actually capture.

One of the earliest successful cameras was the Troika, used in submarine canyons (Cousteau, 1958).

The continental margin of the eastern coast of the United States has been extensively surveyed by use of a number of different camera systems. Rowe and Menzies (1968, 1969), Schoener and Rowe (1970), Menzies et al. (1973), and Rowe (1971) surveyed populations south of Cape Hatteras, using a bottom contact-triggered model on R/V *Eastward*. To the north between Cape Hatteras and Cape Cod, Wigley and Emery (1967) surveyed populations of the brittle star *Ophiomusium lymani* and the polychaete worm *Hyalinoecia tubicola* on the continental slope (Hersey, 1967). Their system consisted of a grab with a single-shot

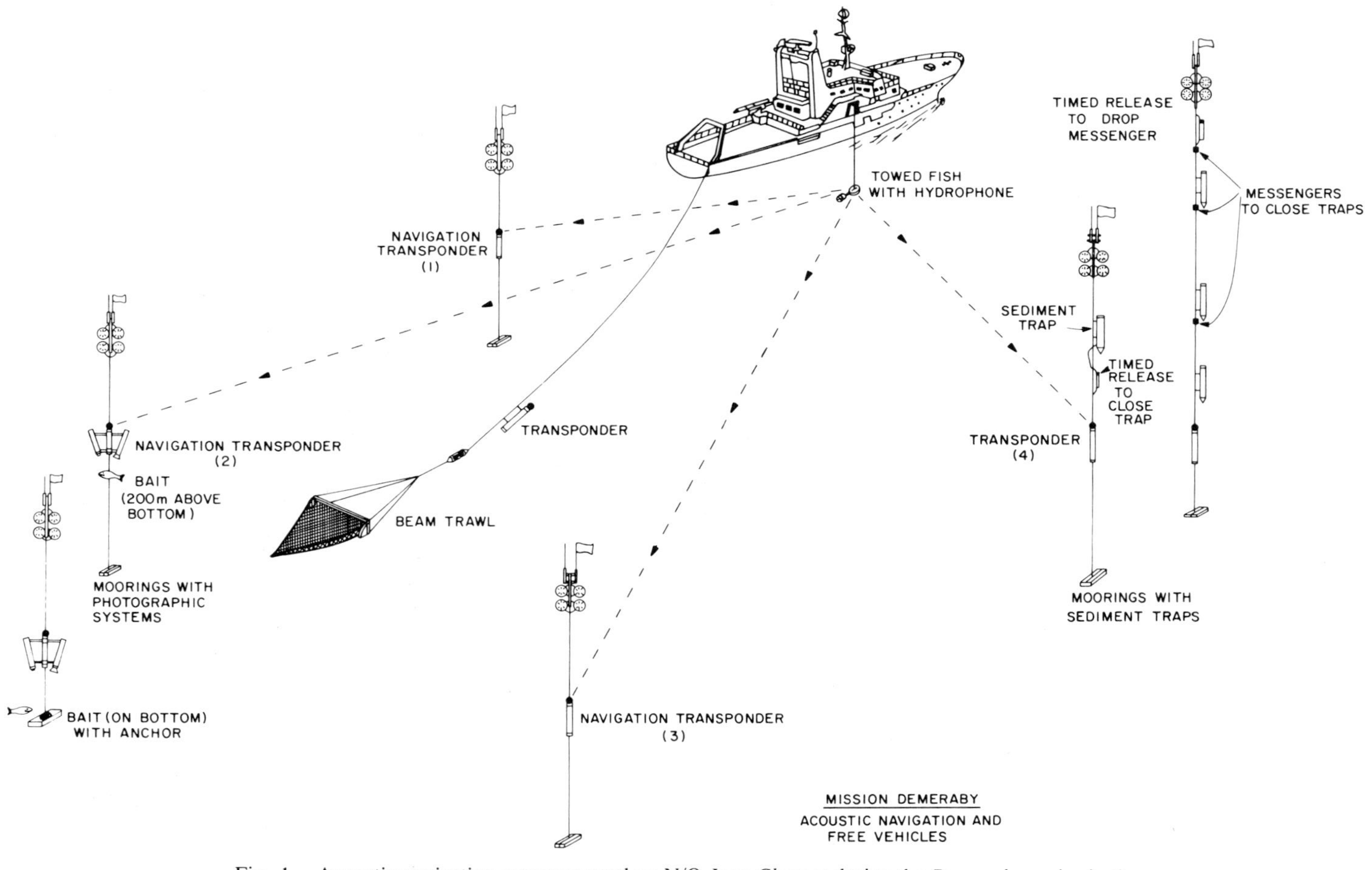

Fig. 1. Acoustic navigation arrays as used on N/O *Jean Charcot* during the *Demeraby* cruise in the Demerara abyssal plain (September 1980). Drawn by J.-C. Cavarec, COB.

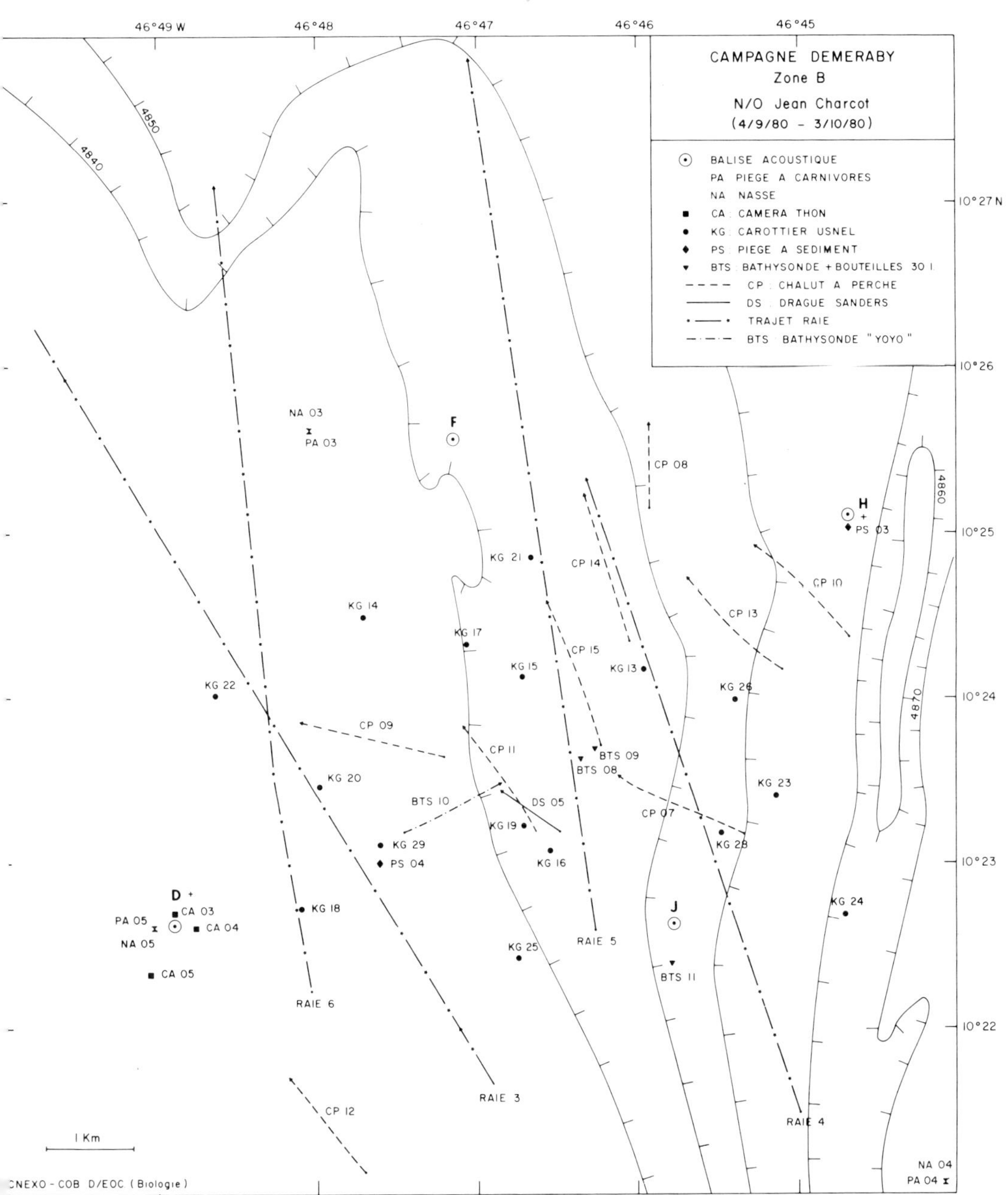

Fig. 2. Map showing an example of biological sampling plotted at a station delimited by four transponders. Bathymetry was obtained by SEABEAM echosounder.

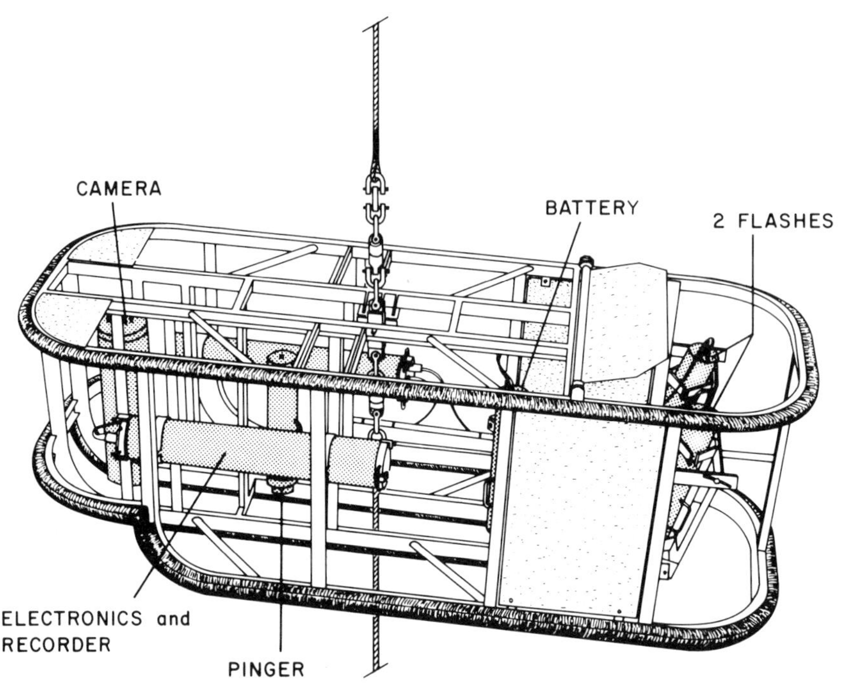

Fig. 3. Bottom photographic system called *Remarquage Abyssal d'Instruments pour l'Eploration* (RAIE) conceived at CNEXO.

camera mounted inside. Grassle et al. (1975), using the DSRV *Alvin* as a camera-carrying platform, made detailed surveys at several depths down the continental slope south of Massachusetts. Haedrich and Rowe (1978) combined the quantitative data accrued from their own bottom camera and trawl data with the Grassle et al. data to estimate biomass of the animals surveyed.

Estimates of holothurian and stalked crinoid abundances in rough topography have been made by use of the Troika in the Bay of Biscay (Sibuet and Lawrence, 1981; Conan et al., 1981).

Cameras with much larger film capacity now being used for general bottom surveying by geologists are also used by biologists. Some survey cameras are dragged as a sled across the bottom (Laban et al., 1963; Gennesseaux, 1966; Thiel, 1970). In other cases a ship's echo sounder is employed to keep the camera several meters above the bottom, while the camera takes in excess of 3000 shots automatically. Such a system (called the *poisson RAIE*) (Fig. 3) is used by biologists at the COB in Brest, France and has been used on U.S. vessels (where it is called *ANGUS*) on the Mid-Atlantic ridge and the Galápagos and Mexican hydrothermal vent regions. The camera system also provides an internal data record on the film that includes time, date, and station number. With the 3000 exposures, the acoustic navigation can be used with the data on the film to reconstruct where each exposure was made within several meters. This approach has proved especially useful on the Demerara Abyssal Plain and in the Cape Verde basin for comparison of large epibenthic forms and sedimentary tracks and trails with animals captured in trawls and box corers (Mauviel, 1982; Sibuet, 1982). The

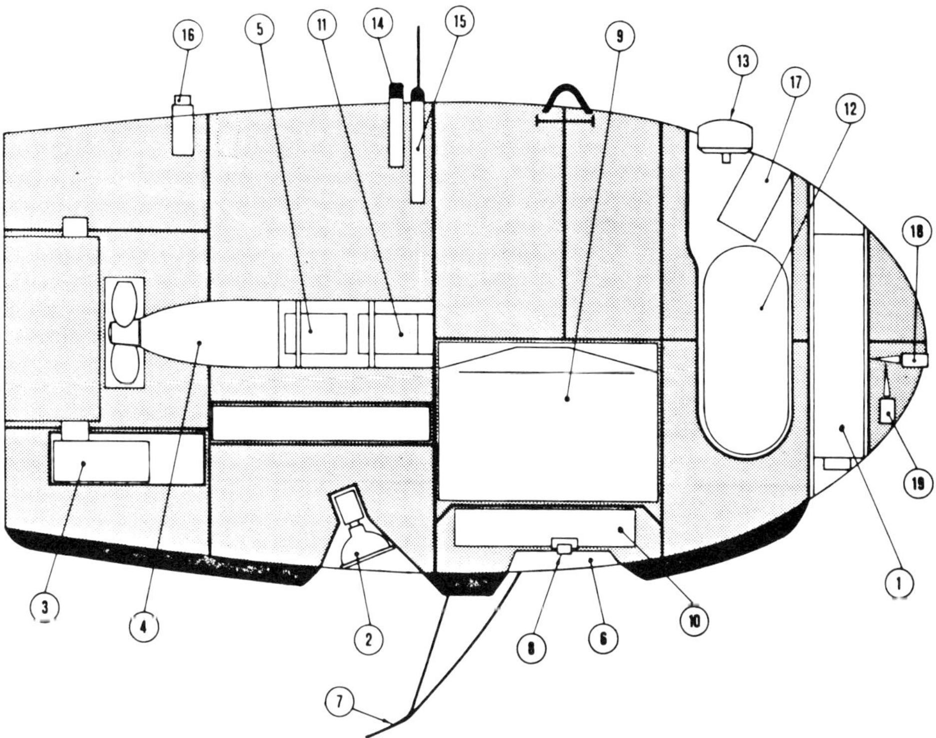

Fig. 4. Free unmanned vehicle Epaulard conceived at CNEXO. It is equipped with camera and stobe (1, 2), propeller and motor (4), rudder (3), transducer (13), and 12-kHz pinger (16), as well as controls to keep it a specific height above bottom and assist in launch and recovery on the surface by the tending ship.

Epaulard system developed by the Centre National pour l'Exploitation des Océans (CNEXO) is an unmanned free autonomous vehicle that can hover close to the bottom at depths of up to 6 km (Fig. 4). It will undoubtedly have extensive future use.

Some cameras have been placed in the entrances of beam trawls to record what the trawl catches, but early trials found that a trawl will bounce and pitch, making big clouds of sediment through which photography is impossible (Menzies et al., 1973). Quantifying trawl catches with cameras has been more successful recently (Rice et al., 1982) (Fig. 5) by use of a trawl that is better able to glide over the bottom.

The same large-capacity multishot automatic cameras used for surveys have been employed as stationary time-lapse cameras. The longest deployment of which we are aware was that by Paul et al. (1978), who photographed a small area in a manganese field in the Pacific for 206 days. This was accomplished by lowering the frequency of the exposures. Generally an exposure is taken every 10 sec or so while surveying; however, delays of several minutes to hours are incorporated in the timing circuit to conserve battery power and allow spreading the film out over weeks to months. Cinematic cameras can also be used if individual exposures are set in synchrony with a strobe (Edgerton et al., 1968). A 16-mm camera

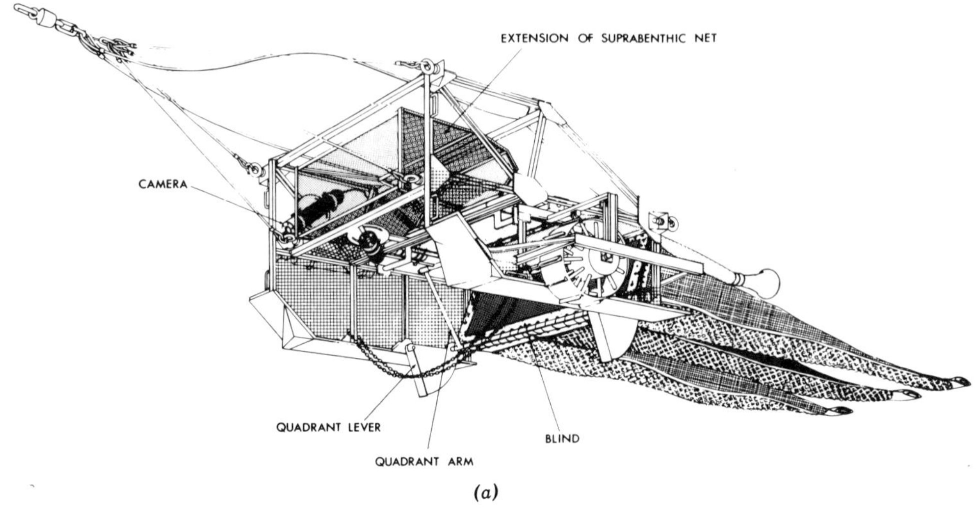

(a)

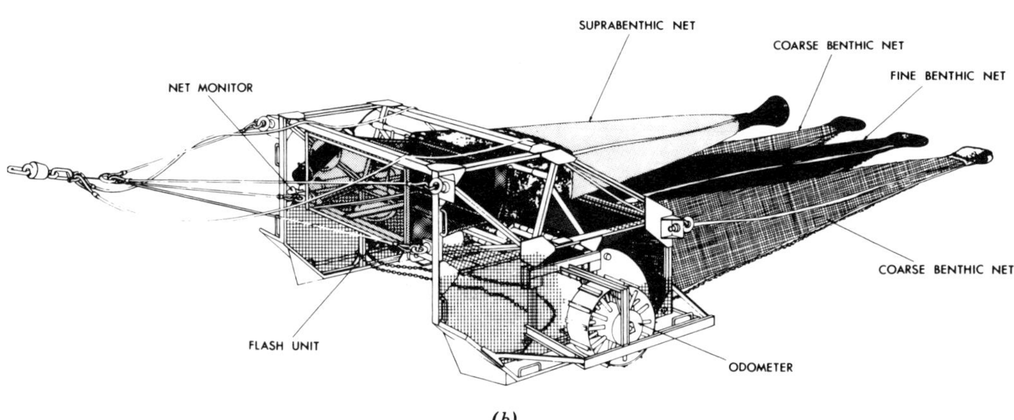

(b)

Fig. 5. Institute for Oceanographic Sciences epibenthic sledge with camera for photographing large organisms before they are captured. (Reprinted with permission from Rice, A. L., R. G. Aldred, E. Darlington, and R. A. Wild, 1982. The quantitative estimation of the deep-sea megabenthos; a new approach to an old problem. *Oceanologia Acta,* **5,** 63–72.)

deployed in the Hudson canyon revealed that extensive amounts of tidal resuspension occur on very short time scales, and in such dynamic areas as a canyon, small organisms rapidly remold sediment surface features etched by the burrowing or crawling of larger organisms (Rowe et al., 1974).

An exciting development was the placement of a can of bait below a moored, short-term, time-lapse camera. First it was discovered that numerous large fish and invertebrates were attracted to such material (Isaacs and Schwartzlose, 1975; Hessler et al., 1972). Such baited camera moorings led to the suggestion that amphipods replace fish in deep trenches (Hessler et al., 1978). When large fish were used as baits for longer periods of time, it was learned that such material can last several weeks on the bottom (Rowe and Staresinic, 1979). We have recently found at 4.8 km on the Demerara Abyssal Plain that a sequence of attraction by different species can be followed (Rowe and Sibuet, in preparation).

Twenty-hour deployments of a camera above a baited trap revealed that the presence of fish at 2 to 4.7 km can be periodic and semidiurnal in the Bay of Biscay (Guennegan and Rannou, 1979).

Television might be the best observational gear for use in the deep sea. Events can be seen as they are happening aboard ship and can be easily recorded. Although TV is used widely on shallow vehicles and on submersibles, it has not had wide application yet because of large power drain in the conductor cables between the camera and the surface vessel. The greatest future of TV use may be by experimentalists who need to control remote manipulators and vehicles.

5. Submersibles

Deep submergence research vessels have been widely used as purely observational tools (Barham et al., 1967; Pollio, 1968). The DSRV most widely used in work reported on by the authors in this volume has been the DSRV *Alvin*, managed by the Woods Hole Oceanographic Institution. French scientists have used the DSRV *Cyana* and the bathyscaphe *Archimede* (CNEXO, 1972); a French DSRV capable of working at 6-km depth is under construction. The bathyscaphe, the latest version of which is the FNRS III, was the first submersible to be used for extensive biological work (Pérès, 1959) and has reached depths of 11 km. Its lack of flexibility relative to a submersible such as the *Cyana* or *Alvin* has limited its utilization to an observational platform [Picard, Pérès, and Drach, cited in Houot (1967)]. Submersibles, however, are the most expensive of all purely observational methods, they present risk to the investigators, and their weather limitations all combine to make other methods preferable when possible. (They are, of course, indispensable when investigators must see their prey.)

Experimental manipulation, rather than surveying, is probably a DSRV's best biological use. It allows biologists to respond to events in their experiments as they perform them. The most successful manipulations on the sea floor have been carried out with DSRV *Alvin*. A manipulator operated by the pilot from inside the DSRV *Alvin*'s sphere, for example, can grasp objects, turn 360° at the wrist, and move vertically and laterally a meter or so. It has a payload of about 25 kg. Typically experimental gear such as box cores, incubation chambers, bell jars, and current meters are carried to the bottom in a basket mounted below the pilot's forward porthole. The pilot then lifts the gear out with the manipulator(s) and operates it for an observer. Most gear is designed to be operated by the manipulator's two opposing fingers (Winget, 1969; Rowe, 1979).

Devices that can be attached to or carried in the *Alvin* basket or that have been sent to the sea floor as an autonomous free vehicle to be operated by a manipulator(s) include (1) fall-cone penetrometer (Rowe, 1979); (2) sheer vane device (Winget, 1969); (3) dye pellet array for boundary-layer current measures (Cacchione et al., 1978); (4) odometer wheel (Rowe, 1979); (5) Ekman box corer (Rowe and Clifford, 1973; Thiel and Hessler, 1974); (6) bead spreader (to measure sediment mixing); (7) close-up camera(s), stereographic pairs to estimate subject size(s); (8) hypodermic syringes, in various configurations for subsampling experimental chambers and so on; (9) incubation chambers to measure metabolic processes, with (a) slurp gun for pelagic organisms (Smith, in press), (b) traps for fish (Smith and Hessler, 1974), and (c) bell jars to enclose a substrate (Smith and Teal,

1973); and (10) particle traps (Rowe and Gardner, 1979). Winget (1969) presents an extensive account of possible tools that the *Alvin* can use. A universal feature is their T-shaped handles.

6. Free Vehicles

A popular alternative to risky, expensive, weather-limited DSRVs has been free vehicles. Many of the devices first used successfully with a DSRV have later been converted to use as part of an autonomous free package. Their basic components include flotation; an anchor release system; a disposable anchor; and, of course, the experimental gear itself. The simplest of the free vehicles have been those that use corrosable links, usually made of magnesium, to drop an anchor after some predetermined length of time (Shutts, 1975). Others use timed releases that employ a clock-controlled switch in a deep-sea housing. The switch activates a solenoid that drops the anchor or puts a charge through a dissolving wire holding the anchor (Williams and Fairhurst, 1977). Timing circuits can also be used to sample sequentially over seasons (Deuser et al., 1981). The best control over free vehicles, however, is with acoustic releases, wherein an anchor-release, solenoid, or "burn wire" is activated when a sound of a particular frequency is sent to the device from a surface vessel (Smith et al., 1976). Several commands can be received by some models so that a series of mechanical tasks can be carried out (Wiebe et al., 1976; Smith et al., 1978). This could involve injection of a syringe full of poison or labeled nutrients into a container, taking a syringe sample from a container at the end of an incubation (Smith et al., 1978), closing the lid on a trap (Hinga et al., 1979; Rowe and Gardner, 1979; Smith and Hessler, 1974), or recovering a recolonized azoic mud experiment (Desbruyérès et al., 1980). The authors have routinely used a Williams–Fairhurst timed release to close sediment traps several hours before a mooring is recalled from the bottom with an acoustic release.

7. Remotely Operated Tethered Unmanned Vehicles

A number of remote tethered vehicles have been developed for engineering uses (NOAA Report, 1979). One of these, the SIO remote underwater manipulator (RUM), for example, has been used by deep-sea biologists in the project "Quagmire" off the coast of San Diego, California (Thiel and Hessler, 1974). The RUM is a vehicle that carries a multishot camera and strobe, television and lights, and a sensitive manipulator similar to that described above on the *Alvin*. These are all mounted on tank treads that can be driven along the bottom. And everything is controlled at the surface within a boxlike vessel called the *ocean research barge* (ORB). A center well within the ORB contains a winch that pulls RUM up by an umbilicus into the well when not in use. The ORB is anchored at three points, and positioning of the RUM involves moving the ORB around on the three anchors and driving the RUM over the bottom with the tank tracks. Neither is particularly easy or precise. The RUM has been used to investigate spatial pattern (Jumars and Eckman, Chapter 10 of this book) and metabolism (Smith, 1974; Smith and Hessler, 1974).

Presumably, remote manipulators will gain wider use in the future because their use is not life threatening, they are much less expensive to build or use than

DSRVs, and they are not as weather limited (Ballard, 1982). They are now widely used by industry at moderate depths.

8. Pressure Retrieval Systems

All organisms brought to the surface from great depths are rapidly decompressed and depending on the season may experience radical temperature change (Brauer, 1972). Multicellular organisms have only rarely been brought to the surface from depths of 2 km and kept alive (Paul, 1973). While microbial assemblages from great depths have been cultured at atmospheric pressure or returned to deep-sea pressures in pressure vessels, it is not known how the composition and metabolism of the original assemblage has been altered by the decompression.

In order to overcome the effects of decompression, a number of pressure-retaining samplers have been constructed recently (Macdonald and Gilchrist, 1969). Such samplers can be lowered to great depths on a wire where they can be closed mechanically. When hawled in, they retain their contents of water at ambient pressure (Jannasch and Wirsen, 1973). A device of similar design and small internal volume can be taken to the sea floor in DSRV *Alvin*'s basket, protected from thermal changes by an insulated container. The advantage to this sampler is that it can be opened and shut with the DSRV's manipulator, thereby controlling the proximity of the pressure vessel to the bottom (Jannasch et al., 1973). An important adjunct to the above devices is the capability to take samples from the vessel, add labeled medium to it, and to monitor oxygen consumption or CO_2 production (Tabor et al., 1981).

Greater internal volume allows such a pressure retaining device to be used as a trap (Yayanos, 1977). The numerous mobile scavenging amphipods that come to bait have been brought to the surface from 5.7 km with only minor alterations in pressure (Yayanos, 1978). Incubation of remains of an amphipod that had been recovered without decompression led to the isolation of a barophilic bacterium that had apparently lived on or in the amphipod. This was the first time that a barophile, an organism that grows more rapidly under high pressure than at atmospheric pressure, had been isolated since the original description of such organisms by ZoBell and Oppenheimer (1950).

9. Summary and Forecast

The remarkable strides taken in deep-sea biology have been due to the utilization of (1) submersibles and remote vehicles with manipulators, (2) free vehicles with timed and acoustic releases, (3) seafloor acoustic navigation systems, and (4) pressure-retrieval systems. Combinations of these systems allow experiments on rates of processes and small-scale dispersion to be conducted successfully. Initial progress of this nature was due almost solely to the use of submersibles (Rowe, 1979); however, a trend toward greater use of free vehicles and remote manipulators can be expected because of the expense, weather limitations, and risk associated with the use of manned submersibles. Submersibles will continue to be used to conduct many complex biological experiments that cannot be conducted by any other method.

Acknowledgments

We would like to thank our colleagues Lucien Laubier and Daniel Desbruyérès for comments on an early draft of this manuscript and Jean Claude Cavarec, who was in charge of the acoustic navigation on the Demeraby expedition, for drawing Figure 1. The work was supported by CNEXO in France and the DOE in the United States. It is contribution No. 767 from the Centre Océanologique de Bretagne, France.

References

Anonymous, 1977. *ATNAV II, the Expanded Capacity Acoustic Transponder Navigation System.* Technical Bulletin, AMF Sea-Link Systems, Herndon, VA, 19 pp.

Backus, R. H., 1966. The "pinger" as an aid in deep trawling. *J. Cons. Perm. Int. Explor. Mer.,* **30,** 270–277.

Ballard, R. D., 1982. Argo and Jason. *Oceanus,* **25,** 30–35.

Bandy, O., 1965. The pinger as a deep water grab control. *Undersea Technol.,* **6,** 36.

Barham, E. G., N. J. Ayer, Jr., and R. E. Boyce, 1967. Macrobenthos of the San Diego Trough: Photographic census and observations from bathyscaphe, *Trieste. Deep-Sea Res.,* **14** (6), 773–784.

Brauer, R. W., ed., 1972. *Barobiology and the Experimental Biology of the Deep Sea.* University of North Carolina Sea Grant Program, Chapel Hill, NC, 428 pp.

Cacchione, D. A., G. T. Rowe, and A. Malahoff, 1978. Submersible investigation of outer Hudson Submarine Canyon. In *Sedimentation in Submarine Canyons, Fans, and Trenches.* D. J. Stanley and G. Kelling, eds., Dowden, Hutchinson and Ross, Stroudsburg, PA, pp. 42–50.

Centre National pour l'Exploitation des Océans, 1972. *Bathyscaphe "Archimede" Campagne 1966 a Madere, Campagne 1969 aux Açores.* Rapport CNEXO No. 3, 125 pp.

Conan, G., M. Roux, and M. Sibuet, 1981. A photographic survey of a population of the stalked crinoid *Diplocrinus* (Annacrynus) *wyvillethomsoni* (Echinodermata) from the bathyal slope of the Bay of Biscay. *Deep-Sea Res.,* **28A,** 441–453.

Cousteau, J. Y., 1958. *Calypso* explores an undersea canyon. *Natl. Geogr. Mag.,* **113,** 373–396.

Desbruyérès, D., J. Bervas, and A. Khripounoff, 1980. Un cas de colonisation rapide d'un sediment profond. *Oceanologica Acta,* **3,** 285–291.

Deuser, W. G., E. H. Ross, and R. F. Anderson, 1981. Seasonality in the supply of sediment to the deep Sargasso Sea and implications for the rapid transfer of matter to the deep ocean. *Deep-Sea Res.,* **28A,** 495–505.

Edgerton, H. E., V. E. MacRoberts, and K. Read, 1968. An elapsed time photographic system for underwater use. *Eighth International Congress on High Speed Photography,* 52 pp.

Gennesseaux, M., 1966. Prospection photographique des canyons sous-marins du Var et du Var et du Paillon (Alpes Maritimes) au moyen de la Troika. *Rev. Geogr. Phys. et Geol. Dyn.,* **8** (1), 3–38.

Grassle, J. F., 1977. Slow recolonization of deep-sea sediment. *Nature,* **265** (5595), 618–619.

Grassle, J. F., H. L. Sanders, R. R. Hessler, G. T. Rowe, and T. McClellan, 1975. Pattern and zonation: A study of the bathyal megafauna using the research submersible *ALVIN. Deep-Sea Res.,* **22,** 457–481.

Guennegan, Y., and M. Rannou, 1979. Semi-diurnal rhythmic activity in deep sea benthic fishes in the Bay of Biscay. *Sarsia,* **64** (1–2), 113–116.

Haedrich, R. L. and G. Rowe, 1978. Megafaunal biomass in the deep-sea. *Nature,* **269,** 141–142.

Heezen, B. C. and C. D. Hollister, 1971. *The Face of the Deep.* Oxford University Press, London, 659 pp.

Hersey, J. Brackett, ed., 1967. *Deep-Sea Photography.* Johns Hopkins Press, Baltimore, 310 pp.

Hessler, R. R., C. L. Ingram, A. Aristides Yayanos, and B. Burnett, 1978. Scavenging amphipods from the floor of the Phillipine Trench. *Deep-Sea Res.,* **25,** 1029–1047.

Hessler, R. R., J. Isaacs, and E. Mills, 1972. Giant amphipod from the abyssal Pacific Ocean. *Science,* **175,** 636–637.

Hessler, R. R. and P. A. Jumars, 1974. Abyssal community analysis from replicate box cores in the central north Pacific. *Deep-Sea Res.*, **21**, 185–209.

Hinga, K. R., J. McN. Sieberth, and G. R. Heath, 1979. The supply and use of organic material at the deep sea floor. *J. Mar. Res.*, **37**, 557–579.

Holme, N. A. and A. D. McIntyre, 1972. *Methods for the Study of Marine Benthos.* IBP Handbook No. 16. Blackwell, Oxford, 334 pp.

Honjo, S., 1978. Sedimentation of materials in the Sargasso Sea at 5,367 m deep station. *J. Mar. Res.*, **36**, 469–4192.

Honjo, S., 1980. Material fluxes and modes of sedimentation in the mesopelagic and bathypelagic zones. *J. Mar. Res.*, **38**, 53–97.

Houot, A. 1967. *Campagnes de l'Archimede Rapport*, 90 pp.

Isaacs, J. D. and R. A. Schwartzlose, 1975. Active animals of the deep-sea floor. *Sci. Am.*, **233** (4), 85–91.

Jannasch, H. W. and C. O. Wirsen, 1973. *In situ* response of deep sea microorganisms to nutrient enrichment. *Science*, **180**, 641–643.

Jannasch, H. W., C. O. Wirsen, and C. L. Winget, 1973. A bacteriological pressure retaining deep-sea sampler and culture vessel. *Deep-Sea Res.*, **20**, 661–664.

Kripounoff, A., D. Desbruyérès, and P. Chardy, 1980. Les peuplements benthiques de la faille Vema: Donnees quantitatives et bilan d'energie en milieu abyssal. *Oceanologica Acta*, **3**, 187–198.

Laban, A., J. Pérès, and J. Piccard, 1963. La photographie sous-marine profonde et son exploration scientifique. *Bull. Inst. Oceanogr. Monaco*, **60** (1258): 1–32.

Laubier, L., J. Martinais, and D. Reyss, 1972. Deep sea trawling and dredging using ultrasonic techniques. In *Barobiology and the Experimental Biology of the Deep Sea*, R. Brauer, ed., op. cit.

Laubier, L. and M. Sibuet, 1977. Campagnes BIOGAS. *Résultats des Campagne à la Mer.* Number 11, CNEXO Publication, 57 pp.

Laubier, L. and M. Sibuet, 1979. Ecology of the benthic communities of the deep Northeast Atlantic. *Ambio*, **6**, 37–42.

Macdonald, A. G. and I. Gilchrist, 1969. Recovery of deep seawater at constant pressure. *Nature*, **222**, 71–72.

Mauviel, A., 1982. *La Bioturbation Actuelle dans le Milieu Abyssal de l'Océan Atlantique Nord.* Thèse de 3ème Cycle, U. B. O., 114 pp.

Menzies, R. J. and G. T. Rowe, 1968. The LUBS, a large undisturbed bottom sampler. *Limnol. Oceanogr.*, **13**, 708–714.

Menzies, R. J. and G. T. Rowe, 1969. The distribution and significance of detrital turtle grass, *Thalassia testudinum*, on the deep-sea floor off North Carolina. *Int. Revue ges. Hydrobiol.*, **54**, 219–222.

Menzies, R. J., R. Y. George, and G. T. Rowe, 1973. *Abyssal Environment and Ecology of the World Oceans.* Wiley, New York, 488 pp.

NOAA, 1979. Remotely operated vehicles. U.S. Department of Commerce, Washington, DC, 150 pp. plus appendices.

Paul, A. Z., 1973. Trapping and recovery of living deep-sea amphipods from the Arctic Ocean floor. *Deep-Sea Res.*, **20**, 289–290.

Paul, A. Z., E. Thorndike, L. G. Sullivan, B. Heezen, and R. Gerard, 1978. Observations of the deep-sea floor from 206 days of time-lapse photography. *Nature*, **272**, 812–814.

Pérès, J. M., 1959. Remarques génèrales sur une ensemble de 15 plongeés effectués avec le bathyscaphe F.N.R.S. III. *Ann. Inst. Oceanogr.*, **35**, 259–285.

Pollio, J., 1968. Stereo-photographic mapping from submersibles. *Proceedings of the SPIE Seminar on Underwater Photo-optical Instrumentation Applications, San Diego, Calif., February 5–6, 1968.* Society of Photo-optical Instrumentation Engineers, Redondo Beach, CA, 1968, pp. 67–71 (*SPIE Seminar Proceedings*, Vol. 12).

Renard, V. and J. P. Allenou, 1979. SEABEAM, multi-beam echo-sounding in JEAN CHARCOT. *Int. Hydrogr. Rev.*, **56**, 35–67.

Rice, A. L., R. G. Aldred, E. Darlington, and R. A. Wild, 1982. The quantitative estimation of the deep-sea megabenthos; a new approach to an old problem. *Oceanologica Acta*, **5**, 63–72.

Rowe, G. T., 1971. Observations on bottom currents and epibenthic populations in Hatteras Submarine Canyon. *Deep-Sea Res.*, **18,** 569–581.

Rowe, G. T., 1979. Monitoring with deep submersibles. In *Monitoring the Marine Environment*, D. Nichols, ed. *Biol. Soc. London*, pp. 75–85.

Rowe, G. T. and C. H. Clifford, 1973. Modifications of the Birge–Ekman box corer for use with SCUBA or deep submergence research vessels. *Limnol. Oceanogr.*, **18** (1), 172–175.

Rowe, G. T. and W. Gardner, 1979. Sedimentation rates in the slope water of the northwest Atlantic Ocean measured directly with sediment traps. *J. Mar. Res.*, **37,** 581–600.

Rowe, G. and R. L. Haedrich, 1979. The biota and biological processes of the continental slope. In *Continental Slopes,* O. Pilkey and L. Doyle, eds. Society of Economic Petrologists and Mineralogists Special Publication No. 27, Tulsa, OK, pp. 49–59.

Rowe, G. T. and R. J. Menzies, 1967. Use of sonic techniques and tension recordings as improvements in abyssal trawling. *Deep-Sea Res.*, **14,** 271–274.

Rowe, G. T. and R. J. Menzies, 1968. Orientation in two bathyal, benthic decapods, *Munida valida* Smith and *Parapagurus pilosimanus* Smith. *Limnol. Oceanogr.*, **13,** 549–552.

Rowe, G. T. and R. J. Menzies, 1969. Zonation of large benthic invertebrates in the deep-sea off the Carolinas. *Deep-Sea Res.*, **16,** 531–537.

Rowe, G. T. and N. Staresinic, 1979. Sources of organic matter to the deep-sea benthos. *Ambio Special Report, The Deep Sea—Ecology and Exploitation,* **6,** 19–23.

Rowe, G. T., G. Keller, H. Edgerton, N. Staresinic, and J. Mac Ilvaine, 1974. Time-lapse photography of the biological reworking of sediments in Hudson Submarine Canyon. *J. Sed. Pet.*, **44** (2), 549–552.

Schoener, Amy and G. T. Rowe, 1970. Pelagic *Sargassum* and its presence among the deep-sea benthos. *Deep-Sea Res.*, **17,** 923–925.

Shutts, R., 1975. *Unmanned Deep-Sea Free-Vehicle System.* Marine Technicians Handbook Series, Inst. Marine Resources, Scripps. Inst. Oceanogr., 50 pp.

Sibuet, M., 1982. *Résultats de la Campagne de Biologie Benthique dans le Bassin du Cap Vert.* VII—*Mégafaune Benthique.* Report CEA, TMC/14 575 Biologie.

Sibuet, M. and J. Lawrence, 1981. Organic content and biomass of abyssal Holothuroids (Echinodermata) from the Bay of Biscay. *Mar. Biol.*, **65,** 143–147.

Smith, K. L., Jr., 1974. Oxygen demand of San Diego Trough sediments: An *in situ* study. *Limnol. Oceanogr.*, **19,** 939–944.

Smith, K. L., Jr., 1978a. Benthic community respiration in the N.W. Atlantic Ocean: *In situ* measurements from 40 to 5200 m. *Mar. Biol.*, **47,** 337–347.

Smith, K. L., Jr., 1978b. Metabolism of the abyssopelagic rattail *Coryphaenoides armatus* measured *in situ*. *Nature,* **274,** 362–364.

Smith, K. L., Jr., in press. Metabolism of deep-sea zooplankton measurements *in situ*. *Limnol. Oceanogr.*

Smith, K. L., Jr., and R. R. Hessler, 1974. Respiration of benthopelagic fishes: *In situ* measurements at 1230 meters. *Science,* **184,** 72–73.

Smith, K. L., Jr. and J. M. Teal, 1973. Deep-sea benthic community respiration: An *in situ* study at 1850 meters. *Science,* **179,** 282–283.

Smith, K. L., Jr., C. H. Clifford, A. Eliason, B. Walden, G. T. Rowe, and J. M. Teal, 1976. A free vehicle for measuring benthic community metabolism. *Limnol. Oceanogr.*, **21,** 104–170.

Smith, K. L., Jr., G. A. White, M. B. Laver, and J. A. Haugsness, 1978. Nutrient exchange and oxygen consumption by deep-sea benthic communities: Preliminary *in situ* measurements. *Limnol. Oceanogr.*, **23** (5), 997–1005.

Spärck, R., 1951. Density of bottom animals on the ocean floor. *Nature,* **168** (4264), 112–113.

Spencer, D. P., P. Brewer, A. Fleer, S. Honjo, S. Krishnaswami, and Y. Nozaki, 1978. Chemical flux from a sediment trap experiment in the deep Sargasso Sea. *J. Mar. Res.*, **36,** 493–523.

Tabor, P. S., J. W. Deming, K. Ohwada, H. Davis, M. Waxman, and R. Colwell, 1981. A pressure retaining deep ocean sampler and transfer system for measurement of microbial activity on the deep sea. *Microb. Ecol.*, **7,** 51–65.

Thiel, H., 1970. Ein fotoschlitten für biologische und geologische karterungen des meeresbotens. *Mar. Biol.*, **1**, 223–229.

Thiel, H. and R. R. Hessler, 1974. Ferngesteuertes Unterwasser—fahrzeug erforscht Tiefseeboden. *UMSHAU in Wissenschaft und Technik*, **14**, 451–453.

Wiebe, P., S. Boyd, and C. Winget, 1976. Particulate matter sinking to the deep-sea floor at 2000 m in the tongue of the Ocean, Bahamas with a description of a new sedimentation trap. *J. Mar. Res.*, **34**, 341–354.

Wigley, R. L. and K. O. Emery, 1967. Benthic animals, particularly *Hyalinoecia* (Annelida) and *Ophiomusium* (Echinodermata), in sea-bottom photographs from the continental slope. In *Deep-Sea Photography*, J. B. Hersey, ed. The Johns Hopkins Press, Baltimore, pp. 235–250.

Williams, A. J., III and K. Fairhurst, 1977. Two oceanographic releases. *Ocean Eng.*, 205–210.

Winget, C. L., 1969. *Hand Tools and Mechanical Accessories for a Deep Submersible*. WHOI Technical Report 69-32, 179 pp.

Yayanos, A. A., 1977. Simply actuated closure for a pressure vessel: Design for use to trap deep-sea animals. *Rev. Sci. Instrum.*, **48**, 786–789.

Yayanos, A. A., 1978. Recovery and maintenance of live amphipods at a pressure of 580 bars from an ocean depth of 5700 meters. *Science*, **200**, 1056–1059.

Yayanos, A. A., A. S. Dietz, and R. van Boxtel, 1979. Isolation of a deep-sea barophilic bacterium and some of its growth characteristics. *Science*, **205**, 808–810.

ZoBell, C. E. and C. H. Oppenheimer, 1950. Some effects of hydrostatic pressure on the multiplication and morphology of marine bacteria. *J. Bacteriol.*, **60**, 771–781.

3. BIOMASS AND PRODUCTION OF THE DEEP-SEA MACROBENTHOS

GILBERT T. ROWE

1. Introduction and Historical Perspective

The investigation of the numbers of organisms per unit area of bottom began when quantitative sampling devices were used on the *Galathea* expedition (Spärck, 1956). A number of Soviet studies began in the same period, with the use of an OKEAN grab (Zenkevitch, 1961).

The purpose of quantifying the abundance of life in the deep sea followed the rationale that early naturalists used in quantifying abundance of life in shallow water. They were searching for exploitable fisheries or the importance of the benthos in nourishing exploitable bottom fish populations. The Soviets, led by Len Zenkevitch, concluded that the fewest organisms were found on deep abyssal plains far from shore because such environments were (1) least affected by the flow of detrital matter from continents and (2) pelagic productivity was lowest there (Zenkevitch, 1961). Separation of the importance of the two sources of organic matter was not possible. The Soviet information (Fig. 1), viewed as worldwide contours of biomass, illustrated a high biomass in the equatorial Pacific or a positive relationship between biomass and pelagic production (Zenkevitch, 1961). The early Soviet work allowed a good general view of the deep ocean. Belyaev (1966) stated: "After many years of research . . . it is evident that the density of the oceanic benthic fauna decreases with depth and distance from shore, as well as from polar and temperate to tropical waters . . . . All these changes are related to one crucial factor—the dependence of the benthic fauna on food resources."

Quantification of stocks for the purpose of understanding productivity and energy flow requires knowledge of the relationships between biomass and respiration, growth, predation, gonad production, and excretion. Although such information on shallow species is available, reliable information on deep-sea organisms has been gathered only recently on just a few of the larger more common species. In his early synthesis Zenkevitch presumed that such information could be inferred from standing stocks.

In this chapter, after a brief review of methods of sampling and sample processing, the findings of Zenkevitch and a review of some of the work on which his broad generalizations were based are presented. That review is extended to specific contemporary studies designed to answer questions about quantitative variations that are related to depth, surface production, temperature, and bottom topography. Organisms smaller than the macrofauna are not considered here. The

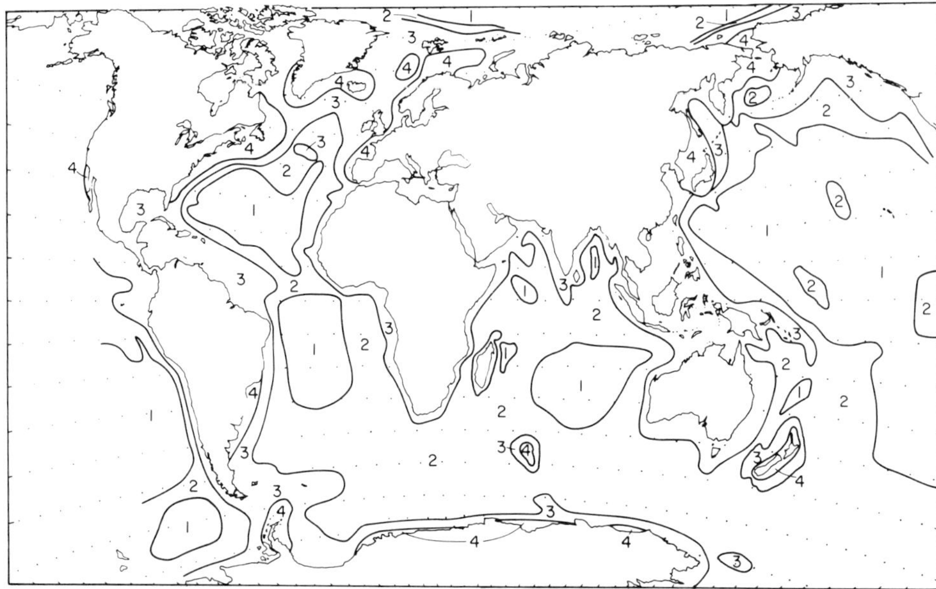

Fig. 1. Worldwide wet weight biomass of the macrobenthos ($1 \leq 0.1$ g/m^2, $2 = 0.1$ to 1 g/m^2, $3 = 1$ to 50 g/m^2, and $4 \geq 50$ g/m^2). [Modified from Vinogradova, (1976), after Zenkevitch, (1961).]

meiofauna and microbiota are covered in other chapters of this volume (Thiel, Chapter 5; Jannasch and Wirsen, Chapter 6) according to definitions presented below. The different measurements of biomass are considered to the extent that deep-sea biologists have studied the subject. Included with this are approaches to conversion of wet weights to variables such as dry weight, carbon, and nitrogen. The final section is devoted to the recent measurements of growth and gonad development and how these are influenced by sources of organic matter and how such information, when coupled with abundance, can be used to construct material or energy budgets.

2. Sampling Deep-Sea Fauna

Early quantitative deep-sea benthic studies used shallow-water techniques. Clam shell grabs were lowered to the bottom, closed securely, and then returned to the surface. Aboard ship the animals were separated from the sediment with sieves or screens on a table or in a hopper. A cleanly washed sample of animals was fixed in formalin and preserved in alcohol, and back in the museum or laboratory the animals were separated to species, weighed, measured, and identified as accurately as possible. The kinds of devices used have been reviewed by Holme and McIntyre (1973).

The approach used in shallow water has a number of problems in the deep sea. The grabs are too small. Instead of sinking vertically, small grabs are carried far away from the ship horizontally as the ship drifts. They hit on their sides, fail to close, or take a sample only partially full. This "kiting" on the long transit can result in the wire sinking faster than the grab. Small grabs can also close prematurely because of the rolling of the ship. Failure to detect the grab's bottom contact results in piling wire on the bottom. The wire can become kinked and knotted. Small gear also covers too little area to adequately sample the sparse fauna. The jaws or covers often leak, losing smaller forms on the long return to the ship. Whereas such losses in shallow water might result in a loss of a few percent of the total abundant fauna, in the deep sea such a loss could be of far greater significance.

Successful deep-sea samplers thus have had to be much bigger and heavier, cover a larger area, and be virtually watertight. Most have been enlarged versions of shallow water models. Menzies et al. (1973) review such instruments and discuss the criteria for a good deep-sea sampler. More recent gear is also presented in Chapter 2 of this volume (Rowe and Sibuet).

The present-day literature in this review is based on a few kinds of gear that deserve mention. The OKEAN grab has always been used by the Soviets; it covers 0.25 m^2. An anchor dredge (Carey and Hancock, 1965) gained wide use because it usually works and sample size could be several times that of even the largest grab or box corer (Sanders et al., 1965). This dredge slices off a 10-cm-thick layer of surface sediment until it fills up and then it rejects additional mud. Area covered can be estimated with the following equation:

$$\text{Area covered} = \frac{\text{dredge width} \times \text{dredge depth}}{\text{sample volume}}$$

Unfortunately, if it fails to penetrate the full 10 cm, an overestimate could be made or if the forward plate plows into the surface mud an underestimate would result, with a loss mostly of surface-living species. The open front allows winnowing or loss of sample on the return to the surface. We can do no more than guess at how these biases have affected published data.

Considerable data have been collected with a large spade or box corer (Fig. 2), often referred to as the *US-NEL box corer* (for U.S. Naval Electronics Laboratory). Early models used by geologists were enlarged by biologists to cover a greater area, flapper valves were added at the top of the box to minimize a bow wave, and inserts were sometimes added inside the box to allow the 0.25-m^2 area to be subsampled evenly (Jumars, 1976; Jumars and Eckman, this volume, Chapter 10). The history of its development and use are covered by Hessler and Jumars (1974).

Manned and unmanned submersibles, too, have been used to take deep-sea samples, allowing sampling in specific patterns. This has been accomplished with small Ekman box corers (see Chapter 2) handled with a submersible's mechanical arm (Rowe and Clifford, 1973; Rowe et al., 1982; Thiel and Hessler, 1974).

3. Sample Processing

An important difference between shallow and deep-sea sampling has been the use over the years of progressively finer sieves for washing the animals from the sediment. Because the fauna is sparse and individuals are small, samples washed through shaker tables lined with conventional screens retained very few organisms. Invertebrates have been categorized by sieve size (Mare, 1942) (Table I), but the categories designated by Mare for shallow water have been abandoned by deep-sea biologists (Table II). Many shallow-water invertebrate taxa considered

Fig. 2. The US-NEL spade (box) corer, with spare box in foreground removed from corer and resting on dolly used to move full box around deck. (Photograph courtesy of Ocean Instruments, San Diego, California.)

TABLE I
Size Categories in Benthic Invertebrate Communities

Microbenthos	<0.04 mm (=40 μm) (Burnett, 1978)
Meiobenthos	>0.1 mm, 0.05–0.001 mg/organism (Mare, 1942)
	>30–60 μm (=0.03–0.06 mm) (Hulings and Gray, 1971), excluding macrofauna
Macrobenthos	>1-mm sieve, >1 mg (Mare, 1942)
	>0.42-mm sieve (Sanders et al., 1965, etc.)
	>0.5-mm sieve (Frankenberg and Menzies, 1968)
	>0.297-mm sieve (Hessler and Jumars, 1974)
	>0.250 mm (Khripounoff et al., 1980)
	>0.12 mm (Paul and Menzies, 1974)
Megabenthos	Visible on the sea floor in bottom photographs (Barham et al., 1967; Hessler and Jumars, 1974; Rowe and Menzies, 1969; Grassle et al., 1975; Rice et al., 1982); "trawl-caught" organisms > 3-cm stretch mesh (Haedrich and Rowe, 1978)

TABLE II
Macrobenthos Sieve Sizes as Used by Various Authors

Spärck, 1956	Not stated; presumed to be 1.75 mm by Jumars and Hessler (1976)
Soviet studies	Early studies, 1955–1970, unstated, but presumed to be 0.5 mm, as in later work according to N. Vinogradova, personal communication
Frankenberg and Menzies, 1968	0.5 mm
Sanders et al., 1965	0.42 mm
Griggs et al., 1969	0.42 mm
Carey and Ruff, 1974	0.42 mm
Rowe and Menzel, 1971	0.42 mm
Rowe, 1971a, 1971b	0.42 mm
Rowe et al., 1974	0.42 mm
Rowe et al., 1975	0.42 mm
Rowe et al., 1982	0.42 mm
Gage, 1977	0.42 mm
Dahl et al., 1977	0.42 mm
Hecker and Paul, 1979	0.42 mm
Carey, 1981	0.42 mm
Hessler and Jumars, 1974	0.297 mm
Jumars and Hessler, 1976	0.297 mm
Smith, 1978	0.297 mm
Khripounoff et al., 1980	0.250 mm
Desbruyérès et al., 1980	0.250 mm
Paul and Menzies, 1974	0.12 mm

to be macrofauna by Mare are lost through sieves with openings of 1 or 2 mm; thus Sanders et al. (1965) used a much smaller sieve (0.42 mm or 420 μm), believing that it captures most of those groups.

The abandonment of Mare's scheme was provoked by the need to obtain a better representation of what lived in the deep sea. Mare intended to differentiate between the macrofauna (polychaetes, crustaceans, mollusks, and echinoderms mostly) and a "meiofauna," which she defined as nematodes, benthic copepods, kinorhinchs, forams, and so on, and small polychaetes and bivalves, generally being much smaller than 2 mm. The work of Spärck (1956) somewhat earlier varied only slightly from this [using a 1.76-mm sieve "presumably," according to Jumars and Hessler (1976)]. Soviet investigations used 0.5-mm sieves with grab samples but have called the captured organisms a fraction of the "meiofauna" as Mare would have intended. The decrease in the size of the sieves appears to have been a competitive chronological progression (Table II). Some authors have steadfastly retained the principle that everything above a certain sieve is macrofauna, no matter what the taxa involved. The first variation from this practice appears to have been that of Sanders et al. (1965), who reasoned that those taxonomic components of the fauna that are caught on 0.42-mm sieves along with the macrofauna but are generally regarded as meiofaunal taxa, should not be counted in assessments of animal density or diversity. In their studies off the

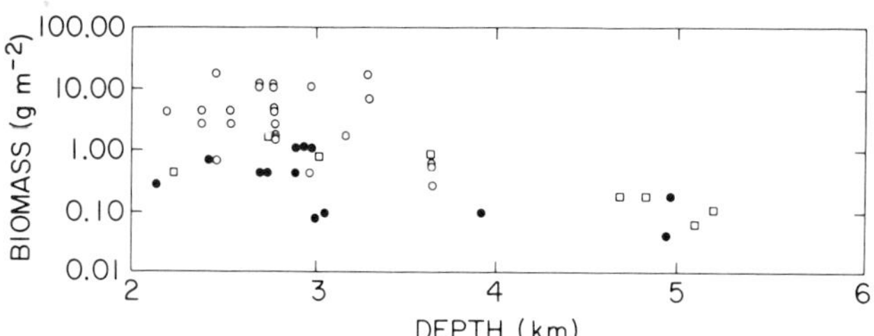

Fig. 3. Comparison of wet weight biomass in the northwestern Atlantic using 0.297-mm sieve [□, Smith (1978)] and 0.42-mm sieve [●, Rowe et al. (1974); ○, (1982)].

eastern coast of the United States, off the southwestern coast of Africa, and off the northeastern coast of South America, using a 0.42-mm sieve, they thus excluded nematode worms, harpacticoid copepods, and ostracods. An effect of this precedent has been an arbitrary variation between investigators in which taxa are counted when animal density is estimated, thus making comparisons of ocean basins based on abundance very difficult.

Screen size has a distinct influence on how much is caught, of course. Smith (1978) pointed this out when comparing animal density in the northwestern Atlantic. Using a mesh of 297 μm, he caught more individuals than did Rowe et al. (1974) and Sanders et al. (1965), who used 420 μm in the same region. This effect of sieve mesh size is far less apparent when biomass from the two sieves is compared. Wet weight biomass from samples sieved through 420-μm mesh (Rowe et al., 1982; Rowe et al., 1974) are nestled indistinguishably within biomass caught by Smith by use of a 297-μm sieve (Fig. 3), for example. Jumars and Hessler (1976) concluded that only about 15% of the macrofaunal individuals are lost from using a 420-μm rather than a 297-μm sieve.

Vital stains, typically Rose Bengal, aid in the sorting of benthos from sediments. This helps the sorter to locate very small organisms and also facilitates recognition of those organisms such as forams that were living when captured, as opposed to merely empty shells or tests. Only a few drops are necessary and should be added to the fresh or foramalin-preserved samples rather than after transfer to alcohol (H. Thiel, personal communication).

4. Measurements of Biomass

The density of organisms per unit area has been a variable often measured; however, biomass, the weight of living organisms, is a more meaningful measure for quantification of life on the sea floor (Table III). The preferable method of determining a wet-weight biomass is to weigh all the animals freshly caught, but generally this weight is taken as the animals are being sorted out and identified back in the laboratory after fixation and preservation. The latter is "a preserved wet weight" and may differ slightly from a fresh weight, especially if preservation has been in alcohol. To my knowledge, all such data from the deep sea are in "preserved wet weight." The amount of water in the tissues of different taxa can vary, and thus a number of investigators have dried their samples to a constant weight to provide more equitable comparison between taxa (Khripounoff et al., 1980). Shells, too, are problems; therefore, some investigators either remove the

TABLE III
Measures of Biomass

Fresh wet weight	Weight after blotting excess liquid
Preserved wet weight	Weight as above after preservation (Rowe et al., 1974, 1982; Hecker and Paul, 1979; Carey, 1981)
Dry weight	Weight after constant weight at 60–90°C
Dry weight with shells removed	As above, but with shells cut away or dissolved with acid
Ash-free dry weight	Weight loss on combustion at 500°C
Elemental analysis of carbon or nitrogen	Elemental analyzer on combustion (Rowe and Menzel, 1971; Rowe, 1971b)
Partitioning of organic matter into lipid, carbohydrate, and protein	Chemical analysis of individual component compounds (Sibuet and Lawrence, 1981, Barnes et al., 1976)
Caloric content	Heat production on ignition (Khripounoff et al., 1980)

tissues from shells or dissolve the shells with a dilute acid (Rowe and Menzel, 1971; Rowe, 1971b). Organic matter combusts at 500°C, but $CaCO_3$ (calcium carbonate) and other inert material does not, so some investigators have determined an ash-free dry weight that is an estimate of the weight of all organic tissue. These methods are detailed in Holme and McIntyre (1973).

Biomass can also be estimated by analyzing the samples for individual elements, such as carbon or nitrogen, again with separate removal of carbonate carbon (Rowe and Menzel, 1971), for caloric energy content determined by heat production (calorimetry) on rapid combustion (Khripounoff et al., 1980) or categories of organic compounds such as lipids, carbohydrates, and proteins (Sibuet and Lawrence, 1981). It was concluded that the patterns of variations in mass with depth were the same with the use of wet, dry, or organic carbon weights (Rowe and Menzel, 1971) (Fig. 4). Unpublished data from Peru follow the same trends (Fig. 5).

Numbers of animals, on the other hand, could not be substituted for measures of mass exactly, even though the measures of mass and abundance followed the same trends, as Sanders et al. (1965) indicated in their justification for not measuring biomass. This is principally because average size of the macrofauna appears to decrease with depth in some basins (Thiel, 1975; Polloni et al., 1979; Gage, 1977).

A primary use of biomass measures in the past has been comparison between studies. Numbers of individuals are more difficult to use because such data are more sensitive to (1) screen size, (2) fragmentation during sieving, (3) loss from a sampling apparatus, (4) variable inclusion or exclusion of different meiofaunal groups, and (5) overlooking the smaller individuals as the animals are being separated from the inanimate residue (sands and silts) that is retained by the sieves. Because most of the total biomass tends to be found in less abundant, large individuals rather than in numerous smaller ones, biomass is less sensitive to such artifacts; thus biomass is more dependable for comparing studies and estimating production.

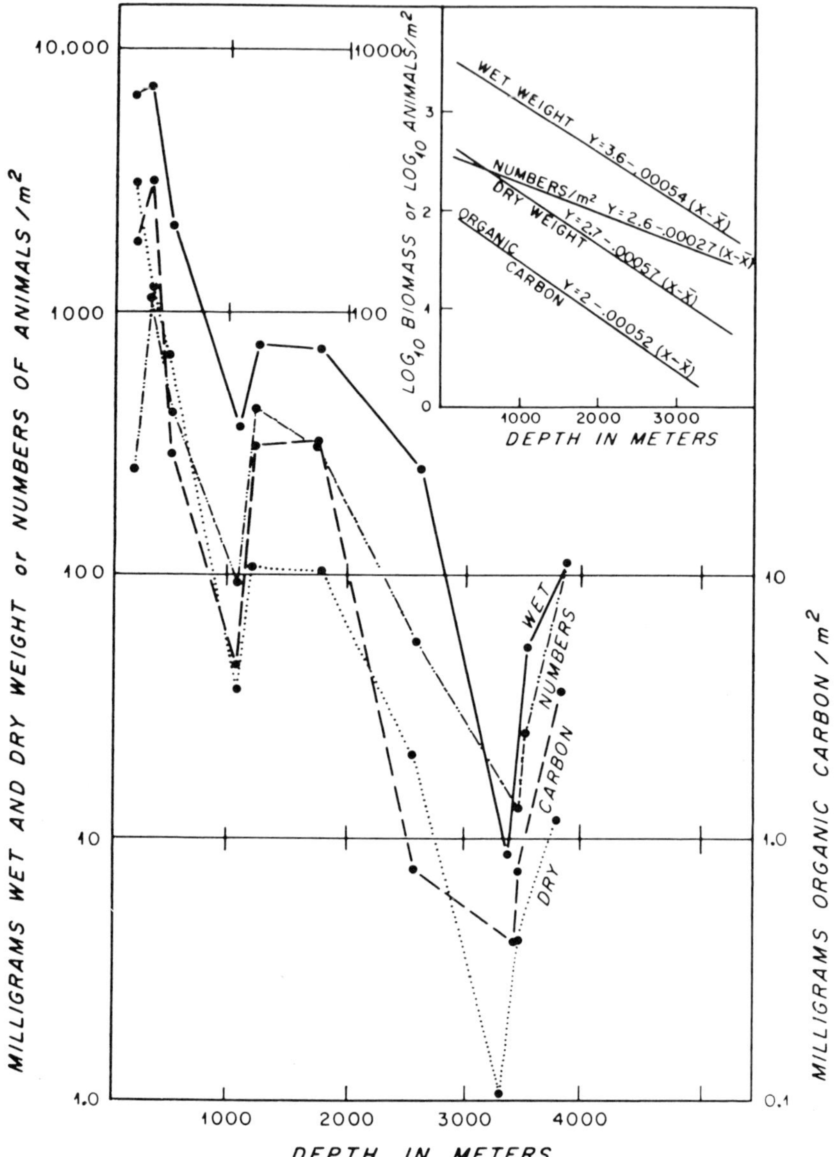

Fig. 4. Comparison of density, wet, dry, and carbon weights in the Gulf of Mexico. [From Rowe and Menzel (1971).]

An idea of how mass is distributed qualitatively among different groups reveals some interesting features of deep-sea animals (Table IV). The echinoderms are among the largest of the benthic invertebrates, but much of their tissue is $CaCO_3$ or water. A trawl thus may contain mainly holothurians by wet weight; on drying, however, most of the organic carbon might be found in a few macrourid fish. Some of the other groups are just the reverse. Bivalve mollusks grow to great size with massive shells in shallow water, but deep-sea shelled mollusks have light, fragile shells. As noted in the section on abyssal productivity, echinoderms have a

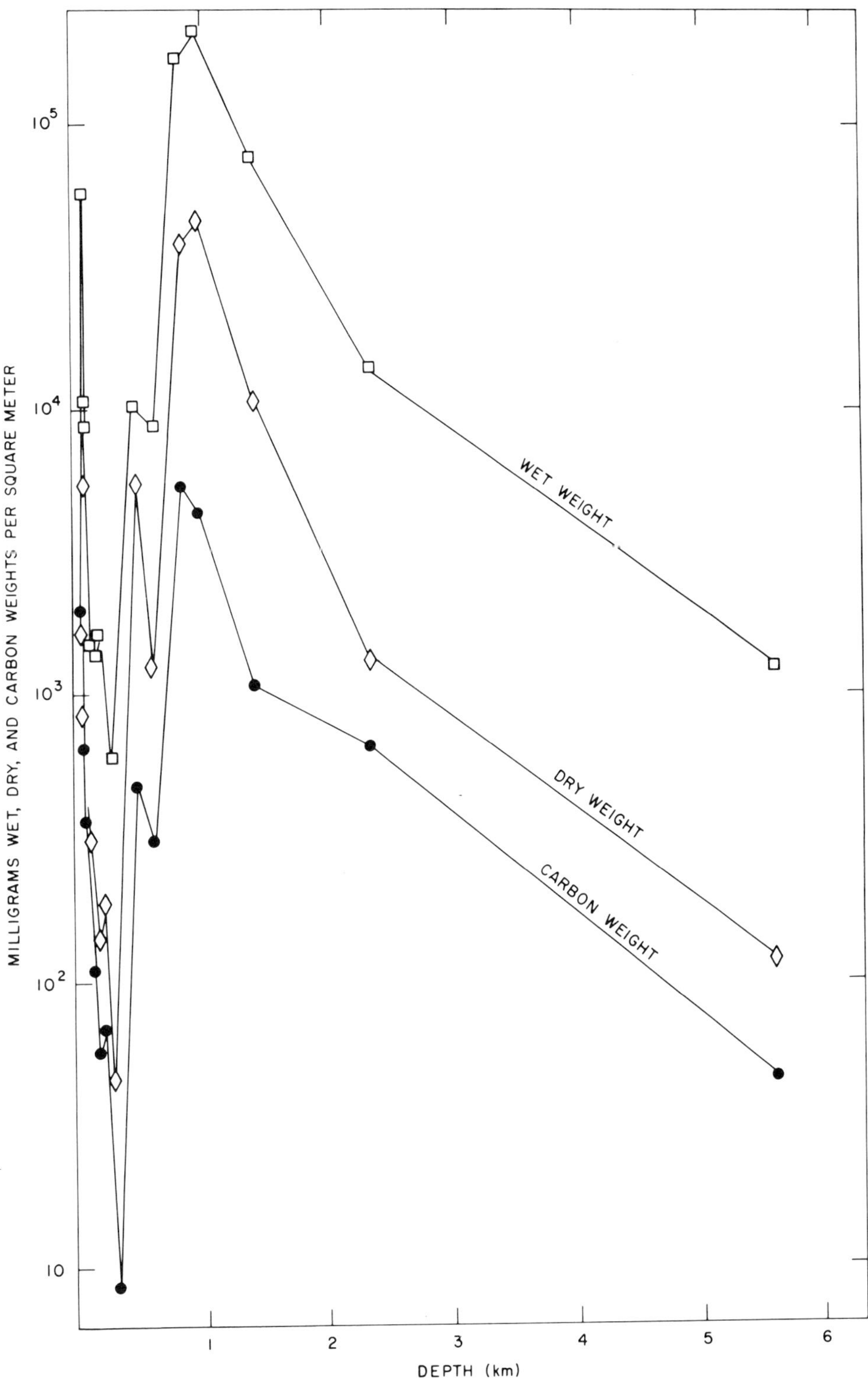

Fig. 5. Comparison of wet, dry, and carbon weights in samples from off the Peruvian coast.

TABLE IV
Comparison of Measures of Biomass in Different Taxa

Taxon	Total Numbers of Individuals	Total Wet Weight[a] (mg)	Shell-Free Total Dry Weight[b] (mg)	Organic Carbon as Percent of Wet Weight	Percent Carbon of $CaCO_3$ Free Dry Weight	Number of Analyses
Polychaete annelid worms	2695	27,330	3580	5.1% (±1.28)[c]	40.6%	21
Amphipod crustaceans	113	336	44.6	4.5% (±1.08)	47.2% (±17.4)	9
Tanaid crustaceans	58	31	3.2	2.9% (±2.04)	43.9% (±16.3)	5
Cumacean crustaceans	53	128	9.9	2.7% (±.2)	34.0% (±7.4)	2
Decapod crustaceans (crabs)	1	21	4.3	4.9%	24.1	1
Cirripede crustaceans (goose neck barnacles)	21	585	56.2	3.7%	38.4	1
Bivalve mollusks	149	10,302	4957.1	3.4% (±1.51)	24.8% (±13)	10
Gastropod mollusks	14	1,488	151.9	3.4% (±1.45)	29.8% (±8.9)	5
Aplacophoran mollusks	18	1,375	219.0	5.7% (±2.79)	37.5% (±7.6)	6
Scaphopod mollusks	1	58	5.6	4	42	1
Sipunculid worm	94	555	76.0	5.2% (±1.21)	39.9% (±7.9)	7
Anthozoan coelenterate (anemone)	1	228	11.9	2.2	42.9	1
Hydrozoan coelenterate	10	36.3	8.9	2.3% (±0)	23.1% (±14)	3
Ophiuroid echinoderms (brittlestars)	32	42,369	9563	3.16% (±0.98)	36.4% (±14.2)	5

Echinoid echinoderms (urchins)	11	30,604	3821	1.3% (±.28)	13.3% (±5.6)	3
Asteroid echinoderms (sea stars)	1	2,310	200	4	46	1
Holothuroid echinoderms (sea cucumbers)	4	675	146	1.9	8.7	1
Porifera (sponges)	2	253	24.8	0.8% (±.2)	7.4% (±4.7)	2
Nematode[d] worms	2763	451	59	3.2% (±2.05)	46.5% (±21.6)	12

Organisms	Percent Carbon of Wet Weight	Percent Carbon of Dry Weight
Polychaetes	5.1%	40.6%
Crustaceans	3.7%	37.5%
Mollusks	4.1%	33.5%
Echinoderms	2.6%	26.1%
Nematodes	3.2%	46.5%
Macrobenthic infauna (polychaetes, amphipods, and bivalves)	4.3%	37.5%
Average of all categories (n = 19)	3.4%	33.0%
Total number of animals 5983	Total number of analyses 99	

[a] Formalin-preserved weight, after 2 min of blotting on paper towel.
[b] Drying to constant weight at 60 to 90°C after acid treatment to remove hard parts.
[c] Standard deviation.
[d] Usually considered meiofauna.

relatively fast rate of growth, whereas some mollusks grow slowly; this may be related to their abilities to deposit $CaCO_3$ shell matter. In any case, good measures of biomass are invaluable in any attempt to assess the relative importance of species in the biological dynamics of an ecosystem, and the deep sea is no exception. Comprehension of the precise partitioning of energy within the community will require assessment of the relative distribution of fat, carbohydrate, and protein among the more important species, as has been initiated for holothurians (Barnes et al., 1976; Sibuet and Lawrence, 1981).

A comparison of the measures of mass from one set of samples [off the coast of Pisco, Peru (Rowe, 1971b)] illustrates both the general consistencies and the unpredictable variance that results from lumping different mass weights of whole samples of macrofauna (Fig. 5). In the deeper samples (2 to 5.7 km), wet is about 10 times dry weight and 1.5 $\log_{10}$ units carbon weight. At lesser depths, the unpredictable nature of the Peruvian coast (low oxygen at ca. 75 to 750 m) is imposed on the measurements, but this is evident in all the sets in a more or less equivalent fashion. The same conclusions would be drawn from each set of data: a low oxygen concentration retards the development of biomass (Frankenberg and Menzies, 1968; Rowe, 1971a). Nonetheless, for use in a budget or model of the system, the carbon values would be most useful. In lieu of these, conversion from wet or dry weights to carbon would suffice (Table IV); or, for a nitrogen model, the carbon values could be divided by 5 (from Table V) to give values in terms of nitrogen.

In this set of samples from the Peruvian coast, 5983 animals were analyzed for organic carbon in 99 separate analyses. The analyses were made on samples separated among taxonomic groups that were as specific as possible. A summary of these analyses (Table IV) allows presentation of the relationships between measures of wet, dry, and carbon mass in major taxa. This allows conversion from wet or dry weights into units of carbon, but it must be used with caution. The major components of the infauna are polychaetes, crustaceans, and bivalves; hence the average of these groups (wet weight is 4.3% carbon, and dry weight is 37.5% carbon) would be most appropriate for grab sample wet or dry weight conversion to carbon, rather than the overall animal average.

The echini were also analyzed by using samples from the northwestern Atlantic as an estimate of the epifaunal carbon and nitrogen content. For determination of organic carbon in the wet weight, the following relationship should be used (Table V):

$$\text{Organic carbon weight} = \text{g wet weight (0.053 g carbon/g dry weight)} \times 0.36 \frac{\text{g dry weight}}{\text{g wet weight}}$$

$$\text{Organic carbon} = 0.019 \text{ g carbon/g wet weight}$$

These values for organic carbon are considerably lower than for the fauna in general in Table IV or, for that matter, for the echini in Table IV. This is because a large fraction of the echini weight is carbonate endoskeletal plates and spicules (71%). The higher values *for echini* in Table IV are suspected to be an artifact resulting from incomplete dissolution of the carbonate during acidification.

TABLE V
Epifaunal Organisms

Taxon	H_2O (%)[a]	CO_3 (%)[b]	Organic Carbon (%)[c]	Nitrogen (%)[d]
Ophiuroid echinoderms (brittlestars)				
Amphiophiura bullata	38.4	87.7	5.2	1.07
Ophiomusium lymani	37.3	90.7	3.60	0.80
O. armigerum	48.9	88.4	4.48	1.06
Ophiura ljungmani	55.0	86.0	5.86	1.22
Asteroid echinoderms (sea stars)				
Zoroaster fulgens	66.1	79.5	7.91	1.55
Paragonaster subtilis	49.1	77.3	3.79	0.94
Dytaster insignis	66.9	73.2	7.89	2.04
Echinoid echinoderms (urchins)				
Echinus sp.	70.9	71.2	7.56	0.99
Holothuroid echinoderms (sea cucumbers)				
Benthodytes sp. (epifaunal)	91.8	6.3	—	3.8
Euphronides cornuta (epifaunal)	91.4	53.7	—	0.98
Molpadia blakei (burrowing)	70.8	36.7	2.86	0.60
Molpadia sp.	85.4	41.8	3.38	0.57
Mean	64.3	70.7	5.25	1.30

[a] Percent H_2O = [(wet wt − dry weight)/wet weight] (100).
[b] Percent CO_3 = [(dry weight − dry weight after acid treatment)/dry weight] (100).
[c] Percent organic carbon = {1 − [(dry weight − carbon weight after acid)/dry weight]} (100).
[d] Percent nitrogen = {1 − [(dry weight − nitrogen weight after acid)/dry weight]} (100).

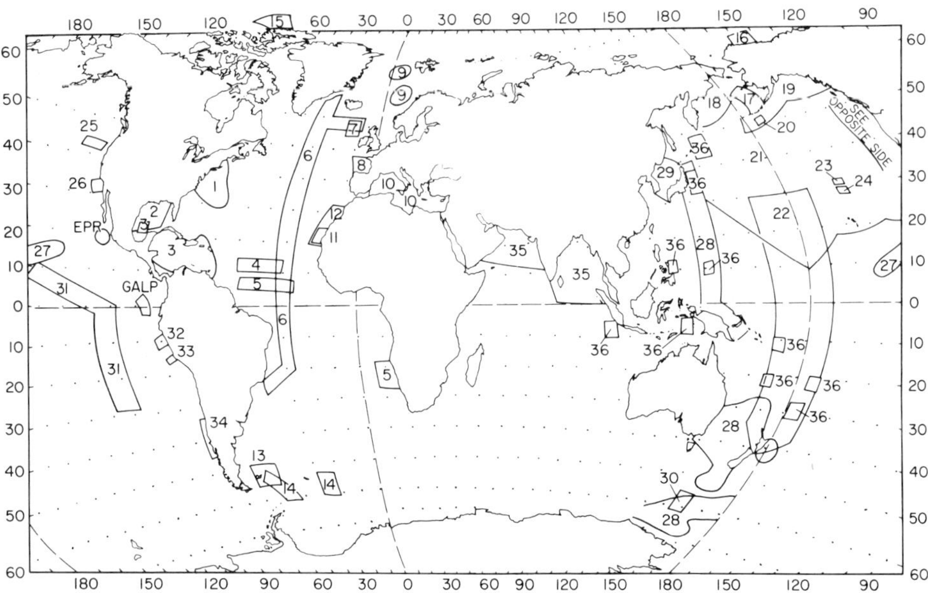

Fig. 6. Worldwide distribution of deep-sea quantitative sampling. (1) Sanders et al. (1965), Rowe et al. (1974), Smith (1978), Rowe et al. (1975), Rowe et al. (1982) (in area, but no data included from Menzies and Rowe (1968), Wigley and Theroux (in press); (2) Rowe and Menzel (1971), Rowe (1971b), Rowe et al. (1974); (3) Pasternak et al. (1975), Muskalev and Pasternak (1970); (4) Khripounoff et al. (1980); (5) Sanders (1969); (6) Kuznetsov (1960); (7) Gage (1977); (8) Laubier and Sibuet (1979), Desbruyérès et al. (1980); (9) Dahl et al. (1977); (10) Rowe (unpublished data), Chukchin (1963); (11) Nichols and Rowe (1977); (12) Thiel (1978), also this volume, Chapter 5; (13) Savilova (1978); (14) Vinogradova et al. (1974); (15) Paul and Menzies (1974); (16) Carey and Ruff (1974); (17) Lus (1970); (18) Belyaev (1960c); (19) Muskalev et al. (1973); (20) Jumars and Hessler (1976); (21) Zenkevitch et al. (1960); (22) Filatova (1960); (23) Hessler (1974), Hessler and Jumars (1974); (24) Filatova and Levenstein (1961); (25) Carey (1981); (26) Jumars and Hessler (1974), Smith and Hinga, this volume, Chapter 8; Barnard and Hartman (1959); (27) Hecker and Paul (1979); (28) Belyaev (1960a,b); (29) Levenstein and Pasternak (1976); (30) Vinogradova et al. (1978); (31) Zezina (1978); (32) Frankenberg and Menzies (1968); (33) Rowe (1971a,b); (34) Gallardo (1963); (35) Belyaev and Vinogradova (1961); Sokolova and Pasternak (1962), Neyman et al. (1973) (includes all Indian Ocean); (36) Belyaev (1966). Trenches East Pacific Rise hydrothermal vent communities (EPR); Galápagos spreading center hydrothermal vent communities (GALP).

5. Worldwide Distribution of Quantitative Deep-Sea Sampling

Many regions have been quantitatively sampled (Fig. 6), but the total number of samples and the area they cover is small relative to the immense area they are intended to represent. Biomass measurements considered in this paper totaled 709. These included Soviet work with the OKEAN grab (0.25 m^2), several studies in which the US-NEL spade corer was used (0.25 m^2 usually), the *Alvin*-operated Ekman box corer (0.04 or 0.023 m^2), a van Veen grab (0.2 or 0.05 m^2), or an anchor dredge (variable area). The analysis does not include many studies that surveyed coastal shelves as a prime goal, but extended into the deep sea (e.g., Wigley and Theroux, in press; Barnard and Hartman, 1959; Gallardo, 1963, Nichols and Rowe, 1977). It does include the shallow data from those studies where transects were made from shallow shelves into the deep sea, where the author was motivated to sample a depth gradient.

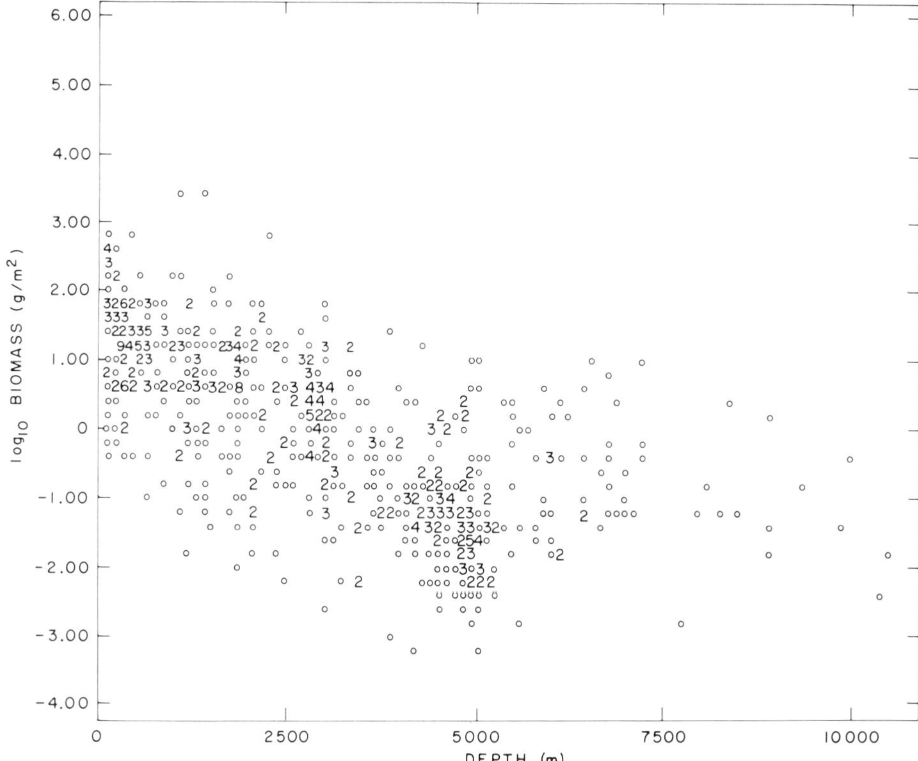

Fig. 7. Biomass from 709 deep-sea quantitative samples. See Fig. 6 for sources of data, with restrictions as enumerated in text.

The locations in the world ocean where samples have been taken (Fig. 6) are skewed by proximity to the laboratories where deep-sea biologists have worked. Nonetheless, thanks to the Soviets, few broad areas remain where no samples have been taken. Several studies referred to in Fig. 6 in addition to those named above are not included in the total 709 samples because raw data on individual samples or even averages for sets of samples were not presented in publications available to me; for example, there are no data from the Indian Ocean. There is no reason to believe, however, that those reports would alter the patterns I present. Readers with better libraries and more patience than I, nonetheless, are encouraged to find the data and make those judgments for themselves.

Semilog plots of biomass on depth (Fig. 7) display a trend that is similar to that published on far fewer data (Rowe, 1971b; Rowe et al., 1974, 1982; Haedrich and Rowe, 1978):

$$\log_{10} \text{biomass} = 1.25 - 0.00039 \text{ (depth)}$$

where Standard error of the estimate $= 1.06$
Standard error of the slope $= 0.0002$
Standard error of the intercept $= 0.069$
Standard error of $r = -0.7$.

Biomass is wet preserved weight in grams and depth is in meters.

Of the total 709 samples, 408 were from Soviet investigations. The Soviet data alone can be represented by

$$\log_{10} \text{biomass} = 1.18 - 0.00033 \text{ (depth)}, \quad \text{with } r = -0.64$$

The slope of the Soviet data (−0.00033) is slightly less than the total (−0.00039), probably because the Soviets have made extensive studies of trenches (see Section 6) rather than from the abyssal plains below unproductive central gyres. As shown in Fig. 7, the emphasis biases the general pattern.

6. High Biomass at High Latitudes

Biomass is higher at high latitudes than near the equator. Compare, for example, Soviet data from the Bay of Alaska (Muskalev et al., 1973) with that from the Caribbean (Pasternak et al., 1975) (Fig. 8). If deep-sea biomass is closely related to surface production (Rowe, 1971b; Rowe et al., 1974), why should latitude make a difference? In fact, although regions of the Arctic and Antarctica do have high seasonal spurts of primary production, their average productivity over the year is relatively low as a result of ice cover and light limitation compared to temperate latitudes, and one would expect this to be reflected in below-average benthic biomass. The answer probably lies in the reasoning expressed by Vinogradov and Tseitlin (Chapter 4, this volume). Seasonal spurts of primary productivity are poorly coupled with pelagic consumption at high latitudes, thus allowing more organic matter to escape from the epipelagic zone without remineralization. The mechanisms responsible for export from temperate coastal zones are probably similar but less pronounced (Walsh et al., 1981). The latter authors also hypothesize that this export has been increasing since the industrial revolution in areas subject to coastal nutrient loading.

7. Trenches

Trenches appear to have higher biomass than adjacent lesser abyssal depths. First suggested by the Soviets (Belyaev, 1966), this has been confirmed in many

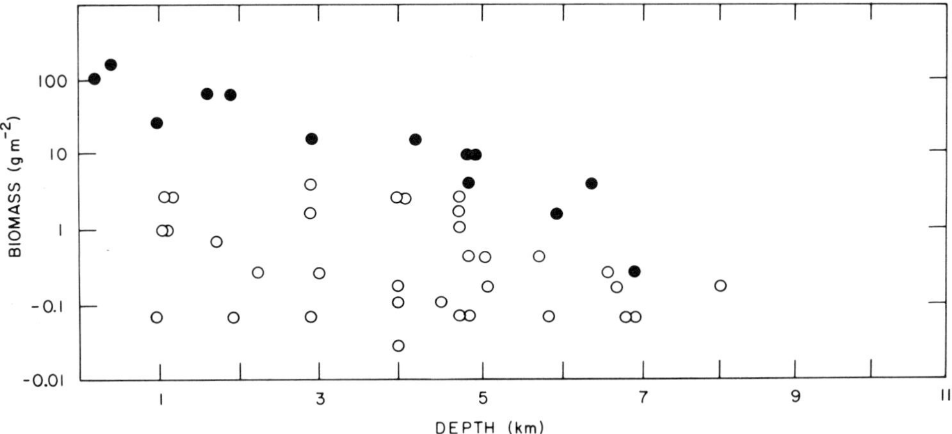

Fig. 8. Comparison of biomass in the Caribbean at low latitudes in tropical waters [○, Pasternak et al. (1975)] with that in the Bay of Alaska at higher latitudes [●, Muskalev et al. (1973)].

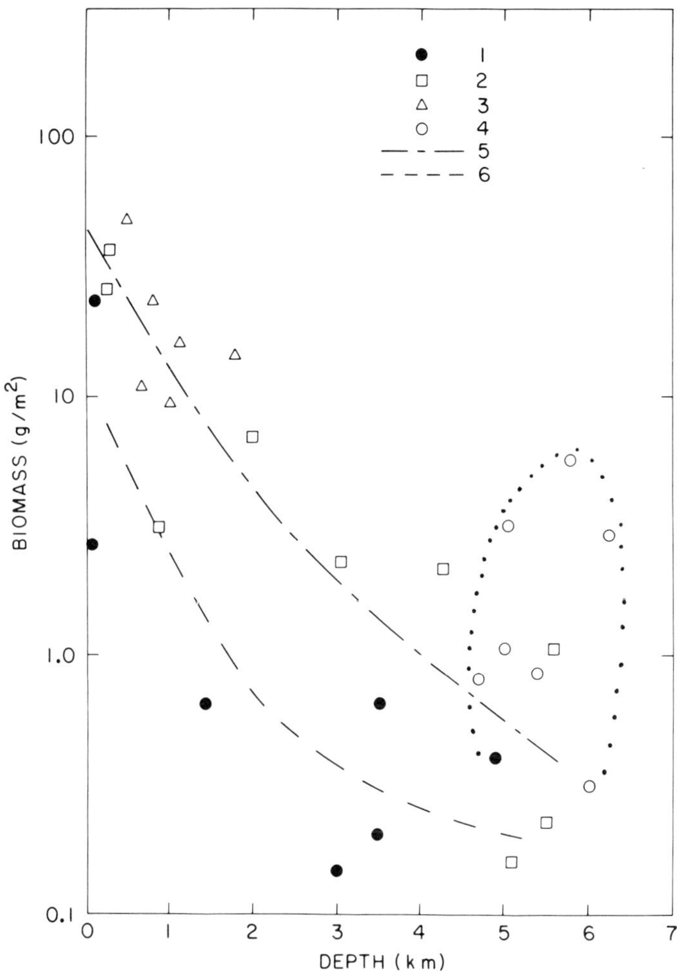

Fig. 9. Biomass between Australia and the Antarctic (Vinogradova et al., 1978); see Fig. 6.

regions, including the area between Australia and Antarctica (Vinogradova et al., 1978) (Fig. 9). In all the data on biomass from depths of 4 to 11 km, amounting to 242 samples, the usual decline of stocks with depth is not seen. The rate of change, although not statistically different from zero, is positive rather than negative (Fig. 10).

On the other hand, within trenches the stocks of macrofauna decrease with depth at a rate that is similar to the general declining pattern. Taking just the data from greater than 6 km, the accepted shallow boundary of trenches, we can see total stocks decrease with depth according to the following relationship:

$$\log_{10} \text{biomass} = 1.105 - 0.00023\,(\text{depth}),$$

based on 57 samples, with the slope different from zero at a significance level of 2%.

The higher biomass in trenches compared to adjacent regions of lesser depth probably reflects net accumulation of sediment from the adjacent shallow continental margin. This is a mechanism similar to that inferred from work on the

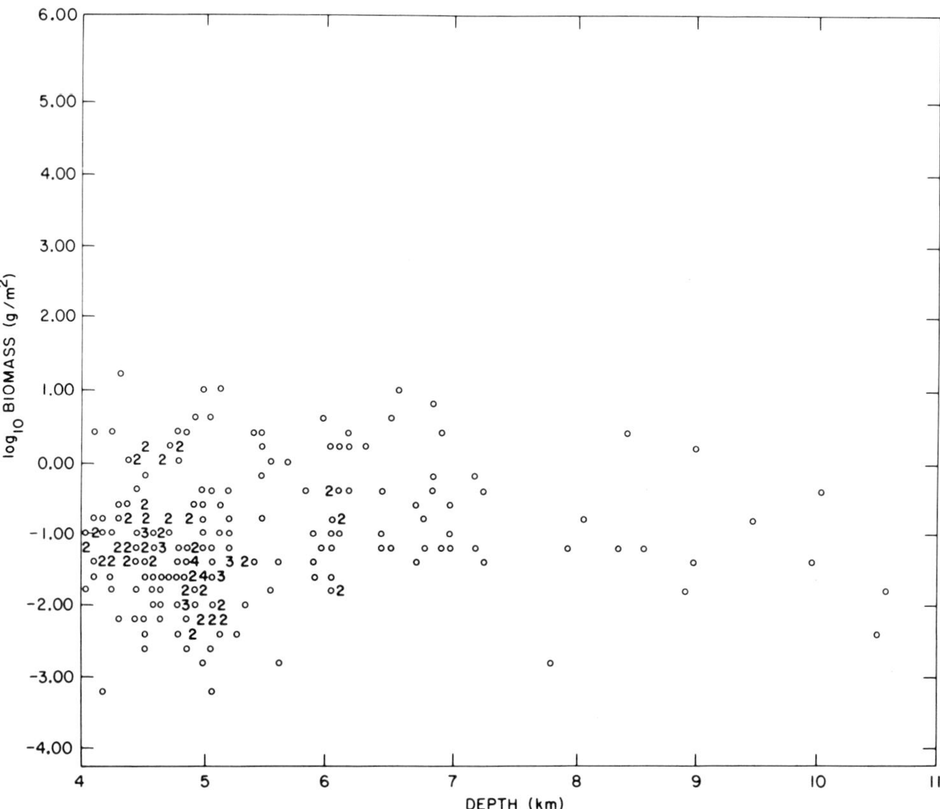

Fig. 10. Biomass (wet preserved weight) at 4- to 11-km depth in the world ocean. Data sources enumerated in Fig. 6; see text.

continental margin off the Oregon coast (Carey, 1981) or in sections of submarine canyons where sediments are accumulating (Rowe et al., 1982).

The decrease in biomass within trenches suggests that the mechanisms that deplete organic energy supplies at the lesser depths of the meso- and bathypelagic zones also act at a depth between 6 and 11 km in trenches. Indeed, although species composition of the "Hadal" zone is different from the abyss (Belyaev, 1966; Wolff, 1960; Menzies et al., 1973), it appears that consumption of food supplies can be just as rapid or more so (Hessler et al., 1978).

8. Biomass Relative to Depth

Standing stocks decrease exponentially with depth when stocks are compared along the depth gradient of a single continental margin (Figs. 4 and 5). The rate of decrease b in stocks with depth x varies on different continental margins, as does the overall level of biomass $\bar{X}$, depending on surface production (Rowe, 1971b), the width of the continental shelf (Rowe and Haedrich, 1979), and the latitude (Belyaev, 1966):

$$y = Ae^{-bx}$$

where y is biomass, A is intercept, b is slope, and x is the depth (Rowe, 1971b; Rowe et al., 1974).

As zooplankton biomass also decreases in a similar exponential fashion (Vinogradov, 1968), it has been suggested that the deep-sea benthos and zooplankton have the same food source, namely, surface-derived organic matter (Rowe et al., 1974). The greater the utilization of the organic matter as it sinks to the bottom, the less is available to the benthos. Packaging by pelagic organisms into fecal matter increases the sinking velocity of the material, but the only particulate organic carbon (POC) that escapes is *that fraction that is not utilized* for energy and growth, for example, the feces. Hence the more efficient the pelagic fauna, the less energy available to the bottom community.

The argument has been made that biomass is a function less of depth than of distance from land, which by some unspecified mechanisms controls the amount of terrestrial runoff and coastal zone productivity that reaches the benthos (Bruun, 1957; Belyaev, 1966). Often trenches at Hadal depths (6 to 11 km) very close to shore have a biomass of several grams per square meter, rather than the few milligrams that would be predicted from regressions of biomass on depth. Undoubtedly, trenches are exceptional because they trap sediment. It is in comparison of trenches to abyssal plains that a biomass versus depth relationship clearly breaks down (see Section 7).

The simple semilogarithmic linear relationship is an oversimplification, and a number of exceptions to this "rule" have been described. Khripounoff et al. (1980) found that the Vema Channel in the equatorial Atlantic was exceptional, but the reasons for it could not be discerned. Most exceptions, as in trenches, probably result from topographic anomalies that trap detrital material. These include submarine canyons (Rowe et al., 1982; Griggs et al., 1969) and physiographic boundaries where the bottom slope changes sharply, such as the continental slope–rise junction at about 2 km depth (Carey, 1981).

9. Regions of Lowest Biomass

Regions of lowest biomass are characterized by macrofaunal abundances ≤ 100 animals m^{-2} that weigh on the order of 0.05 g/m^2 (Fig. 6). Converted to carbon (Table IV), that amounts to 2.0 mg C/m^2. Such communities were found on the abyssal plains of the central gyres of ocean basins far from shore at depths of 4.5 and 6 km (Fig. 10).

Low stocks can be found at lesser depths as well. I found them in the Gulf of Mexico on the Sigsbee Abyssal Plain (3.3- to 3.7-km depth) and in the eastern Mediterranean at 2 to 3 km. This may be a function of low surface production or an efficient utilization of energy by the pelagic biota. These basins also have temperatures that are higher than most deep basins (4 and 13.5°C) that could increase the rate at which organic matter is turned over, thereby resulting in lower standing stocks. This possibility warrants consideration.

10. Growth Rates and Productivity

As originally explained by Zenkevitch (1961), one purpose for studying biomass on the deep-sea floor was to infer the levels of secondary (heterotrophic) productivity there relative to the rest of the ocean. Most of the communities have an extremely low biomass; therefore, all rates of biological processes can be presumed to be low relative to shallow water. The absolute rates of those processes however have only recently been measured (Jannasch and Wirsen, Chapter

6; Smith and Hinga, Chapter 8; Vinogradov and Tseitlin, Chapter 4, all this volume). If growth or production were directly proportional to biomass, then Fig. 6 could be used to estimate production at any depth. One would infer from this that production decreases exponentially down to about 5 km and then may increase in some trenches. What information is available, however, does not confirm that we are yet at liberty to make such generalizations.

Estimates of growth rates in the deep sea present a confusing, often contradictory, picture. The classic approach of estimating production is to measure growth of cohorts identified from size–frequency data over time. This requires taking a series of samples from the same population over periods of time that are long enough for growth to be observable, and the animals must undergo periodic spawning in order to separate the cohorts.

Because of the difficulties presented by these requirements, indirect methods have been employed. Growth rings on the tiny cosmopolitan brachiopod *Pelagodiscus atlanticus* suggest that this species has a production : biomass ratio of 0.3 to 0.4 (Zezina, 1975). It must be assumed that the growth rings are annual responses to the seasonal input of organic matter, a phenomenon regarded as likely (Deuser et al., 1981). They appear to live for 3 to 6 years. Continental slope specimens grow faster than do ones on the abyssal plain, not surprisingly. A length–weight relationship indicates that weight = 0.036 length$^{3.41}$, where weight is alcohol preserved after removal of liquid from the mantle space by air drying. (This is probably slightly less than the "preserved wet weight" described in Section 4 because of the drying nature of the alcohol.) Most specimens grow about 1 mg/yr, or using the general wet weight to carbon conversion (4%), about 40 μg of carbon per specimen per year.

The pentacrinid crinoid *Annacrinus wyville–thomsoni* reach an age of at least 15 to 20 years (Roux, 1976), based on their presumed skeletal regeneration rate. If their mass were known, their growth rates could be calculated.

Startlingly slow growth was inferred from using natural radionuclides (^{228}Ra chronology) to estimate the age of a small but common bivalve, *Tindaria callistiformis,* in the deep northwestern Atlantic (Turekian et al., 1975). Sizes of 8 mm appeared to be about 100 years old, and it took 50 years to reach sexual maturity. The statistical error in the estimate was very wide, however.

Measurements of growth on megafaunal species, made with traditional size–frequency data, suggest rates of growth are slower than in shallow water, but somewhat greater than *Tindaria callistiformis*. A permanent station in the southern Rockall Trough in the northeastern Atlantic sampled for 3 years (Gage et al., 1980) has allowed estimates of growth of some megafauna by standard methods (Tyler and Gage, 1980). A pattern typical of shallow species was observed in the brittle star *Ophiura ljungmani,* with cohorts being added on a yearly basis and maximum age appearing to be about 10 years. Growth rates of about 1 mm/yr (disk diameter) were inferred for specimens of greater than 1.5 mm. Reproductive maturity was reached at 3.5 to 4 mm.

The same group of researchers has also reported seasonally periodic reproduction in two common deep-sea bivalves, *Ledella messanensis* and *Yoldiella jeffreysi* (Lightfoot et al., 1979). *Ledella messanensis* becomes sexually mature at about 1.8- to 2.0-mm shell length but continues to increase in fecundity up to 4-mm shell length, where the mean egg number was 174 and the mean egg size was

109 μm. No information was given on growth rates, only relative maturity of the gonads.

It may seem contradictory that on one side of the Atlantic common bivalves would reproduce seasonally and contribute a large fraction of their production to gonads, whereas on the western side, at just 1 km deeper, another species would grow at exceptionally low rates and spawn continuously (Sanders and Hessler, 1969; Grassle and Sanders, 1973). Unfortunately, no quantitative growth information is available from the western Atlantic, but the valuable data on the gonad production rate for the two eastern Atlantic species can be used to estimate indirectly where they are investing their resources. Using the egg size (d = 109 μm), mean egg number (174), and a presumed egg density of 1.05 g/cm^3, we can estimate the carbon invested in gonads:

$$\text{Egg volume} = (4/3)\ \pi\ (54\ \mu\text{m})^3 = 6.6 \times 10^5\ \mu\text{m}^3$$

$$\text{Egg wet weight} = (1.05 \times 10^{-6}\ \mu\text{g wet weight}/\mu\text{m}^3)\ (6.6 \times 10^5\ \mu\text{m}^3/\text{egg})$$

$$\begin{aligned}\text{Egg weight/animal} &= (1.05 \times 6.6 \times 10^{-1})\ (174\ \text{eggs/animal}) \\ &= 120\ \mu\text{g wet weight eggs/individual}\end{aligned}$$

If half of all the macrobenthos makes a similar investment [and at this location and depth in the northeastern Atlantic abundance is on the order of 1800 individuals/m^2 (Gage, 1977)], the average gonad investment would be about 100 mg/m^2 · year. This amounts to only about 5 mg carbon/m^2 · yr. Wet weight of the macrofauna at this locale was 3.7 g/m^2 (Gage, 1977); therefore, the reproduction: biomass ratio would be only 100 mg/m^2·yr per 3.7 g/m^2 = 0.03.

Although there appears to be great disparity between growth rates at approximately 3-km depth in the eastern Atlantic versus 4-km depth in the western Atlantic, I think it is premature to conclude that either estimate is or is not representative of its own environment, as each environment is characterized by quite different levels of energy flow. The purpose of this comparison is to emphasize that much more work needs to be oriented to production and growth. Comparisons utilizing both radiometric or cohort-identified time series are needed in contrasting deep-sea environments. All size categories of the deep-sea benthos should be included. Clearly, quantification of growth must be put in terms of size, mass, or energy to facilitate intercomparisons.

Acknowledgments

This work was supported by grants or contracts from the National Science Foundation between 1970 and 1978 and the Department of Energy (U.S.) between 1978 and 1982. Special thanks are due Dr. Nina Vinogradova for helping to furnish me with difficult-to-find Soviet articles. I will always be obliged to my former colleagues Pam Polloni, Dick Haedrich, and Hovey Clifford for their help in the study of deep-sea life.

References

Barnard, J. L. and O. Hartman, 1959. The sea bottom off Santa Barbara, California: Biomass and community structure. *Pacif. Nat.*, **1**, 1–16.

Barnes, A. T., L. B. Quetin, J. J. Childress, and D. Pawson, 1976. Deep-sea macroplanktonic sea cucumbers: Suspended sediment feeders captured from deep submergence vehicle. *Science*, **194,** 1083–1085.

Barnham, E. G., N. J. Ayer, and R. E. Boyce, 1967. Megabenthos of the San Diego Trough: Photographic census and observations from bathyscaphe *Trieste*. *Deep-Sea Res.*, **14,** 773–784.

Belyaev, G. M., 1960a. Some regularities in the quantitative distribution of the bottom fauna in the western Pacific. *Trudy Inst. Okeanol., Akad. Nauk SSSR* (journal of USSR Academy of Science, Oceanology Institute), **41,** 98–105 (in Russian).

Belyaev, G. M., 1960b. Quantitative distribution of benthos in the Tasmanian Sea and in the Antarctic waters south of New Zealand. *Doklady Akad. Nauk SSSR*, **130** (4), 875–878 (in Russian).

Belyaev, G. M., 1960c. The quantitative distribution of the bottom fauna in the Western Bering Sea. *Trudy Inst. Okeanol.*, **34,** 85–104.

Belyaev, G. M., 1966. *Bottom Fauna of the Ultraabyssal of the World Ocean*. Institute of Oceanology, USSR Academy of Science, Moscow, 247 pp.

Belyaev, G. M. and N. V. Vinogradova, 1961. Quantitative distribution of the bottom fauna in the northern part of the Indian Ocean. *Doklady Acad. Nauk SSSR*, **138,** 1191–1194 (in Russian).

Bruun, A. F., 1957. Deep-sea and abyssal depths. In *Treatise on Marine Ecology and Paleoecology*, Vol. 1. *Ecology*, J. Hedgpeth, ed. (reprinted in *Geol. Soc. Am., Mem.*, **67,** 641–672).

Burnett, Bryan R., 1978. Quantitative sampling of microbiota of the deep-sea benthos. I. Sampling techniques and some data from the abyssal central North Pacific. *Deep-Sea Res.*, **24,** 781–789.

Carey, A. G., Jr., 1981. A comparison of benthic infaunal abundance on two abyssal plains in the northeast Pacific Ocean with comments on deep-sea food sources. *Deep-Sea Res.*, **28,** 467–479.

Carey, A. G., Jr. and D. R. Hancock, 1965. An anchor-box dredge for deep-sea sampling. *Deep-Sea Res.*, **12,** 983–984.

Carey, A. G., Jr. and R. E. Ruff, 1974. Ecological studies of the benthos in the western Beaufort Sea with special reference to bivalve molluscs. In *Polar Oceans*, M. J. Dunbar, ed. pp. 505–528. discussed on pp. 528–530.

Chukhchin, V. D., 1963. Quantitative distribution of benthos in the eastern part of the Mediterranean Sea. *Trudy Sevastopol Biol. Sta.*, **16,** 215–233 (in Russian).

Dahl, E., L. Laubier, M. Sibuet, and O. Strömberg, 1977. Some quantitative results on benthic communities of the deep Norwegian Sea. *Astarte*, **5,** 61–79.

Desbruyérès, D., J. Bervas, and A. Khripounoff, 1980. Un cas de colonisation rapide d'un sediment profond. *Oceanologica Acta*, **3,** 285–291.

Deuser, W. G., E. H. Ross, and R. F. Anderson, 1981. Seasonality in the supply of sediment to the deep Sargasso Sea and implications for the rapid transfer of matter to the deep ocean. *Deep-Sea Res.*, **28A,** 495–505.

Dickinson, J. J. and A. G. Carey, 1978. Distribution of gammarid Amphipoda (Crustacea) in Cascadia Abyssal Plain (Oregon). *Deep-Sea Res.*, **25,** 97–106.

Filatova, Z. A., 1960. On the quantitative distribution of the bottom fauna in the Central Pacific. *Trudy Inst. Okeanol., Akad. Nauk SSSR*, **41,** 85–105 (in Russian).

Filatova, Z. A. and R. J. Levenstein, 1961. The quantitative distribution of the deep-sea Pacific. *Trudy Inst. Okeanol., Akad. Nauk SSSR*, **45,** 190–213 (in Russian).

Frankenberg, D. and R. J. Menzies, 1968. Some quantitative analyses of deep-sea benthos off Peru. *Deep-Sea Res.*, **15,** 623–626.

Gage, J. D., 1977. Structure of the abyssal macrobenthic community in the Rockall Trough. In *Biology of Benthic Organisms*, B. E. Keegan, P. O. Ceidigh, and P. J. S. Broaden, eds. Pergamon Press, New York, pp. 247–260.

Gage, J. D., R. H. Lightfoot, M. Pearson, and P. A. Tyler, 1980. An introduction to a sample time-series of abyssal macrobenthos: Methods and principle sources of variability. *Oceanologica Acta*, **3,** 159–176.

Gallardo, A., 1963. Notas sobre la densidad de la fauna bentonica, en el sublitoral del norte de Chile. *Gayana Zool.*, **8,** 3–15.

Grassle, J. F. and H. L. Sanders, 1973. Life histories and the role of disturbance. *Deep-Sea Res.*, **20,** 643–659.

Grassle, J. F., H. L. Sanders, R. R. Hessler, G. T. Rowe, and T. McClellan, 1975. Pattern and zonation: A study of the bathyal megafauna using the research submersible *ALVIN*. *Deep-Sea Res.*, **22,** 457–481.

Griggs, G. B., A. G. Carey, Jr., and L. D. Kulm, 1969. Deep-sea sedimentation and sediment fauna interaction in Cascadia Channel and on Cascadia Abyssal Plain. *Deep-Sea Res.*, **16,** 157–170.

Haedrich, R. L. and G. T. Rowe, 1978. Megafaunal biomass in the deep-sea. *Nature*, **269,** 141–142.

Hecker, B. and A. Z. Paul, 1979. Abyssal community structure of the benthic infauna of the eastern equatorial Pacific: DOMES sites A, B, and C. In *Marine Geology and Oceanography of the Pacific Manganese Module Province*, J. L. Bischoff and D. Z. Piper, eds. Plenum Press, New York, pp. 287–308.

Hessler, R. R., 1974. The structure of deep benthic communities from central oceanic waters. In *The Biology of the Oceanic Pacific*, C. B. Miller, ed. Oregon State University Press, Corvallis, pp. 79–93.

Hessler, R. R., C. L. Ingram, A. Aristides Yayanos, and B. Burnett, 1978. Scavenging amphipods from the floor of the Philippine Trench. *Deep-Sea Res.*, **25,** 1029–1047.

Hessler, R. R. and P. A. Jumars, 1974. Abyssal community analyses from replicate box cores in the central North Pacific. *Deep-Sea Res.*, **21,** 185–209.

Hinga, K., J. M. N. Sieburth, and G. R. Heath, 1979. The supply and use of organic material at the deep-sea benthos. *J. Mar. Res.*, **37,** 557–579.

Holme, N. A. and A. D. McIntyre, 1973. *Methods for Study of Marine Benthos*. IBP Handbook No. 16, Blackwell, Oxford, 334 pp.

Honjo, S., 1978. Sedimentation of materials in the Sargasso Sea at 5,367 m deep station. *J. Mar. Res.*, **36,** 469–492.

Hulings, N. C. and John S. Gray, 1971. *A Manual for the Study of Meiofauna*. Smithsonian Contributions to Zoology, No. 78. U.S. Government Printing Office, Washington, 84 pp.

Jumars, P. A., 1976. Deep-sea species diversity: Does it have a characteristic scale? *J. Mar. Res.*, **34,** 217–246.

Jumars, P. A. and R. R. Hessler, 1976. Hadal community structure: Implications from the Aleutian Trench. *J. Mar. Res.*, **34,** 547–560.

Khripounoff, Alexis, D. Desbruyérès, and P. Chardy, 1980. Les peuplements benthiques de la faille VEMA: Donnees quantitatives et bilan d'energie en milieu abyssal. *Oceanologica Acta*, **3,** 187–198.

Kuznetsov, A. P., 1960. Data concerning quantitative distribution of bottom fauna of the bed of the Atlantic. *Dokl. Akad. Nauk SSSR*, **130,** 1345–1348 (in Russian).

Laubier, L. and M. Sibuet, 1979. Ecology of the benthic communities of the deep North East Atlantic. *Ambio Spec. Rep.*, **6,** 37–42.

Levenstein, R. and F. A. Pasternak, 1976. Quantitative distribution of the bottom fauna in Japan Sea. *Akad. Nauk, SSSR*, **99,** 197–210 (in Russian).

Lightfoot, R. H., P. Tyler, and J. D. Gage, 1979. Seasonal reproduction in deep-sea bivalves and brittlestars. *Deep-Sea Res.*, **26A,** 967–973.

Lus, V. Ya., 1970. Quantitative distribution of benthos on the continental slope of the eastern part of the Bering Sea. In *Soviet Fisheries Investigations in the Northeastern Pacific*, P. A. Moiseev, ed., pp. 116–124 (reprinted in *Izvestiya*, **72**) (in Russian).

Mare, M. F., 1942. A study of a marine benthic community with special references to the micro-organisms. *J. Mar. Biol. Assoc. U.K.*, **25,** 517–554.

Menzies, R. J. and G. Rowe, 1968. The LUBS, a large undisturbed bottom sampler. *Limnol. Oceanogr.*, **13,** 708–714.

Menzies, R. J., R. Y. George, and G. T. Rowe, 1973. *Abyssal Environment and Ecology of the World Ocean*. Wiley, New York, 488 pp.

Muskalev, L. I. and F. Pasternak, 1970. Over the quantitative distribution of the benthos in the southeast Gulf of Mexico and near Cuba. *Oceanological Res.*, IGY Section X, 128–143 (in Russian).

Muskalev, L. I., O. Zezina, R. K. Kudinova-Pasternak, and T. L. Muromtseva, 1973. Quantitative and ecological characteristics of bottom assemblages in bathyal depths of the Bay of Alaska. *Trans. Shirshov Inst. Oceanol.*, **91,** 73–79 (in Russian).

Neyman, A. A., 1969. Some data on the bottom fauna of the northern Indian Ocean shelves. *Oceanology,* **6,** 1071–1077.

Neyman, A. A., M. Sokolova, N. Vinogradova, and F. Pasternak, 1973. Some patterns of the distribution of bottom fauna in the Indian Ocean. In *The Biology of the Indian Ocean, Ecology Studies 3,* B. Zeitzschel, ed. Springer-Verlag, pp. 467–473.

Nichols, J. A. and G. T. Rowe, 1977. Infaunal macrobenthos off Cap Blanc, Spanish Sahara. *J. Mar. Res.,* **35,** 525–536.

Pasternak, F. A., L. I. Muskalev, N. F. Fedikov, 1975. Some peculiarities in distributional patterns of the deep-sea bottom fauna of the Caribbean Sea and Gulf of Mexico. *Trans. Shirshov Inst. Oceanol.,* **101,** 52–64 (in Russian).

Paul, A. Z. and R. J. Menzies, 1974. Benthic ecology of the high Arctic deep sea. *Mar. Biol.,* **2,** 7251–7262.

Polloni, P. T., R. L. Haedrich, G. T. Rowe, and C. H. Clifford, 1979. The size–depth relationship in deep ocean animals. *Internationale Revue der gesamten Hydrobiologie,* **64,** 39–46.

Rice, A. L., R. G. Aldred, E. Darlington, and R. A. Wild, 1982. The quantitative estimation of the deep-sea megabenthos; a new approach to an old problem. *Oceanologica Acta,* **5,** 63–72.

Roux, M., 1976. Aspects de la variabilite, et de la croissance au sein d'une population de la Pentacrine actuelle: *Annacrinus wyville thomsoni* Jeffreys (Crinoidea). *Thalassia Yugoslavica,* **12,** 307–320.

Rowe, G. T., 1971a. Benthic biomass in the Pisco, Peru upwelling. *Inv. Pesq.,* **35** (1), 127–135.

Rowe, G. T., 1971b. Benthic Biomass and surface productivity. In *Fertility of the Sea,* Vol. 2, J. D. Costlow, Jr., ed. Gordon and Breach, New York, pp. 441–454.

Rowe, G. T., 1971c. Observations on bottom currents and epibenthic populations in Hatteras Submarine Canyon. *Deep-Sea Res.,* **18,** 569–581.

Rowe, G. T. and R. J. Menzies, 1969. Zonation of large benthic invertebrates in the deep-sea off the Carolinas. *Deep-Sea Res.,* **16,** 531–537.

Rowe, G. T. and D. W. Menzel, 1971. Quantitative benthic samples from the Deep Gulf of Mexico with some comments on the measurement of deep-sea biomass. *Bull. Mar. Sci.,* **21,** 556–566.

Rowe, G. T. and C. H. Clifford, 1973. Modifications of the Birge-Ekman box corer for use with SCUBA or deep submergence reseach vessels. *Limnol. Oceanogr.,* **18** (1), 172–175.

Rowe, G. T. and W. Gardner, 1979. Sedimentation rates in the slope water of the northwest Atlantic Ocean measured directly with sediment traps. *J. Mar. Res.,* **37,** 581–600.

Rowe, G. T. and R. L. Haedrich, 1979. The biota and biological processes of the continental slope. In *Continental Slopes,* O. Pilkey and L. Doyle, eds. Society of Economic Petrologists and Mineralogists Special Publication No. 27, Tulsa, OK, pp. 49–59.

Rowe, G. T., P. T. Polloni, and S. G. Hornor, 1974. Benthic biomass estimates from the northwestern Atlantic Ocean and the northern Gulf of Mexico. *Deep-Sea Res.,* **21,** 641–650.

Rowe, G. T., P. T. Polloni, and R. L. Haedrich, 1975. Quantitative biological assessment of the benthic fauna in deep basins of the Gulf of Maine. *J. Fish. Res. Board Can.,* **32** (10), 1805–1812.

Rowe, G. T., P. T. Polloni, and R. L. Haedrich, 1982. The deep-sea macrobenthos on the continental margin of the northwest Atlantic Ocean. *Deep-Sea Res.,* **29,** 257–278.

Sanders, H. L., 1969. Benthic marine diversity and the stability time hypothesis. In *Diversity and Stability in Ecological Systems,* Vol. 22. Brookhaven Symposium in Biology, pp. 71–81.

Sanders, H. L. and R. R. Hessler, 1969. Ecology of the deep-sea benthos. *Science,* **163,** 1419–1424.

Sanders, H. L., R. R. Hessler, and G. R. Hampson, 1965. An introduction to the study of deep-sea benthic faunal assemblages along the Gay Head—Bermuda transect. *Deep-Sea Res.,* **12,** 645–867.

Savilova, T. A., 1978. On the quantity distribution of bottom fauna of Falklands Islands. *Akad. Nauk SSSR,* **113,** 237–241 (in Russian).

Sibuet, M., and J. M. Lawrence, 1981. Organic content and biomass of abyssal holothuroids (Echinodermata) from the Bay of Biscay. *Mar. Biol.,* **65,** 143–147.

Smith, K. L., Jr., 1978. Benthic community respiration in the N.W. Atlantic Ocean: *In situ* measurements from 40 to 5200 m. *Mar. Biol.,* **47,** 337–347.

Sokolova, M. N. and E. A. Pasternak, 1962. The quantitative distribution of the bottom fauna in the northern part of the Arabian Sea and Bengal Bay. *Dokl. Akad. Nauk SSSR,* **144** (3), 645–648 (in Russian).

Spärck, R., 1951. Density of bottom animals on the ocean floor. *Nature,* **168** (4264), 112–113.

Spärck, R., 1956. The density of animals on the ocean floor. In *The Galathea Deep-Sea Expedition 1950–1952.* Allen and Unwin, London, pp. 196–201.

Thiel, H., 1975. The size structure of the deep-sea benthos. *Int. Rev. Ges. Hydrobiol.,* **60** (5), 575–606.

Thiel, H., 1978. Benthos in upwelling regions. In *Upwelling Ecosystems,* R. Boje and M. Tomczak, eds. Springer, Berlin, pp. 124–138.

Thiel, H. and R. R. Hessler, 1974. Ferngesteuertes Unterwasser-fahrzeug erforscht Tiefseeboden. *UMSHAU in Wissenschaft und Technik,* **14,** 451–453.

Turekian, K. K., J. K. Cochran, D. P. Kharkar, R. Cerrato, J. Vaisnys, H. L. Sanders, J. F. Grassle, and J. Allen, 1975. The slow growth rate of a deep-sea clam determined by 228 Ra chronology. *Proc. Natl. Acad. Sci. (USA),* **72,** 2829–2832.

Tyler, P. and J. Gage, 1980. Reproduction and growth of the deep-sea brittlestar *Ophuira ljungmani* (Lyman). *Oceanologia Acta,* **3,** 177–185.

Vinogradov, M. E., 1968. *Vertical Distribution of the Oceanic Zooplankton.* Academy of Sciences of the USSR, Institute of Oceanography, Moscow, 339 pp.

Vinogradova, N., 1976. The large view of the ocean. *Oceanology, Priroda,* **11,** 94–105 (in Russian).

Vinogradova, N. G., R. K. Kudinova-Pasternak, L. Muskalev, T. Muromtseva, and N. Fedikov, 1974. Some regularities of the quantitative distribution of the bottom fauna of the Scotia Sea and the deep-sea trenches of the Atlantic Sector of the Antarctic. *Trans. Shirshov Inst. Oceanol.* **98,** 157–182 (in Russian).

Vinogradova, N. G., O. N. Zezina, and R. J. Levenstein, 1978. Bottom fauna of deep-sea trenches of the Macquarie complex. *Trudi Inst. Okeanol., Dokl. Akad. Nauk SSSR,* **112,** 174–192 (in Russian).

Walsh, J. W., G. T. Rowe, R. Iverson, and C. P. McRoy, 1981. Biological export of shelf carbon is a sink of the global CO_2 cycle. *Nature,* **291,** 196–201.

Wigley, R. L. and R. B. Theroux, in press. Macrobenthic inventebrate fauna of the middle Atlantic Bight region: Part II. Faunal composition and quantitative distribution. U.S. Geological Survey Professional Paper.

Wolff, T., 1960. The Hadal community, introduction. *Deep-Sea Res.,* **6,** 95–124.

Zenkevitch, L. A., 1961. Certain quantitative characteristics of the pelagic and bottom life of the ocean. In *Oceanography,* M. Sears, ed. Publication No. 67, AAAS, Washington DC, pp. 323–335.

Zenkevitch, L. A., N. G. Barsanova, and G. M. Belyaev, 1960. Quantitative distribution of bottom fauna in the abyssal area of the world ocean. *Dokl. Akad. Nauk SSSR,* **130** (1), 183–186 (in Russian).

Zezina, O. N., 1975. On some deep-sea brachiopads from the Gay Head-Bermuda transect. *Deep-Sea Res.,* **22,** 903–912.

Zezina, O. N., 1978. On benthos research in the west part of the Peruvian basin and at nearby regions of the east Pacific Rise. *Akad. Nauk SSSR,* **113,** 36–43.

4. DEEP-SEA PELAGIC DOMAIN (ASPECTS OF BIOENERGETICS)

M. E. VINOGRADOV AND V. B. TSEITLIN

1. Introduction

The primary organic matter is produced almost exclusively in the sunlit layers, and from there by various ways it penetrates to the depths where it constitutes the basis of existence of the whole population of the underlying water layers and the bottom. While it sinks down, its quantity steadily decreases due to consumption and mineralization. All other sources of organic matter (chemosynthesis by the deep-sea bacteria and fragments of terrestrial organisms) have a subordinate significance. Thus the plankton assemblage of the open sea represents a unique formation characterized by a spatial disconnection between the layers where primary organic matter is produced and the underlining layers where a considerable share of this matter is consumed.

The plankton community inhabiting the surface production zone includes autotrophic organisms and does not depend on a food influx from outside. Communities of the deeper layers and the deep-sea bottom exist on the energy originating from the surface population and not totally consumed by the latter. Thus this energy incorporated in organic matter, living or dead, descends to the depths.

The deep-sea communities energetically depend on the surface communities. This indissoluble dependence substantially affects the structure and functioning and the ecological peculiarities of these communities and the morphology and physiology of their components more than does the change of the physicochemical environmental parameters with depth (e.g., Marshall, 1960, 1965; Vinogradov, 1962, 1968; Birstein and Vinogradov, 1971; Mauchline, 1972; Rowe, 1980).

It is obvious, therefore, that the existence of the inhabitants of the ocean depths is rigidly determined by the quantity of the energy that is produced in the surface layers and remains unconsumed. This quantity of energy also depends on the ways in which this energy contained in organic matter penetrates into the deep waters. Consequently, it is extremely important to measure the energy flow from the surface to the deep layers and the degree of its dissipation at various depths and to disclose various modifications in which energy and matter are transferred.

The communities of different depths are connected by the flows of dead organic matter and by constant and fairly intensive replacement of organisms. The interzonal herbivorous migrators feeding on the surface layers constitute in the temperate regions up to 40% of the biomass of the net plankton of the bathypelagic zone (1000 to 3000 m), and in some regions and in certain seasons their share increases up to 60% (Vinogradov, 1968; Vinogradov and Arashkevich, 1969).

Naturally, they are components of the surface as well as deep-water communities. Leavitt (1938) even proposed to name the mesopelagic depths (200 to 750 m) as the "dynamic photic zone," in consideration of the fact that the basic portion of its community consists of migrators feeding on the surface layers. From this standpoint, the energetically autonomous community of the surface layers and the deep-water communities that energetically depend on the former should be regarded as parts of a united assemblage inhabiting the entire water column and thus constituting a joint ecosystem.

The vertical stratification of the biota in the whole water column cannot be fully explained by the gradients of the physical environmental patterns (Banse, 1964; Paxton, 1967; Vinogradov, 1970a,b, 1972). It is preserved despite the affecting of turbulent vortices that intermix separate parts of the ecosystem. It is probable that in the community itself certain inherent peculiarities favor the maintenance of a sharp stratification in population distribution in defiance of external disturbing factors. Members of a community must spend the energy to maintain distributional stratification, which certainly has a definite although as yet not always clear biological meaning.

Dead organic matter (animal and plant fragments, fecal pellets, exuviae, etc.), while sinking down, may be consumed repeatedly by the inhabitants of the depths. At the same time changes take place in the abundance and size of organic particles (McCave, 1975) and in the contents of labile organic matter (Parsons, 1963; Bogdanov et al., 1968; Bogdanov and Shaposhnikova, 1971; Conover, 1978) and, consequently, their nutritious value (Vinogradov, 1961; Krause, 1962; Wheeler, 1967).

Recently attempts were made to measure the metabolism of the deep-sea pelagic animals (Childress, 1971; Childress and Nygaard, 1973; Meek and Childress, 1973; Smith and Hessler, 1974; Smith, 1978) and their need for food and their actual diet (MacDonald, 1975; Gorelova and Tseitlin, 1979) and the share of organic matter originating in the overlying layers and utilized by those, as well as the share of this organic matter in the sedimentation and food supply of benthic organisms (Griggs et al., 1969; Sanders and Hessler, 1969; Rowe, 1971; Wiebe et al., 1976; Rowe and Gardner, 1979).

2. Variations in Abundance with Depth

The production of the pelagic animals in deep waters is determined by the supply of organic matter from the overlying layers and, above all, depends on the productivity of the euphotic zone. Therefore, it is obvious that in areas with a high production of surface plankton, deep-sea plankton will be more abundant, too. The adjustment between the life cycles of producers and consumers in surface communities is also a contributing factor here. In some degree this adjustment also determines the utilization of production within surface communities themselves and the share of production penetrating into deep water (Banse, 1964). However, in areas where the adjustment of life cycles of producers and consumers of surface communities differs in various degrees (e.g., boreal and tropical regions) and where the trophic structure of deep-sea communities also differs, the ratio between abundance of surface and deep-sea plankton may be determined by more complicated interrelationships (Vinogradov, 1968).

The quantity of the organic matter reaching to a certain depth just determines the production of the plankton, whereas the ratio between the production and the biomass may differ at various depths. If we agree with the suggestion made by Mauchline (1972) that the relation between body length and lifetime in the cold-water epipelagic and in bathypelagic animals is similar, we may suppose that the bathypelagic species from such groups as Euphausicea, Mysidacea, and possibly Chaetognatha may live two to seven times longer than the epipelagic ones and hence that their biomass should be much larger at the same production rate.

A. *Microzooplankton (Protozoa)*

A general pattern of vertical distribution of unicellular heterotrophic organisms is poorly studied, since, above all, it is related to great methodological difficulties.

Observations made in the Mediterranean Sea, off the Atlantic African coast, and in the northwestern Indian Ocean (Lecal, 1952; Bernard and Lecal, 1960; Bernard, 1963) have shown that the deep waters of these regions down to 1000 to 3000 m are rich in single-cell heterotrophic dinoflagellates and other organisms. It remains still unknown whether the heterotrophic unicellular organisms are as abundant in other regions. In particular, in "Vityaz" collections made in the Pacific and Indian Oceans, these organisms are not encountered in such large quantities. Moreover, the number of the nannoplankton at various depths of the Ionian Sea according to Grezė (1963) is by 50 to 200 times less than that given by Bernard (1961) for the same region, not withstanding the fact of nearly full coincidence in techniques and in the season of the studies carried out by both researchers.

As Bernard (1963, 1966) and others suggest, the unicellular organisms play a substantial role in the nutrition of the deep-sea zooplankton. Bernard believes that, in contrast to the diatoms and peridinians, which rapidly decompose, when shifting downward, the palmelloid stages of *Cyclococcolithus fragilis* reach great depths of several thousand meters nearly intact, supplying the animals living there with food and vitamins.

Another group of nannoplankton, so-called olive–green cells having round bodies 1 to 15 μm in size, by some authors (e.g., Fournier, 1970) are assumed to belong to the flagellates. According to many authors (e.g., Schiller, 1931; Hasle, 1959; Fournier, 1966, 1970; Hamilton et al., 1968; Melnikov, 1975), they are found mainly below the euphotic zone, with a maximum concentration in the 500- to 1000-m layer, where their population density reaches 100,000 to 200,000 cells per liter. At further depths the number of olive–green cells decreases somewhat, but even at 3000 to 4000 m it is usually not less than 25,000 to 50,000/liter. They are found in large quantities in the gut of the deep-sea isopods (Menzies, 1962), prawns (Wheeler, 1970), and copepods (Fournier, 1973; Harding, 1974).

According to Bernard (1964, 1966), the maximum population density of the unicellar heterotrophic organisms is at 700 to 1000 m and lower. Their abundance at these depths does not correlate with the abundance of plankton in the surface layers. If the unicellular organisms actually served as the main food source of the deep-sea fauna, the quantitative distribution of the latter would be similar to the distribution of these unicellular organisms. However, in contrast to these organisms, the quantity of zooplankton decreases very rapidly with depth, so that at the depths greater than 1000 to 2000 m, it becomes hundreds of times lower than in the surface layers. Besides, there is a practically universal and constant ratio between the quantity of surface and deep-sea mesoplankton. This pattern of vertical distribution of mesoplankton would be easily explained by assuming that the food supply of the deep-sea fauna comes from the surface layers, but it becomes wholly inexplicable by assuming that the autochthonous heterotrophic unicellular organisms serve as such a food source.

Most of the deep-sea pelagic animals have large lecithotrophic larvae, a fact that also contraindicates the nutritional importance of unicellular organisms that might be a constant and easily available food supply for usual planktotrophic larvae. Thus the quantitative distribution of unicellular heterotrophic organisms still requires special and painstaking investigations.

B. *Mesoplankton (Net Plankton)*

The maximum concentration (population density, biomass) of mesoplankton is usually confined to the upper 100-m layer. Then it sharply and often unevenly decreases down to the lower layers of the mesopelagic zone. Depending on the hydrologic regime, seasonal cycles of the communities, and even the time of the day, the layers of maximal concentrations may be confined to various depths. In temperate regions in summer they are confined to the upper 50-m layer, and in winter, to the 200- to 500-m or even 500- to 700-m layer (Wiborg, 1954; Foxton, 1956; Vinogradov, 1968). In the tropics, the maximum concentration is usually confined to the 25- to 50-m or 50- to 100-m layer. In the dichothermal waters of the northern Pacific, even in summer, there is a maximum of biomass in the 200- to 300-m layer, which is almost equal to the maximum in the 0- to 50-m layer, and a sharp minimum in the cold 75- to 200-m layer (Bogorov and Vinogradov, 1955; Vinogradov, 1968).

However, despite the heterogeneity of vertical distributions in the greatest portion of the open ocean in the 0- to 500-m layer, approximately two-thirds (60 to 65%) of the biomass of mesoplankton inhabits the entire water column from 0 to 4000 m, whereas within the 500- to 4000-m layer, one third (35 to 40%) (Vinogradov, 1960, 1962, 1968)—that is, the total quantity of plankton in the bathy- and abyssopelagic zones is directly dependent on the quantity of epipelagic mesoplankton.

In the bathy- and abyssopelagic zones down to a depth of 6000 to 7000 m, the quantity of plankton decreases with depth more slowly and more homogenously, generally conforming to the exponential equation $y = ae^{-kx}$, where y is the biomass of the mesoplankton, a, is a coefficient related to the quantity of plankton in the overlying layers, and k is the coefficient of the rate of decreasing of the plankton with depth (Vinogradov, 1958, 1960, 1968; Johnston, 1962). If the biomass of the plankton is expressed in milligrams per cubic meter wet weight and the depth in meters, then for the highly productive Kurile–Kamchatka region of the Pacific (in the spring), this relationship assumes the form $y = 56.2 \exp(-6.5 \times 10^{-4}x)$, for the oligotrophic tropical region of the Mariana Trench $y = 1.82 \exp(-8.5 \times 10^{-4}x)$, for the mesotrophic tropical coastal region of the Bougainville Trench $y = 5.74 \exp(-8.5 \times 10^{-4}x)$, and for the mesotrophic subtropical waters of the Southern Hemisphere (region of the Kermadec Trench) $y =$

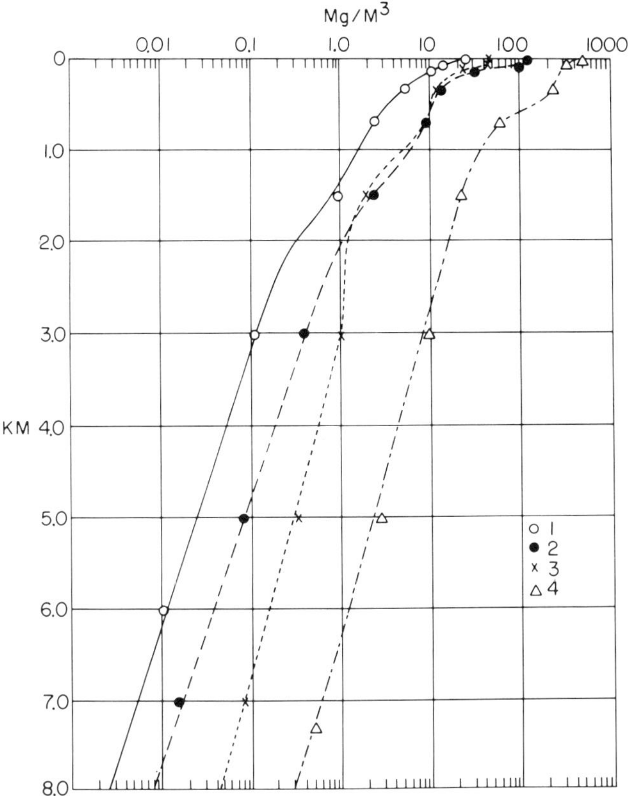

Fig. 1. Vertical distribution of wet weight biomass of mesoplankton (mg/m^3) in different regions of the ocean: (1) Mariana Trench; (2) Bougainville Trench; (3) Kermadec Trench; (4) Kurile–Kamchatka Trench [according to Vinogradov (1968)].

$7.75 \exp(-6.5 \times 10^{-4}x)$. The curves described by these equations are presented in Fig. 1.

The variations of the coefficient correspond to the differences in the biomass of the surface mesoplankton and testify to a significant poverty of deep-sea plankton of the oligotrophic regions of the tropical ocean in comparison with the marginal tropical seas (Bougainville Trench), and especially in comparison with the eutrophic subpolar regions (northwestern Pacific). This relationship is very clearly manifest on the meridional cross section through the Pacific Ocean (Fig. 2).

The rates of mesoplankton biomass decrease with depth in the tropical regions (Mariana and Bougainville Trenches) are higher ($K = 8.5 \times 10^{-4}$) than in the subpolar region of the northwestern Pacific or in the southern regions of this ocean, e.g. in the Kermadec Trench $K = 6.5 \times 10^{-4}$. Thanks to this, the biomass of the net plankton within the 4000- to 8000-m layer in the region of the Kurile–Kamchatka Trench is 160 times lower than in its 0- to 500-m layer, and in the Bougainville and Mariana Trenches it is even 760 to 780 times lower when corresponding layers are compared here. In the Mariana trench the plankton biomass within the 4000- to 8000-m layer does not amount to even one-thousandth of its biomass in the 0- to 50-m layer of the epipelagic zone.

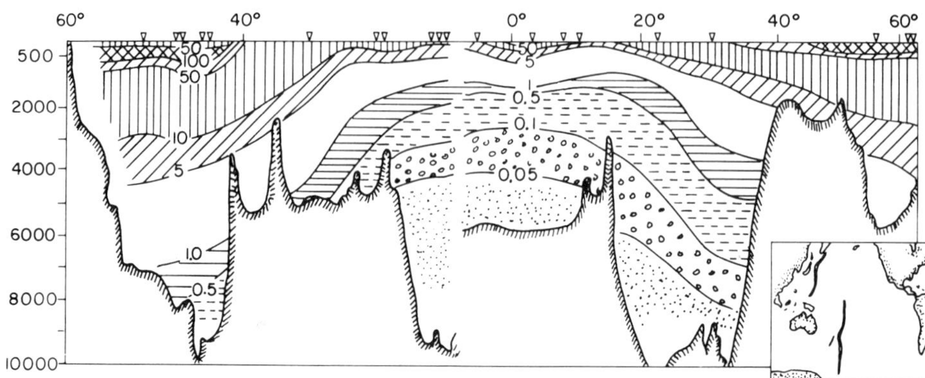

Fig. 2. Distribution of wet weight mesoplankton biomass (mg/m^3) on a meridional section across the Pacific Ocean. Triangles indicate the position of the plankton stations. [After Vinogradov (1968).]

However, the rates of decrease of plankton biomass with depth lower than 750 to 1000 m conform exponentially only in the first approximation to the equations given above. Indeed, comparatively narrow layers with sharply decreasing biomass alternate with extensive layers where biomass decreases considerably slower (Fig. 3), and where supposedly there may be layers of increased concentration of mesoplankton. For instance, in the northwestern Pacific Vinogradov (1970b) found that the quantity of plankton decreases most sharply at mesopelagic depths from 500 to 1000 m in summer. This decrease occurs as a result of a sharp decrease in the biomass of the upper-interzonal filter–feeding copepods of the genera Calanus and Eucalanus. In the bathypelagic zone from the depth of 750 to 1000 m to a depth of 2500 m, the rates of decrease of biomass are low. Throughout this entire depth range, the biomass decreases only two to three times. In turn, at the boundary between the bathy- and abyssopelagic zones, within the 2500- to 3500-m layer, it decreases by about 10 times, although the rates of decrease of copepod biomass do not undergo significant changes here. The overall decrease of biomass occurs mainly as a result of the almost total disappearance of planktophagous hunters primarily by the disappearance of chaetognaths and small fish, so characteristic for the bathypelagic zone.

At greater depths, in the abyssopelagic zone, the biomass decreases again, but rather slowly, changing from the 3000- to 4000-m layer to the 5000- to 6000-m layer by only 1.5 to 2 times overall. It is difficult to judge the variability of biomass in the ultraabyssal waters of the trench because of the paucity of material.

Judging from these data obtained in the northwestern Pacific, the change with depth of the rates of decrease of mesoplankton biomass (Fig. 4) makes it possible to mark out three major layers corresponding to (1) the epi- and mesopelagic zones, (2) the bathypelagic zone, and (3) the abyssopelagic zone. They are defined by a sharp change in the species and trophic composition of the communities, as well as in the morphological and ecological adaptations of the animals.

In the tropics the same three major layers may be distinguished. However, the boundaries between them in both oligotrophic and mesotrophic areas lie higher than in the subarctic region. The change from the surface plankton to the midwater plankton takes place at 200 to 500 m or even at a 100- to 200-m depth and the change from the midwater plankton to the abyssal one, at a 1500- to 2000-m depth.

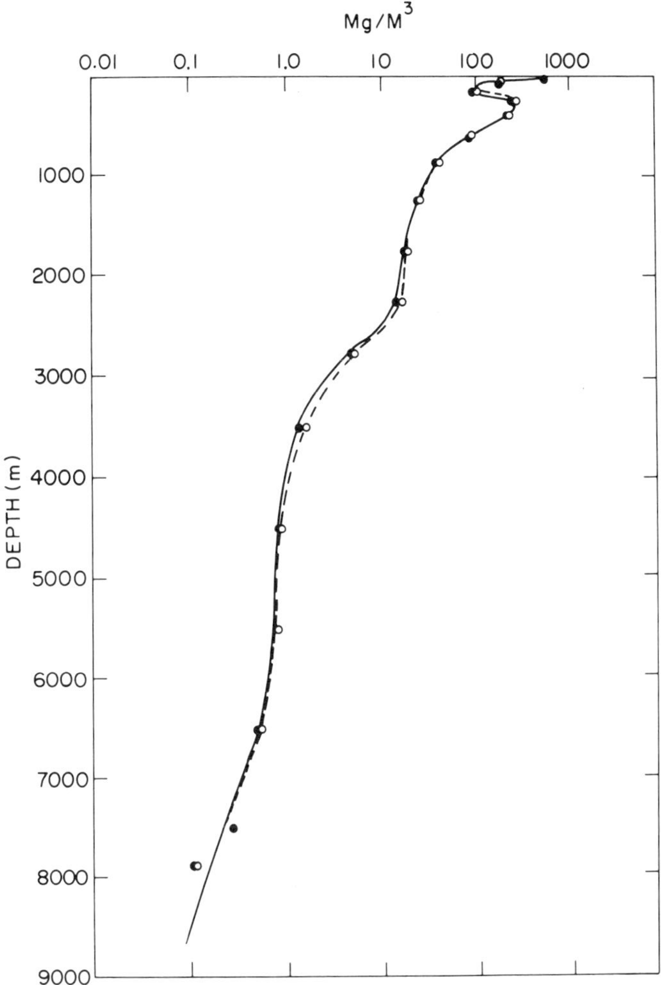

Fig. 3. Vertical distribution of wet weight mesoplankton biomass (mg/m^3) in Kurile–Kamchatka region (mean for nine stations with fractional vertical hauls, July/August 1966). [After Vinogradov (1970b).]

These differences result from the fact that in the subpolar region the upper-interzonal herbivores—performing seasonal migrations down to 1000- to 2000-m depth and even lower—play an important role, whereas in the tropics, the interzonal species perform mainly diurnal migrations with a much narrower range: from the surface down to 400 to 800 m. Hence the midwater pattern of plankton distribution in both regions, related to large quantities of migrators from the surface layers differs primarily in the vertical scope: the latter in the subpolar regions is much more extensive than that in the tropics.

C. *Macroplankton*

The basic information on the vertical changes of the species composition, biomass, and population density of macroplankton was obtained with midwater

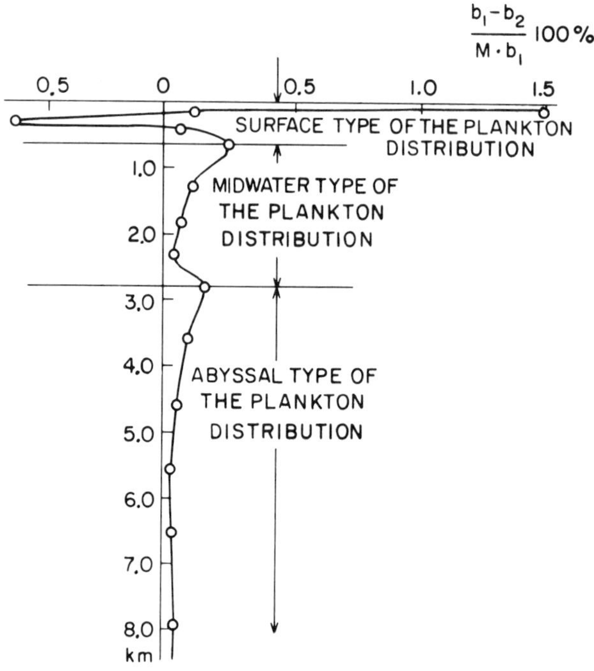

Fig. 4. Intensity of change in mesoplankton biomass with depth in Kurile–Kamchatka region [average for nine stations, July/August 1966; b_1—mean biomass of the overlying layer, b_2—mean biomass of the underlying layer; M—distance (m) between centers of the layers compared].

trawls. To evaluate these trawls as a tool capable of providing quantitative assessment, it is necessary to know the size limitations of the animals captured. The most frequently used Isaacs–Kidd midwater trawl captures the animals not exceeding 20 to 100 mm (Foxton et al.,1968). Of course, much smaller, as well as much larger, individuals are often caught, but the main size group is this, which we call *macroplankton*. It is easy to imagine that there may be various classifications; for instance, Sieburth et al. (1978) consider as macroplankton animals ranging from 20 to 200 mm, whereas Cushing et al. (1958) assign to macroplankton the animals ranging from 1 mm to 1 cm in length. However, the quantity of animals actually caught by the trawl depends also on the speed of towing, the size of its opening, and the mobility of creatures, in other words, on the species composition of macroplankton. Besides, there is no reason to believe that with the same efficiency the trawl catches the same animals at different times of day or at various depths; depending on the illumination, they may evade the trawl with intermittent success. Because of these methodical difficulties, the vertical distribution of macroplankton has not been studied in such a detail as that of mesoplankton.

Furthermore, the distribution of macroplankton with depth is stratified much more sharply than that of mesoplankton. All data available show the quantitative heterogeneity in vertical distribution of the main taxonomic groups of macroplankton, characterized by two or three maxima of its concentration. Distributional patterns may be substantially influenced by local hydrologic regimes and nutritive content of waters.

When conclusions are drawn from the catches of horizontal nets, including midwater trawls, the variation of the layers between high and low population densities of macroplankton hampers adequate evaluation of the quantities within the water column.

On the other hand, the sampling made by vertical or oblique nets is insufficiently representative because of a small volume of filtered water and a low speed of hauling of the net (ca. 1 m/sec). Undoubtedly, the plankton nets better capture the passive bathypelagic macroplankton than the active epi- and mesopelagic members of this group. Nevertheless, Vinogradov (1968, 1977), using his own data on vertical hauls of plankton nets with an opening 1 m^2, and having averaged the data on vast series of samples made in each region where more than 30 to 60 $\times$ 10^3 m of water was filtered, showed that in the cold eutrophic Kurile-Kamchatka trench mean macroplankton biomass[1] within the 0- to 3000-m layer comprises 1.5 mg/m^3, or 2.3% of the mesoplankton biomass. In the oligotrophic tropical regions of the Pacific and Indian Oceans in the 0- to 2000-m layer the macroplankton biomass constitutes 0.7 mg/m^3, or 13% of the mesoplankton biomass; and in mesotrophic equatorial regions (from 12°N to 12°S) it comprises 4.3 mg/m^3, or 63%.

In the northwestern Pacific within the layers of maximal prevalence of macroplankton at the depths of 750 to 1000 m its biomass comprises 3.4 mg/m^3, or 4.4%; and within the 2000- to 3000-m layer it comprises 1.3 mg/m^3, or 12% of the total net plankton biomass. In the oligotrophic tropical regions at 500- to 1000-m depth, macroplankton biomass constitutes 1.6 mg/m^3, or 25%; and in the equatorial region at the depth of 1000 to 2000 m, it is 5.9 mg/m^3, or 83% of the same.

And now let us turn to midwater trawls data. According to Legand et al. (1972), the maximum of the biomass of macroplankton (gelatinous animals not considered) in the equatorial waters of the central Pacific lies at 450 to 950 m, attaining 5.5 mg/m^3. By night, the pattern of distribution smoothes since the interzonal species climb up, and within the 0- to 900-m layer the mean biomass of macroplankton comprises only 3 to 4 mg/m^3. In the 0- to 1200-m layer in the equatorial Pacific its biomass constitutes about 20% of the total plankton biomass.

Similar results were obtained by Vinogradov and Parin (1973) in the tropical Pacific, where macroplankton biomass in productive regions at depths from 60 to 100 down to 400 to 500 m decreases on the average from 12 to 4 mg/m^3. In the oligotrophic regions the distribution was more homogenous, with mean biomass comprising about 2 mg/m^3. In the regions studied, macroplankton biomass within the 0- to 500-m layer constitutes 15 to 35% of the biomass of mesoplankton. These data are similar to those of Legand et al. (1972). They may be considered as the usual ratio between macroplankton and mesoplankton in the tropical zone in the entire upper 1000-m layer (Parin et al., 1977).

Let us compare these results with a tentative theoretical evaluation of the biomass ratio between macroplankton and mesoplankton within the same depth range. The biomass B of the group of animals whose weights W range from W_{min} to W_{max} or whose sizes L range from L_{min} to L_{max} may be estimated from the equation (Tseitlin, 1981a).

[1] Vinogradov (1968) assigns the animals measuring from 30 to 100 mm in length to macroplankton.

$$B = \frac{B_0}{2.7} \ln \frac{W_{max}}{W_{min}} \quad \text{or} \quad B = B_0 \ln \frac{L_{max}}{L_{min}} \tag{1}$$

where B_0 is a value characterizing the plankton abundance in a definite region at a definite depth. Defining mesoplankton as the size group from $L_{min}^{mez} = 0.2$ mm to $L_{max}^{mez} = 20$ mm (Sieburth et al., 1978) and obtaining the magnitude of the biomass of mesoplankton B_{mez} from equation 1, we obtain $B_0 = B_{mez}/4.6$. Since macroplankton captured with midwater trawl measures $L_{min}^{macro} = 20$ to $L_{max}^{macro} = 100$ mm, then $B_{macro} = 1.6\ B_0$. Hence, it follows that $B_{macro} = 0.35\ B_{mez}$; thus this evaluation nearly conforms with the observed data presented above. It is necessary to keep in mind that a calculated estimate takes into account the macroplanktonic animals of all groups, including all gelatinous creatures. The latter, as is well known, can sometimes be extremely abundant.

3. Energy Flow to the Depths

The means by which the nutritious organic matter penetrates into the depths and its form there have been discussed repeatedly and in detail by many authors (e.g., Vinogradov, 1962, 1968; Menzies, 1962; Fournier, 1972; MacDonald, 1975; McCave, 1975).

In general, it reaches the depths by two basic routes: (1) by sinking down as dead organic matter that comes as a result of living activity of the surface, littoral, or terrestrial organisms. Besides, as suggested by Rudjakov and Tseitlin (1980), a passive sinking of animals from overlying layers whose movable activity for certain reasons is weakened may be of some significance and (2) by active downward migration of herbivorous animals grazing in subsurface layers rich in food or active upward migration of the deep-sea carnivorous animals.

Some quantitative data characterizing these routes are of interest here.

A. *Vertical Migrations of Zooplankton*

Unquestionably an important route of food supply for oceanic depths is the active transport of organic matter from the surface zone by migrating animals (e.g., Vinogradov, 1953, 1955, 1959, 1962; Tchindonova, 1959). This is of paramount significance in highly productive temperate regions. This transport here seems to be achieved mainly by seasonal or ontogenetic migrations.

The grazing of the populations of the upper-interzonal phytophagous animals in high latitudes occurs during a relatively short spring–summer period of development of phytoplankton and microzooplankton in surface layers. In the autumn, the filter–feeders travel to the depths and rise again to the surface in the spring as themselves or their offspring. Because of the peculiarities of their life cycles, a considerable share of their populations remains at the depths in the spring–summer period, too.

As an example the data obtained in the northwestern Pacific may be considered. Here filter–feeding copepods grazing mainly on phytoplankton (the most abundant species of which are *Calanus cristatus, C. plumchrus*, and *Eucalanus bungii*) amount to 52% of the total biomass of the plankton in the layer at 0 to 4000 m. The share of *C. cristatus* comprises 23%, that of *C. plumchrus* 11%, and that of *E. bungii* 17% (Vinogradov and Arashkevich, 1969). In upper layers of the epipelagic zone where they graze, they usually amount to 70 to 90% of the total biomass of plankton or even more. Late in summer and in the autumn, the elder Calanus copepodites (fifth) descend to mesopelagic and bathypelagic depths, where they reach maturity, while not feeding at all. Moreover, the mandibulae in such mature specimens (sixth) reduce.

Among the species mentioned above, only a few individuals penetrate to the greatest depths, and their occurrence there does not affect the existence of the population. The vertical range of the distribution of the main biomass of the populations is more limited (Table I). In *C. cristatus* 90% of the whole population inhabit the layers above 1500 m; however, 90% of the mature males inhabit the 500- to 2000-m layer, and the mature females the 500- to 3000-m layer; 90% of the population of *C. plumchrus* inhabits the 0- to 750-m layer, whereas 90% of males inhabit the 300- to 750-m layer and females, the 200- to 1000-m layer. In *E. bungii* more than 90% of its population live above 750 m. However, an insignificant share of the population occurring at greater depths may constitute an important component in the diet of the low-density population of the bathypelagic and abyssopelagic zones.

In other cold-water and temperate regions, similar species of the interzonal copepods such as *Calanus finmarchicus* in the North Atlantic; *C. glacialis* and *C. hyperboreus* in the Arctic; and *C. propinquus, Calanoides acutus*, and *Rhincalanus gigas* in the Antarctic play the same role in active transport of organic matter to the great depths (e.g., Somme, 1934; Mackintosh, 1937; Wiborg, 1954; Ostvedt, 1955; Foxton, 1956; Kashkin, 1962; Andrews, 1966; Vinogradov, 1968). Active transport of organic matter from the subsurface layers to the bathypelagic and abyssopelagic depths (down to 6000 to 7000 m) is likewise carried out by certain macroplanktonic mysids of the genera *Boreomysis (B. plebeja, B. curtirostris, B. incisa), Hyperamblyops*, and *Dactylamblyops* by such euphausiids as *Benteuphausia amblyops* and certain other crustaceans (Tchindonova, 1959; Vinogradov, 1959, 1962, 1968, 1970).

These crustaceans probably rise up irregularly to the subsurface layers where they graze on phytoplankton and then descend to greater depths so rapidly that the phytoplankton does not have time to be digested, so that diatomic cells appear to be almost intact, full of protoplasts, and "fresh." Crustaceans with stomachs packed with such cells were found from 200- to 500-m to 4000- to 6000-m depths and once were found in a catch from 6000 to 7000 m. At the abyssal depths, on several occasions, specimens of *Boreomysis incisa* were encountered whose weight ranged from 20 to 55% of the total biomass of the plankton taken in the 4000- to 6000-m layer (Vinogradov, 1970).

The active transport of organic matter during the diurnal migrations of the herbivores and omnivores such as copepods of the genus *Metridia* and certain euphausiids reaches only the mesopelagic zone (400 to 700 m). Usually its daily capacity is not large.

However, for the carnivorous mesopelagic population, even a scarce diurnal transfer of organic matter, because of its regularity, may play a significant role. The number of animals migrating in this way and their share in the total biomass of plankton varies substantially with season and area. The rates of daily migrations increase with animal age (Table II); these migrations are especially intensive in the autumn. Where in the plankton prevail intensively migrating species, up to 50% of

TABLE I
Upper-Interzonal Copepods[a]

Species	0 to 50 m	50 to 100 m	100 to 200 m	200 to 500 m	500 to 750 m	750 to 1000 m	1000 to 1500 m	1500 to 2000 m	2000 to 2500 m	2500 to 3000 m	3000 to 4000 m
C. cristatus	53.4	15.6	4.3	15.5	4.4	22.3	30.3	36.8	14.3	9.9	7.3
C. plumchrus	16.9	10.7	5.8	9.9	17.6	12.6	8.4	4.6	1.1	2.6	0.6
E. bungii	4.4	20.7	23.9	31.0	14.4	2.4	1.4	0.4	0.4	0.3	0.1
Total biomass of these copepods (mg/m^3)	512	62.6	37.7	137	38.8	15.5	10.7	8.13	2.49	0.62	0.1

[a]The share of the most abundant upper-interzonal copepods in the total biomass of mesoplankton measured at various depths in the Kurile–Kamchatka region in percentages, and their total biomass measured in milligrams per square meter. Average values from six stations made during July to August, 1966 are given (After Vinogradov and Arashkevich, 1969).

TABLE II
Intensity of *Metridia pacifica* Migrations in Northwestern Pacific in May (in percent of Maximum Possible Intensity in the 0- to 500-m Layer) (Vinogradov, 1968)

Copepodite Stages	I	II	III	IV	V	VIq
K_{500}, %	15.6	15.6	19.8	22.8	71.9	98.8

the total biomass of plankton may shift from the surface layers into the mesopelagic zone (e.g., in the autumn in the northern Okhotsk Sea, where *Metridia ochotensis* dominates in the plankton communities).

In productive temperate regions the migrating upper-interzonal herbivores and omnivores shifting down from the epipelagic zone into the meso- and bathypelagic depths thus actively transfer a large quantity of organic matter, which provides food for carnivorous bathypelagic plankton. The vertical climb of the carnivorous invertebrates and planktophagous fishes for grazing in epipelagic zone is of subordinate significance here, since such migration is characteristic for a small portion of plankton (a few copepods of the genus *Pleuromamma* and some other genera, certain pelagic gammarids, a few lantern fish, and some other groups of mesopelagic animals).

Thus it turns out that the interzonal phytophagous plankters, which are the principal consumers of the primary organic matter of the surface layers, spend a part of their life in the mesopelagic and bathypelagic depths here as components of the deep-sea fauna and, at the same time, as its food source. In the deep-sea pelagic communities they play the role of distinctive "pseudoproducers." The pelagic communities of the ocean depths, which are energy-dependent on the surface communities, develop principally on the basis of utilization of the organic resources of these pseudoproducers. Such communities must be of a higher "maturity" and be more stabile than the surface communities where these pseudoproducers enter as main consumers (Margalef, 1968).

The juveniles of the overwhelming majority of the bathypelagic carnivores live at the depths that are the shallowest possible and richest in food, whereas with increasing size and maturation, they descend lower (Fig. 5).

In the tropical regions the pattern of active transport of organic matter substantially differs from the pattern just discussed. Because of the lack of a sharp seasonal cycling and a relatively steady development of the phytoplankton throughout the year, there is nothing to stimulate the plant-eating plankton to abandon the surface layers at unfavorable periods. The seasonal migrations do not occur at all or are very weakly manifest. Most of the herbivores live constantly in the epipelagic zone. The eurybathic upper-interzonal herbivore and omnivore species inhabit a much narrower vertical range than in the cold-water regions, shifting actually no deeper than to the 500- to 1000-m layer (Tables III and IV).

The lack in the tropics of abundant upper-interzonal species, forces most of the mesopelagic macroplankton species to undertake regular diurnal migrations into the surface layers. Therefore, the share of the carnivorous lower-interzonal species in the tropical plankton exceeds by far that in the high latitudes.

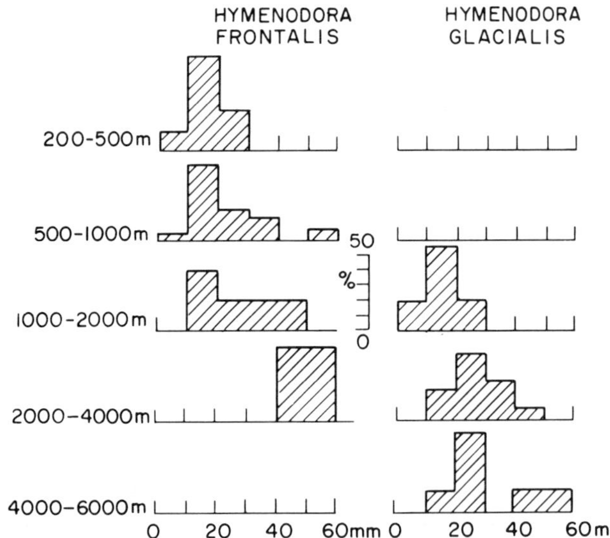

Fig. 5. Change with depths in size distribution of populations of two species of prawns in Kurile–Kamchatka Trench region.

The patterns of diurnal migration of the mesopelagic (lower-interzonal) planktophagous macroplankton represented by different taxonomic groups (e.g., siphonophores, euphausiids, decapods, cephalopods, and fish) were examined by Parin et al. (1977). In many regions the lantern fish are the most abundant migrators of this type.

Because the caloricity in the fish migrating higher upward to the surface layers is greater than that in the fish constantly dwelling in the bathypelagic waters, the role of the former as energy transmitters to the deep-sea communities is particularly great. The bathypelagic predator, having gulped a migrating lantern fish, will obtain more energy than in the case when it would swallow a bathypelagic fish of the same size (Childress and Nygaard, 1973). The catching of active migrators by passive deep-sea predators is considerably facilitated by the fact that many active migrators—including some lantern fish sojourning at the depth in the daytime in a sluggish, almost lethargic state—are an easy prey of those predators (Barnham, 1971).

Probably the most complete data on the patterns of macroplankton migrations are provided by the observations of deep-scattering layers (DSL). Concentrations of scatterers that form the migrating DSL (their concentrations at frequencies of about 20 kHz are rarely lower than 10^{-2} specimens/m^3) in the daytime usually hold at the depths of 200 to 700 m, more rarely at 800 to 1000 m, and sometimes even at 2000 m, whereas by night they climb up into the epipelagic zone, to the thermocline, or—passing through it—to the surface layers (Tchindonova and Kashkin, 1969; Kashkin and Tchindonova, 1971; Kashkin, 1977; Vinogradov, 1974). There are also the migrators inhabiting the deeper waters that by night do not climb up to the surface layers and concentrate at the depths of approximately 400 m (Kanwisher and Ebeling, 1957). However, at present it is not possible to estimate quantitatively the share of plankton migrating in such manner.

Thus, between the cold-water regions where the surface communities usually

TABLE III

Relative Population Density of Upper-Interzonal Herbivorous Copepods of Genus Calanus at Various Depths in Sea of Norway and Northwestern Pacific[a] According to Ostwedt (1955) and Vinogradov and Arashkevich (1969)

Region	Species	0 to 100 m	100 to 200 m	200 to 500 m	500 to 1000 m	1000 to 2000 m	Exceeding 2000 m
The Sea of Norway, annual mean	*C. finmarchicus*	28.4	21.9	29.8	19.9		
	C. hyperboreus	4.7	7.8	8.9	78.6		
Northwestern Pacific, May to June	*C. cristatus*	48.4	2.0	4.6	24.3	15.7	5.0
Northwestern Pacific, July to August	*C. cristatus*	45.5	1.7	28.5	6.4	15.1	2.8
	C. plumchrus	43.6	5.5	26.2	19.2	5.0	0.7

[a] In percentage of the whole population, and for the Sea of Norway, the data are given for the 0- to 2000-m layer.

TABLE IV

Relative Population Density of Upper-Interzonal Herbivorous and Omnivorous Copepods at Various Depths in the Tropical Pacific (in percent of Density of Whole Population) According to Vinogradov and Voronina (1964)

Species	0 to 200 m	200 to 500 m	500 to 1000 m	1000 to 2000 m	Exceeding 2000 m
Neocalanus gracilis	86.4	5.8	7.8	0	0
Rhincalanus cornutus	7.8	84.1	7.7	0.4	0
Gaetanus miles	4.2	86.8	9.0	0	0
Pleuromamma gracilis	64.2	34.5	1.3	0	0
P. xiphias	25.0	52.4	22.6	0	0

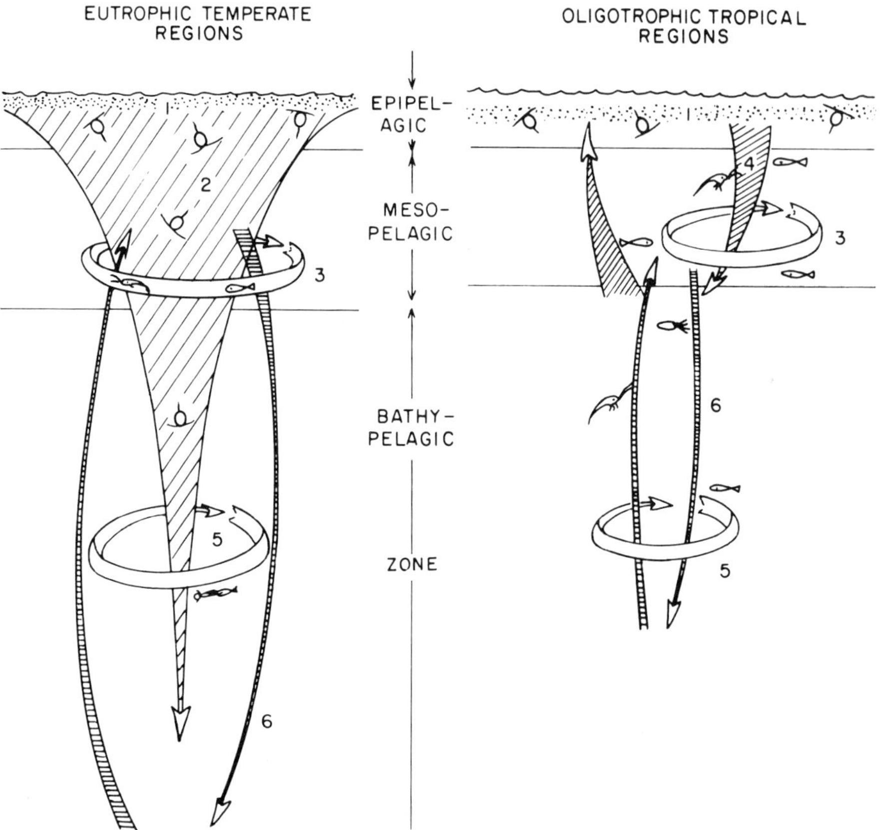

Fig. 6. Main pathways of active transfer of organic matter to meso- and bathypelagic depths in cold-water and tropical regions: (1) zone of development of phytoplankton and that of feeding herbivores; (2) ontogenetic migrations of phytophagous upper-interzonal copepods down to meso- and bathypelagic depths; (3) mesopelagic carnivorous species living on the interzonal species migrating to mesopelagic waters; (4) migrations (mainly diurnal) of carnivorous mesopelagic macroplankton; (5) bathypelagic carnivorous species living on interzonal species migrating to bathypelagic waters; (6) migrations (mainly ontogenetical) of bathypelagic species to overlying layers richer in food.

have an unbalanced annual cycle of production and consumption, and those of the tropics, where the production cycle of surface communities is usually more balanced, there are substantial differences in means by which active transport of organic matter is carried out to the depths (Fig. 6). These differences materially affect structural and functional variability in the deep-sea communities of both regions and, supposedly, also affect diverse rates of decrease of plankton with depth, as was pointed out above.

B. The Flow of Detritus

In addition to the active transfer of organic matter by means of vertical migration, there is another route to the depths, the sinking of detritus. Only a portion of the detritus whose basic components are dead phytoplankton cells, carcasses, and fecal material of animals are available as food to the deep-water inhabitants. A considerable part of it is consumed near the surface. Precisely what portion is determined by three factors—the particle sinking rate, the particle decomposition rate, and the number of consumers.

We calculate the vertical flow of organic matter from the producing zone for the open waters of the tropical zone of the ocean. The proportion of the primary production utilized by the heterotrophic part of the association ranges from 20 to 30% in the zones of rich-plankton development, such as in "young" upwelling waters, to 80 to 100% in "older" waters. In certain situations, the primary production does not cover the demands of the heterotrophs, and the community can exist only on allochthonous influx of organic matter (Vinogradov and Shushkina, 1978; Vinogradov et al., 1976). However, short-term observations do not make it possible to estimate the average flow of detritus from the producing zone. Such data can be obtained from the size distribution of pelagic organisms (Tseitlin, 1981d). We begin the calculations by determining the proportion of the primary production consumed by the heterotrophs of the surface zone that are larger than phytoplankton cells and by the protozoa that have approximately the same dimensions. We assume that the assimilated food is expended on respiration Q and somatic production P. Then the diet of the animals is equal to $C = (P + Q)/U$, where U is the coefficient of assimilation. To calculate the respiration and production, we use the following expressions (Tseitlin, 1981c):

$$Q(W_{min}, W_{max}) = \frac{0.05\ akB_0\ (W_{min}^{\alpha-1} - W_{max}^{\alpha-1}),\ \text{mg/m}^3 \cdot \text{day}}{w_0 \tau q(1 - \alpha)} \tag{2}$$

$$P(W_{min}, W_{max}) = \frac{B_0\ (W_{min}^{\alpha-1} - W_{max}^{\alpha-1},\ \text{mg/m}^3 \cdot \text{day}}{2.7\ T_0 \tau(1 - \alpha)} \tag{3}$$

Equations 2 and 3 define, respectively, the respiration and production (in milligrams wet weight) of a group of pelagic organisms whose weight is found between the minimum weight W_{min} and the maximum weight W_{max}. In these formulas a (milliliters of O_2 per hour) and $\alpha = 0.75$ (Hemmingsen, 1960; Winberg, 1976) are parameters that relate respiration to weight ($Q = aW^{\alpha}$); T_0 is a parameter for the duration of the growth of the animals $T = T_0 W^{1-\alpha}\tau$ (Tseitlin, 1981b); T_0 is equal to 33 days for multicellular and 13 days for unicellular organisms; $w_0 = 1$ mg; W_{min} and W_{max} are dimensionless values numerically equal to the weight expressed in milligrams; τ is the temperature correction; B_0 (in mg/m^3) is proportional to the biomass of the mesoplankton; $k = 4.86$ cal/ml O_2 is the oxycalorific conversion factor; and q (cal/mg) is the caloric value of the animals.

Equation (3) is applicable for limited weight groups of organisms $W_{max}/W_{min} \leq 10^4$. In this case it is supposed that intragroup consumption is absent. To calculate the energy required by the communities, one should use the formula

$$C = \frac{(P + Q)\,(1 - \eta)}{U} \tag{4}$$

where η is the coefficient of ecological efficiency, which is equal to the ratio of the production of two consecutive trophic levels. We assume that $\eta = 0.1$. It has been shown (Tseitlin, 1981b) that in the calculations, one should use the values of the parameters a and q that are peculiar to the smallest dimensions of the animals in question. In our case these are the small crustaceans (we evaluate the consumption by protozoa separately). Therefore, we suppose that $a = 0.125$ ml O_2/hr (Sushchenya, 1972), and $q = 0.7$ cal/mg (Shushkina and Sokolova, 1972). The coefficient of assimilation (U) for the herbivorous animals is assumed to be equal to 0.6 to 0.7 (Conover, 1966; Sushchenya, 1975; Pechen-Finenko, 1979). Here we consider that $U = 0.6$. Suppose $W_{min} = 4 \times 10^{-4}$ mg, $W_{max} = 10^9$ mg and $\tau = 1$ (corresponding to 20°C). Then for the animals that are larger than phytoplankton cells, we obtain $Q_h = 1.23\ B_0$ mg/m^3 · day, $P_h = 0.32\ B_0$ mg/m^3 · day, and according to equation 4, $C_h = 2.32\ B_0$ mg/m^3 · day.

Let us estimate the primary production by equation 3, assuming $W_{min} = 10^{-7}$ mg, $W_{max} = 2 \times 10^{-4}$ mg, $T_0 = 13$ days, and $\tau = 1$. We obtain $P'_p = 5.45B$ mg/m^3· day. But since the estimate of the energy characteristics is made here on the basis of the size structure of the communities (Tseitlin, 1981b), the value found includes the production of nonpredatory protozoa P_a that are found in the same size group. Studies in recent years (Beers and Stewart, 1970; Tumantzeva and Sorokin, 1975) have shown that P_a amounts to approximately 5% of the phytoplankton production P_p. Then $P_a = 0.27B_0$ and $P_p = P'_p/1.05 = 5.19B_0$ mg/m^3 · day. Let us determine the respiration of the protozoa. From equations 2 and 3 it follows that

$$Q = \frac{0.135\ akT_0}{q} P$$

Assuming here that $a = 0.107$ ml O_2/hr—a value found for infusoria (Kchlebovich, 1979), $T_0 = 13$ days, and $q = 0.8$ cal/mg, we obtain $Q_a \simeq 1.13P_a = 0.3B_0$. The total consumption of energy by nonpredatory protozoa is thus equal to $C_a = 0.57B_0/0.6 = 0.95B_0$ mg/m^3 · day. The total consumption of energy by protozoa and the larger heterotrophs is $C = C_h + C_a = 3.29B_0$ mg/m^3 · day; that is, the heterotrophs of the producing layer of the tropical ocean (disregarding bacteria!) consume on an average 63% of the primary production.

However, the rate of descent of the phytoplankton cells (Fig. 7) is important. The large cells require several days to sink through the surface zone, and the cells of average size require weeks. Therefore, in tropical regions the primary production not utilized by the animals is almost totally consumed in the surface zone by the bacteria and makes no direct contribution to the diet of the animals of deep-lying layers. As direct determinations in the tropics have shown (Sorokin et al., 1975; Vinogradov and Shushkina, 1978), only in eutrophic and, in part, mesotrophic waters—upwellings and similar regions—may a substantial part of the phytoplankton not be utilized in the surface zone. In temperate regions, on the other hand, with a discontinuous production cycle by the surface communities, the "dying bloom" phytoplankton obviously makes up a significant part of the detritus descending into the deep layers.

Now let us consider the other components of the detritus—the carcasses and fecal matter of animals. One can point to situations in which one of these two components has dominant significance. For instance, in the case of mass mortality of animals in the zones of fronts or for some other causes, obviously, corpses may predominate. But if we exclude such cases, the predominant component will be the fecal material of animals. The latter may be shown as follows. The unassimilated part of the diet of heterotrophs $C_{na} = (1 - U)C_h = 0.4 \times 2.32B_0 = 0.93B_0$ mg/m^3 · day (here we are not considering the consumption of energy by protozoa—as becomes clear later—their corpses and excretions cannot play a

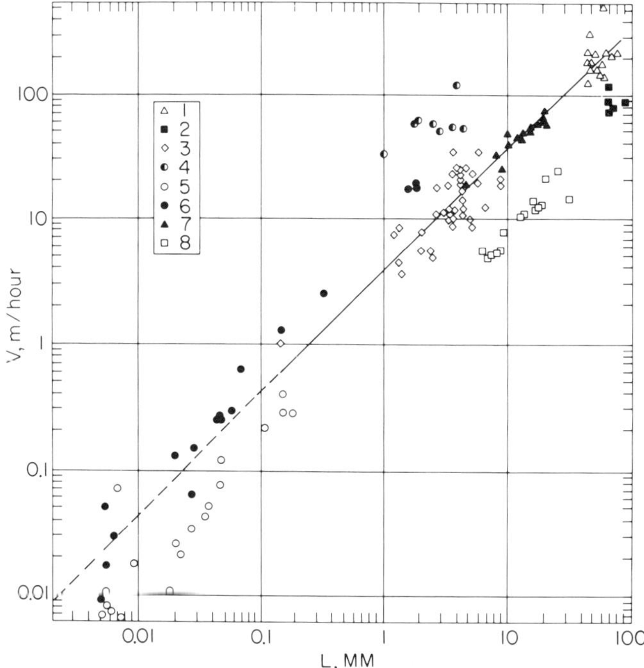

Fig. 7. Dependence of the rate of passive sinking (v, m/hr) on the size of organisms (L—body length of pelagic animals or diameter of phytoplankton cells): (1) decapods; (2) fish (cyclotone); (3) nauplii of copepods; (4) ostracods; (5) active cells of phytoplankton; (6) aging and dead phytoplankton cells; (7) euphausiids; (8) chaetognaths.

significant part in the diet of the deep-sea animals). Since the elimination of the animals E in steady state is equal to the pure production, then $E_h \simeq (\frac{1}{2} - \frac{2}{3})P$ (Tseitlin, 1981d). From this it follows that the ratio of the rate of production of feces C_{na} to the rate of production of corpses E_h is equal to 3–6; feces are produced at a rate three to six times higher.

At present it is difficult to determine what proportion of the detritus produced in the surface zone reaches great depth. The processes of decomposition of detritus, the rate of sinking, and the rate of its consumption as it descends have been little studied. But on the basis of the available data, certain estimates can be made.

The sinking rate is important. Since the rate of the processes of decomposition of organic matter depends strongly on temperature, the more rapidly the warm surface zone is passed through, the greater the quantity of organic matter that will avoid decomposition and reach great depths. Figure 7 shows the sinking rates of various plankton organisms (Rudjakov and Tseitlin, 1980). The regression equation found for copepods and their nauplius larvae and decapods has the form $V = 4L$, where V is the rate of sinking in meters per hour and L is the body length in millimeters. One easily sees that this corresponds to the sinking rate approximately equal to one body length per second. Hence it follows that small crustaceans (L = 0.5 to 1 mm) sink approximately 50 to 100 m in 1 day. In tropical regions, the zooplankton maximum is several tens of meters above the upper thermocline, so that one day is sufficient to pass into the zone of lower temperature. According to Skopintsev (1949), the rate constant of the decomposition reaction of dead plankton at 25°C has the value 0.044 to 0.114 day^{-1}. Other investigators quote smaller values for the constants: 0.038 (Grill and Richards, 1964); 0.0296 (for nitrogen); and 0.0156 (for carbon) (Otsukia and Hanya, 1972). Pavlova (1968) studied the rate of decomposition of zooplanktonic organisms and found that at the temperature of 17 to 23°C, total decomposition of copepods occurs in 30 to 40 days, for small radiolaria in 5 days, and that *Sagitta*, which has weaker outer covers, decomposes into parts in 5 days and totally decomposes in 13 days. Harding (1973) lists lower rates of decomposition for copepods: 11 days at 4°C and 3 days at 22°C. We note that the decomposition rate of 30 days corresponds to the constant 0.04 day^{-1}, which is close to the value listed above. Evidently, the infestation of the corpse by bacteria and protozoa in the surface zone may accelerate to a significant degree the subsequent process of decomposition. Suppose that the carcasses of the animals sink at the same rate as live individuals.[2] Then the corpses of crustaceans 0.5 to 1 mm in size in 10 days will be at depths of 500 to 1000 m (corresponding to temperatures of 5 to 10°C in tropical waters), and in 30 days, at 1500 to 3000 m.

In brief, the corpse of even a small crustacean, under the appropriate conditions, can reach very significant depths. This is all the more valid for the large crustaceans, whose rates of descent are greater. But we are still uncertain as to (1) what extent the rate of decomposition as defined in laboratories corresponds to the rate *in situ* and (2) how rapidly the rate of descent of the decomposing animal changes. Until experiments answer this question more accurately, it may be considered that carcasses are components of the diet of the deep-sea animals.

The results of measurements of the sinking rates of the feces of certain crustaceans are shown in Fig. 8 (Small et al., 1979). As in the case of the sinking rates of the plankton organisms, the point spread of the data is very large. The sinking rates of feces depend not only on their size, but are also determined to a significant degree by the character of the food of the animal. This was demonstrated in laboratory experiments, where feces obtained by feeding animals phytoplankton with the addition of a mineral suspension had much greater rates of sinking than feces obtained when the animals were fed pure phytoplankton (Turner, 1977; Fowler and Small, 1972). This same circumstance—of mineral particles in the food—apparently explains the high sinking rates of feces in experiments by Smayda (1969) (see Fig. 8). Small et al. (1979) emphasize the importance of studying the characteristics of the feces of animals feeding on natural food—in this case the sinking rate may be substantially greater than in the case of a laboratory diet. In view of the large spread of the data on the sinking rates of feces, it can be said that they exceed by three to five times the average sinking rates of planktonic organisms of the same dimensions.

The data on the decomposition rate of feces are even fewer than those on the rates of sinking. According to Turner (1979), the membranes of the feces are destroyed at 22°C in 3 to 11 days; at 5°C they are preserved intact for 35 days. Honjo and Roman (1978) observed the decomposition of the membranes within 24 hours at the temperatures of 10 and 15°C for copepods fed laboratory phytoplank-

[2]Figure 7 presents data for animals that have a neutral or positive buoyancy. Their corpses, respectively, will remain at a constant depth or ascend until the moment when the body tissues providing buoyancy decompose.

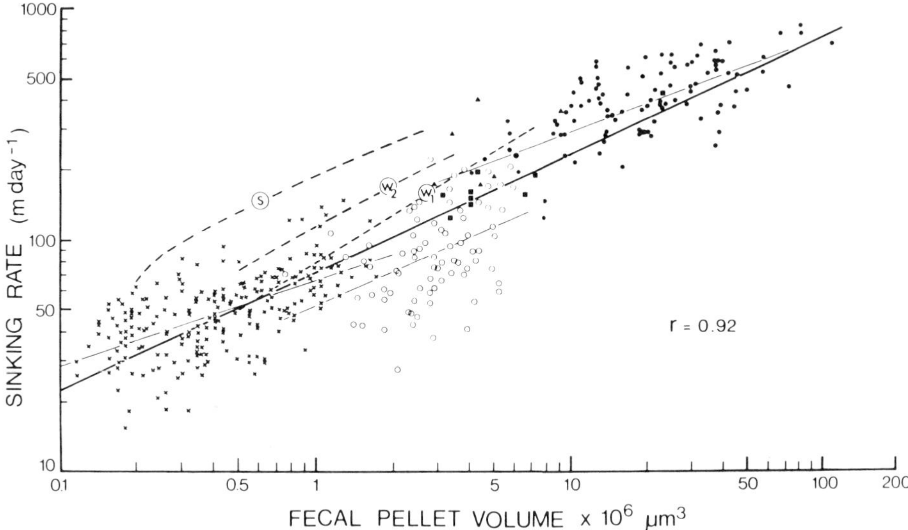

Fig. 8. Composite relationship (thick regression line) between fecal pellet sinking rate (S = m/day) and pellet volume (V = $\mu m^3 \times 10^6$), for mixed small copepods (crosses), *Anomalocera patersoni* (open circles), and various euphausiid species (filled circles, *Meganyctiphanes norvegica*; triangles, *Euphausia krohnii*; squares, *Nematoscelis megalops* [from Fowler and Small (1972)]. Composite regression equation is $\log S - 0.513 \log V - 1.214$. Temperature was 14°C for all experiments. Mean and range of values from Smayda (1969, 1971) are given for comparison (dashed line, S), as are mean and range of values from Wiebe et al. (1976) for their 5°C experiments (dashed line, W_1) and 21.7°C experiments (dashed line, W_2). Fine lines are individual regression lines for small copepods, *A. patersoni*, and euphausiids [based on information provided by Small et al. (1979)]. [Reprinted with permission of Springer-Verlag, New York, Inc. (*Marine Biology*) from Small, L. F., S. W. Fowler, and M. J. Ünlü, 1979. Sinking of natural copepod fecal pellets. *Mar. Biol.*, **51**, 239.]

ton. The same was observed by Small et al. (1979), with the total decomposition on such a diet occurring in several days. But the decomposition process of feces formed on natural food, according to their observations, occurred slowly. In deep-water traps, the feces do not undergo decomposition for the course of many weeks. Evidently it can be assumed that feces are considerably more resistant to decomposition than the corpses of animals. As Fowler and Small (1972) noted, the chances of dead euphausiids reaching the bottom are great because of the high rate of sinking whereas the chances of feces are great because of slow decomposition. A critical factor for preserving the fecal membrane intact is apparently the time necessary to pass through the warm upper layer.

From Fig. 8, it follows that the feces of small copepods ($5 \times 10^5\ \mu m^3$ in volume) pass through the 100-m layer of water in approximately 2 days. It can be assumed that the feces of these copepods will pass through the warm surface layer without being destroyed or else by being destroyed insignificantly. But the entire process must depend strongly on the level of bacterial activity and consumption by surface detritophages. If the latter are sufficiently large, the small feces will participate in the material cycle occurring in the surface layer. In the case of partial destruction, pellets sink, not in the form of compact fragments, but as a more amorphous fecal substance, whose quantity, as noted by Bishop et al. (1977), exceeds the dense quantity of fragments. All these factors, whose role requires further investigation, contribute considerable uncertainty to the estimate of the flow of small fecal material. Therefore, only with reservations can we consider that all the unassimilated matter by the inhabitants of the surface zone (C_{na}) is food for the deep-sea animals.

Assume that the producing layer occupies the upper 125 m. Then the primary production $P_p = 650B_0$ mg/m^2 · day, and the flow of corpses and feces downward $C_{na} + E_h = 143B_0$ mg/m^2 · day and amounts to approximately 22% of the primary production. Taking (with $B_{mez} = 27$ mg/m^3)$B_0 = 5.9$, we obtain $P_p = 3800$ mg/m^2 · day = 380 mg C/m^2 · day and $C_{na} + E_h = 840$ mg/m^2 · day = 84 mg C/m^2 · day. We obtain this number by using an assimilation coefficient $U = 0.6$. When $U = 0.7$, we obtain $C_{na} + E = 60$ mg C/m^2 · day. Therefore, when characteristic values of U differ by only approximately 15%, the downward flow of detritus varies by 40%.

In addition to the corpses and feces of animals, one should also mention exuvia and the suspension formed from dissolved organic matter. The quantity of exuvia can be estimated from the results of the study by Golubev (1977) to be approximately 15% of somatic production. If one considers that this production is about 20% of the assimilable energy, about 3% of the latter is spent on exuviae—a value that is too small to be considered in our rough calculations. Regarding the formation of the suspension of dissolved organic matter, this process occurs not only close to the surface, but also at considerable depth. First, it is possible to raise objections to its source (Riley, 1970). Evidently this component of the detritus cannot be subjected to a quantitative accounting.

Finally, there remains the feces and corpses of the animals living below the production zone. Honjo (1978) studied feces at the depth of 5367 m. He found that they can be classified into "green" and "red" feces. The "green" feces belong to animals living near the surface; they contain relatively fresh plant cells, about 15% organic carbon, and 25% mineral particles. The "red" feces belong, apparently, to the deep-sea animals. They contain more mineral particles (≤85%) and are poor in organic carbon (~5%). They have no peritrophic membrane. Honjo suggests that the "red" feces are the product of mesopelagial and bathypelagial coprophages that consume the "green" feces. If the surface animal feeds intensively but poorly excretes the nutrients, the deep-water coprophages have a higher assimilation efficiency, limiting their requirements (Honjo, 1978).

4. Trophic Relationships in Deep-Sea Communities

A. *Change of Trophic Relationships in Communities with Depth*

As stated above, while the organic matter formed in the surface producing zone gradually penetrates into greater depths, its quantity must inevitably decrease as a result of consumption by organisms of the intermediate layers and its mineralization. The form of this organic matter also changes. At some depths the organic matter prevails as such, represented by migrating organisms themselves, and at the other depths, as detritus at various stages of mineralization. All this results in variation in trophic structure of the communities with depth.

Perhaps while scrutinizing this problem it would be better to begin with communities of the highly productive northwestern Pacific, fairly well studied regarding quantitative data on the vertical distribution of the total plankton biomass (see above), as well as the distribution of the main taxonomic groups from the surface down to the depth of 6000 to 8000 m (Vinogradov, 1968) (Table V).

The animal population of the upper euphotic zone rich in phytoplankton is characterized here by its predominance of herbivores, represented primarily by *Calanus plumchrus* and *C. cristatus* and, to a lesser degree, by that of omnivores, such as *Eucalanus bungii* and *Metridia pacifica*, whereas the last mentioned group prevails at somewhat greater depths than the more obligatory herbivores (Table I). The predators, among which one of the most important is *Sagitta elegans* s.l., play particularly subordinate roles in the surface layers (Fig. 9).

In the lower epipelagic zone at a depth of 80 to 200 m, in the cold intermediate layer peculiar for the northwestern Pacific, the number of the herbivores of the genus *Calanus* diminishes sharply (by an order of magnitude). The biomass of *Eucalanus* within the same depth range changes less. At the same time, the population density of chaetognaths undergoes no significant variations, which results in an increase in their share of the plankton.

At mesopelagic depths the biomass and share of interzonal herbivorous copepods increase again, but below 500 m they in any case are represented by the "deep-water aphagous" members of populations. The number of chaetognaths remains here nearly the same as in the overlying layers, but because of the increase in the quantity of the total plankton biomass, their share lessens here. Their species composition also changes. In the 200- to 300-m layer a significant number of *Sagitta elegans* s.l. is still present, but along with it emerge fairly large numbers of *Eukrohnia hamata*, which dominates the 300- to 500-m layer.

At the depth of 300 to 750 (1000) m omnivorous copepods have a great importance; this fact is supposedly related to the development at these depths of radiolarian families Aulacanthidae and Aulasphaeridae, which are sometimes encountered in great numbers, amounting up to 10 to 20% of the total sample weight. Below 500 to 750 m down to the lowermost borders of the bathypelagic zone, a major share of the plankton biomass (35 to 45%) (as stated above) is made up of populations of interzonal filter–feeding copepods (*C. cristatus, C. plumchrus*), which are a food reserve for bathypelagic predators. In the lower mesopelagic layers carnivorous prawns, such as *Gennadas borealis* and *Hymenodora frontalis*, are regularly met with and sometimes play an important role there. In the 500- to 1000-m layer lives a bulk of population of a jelly fish, *Crossota brunnea*, and some other coelenterate species.

Below 1000 m, the structure of the community changes substantially compared with that of overlying layers, staying nearly unaltered throughout the entire bathypelagic zone down to 3000 m. The highest biomass here are the carnivores, where chaetognaths; prawns; and planktophagous fish, such as some Cyclothones, Melamphaids, and Macrourids, prevail. Copepods here make up one-third to two-thirds of the biomass of the net plankton, whereas more than a half of them are represented by aphagous V and VI copepodite stages of the genus *Calanus*. The biomass of prawns reaches its maxima within the 1000- to 1500-m layer, where *Hymenodora frontalis* prevails and within the 2000- to 2500-m layer, where *H. glacialis* predominates. Chaetognaths (*Eukrohnia fowleri*) dominate within the depth range stretching out from 1500 to 3000 m, where they give on average more than 30% of the total plankton biomass and at definite stations in the 2000- to 3000-m layer, even up to 55 to 60%. Because of the presence of both these groups and,

TABLE V

Share of Animals of Main Taxonomic Groups in Mesoplankton of the Northwestern Pacific[a]

Depth, m	Herbivorous Copepods *C. plumchrus*, *C. cristatus*	Omnivorous Copepods *E. bungii*, *M. ochotensis*, *M. pacifica*	All Copepoda	Mysidacea	Amphipoda	Euphausiacea	Decapoda	Chaetognatha	Total Biomass, Wet Weight, mg/m^3 (Coelenterates Excluded)
0 to 50	70.3	5.4	84.9	<0.1	2.3	3.7	<0.1	8.9	626
50 to 100	26.3	24.7	59.1	<0.1	3.6	7.2	0.2	29.4	109
100 to 200	10.1	27.5	41.4	<0.1	5.4	5.2	1.1	45.7	89
200 to 500	25.4	36.7	78.3	0.2	2.9	1.2	0.1	14.8	247
500 to 750	22.0	20.6	69.0	4.2	1.1	0.2	1.4	17.3	103
750 to 1000	34.9	3.3	74.4	6.4	0.6	<0.1	1.8	14.3	51
1000 to 1500	38.7	3.3	73.2	7.7	1.1	0.1	8.6	6.0	27
1500 to 2000	41.4	0.9	53.8	2.1	1.3	0.3	4.7	33.1	19
2000 to 2500	15.4	1.2	33.4	7.6	1.3	<0.1	8.3	45.1	17
2500 to 3000	12.5	1.2	36.5	0.4	0.6	11.4	8.1	40.3	5.2
3000 to 4000	7.9	0.2	70.6	4.0	1.7	<0.1	3.4	5.5	1.6
4000 to 5000	<0.1	0	55.4	36.6	1.9	3.0	<0.1	1.2	0.91
5000 to 6000	0	0	40.7	29.1	23.7	<0.1	<0.1	0.9	0.86
6000 to 7000	0	0	48.4	13.3	30.1	0	0	0.8	0.60
7000 to 8700	0	0	57.1	0	21.9	0	0	<0.1	0.32

[a] In percent of biomass of net plankton in each layer. Taken during July and August 1966, an average from nine stations.

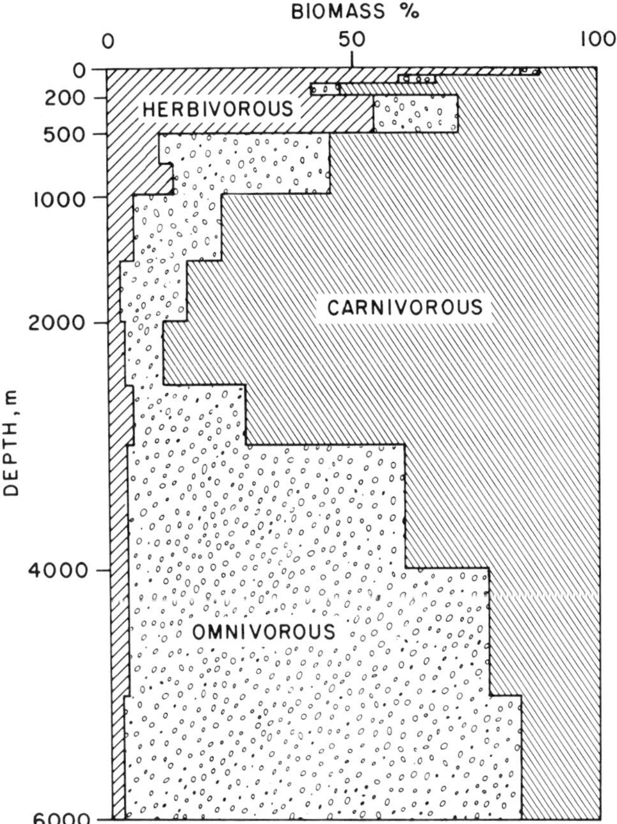

Fig. 9. Change with depth of the share of main trophic groups of mesoplankton in Kurile–Kamchatka region (July/August 1966), disregarding the interzonal aphagous V to VI copepodite stages of *Calanus cristatus* and *C. plumchrus*, which do not feed below 500 m.

to a lesser extent, the carnivorous copepods, mysids, and euphausiids, as well as fish, the carnivorous members of the community dominate at bathypelagic depths, comprising up to 55 to 80% of the total net plankton biomass here.

Below 3000 m the proportion of predators diminishes. It seems likely that because of extremely low population densities of prey animals and their scarce biomass, the predation as such (even the passive one) becomes unreasonable from the viewpoint of energetics. Chaetognaths and fish almost wholly disappear out of plankton below 3000 m, and so do decapods and coelenterates below 4000 to 5000 m. Here there is a prevalence of small euryphagous animals (chiefly copepods) able to utilize a wide variety of food. However, at the depths of 4000 to 6000 m at the expense of the migrating mysids (see above), the role of the phytophagous crustaceans somewhat increases.

Deeper than 6000 m, that is, within deep-sea trenches, detritophagous animals predominate, and only in the near-bottom layer does a significant number of carnivorous polychaetes and gammarids emerge. These seem to feed off the bottom. It is noteworthy that the rearrangement of the trophic structure of the communities between the epi- and mesopelagic layers, as well as between the bathy- and abyssopelagic ones (the 500- to 750- and 2500- to 3000-m layers) occurs rather sharply, accompanied by particularly intensive decrease in the total plankton biomass, as already mentioned (Section 2).

In the tropics, as a result of a nearly unvariable and low primary production throughout the year even in the epipelagic communities, the role of herbivores considerably lessens; however, in turn, as it seems, the relative role of the detritus food chain becomes more important. As already stated, the active downward transport of organic matter from the surface layers in the tropics proceeds mainly through the migration of carnivorous animals that are larger than herbivorous copepods. Therefore, in the meso- and bathypelagic tropical communities greater significance is attributed not the relatively small chaetognaths, as in the northwestern Pacific, but to the macroplanktonic, often large, carnivorous forms, primarily fish, copepods, and prawns (Section 2.C). It is obvious that the average animal sizes of the meso- and bathypelagic populations in the productive tropical regions will exceed those in the productive cold-water regions.

Along with large carnivores a significant role is also played by euryphagous animals that feed on detritus shifted downward, and on unassimilated residues of the food taken by daily migrators in upper layers and defecated by them at the depth. The lessening of the role of predators in abyssal communities and the increase in importance of euryphagous animals occurs in the tropics at lower depths than in the cold-water regions (Vinogradov, 1968).

Before discussing trophic relationships within the deep-sea communities, it is necessary to mention the question of the cyclic nature of their development related to the seasonal variability in food supply from the surface layers, which is poorly known as yet.

Mauchline (1972) made a special analysis of the information contained in the literature concerning the ratio between the adult males and females and the maturity of the males in the populations of the deep-sea crustaceans. He found that usually the females prevail and that in the populations there occur almost always adult males. This permitted him to suggest that there is a prolonged or even year-round reproduction period. However, Mednikov (1961) thinks that the prevalence of females in the deep-sea species and the increase of this predominance with depth have adaptive significance for increasing fertility of the populations under limited feeding conditions.

Unfortunately, direct observations of the seasonal variations of the age composition within populations of the deep-sea animals are very scarce. However, in the northern Pacific it was noticed that in the region of the Kurile–Kamchatka trench in the abyss at the depths of 5000 to 6000 m late in summer, in certain populations of such pelagic gammarids as *Scopelocheirus schellenbergi, Vitjaziana gurjanovae, Parargissa arquata* juveniles predominated, whereas in the spring, the share of young individuals in these species was less. Apparently this is related to the sinking late in summer of senescent radiolarians Aulacanthidae from the 200- to 500-m layer, where they are very abundant, into the abyssopelagic zone. The phaeodia of these radiolarians in this season serve as basic food of the abyssal gammarids and copepods and provide a marked intensification of their reproduction (Vinogradov, 1970a,b). Therefore, at least in some temperate regions, the seasonal cycles of development of the surface communities and food transport to the depths also lead to seasonal variations in the existence not only of the mesopelagic and bathypelagic, but also of the abyssopelagic communities.

B. *Adaptation to Existence Under Constant Food Scarcity*

In the surface layers with rich food supplies, even significant energy losses can be easily recovered by increasing the intensity of feeding. As Schmalgauzen (1968) points out, when food is abundant, the most active and fertile species attain advantage in evolutionary development. On the contrary, in the case of the shortage of energy resources, the advantage should fall to the narrowly specialized species, whose metabolism is more economical, as it is lowered even at the expense of the animal activities. Indeed, when low population density and scarcity of prey animals prevail, the predator's energy expenditures will exceed consumption when it searches for and pursues its prey. Therefore, the hatching by deep-water predators is done passively by watching for and luring their prey. At this time various appendages are used, with light organs in fish and cephalopods, elongated grasping limbs in crustaceans, and so on. At bathypelagic depths there are no nectonic animals. The entire authochthonic population of the depths, even the large predators, are planktonic organisms passively hovering in the water (Parin, 1979).

The scarcity of food resources and its resulting need of reducing energy expenditures, bring substantial morphological modifications of the organism in many groups and, thus, the emergence of new taxa (Birstein and Vinogradov, 1971). Adaptations to the extremely scarce food supply in the deep-sea animals occur in various ways in different taxonomic groups. They involve physiological, morphological, and ecological characteristics of organisms and considerably affect the structure of the deep-sea communities. This aspect was especially studied by many students (e.g., Denton and Marshall, 1958; Marshall, 1960, 1971; Walters, 1961; Vinogradov, 1968; Birstein and Vinogradov, 1971; Mauchline, 1972; Childress and Nygaard, 1973). It seems likely that the trend to reduce energy expenditures is inherent to the entire deep-sea population and that it changes the physiognomy of the deep-water animals more deeply than the adaptations to other environmental factors.

C. *Quantitative Evaluation of the Energy Expenditures of Deep-Water Animals*

There are two direct methods for estimating the energy expenditures: by measuring the rate of respiration or the 24-hr diet of the animals. When one speaks of deep-sea animals, both methods encounter tremendous difficulties. This scarcely requires a detailed explanation—it is sufficient to recall that the role of all factors determining the respiration and the diet of marine epipelagic (and freshwater) animals is far from fully explained. Also, they are far more accessible to observation and experiment than the deep-water animals. Since the results of measurement of the respiration rate usually have a very significant variance, the acquisition of reliable data requires numerous repetitious experiments. It is precisely this condition that is difficult to realize in the case of deep-sea animals—the investigators rarely deal with a sufficiently large number of specimens. Moreover, under laboratory conditions, it is relatively simple to maintain the temperature and concentration of oxygen specific to the living depth of the animals being studied, but not the pressure of these depths.

The results of laboratory experimental studies of the effect of pressure on the respiratory rate show that this effect may vary extremely. In particular, it may differ among the migrating and nonmigrating animals. Among the migrating decapods, the respiration rate remains essentially constant under combined exposure to elevated pressure and lower temperature, but for the nonmigrating epipelagic euphausiids, an increase in pressure reduces the rate of respiration (Teal, 1971). At the same time, the pressure insignificantly affects the respiration intensity of the mesopelagic fish *Anoplogaster cornuta* (Meek and Childress, 1973). Measurements of the rate of respiration of mollusks have shown that it remains unchanged throughout the entire range of pressure at which the animals live and depends only on temperature (Smith and Teal, 1973). The combination of the variations of temperature, pressure, and oxygen concentration may cause significant circadian fluctuations in the rate of respiration (Childress, 1977). King and Packard (1975), while evaluating the rate of respiration of plankton by measuring electron-transfer activity, found that pressure variations within the depth limits of the animals studied have no effect on its magnitude. Hochachka (1971), on the basis of biochemical consideration, suggests that the adaptation of animals to living at great depths must consist of achieving pressure independence for the rate of the key metabolic reactions. If one accepts the latter viewpoint, the results of measurements of the rate of respiration of deep-sea animals at atmospheric pressure should not evoke serious doubts.

A comparison of the results of such experiments at the depths at which animals live shows that the respiration rate declines with increasing depth (Childress, 1971, 1975; Torres et al., 1979). On the basis of simple model representation, its dependence on depth can be found. We shall do this for the pelagic zooplanktophages—the group of animals that feeds on the mesoplankton (Tseitlin, 1981c). The dependence of the biomass of the latter, $B(z)$, on depth is considered to be known.

The respiration intensity Q_s is defined by the formula

$$Q_s(z) = \frac{Q}{W\tau(z)} = \frac{aW^{\alpha-1}}{\tau(z)} \qquad (5)$$

where Q is the rate of respiration at 20°C (in milliliters of oxygen per hour); W is the weight of the animal (in grams), a (in milliliters of oxygen per gram per hour) and α are parameters, and the temperature correction

$$\tau = \exp\,[0.085(20 - t)] \qquad (6)$$

where t is the temperature at the depth of habitation; it corresponds approximately to $Q_{10} = 2.35$. Assuming that the respiration makes up the basic part of the energy expenditures, the energy balance equation for nonmigrating pelagic zooplanktophages can be written in the form (Tseitlin, 1976):

$$k_0 B(z) L^2 \xi(L) f[V(L)] = \frac{aW^{\alpha}}{\tau(z)} \qquad (7)$$

where k_0 is a constant, L is the body length of the animal, $\xi(L)$ is a function of the size of the predator characterizing the proportion of the mesoplankton which can serve it as food, and $f[V(L)]$ is a function of the velocity of motion of the predator (increasing as its size increases); for passive predators $f = 1$. It is convenient to write equation 7 by substituting values dependent on depth z into the left-hand part:

$$B(z)\tau(z) = \frac{aW^{\alpha}}{k_0L^2\xi(L)f[V(L)]} \tag{8}$$

The product $B(z)\tau(z)$ declines with depth; the variation of its magnitude in tropical regions of the ocean is shown in Fig. 10. But then the right-hand part of equation 8 must also decline with depth. We assume that the exponent α is a constant and that the form of the function ξ and f does not depend on depth. Reduction of the right-hand part with increasing depths can be accomplished by either (1) reducing the magnitude of a that characterizes the level of metabolism, (2) increasing the size of the animals, or (3) both. Since zooplanktophages of the same size are encountered at all depths, the diminution of the right-hand part can

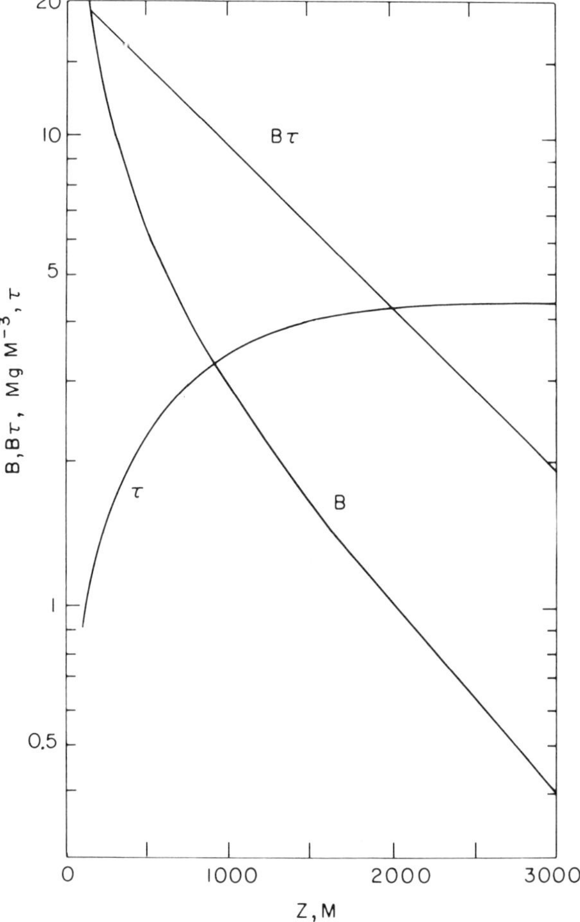

Fig. 10. Dependence of mean values of biomass of mesoplankton B, temperature correction τ, and their products ($B\tau$) on depth z.

be caused only by a reduction in the level of metabolism—the value a, that is, $a = a(z)$. From equation 8, it then follows that

$$a(z) = C_0 B(z) \tau(z) \tag{9}$$

Substituting equation 9 into equation 5, we find

$$Q_s(z) = C_0 B(z) W^{\alpha - 1} \tag{10}$$

Let us determine the constant multiplier C_0. Considering the values a, B, and τ to be known for the upper 100-m layer, we obtain

$$C_0 = \frac{a(100)}{B(100)\tau(100)} \tag{11}$$

Thus the respiration intensity is proportional to the magnitude of C_0 that is constant for the trophic and temperature conditions prevailing in the surface layer and to the biomass of the food (mesoplankton) at the depth of habitation $B(z)$. Therefore, the effect of temperature in equation 10 is manifest through the vertical distribution of the mesoplankton.

Table VI presents experimental data on the respiration intensity of some non-migrating pelagic zooplankters as well as these values as calculated by equation 10. In the calculation, we used the average curve of vertical distribution of meso-plankton (Fig. 10) for tropical regions. The value $a(100) = 0.125$ ml O_2/g · hr is the average for crustaceans at 20°C (Sushchenya, 1972); $\tau(100) = 0.9$, which corresponds to approximately 19°C when $Q_{10} = 2.35$; and $\alpha = 0.75$. Only in two cases of 19 do the data calculated diverge by more than a factor of 2—a very good result if one considers that such a difference is hardly rare when one makes intraspecific determinations of the respiration of animals of the same size. The regression equations found from the experimental (Q_s^{exp}) and calculated (Q_s^{cal}) data are as follows:

$$Q_s^{\text{exp}} = 29.7 z^{-1.0 \pm 0.2}$$

$$Q_s^{\text{cal}} = 63.9 z^{-1.1 \pm 0.1} \tag{12}$$

where Q_s is expressed in milliliters of oxygen per gram per hour and z, in meters. The range of application of the equations is limited to the depths greater than 100 m, since in the upper layer, the temperature is practically constant, and the dependence of the biomass of the mesoplankton on depth has a complex character; that is, the magnitude $B(z)\tau(z)$ does not diminish in this case. In the range of depths from 100 to 5000 m, the discrepancy of the results of the calculations (equation 12) does not exceed 30%.

The pathway just discussed—a reduction in the level of metabolism a—can be regarded as a physiological adaptation to great depths. The other possibility is the variation of size at a constant value of a. Evidently it is possible to follow the variation of body size with increasing depth of habitation only for animals belonging to related species (Birstein, 1963). An example of such an increase in size ("deep-water gigantism") is the genus *Styloheiron* presented in Table VI. In this case, the diminution of the right-hand part of equation 8 with depth is assured only by functions dependent on the weight and size of the animal, if one assumes that a will remain constant. The advantages related to increasing size with depth are determined by many factors, such as the increase in the number of potential prey and the increase in the volume caught. But deep-water gigantism is a rare phenomenon that is characteristic of animals that are usually of small size. Therefore, a common means for reducing energy expenditures with depth is to reduce the metabolic level.

Attempts to determine the energy expenditures of deep-sea animals from their daily diet involve even greater difficulties. The possibility of determining the diets under laboratory conditions is practically excluded, and the only method remaining is to study the stomach contents of animals that have been captured. Studies of such a type (Holton, 1969; Legand and Rivaton, 1969; Hopkins and Baird, 1973, 1977; Merrett and Roe, 1974; Baird et al., 1975; Donaldson, 1975; Tyler and Pearcy, 1975) have been performed on macroplankton animals—fish and shrimp. They explained many of the important feeding characteristics of some species—the composition of the food and the rhythm of feeding, selectivity by species composition of food objects and their dimensions. However, regarding the magnitude of the daily diet of interest to us here, there are practically no data available. An attempt to estimate the daily diet by making specific assumptions concerning the dynamics of digestion of the food was made for the myctophids *Myctophum nitidulum* that migrate to the surface (Tseitlin and Gorelova, 1978). For the young of this species (size 11 to 20 mm), the magnitude of the specific daily diet (weight of the food referred to the weight of the fish) was found to be equal to 13%; for the adult fish (50 to 60 mm), it was approximately 3%. The latter value is possibly somewhat too high. An estimate of the specific daily diet C/W of nonmigrating mesopelagic fish was made for the genus *Cyclothone* (Gorelova and Tseitlin, 1979). To calculate this value, Gorelova and Tseitlin used the formula

$$\frac{C}{W} = \frac{24\,(N_f/N)\,Y_m^{0.38}}{57.9\,W^{0.182}\tau}$$

where N_f/N is the proportion of fish having food in their stomachs; $Y_m = M_p/W$ is the average value of the "stomach-filling" index, M_p is the weight of the prey, and W is the weight of the fish in grams. The values of Y_m are approximately equal for all species of *Cyclothone* ($\simeq 0.03$). The temperature correction τ was found from the graph in Fig. 10. The results of the calculations are shown in Table VII.

The value of N_f/N, which characterizes the feeding frequency, exercises a substantial influence on the specific daily diet. A high proportion of empty stomachs among cyclothones (~80 to 90% in Table VII) was noted earlier. Birstein and Vinogradov (1955) studied the stomachs of 35 specimens of *C. microdon* (one of the most well-studied deep-sea species) and found all of them were empty. From the results obtained, the feeding rates of the various species were estimated: *C. alba* receives one victim of average weight approximately once every 2 days, *C. pseudopallida* once every 3 days, and *C. pallida* once every 10 days. The possibility of such infrequent acquisition of food by deep-water animals was noted by MacDonald (1975): According to the estimates, the monthly food requirements of the ostracods *Gigantocypris* may be satisfied by catching one large copepod.

The data presented in Table VI on the respiration rates of cyclothones were evaluated with respect to their daily diets on the assumption that the assimilated diet is expended exclusively on metabolic processes. We note that they do not

TABLE VI

Respiration Intensity of Some Pelagic Planktophagous Animals

Species	Weight (g)	Average Habitation Depth (m)	Sources of Data on Habitation Depths	Biomass of Zooplankton at Habitation Depth (mg/m^3)	Respiration Rate in Experiment $Q_{s,exp}$ (ml O_2/g · hr)	Sources of Data on $Q_{s,exp}$ (O_2/g · hr)	Calculated Respiration Rate $Q_{s,calc}$ (ml/O_2/g · hr)	$\frac{Q_{s,exp}}{Q_{s,calc}}$
Euphausiacea								
Styloheiron suhmii	0.0037	25	Baker, 1970	26	0.51	Shushkina and Pavlova, 1973	0.66	0.77
S. affine	0.0051	80	Baker, 1970	25	0.44	Shushkina and Pavlova, 1973	0.59	0.75
S. longicorne	0.0155	200	Baker, 1970	13	0.25	Shushkina and Pavlova, 1973	0.23	1.09
S. elongatum	0.0249	300	Baker, 1970	9.4	0.19	Shushkina and Pavlova, 1973	0.15	1.24
S. maximum	0.106	550	Baker, 1970	5.3	0.11	Shushkina and Pavlova, 1973	0.06	1.83
Pisces								
Cyclothone alba	0.056	550	Badcock and Merrett, 1976, 1977	5.3	0.10	Gorelova and Tseitlin, 1979[b]	0.069	1.49
C. pseudopallida	0.145	750	Badcock and Merrett, 1976, 1977	4.0	0.072	Gorelova and Tseitlin, 1979[b]	0.041	1.75
C. pallida	0.289	800	Badcock and Merrett, 1976, 1977	3.5	0.022	Gorelova and Tseitlin, 1979[b]	0.030	0.67
C. acclinidens	0.061	550	Badcock and Merrett, 1976, 1977	5.3	0.037	Gorelova and Tseitlin, 1979[b]	0.067	0.55

Melanostigma pammelus	1.65	600	Childress, 1975[a]	4.9	0.016	Childress, 1975[a]	0.027	0.59
Anoplogaster cornuta	40.00	700	Childress, 1975[a]	4.1	0.014	Childress, 1975[a]	0.010	1.40
Mysidacea								
Gnathophausia zoea	5.2	700	Childress, 1975[a]	4.1	0.025	Childress, 1975[a]	0.017	1.46
G. gracilis	1.79	1500	Childress, 1975[a]	1.8	0.021	Childress, 1975[a]	0.010	2.10
Boreomysis californica	0.033	800	Childress, 1975[a]	3.5	0.036	Childress, 1975[a]	0.051	0.71
Decapoda								
Paciphaea emarginata	6.7	750	Childress, 1975[a]	4.0	0.016	Childress, 1975[a]	0.016	1.00
Notostomus sp.	15.3	1200	Childress, 1975[a]; Chace, 1940	2.2	0.006	Childress, 1975[a]	0.007	0.86
Acanthephyra curtirostris	1.77	750	Childress, 1975[a]	4.0	0.033	Childress, 1975[a]	0.022	1.50
Copepoda								
Bathycalanus princeps	0.033	800	Childress, 1975[a]; Grice and Hulsemann, 1977; Deevey and Brooks, 1977	3.5	0.020	Childress, 1975[a]	0.051	0.39
Paraeuchaeta norvegica	0.038	1000	Conover, 1960	2.7	0.080	Conover, 1960	0.050	1.60

[a] Average depths for the range indicated in the article.
[b] Estimate based on nutrition data.

Table VII

Calculation of Specific Daily Diets of Mesopelagic Fishes of Genus *Cyclothone*

Species	Size (mm)	Weight (g)	Depth of Habitation (m)	Temperature at Habitation Depth (°C)	Temperature Correction	Proportion of Fish Having Food in Stomach	Specific Daily Diet (Percent of Body Weight)
Cyclothone alba	24.7	0.056	550	9.0	2.4	0.21	1.6
C. pseudopallida	35.5	0.145	750	6.8	2.9	0.22	1.2
C. pallida	46.0	0.289	800	6.5	3.0	0.08	0.4
C. acclinidens	25.5	0.061	550	9.0	2.4	0.08	0.6

deviate from the series of data obtained by direct measurement of the rate of oxygen consumption.

The basic difficulty in determining the energy expenditures of animals by their daily diets is the decreasing number of filled stomachs with depth. The deeper the animals live, the more specimens are required to obtain reliable data on the diets, and the fewer animals it is possible to acquire for this purpose. The other problem that pertains equally to both methods of estimating the energy expenditures is the determination of the habitation depth of the animals. The information that we presently possess is usually sufficient only for the roughest estimates of the relationship between the energy characteristics of the animals and the depth at which they live.

D. *Utilization of Descending Organic Matter in Water Column*

For estimation of the proportion of descending detritus that is consumed by pelagic organisms, it is necessary to know the energy requirements of all of these organisms and their vertical distribution. As is clear from the aforestated, our information on both is far from complete. The best data on the vertical distribution are found for the mesoplankton, whereas data on energy requirements give us some idea of expenditures for respiration, but they are not adequate as a basis for calculation. Therefore, it is advisable to use the method based on the size distribution of the organisms (Tseitlin, 1981e) and to use equations 2 and 3 for the calculations. Their use means that we consider the shape of the dimensional spectrum of the organisms to be independent of depth. For small particles, an approximate constancy of the spectrum down to the depth of about 4000 m was noted by Sheldon et al. (1972) and Lal (1977). Tseitlin (1981a) showed that the shape of the size spectrum of mesoplankton is approximately the same down to the depth of 500 m. Since it is scarcely possible to determine the size structure of the deep-sea community, the hypothesis of its invariance with depth remains to be judged. As in all calculations based on the size distribution of the organisms, to obtain a result in numerical form, one must know the vertical distribution of some size group. As before, we choose mesoplankton, as its vertical distribution has been studied better than that of any other group of animals. Moreover, it is precisely the mesoplankton that determines the basic proportion of energy consumed by the deep-water animals.

The quantity of detritus consumed in the layer occupying the depths from z_{i-1} to z_i is defined as

$$C_i(z_{i-1}, z_i) = \frac{1-\eta}{U}\int_{z_{i-1}}^{z_i} (Q + P)dz, \qquad z_i > z_{i-1} \tag{13}$$

where Q and P are the rates of respiration and production of the animals inhabiting this layer; U is the assimilation coefficient of the food. The magnitude of the flow of detritus $\Phi(z_i)$ to the lower boundary of the layer depends on the flow to the upper boundary $\Phi(z_{i-1})$, the absorption in the layer $C(z_{i-1}, z_i)$, the quantity of food not assimilated by the animals inhabiting the layer $(1 - U)\ C(z_{i-1}, z_i)$ and elimination $E(z_{i-1}, z_i)$:

$$\Phi(z_i) = \Phi(z_{i-1}) - C(z_{i-1}, z_i) + (1 - U)C(z_{i-1}, z_i) + E(z_{i-1}, z_i)$$

or, taking equation 13 into consideration,

$$\Phi(z_i) = \Phi(z_{i-1}) - (1 - \eta)\int_{z_{i-1}}^{z_i} (Q + P)dz + E(z_{i-1}, z_i) \tag{14}$$

The magnitude of elimination is equal to the pure production in the layer in question. We note that the magnitude of the absorption in the layer (z_{i-1}, z_i) does not depend on the coefficient of assimilation.

Since we are concerned with the magnitudes of the flow throughout the entire layer of water lying below the upper-producing zone that enters into this layer, the flow $\Phi(z_{i-1})$ will be assumed to be equal to the flow of carcasses and fecal material from the producing zone as defined above. Equation 14 can be reduced to the form (Tseitlin, 1981e):

$$\Phi(z_i) = C\left\{\frac{[(1 - \eta) - U(1 - K_2)]\Delta z}{U(1 - K_2)} - I(z_{i-1}, z_i)\right\}, \tag{15}$$

where

$$C = \frac{B_0(W_{\min}^{\alpha-1} - W_{\max}^{\alpha-1})(1 - K_2)}{2.7(1 - \alpha)T_0\tau K_2}$$

$$I(z_{i-1}, z_i) = \begin{cases} 10^4(z_{i-1}^{-1} - z_i^{-1}) \quad \text{when} \quad z_{i-1} \geq 100 \text{ m},\ z_i \leq 800 \text{ m} \\ 37.3\,[\exp(-1.88 \times 10^3 z_{i-1}) - \exp(-1.88 \times 10^3 z_i)] \\ \text{when} \quad z_{i-1} \geq 800 \text{ m} \end{cases}$$

where Δz is the magnitude of the upper producing zone (in meters) and the value $K_2 = 0.2$ to 0.3. In deriving equation 15 it was assumed that the biomass of the mesoplankton B (and consequently B_0), the temperature correction τ, and the parameters a and T_0, which determine (besides other characteristics) the level of respiration and the duration of growth of the animals, respectively, were dependent on depth. In calculating the function $I(z_{i-1}, z_i)$, we used averaged data from the vertical distribution of the mesoplankton and temperature in the tropical regions of the ocean. All the values entering into equation 15 are the same as those used for calculating the production and respiration of heterotrophs.

The graphs showing the ratio of magnitude of the detritus flow $\Phi(z)$ to the primary production P_p at different values of Δz and K_2 are presented in Fig. 11. In the same figures, we show the values observed for this ratio according to the data of a number of researchers [cited by Smith and Hinga, this volume, Chapter 8, and Bishop et al. (1977)]. One sees that the rate of variation of the detritus flow rapidly declines with depth. Since the differences of the flows are determined according to equation 14 by consumption and production of detritus by the deep-water animals, the curve $\Phi(z)$ reflects not only the decrease in the abundance of the animals with depth, but also the decrease in their energy requirements. This is quite strongly manifest at depths below 1500 m. The values Δz and K_2 very substantially influence the relative magnitude of the flow Φ/P_p at great depths. For depths of less than 500 m, the results of the calculations with the real values of these parameters differ little.

It may be of interest to compare the results obtained here with the estimates made by Riley (1970) according to data for the Sargasso Sea. The magnitude of the primary production there, according to his calculations, is equal to 320 to 380 mg

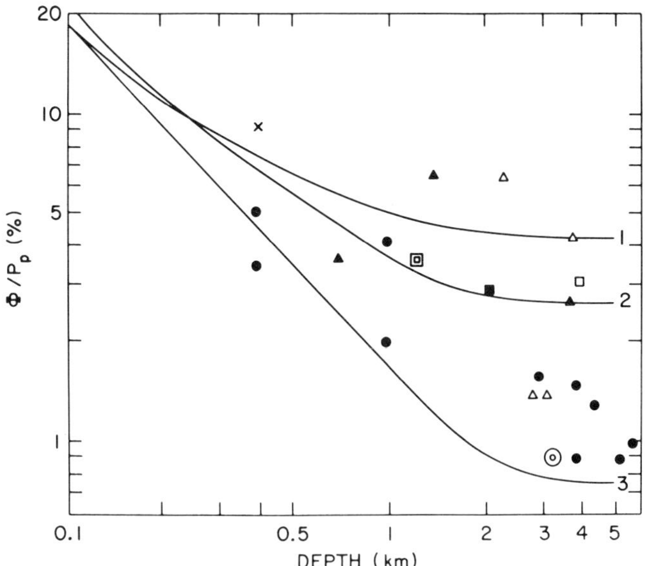

Fig. 11. Dependence of ratio of downward flow of organic matter (Φ) to primary production (P_p) on depth. Solid curves are calculated by equation 15: (1) Δz = 125 m, K_2 = 0.25; (2) Δz = 125 m, K_2 = 0.2; (3) Δz = 100 m, K_2 = 0.25; △—Rowe and Gardner (1979), ▲—Hinga et al. (1979), □—Soutar et al. (cited by Smith and Hinga, this volume, Chapter 8), ●—Honjo (1980), ○—Deuser and Ross (1980), ■—Wiebe et al. (1976), ×—Bishop et al. (1977). Double signs indicate overlapping data.

C/m^2 · day; that is, it corresponds to the value calculated by us (300 to 380 mg C/m^2 · day). Riley suggests that approximately 80% of the primary production is utilized in the surface layer and upper part of the mesopelagial. According to Fig. 11, 90% and more of the primary production is consumed above 300 m. The size of the flow at the depth of 300 m is estimated by Riley at 50 mg/m^2 · day; according to our data, it is equal to 27 to 38 mg/m^2 · day. The calculations made by us indicate a steeper decline of the flow in the upper layers of the pelagial. One possible reason for this phenomenon may be our assumption that the entire flow of organic matter from the producing zone consists of carcasses and feces of animals. In reality, part of the unutilized phytoplankton also descends (if only to the 100- to 200-m layer). And since a significant part of the consumption falls on the 100- to 200-m layer, the greater part of the detritus would reach the deep-lying layers.

The satisfactory coincidence between the calculations and the results of observations may appear strange. It would appear that calculations of this type cannot serve for the determination of the quantity of organic matter descending to the bottom. Actually, this quantity is defined as the difference between the primary production and the absorption in the column of water. But both of these values are determined with a very low degree of accuracy. It is clear that their difference can be significant only in the case when the real value of the flow at the bottom is sufficiently large. According to the data of Smith and Hinga (this volume, Chapter 8) and Honjo (1978), the latter amounts to 1 to 3% of the primary production at 5000 m. If this estimate is close to the real value of the bottom flow of detritus, the difficulty with finding its value with our method is evident. Neither primary production nor absorption can be accurately determined by performing even several determinations of them during a small time interval. Calculations should be based on yearly averages, and even better, on the average estimates of the primary production and absorption over several years. With these conditions, it may be possible to rely on the results.

Acknowledgments

The authors thank Dr. G. T. Rowe, who has looked through the original literal translation and has made critical remarks, mostly concerning the improvement of terminology and Dr. V. M. Makushok, who took on his shoulders the ungrateful task of translating parts of the text.

References

Andrews, K. I., 1966. The distribution and life-history of *Calanoides acutus* (Giesbrecht). *Discov. Rep.*, **34,** 119–161.

Arashkevich, E. G., 1969. The character of feeding of copepods in the Northwestern Pacific. *Oceanology*, **9,** N(5), 857–873.

Badcock, J. and N. R. Merrett, 1976. Midwater fishes in the eastern North Atlantic. 1. Vertical distribution and associated biology in 30°N, 23°W, with developmental notes on certain myctophids. *Progr. Oceanogr.* **7** (1).

Badcock, J. and N. R. Merrett, 1977. On the distribution of midwater fishes in the eastern North Atlantic. In *Oceanic Sound Scattering Prediction, Marine Studies Series*. Plenum Press, New York.

Baird, R. C., T. L. Hopkins, and D. F. Wilson, 1975. Diet and feeding chronology of *Diaphus taaningi (Myctophidae)* in the Cariaco Trench. *Copeia*, **1975,** 356–365.

Baker, A. de C., 1970. The vertical distribution of euphausiids near Fuerte-ventura, Canary Islands. *J. Marine Biol. Assoc. U. K.*, **50** (2), 301–342.

Banse, K., 1964. On the vertical distribution of zooplankton in the sea. *Progr. Oceanogr.*, **2,** 53–125.

Barham, E. G., 1971. Deep-sea fishes: Lethargy and vertical orientation. *Sci. Rep. Maury Center Ocean*, **5,** 100–118.

Beers, I. R. and G. H. Stewart, 1970. Numerical abundance and estimated biomass in microzooplankton. In *The Ecology of the Plankton off La Jolla, California, in the Period April Through September, 1967*, Part VI, D. Strickland, ed. (Reprinted in *Bull. Scripps Inst. Oceanogr.*, **17,** 67–87.)

Bernard, F., 1961. Problems of elementary fertility in the Mediterranean from 0 to 3000 meters depth. *Res. Sci. Campagn. "Calypso,"* **5** (16), 61–159.

Bernard, F., 1963. Sinking rates of *Cycloccolithus fragilis* (Lohm.). Consequences for the life cycle of the tropical sea. *Pelagos*, **1** (2).

Bernard, F., 1964. The nannoplankton in the aphotic zone of the tropical sea. *Pelagos*, **2** (2), 1–32.

Bernard, F., 1966. Abundance of nannoplankton in the aphotic layers of the ocean—probable consequences for the deep productivity of the tropical ocean. In *Second International Oceanography Congress Abstracts*, papers No. 37-SII, A. P. Vinogradov, ed. Nauka, Moscow, pp. 38–39.

Bernard, F. and J. Lecal, 1960. Unicellular plankton caught in the Indian Ocean by the Charcot (1950) and the Norsel (1955–1956). *Bull. Inst. Oceanogr.*, No. 1166.

Birstein, J. A., 1963. *Deep-Water Isopod Crustaceans of the Northwestern Part of the Pacific Ocean*. ISD. AN USSR, Moscow.

Birstein, J. A. and M. E. Vinogradov, 1955. Notes on the feeding of deep-sea fishes in the Kurile-Kamchatka region. *Zool. J.*, **34,** 842–849.

Birstein, J. A. and M. E. Vinogradov, 1971. Role of the trophic factor in the taxonomic discreteness of marine deep-sea fauna. *Bull. Mosk. Obsh. Ispyt. Prirodi, Ser. Biol.*, **76** (3), 59–92.

Bishop, J. K., J. M. Edmond, D. R. Ketten, M. P. Bacon, and W. B. Silker, 1977. The chemistry, biology and vertical flux of particulate matter from the upper 400 m of the equatorial Atlantic Ocean. *Deep-Sea Res.*, **24,** 511–548.

Bogdanov, J. A., Y. A. Grigorovich, and M. G. Shaposhnikova, 1968. Protein determination in the water suspended matter. *Oceanology*, **8** (6), 1087–1090.

Bogdanov, J. A. and M. G. Shaposhnikova, 1971. Concentration of suspended organic matter in the tropical Pacific water. *Oceanology*, **11** (4), 668–673.

Bogorov, B. G. and M. E. Vinogradov, 1955. Some essential features of zooplankton distribution in the northwestern Pacific. *Trans. Inst. Oceanol.*, **18,** 113–123.

Chace, F. A., Jr., 1940. Plankton of the Bermuda Oceanographic Expedition. IX. The bathypelagic Caridean Crustacea. *Zoologica*, **25** (11).

Childress, J. J., 1971. Respiratory rate and depth of occurrence of midwater animals. *Limn. Oceanogr.*, **16,** 104–106.

Childress, J. J., 1975. The respiratory rates of midwater crustaceans as a function of depth of occurrence and relation to the oxygen minimum layer off southern California. *Comp. Biochem. Physiol.*, **50A,** 787–799.

Childress, J. J., 1977. Effects of pressure, temperature and oxygen on the oxygen consumption rate of the midwater copepod *Gaussia princeps*. *Mar. Biol.*, **39,** 19–24.

Childress, J. J. and M. H. Nygaard, 1973. The chemical composition of midwater fishes as a function of depth of occurrence off southern California. *Deep-Sea Res.*, **20** (12), 1093–1109.

Conover, R. J., 1960. The feeding behavior and respiration of some marine planktonic crustacea. *Biol. Bull.*, **119,** 399–415.

Conover, R. J., 1966. Assimilation of organic matter by zooplankton. *Limnol. Oceanogr.*, **11,** 338–345.

Conover, R. J., 1978. Transformation of organic matter. In *Marine Ecology*, Vol. 4, O. Kinne, ed. Wiley, New York, pp. 221–499.

Cushing, D. H., G. F. Humphrey, K. Banse and T. Leavastu, 1958. Report of Committee on terms and equivalents. Rapp. et procès-verbaux réunions. *Cons. perm. int. explor. mer.*, **144,** 15–16.

Deevey, G. B. and Brooks, A. L., 1977. Copepods of the Sargasso Sea off Bermuda: Species composition, and vertical and seasonal distribution between the surface and 2000 m. *Bull. Mar. Sci.*, **27** (2), 256–291.

Denton, E. J. and N. B. Marshall, 1958. The buoyancy of bathypelagic fishes without a gas-filled swimbladder. *J. Mar. Biol. Assoc. U. K.*, **37,** 753–767.

Deuser, W. G. and E. H. Ross, 1980. Seasonal changes in the flux of organic carbon to the deep Sargasso Sea. *Nature*, **283,** 364–365.

Donaldson, H. A., 1975. Vertical distribution and feeding of sergestid shrimps (Decapoda: *Natantia*) collected near Bermuda. *Mar. Biol.*, **31,** 37–50.

Fournier, R. O., 1966. North Atlantic deep-sea fertility. *Science*, **153** (3741), 1250–1252.

Fournier, R. O., 1970. Studies on pigmented microorganisms from aphotic marine environments. *Limnol. Oceanogr.*, **15** (5), 657–682.

Fournier, R. O., 1972. The transport of organic carbon to organisms living in the deep oceans. *Proc. Roy. Soc. Edinburgh*, **73,** 203–211.

Fournier, R. O., 1973. Studies on pigmented microorganisms from aphotic marine environments. III. Evidence of apparent utilization by benthic and pelagic Tunicata. *Limnol. Oceanogr.*, **18,** 38–43.

Fowler, S. W. and L. F. Small, 1972. Sinking rates of euphausiid fecal pellets. *Limnol. Oceanogr.*, **17,** 293–296.

Foxton, P., 1956. The distribution of the standing crop of zooplankton in the Southern ocean. *"Discovery" Rep.*, **28,** 191–236.

Foxton, P., W. Aron, M. Legand, and T. Nemoto, 1968. Micronecton. Report of working party No. 4. In *Monographs on Oceanographic Methodology*, Vol. 2. UNESCO Press, Geneva, pp. 164–167.

Golubev, A. P., 1977. Role of exuvial detachments in the energy balance of the mysid *Paramysis lacustris* of the Kaunas reservoir. In Nineteenth Scientific Conference on the Study and Adaptation of the Water Bodies of the Baltic States and Byelorussia. Themes of Reports, Nos. 39–40, Minsk.

Gorelova, T. A. and V. B. Tseitlin, 1979. Feeding of the mesopelagic fishes of the genus *Cyclothone*. *Oceanology*, **19,** 1110–1115.

Greze, V. N., 1963. Specific traits noted in the structure of the pelagial in the Ionian Sea. *Oceanology*, **3** (1), 100–109.

Grice, G. D. and R. Hulsemann, 1967. Bathypelagic calanoid copepods of the western Indian Ocean. *Proc. U.S. Nat. Museum*, **122** (3593), 1–67.

Griggs, G. B., A. G. Carey, and L. D. Kulm, 1969. Deep-sea sedimentation and sediment fauna interaction in Cascadia Channel and on Cascadia abyssal plain. *Deep-Sea Res.*, **16,** 157–170.

Grill, E. V. and F. A. Richards, 1964. Nutrient regeneration from phytoplankton decomposition in sea water. *J. Mar. Res.*, **22,** 51–69.

Hamilton, R. D., O. Holm-Hansen, and J. D. H. Strickland, 1968. Notes on the occurrence of living microorganisms in deep water. *Deep-Sea Res.*, **15,** 651–656.

Harding, G. C. H., 1973. Decomposition of marine copepods. *Limnol. Oceanogr.*, **18,** 670–673.

Harding, G. C. H., 1974. The food of deep-sea copepods. *J. Mar. Biol. Assoc. U. K.*, **54** (1), 141–156.

Hasle, G. R., 1959. A quantitative study of phytoplankton from the Equatorial Pacific. *Deep-Sea Res.*, **6** (1), 38–59.

Hemmingsen, A. M., 1960. Energy metabolism as related to body size and respiratory surfaces, and its evolution. *Rep. Steno Mem. Hosp. and Nordisk Insulinlaboratorium*, **9,** 11.

Hochachka, P., 1971. Enzyme mechanisms in temperature and pressure adaptation of off-shore benthic organisms: The basic problem. *Am. Zoolog.*, **11,** 425–435.

Holton, A. A., 1969. Feeding behavior of a vertically migrating lantern fish. *Pacif. Sci.*, **23,** 325–331.

Honjo, A., 1978. Sedimentation of materials in the Sargasso Sea at a 5367 m deep station. *J. Mar. Res.*, **36,** 469–492.

Honjo, S., 1980. Material fluxes and modes of sedimentation in the mesopelagic and bathypelagic zones. *J. Mar. Res.*, **38** (1), 53–97.

Honjo, S. and M. R. Roman, 1978. Marine copepod fecal pellets: Production, preservation and sedimentation. *J. Mat. Res.*, **36,** 45–57.

Hopkins, T. L. and R. C. Baird, 1973. Diet of hatchetfish *Sternoptyx diaphana*. *Mar. Biol.*, **21,** 34–46.

Hopkins, T. L. and R. C. Baird, 1977. Aspects of the feeding ecology of oceanic midwater fishes. In *Oceanic Sound Scattering Prediction*, N. R. Anderson and B. J. Zahuranec, eds. Plenum Press, New York, pp. 325–360.

Johnston, R., 1962. An equation for the depth distribution of deep-sea zooplankton and fishes. *Rapp. et Proces-Verb. Cons. Internat. Explor. Mer.*, **153** (38).

Kanwisher, I. and A. Ebeling, 1957. Composition of swim-bladder gas in bathypelagic fishes. *Deep-Sea Res.*, **4** (3).

Kashkin, N. I., 1962. On the adaptive value of seasonal migrations of *Calanus finmarchicus* (Gunnerus, 1770). *Zool. J.*, **41** (3), 342–357.

Kashkin, N. I., 1977. Fauna of DSL. In *Oceanology, Biology of the Ocean*, Vol. 1, M. E. Vinogradov, ed. Nauka, Moscow, pp. 299–318.

Kashkin, N. I., and J. G. Tchindonova, 1971. Mesopelagic fishes as resonant scatterers in the DSL of the Atlantic Ocean. *Oceanology*, **11,** 482–493.

Kchlebovitch, T. V., 1979. Rate of oxygen consumption by infusoria. In *General Principles of the Study of Aquatic Ecosystems*, G. G. Winberg, ed. Nauka, Leningrad.

King, F. D. and T. T. Packard, 1975. The effect of hydrostatic pressure on respiratory electron transport system activity in marine zooplankton. *Deep-Sea Res.*, **22,** 99–105.

Krause, H. R., 1962. Investigation of the decomposition of organic matter in natural waters. *FAO Fisher. Biol.*, Report No. 34.

Lal, D., 1977. The oceanic microcosm of particles. *Science*, **198,** 997–1009.

Leavitt, B. B., 1938. The quantitative vertical distribution of macrozooplankton in the Atlantic Ocean basin. *Biol. Bull.*, **74** (3), 376–394.

Lecal, I., 1952. Depth separation of the Coccolithophorides at some western Mediterranean stations. *Bull. Inst. Oceanogr.* (1018), 1–13.

Legand, M. and J. Rivaton, 1969. Biological cycles of the mesopelagic fish in the eastern Indian Ocean. III. Predatory activity of micronectonic fish. *Cahiers ORSTOM Oceanographic*, **7,** 29–45.

Legand, M., P. Bourret, P. Fourmanoir, P. Grandperrin, J. A. Guéredrat, A. Michel, P. Rancurel, R. Repelin and C. Roger, 1972. Trophic relationships and vertical distribution of the mesopelagic zone of the intertropical Pacific Ocean. *Cahiers ORSTOM Oceanographic*, **10** (4).

MacDonald, A. G., 1975. *Physiological Aspects of Deep-Sea Biology*. Cambridge University Press, Cambridge, 450 pp.

Mackintosh, N. A., 1937. The seasonal circulation of the Antarctic macroplankton. *Discov. Rep.*, **16,** 367–412.

Margalef, R., 1968. *Perspectives in Ecological Theory*. University of Chicago Press, 111 pp.

Marshall, N. B., 1960. Swim bladder structure of deep-sea fishes in relation to their systematics and biology. *Discov. Rep.*, **31,** 121.

Marshall, N. B., 1965. *The Life of Fishes*. Weid and Nicolson, London.

Marshall, N. B., 1971. *Explorations in the Life of Fishes*. Cambridge, Mass.: Harvard Univ. Press.

Mauchline, J., 1972. The biology of bathypelagic organisms, especially Crustacea. *Deep-Sea Res.*, **19** (11), 753–780.

McCave, I. N., 1975. Vertical flux of particles in the ocean. *Deep-Sea Res.*, **22,** 491–502.

Mednikov, B. M., 1961. On the sex ratio in deep water Calanoida. *Crustaceana*, **3,** 105–109.

Meek, R. P. and J. J. Childress, 1973. Respiration and the effect of pressure in the mesopelagic fish *Anoplogaster cornuta* (Beryciformes). *Deep-Sea Res.*, **20** (12), 1111–1118.

Melnikov, J. A., 1975. Microplankton and organic detritus in the water of the southeast Pacific. *Oceanology*, **15** (1), 146–156.

Menzies, R. J., 1962. On the food and feeding habits of abyssal organisms as exemplified by the Isopoda. *Int. Rev. Ges. Hydrobiol.*, **47** (3), 339–358.

Merrett, N. R., and H. S. J. Roe, 1974. Patterns and selectivity in the feeding of certain mesopelagic fish. *Mar. Biol.*, **28,** 115–126.

Ostvedt, O. I., 1955. Zooplankton investigation from weather ship "M" in the Norwegian Sea, 1948–49. *Hvalradest Skr.*, **40,** 93 pp.

Otsukia, A. and T. Hanya, 1972. Production of dissolved organic matter from dead green algal cells. I. Aerobic microbial decomposition. *Limnol. Oceanogr.*, **17,** 248–257.

Parin, N. V., 1979. Some characteristics of the spatial distribution of ocean pelagic fish. In *Biological Resources of the Ocean*, S. A. Studenetzkiy, ed. Nauka, Moscow, pp. 102–112.

Parin, N. V., K. N. Nesis, and N. I. Kashkin, 1977. Macroplankton and nekton, in Chapter 3, Vertical distribution of plants and animals in the ocean. In *Oceanology, Biology of the Ocean*, Vol. 1, M. E. Vinogradov, ed. Nauka, Moscow, pp. 159–173.

Parsons, T. R., 1963. Suspended organic matter in sea water. *Progr. Oceanogr.*, **1,** 203–239.

Pavlova, E. V., 1968. Destruction of zooplankton organisms of the Mediterranean Sea. In *Expeditionary Studies in the Mediterranean in May–June 1968*, V. N. Greze, ed. Naukova Dumka, Kiev.

Paxton, I. R., 1967. A distributional analysis for the lantern fishes (family *Myctophidae*) of the San Pedro Basin, California. *Copeia* (2), 422–443.

Pechen-Finenko, G. A., 1979. Assimilability of food by planktonic crustaceans. In *General Principles of the Study of Aquatic Ecosystems*, G. G. Winberg, ed. Nauka, Leningrad.

Riley, G. A., 1970. Particulate and organic matter in seawater. *Adv. Mar. Biol.*, **8,** 1–118.

Rowe, G. T., 1971. Benthic biomass and surface productivity. In *Fertility of the Sea*, J. Costlov, ed. Gordon and Brench, New York, pp. 441–454.

Rowe, G. T., 1981. The deep-sea ecosystem. In *Analysis of Marine Ecosystems*, A. Longhurst, ed. Pergamon Press, London, 235–267.

Rowe, G. T. and W. Gardner, 1979. Sedimentation rates in the slope water of the northwest Atlantic Ocean measured directly with sediment traps. *J. Mar. Res.*, **35,** 581–600.

Rudjakov, Yu. A., and V. B. Tseitlin, 1980. Rate of passive sinking of planktonic organisms, *Oceanology*, **20** (5), 931–936.

Sanders, H. L. and R. R. Hessler, 1969. Ecology of the deep-sea benthos. *Science*, **163,** 1419–1424.

Schiller, I., 1931. Autochthonic plant organisms in the deep-sea. *Biol. Zbl.*, **51.**

Schoener, A., 1968. Evidence for reproductive periodicity in the deep-sea. *Ecology*, **49,** 81–87.

Sheldon, R. W., A. Prakash, W. H. Sutcliff Jr., 1972. The size distribution of particles in the ocean. *Limnol. Oceanogr.*, **17** (3), 327–340.

Shmalgauzen, I. I., 1968. Control and regulation in evolution. In *Kiberneticheskie Voprosi Biologii*, R. L. Berg and A. A. Lyapunov, eds. Nauka, Novosibirsk, pp. 34–73.

Shushkina, E. A. and E. P. Pavolova, 1973. Rates of metabolism and production of zooplankton in the Equatorial Pacific. *Oceanology*, **13** (2), 339–345.

Shushkina, E. A. and I. A. Sokolova, 1972. Caloric equivalents of the body mass of the tropical organisms from the pelagic part of the ocean. *Oceanology*, **12** (5), 860–867.

Sieburth, J. McN., V. Smetacek, and J. Lenz, 1978. Pelagic ecosystem structure: Heterotrophic compartments of plankton size fractions. *Limnol. Oceanogr.*, **23** (6), 1256–1263.

Skopintsev, B. A., 1949. Dynamics of organic matter in a water reservoir; the rate of decomposition of organic matter from dead plankton. *Tr. Vses. Gidrobiol. Obshchest.*, **1**, 34–43.

Small, L. F., S. W. Fowler, and M. J. Ünlü, 1979. Sinking rates of natural copepod fecal pellets. *Mar. Biol.*, **51**, 239–241.

Smayda, T. J., 1969. Some measurements of the sinking rate of fecal pellets. *Limnol. Oceanogr.*, **14**, 621–625.

Smayda, T. J., 1971. Normal and accelerated sinking of phytoplankton in the sea. *Mar. Geol.*, **11**, 105–122.

Smith, K. L., 1978. Metabolism of the abyssopelagic rattail *Coryphaenoides armatus* measured *in situ*. *Nature*, **274**, 362–364.

Smith, K. L. and R. R. Hessler, 1974. Respiration of benthopelagic fishes: *In situ* measurements at 1230 meters. *Science*, **184**, 72–73.

Smith, K. L., Jr., and J. M. Teal, 1973. Temperature and pressure effects on respiration of the ecosomatous pteropods. *Deep-Sea Res.*, **20**, 853–858.

Somme, J. D., 1934. Animal plankton of the Norwegian coast water and open sea. I. Production of *Calanus finmarchicus* and *C. hyperboreus* in the Lofoten Area. *Fiskeridirekt. Skrift. Havund.* **4** (9), 1–163.

Sorokin, Yu. I., E. B. Pavelyeva, and M. I. Vasilyeca, 1975. Productivity and trophic role of bacterioplankton in the area of equatorial divergence. *Trans. Inst. Oceanol.*, **102**, 184–198.

Sushchenya, L. M., 1972. *Respiration Intensity of Crustaceans*, Naukova Dumka, Kiev.

Sushchenya, L. M., 1975. *Quantitative Dietary Laws of Crustaceans*, Nauka i Tekhnika, Minsk.

Tchindonova, Ju. G., 1959. Feeding of some groups of deep-sea macroplankton in the northwestern Pacific. *Trans. Inst. Oceanol.*, **30**, 166–189.

Tchindonova, Ju. G. and N. I. Kashkin, 1969. Comparison of the biological and acoustical methods for estimating the deep scattering layers. *Oceanology*, **9**, 528–539.

Teal, J. M., 1971. Pressure effects on the respiration of vertically migrating decapod crustacea. *Am. Zool.*, **11**, 571–576.

Torres, J. J., B. W. Belman, and J. J. Childress, 1979. Oxygen consumption rate of midwater fishes as a function of depth of occurrence. *Deep-Sea Res.*, **26A**, 185–197.

Tseitlin, V. B., 1976. Strategy of hunting and vertical distribution of pelagic zooplankton feeders in a tropical ocean. *Oceanology*, **16** (5), 883–894.

Tseitlin, V. B., 1981a. Size structure of the pelagic population in the tropical oceanic regions. *Oceanology*, **21** (1), 125–131.

Tseitlin, V. B., 1981b. Energetical characteristics and size distribution of pelagic organisms in the tropical regions of the ocean. *Oceanology*, **21** (3), 529–536.

Tseitlin, V. B., 1981c. Dependence of weight-specific respiration rate on the depth of habitat. *Doklady Acad. Nauk USSR*, **253** (5), 1224–1226.

Tseitlin, V. B., 1981d. Estimation of the vertical detritus flow from the surface zone in the tropical ocean regions. *Oceanology*, **21** (4), 713–718.

Tseitlin, V. B., 1981e. Consumption of the sinking detritus by pelagic animals. *Oceanology*, **21** (5), 889–893.

Tseitlin, V. B. and T. A. Gorelova, 1978. Studies of feeding of the lanternfish *Myctophum nitidulum* (Myctophidae, Pisces). *Oceanology*, **18** (4), 742–748.

Tumantseva, N. I., Sorokin, Yu. I., 1975. Microzooplankton of the area of equatorial divergence in the eastern part of the Pacific Ocean. *Trans. Inst. Oceanology*, **102**, 206–212.

Turner, J. T., 1977. Sinking rates of fecal pellets from the marine copepod *Pontella meadia*. *Mar. Biol.*, **40**, 249–250.

Turner, J. T., 1979. Microbial attachment to copepod fecal pellets and its possible ecological significance. *Trans. Am. Microsc. Soc.*, **98** (1), 131–135.

Tyler, H. A. and W. G. Pearcy, 1975. The feeding habits of three species of lanternfishes (family Myctophidae) off Oregon, USA. *Mar. Biol.*, **32,** 7–11.

Vinogradov, M. E., 1953. The role of vertical migration of zooplankton in the feeding of deep-sea animals. *Priroda* (6), 95–96.

Vinogradov, M. E., 1955. The vertical migration of the zooplankton and its importance for the feeding of deep-sea pelagic fauna. *Trans. Inst. Oceanol.*, **13,** 71–76.

Vinogradov, M. E., 1958. On the vertical distribution of deep-sea plankton in the west part of the Pacific Ocean. Fifteenth International Congress Zoology, Section III, Paper 31, London.

Vinogradov, M. E., 1959. The vertical migrations of the deep-sea plankton. In *Itogi Nauki*, Vol. 1, *Advances in Oceanology*, L. A. Zenkevitch, ed. Acad. Sci. USSR Press, Moscow, pp. 204–220.

Vinogradov, M. E., 1960. Quantitative distribution of the deep-sea plankton in the western and central Pacific. *Trans. Inst. Oceanol.*, **41,** 55–84.

Vinogradov, M. E., 1961. Food sources of deep-water fauna. Speed of decomposition of dead Pteropoda. *Doklady Acad. Nauk USSR* (Biol. Sci.), **138** (6), 39–42.

Vinogradov, M. E., 1962. Feeding of the deep-sea zooplankton. *Rapp. et Proc.-Verb. Cons. Internat. Explor. de la Mer.*, **153,** 114–120.

Vinogradov, M. E., 1968. *Vertical Distribution of the Oceanic Zooplankton.* Nauka, Moscow, 320 pp.

Vinogradov, M. E., 1970a. Some peculiarity of the change in ocean pelagical communities with the change ot the depth. In *Program and Method of Investigation of Biogeooceanological Water's Surroundings*, L. A. Zenkevitch, ed. Moscow, Nauka, pp. 84–96.

Vinogradov, M. E., 1970b. The vertical distribution of zooplankton in the Kurile Kamchatka region of the Pacific Ocean (based on the data of the 39th cruise of r/v *Vityaz*). *Trans. Inst. Oceanol.*, **86,** 99–116.

Vinogradov, M. E., 1972. Vertical stratification of zooplankton in the Kurile Kamchatka trench. In *Biological Oceanography of the Northern North Pacific Ocean,* A. Y. Takenouti, ed. Idemitus Shoten, Tokyo, pp. 333–340.

Vinogradov, M. E., 1977. Zooplankton, in Chapter 3, Vertical distribution of plants and animals in the Ocean. In *Oceanology, Biology of the Ocean*, Vol. 1, M. E. Vinogradov, ed. Nauka, Moscow, pp. 132–151.

Vinogradov, M. E. and E. G. Arashkevich, 1969. The vertical distribution of interzonal filter-feeding copepods and their importance in the communities of different depths in the northwest Pacific. *Oceanology*, **9,** 488–499.

Vinogradov, M. E. and N. V. Parin, 1973. Some dates of vertical distribution of macroplankton in tropical regions of Pacific Ocean. *Oceanology*, **13,** 137–148.

Vinogradov, M. E., Shushkina E. A., I. N. Kukina, 1976. Functional characteristics of a planktonic community in the equatorial upwelling. *Oceanology,* **16** (1), 122–137.

Vinogradov, M. E. and E. A. Shushkina, 1978. Some development patterns of plankton communities in the upwelling areas of the Pacific Ocean. *Mar. Biol.*, **48,** 357–366.

Vinogradov, M. E. and N. M. Voronina, 1964. Quantitative distribution of plankton in the upper layers of the Pacific equatorial currents area. II. The vertical distribution of some species. *Trans. Inst. Oceanol.*, **65,** 58–76.

Walters, V., 1961. A contribution to the biology of the *Giganturidae*, with description of a new genus and species. *Bull. Museum. Compar. Zool.*, **125** (10).

Wheeler, E. H., 1967. Copepod detritus in the deep-sea. *Limnol. Oceanogr.*, **12** (4), 679–702.

Wheeler, E. H., 1970. Atlantic deep-sea Copepoda. *Smithsonian Contrib. Zool.*, **55,** 1–31.

Wiborg, K. F., 1954. Zooplankton in relation to hydrography in the Norwegian Sea. *Fiskeri Tidsskr., Ser. Havunders*, **11** (4), 66 pp.

Wiebe, P., S. Boyd, and C. Winget, 1976. Particular matter sinking to the deep-sea floor at 2000 m in the Tongue of the Ocean, Bahamas with description of a new sedimentation trap. *J. Mar. Res.*, **34,** 341–354.

Winberg, G. G., 1976. Dependence of energy metabolism on the body mass in aquatic poikilotherms. *Zh. Obsch. Biol.*, **37,** 56–70.

5. MEIOBENTHOS AND NANOBENTHOS OF THE DEEP SEA[1]

HJALMAR THIEL

1. Definitions

Benthic organisms are divided according to feeding types, habitats occupied, respiratory requirements, and size. The first groups are ecological, and even though size is related to metabolism and turnover rates, size groups originated primarily in conjunction with the methods used for sampling and processing, and these have changed historically. Meio- and nanobenthos have to be viewed together with mega- and macrobenthos, altogether comprising the endo- (or in-) fauna and the epifauna and epiflora of the benthic communities.

The term "meiofauna" was coined by Mare (1942). "Meio" is derived from Greek and means "the smaller" fauna, seen in context with macrofauna. Early quantitative studies on the benthos were restricted to those faunal components that were retained on sieves used to wash off the finer sediment particles. The mesh size most frequently employed as a lower limit was a 1 × 1-mm mesh (Thorson, 1957; Holme and McIntyre, 1971). The upper size limit for meiofauna and the lower one for macrofauna can thus be defined as given by the 1-mm mesh.

This 1-mm limit is not generally agreed on. Mare (1942) suggested 2 mm, and Sanders et al. (1965) in their study on macrofauna of the Gay Head–Bermuda transect realized that a good portion of fauna was lost by using the 1-mm sieve, and they introduced the 420-μm mesh size. Later on, Menzies and Rowe (1968) applied 0.5 mm, and Hessler (1974) washed his samples from the northern Pacific deep sea through 297 μm to collect *in toto* those taxa that were generally believed to belong to the macrofauna.

The lower size limit for meiofauna differed strongly between researchers, at least partly dictated by the amount of time that one was willing to spend on sorting organisms from the sediment; the time required increases considerably with decreasing lower size limit. But this problem pertains mainly to shallow water investigations. Deep-sea meiofauna has been studied only since the mid-1960's, and because the number of workers has been limited, methods are more comparable. Wigley and McIntyre (1964) were the first to leave the continental shelf for meiofauna sampling, taking five samples between 99- and 567-m depth in the northwestern Atlantic. The lower mesh size they used was 74 μm because of insignificant faunal weights in the smaller fractions. Thiel (1966) sorted the meiofauna from samples taken in the northwestern Indian Ocean down to a depth

[1] Submitted April, 1980.

of 5030 m with the lower limit of 65 μm, and later Dinet et al. (1973) and Thiel (1971) set the lower limit to 50 and 42 μm, respectively, in order to collect smaller abundant specimens as well. Sixty-three samples from different depths beyond the shelf edge resulted on the average in the following percentages in distribution between the size fractions:

9 to 12%	for	150 to 1000 μm
16 to 21%	for	100 to 150 μm
44 to 48%	for	65 to 100 μm
21 to 27%	for	42 to 65 μm

This demonstrates that the smallest fraction between 42 and 65 μm, although less important in terms of biomass, still contains 21 to 27% of the animals by number, and it is important to include this size fraction.

Experience and discussion on the lower limit for the meiofauna had somewhat settled when Burnett (1973) started his studies on what he termed the *microfauna* and the *microbiota* (Burnett 1977 to 1981), that is, those organisms passing the 42-μm meshes but being not as small as bacteria. This size group is composed of unicellular organisms such as prokaryotes, yeastlike cells, different protozoan taxa, and more rarely of small individuals of taxa normally found in the meiofauna such as nematodes and nauplii.

Although the faunal components of the benthos classified by the prefix "macro-" and "meio-" are generally well understood (for deviations, see below), "micro-" does not sufficiently characterize a definite size group. "Micro-" is in use together with "biota" and "organisms" often, but not always, including bacteria and fungi or even containing only these groups. It is combined with "flora" or "algae" for the unicellular plants in sediments. It is connected with "fauna," synonymous with meiofauna, or by geologists it is applied to organisms with skeletal structures allowing fossil records useful for sedimentary stratigraphic purposes. Thus "micro-" seems to be an overexploited prefix, leading to confusion of terms and organism groups. Therefore, I prefer to avoid "micro-" and propose instead "nano-," in the compositions "nanofauna," "nanoflora," and "nanobenthos," or "nanobiota," as used already by Burnett (1981). "Nano" derives from the Latin word for "dwarf" and in its combination with "benthos" linguistically the Greek "nanno" would have been preferable. "Nano," however, has long been applied to classify a definite plankton size group (Sieburth et al., 1978), it is consistent with the expressions "fauna" and "flora." Finally, it is widely in use in the metric system, as in "nanogram." The nano size group then covers metazoans, protozoans, and fungilike organisms smaller than 42 μm, but it excludes bacteria. According to Burnett (1973), his method allows for the quantitative evaluation of the nanobenthos down to 2 μm.

It would be useful for reasons of comparison, as, for example, of metabolic rates, to adjust the benthos size groups to the spectrum of plankton size fractions given by Sieburth et al. (1978), ascribing the taxonomic–trophic compartments of the plankton to logarithmic size groups on the base 2, which roughly couples with the live weights in the metric system. Yet such a classification would not agree with the historically developed size groups erected for the benthos, and at this time that would create considerable confusion.

TABLE I
Benthos Size Groups Defined by Mesh Size, Sample Size, and Processing Methods

Size Group	Mesh Size	Sample Size	Processing	
			Shipboard	Laboratory
Macrofauna	>1 mm (0.5, 0.42, 0.3 mm, other authors)	600 to 2500 cm^2	Sieving (sorting), preservation	(Staining), sorting with lens or stereomicroscope
Meiofauna	42 to 1000 μm (50, 63 μm other authors)	Subsample 2 to 10 (25) cm^2	Preserved totally	Sieving, staining, sorting with stereomicroscope
Nanobenthos[a]	2 to 42 (50) μm	Subsample 0.71 or 2.84 cm^2	Preserved totally	Slide preparation using subvolumes of 0.25 ml, staining, microscopic scanning

[a] Microbiota in Burnett (1977 to 1981).

A definite requirement in ecological research is the quantitative assessment of organism densities in space and time. Thus it is a prerequisite for the methods applied that they be suitable for quantitative assessments of the chosen organism size groups. In a simple scheme the benthic size groups can be defined as given in Table I. These size categories are additionally divided by their processing methods (see Section 2), which, in turn, are related to sample size.

Problems become apparent when faunal size groups are coupled with definite taxa. A number of taxa are predominantly meiofaunal size (Nematoda, Copepoda, Harpacticoidea, Ostracoda, Tardigrada, Kinorhyncha), whereas others are regarded as macrofauna. However, some species of typical meiofaunal size become larger than 1 mm, such as in the Nematoda and Ostracoda. Juveniles of predominantly macrofaunal taxa start off with larvae of meiofaunal size, as in the Mollusca and Polychaeta, and they grow into macrofauna size. These are termed "temporary" meiofauna as opposed to the "permanent" meiofauna. In the deep sea, traditionally macrofaunal taxa have a high number of representatives exhibiting meiofaunal size, giving the impression that size on an average is smaller than in shallow waters [Thiel (1975); cf. Section 5, this chapter].

One taxon is especially difficult, the Protozoa, and within these, the Foraminifera. Most known species are meiofauna, but there are likely a large number of nanofaunal size yet to be described. Foraminifera, such as the Schizamminidae, may grow several millimeters or even centimeters in length (Nørvang, 1961; unpublished observations). Xenophyophoria (Tendal, 1972; Gooday, 1978; Rice et al., 1979) may stretch their pseudopodia several centimeters. They collect and stabilize large lumps of sediments, up to 6 cm high as photographed by Rice et al. (1979). The Komokiacea are branched or brushlike Foraminifera of a few millimeters across (Tendal and Hessler, 1977) (see Addendum 1).

To avoid complication of terms, the benthos size groups should purely be defined by size, indicating simultaneously the processing methods. Under biological aspects the defined size categories and their limits or any other size classification have little importance and are thus sufficiently well established by the sampling and processing methods.

Considering the size of benthic organisms in metabolic units, some general relations become apparent (Fenchel, 1968, 1974, 1978; Gerlach, 1971). Oxygen consumption rates per unit mass are higher in small organisms. However, it is not clear how Foraminifera follow such a relationship. Although they are regarded as small, having a small plasma volume, by the methods used they appear in all the three size groups, depending to a large extent on their tests of secreted carbonates or agglutinated sediment particles. In relation to their actively living part, the protoplasm, many or most Foraminifera obviously show up in too large a size fraction. This holds as well for other shell-bearing taxa. Taxonomic position, size, and metabolic rate all call for a definition of macrofauna, meiofauna, and nanobenthos by size or sieve mesh width, by sampling and processing methods.

Although size categories and limits of size groups are decided on by the investigator for each study, it must be stressed that generally accepted limits or sieves favor the comparability of results. This was one of the guiding ideas, when the Baltic Marine Biologists discussed their "Recommendations on Methods for Marine Biological Studies in the Baltic Sea" (Dybern et al., 1976).

However, rigid limits may as well result in severe limitations. An overlap in size groups seems to be useful when the total fauna is to be assessed quantitatively. The small samples taken in meiofauna research may reveal not enough of the large specimens in this size class for quantitative assessment. The size fraction of 0.5 to 1.0 mm of a larger (i.e., macrofaunal) sample may better represent the animals in this size order. If both the sieves, 0.5 and 1.0 mm for both size classes, are employed for the macrofauna as well as for the meiofauna, this overlap would allow for an appropriate estimate of faunal densities. A similar overlap should be introduced between meio- and nanobenthos. Burnett (1977) was aware of the need for an overlap when he used a 50-μm mesh size as the upper limit for his microbiota. However, in addition, a 42-μm sieve is obligatory to allow for a full comparability with the meiofauna data.

It is important to state in each report which limits and which sieves had been adopted. Mesh sizes that have been employed regularly in other studies should be chosen by future investigators to achieve widest comparability of results.

In this context, the deviating size limits for the faunal components applied by Soviet researchers (Sokolova, 1970, 1972) must be noted. Their trawls have taken macrofauna down to only 5 mm; smaller fauna, with limits of 0.5 to 5 mm, has been sampled with grabs and is termed *meiofauna*. These size groups fall into the macrofaunal category of western Europeans and Americans.

2. Methods for Study of Deep-Sea Meio- and Nanobenthos

A. Sampling

Deep-sea research in general is time-consuming, and sampling methods must be adapted to this. Great care should be taken to apply methods that warrant comparable data sets [see Gage (1979) for macrofauna]. In spite of shortcomings well known to deep-sea researchers (Menzies et al., 1973), sampling is mostly performed with a grab, and actual meiofauna and nanobenthos samples are isolated from the grab samples by subsampling (e.g., Thiel, 1966; Hessler, 1971; Douglas et al., 1978). In shallow waters Elmgren (1973) has demonstrated that sampling by a diver is superior to all other methods. The diver in shallow waters, and comparably well a remote underwater manipulator in the deep sea (Thiel and Hessler, 1974; Thistle, 1978; Jumars, 1978; Burnett, 1981) or a submersible (Rowe et al., 1975; Haedrich and Rowe, 1977) can place a corer on the sediment surface with virtually no disturbance. But under most circumstances those highly developed methods are not available and blind sampling cannot be avoided. Barnett, Watson, and Connelly (see Scottish Marine Biological Association, 1977) constructed a multiple frame corer for careful sediment penetration. After settlement of its frame on the bottom, four core tubes descend slowly to the sediment surface and into the sediment under the control of a piston in a water-filled cylinder (Craib, 1965). The area enclosed by each core tube is 25 cm^2, which provides samples large enough for detailed studies of harpacticoid copepods and other meiofauna. A later development of the multiple corer has been the successful introduction of a 12 core unit (see Scottish Marine Biological Association, 1979). A camera and strobe flash mounted on the corer framework photographs the cores being pushed into the sediment, allowing subsequent analyses of sample disturbance, and provides photographs of the sediment surface prior to coring.

A pressure wave in front of a grab can blow light materials away from the sediment surface. This effect was demonstrated by Ankar (1977), who filmed the arrival of a Van Veen-type grab on the bottom in different sediments of shallow waters, and it was investigated by Douglas et al. (1978), comparing different types of corer and grab employed for the collection of Foraminifera. The pressure wave is weaker in samplers such as the Reineck box corer or the large undisturbed bottom sampler (LUBS) (Menzies and Rowe, 1968), which allow a free flow through the device on its way to the bottom. Disturbances such as a lateral shift during the process of sample cutting and winnowing may occur, but these effects have not been quantified. From a sample taken in the Mediterranean at a depth of 2701 m, 10 subsamples of 25-cm^2 surface area from a total sample size of 600 cm^2 were cored. These subsamples were arranged in two rows, Nos. 1 to 5 and 6 to 10 side by side. The sample was slightly tilted, with subsamples 1 and 6 somewhat higher than the subsequent ones. The first and the second centimeter were sorted independently. Both layers show a high degree of variability for the two most abundant meiofauna groups, the nematodes and the harpacticoids (Table II).

From Table II it becomes apparent that the deviation from the mean value found in the meiofauna distribution for the first centimeter is higher than that for the second, a difference that cannot be fully explained. Artificial deviation may be introduced to the upper layer by the sampling process; however, it may as well be

TABLE II
Distribution of Nematodes and Harpacticoids (>42 μm) in a Reineck Box Core Sample[a]

	Number of Nematodes		Number of Harpacticoids	
Number of Subsample	First Centimeter	Second Centimeter	First Centimeter	Second Centimeter
1	241	172	7	6
2	346	147	19	9
3	264	209	13	7
4	256	192	6	10
5	207	174	4	2
6	368	210	10	2
7	435	190	9	8
8	421	203	7	7
9	306	204	28	5
10	307	146	7	3
Mean	315	184	11	6
Standard deviation	±69	±21	±6.6	±2.6
Standard deviation (%)	±22	±12	±60	±43

[a]Taken from 2701-m depth in the Mediterranean. Subsamples cover 25 cm^2.

due to a more complex microhabitat structure of the sediment surface and to stronger bioturbation caused by some larger fauna. From Table II it becomes evident that the sampling bias at least seems to be small. No general trend of animal distribution emerges in the upper layer, which might be related to the tilting of the sediment in the box or to other outer influences. Although the extent of bias in each single sample is not known, the box grab seems to be suitable for sampling deep-sea habitats and for subsampling. However, one should keep in mind that the quality of a sample as well depends on sea conditions.

A more detailed study was performed by Jumars (1975a). He had used the USNEL box corer (Hessler and Jumars, 1974) covering an area of 2500 cm^2 and being divided by Jumars' vegematic subcorer into 25 fields of 100 cm^2. For sediment-surface-living polychaetes, he could demonstrate a lower abundance in the outer 16 subcores than in the inner 9, indicating a stronger bow wave effect along the rim of the box. This clearly indicates the effect of restricted water flow through a grab and that it is essential to have the top as wide open as possible for optimal flow through the sampling chamber. A large modified USNEL-type grab was used by Thiel (1980a, 1981) in which the top is 52% open. On penetration of the bottom, the open areas are covered with lids, triggered through the relative movements of the column and the frame. A more recent version of this grab type in which the lids are triggered by the moving spade is pictured in Rowe and Sibuet (this volume, Chapter 2).

Another possible shortcoming of grab sampling is the disturbance of natural stratification from the sideways movement of the grab due to its swinging or to ship's drift at the moment that the box is penetrating the sediment. One way to diminish the risk of disturbance is to lower the grab down to 20 or 30 m above the bottom relatively quickly, and then stop the winch. When the grab has come to some rest, which can be seen on the pinger record (Rowe and Sibuet, this volume, Chapter 2), the grab is lowered on down to the bottom at the lowest speed, guaranteeing gentle but good penetration into the sediment.

Further alterations in the distribution of organisms may be introduced by the sloshing of the water in the box above the sample, and this effect increases with increasing wave height. This is strongest when the grab is removed from the water, coming over the ship's side, and is resting on deck. Water may run out of the box, possibly carrying small organisms with it, and sloshing may resuspend sediment and simultaneously redistribute the fauna. This can be reduced with baffles or built-in subcores such as the vegematic subdivisions used by Jumars (1975a,b), but subsampling becomes difficult when the subcores are too narrow. The smearing on the sides of the subcores and their compression may be unwanted additional effects, increasing with the thickness of the wall of the coring box plus the subcores.

Despite the aforementioned shortcomings and arguments against such subsampling, grab sampling with commensurate subsampling seems to be the best compromise between sampling time and sample quality, especially considering that it is rare that only the fauna is wanted. Usually geologic and chemical properties and other live elements are to be sampled in addition. Subsampling the same grab sample is possible for all these components, and a large grab area assures closely spaced subsamples and the opportunity to correlate results.

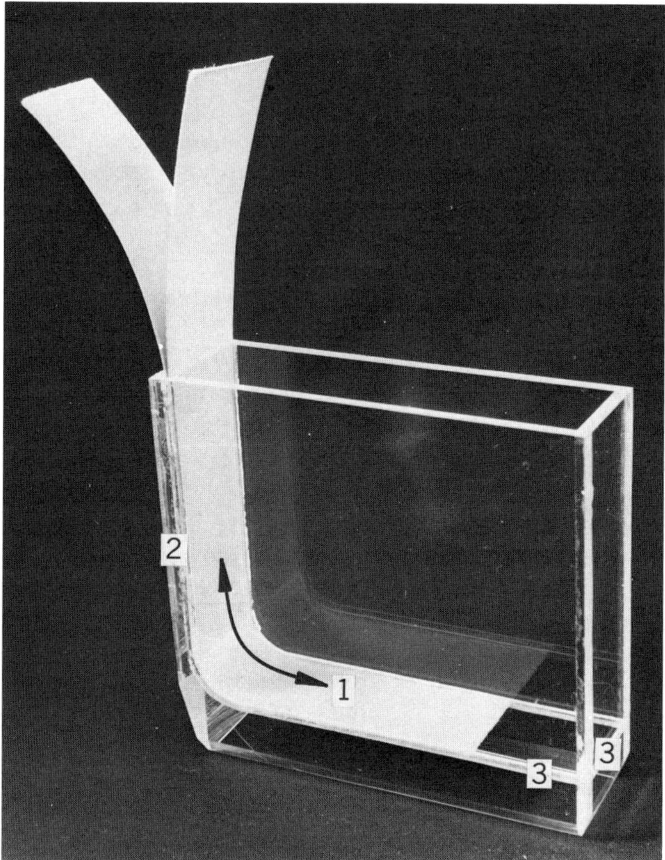

Fig. 1. The meiostecher for undisturbed subsampling an area of 2.5 × 10 cm: (1) plastic tongue to cut off the subsample from the deeper sediment layers; (2) double wall; (3) grooves for guiding the tongue.

B. Subsampling

Water in most cases covers the grab sample. By mechanical disturbance during sampling and retrieving, it may contain floating meiofauna. This water can be drained off first through an adjustable outlet (Hessler and Jumars, 1974) or carefully siphoned into a sieve of the smallest faunal size fraction wanted. What few meiofauna are caught are later added to those of the top sediment layer.

When studies of abyssal meiofauna began, it was believed that the expected low faunal density would necessitate a subsample size on the order of 25 cm^2. Therefore, the meiostecher (Fig. 1) was constructed (Thiel, 1966). Later its size was reduced to 10 cm^2. The meiostecher is a rectangular tube of 2.5 × 10 (or 4) cm inner dimensions. One of the narrow sides has a double wall, and the space between the two walls is prolonged by a horizontal groove in both the wide sides. After inserting the meiostecher into the sediment, a plastic tongue is pushed through the double wall, and, gliding along the groove, it divides the sediment inside the meiostecher from the deeper core. This subsampler leaves the unwanted sample area undisturbed for further investigations. It is convenient to

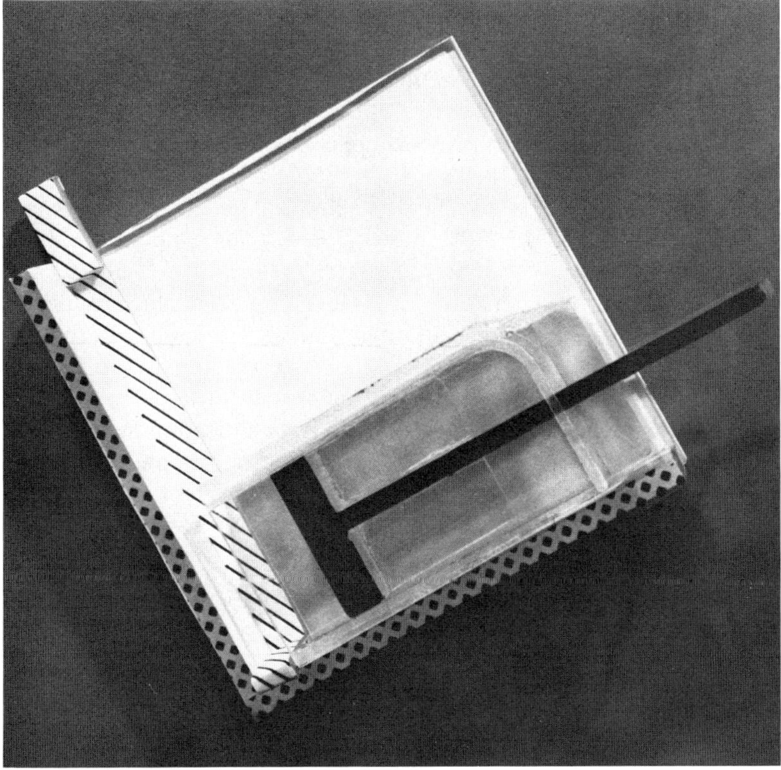

Fig. 2. Plexiglass support for slicing the meiostecher subsamples: [black box] plunger inside the 10 cm^2-meiostecher; [white box] support basic plate; [dotted box] support vertical sides; [hatched box] 1-cm-distance bar.

construct a small plexiglass support for the meiostecher for exact, rapid slicing of the subsample (Fig. 2). It is pushed out off the meiostecher with a plunger and glides over a 1-cm-distance bar until it hits the frame. The layer then can be cut off easily and preserved.

Tubes of different diameters have been used for subsampling (Wigley and McIntyre, 1964; Dinet, 1973, Dinet and Vivier, 1977), and a tube, stoppered at the top, allows the subcore to be isolated provided the sediment is not too soft, sandy, or stiff. Similarly, the subsamples can be taken with hypodermic syringes, whose anterior ends have been cut off.

The size of a subsample depends on the local variability of the fauna and its distributional scales (Jumars and Ekman, this volume, Chapter 10), and last but not least on the time one wants to spend on separating the organisms from the sediment. In most instances it would be too time consuming to sort more than 10 cm^2 per grab sample. To obtain a realistic average abundance and to learn about spatial distribution, this area should not be subcored as one unit. Subsampling with five syringes covering areas of 2 cm^2 or more (Thiel, 1979b) seems to be an effective method. Evaluation of the faunal density within a single grab sample by 5×2 cm^2 and projection of this result into a larger area is clearly a compromise between precision and time; in other words, the results must be evaluated with care.

For the study of Foraminifera abundance and diversity, Douglas et al. (1978) regard 10-cm^2 samples as adequate in areas with high densities but collect 100 cm^2 or more because 300 specimens provide a reasonable level of precision in estimating species proportions.

Nanobenthos sampling is principally achieved in the same manner as for meiofauna. Burnett (1973 to 1981) developed his subsampling method, testing different types of tubings with inner diameters of 0.95 cm and of 1.9 cm (areas of 0.71 cm^2 and 2.84 cm^2, respectively). Soft plastic tubes displaced sediment extensively as a result of side friction, although this may be reduced through silicone coating. Acrylic plastic tubes seem best because of their hydrophobic surface, allowing the tubes to slip through the sediment with little or no displacement of particles and fauna (Burnett, 1978, 1979b). In a recent sampling program by Burnett (1978, 1979b) these microcorers were mounted to a frame inside a box of the box grab, separating the subsamples from the sample right at the sampling site. This avoids disturbance later but it is feasible only in easily penetrated muddy sediments.

C. Preservation

Meiofauna is preserved at sea in 4% seawater formalin, but this should be buffered with hexamethylenetetramin (urotropin), about 10 g/liter, or with sodium borate in excess, to avoid dissolution of carbonate structures. A high concentration of foraminiferan carbonate shells cannot sufficiently buffer the sample. Ostracode shells seem to be especially susceptible to dissolution, even when in buffered preservative. Alcohol preservation should be used for special investigations of this group. This preservation method is adequate for all the hard meiofauna. The soft meiofauna has been processed properly only by Coull et al. (1977), with collections close to shore. The work force they used allowed fast, live-sample sorting immediately after a short cruise. Only the numbers of Turbellaria (maximum 3.4%) and Oligochaeta (up to 1%) from 400-m depth were significantly higher in replicates with living organisms in comparison to samples sorted after preservation. Gastrotriches were encountered equally well from life and preserved samples in these tests, but often they were rather badly preserved, and some may have been totally destroyed. Special methods must be used if these groups are to be investigated. Formalin with a concentration of 8% may better preserve this soft fauna, as it is used for the similarly soft and even more fragile nanobenthos.

Burnett (1977) preserved the nanobenthos in sediment layers of 5-mm thickness and included the water above the microcore with the top layer. The formalin concentration in the preserved sample is 8%, and its volume is adjusted to 15 ml for later quantitative evaluation. Buffering the sample with hexamethylenetetramin allowed staining (see below) only after repeated washing; Burnett thus suggests buffering the preservative with sodium borate.

For long-term storage of samples, the pH of the preservative should be checked every few months and further buffer should be added when the solution becomes acidic (see Addendum 2).

D. Meiofauna Sieving and Sorting

The methods for the sorting of meiofauna are generally the same for shallow and deep sediment samples. Uhlig et al. (1973) compared the efficiencies of different methods and their combinations. The most effective sorting is achieved by use of a stereomicroscope with magnifications of up to 50. The hydrodynamically rather light particles of many deep-sea sediments often do not allow presorting through water agitation methods such as elutriation. For comparability of results, all my samples were hand sorted. Rachor (1975) applied mechanical disturbance and decantation at least eight times to every sample before subsequent preservation and later sorting. However, his low meiofaunal counts (see Section 4.A) may be due partly to this procedure. Centrifugation in a density gradient by use of Ludox-TM for dividing the lighter organisms from the heavier sediment particles should be tested for deep-sea samples.

For a better identification of the few organisms between the bulk of similarly colored sediment grains, a selective staining improves the sorting efficiency considerably. Rose Bengal is used by most investigators, but eosin and other homologues may be applied as well. Thiel (1966) studied the staining properties of Rose Bengal and eosin in a wide range of pH values. Whereas staining with Rose Bengal was good at pH above 6, eosin was good only above 5. Good staining can be achieved easily through the addition of phenol to the staining solution to give it a 5% phenol concentration. Good coloring is reached after 10 to 20 minutes, and the excess stain is washed away with tap water.

The protoplasm of Foraminifera stains equally well, but in many specimens it is difficult to decide whether it was alive at the time of collection. Rose Bengal may stain the inner organic lining of the test, which is rather stable, but this would not indicate a "live" specimen. In thick-walled agglutinating Foraminifera, the stained protoplasm may be seen only faintly or may be totally obscured. For better distinction of living from nonliving Foraminifera, Walker et al. (1974) proposed Sudan Black B, a blue–black color, to be superior to Rose Bengal. This chemical can be used acetylated or in heated (40°C) saturated solution. In specimens with a dark-colored test, with a dense agglutinated or even calcareous test, the stained protoplasm may not be discernible. From my experience, wet sorting of Foraminifera together with the meiofauna after Rose Bengal staining is superior to dry sorting as used by geologists.

Principally, the total sample could be sorted in one unit with its particle sizes of 42 to 1000 μm; however, it is easier and less exhaustive for the sorter's eyes to work on more narrow size fractions with definite magnifications at a time. For the larger particles, a lower optic magnification is sufficient, whereas the smaller ones need higher magnification. The fractionation by sieves takes a little extra time, but it saves sorting time.

Before sieving, this process can be improved by an ultrasonic treatment (Thiel et al., 1975) to break up fecal pellets and sediment aggregations. If these contribute to the sediment in considerable amounts, sorting will be enhanced through the reduction in sample size. Foraminifera stain better, and fauna associated with mud balls and fecal pellets become readily detectable after sonification. Some damage may occur to fragile taxa. Crustaceans may loose antennae, extremities, or furca, and agglutinating Foraminifera may be destroyed after 10 sec of ultrasound application, and the most fragile Xenophyophoria and Komokiacea, difficult to discern in their outer appearance from fecal pellets and mud balls, may disappear altogether.

Sieving processes are difficult to standardize, and errors may be introduced into the number of meiofauna per size fraction without a change in the total number. The mesh sizes of the sieves employed in fractionation should be chosen according to those sizes abundantly used by other researchers and according to the lower limit of organism size studied. As pointed out above, 42, 63, and 100 μm were often taken as the lower limit for meiofauna, and fractioning with these mesh sizes allows for a broader comparison with published data. During early studies on the deep-sea meiofauna, Thiel (1966) recommended the use of an additional sieve of 150 μm to keep all larger particles separated, and later he introduced 420 μm (or 500 μm) for better comparison with those macrofaunal studies in which this mesh size is used as a lower limit. For proper separation of meio- and macrofauna, the latter should be isolated by employing a 1-mm sieve. An agreement on generally accepted mesh sizes could well improve the comparability of results.

Processing biases almost always reduce the actual number of organisms per sediment area or volume because it is practically impossible to sort all the animals out of the samples. Double countings of samples revealed a loss of 5 to 10% on the average, which would not normally justify regular resorting. Quantitative meiofauna counts have to be regarded as low values.

E. Nanobenthos Processing

Burnett (1973 to 1981) alone has isolated nanobenthos quantitatively from deep-sea sediments. Because microscopic sample inspection is necessary, slides must be prepared. A sediment slurry of the preserved sample is produced by gentle mixing and 0.25 ml are pipetted onto a cover glass, previously coated thinly with Meyer's albumen for fixing the particles to it through drying at 80°C. This is immediately followed by staining with hematoxylin and mercuric bromophenol blue, followed finally by mounting the cover glass on a slide. Evaluation is done in two steps: for organisms larger than 10 μm, four entire slides prepared from the same sample are scanned at a magnification of 300; and for the organisms smaller than 10 μm, about one-eighth of the four slides is scanned with a magnification of 600. From these counts quantitative values of nanobenthos in the sediment are estimated. This type of quantitative evaluation introduces further subsampling variance in that subunits are taken from subsamples and the slide areas are not totally scanned and counted.

Nanobenthos is fragile and easily destroyed by preservation fluids, drying, and chemical treatment. Burnett (1977) made a number of tests with some species, but these cannot be applied in general. Species react differently to the various steps in the procedure. In strong fixative concentration Burnett (1979a) found limited disruption of protozoa. Shortcomings listed by Burnett (1977) in addition to box corer sampling bias (see above) are smearing of surface material along the inner wall of the microcorer and compression of the sample due to friction, which becomes more effective in small-diameter tubes; disruption by fixation, temperature, and mechanical agitation; nonstaining and understaining; and volumetric subsample deviation and fragmentation of colonial organisms. Most biases reduce nanobenthos counts and will be underestimates more likely than overestimates.

F. Biomass Determination

The determination of weights for meio- and nanobenthos are tedious and rarely done in deep-sea studies. Average weights are used from shallow water organisms or weights are extrapolated from subsamples. Rachor (1975) alone measured all the isolated nematodes found, according to a formula presented by Andrássy (1956) for the weight $W = l \times d^2 \times 0.665$, in which l is the length of the nematode without thin tips or tails, d is its largest diameter, and 0.665 derives from the specific weight for nematodes of 1.13 (Wieser, 1960a) divided by an empirical constant of 1.7. Harpacticoid and nauplii weights were determined similarly. Average values in Rachor's study are 0.51 μg for nematodes, 1.03 μg for harpacticoids, and 0.28 μg for nauplii. Extremely large nematodes are excluded from the average. Rachor lists a number of weights arrived at by other authors mainly from shallow waters and describes data given by Thiel (1971, 1972b) for the deep sea as too high. This may well be the case. Vivier (1978b) points to the wide variability in nematode weights and the inherent error in biomass determination. As long as standard weights for all individuals in a sample or in a size fraction are used and the weights for total samples are calculated through multiplication with faunal numbers, the weights are inferior to counts. Therefore, during the course of this chapter only numbers of individuals are presented and weights are discussed only briefly.

Burnett (1979a) took size measurements of some ciliates, flagellates, and a blue–green bacterium and calculated their biomass. The various species react differently to the mounting and staining process. Shrinkage was found to range between 46 and 92%, with an average of 70%.

3. Distributional Patterns

Investigations on the meiofauna of the bathyal and abyssal environments are as old as biological deep-sea research, without being specified as such. Among the first publications on the voyage of the British *Challenger* during the years 1872 to 1876 were the extensive reports on Ostracoda by G. S. Brady in 1880 (Brady, 1880) and on Foraminifera by H. B. Brady (Brady, 1884) in 1884. Questions addressed morphology and taxonomy in those early days. Later collections of meiofauna were occasionally made, mainly as a by-product with macrofauna. As long as sieving was done with the 1-mm mesh sieve, however, organisms belonging to groups of predominantly meiofaunal size were mostly washed away. The Danish *Galathea* expedition with magnificent macrofauna collections from all ocean depths revealed, as far as published, very little meiofauna. Wieser (1956) described 3 nematode species, with 24 specimens in total.

A. Depth Distribution

Wieser (1960b) was the first to compile data on deep-sea meiofauna. Special research directed to bathyal meiofauna dates back to Wigley and McIntyre (1964)

TABLE III
Some Early Depth Records Demonstrating the Increase of Knowledge Regarding Distribution of Nano- and Meiobenthic Taxa in Bathyal, Abyssal, and Hadal Habitats

	Depth (m) (Reference)
Nanobenthos	
Fungi	4610 (Höhnk, 1961); 5315 (Kohlmeyer, 1977)
Flexibacteria	1200 (Burnett, 1979a)
Yeastlike cells	5498 (Burnett, 1977)
Sarcodinia	1200 (Burnett, 1973); 5498 (Burnett, 1977)
Suctoria	1130 (Burnett, 1973)
Ciliata	3920 (Uhlig, 1970)[a]; 5498 (Burnett, 1977)
Foraminifera	1200 (Burnett, 1979a)
Meiobenthos	
Foraminifera	7220 (Brady, 1884)
Halammohydra (Coel.)	1480 (Uhlig, 1970)[a]
Turbellaria	1275 (Wieser, 1956); 3828 (Uhlig, 1971)[a]; 7298 (Jumars and Hessler, 1976)
Nematoda	3311 (Ditlefsen, 1926); 4540 (Wieser, 1956); 10,414 to 10,687 (Wolff, 1960)
Gastrotricha	1480 (Uhlig, 1970)[a]
Kinorhyncha	385 (Wieser, 1956); 4690 (Thiel, 1966)
Tardigrada	385 (Marcus, 1934); 4690 (Thiel, 1966)
Oligochaeta	2000 (Cook, 1969); 4850 (Cook, 1970); 7298 (Jumars and Hessler, 1976)
Archiannelida	3039 (Uhlig, 1970)[a]
Halacarida	5320 (Thiel, 1972b); 9807 (Thiel, unpublished)
Ostracoda	5030 (Brady, 1880); 7657 (Wolff, 1960)
Harpacticoidea	9995 to 10,002 (Wolff, 1960)

[a] Rough depth limits were published in Uhlig (1970, 1971); depths are presented in this table according to personal communication.

with samples down to 576 m and to Thiel (1966) reaching abyssal depths of 5030 m. Hadal meiofauna is reported on by George and Higgins (1979) from the Puerto Rico Trench. Thiel (1979a) included data on samples received from Dr. R. R. Hessler taken in the Philippine Trench at 9800 m depth, the deepest so far, whereas Jumars and Hessler (1976) reported on some meiofaunal specimens down to 0.297-mm mesh size in their study of the macrofauna of a core from 7298 m in the Aleutian Trench. Maximum depth for meiofaunal taxa (Table III) suggests that numerous taxa will be found in deep-sea trenches in the future.

The nanobenthos have been even more ignored in deep-sea studies. Uhlig (1970, 1971) employed his seawater ice technique on deep-sea samples and isolated ciliates from down to 3920 m for the first time, together with some soft meiofauna (Table III). Aware that many deep-sea organisms do not withstand transport from great depth to the surface or sample processing, and that most of them exhibit a sessile or very slow moving mode of life, Burnett (1973 to 1981) successfully preserved nanobenthal organisms. The taxa he discovered on the slides and the depths from which they were isolated are presented in Table III. Further fascinating results on depth records and taxonomic composition can be expected for the next decades.

B. Sediment Penetration

Benthic organisms inhabit a thin layer of the sedimentary regime; however, they all ultimately depend on the sediment surface. The two most important conditions indispensable to benthos life are food and oxygen. Both are delivered to the habitats through the overlying water column, arriving at the bottom water layer and at the sediment by diffusion in the case of oxygen, by sinking in the case of food particles; or both are delivered by being moved passively through water transport and animal activity.

For the deep-sea environment, oxygen becomes the limiting parameter in only rare cases. Anoxic conditions result from very low rates of water exchange, such as in the isolated bathyal Santa Barbara Basin off the California coast and in the Black Sea, and/or on high inputs of organic matter. This occurs in upwelling areas off Peru, Chile, southwestern Africa, and to a limited extent off northwestern Africa related to sediment contents of organic matter higher than about 5%. Whereas the steadily anoxic environments such as the Santa Barbara Basin and the deep Black Sea are devoid of metazoa and protozoa, in other areas a thin surface layer of the sediment may be oxygenated and inhabited by meiofauna. This is exemplified by nematodes living under low oxygen conditions in the irregularly flushed deep basins of the Baltic Sea (Elmgren, 1975, 1978). Meiofauna such as Nematoda, Turbellaria, Gnathostomulida, and Rotifera may penetrate deeper into anoxic sediment layers. As this is known from shallow waters (Fenchel, 1969, 1971; Fenchel and Riedl, 1970, Fenchel, 1978, Powell et al., 1979, 1980; Revsbeck et al., 1980), it can be expected as well for deep-sea environments such as the transition zone of oxic to anoxic conditions. Reise and Ax (1979, 1980) doubt the existence of the sulfide biome because the organisms are believed to use the oxygen available in small quantities from isolated microhabitats, and no species is known that does not require aerobic respiration during parts of its life cycle. But Powell et al. (1979, 1980) discovered sulfide detoxification mechanisms that allow the organisms to tolerate continuous sulfide stress at least for parts of their lives.

No information is available on the nanobenthos from oxygen-poor or oxygen-deficient deep habitats, although in shallow waters species of these taxa were discovered to be more abundant in some anoxic sediments than meiofauna. Fenchel (1969) mentions this for ciliates, and Gallardo (1977) reports that many kinds of prokaryotes occur off the coasts of Peru and Chile in thick, interwoven mats in the sediment surface, where anoxic water covers subtidal to bathyal depths. Other nanobenthal and meiofaunal components are to be expected in such organic-rich habitats or transition zones between oxic and anoxic environments. Thane and Perry (1975) described shallow-water ciliates lacking cytochrome oxidase and mitochondria and being adapted to anaerobic marine habitats.

Most deep-sea sediments are well oxygenated and show high positive redox potentials (unpublished data from measurements in box grab samples). If oxygen becomes rare at all, this happens in sediment depths much deeper than the inhabited thin surface zone (Revsbeck et al., 1980). The penetration of meiofauna and nanobenthos seems to be correlated with the input of organic matter to the sediment, and this is reflected to some extent by the content of organic matter in the sediments, although this is not equivalent to food materials. In nearshore shallow waters silt and clay sediments contain up to around 2% organic carbon, but this generally drops to 1% in offshore areas, to 0.5% in bathyal and abyssal zones under the influence of high surface productivity, and down to 0.25% or less in central oceanic regions. [For deviations, see Aldred et al. (1979), Thiel (1978a and b, 1979a), Walsh (1981), and this Chapter, Section 4.B, subsection on northwestern African upwelling region in eastern Atlantic depth transects and Section 5.1.] Only small amounts of organic matter are accumulated in most abyssal and hadal sediments (Müller and Suess, 1979). It is oxidized and degraded through organism metabolism, as suggested by Pamatmat (1973) and Smith (1978) by the measurement of seabed oxygen demand (Smith and Hinga, this volume, Chapter 8). Chemical oxidation is seemingly very low in the deep sea, and in most cases too low to be measured with *in situ* techniques.

In most investigations meiofauna sorting is restricted to the upper 4 or 5 cm of the sediment because of the low numbers of individuals deeper down. From the Iberian deep sea, Thiel (1972b) sorted 6 out of 11 samples to 7-cm sediment depth (Table IV). At a sediment depth deeper than 4 cm, 14.5% of the fauna were encountered and 8.1%, below 5 cm. Subsampling procedure may have dislocated meiofaunal individuals into deeper layers, possibly indicated by the ostracodes found deep down (Table IV), yet no data are available to assure this assumption.

Rachor (1975) arrived at roughly similar values; however, deeper than 4 cm he found only 3% of the nematodes, 6% of the copepods, and 7 to 8% of the nauplii. Differences may be due partly to varying methods. From the Mediterranean samples between 200- and 800-m depth, Vivier (1978a) arrived at 12.8 and 9.3% together for the fourth and fifth layer and at 7.2 and 2.4% for the two subsequent layers.

This general decrease (Coull et al., 1977; Dinet, 1973, 1976; Dinet and Vivier, 1977, Thiel, 1971, 1972b) is not encountered in all samples, and even intermediate reversals in faunal densities occur. For the nematodes of 11 samples from the Iberian deep sea (Thiel, 1972b), the numbers and percent ranges were found as given in Table V. Only once in 44 1-cm sections did a layer contain more specimens than the layer above it, and deviations from the expected decrease with sediment depth were found in 17 layers out of 76 in water depths between 80 and 1000 m from the northwestern African upwelling region. Further examples can be given from the central northern Pacific Ocean [from 5100 m (unpublished data), and 5850 m (Burnett, 1979)] and the Red Sea [500 to 1500 m (unpublished data)], where a regular decrease in faunal numbers was encountered. Coull et al. (1977) presented an unexpected example from 4000-m depth in which the Foraminifera had their highest density in the second layer, whereas for the Nematoda and the other faunal components, it was found in the third layer. There was some indication that sample disturbance caused this distribution, but it may as well have had biotic causes. In the deeper waters of the North Sea (150 m) a higher number of samples exhibited irregularly decreasing meiofaunal densities.

The nanobenthos inhabits similar depths, yet our only knowledge is from the work of Burnett (1973 to 1981). He had to cope with many difficulties in collecting quantitative data on these small organisms. In a core from 5498 m in the central

TABLE IV

Depth Distribution of Meiofaunal Taxa (>42 μm) in Sediments of Iberian Deep Sea[a]

Layer (cm)	Nematoda	Polychaeta	Copepoda	Nauplii	Ostracoda	Total	Total (%)
0 to 1	1370	7	31	7	8	1423	42.6
1 to 2	679	4	12	5	8	708	21.2
2 to 3	467	2	6	2		477	14.3
3 to 4	238		6	2	1	247	7.4
4 to 5	209		5			214	6.4
5 to 6	145		1			146	4.4
6 to 7	122		1		2	125	3.7

[a] Number and percentage averaged from six samples (25 cm^2) collected at around 5300-m depth.

TABLE V
Ranges of Numbers and Percentages of Nematodes (>42 μm)

Layer (cm)	Range		Mean and Standard Deviation	Percent in Layer
	Numbers	Percentage		
0 to 1	97 to 371	40 to 70	245 ± 82	52
1 to 2	38 to 196	15 to 37	117 ± 54	25
2 to 3	27 to 116	8 to 24	71 ± 26	15
3 to 4	15 to 71	3.4 to 18	37 ± 15	8

[a]Mean, percentage, and standard deviation for the four layers calculated from 11 subsamples (25 cm^2) collected in the Iberian deep sea.

northern Pacific (Burnett, 1977), he discovered the majority of ciliates, flagellates, amoebas, testate cells, yeastlike cells, and other cells in an upper layer of probably not more than 5-mm thickness. In San Diego Trough sediments of 1230-m depth, nanobenthal populations reached 10 cm down into the sediment. Although a general decline in nanobenthal densities was described, this did not hold for all the subsamples, and some even showed a dramatic increase in deeper layers.

Thus for both meio- and nanobenthos, the vertical distribution in the sediment column is similar: a more regular decrease is correlated with low standing stock of organisms, and deviations from this show up where standing stock is higher.

As pointed out above, in most deep-sea sediments there is no oxygen–hydrogen sulfide discontinuity layer present in the range of faunal penetration, and the chemical properties measured thus far show no or very small and gradual change. Thus the food supply to the sediment surface should be what concentrates the organisms to this interface. This pattern would be expected where populations are limited by low food concentrations, specifically, in deep and central oceanic waters. On the other hand, it should be less developed where food concentrations are higher, allowing for higher organism densities in the top layer and deeper down in the sediment. Figure 3 is an attempt to demonstrate graphically the different types of meiofaunal depth distribution. This pattern is regulated through macrofauna predation and by competition for the best-suited microhabitat or adaptation to different sediment zones. The deeper zones are often more compact, with smaller interstitial spaces or closer-packed particles. Penetration of the deeper layers needs morphological and possibly energetical adaptations, allowing for avoidance reactions to predators and leading to species' spatial separation. This species interaction primarily argues for the expected gradual decrease in organism density with sediment depth.

However, high food concentration favors not only small organisms, but also the macrofauna of various feeding habits. Filter feeders may concentrate their fecal pellets on the sediment surface, or deeper down, burrowers mix the sediment, including food sources for the smaller ones or even pushing the less vagile organisms around. Predators like sipunculids and holothurians feed in various layers (Hansen, 1978) and the vagile surface fauna feeds in the richest surface layer. Greater abundance and biomass (Rowe, this volume, Chapter 3) of the macrofauna would have two foreseeable effects. Bioturbation rates would be

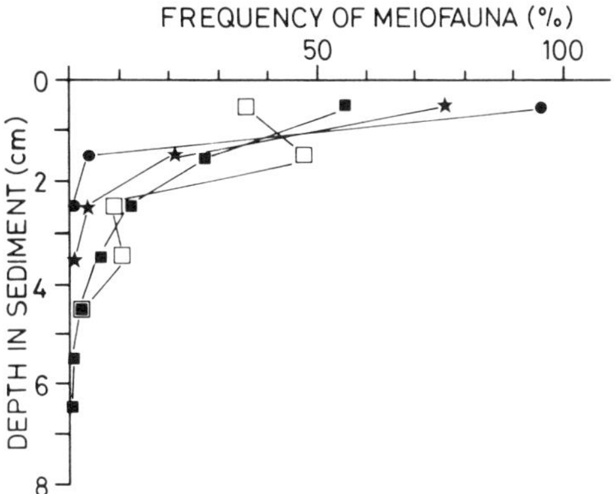

Fig. 3. Generalized scheme of meiofaunal depth distribution in the sediment under different conditions of organic matter input: ■ low to medium organic matter input, sediment well oxygenated, slow and mostly regular decrease of faunal densities, deep penetration; □ medium to high organic matter input, sediment well oxygenated; slow but frequently irregular decrease of faunal densities, deep penetration; ★ very low organic matter input, most of the fauna concentrates near the sediment surface, strong decrease with depth, penetration limited; ● very high organic matter input, redox discontinuity layer between 1 and 2 cm. Most of the fauna concentrates in the oxic layer, with strong decrease within the discontinuity layer and the anoxic zone.

greater, thereby moving both small organisms and food material deeper, erasing to some degree the general pattern. Also, more macrofauna would increase the likelihood of alterations in the distribution of the small biota due to predation. Thus the distribution of organisms in the sediment column can consistently be explained for the food-poor deep and central oceanic environments and for coastal areas with a higher food input.

C. Variation in Abundance with Time

Series of samples collected from some permanent stations are available only from the French Biogas program in the Golfe of Gascogne (Bay of Biscay). Six stations were occupied up to six times within 2 years (Dinet and Vivier, 1977). The results (Table VI) are based on one to six subsamples per box grab.

The means in faunal densities decrease from shallower to deeper stations (1 to 4 and 6 to 5), and the standard deviations (26 and 40%) are not exceptionally high if compared with other collections (this Chapter, Section 3.D). But the figures are not sufficient to express any change in meiofauna numbers during the course of the year (Dinet and Vivier, 1977).

One explanation is that the density may actually not change strongly during the year. Also, the grab samples taken from a single station are scattered over distances up to 10 nautical miles and depth within a single station varies considerably (Table VI). Thus the high, small-scale variability observed between subsamples and samples (this Chapter, Section 3.D) may mask any difference that could be related to the time of year.

Douglas et al. (1979) report an extensive sampling program for Foraminifera in the borderland of the Southern California Bight down to 1870 m. Sampling was achieved during March (and February) and September (and August). No clear seasonal pattern was detected in foraminiferal standing stocks.

TABLE VI
Time Series of Samples from Six Biogas Stations[a]

Station	1	2	3	4	5	6
Biogas 1: October 1972	—	419[b]	—	—	171	—
Biogas 2: April 1973	478	536	—	—	—	—
Biogas 3: October 1973	709	367	395	213	105	379
Biogas 4: February 1974	407	142	258	—	—	531
Biogas 5: June 1974	671	328	162	(16)	—	464
Biogas 6: October 1974	—	584	373	264	(7)	706
Mean	566	396	297	238	138	520
Standard deviation	±147	±158	±108			±139
Standard deviation (%)	26	40	36			27
Depth range (m)	2035 to 2235	2690 to 3370	4096 to 4300	4550 to 4725	4315 to 4464	1920 to 2480
Depth difference (m)	200	680	204	175	149	560

[a] From Dinet and Vivier (1977) with mean, standard deviation, and depth ranges added (10-cm^2 subsamples, 50-μm sieve).
[b] Corrected according to data list on page 95 in Dinet and Vivier (1977).

In Sections 4 and 5 quantitative data are compared as though fluctuations within a year or between years do not exist.

D. *Faunal Density and Diversity*

Most publications on deep-sea environments discuss meiofauna abundance on the level of higher taxa. A few papers consider diversity of Harpacticoidea (Coull, 1972; Hessler and Jumars, 1974; Thistle, 1978, 1979), Nematoda (Tietjen, 1976; Dinet and Vivier, 1977), and live Foraminifera [Buzas and Gibson (1969), "living assemblages give the same pattern" as "living and dead" (Buzas, 1972; Bernstein et al., 1978, Bernstein and Meador, 1979, Douglas et al., 1979, Douglas and Woodruff, 1981)] on the species level. The general finding of these authors is an increase in diversity with water depth, with diversity defined as number of species relative to number of individuals and sample area. Food limitation, as expected, decreases the diversity through competition for energy resources (Thiel, 1975), and a reduction was found in abyssal plains for cumaceans (Jones and Sanders, 1972) and for bivalves and gastropods (Rex, 1973). Along two transects off North Carolina, Tietjen (1976) discovered a decrease of diversity with increasing depth, which he attributed to sediment types and species richness. Sandy habitats are regarded to have a higher number of microhabitats, allowing for a higher diversity compared to clay–silt sediments at greater depth. But within-habitat diversity off North Carolina is higher than in the shallow mud sediments of Long Island Sound.

Diversity in the deep sea is discussed by Rex (this volume, Chapter 11).

Species diversity is thought to develop and be maintained through a high degree of physical stability (Hessler and Sanders, 1967; Sanders, 1968, 1969), small-scale biological disturbances (Dayton and Hessler, 1972), predation and productivity (Rex, 1976), and microhabitat structuring (Jumars, 1976; Thistle, 1978, 1979). The

Fig. 4. Intensively structured sediment surface demonstrating variability in space and time (Iberian Deep Sea Basin, 5170-m depth, photosled transect 58 run by Dr. Gerd Schriever, Kiel, 1980). Starfish imprint, holothurian walking track, feeding traces, lumps, and tubes.

scales of diversity are discussed by the last two authors, who conclude that influences are discernible on distances of hundred meters, meters, and in the centimeter range. Thistle (1978, 1979) measured structural components on the 10-cm scale and found distances or sample size on the order of 10 × 10 cm to be too large. This indicates for the meiofauna a small-scale system, which corresponds to their small size and even to their presumably small ambit (Jumars, 1975b). Deep-sea photographs (Figs. 4 and 5) show a highly diverse small-scale structured sediment surface. The relative stability of the physical environment, the low density of organisms, and the slow biological rates together set time scales that allow the meiofauna to adapt to the high small-scale habitat structure. Species density cannot decrease beyond a level that allows for occasional meetings of

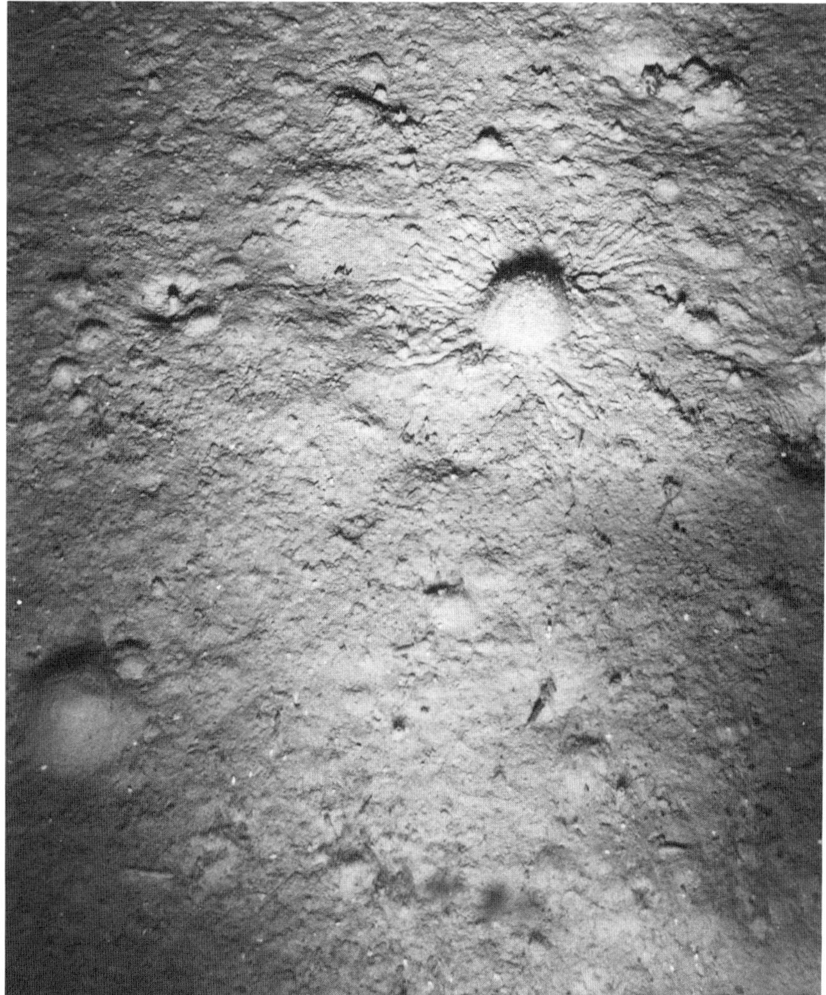

Fig. 5. Intensively structured sediment surface demonstrating variability in space and time (Iberian Deep Sea Basin, 5170-m depth, photosled transect 58 run by Dr. Gerd Schriever, Kiel, 1980). A xenophyophorian (?) mound and starlike feeding tracks with lumps and tubes.

individuals for reproductive purposes, and community density is limited by energy availability. Thus the low standing stock, the high diversity, and the deep-sea time scales seem to predict that variability of faunal densities would be high on rather small scales of space.

E. *Horizontal, Small-, and Medium-Scale Distribution*

The evaluation of quantitative data on meio- and nanobenthos densities (this Chapter, Section 4) deserves discussion relative to the value of single and replicate subsamples and samples or to the variability within grabs (between subsamples) and between grabs, respectively. Small- and medium-scale variability in faunal densities must be seen in relation to the large-scale ecological understanding at which this research aims.

Although the amount of time necessary for sample sorting generally does not allow replication, some replication has been attempted. Whereas Dinet (1973, 1979) sorted three replicates from each sample, Coull et al. (1977) worked up six 10-cm^2 meiostecher subsamples per sample. Data on ten subsamples of 25 cm^2 from one box grab (Thiel, unpublished data) and subsamples of 0.9 cm^2 for nematodes (Burnett, 1979b) are also available.

Standard deviations (Table VII) range between 4.6 and 68% of the mean. On a small scale within areas of 600 cm^2 (see works by Dinet and Thiel) and 200 cm^2 (Coull et al., 1977), that is, between the subsamples, no regularities in standard deviation according to faunal densities or water depth become apparent.

Coull et al. (1977) compared the variability found within grabs with that encountered between grabs of the same station. Of the total station variance, 85.7% was due to within grab variance, whereas 14.3% was due to between-grab variance.

No other data for comparable calculations are available because of the lack of replicate samples from a single station and replicate subsamples isolated from these samples. Most sample series were taken along depth transects and thus are not suitable for a comparable analysis. However, the same statistics can be applied to a series of six box grab samples collected from the Iberian deep sea (Thiel, 1972b) evaluating the deep-sea plain as a rather uniform region. The most distant grabs were about 60 nautical miles apart, but the depth change was only 68 m, or less than 1%. Two subsamples were taken from five of the grabs, with the last sample cored just once (Table VIII). After $\log_{10}$ transformation the variance components are 66% for within-grab variance and 34% for between-grab variance. Thus small-scale variance is about twice the medium-scale variance on the total faunal level.

The patchy distribution of Foraminifera was studied in five 0.25-m^2 cores from the abyssal (5500- to 5800-m) central northern Pacific by Bernstein et al. (1978) for live and dead components combined; from one of these grab samples, the live unit Foraminifera were counted (Table VII) (Bernstein and Meador, 1979). Both these investigations exhibit strong patchiness on the within-core scale of centimeters and on the between-core scale in the order of kilometers. For 12 of 22 species, a positive within-species correlation was found for live and dead specimens, suggesting that patchiness persists for at least more than one generation. Additionally, the Foraminifera can be regarded as structuring the habitat of meio- and nanobenthal organisms, partly causing their small-scale distributional patterns.

Other sample series are not suitable for variance analysis because environmental differences do not allow pooling of the samples or because of the lack of replicate subsamples. Results so far indicate a higher small-scale or within-grab variance than the medium-scale or between-grab variance. Douglas et al. (1979) have taken replicate box cores (southern California Borderland to 1870 m). From each grab one sample of 60 to 120 cm^2 was sorted for live Foraminifera, and the replicates showed a high degree of variability through all the investigated depth range.

Comparing quantitative meiofauna data on ocean-wide scales, one must consider these limitations and the fact that variances are rather high. Data evaluation, therefore, is a challenge, but effective explanations of differences in faunal densities within and between most sample series can be made with some confidence.

TABLE VII

Arithmetic Mean ($\bar{x}$), Standard Deviation (s), and Coefficient of Variation [$V = (s/\bar{x})100$] for Replicate Subsamples from Each One-Box Core

Station or Grab Number	Depth (m)	Number of Subsamples	Arithmetic Mean $\bar{x}$	Standard Deviation s	Percentage of Standard Deviation V
Dinet (1973), Southern Atlantic—Walvis Ridge: Area 10 cm², Depth 4 cm					
04	4100	3	496	±192	±39
06	2800	3	909	±146	±16
08	3690	3	646	±175	±27
09	4660	3	302	±55	±18
10	1140	3	366	±75	±21
Dinet (1976), Mediterranean: Area 10 cm², Depth 4 cm					
KR 20	1069	3	86	±12	±14
KR 21	1209	3	98	±12	±12
KR 25	459	3	204	±9	±5
Coull et al. (1977), Northern Atlantic—North Carolina: Area 10 cm², Depth 6 cm					
301	400	6	581	±208	±36
302	400	5	322	±101	±31

303	400	6	402	± 140	± 35
305	400	4	446	± 123	± 28
326	800	3	593	± 142	± 24
329	800	4	1116	± 192	± 17
316	4000	4	78	± 27	± 34
320	4000	4	70	± 47	± 68
Burnett (1979b), Central Northern Pacific: Area 2.8 cm^2, Depth 4 to 6 cm					
H 288	5853	5	20	± 5	± 25
Thiel (unpublished), Mediterranean: Area 25 cm^2, Depth 2 cm					
KG 456	2701	10	550	± 93	± 17
Thiel (unpublished), Central Pacific: Area 10 cm^2, Depth 2 cm					
KG 482	5120	5	50	± 20	± 40
Thiel (unpublished), Philippine Trench: Area 5 cm^2, Depth 4 cm					
M 210	9807	4	79	± 31	± 39
Bernstein and Meador (1979), North Pacific Gyre: Area 100 cm^2, Depth 2 cm					
H 153	5300	24	15	± 6	± 39

TABLE VIII
Meiofauna Counts (25 cm^2, 4-cm Depth) from 11 Subsamples of Six Box Grabs from Iberian Deep Sea

Box Grab Number	Depth (m)	Meiofauna Counts	
		Replicate 1	Replicate 2
89	5325	340	635
91	5335	363	—
92	5340	405	209
93	5305	529	525
94	5320	579	450
95	5272	655	665

4. Meiobenthos and Nanobenthos Abundances

Knowledge of the quantitative distribution of meiofauna in deep-sea environments, summarized by Thiel (1975, 1979a) allows some general conclusions on the dependence of meiofauna on ecological factors and on the importance of the meiofauna in the deep benthic ecosystem. So little is known of the nanobenthos that their evaluation must remain mostly speculative.

A. *Comparability of Quantitative Data*

Ecological evaluation of the results obtained on an ocean-wide scale depends on the comparability of the available data. The methods of different authors are often not described well enough, but we must accept published values as comparative data. In some cases differences are apparent that may be related to sample processing. For example, the results reported by Dinet et al. (1973) on nine samples from the deep western Mediterranean show meiofaunal densities a factor of 3 to 4 lower than my own results (Fig. 10) from an area close to their sampling site. Such a difference is not to be expected for a distance of some 50 nautical miles and a minor depth increase. According to Dinet (personal communication), sample processing was not thorough during their earlier deep-sea meiofauna studies.

Another collection of meiofauna samples (Rachor, 1975; Thiel, 1975) allows a better comparison. Some of the subsamples were actually taken from the same five grabs with the same type of subcorer, the meiostecher (Table IX). Differences of such an order of magnitude by far exceed natural variability in faunal densities (see Section 3.E) and can be explained only by different processing methods since sampling and subsampling were identical. Whereas Rachor used 50-μm mesh gauze as a lower limit, I used 42 μm, yet 30 to 90% of the fauna would not be washed through the 50-μm mesh but retained on the 42 μm mesh. The main loss in Rachor's subsamples probably occurred during its dilution with seawater and subsequent decanting, a process that was repeated at least eight times for each single subsample. A further loss may have occurred during the sorting process, which almost always happens. As pointed out above, all meiofaunal counts must

TABLE IX

Meiofauna Densities per 10 cm^2 and 4-cm Sediment Depth from Subsamples Isolated from the Same Grabs During *Meteor* Cruise 19, 1970 by Rachor and Thiel

Station Number	Depth (m)	Meiofaunal Counts 10 cm^{-2}			Difference Factor
		Rachor	Thiel		
19-238	1469	86	468 674	571	6.6
19-234	1689	29		304	10.5
19-232	2140	56	574 345	459	8.2
19-263	3809	62		192	3.1
19-223	5111	9		163	11.6

be regarded as low figures, as well those presented in Table IX from my own studies. Evaluation of data in this manner is obligatory to the understanding and interpretation of faunal distribution.

B. *Meiofauna*

Meiofauna density data can be grouped into two categories: samples from locations in central oceanic regions and from those transects that cross continental margins down onto abyssal plains (see Addendum 3).

Western Atlantic Depth Transects (Fig. 6)

The first quantitative data published from beyond the shelf were those due to Wigley and McIntyre (1964) from a transect off Martha's Vineyard (island off Massachusetts coast) about 40°N. These can be compared with two transects in 33 to 34°N worked by Tietjen (1971) and another one in the same general area by Coull et al. (1977). Differences in methods make full comparison difficult. Wigley and McIntyre washed their samples through a lower screen mesh size of 74 μm, whereas Tietjen and Coull et al. deployed 44 and 42 μm, respectively. The low figures reported by Wigley and McIntyre may be explained at least partly by this difference, but sediment type and food availability may also play a role. Whereas Tietjen's meiofaunal counts are not much higher than those of Wigley and McIntyre, those by Coull et al. resulted in much higher values at 400 and 800 m. This may be due to the sampling methods. Coull et al. used a small box grab and the meiostecher for subsampling, whereas Tietjen took his samples with a gravity corer and a box dredge. This may as well explain his much higher values for the Foraminifera above all other meiofauna, but these are not out of the order of magnitude for total meiofauna given by Coull et al. Both papers exhibit rather low densities in 400- to 500-m depth, and Tietjen correlates this with sediment type and food income. The Gulf Stream occurs at this water depth (Newton et al., 1971), causing sediment erosion and possibly resulting in less food particle sedimentation. Coull et al. describe the sediment in 400-m depth as fine sand with

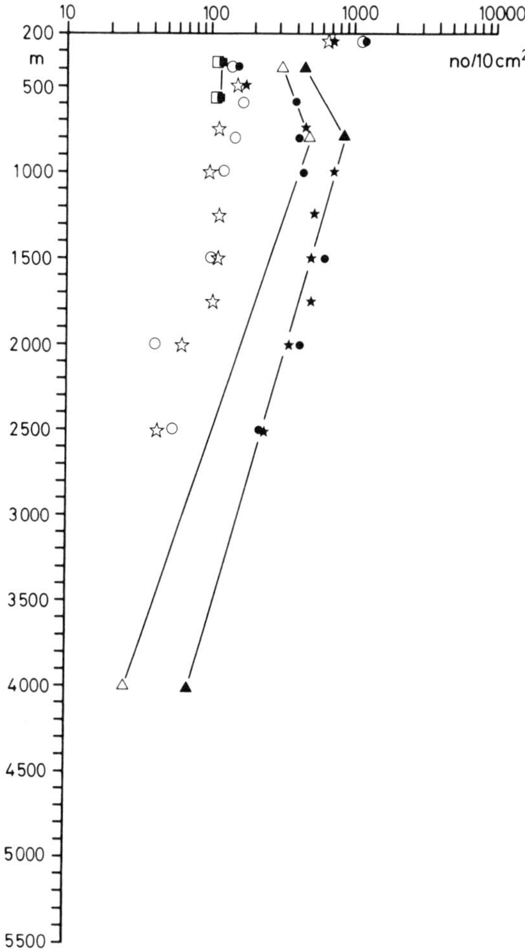

Fig. 6. Western Atlantic depth transects: number of individuals/10 cm^2 sediment surface: □—□ meiofauna and ■—■ meiofauna plus Foraminifera from off Massachusetts (Wigley and McIntyre, 1964); ☆ ☆, ○ ○ meiofauna and ★ ★, ● ● meiofauna plus Foraminifera from two transects off North Carolina (Tietjen, 1971); △—△ meiofauna and ▲—▲ meiofauna and Foraminifera off North Carolina (Coull et al., 1977).

17.7% silt–clay, whereas at 800 m a fine silt with 89.8% silt–clay components indicates a strong difference in the sedimentary regime (see Rowe, this volume, Chapter 3, for upper-slope mud–sand line).

Eastern Atlantic Depth Transects: Portugal, Morocco, and Bay of Biscay (Fig. 7)

Three transects were sampled off the coasts of Morocco (33°30′N) and Portugal (37°N) during *Meteor* cruise 8 (1967) and *Meteor* cruise 19 (1970). The latter was independently processed by Rachor (1975) and Thiel (1975). Differences in those *Meteor* 19 results are presumably due to processing methods applied (see Section 4.A). I regard values presented by Rachor as too low. In deep water the two

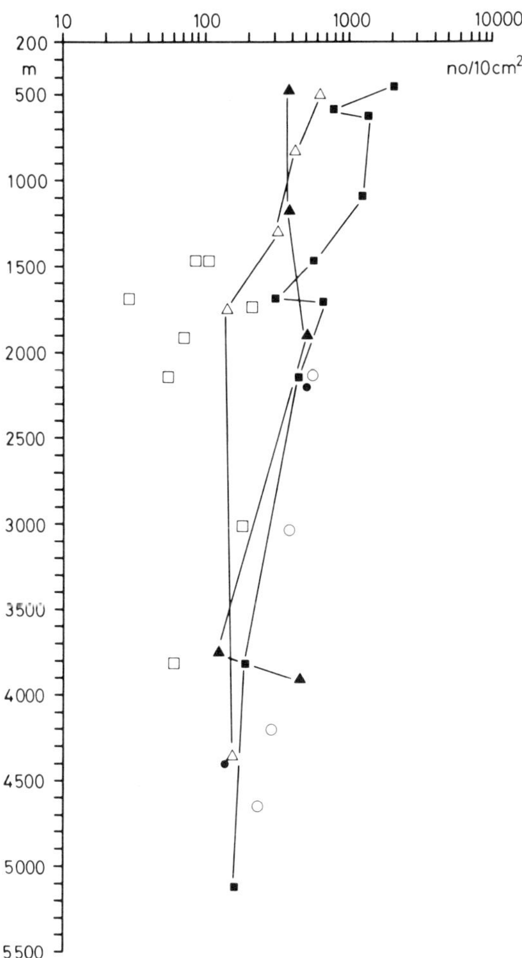

Fig. 7. East Atlantic depth transects: number of meiofauna/10 cm^2 sediment surface: ○ ○ Golfe de Gascogne, Stations 1 through 4 (Dinet and Vivier, 1977); ● ● Golfe de Gascogne, Stations 5 through 6 (Dinet and Vivier, 1977); □ □ off Portugal (Rachor, 1975); ▲—▲ off Portugal, 1967; ■—■ off Portugal, 1970; △—△ off Morocco, 1967.

Meteor 8 and 19 transects off Portugal were about the same, but differences in the upper 1200 m are difficult to explain.

Excluding the two stations in 580-m and 1160-m depth from off Portugal (*Meteor* 8 cruise), data from off Morocco are much lower than those from Portugal. One explanation for this discrepancy is that primary production off Portugal is higher as a result of upwelling and river runoff, as the latter is a constant and powerful source of nutrients off Portugal by the River Tejo (Thiel, 1978a). The question as to whether within this transect the two minima are real must remain open.

Dinet and Vivier (1977) present data from repeated sampling over 2 years in the Bay of Biscay (Golfe de Gascogne). Because seasonal variation was not discovered so far (see Section 3.C), two to six values (Dinet and Vivier, 1977, Table 1) based on several subsamples from each single station were averaged, but omitting two very low, obviously biased figures. The results fall well within the range of the respective depths off Portugal, with the four northern stations somewhat higher than the two southern ones.

Comparing the eastern Atlantic with the western Atlantic (Fig. 6), one must consider the fact that the Foraminifera are not included in Fig. 7. Wigley and McIntyre's values for the meiofauna are very low, whereas Tietjen's counts, although high at 250 m depth, are in general rather low. The numbers for *Meteor* 19 are high, with the maximum of 2046/10 cm^2 surface area at 461 m depth. If the Foraminifera (not included in Fig. 7) are counted together with the meiofauna, they reach more than 20% of the total number in the depths of 461 and 481 m off Portugal and their percentage is less than 10% in deeper waters. This does not change the general trend in meiofaunal distribution for these transects.

Differences in primary production of the northern Atlantic are the likely sources for the differing standing stocks. This is difficult to assess, however. Data are available in a computerized outprint (Bunt, 1975), giving 50 to 100 g C/m^2 · yr for off North Carolina, 100 to 200 g C/m^2 · yr for off Portugal and the Bay of Biscay, and the same with some indication of 200 to 400 g C/m^2 · yr for the northwestern African upwelling zone. Regardless of whether these figures are correct, the trend in productivity levels is the same as that for meiofaunal standing stocks and may help to explain the differences.

Eastern Atlantic Depth Transects: Northwestern African Upwelling Region (Fig. 8)

Upwelling areas are characterized by a high primary production, and high benthos standing stocks (and production) are to be expected (Thiel, 1978a, 1982). For the central northwestern African upwelling area, primary production is 300 g carbon/m^2 · yr^{-1} (Schemainda and Nehring, 1975). Comparing Figs. 7 and 8 (*Meteor* cruise 26, 1972), we see that the standing stock of meiofauna in the upwelling region (21°N) is not much higher than in the Portugal transect from *Meteor* 19 and that not much change is introduced when Foraminifera are included, accounting for 10 to 20% of total counts. This similarity may be due to local food income off Portugal or to high competition for food resources with macrofauna off Mauritania (this Chapter, Section 5.C, subsection on biotic interactions). Between the transects off northwestern Africa a decrease is observed from the upwelling area to the Morocco transect (Fig. 7).

A striking feature in the upwelling transects is the increase of meiofaunal numbers with depth. Values higher than those figured for 200 m (Fig. 8) were found on the shelf (Thiel, 1978a, 1979a), and for the upper continental slope a pronounced minimum was discovered with a subsequent maximum in a depth of 445 m, including the Foraminifera or in 810 m without them. Similarly, the other transect exhibits a maximum in depths of 800 to 1175 m. This pattern is a deviation from the generally accepted decrease of standing stocks with depth. Thiel (1978a) related the upper slope minimum to the undercurrent with a maximum of 10 cm/sec (Mittelstaedt, 1976). Other processes involved are discussed by Thiel (1978b, 1979a). Fahrbach and Meincke (1978) describe internal waves, and tidal current

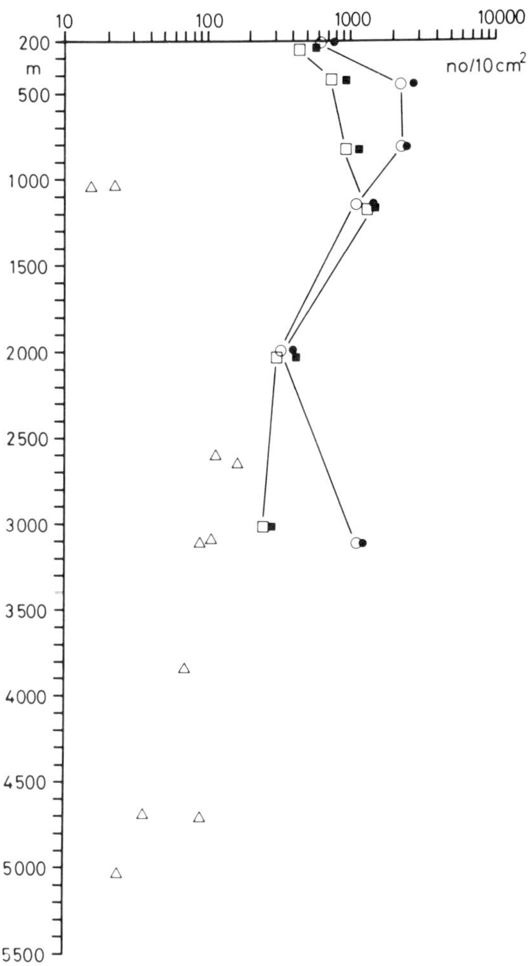

Fig. 8. Eastern Atlantic and Indian Ocean depth transects, upwelling areas; number of individuals/10 cm^2 sediment surface: □—□, ○—○ meiofauna and ■ ■, ● ● meiofauna plus Foraminifera from two transects off Mauritania; △ △ off Somalia, Indian Ocean.

velocities of 30 cm/sec and up to 50 cm/sec are recorded 8 to 10 m above the bottom in the data reports by Meincke et al. (1975) and Brockmann et al. (1977).

Shear stress at such current velocities is probably sufficient to cause erosion. The light organic particles, food material for the benthos, are partly prevented from settling to the bottom and others are resuspended from the seabed. An additional indication for restricted particle settlement is the high standing stock of macro- and megafaunal filter feeders, where the meiofauna is at its minimum (Thiel, 1978b).

Further evidence is presented by the downslope water transport reported by Brockmann et al. (1977) and Mittelstaedt (1976). Bein and Fütterer (1977) and Diester-Haass and Müller (1979) describe geologic indications of a downslope sediment transport, and Kullenberg (1978) measured a higher turbidity in the bottom water layer than in the overlying water masses. All these observations help to explain the high concentrations of biologically produced matter such as organic carbon (Diester-Haass, 1978) and chloroplastic pigment equivalents determined in the sediments (Thiel, 1978a, 1982).

Consistency in physical, geologic, and biological results allows the conclusion that meiofaunal densities again are closely related to the food conditions. Organic matter is transported downslope away from the upper zone, nourishing all types of organism in deeper waters. An example is the large actinian *Actinoscyphia aurelia,* which was dredged in large amounts between 1000- and 2000-m depth (Aldred et al., 1979) beyond the central upwelling off Mauritania.

Northern Atlantic and Norwegian Sea Depth Transects (Fig. 9)

Four transects were worked out in the northeastern North Atlantic and the Norwegian Sea. Not readily understood are the transects off northern Norway and south of Spitzbergen (*Anton Dohrn* cruise 91, 1965), covering only four and five samples, respectively. Both series exhibit high faunal densities with a maximum in 1200- to 1500-m depth. A relation to current activities and food input is not apparent, possibly because of our limited knowledge. Relatively high plankton death rates and high detritus concentrations can be expected at the polar front between the Atlantic and the Arctic waters in the Barents Sea, which may result in a good food source for slope organisms, and the low temperatures of the deep water may effect food preservation and its long-term availability. High benthic standing stock in the Antarctic Ocean is reported by Dayton and Oliver (1977) from shallow waters rich in food materials.

The other two transects combined cross the Iceland–Faroe Ridge from 2500 m in the northern Atlantic to 1800 m in the Norwegian Sea (*Anton Dohrn* cruise 98, 1966). They are separated into two units because they run from the sill of the ridge down to greater depth in different directions, exhibit highly different faunal densities, and are situated under different hydrographic and sedimentary regimes (Thiel, 1971, 1972a).

Both slopes of the Iceland–Faroe Ridge are governed by the interoceanic water transport, resulting in the overflow of cold water with a higher density from the Norwegian Sea into the Atlantic Ocean (Lynn and Reid, 1968). Water masses originating from the northern Atlantic and cooling off in the Greenland Sea or further south (Peterson and Rooth, 1976) sink into the deep basin, which is bordered by the Scandinavian and the Greenlandic land masses and sealed off from the Atlantic by the Greenland–Iceland–Faroe–Shetland sill. The slowly accumulating water masses in the Greenland and Norwegian Seas deep layers spill over the sill into the Atlantic Ocean. The Norwegian Sea slope is governed by this water mass, characterized by temperatures of less than 0°C beyond 500-m depth (Dietrich, 1967) and by very little current activities; thus the Norwegian Sea slope belongs to a quiet sedimentary regime. On the contrary, the Atlantic slope is strongly influenced by water masses mixed from deep Norwegian Sea water and from Atlantic Ocean surface water moving north from the Atlantic Ocean into the Norwegian Sea. The near-bottom overflow is an intermittent process, and current velocities reaching 50 cm/sec do not allow deposition of food particles or resuspension of newly settled materials will occur.

Observations, with the exception of meiofauna, made during the *Anton Dohrn*

TABLE X
Trends in Organism Densities and Abiotic Conditions for Slopes of Iceland–Faroe Ridge According to Different Authors

	Atlantic Slope	Norwegian Sea Slope	Source[a]
Number of macrofauna per square meter from 10 replicates at			
600 m	590	1208	K, T
1000 m	280	988	
1500 m	253	644	
1800 m	218	1117	
Range of macrofauna wet weight per square meter in (g)	1.4 to 3.0	3.7 to 18.0	K, T
Fungi	More	Less	G
Bacteria	Less	More	B
pH in sediment surface	8.0 to 8.2	7.7 to 7.9	U
Sorting coefficient $\sqrt{75/25}$	1.27 to 1.84	2.21 to 3.67	T
Sediment water content (%)	25 to 35	45 to 70	U

[a]Code: B = Bansemir (1969); G = Gaertner (1968); K = Knorr (1969); T = Thiel (1971); U = unpublished.

cruise 98 in 1966 are not numerous (Table X), and their value should not be overestimated. Plate counts for bacteria have their limitations, and pH measurements made in box grab samples are relative values only, although general trends are well exhibited and allow some generalization. Sediment properties and faunal densities characterize the Norwegian Sea slope as a high sedimentary regime. Organism remains, which are a good food source, originate in the southeast of Iceland, where several water bodies meet (Dietrich, 1967; Meincke, 1978): the deep water of the Norwegian Sea; the northern Iceland winter water with arctic intermediate water; the eastern Iceland water; and the northern Atlantic surface water. The mixing of these water masses creates considerable stress on organisms, and many of them are not able to withstand the changing conditions and die, which results in high detritus concentrations (Körte, 1966). A high standing stock of macro- and meiofauna is supported by the organic materials settling out. Low metabolic rates resulting from the very low temperatures may tend to preserve organic matter and also result indirectly in high faunal densities.

The Atlantic slope is adversely affected by intermittent strong currents, net sediment deposition is low, and it supports much less fauna.

According to Hansen and Meincke (1979), meandering and tidal currents influence the sill of the Iceland–Faroe Ridge. These may prohibit settling of food materials, and thus animal populations at sill depth are sparse.

In addition to the low faunal numbers on the Atlantic slope, a pronounced minimum is found in 1000 m (Thiel, 1971), which is the depth of the flow of cold Norwegian Sea water through the Faroe Bank Channel, passing east and south around the Faroe Islands. Leaving this channel, this water mass can be traced

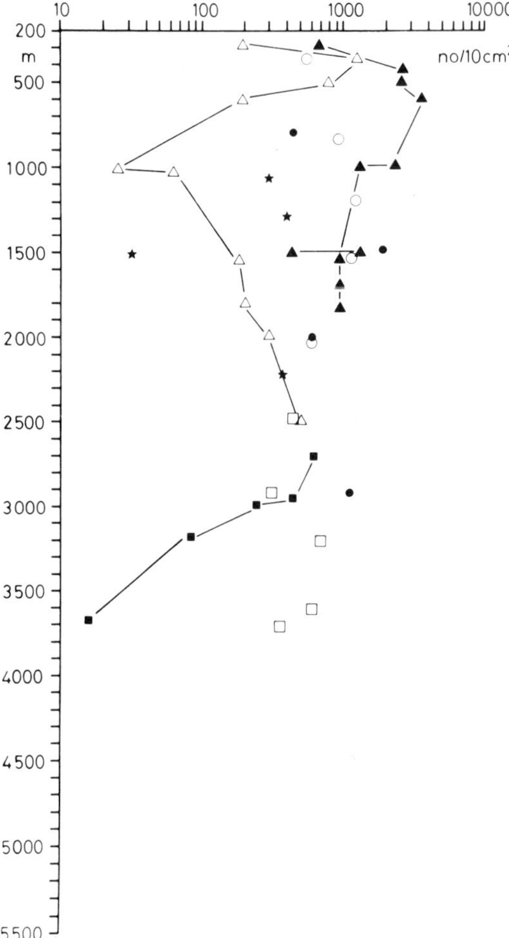

Fig. 9. Northern Atlantic and Norwegian Sea depth transects; number of meiofauna/10 cm² sediment surface: △—△ Iceland–Faroe Ridge, Atlantic slope transect; ▲—▲ Iceland–Faroe Ridge, Norwegian Sea slope transect; ★ ★ Iceland–Faroe Ridge, north of Scotland (Heip, Herman, Vanosmael, and Vincx, personal communication); ● ● northern Norway transect; ○ ○ southern Svalboard transect; ■—■ Norwegian Sea (Dinet, 1979); □ □ Greenland Sea (Dinet, 1979).

along the 1000-m isobath of the ridge (Dietrich, 1967). However, the minimum in faunal numbers and the water's path along the Atlantic slope may be unrelated. The subsequent gradual increase in meiofauna densities down to a depth of 2500 m indicates that sedimentation and food input to the sediment become more favorable for the fauna with increasing depth.

Unpublished counts from four deep stations sampled between Scotland and Iceland (Fig. 9) were provided by Heip, Herman, Vanosmael and Vinex (personal communication). While the station in 1064 m was taken between Scotland and the Faroe Islands, the other three were sampled from the Atlantic slope of the Iceland–Faroe Ridge. In two of the samples, faunal densities deviate strongly from those of Thiel (1971). This may be due to local influence of the current and sedimentary regimes.

Dinet (1979) discusses the results from 17 samples collected at 10 stations in the Norwegian and Greenland Seas. From each sample two subsamples were sorted, and the mean from each two or four subsamples per station is presented in Fig. 9. The data for the Norwegian–Lofoten and the Spitsbergen–Greenland Basins give the impression of depth transects, but they cover the length of the four basins, respectively (Dahl et al., 1976). Very low meiofaunal values were discovered in the two stations of the Norwegian–Lofoten Basins deeper than 3000 m. Compared to the deep stations of the Spitsbergen and Greenland Basins, the difference appears to be extremely high. Dinet (1979) attributed the low values to a higher degradation of organic matter in the water column and on the seabed due to the highly oxygenated water of the Norwegian Basin, but the same should apply to the Spitsbergen–Greenland Basins. It seems to be more likely that the downward transport of the cooled water in the Greenland Sea is responsible for a higher food concentration at the sediment–water interface.

Mediterranean Sea Depth Transects and Deep Basin (Fig. 10)

Six successful grab samples were taken across the Gibraltar Ridge during *Meteor* cruise 21 (1970) for a comparison to the Iceland–Faroe Ridge transects. Two samples were collected in the Atlantic Ocean with low densities, whereas those four from the Mediterranean Sea are either much higher or deeper (Thiel, 1975). The likely explanation is similar to that for the Iceland–Faroe Ridge. Mediterranean Sea water, which has a higher density, crosses the Gibraltar Ridge, leaving a quiet sedimentary basin behind. Because of its high density, it rushes down the Atlantic slope at up to 100 cm/sec, creating eroding conditions (Heezen and Hollister, 1971; Zenk, 1971).

The meiofaunal counts presented by Dinet et al. (1973) seem to be rather low and are regarded as biased by the methods (Dinet, personal communication). Dinet (1976) presents some data for the Aegean Sea from bottoms between 459- and 1209-m depth. These are rather low in comparison to Atlantic samples but generally higher than in the deep samples from the Mediterranean. Primary production is low throughout this sea, although the Black Sea water is believed to transport organic matter into this area. However, the samples were also taken close to land masses, and the discharge of river water and land runoff may be of additional importance for food delivery into this environment.

A further series of 21 samples was reported by Vitiello and Vivier (1974) and Vivier (1975, 1978a) in the Canyon de Cassidaigne to study the influence of industrial red mud discharged from an outflow pipe in 330-m depth. Density of meiofauna ranged between 117 and 801 individuals/10 cm^2 down to a sediment depth of 7 cm between water depths of 245 and 810 m. Obviously, the red mud settles out in the nearby deeper parts of the canyon, where two stations (not included in Fig. 10) are devoid and depauperate in meiofauna, whereas numbers around 100 individuals/10 cm^2 are only partly influenced by the discharge.

Four rather low values are given by Soyer (1971) for depths between 250 and 500 m with only 43 to 60 specimens per 10 cm^2. His samples were washed through 88-μm mesh sieves, and this may partly account for the low contents.

In summary, the densities of meiofauna from deep-sea stations in the Mediterranean are lower than those in other areas. This was to be expected because of the generally low primary productivity (50 to 100 g $C/m^2 \cdot yr$) of Mediterranean waters

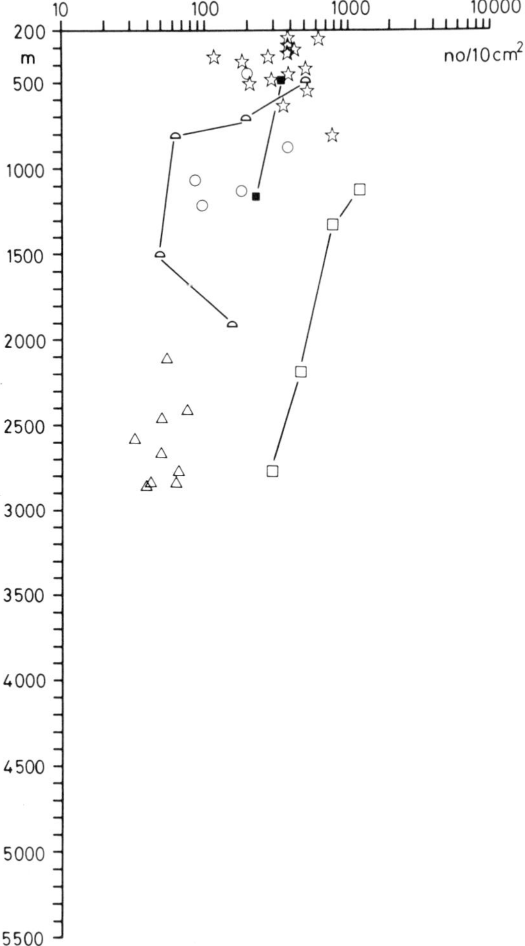

Fig. 10. Mediterranean Sea and Red Sea; meiofauna/10 cm²: △ △ western Mediterranean Sea (Dinet et al., 1973); ○ ○ Egean Sea (Dinet, 1976); ☆ ☆ northwestern Mediterranean Sea (Vivier, 1975, 1978a); □—□ Gibraltar–western Mediterranean slope transect; ■—■ Gibraltar–Atlantic slope transect; ⌓—⌓ central Red Sea depth transect.

(Sournia, 1973; Bunt, 1975; Finenko, 1978) and because of presumably higher degradation and regeneration rates that result from the high temperatures of 13.2°C down to the seabed (this Chapter, Section 5.B).

Red Sea Depth Transect (Fig. 10)

A few quantitative box grab samples were taken in the Red Sea during *Sonne* cruise 2, 1977 (Thiel, 1980a), and first results are presented by Thiel (1979b). The meiofaunal densities encountered in depths between 507 and 1977 m range between those of the Mediterranean, and at intermediate depths they are lower and can be compared with central oceanic regions, although the depth is much less. My explanation (Thiel, 1979b) is again that the fertility of the sea is extremely low, limited through nutrient availability for phytoplankton growth (Halim, 1969). The Red Sea is a narrow basin, and no position is farther from the continents than 135 km. However, organic matter production on the bordering land masses is extremely low, land runoff is limited and aperiodic, and it could deliver little suitable foodstuff from the desert regions. The higher faunal densities were encountered near to the coast, where a steep slope descends from the highly productive coral reefs down to a depth of 500 m. Some organic remains from the reef biocoenoses will be transported downslope and can be expected to concentrate at the slope base but will not accumulate far out on the broad terrace located between about 500- and 900-m depth. This is demonstrated by a strong decrease in meiofaunal densities in the samples collected from this terrace. A comparatively high value was found again in the deepest sample from the central graben of 1977 m. I assume that more organic matter becomes trapped there, as in oceanic trenches, than on the terrace or on the steep graben slopes. Even debris from the ship traffic concentrating along the central axis of the Red Sea (Thiel, personal observation) could be a food source, as evidenced by the abundance of coal and cinders in all trawl samples (Thiel, 1979b).

In addition, degradation of organic matter would be high in the warm surface (28 to 38°C) and deep (21.5°C) waters. Small organic particles will not sink deep before advanced or even complete disintegration. The Red Sea, with its high temperatures, can be considered as an "energy-wasting" environment. It stands in contrast with the "energy-saving" environment of the Norwegian Sea, which has low temperatures. This results from metabolic activities adversely directed by warm and cold environmental conditions, respectively (this Chapter, Section 5.B).

Indian Ocean Depth Transect (Fig. 8)

The results from the first abyssal meiofauna study, which happened to be done in the Indian Ocean on *Meteor* cruise 1 1964 to 1965 (Thiel, 1966), are added to ensure thoroughness, but I do this with some hesitation. Comparability with other results is limited by possible sampling and sorting effects, realized only after later experiences. The lower size limit was 65 μm, reducing the standing stock by about 10 to 25%. Regarding the values as comparable within this series, the low counts at 1045 and 1050 m do not fall into the otherwise decreasing sequence in this transect. It was suggested (Thiel, 1966) that low oxygen levels may occur in the sediment surface layer in corresponding depths with the oxygen minimum zone in the water column. Dietrich et al. (1966) report oxygen values of 0.65 to 0.93 ml/liter for 1000-m depth and of 0.96 to 1.13 ml/liter for 1100-m depth. These oxygen minimum values are located near the station where the low meiofauna densities were encountered. Deeper down the oxygen content increases to 2.4 to 2.6 ml/liter in 1700-m depth, obviously not influencing meiofaunal densities anymore.

The Indian Ocean off Somalia is governed by the monsoon winds, creating upwelling conditions. This prevails for only 6 months, and after the monsoon season the wind blows from the sea and productivity diminishes. Under these conditions standing stocks are expected to be lower than off northwestern Africa, where the winds go offshore and upwelling prevails for longer periods of the year (Thiel, 1978a, 1982).

Central Oceanic Regions: Abyssal (Fig. 11)

Whereas the depth transects partly reach down into the abyssal plains, some special sampling programs have considered the deep basin areas away from conti-

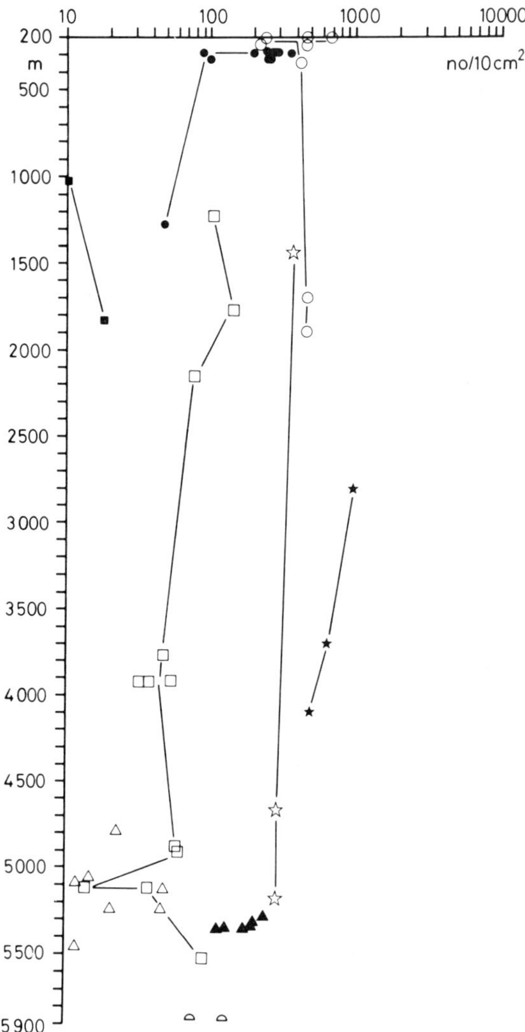

Fig. 11. Central oceanic regions; numbers of meiofauna/10 cm^2: ●—● plateau and slope of Great Meteor Seamount; ○—○ plateau and slope of Josephine Seamount; ■—■ steep southern slope of Josephine Seamount (Rachor, 1975); □—□ Tejo-, Horseshoe, and other eastern Atlantic deep-sea areas (Rachor, 1975); ☆—☆ southern Atlantic north of Walvis Ridge (Dinet, 1973); ★—★ southern Atlantic south of Walvis Ridge (Dinet, 1973); ▲ ▲ Iberian deep-sea basin (Thiel, 1972b); △ △ central Pacific Ocean; ⌓ ⌓ central northern Pacific gyre area (Burnett, 1979b).

nental margins. Most samples were collected from the northeastern Atlantic. During *Meteor* 3 (1966), the present author (Thiel, 1972b) obtained six grab samples and sorted 11 subsamples from the Iberian Deep Sea plain. The overall mean is 184/10 cm^2, with a standard deviation of ± 41, which again is higher than the data due to Rachor (1975) received from samples collected farther south in the Tejo and Horseshoe plains, and in the deep-sea areas between Josephine Seamount and Madeira and west of this line in the Canary Basin. All of his values seem to be too low (see Section 4.A), except the deepest one from the Canary

Basin, which, however, is the most central Atlantic and a low density was to be expected.

Dinet (1973) has sampled on both sides of the Walvis Ridge in depths between 1400 and 5200 m. The densities of meiofauna he encountered in the southern three stations are about two times higher than north of the ridge and show a decline with depth, whereas the three northern samples are rather similar to each other. Dinet explains the differences found for the slopes of the Walvis Ridge in analogy to the results from the Iceland–Faroe Ridge (Thiel, 1971). Indeed, the Antarctic Bottom Water penetrates the Cape Basin and reaches north of the ridge, but this water does not spill over the ridge. Actually, the ridge separates the water masses in the Cape and the Angola Basins—the latter was Antarctic Bottom Water as well—which moved north between the mid-Atlantic Ridge and the South American continent and penetrated the Angola Basin through the Romanche Deep. In addition, station 10 at 1440-m depth seems to be too shallow to be influenced by any overflow, and station 02 in 5170-m depth should be out of the range of influence of such a phenomenon. Another explanation can be offered. The sea off southwestern Africa is governed by strong upwelling, which centers in offshore regions, somewhat south of Dinet's sampling area. Primary production is high, and organic matter is transported by the Angola Current to the north and by the prevailing winds to the northwest into the sampling area. It seems more likely that the southern stations are nearer to a rich food source. Relatively high faunal densities were found off northwestern Africa under the influence of upwelling at 3000 m (Thiel, 1978a, 1982), and the same must be assumed off southwestern Africa, with station 02 having the least faunal numbers and being the deepest as well as the most distant station (Dinet, 1973).

Two other sampling sets are included in Fig. 11, both from the Pacific Ocean and both far away from land masses. Burnett (1979b) worked beyond the central North Pacific gyre in an area chosen for its low productivity. However, his values for faunal densities (40 to 147/10 cm^2) are higher by a factor of up to 12 compared to values (12 to 50/10 cm^2) found by Thiel (1975) in the central Pacific south of Hawaii. A likely explanation for some of the differences is the sampling procedure. Burnett used a grab with a better water flow through the box, thus probably avoiding pushing away the sediment surface just before sampling. My own data are based on samples that were taken less carefully for the study of manganese nodules during *Valdivia* cruise 4 (1972). Furthermore, it should be mentioned that Burnett's data are nematode counts plus harpacticoid estimates, whereas my data additionally include single specimens of nauplii, ostracods, polychaetes, and a tardigrade.

Central Oceanic Regions: Seamount Plateaus (Fig. 11)

The meiofauna density was studied from two seamounts on *Meteor* cruise 9, 1967 (Thiel, 1970, 1975). Josephine Seamount is situated some 300 nautical miles west of southern Portugal. Its top is 170 m beyond the sea surface, and the plateau is oval and slightly sloping to the north–northwest (Rad, 1974; Thiel, 1970). Whereas this seamount is near the continent, Great Meteor Seamount is 1600 km from the African coast, 1100 km west of the Canary Islands, and 1100 km south of the Azores. Its plateau has a minimum depth of 275 m in the south, and gentle slopes extend to about 350-m depth (Rad, 1974; Thiel, 1970). Plateau values for the meiofauna from six grabs from Josephine Seamount in depths between 206 and 355 m ranged from 210 to 670 individuals/10 cm^2 without any definite depth gradient, having a mean and standard deviation of 403 ± 140/10 cm^2 (subsamples with less than 4-cm penetration depth were excluded). Two grabs were taken at 1700- and 1900-m depth to the north, where the slope is low. With 470 and 500 individuals/10 cm^2, these deep samples are as high as the plateau values, which could possibly be explained by downslope organic matter transport from the plateau. Four samples taken by Rachor (1975) to the southwest of the Josephine Seamount are much lower in faunal densities, a fact that holds for all his samples. As the southwestern slope is much steeper than the northern slope, the differences may be partly explained by a lesser food particle concentration along the steeper slope.

Subsamples from ten grabs were taken on the plateau of Great Meteor Seamount between 292- and 340-m depth and were sorted to 4-cm sediment depth. The meiofauna ranged from 87 to 346, with a mean and standard deviation of 227 ± 74 individuals/10 cm^2. Again, no gradient in plateau depths was visible. One subsample from the steep slope at 1280 m contained 48 specimens.

Mean values for the plateaus of the two seamounts are well separated, yet the single values in faunal densities show some overlap. This can possibly be explained by the position of the seamounts, as the Josephine Seamount is closer to the continent and in an area of higher primary and secondary production than the Great Meteor Seamount, which is well isolated in the central Atlantic region.

In the depth range of the two plateaus, between 170 and 350 m, only a few other data are available. On the Iceland–Faroe Ridge (Fig. 9), the values are higher or in the same range in the immediate overflow region. In the West African upwelling region (Fig. 8), the 200-m minimum of the northern transect reaches down into the plateau values, whereas others and almost all from Morocco and Portugal (Fig. 7), even those from deeper stations, exhibit higher densities. This distribution is in good agreement with the assumed general dependence of meiofauna densities on food sedimentation.

Somewhat unexpected is the relatively small difference between the shallow and the deep central oceanic regions, the seamount plateaus, and the deep-sea plains. If the means from sample groups are used, from the plateau of Great Meteor Seamount with 227 ± 74 specimens/10 cm^2 and from the Iberian deep-sea plain with 184 ± 41 specimens/10 cm^2, not much difference is exhibited for a depth range of 5000 m. The seamount lies in the central ocean in the subtropical region, whereas the deep-sea plain, more than 12° in latitude further north, is a temperate, near-continent area. Primary productivity (Ryther, 1963; Bunt, 1975; Finenko, 1978) is less than 50 g carbon/$m^2 \cdot$ yr for the area of the Great Meteor Seamount and about 100 g C/$m^2 \cdot$ yr for the waters above the Iberian deep sea. Josephine Seamount, which is only 6° in latitude right to the south of the Iberian deep-sea plain, has the same position in relation to the continent as the Iberian deep-sea plain, and the mean values of meiofaunal densities differ by little more than a factor of 2. These small differences are not adequately explained. Sinking and transport of organic matter for about 5000 m should result in its strong degradation and in more pronounced differences between faunal densities, especially since the plateaus of the seamounts are well within the range of the daily vertical plankton migration (Hesthagen, 1970). On the contrary, Kinzer (1972), Boysen et al. (1972), and Nellen (1973) demonstrated through quantitative plankton and nekton sampling above the seamount plateaus that species number and biomass were considerably less than outside the seamount area, which will at least partly explain the low meiofauna densities of the seamount plateaus. Food income through vertical migration seems to be much reduced for the large plateaus of seamounts.

Another difference between shallow and deep central oceanic regions is the stock of macrofauna and fish exerting a high feeding pressure on the meiofaunal stock of shallow waters. Dredging and trawling on the plateau of Great Meteor Seamount resulted in large catches of sea urchins migrating gregariously and of demersal fish (Ehrich, 1977) partly feeding on benthic organisms.

Additionally, anticyclonic currents above the plateau of Great Meteor Seamount (Meincke, 1971a,b) may reduce sedimentation. Horn et al. (1971) give 3 to 5 cm/sec for near-bottom velocities, and Rad (1974) as well as Stackelberg et al. (1979) found evidence for stronger currents from geologic observations, and they expect constant winnowing and stirring up of fine-grained material. This would for sure contain organic matter, which in these processes will be lost from the plateau.

Hadal Regions

A first quantitative sample from hadal regions including meiofauna with the macrofauna was taken by Jumars and Hessler (1976) at 7200-m depth in the Aleutian Trench; but since a sieve with 297-μm meshes was applied, the value of 1.3 individuals/10 cm^2 (without Foraminifera) is not comparable to other meiofauna studies. George and Higgins (1979) sampled the Brownson Deep in the Puerto Rico Trench at 8560-m depth and reported 17 specimens/10 cm^2 retained by a 62-μm mesh. The two deepest samples so far available were collected by R. R. Hessler from Scripps Institution of Oceanography during a cruise on R. V. *Washington* in 1975. Each subsample was isolated from a grab sample taken at 9807-m depth from the floor of the Philippine Trench, and they were placed at my disposal (Thiel, 1979a). Meiofauna counts were 20 and 46 individuals/10 cm^2, the same order of magnitude as found in the other two trenches and in the central Pacific, but only 0.1 to 0.25 of Burnett's figures from beyond the central northern Pacific Gyre. Additionally, although not comparable with other data, the two subsamples from the Philippine Trench contained relatively high amounts of spherical testate, agglutinating forms. With 112 and 251 individuals/10 cm^2, these presumed Foraminifera are 3 to 12 times more abundant than the meiofauna.

C. *Foraminifera*

The Foraminifera have long been a taxon utilized by geologists for stratigraphic purposes, but only 25 years ago Phleger introduced quantitative investigations on living foraminifers into geologic work. It is extremely difficult to evaluate the results from samples collected for geologic purposes because often only the surface was sampled in undefined layers. A summary on live Foraminifera (Thiel, 1975) demonstrates that only 4 to 61% of the living Foraminifera inhabit the surface 1-cm layer compared to the top 4 cm. No regularities in relation to depth are apparent, and any deduction from the 1-cm layer to total standing stock is impossible. Corers seldom adequately sample the top layer, and so only data from grabs are included here.

The most intensive work is that of Douglas et al. (1979) in the Southern Califor-

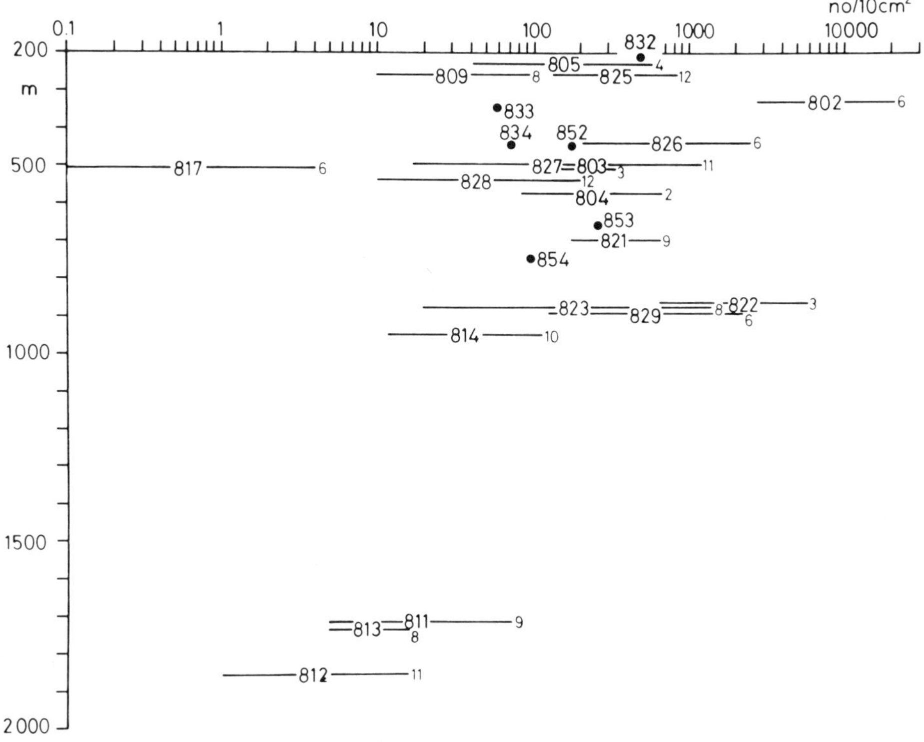

Fig. 12. Density of Foraminifera/10 cm^2 surface area and roughly (Douglas, personal communication) the upper 1-cm layer for the stations deeper than 200 m, Southern California Borderland [from Douglas et al. (1979); values for living Foraminifera (standing crop) from appendix II corrected according to appendix I: sample volume]. The lines connect minimum and maximum values in each station. Station number is given within each line and the number of samples, behind each line. Coastal slope and trough: Stations 832, 833, and 834, off Point Conception; Stations 802, 803, and 804 off Santa Barbara; Stations 821 to 823 and 825 to 829 off south of Los Angeles; Stations 852 to 854 off north of San Diego. Island slopes, outer troughs, and banks: Stations 805 and 809, slope off Santa Rosa Island; Stations 811 and 812, trough off Santa Cruz Island; Station 813, trough off San Clemente Island; Station 814 and 817, Tanner Bank and slope.

nia Borderland. Box grab types were used for sampling, and the subsamples measured around 100 cm^2 and they were 1 cm thick. The abundance in all the samples from deeper than 200 m (Fig. 12) decreases from the shallower to the deeper stations and from coastal to offshore areas. Summer and winter data exhibit no seasonal trend. The offshore Tanner Ridge with low values has a coarse sediment with presumably little food income. The rather high values in some of the deeper basins could be related to very low oxygen concentrations. Below a depth of about 300 m the oxygen values drop to below 1 ml/liter, and below 500 m they are even <0.5 ml/liter (Douglas et al., 1979, Table 2). Under low oxygen tension Foraminifera seem to outcompete the macrofauna, feeding pressure becomes reduced, and food resources are good. Phleger and Soutar (1973) found up to 1175 individuals/10 cm^2 in the Santa Barbara Basin and 2150 specimens/10 cm^2 in a basin (530-m depth) off Baja California, both with oxygen concentrations as low as 0.1 ml/liter.

Walch (1978) sampled Foraminifera with a box grab, crossing the eastern

Pacific rise north of the equator in an area of relatively high fertility. In the surface 1-cm layer she found 40 to 170 specimens in 10-cm^2 samples in the 74-μm fraction [according to Douglas and Woodruff (in press)]. From a sample collected in the area of the central northern Pacific Gyre, Bernstein and Meador (1979) sorted the unit Foraminifera (nonfragmenting species during processing) and recorded on the average 1.5 specimens/10 cm^2, larger than 297 μm. Saidova (1970) reports on a series of samples taken in the Kurile–Kamchatka Trench between about 250- and 9000-m depth (cf. Thiel 1975). The abundance varies from 5 to 160 specimens/10 cm^2, with no definite distributional trend. This is true as well for a transect from 443- to 6875-m depth in the South Sandwich Trench with seven samples exhibiting 6 to 140 specimens/10 cm^2 with its maximum at 4720 m (Basov, 1974). A depth transect near the Falkland Islands (Basov, 1974) exhibited a decrease from 720-m depth with 480 specimens/10 cm^2 down to 48 specimens/10 cm^2 in 1660 m. Further data on Foraminifera standing stock in the upper 1-cm sediment layer are compiled by Thiel (1975).

Although Wigley and McIntyre (1964) included this group in their investigations with the meiofauna (Fig. 6), other authors did not follow this example, partly because Foraminifera have rarely been regarded as meiofauna and partly because of the time it takes to sort them. From the point of view of total standing stock, however, each taxon should be evaluated in the size fraction in which it is best represented. The Foraminifera thus should be processed together with the meiofauna.

Densities of Foraminifera are presented in Figs. 6 and 8 together with the meiofauna. In their two samples (Fig. 6) Wigley and McIntyre (1964) found the Foraminifera to be 4 and 5% of the total meiofauna, whereas Coull et al. (1977) encountered 31, 33, and 65% as average numbers in 400, 800, and 4000 m, respectively (Fig. 6). Exceptionally high densities are reported by Tietjen (1971), if compared to meiofauna. Whereas his upper-slope stations contained 2 to 11% Foraminifera, deeper down the slope and on the continental rise a strong increase with up to 90% of total meiofauna was detected, with some indication of an increase with greater depth. This is exhibited as well in the few depths Coull et al. (1977) sampled, and their figures are well substantiated by two and four replicate samples per station and several subsamples per grab. In the northwestern African upwelling regions (Fig. 8), between 2 and 27% of the total meiofauna was Foraminifera (Thiel, 1975). Protoplasm wet weight of Foraminifera is reported to be 0.1 to 0.9 g/m^2 for the central Arctic Ocean between 1000- and 2600-m depth (Fetter, 1973), up to 7 g/m^2 between 1000 and 4000 m, and 8 g/m^2 in the Kurile–Kamchatka Trench (Saidova, 1967, 1970), but it seems doubtful that the inner test volume of the Foraminifera is equivalent to the plasma volume. The data on the abundance of Foraminifera so far assembled suggest the ecological and metabolic importance these protozoa may have in deep benthic communities (Thiel, 1975), especially under low oxygen conditions, and this should be understood in common view with the nanobenthos.

D. *Nanobenthos*

This size group was quantitatively studied exclusively by Burnett (1977, 1981), who concentrated his efforts on two localities, the bathyal San Diego Trough and the abyssal central North Pacific gyre. Organisms discovered were flagellates, amoebas, ciliates, testacids, foraminifers, yeasts, flexibacteria, and bubblelike prokaryote cells. Severe difficulties concerning the techniques (see Section 2.E) limited results. Because of the intricacy in perceiving them and the number of potential biases inherent with the techniques of quantification, the data are likely minimum counts.

For the central northern Pacific (5498 m), Burnett (1977) made three counts for top water plus surface sediment of 15,370, 25,750, and 26,800 individuals/cm^2. In the next 0.5-cm layer, he encountered on the average 1150 specimens, and in a later paper (1979b) he reports on further organisms in even deeper sediment layers, not seen before. As a first approximation these counts can be estimated to be 24,000 individuals/cm^2, but these must be regarded as minimum values.

Burnett (1981) studied the nanobenthos of the bathyal San Diego Trough, 1230 m depth, down to 6.3 and 9.8 cm. Not all the layers, however, were continuously worked up, so the total cannot be calculated. For the upper 1.6 cm, the range was 7.1 to 20.8 $\times$ 10^4 organisms/3 cm^3, equivalent to an area of about 2 cm^2. The mean and standard deviation per square centimeter was 13.2 $\pm$ 5.7 $\times$ 10^4. This value represents only about 25% of the nanobenthos in the upper 10 cm of sediment (Burnett, personal communication). Thus 50 $\times$ 10^4/cm^2 is only a rough estimate, but it may indicate the order of magnitude.

These first results on nanobenthos must be considered very preliminary. Data from only two environments are available, and these are minimum counts. They were done by one person, assuring comparability. The values of the San Diego Trough exceed those of the central northern Pacific by a factor of 20; yet, for the San Diego Trough samples, the nanobenthos was still abundant in the lowest layer assessed. Surely, even if these figures are accurate only to orders of magnitude, they demonstrate the high abundance and possible ecological importance of the nanobenthos in these communities.

Parallel to Burnett's sampling in the San Diego Trough, I collected quantitative meiofauna samples and found an average of 2600 specimens/10 cm^2 for the upper 5 cm, roughly two orders of magnitude less than Burnett calculated for the nanobenthos.

From the same general area of the Pacific Ocean (5847- and 5853-m depth), from which the nanobenthos was studied, Burnett (1979b) presents average meiofauna densities of about 130 and 73 specimens/10 cm^2. Thus the nanobenthos in the central northern Pacific may exceed the meiofauna by more than three orders of magnitude.

High densities of the nanobenthos relative to the meiofauna were suggested for theoretical reasons (Thiel, 1975) in my discussion of the size structure of the deep-sea benthos. Organisms with asexual reproduction—including the Foraminifera—should have competitive advantages over those with sexual reproduction in a low-energy habitat.

5. General Relationships

A general evaluation of the distributional patterns described for the depth transects and the oceanic and the hadal regions is biased by the low total number of samples, the paucity of investigations of these size groups of small organisms, and the small-scale spatial variability on the order of centimeters and decimeters, that is, between subsamples and within a grab sample. In many cases only one subsample was processed and in others, two or three; this resulted in an unequal reliability for faunal numbers in different localities and depths. The data so far available, however, deserve discussion and allow some generalization.

A. *Energy Availability*

The results obtained by several authors in different ocean regions indicate that food energy availability is the major factor controlling distributional patterns. Deep-sea studies on the macrofauna offered this explanation earlier, observing a decrease in standing stock with increasing depth and increasing distance from the continents (e.g., Belyaev et al., 1973; Sanders and Hessler, 1969; Monniot, 1979; Wolff, 1977; Rowe, 1971b, this volume, Chapter 3). It is not a new argument in the discussion on meiofauna standing stocks in the deep sea (Thiel, 1971 to 1979b; Dinet, 1973), and Morita (1979) proposed it for the bacteria. Within and between localities on a wide scale, it is generally overwhelmingly apparent that food availability is the prime factor explaining the differences in faunal densities.

All the depth transects exhibit a trend of faunal decrease with depth. Exceptions are observed where currents hinder sedimentation of hydrodynamically lighter particles, including the organic matter, or where these currents erode those materials and transport them to quieter habitats [Iceland–Faroe Ridge, Atlantic slope (see Section 4.B, on northern Atlantic and Norwegian Sea depth transects), and off North Carolina (see Section 4.B, on western Atlantic depth transects)]. Resuspension by energy dissipation of tidal and internal waves has the same effect, when low current intensities transport the materials in suspension away [upwelling off northwestern Africa (see Section 4.B, on eastern Atlantic depth transects, northwestern African upwelling region)].Transects in restricted areas would be expected to have similar meiofaunal densities, but currents can alter this pattern through the above mechanisms [Iceland–Faroe Ridge and Gibraltar Ridge (see Section 4.B, on northern Atlantic and Norwegian Sea depth transects)]. Thus physically high-energy habitats simultaneously support biologically low-energy communities.

By comparing different depth transects or oceanic regions, the deviating meiobenthos standing stocks can be partly explained by productivity; that is, surface primary production is reflected by the benthos. Low production levels are encountered in the Red Sea, the Mediterranean, and central oceans, whereas high levels characterize the eastern Atlantic regions with maximal values in the northwestern African upwelling area.

The exact mechanisms behind these seemingly simple relations are obscure, however, and we are far away from really understanding the true connections between the primary production and the benthos, especially the deep-sea benthos.

In general, the benthos mirrors the primary production of the same area, but surface as well as deep currents may redistribute the organic matter, and its path through the food web, packed in several organismic units, may allow for a far distant transport. Vertical sinking is limited in upwelling regions, where offshore and equatorial surface currents and deeper countercurrents may tend to retain slowly sinking material within the system for longer periods of time (Mittelstaedt, 1976; Thiel, 1978a). Southeast of Iceland, organisms carried in surface waters die when water masses are mixed along oceanic fronts. Production, death, and accumulation zones of organic matter may be far away from each other.

Organic matter is degraded in food webs, and their length and efficiency determine the amount reaching the seabed. Nakajima and Nishizawa (1972) encountered a relatively higher heterotrophic bacterial activity in areas of higher production, that is, intensified degradation processes and pelagic recycling, resulting in a relatively lesser food transport to the bottom. A simple linear relationship between primary production and benthic standing stock has statistical significance, but the large variation in such data indicate that the relationships are complex, as they are a function of biological and physical processes in the water column and of depth and topography.

B. Production and Temperature

Longevity, reproduction rates, and production of deep-sea species are not known. Production of shallow water meiofauna is poorly understood from generation times or respiration studies, and extrapolation to deep-sea species is highly speculative. Nonetheless, by assuming that standing stock is proportional to energy turnover or production, stocks can be used to make relative estimates. This relationship may exist within the same general area and depth range, but in its strict sense it should not be applied to distant geographic regions, to areas of different surface production and sedimentation, to different depth zones, and especially not to different temperature regimes. Pressure too may be important (Somero et al., this volume, Chapter 7). Community respiration (Smith, 1978; Smith and Hinga, this volume, Chapter 8) decreases by three orders of magnitude between shallow (40-m) and deep-sea (5200-m) communities in the northwestern Atlantic. For the depth range discussed in this chapter, 200 m and deeper, two orders of magnitude may be assumed as an approximation (cf. Pamatmat, 1973). With the use of these ratios for an estimation of metabolic rates, every increase in the ratio of meio- to macrofauna between shallow and deep stations would be multiplied by this factor. Differences between slope and abyssal stations (Figs. 6 to 11) would become more pronounced.

However, respiration or total metabolism may not depend solely on depth related environmental factors. On a horizontal scale between regions and oceans, it may be altered by concentration of organic matter and by temperature. Energy conversion efficiency may increase with decreasing food concentrations, as this was experimentally demonstrated by Butler et al. (1970), Sushchenya (1970), Taniguchi (1973), and Gaudy (1974).

Temperature is known from many physiological experiments to change respiration rates by a factor of 2 to 4 per 10°C shift (Q_{10}) on the species level (Precht et al., 1973). Comparison of the temperatures for 2000-m depth in some of the regions studied (Table XI) reveals a difference of more than 20°C for the extreme conditions. Under the aspect of benthos growth and production, the temperature conditions in the Norwegian Sea should have a cooling or food energy preserving effect. Conversely, energy seems to be wasted in the Red Sea. The Q_{10} effect for these two environments would amount to a factor of 4 to 8.

An issue to be determined by further studies is whether these factors are applicable to the different species from these far-distant regions. Because of the

TABLE XI
Bottom Temperatures for 2000-m Depth from Four Study Sites

Area	Temperature of Bottom Water (°C)
Norwegian Sea	−0.5 to −1
Atlantic Ocean	2 to 4
Mediterranean Sea	13.2
Red Sea	21.5

contrasting standing stock and presumably production levels and because of the unknown sedimentation rates of food materials in the two environments, the actual temperature effect cannot be deduced. Adaptation of species to the environmental temperature may reduce the seemingly diverging rates of respiration, metabolism, or production of the benthos. An adaptation of enzyme systems to the environmental temperature is to be expected, but adaptation will not rule out the effects of temperature totally.

C. Significance of Meio- and Nanobenthos in Deep-Sea Communities

Whereas energy availability is recognized as the major factor determining the density patterns of the benthos and temperature introduces large-scale modifications, biological interactions through competition and predation are of importance to community structure. In this context "community" should be understood as the total living assemblage of interacting species, all the size classes of the benthos included. Communities normally described in a restricted sense by only their macrobenthic components should not be regarded as true communities. The macrobenthos is most easily sampled, sorted, and identified, but it is far outnumbered by smaller forms and does not necessarily dominate benthic metabolism.

Size Groups of the Deep-Sea Benthos

Size groups as separated in benthic research are artificial units in a biological respect, and their classification is an artifact of sampling and processing methods (see Section 2), but the separation by size can be used to compare the importance of smaller and larger organisms. Such a comparison is restricted by the data sets available; parallel evaluation of size groups has been done rarely.

While summarizing our knowledge on the abundance of meiofauna in the deep sea, I hypothesized (Thiel, 1975, p. 600) "for the deep-sea benthos: With increasing depth and decreasing food concentration small organisms gain importance in total community metabolism," and I defined the deep sea as a "small organism habitat" (Thiel, 1975, p. 593). Sanders et al. (1965), in justifying their decision to use an 0.42-μm mesh size for their macrofauna study instead of the 1 mm sieve, stated (p. 864) that "if anything, benthic animals, on the average, appear to be slightly smaller at greater depth when compared to the slope regions." Fournier (1972, p. 203) reports "a general decrease in size, biomass and numerical abundance appears to occur with increasing depth." Hessler (1974), speaking about "those taxa traditionally regarded to be macrofauna," stated that "in reality most of the animals taken in central gyre waters are actually meiofaunal in size, that is, they would pass through a 1.0 mm screen." Size reduction was also discussed by Rowe and Menzel (1971, p. 558): "animals on the average are smaller in deeper water," but they suggested "that the inclusion of a few large individuals, which dominated the biomass at several of the shallow stations, biased the results." In their discussion (Rowe and Menzel, 1971, p. 563), however, they say "we conclude that where the fauna is sparse (i.e., in deep, depauperate basins) the biomass must be parceled into small packages of equivalent size, but as the influx of energy increases, the community is able to cycle carbon efficiently through larger organisms." Thus Rowe and Menzel (1971) and Thiel (1966, 1975) independently reached the same conclusions on size structure. I based my hypothesis (Thiel, 1975) on meiofauna counts from many deep-sea regions and cited studies on deserts and plankton as possibly supporting this hypothesis. Gage (1977) observed on the average smaller size in the Rockall Trough with 2875 m compared to Scottish sea lochs, and Dayton and Oliver (1977) reported on smaller individual animals from the deep sea than from shallow waters in the area of McMurdo Sound, Antarctica.

Continuing the discussion on size structure, Polloni et al. (1979) present data on macrofauna from a total of 283 species sampled between 400- and 3600-m depth in the northwestern Atlantic. As they found no statistically significant decrease in average size, they suggested that average small size in the deep sea may not necessarily be a general phenomenon. Haedrich and Rowe (1977) agree on small organism size for infaunal communities in fine-grained and food-poor sediments but point out that a number of fish appear to be "bigger–deeper." The two opinions stand against each other, both lacking a broad size spectrum in the observations. However, comparison of the faunal decrease from shallow to deep seabeds reveals a definite difference between macro- and meiofauna. Rowe (1971a,b) and Rowe et al. (1974) present their results on the macrofauna sampled off Peru, from the Gulf of Mexico, and in the northwestern Atlantic. Wet weights and numbers are given [e.g., in Rowe et al. (1974)], and Rowe et al. (1974, p. 646) state that "with numbers of animals diminishing less rapidly" than biomass, average organism size should be smaller. In general, these authors describe the macrofauna decrease as an exponential decay function (Fig. 13) that also is believed to be "proportional to the rate of decrease in primary production in the offshore direction and to the physical and biological efficiency of moving organic matter down to the maximum depth of the basin." The shortcoming in the argument surely is that they considered only the macrofauna (see Addendum 5).

I presented (Thiel, 1979a) a comparable regression for the meiofauna from transects sampled off Portugal, Morocco, and Mauritania (Fig. 13). The decrease in abundance of meiofauna is less pronounced than for the macrofauna (Rowe, 1971b), and the slopes (b values) for meio- and macrofauna show a ratio of 2:1. Again, these two data sets do geographically not belong together and apply to different ecological conditions. These mathematical formulations omit specific abundance, which could be of importance in ecological evaluation of results. But judging from the data presented in Figs. 6 to 12, the decrease in meiofauna abundance is generally smaller than in the macrofauna. Additionally, in the northwestern Atlantic at 1850-m depth, Grassle et al. (1975) estimated the number of

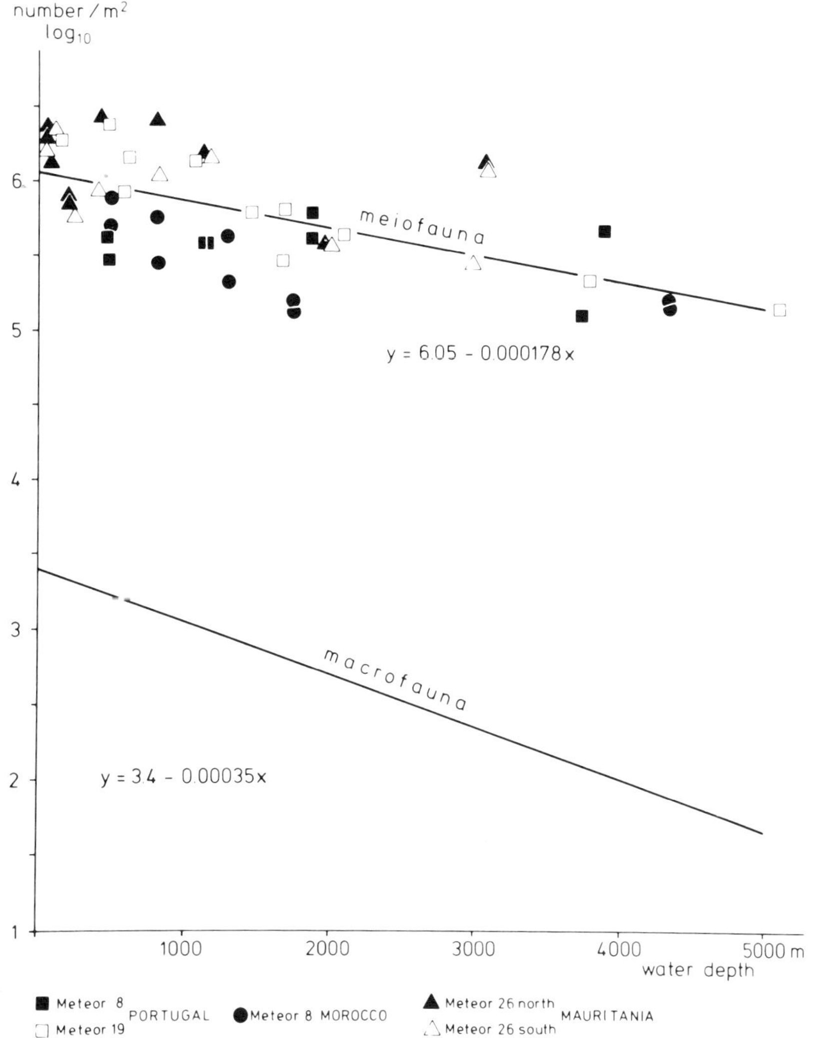

Fig. 13. The decrease of macrofauna with depth according to Rowe (1971b) from western Atlantic, Gulf of Mexico, and eastern Pacific regions, compared to the decrease of meiofauna from different eastern Atlantic regions (Thiel, 1979a).

megafauna to be three orders of magnitude less than the macrofauna, whereas biomass for both the size groups was about equal (Haedrich and Rowe, 1977). This indicates a stronger decrease in metabolic rates of megafauna than in macrofauna. On the basis of new data, Rowe suggests a stronger decrease in macro- than in megafauna (this volume, Chapter 3).

Further arguments against average small organism size in the deep-sea benthos are discussed by Polloni et al. (1979). Gigantism in deep-sea species is a well-known phenomenon in Isopoda and Tanaidacea [Wolff (1960, 1962); see Wolff (1970) for earlier literature] with some indication of it in Cumacea (Jones, 1969). Large species were found among the bivalves, "although in general the abyssal bivalves are not particularly large" (Knudsen, 1970), and Madsen (1961) did not find very large species in the Porcellanasteridae (Echinodermata). In this context Monniot's (1979) results on Ascidia are of importance, because gigantism and stalk constructions were observed in the deep sea along with dwarfism (see below). Polloni et al. (1979) present regression lines of mean weight versus depth of capture and found Echinodermata and Crustacea Decapoda somewhat smaller in deeper waters, but the regressions were not significant, whereas a highly significant size increase with depth was encountered for fish.

These observations do not support my hypothesis, of course, but this approach is different from my original macro–meiofauna comparison, and little in these observations is applicable to this question. The hypothesis does *not* say that all the species in the deep sea are smaller, size decrease must occur in all taxa, and gigantism does not occur. The hypothesis is not based on any taxon level, but on community level in its broad sense (see above). Single large specimens—and even giant ones—may not rule out a high number of small organisms by weight or by metabolic rates.

As I pointed out earlier (Thiel, 1979a), some difficulties arise in defining the benthic community. Many fish species, believed to be demersal according to former records from bottom trawling, have been caught in the pelagic zone and evidently feed on bathypelagic plankton and nekton (Hureau et al., 1979; Polloni et al., 1979; Rowe and Staresinic, 1979; Thiel, 1979a, and earlier papers cited). Evaluation of these species is difficult because of their long-distant migrations and their habit of feeding in part at least in the pelagic zone. From investigations on stomach contents of fish trawled at the continental slope, Sedberry and Musick (1978) predict a higher standing stock of zooplankton and nekton near bottom than in the bathypelagic zone. I suggested (Thiel, 1979a) excluding pelagic feeders from size comparison, or at least considering those fish taking benthic and pelagic prey with caution. Haedrich and Rowe (1977) explain large size among mobile deep-sea fish as a necessity in searching for food. Such an adaptation implies that carcasses falling from shallow water are important in the food economy of the total deep ecosystem or that deep-sea plankton and nekton are not sufficiently well explored.

Baited camera and phototrap lowerings have demonstrated the attraction of quite a number of fishes, decapod crustaceans, and other mobile scavengers or predators to bait (Isaacs, 1969; Dayton and Hessler, 1972; Hessler, 1974; Wolff, 1977; Guennegan and Rannou, 1979; Rowe and Staresinic, 1979; Thiel and Rumohr, 1979; Thiel, 1980b). The many photographs available today convey the impression of high standing stocks of these mobile swimming megafauna species, but no relation to area can be given. The spreading of bait smell with currents may be unidirectional and would not seem to be fast enough for information transfer. I suggested (Thiel, 1979a) vibrations may be a fast stimulus, simultaneously propagating in all directions. Vibration would be caused by gear lowering and its arrival on the bottom and afterward by the swimming and feeding activity of the attracted megafauna. But still, the area or volume from which the organisms are drawn is not known.

Two other explanations have been proposed for large size in the deep sea, one standing in contrast to mobility. Monniot (1979) relates small size in most deep-sea Ascidia to the low-energy income to the seabed, but some species of different families attain a size 3 to 10 times larger than related species. The large species are often stalked and show adaptations in their feeding mechanisms, which characterize them as passive filter feeders. No energy is consumed in creating a current for food particle transport, and they are believed to trap macroscopic prey. A similar explanation may be adequate for large hydroids and gorgonarians. Grasshoff (1972) described the stalked coelenterate *Umbellula thieli* as a final evolutionary step from polypoidal colonies through reduction in polyp numbers to the quasisolitary large headed giant. Grasshoff (1972) and Thiel (1972b) discuss the possibility of food uptake from the sediment surface, but the single-large polyp may have a trapping function like that of the large Ascidia (see Addendum 6).

Other large-size species of the deep sea live together in a special "large-size community" isolated in the small organism habitat. These communities are found around hydrothermal vents in about 2600-m depth at the central Pacific ridge spreading center north of the Galápagos (Corliss and Ballard, 1977; Lonsdale, 1977; Ballard and Grassle, 1979; Galápagos Biology Expedition Participants, 1979; RISE Project Group, 1980; Jannasch and Wirsen, this volume, Chapter 6). The functioning of this community is not well understood so far. Water convection is created by higher temperatures in rock fissures, and a primary carbon source apparently are chemoautotrophic sulfur bacteria. Whereas the water temperature around the hydrothermal vents is elevated by only a few tenths of one degree, or to more than 350°C in the central plume with rapid cooling to the sides, the regular and sufficient food supply apparently allows giant growth under low-temperature deep-sea conditions. Such a food source may apply to the filter–feeding organisms such as the large clams of more than 20 cm in length, the mussels, serpulid worms, and actinians. Brachyuran and galatheid crabs and possibly fish can be regarded as secondary consumers, and the giant vestimentiferans that have no guts are believed to live on dissolved organic matter. The functioning of this extraordinary community is not well understood; however, large organism size under apparently favorable food conditions supports the hypothesis of small organism size under constant food limitation (see Addendum 7).

So far organism size has been given by its actual measurements or by those components of the fauna retained on a sieve of definite mesh width. The quality of size, weight, or organic matter equivalents would allow more meaningful comparisons of production and maintenance, but these data are not available. Plankton and nekton studies have revealed that chemical and morphological changes achieve buoyancy and easier propulsion through the water for decreasing energy expenditure (Marshall, 1979; Denton and Gilpin-Brown, 1973; Childress and Nygaard, 1973; Barnes et al., 1976). Although such adaptations can be assumed for benthopelagic organisms, in benthic species this is not expected. However, Allen and Sanders (1966), Allen (1979), Knudsen (1979), and Oliver (1979) describe a number of morphological adaptations in bivalves to cope with the low food availability in the deep sea. Some deep-living species show a higher water content and thus increasing size, but not through energy-consuming organic matter. The competitive advantages of large size for different taxa may be extended feeding or moving organs, larger food-trapping constructions, or reduction of predation risk in an environment with an admittedly low predation pressure (see Addendum 4).

Similarly, the large Foraminifera, the Komokiacea (Tendal and Hessler, 1977), and particularly the Xenophyophoria (Tendal 1972) exhibit enormous structures (Gooday, 1978; Rice et al., 1979) with only small amounts of protoplasm. Tendal (1979) describes how Xenophyophoria store their fecal material, the stercomata, after initial digestion of organic constituents. He offers the hypothesis that microorganisms grow on the stored fecal matter and then the protozoan redigests the stercomata, harvesting their garden of newly produced microorganisms.

Biotic Interactions

In a qualitative approach Schoener (1969) suggested that food limitation would favor small body size because of competition, and the benthic community of the deep sea seems to support this hypothesis. Examples that may apply to these thoughts can be found in Thiel (1975) from different areas and habitats, in Gage (1977) for comparison of Scottish sea lochs with the deep Rockall Trough, and in Dayton and Oliver (1977) for antarctic subice communities and the deep sea.

Metabolic rates on a weight-specific basis increase with decreasing organism size or weight (Zeuthen, 1953; Hemmingsen, 1960; Fenchel, 1968; Vernberg and Coull, 1974; Fenchel, 1978). The relationship can be compared by the power function $M = aW^{b-1}$, where M is the metabolic rate per unit weight, W is the body weight, and a and b are constants. The value of b varies around 0.75, and it follows that the decrease of body weight by a factor of 10 can be expressed by about $M \times 1.77$, that is, an increase of the metabolic rate by more than 75%. In an energy-limited environment small size would thus be an energetic paradox. Total biomass to be maintained by the available food resources could be larger with larger life units. But Fenchel (1974) demonstrated that in small organisms—and in protozoa in comparison with metazoa—relatively more energy is channeled into growth and reproduction and less in maintenance. Nonreproductive growth could then be higher in small organisms, which fits life conditions in the deep sea, where K strategists with slow growth rate and low reproductive potential predominate. But the question of whether energy loss due to small size and energy gain in favor of production are counterbalanced or together have an energy conserving character and ecological advantage must remain open.

Nevertheless, deep-sea communities are ecologically rather stable and well accommodated; hence the organism size distribution should be explainable. Two examples from shallow waters can be presented for comparison, exhibiting on the average smaller organism size: the sulfide biomes and the northern Baltic Sea.

Elmgren (1978), in reviewing the benthos of the Baltic Sea, demonstrates that in some areas more energy is channeled into the meiofauna than into the macrofauna. In the Bothnian Bay he observed a biomass ratio of 2:1 for meiofauna:macrofauna. Physiological stress of brackish water conditions reduces the species number and the average size of species (Remane and Schlieper, 1971). In addition, the larger species seem to drop out earlier than the smaller ones, favoring the meiofauna.

Physiological stress from oxygen deficiency as well allows only small organisms to live in sulfide biomes (see Section 3.B), where macrofauna no longer exists.

Stress situations, which might influence the faunal components differently according to their size, do not seem to occur in the deep sea. Brackish water and sulfide habitats cannot help to explain the size distribution in the deep sea, although community structure is similar.

The deep sea is a marine environment that does not exert physiological stress differently on organism size groups. The essential factor again seems to be food, whereas in sulfide biomes this is not the limiting factor. The Bothnian Bay of the Baltic, where meiofauna dominates macrofauna, shows very low primary production (Elmgren, 1978) in addition to its low-salinity stress situation.

Further arguments to explain the size structure of communities are given by those offshore communities that exhibit unexpected similar meiofaunal densities under apparently different food income to the benthos. The respective habitats of these communities, grouped as "oceanic regions," are the seamount plateaus and the deep-sea plains (see Section 4.B, subsections on central oceanic regions, abyssal and seamount plateaus and Fig. 12). Meiofauna abundance on the top of the Josephine Seamount (206 to 355 m) was only about twice that in the Iberian deep sea at 5300-m depth. Such a relatively small difference in faunal densities between habitats separated by a vertical range of 5000 m were not expected from the exponential decrease in macrofauna found by Rowe (1971a,b) for the western Atlantic and other transects (Rowe, this volume, Chapter 3). Primary production above the Iberian deep sea (42°N) may be somewhat higher than above Josephine Seamount (36°N), but this would not be sufficient to account for the modest differences in standing stock. Sedimentation in these two environments will differ, and for Josephine Bank some resuspension and transport of organic matter away from the plateau can be expected as it was shown for the Great Meteor Seamount (see Section 4.B, on seamount plateaus). Again, however, a quiet sedimentary regime would not increase the standing stock by an order of magnitude, and the small differences in standing stock are still not understood.

Besides primary production and sedimentation, organismic interactions must be taken into account. It would seem likely that macro- and megafaunal populations are of importance in explaining the relatively high meiofaunal standing stock in the deep sea. Quantitative data on these size groups from the areas, where meiofauna has been studied, are not available. Trawl catches are not suitable for a quantitative comparison, but deep-sea hauls lasting 5 hr regularly bring up much less than 30-min seamount plateau hauls.

Predation by macro- and megafauna on meio- and nanobenthos will be one mode of interaction between these size groups (Thiel, 1982). Not much is known of the effect of predation on meiofauna. McIntyre (1969) argues against this relationship and is of the opinion that meio- and macrofauna are parallel ends of benthic food chains. Counts of meiofaunal organisms in the gut of sipunculids and holothurians indicate that at least part of the smaller size groups is digested (Hansen, 1978). Predation must be seen in close correlation with competition. Under severe food limitation, large infauna can only exist with special adaptations for fast sediment cycling, highly efficient breakdown of organic matter, or intestinal symbionts. For many of the larger invertebrates, food gathering consumes too much energy. The low number of larger sediment feeding organisms in the deep sea exerts a low feeding pressure on the smaller ones, and their competitive advantage is the ability to exploit the rather rare and small particulate food resources. This is supported by observations on macro- and meiofauna in the Bothnian Bay, where meiofauna standing stock exceeds that of macrofauna under food limitation (Elmgren, 1976, 1978).

The opposite is found in shallow waters, where food sedimentation is high and the macrofauna is able to collect or catch sufficiently high food ratios. Sediment feeding macrofauna compete for the same food source as used by the meiofauna, and in addition they swallow the meiofauna, not selecting the particles offered. Feeding pressure and competition are high in shallow waters. On the seamounts large numbers of fish were trawled and echinoids were caught and photographed, migrating gregariously along the bottom, searching for food. Although their rate of meiofauna and nanobenthos uptake is not known, competition for the same food source is evident. However, not only the sediment feeding macrofauna competes for food, but the passive and active particle catchers as well. Whether they trap their food, filter, or browse, they prevent particles from dropping out of the water column to the bottom. Dense populations of gorgonians, sponges, actinians, crinoids, polychaetes, and ascidians were photographically encountered on top of the seamounts. Epibenthos and endobenthos (mega-, macro-, meio-, and nanobenthos) compete for the community's total food source, although not all food components are equally obtainable for all of them. Competitive disadvantage and predation pressure are encountered by meio- and nanobenthos in biologically high-energy environments, and competitive advantage allows them to keep a high standing stock in biologically low-energy environments. While mega-, macro-, and meiofauna fit such a model, when bottom-bound organisms are considered, knowledge on the nanobenthos and the Foraminifera is still too scanty for comparison. High density of smallest metazoans and protozoans was described by Burnett (1977 to 1981) two to four orders of magnitude above the meiofauna, and dense populations of Xenophyophoria were estimated by Rice et al. (1979) up to 50 test structures in an area of 2.6 m^2. Foraminiferan biomass was estimated to range from 15 to 86%, with a mean of 53% from total biomass larger than 149 μm in the deep Arctic Ocean beyond the permanent ice (Paul and Menzies, 1974). Although biomass was calculated from plasma volume—and this was possibly overestimated under the assumption that the plasma filled 90% of the inner test volume (Fetter, 1973), these and the observations cited above demonstrate the significance that the nanobenthos and the Foraminifera may attain in deep benthic associations. Their relationship to the macrofaunal component of the communities may be described as for the meiofauna by predation and competition, and it will be the same again for meiofauna and nanobenthos.

Some macro- and meiofauna species seem to be adapted to low food levels, but where population levels drop to densities that cannot assure for at least occasional meetings of males and females for reproductive purposes, dioecious species drop out. In these environments asexually propagating species have a competitive advantage (Thiel, 1975). Many metazoans reproduce asexually, but many protozoans multiply by fissionary or multifissionary processes, and these organisms may well dominate the deep-sea communities. The partitioning of total energy uptake into growth and maintenance (Fenchel, 1974) must be recalled: the protozoa, with less structural complexity, use relatively less energy for maintenance, but more for growth and reproduction. This is definitely an advantage in low-energy environments.

Organism numbers and calculations of their weights constitute the basis for this discussion, because data on organic matter are not available. This shortcoming should direct quantitative research on deep-sea organisms in the future. Basic questions and applied forces aim at a better understanding of structure and function of deep-sea communities. This will require the concentration of efforts on total community studies, with considerable interdisciplinary input and with extrapolation from results on shallow-sea investigations.

Acknowledgments

I am most grateful to Dr. Gilbert T. Rowe for reviewing the manuscript, for valuable comments and suggestions, and for his language corrections. Thanks are also due to Mrs. H. Heimhold for preparation of all the illustrations, and to Mrs. B. Berghahn for manuscript typing and repeated proofreading.

References

Aldred, R. G., K. Riemann-Zürneck, H. Thiel, and A. L. Rice, 1979. Ecological observations on the deep-sea anemone *Actinoscyphia*. *Oceanol. Acta*, **2**, 389–395.

Allen, J. A., 1979. The adaptations and radiation in deep-sea bivalves. *Sarsia*, **64**, 19–27.

Allen, J. A. and H. L. Sanders, 1966. Adaptations to abyssal life as shown by the bivalve *Abra profundorum* (Smith). *Deep-Sea Res.*, **13**, 1175–1184.

Andrássy, I., 1956. Die Rauminhalts- und Gewichtsbestimmung der Fadenwürmer (Nematoden). *Acta Zool. Acad. Sci. Hungar.*, **2**, 1–15.

Ankar, S., 1977. Digging profile and penetration of the Van Veen grab in different sediment types. *Contr. Askö Lab.*, **16**, 22 pp.

Ballard, R. D. and J. F. Grassle, 1979. Strange world without sun. *Nat. Geogr. Mag.*, **156**, 680–703.

Bansemir, K., 1969. Bakteriologische Untersuchungen von Wasser und Sedimenten aus dem Gebiet der Island-Färöer Schwelle. *Ber. Dt. Wiss. Komm. Meeresforsch.*, **20**, 282–287.

Barnes, A. T., L. P. Quentin, J. J. Childress, and D. L. Pawson, 1976. Deep-sea macroplanktonic seacucumbers: Suspended sediment feeders captured from deep submergence vehicle. *Science*, **194**, 1083–1085.

Basov, I. A., 1974. Biomass of benthic foraminifers in the region of the south Sandwich Trench and Falkland Islands. *Oceanology*, **14**, 277–279.

Bein, A. and D. Fütterer, 1977. Texture and composition of continental shelf to rise sediments off the northwestern coast of Africa. An indication of downslope transportation. *"Meteor" Forsch. Ergebnisse C*, **27**, 46–74.

Belyaev, G. M., N. G. Vinogradova, R. Y. Levenstein, F. A. Pasternak, M. N. Sokolova, and Z. A. Filatova, 1973. Distribution patterns of deep-sea bottom fauna related to the idea of the biological structure of the sea. *Oceanology*, **13**, 114–121.

Bernstein, B. B., R. R. Hessler, R. Smith, and P. A. Jumars, 1978. Spatial dispersion of benthic Foraminifera in the abyssal central North Pacific. *Limnol. Oceanogr.*, **23**, 401–416.

Bernstein, B. B. and J. P. Meador, 1979. Temporal persistence and biological patch structure in an abyssal benthic community. *Mar. Biol.*, **51**, 179–183.

Boysen, O., S. Ehrich, W. Nellen, and G. Hempel, 1972. Fischerei-Biologie. In *Bericht über den Verlauf der Roßbreiten Expedition 1970*, G. Hempel and W. Nellen, eds., pp. 62–63. (Reprinted in *"Meteor" Forsch. Ergebnisse A*, **10**, 51–78.)

Brady, G. S., 1880. Ostracoda. *Rep. scient. res. voyage H.M.S. "Challenger" 1873–76. Zoology*, **1**, 1–184.

Brady, H. B., 1884. Foraminifera. *Rep. scient. res. voyage H.M.S. "Challenger" 1873–76. Zoology*, **9**, 1–814.

Brockmann, C., P. Hughes, and M. Tomczak, 1977. Data report on currents, winds and stratification in the NW African upwelling region during early 1975. *Ber. Inst. Meereskunde Kiel*, **32**, 1–45.

Bunt, J. S., 1975. Primary production of marine ecosystems. In *Primary Productivity of the Biosphere*, H. Lieth and R. H. Whittaker, eds. Springer-Verlag, Berlin, pp. 169–183.

Burnett, B. R., 1973. Observation of the microfauna of the deep-sea benthos using light and scanning electron microscopy. *Deep-Sea Res.*, **20**, 413–417.

Burnett, B. R., 1977. Quantitative sampling of microbiota of the deep-sea benthos. I. Sampling techniques and some data from the abyssal central North Pacific. *Deep-Sea Res.*, **24**, 781–789.

Burnett, B. R., 1978. Microbiota. In *Benthic Biological Studies. Seabed Disposal Program Annual Report, January–December 1977*. Sandia Laboratories, Albuquerque, NM, 16 pp.

Burnett, B. R., 1979a. Quantitative sampling of microbiota of the deep-sea benthos. II. Evaluation of technique and introduction to the biota of the San Diego Trough. *Trans. Am. Microsc. Soc.*, **98**, 233–242.

Burnett, B. R., 1979b. Microbiota and meiofauna. In *Benthic Biological Studies. Seabed Disposal Program Annual Report, January–December 1978*. Sandia Laboratories, Albuquerque, NM, 19 pp.

Burnett, B. R., 1981. Quantitative sampling of microbiota of the deep-sea benthos. III. The bathyal San Diego Trough. *Deep-Sea Res.*, **28**, 649–663.

Butler, E. I., E. D. S. Corner, and S. M. Marshall, 1970. On the nutrition and metabolism of zooplankton. VII. Seasonal survey of nitrogen and phosphorus excretion by Calanus in the Clyde Sea-area. *J. Mar. Biol. Assoc. U. K.*, **50**, 525–560.

Buzas, M. A., 1972. Patterns of species diversity and their explanation. *Taxon* **21**, 275–286.

Buzas, M. A. and T. G. Gibson, 1969. Species diversity: Benthonic Foraminifera in Western North Atlantic. *Science*, **163**, 72–75.

Childress, J. J. and M. H. Nygaard, 1973. The chemical composition of mid water fishes as a function of depth of occurrence off Southern California. *Deep-Sea Res.*, **20**, 1093–1109.

Cook, D. G., 1969. *Peloscolex dukei* n. sp. and *P. aculeatus* n. sp. (Oligochaeta, Tubificidae) from the North-West Atlantic, the latter being from abyssal depth. *Trans. Am. Microsc. Soc.*, **88**, 429–497.

Cook, D. G., 1970. Bathyal and abyssal Tubificidae (Annelida, Oligochaeta) from the Gay Head–Bermuda Transect, with description of new genera and species. *Deep-Sea Res.*, **17**, 973–981.

Corliss, J. B. and R. D. Ballard, 1977. Oases of life in the cold abyss. *Nat. Geogr. Mag.*, **152**, 441–453.

Coull, B. C., 1972. Species diversity and faunal affinities of meiobenthic Copepoda in the deep sea. *Mar. Biol.*, **14**, 48–51.

Coull, B. C., R. L. Ellison, J. W. Fleeger, R. P. Higgins, W. D. Hope, W. D. Hummon, R. M. Rieger, W. E. Sterrer, H. Thiel, and J. H. Tietjen, 1977. Quantitative estimates of the meiofauna from the deep sea off North Carolina, USA. *Mar. Biol.*, **39**, 233–240.

Craib, J. S., 1965. A sampler for taking short undisturbed cores. *J. Cons. Perm. Int. Explor. Mer*, **30**, 34–39.

Dahl, E., L. Laubier, M. Sibuet, and J.-O. Strömberg, 1976. Some quantitative results on benthic communities of the deep Norwegian Sea. *Astarte*, **9**, 61–79.

Dayton, P. K. and R. R. Hessler, 1972. Role of biological disturbance in maintaining diversity in the deep sea. *Deep-Sea Res.*, **19**, 199–208.

Dayton, P. K. and J. S. Oliver, 1977. Antarctic soft-bottom benthos in oligotrophic and eutrophic environments. *Science*, **197**, 55–58.

Denton, E. J. and J. B. Gilpin-Brown, 1973. Floatation mechanisms in modern and fossil cephalopods. *Adv. Mar. Biol.*, **11**, 197–268.

Diester-Haass, L., 1978. Sediments as indicators of upwelling. In *Upwelling Ecosystems*, R. Boje and M. Tomczak, eds. Springer-Verlag, Berlin, pp. 261–281.

Diester-Haass, L. and P. J. Müller, 1979. Processes influencing sand fraction composition and organic matter content in surface sediments off W Africa (12–19°N). *"Meteor" Forsch. Ergebnisse C*, **31**, 21–47.

Dietrich, G., 1967. The international "Overflow" expedition (ICES) of the Iceland-Faroe Ridge, May–June 1960. A review. In *The Iceland–Faroe Ridge International (ICES) "Overflow" Expedition, May–June, 1960*, J. B. Tait, ed. (Reprinted in *Rapp. P.-v. Réun. Cons. perm. int. Expl. Mer*, **157**, 268–274.)

Dietrich, G., W. Düing, K. Grasshoff, and G. Krause, 1966. Physikalische und chemische Daten nach Beobachtungen des Forschungsschiffes "Meteor" im Indischen Ozean 1964/65. *"Meteor" Forsch. Ergebnisse A*, **2**, 1–5 and tables.

Dinet, A., 1973. Distribution quantitative du méiobenthos profond dans la région de la dorsale de Walvis (Sud-Ouest Africain). *Mar. Biol.*, **20**, 20–26.

Dinet, A., 1976. Étude quantitative du méiobenthos dans le secteur nord de la Mer Égée. *Acta Adriatica*, **18**, 83–88.

Dinet, A., 1979. A quantitative survey of meiobenthos in the deep Norwegian Sea. *Ambio, Spec. Rep.*, **6**, 75–77.

Dinet, A., L. Laubier, J. Soyer, and P. Vitiello, 1973. Résultats biologiques de la campagne Polymède. II. Le méiobenthos abyssal. *Rapp. Comm. int. Mer Médit.*, **21,** 701–704.

Dinet, A. and M. H. Vivier, 1977. Le méiobenthos abyssal du Golfe de Gascogne. I. Considérations sur les données quantitatives. *Cah. Biol. Mar.*, **18,** 85–97.

Dinet, A. and M. H. Vivier, 1979. Le méiobenthos abyssal du Golfe de Gascogne. II. Les peuplements de Nématodes et leur diversité spécifique. *Cah. Biol. Mar.*, **20** (1), 9–123.

Ditlefsen, H., 1926. Free-living nematodes. *Danish Ingolf Exp.*, **4,** 1–121.

Douglas, R. G., M. L. Cotton, and L. Wall, 1979. *Distributional and Variability Analysis of Benthic Foraminifera in the Southern California Bight.* BML Technical Report, 21.0, Vol. 2. Bureau of Land Management, Washington DC, 219 pp.

Douglas, R. G., L. Wall, and M. L. Cotton, 1978. *The Effects of Sample Quality on the Recovery of Live Benthic Foraminifera from the Southern California Borderland.* BML Technical Report, 20.0, Vol. 2. Bureau of Land Management, Washington DC, 37 pp.

Douglas, R. G. and F. Woodruff, 1981. Deep-sea benthic Foraminifera. In *The Sea,* Vol. 7, C. Emiliani, Wiley, New York, pp. 1233–1328.

Dybern, B. I., H. Ackefors, and R. Elmgren, eds., 1976. *Recommendations on Methods for Marine Biological Studies in the Baltic.* Baltic Marine Biologists, Publication No. 1, Stockholm, 98 pp.

Ehrich, S., 1977. Die Fischfauna der Großen Meteorbank. *"Meteor" Forsch. Ergebnisse D,* **25,** 1–23.

Elmgren, R., 1973. Methods of sampling sublittoral soft bottom meiofauna. *OIKOS Suppl.,* **15,** 112–120.

Elmgren, R., 1975. Benthic meiofauna as indicator of oxygen conditions in the northern Baltic proper. *Merentutkimuslait. Julk./Havsforskningsinst. Skr.,* **239,** 265–271.

Elmgren, R., 1976. Baltic benthos communities and the role of meiofauna. *Contrib. Askö Lab.,* **14,** 1–31.

Elmgren, R., 1978. Structure and dynamics of Baltic communities, with particular reference to the relationship between macro- and meiofauna. *Kieler Meeresforsch. Sonderh.,* **4,** 1–22.

Fahrbach, E. and J. Meincke, 1978. High frequency velocity fluctuations near the bottom over the continental slope. *"Meteor" Forsch. Ergebnisse A,* **20,** 1–12.

Fenchel, T., 1968. The ecology of marine microbenthos. III. The reproductive potentials of ciliates. *Ophelia,* **5,** 123–136.

Fenchel, T., 1969. The ecology of marine microbenthos. IV. Structure and function of the benthic ecosystem, its chemical and physical factors and the microfauna communities with special reference to the ciliated protozoa. *Ophelia,* **6,** 1–182.

Fenchel, T., 1971. The reduction–oxidation properties of marine sediments and the vertical distribution of the microfauna. *Vie Milieu,* Suppl. 3. *Symp. Eur. Biol. Mar.,* **22,** 509–521.

Fenchel, T., 1974. Intrinsic rate of natural increase: The relationship with body size. *Oecologia* (Berlin), **14,** 317–326.

Fenchel, T., 1978. The ecology of micro- and meiobenthos. *Ann. Rev. Ecol. Syst.,* **9,** 99–121.

Fenchel, T. and R. Riedl, 1970. The sulfide system: A new biotic community underneath the oxidized layer of marine sand bottoms. *Mar. Biol.,* **7,** 255–268.

Fetter, F. C., 1973. Recent deep-sea benthic Foraminifera from the Alpha Ridge Province of the Arctic Ocean. In *Benthic Ecology of the High Arctic Deep Sea,* A. Z. Paul and R. J. Menzies, eds. Fin. Rep. ONR Contr. N000 14-67-A-0235-0005, pp. 296–337.

Filatova, Z. A., 1969. Quantitative distribution of the deep-sea benthic fauna. In *Deep-sea Bottom Fauna, Pleuston,* V. G. Kort, ed. *Biol. Tikh. Okean. Tikii Okeana., Isdated Nauka, Moskwa,* **7,** 353 pp. (in Russian). (Translation publ. by U.S. Naval Hydrographic Office, No. 487, 1970.)

Finenko, Z. Z., 1978. Production in plant populations. In *Marine Ecology,* Vol. 4, *Dynamics,* O. Kinne, ed. Wiley, Chichester, pp. 13–87.

Fournier, R. O., 1972. The transport of organic carbon to organisms living in the deep ocean. *Proc. Roy. Soc. Edinburgh,* **73,** 203–211.

Gaertner, A., 1968. Niedere, mit Pollen köderbare marine Pilze diesseits und jenseits des Island-Färöer-Rückens im Oberflächenwasser und im Sediment. *Veröff. Inst. Meeresf. Bremerh.,* **11,** 65–82.

Gage, J. D., 1977. Structure of the abyssal macrobenthic community in the Rockall Trough. In *Biology of Benthic Organisms,* B. E. Keegan, P. O'Ceidigh, and P. J. S. Boaden, eds. Pergamon Press, Oxford, pp. 247–260.

Gage, J. D., 1979. Macrobenthic community structure in the Rockall Trough. *Ambio, Spec. Rep.,* **6,** 43–46.

Galápagos Biology Expedition Participants, 1979. Galápagos '79: Initial findings of a deep-sea biological quest. *Oceanus,* **22,** 2–10.

Gallardo, V. A., 1977. Large benthic microbial communities in sulphide biota under Peru–Chile Subsurface Countercurrent. *Nature,* **268,** 331–332.

Gaudy, R., 1974. Feeding four species of pelagic copepods under experimental conditions. *Mar. Biol.,* **25,** 125–141.

George, R. Y. and R. P. Higgins, 1979. Eutrophic hadal benthic community in the Puerto Rico Trench. *Ambio, Spec. Rep.,* **6,** 51–58.

Gerlach, S. A., 1971. On the importance of marine meiofauna for benthos communities. *Oecologia* (Berlin), **6,** 176–190.

Gooday, A. J., 1978. Giant testate protozoans (Xenophyophoria) in the abyssal northeastern Atlantic. *J. Geol. Soc. Lond.,* **135,** 478.

Grasshoff, M., 1972. Eine Seefeder mit einem einzigen Polypen: *Umbellula thieli* n. sp. Die von F.S. "Meteor" 1967–1970 im östlichen Nordatlantik gedredschten Pennatularia (Cnidaria : Anthozoa). *"Meteor" Forsch. Ergebnisse D,* **12,** 1–11.

Grassle, J. F., H. L. Sanders, R. R. Hessler, G. T. Rowe, and T. McLellan, 1975. Pattern and zonation: A study of the bathyal megafauna using the research submersible "Alvin." *Deep-Sea Res.,* **22,** 457–481.

Guennegan, Y. and M. Rannou, 1979. Semidiurnal rhythmic activity in deep-sea benthic fishes in the Bay of Biscay. *Sarsia,* **64,** 113–116.

Haedrich, R. L. and G. T. Rowe, 1977. Megafaunal biomass in the deep-sea. *Nature,* **269,** 141–142.

Halim, Y., 1969. Plankton of the Red Sea. *Oceanogr. Mar. Biol. Ann. Rev.,* **7,** 231–275.

Hansen, B. and J. Meincke, 1979. Eddies and meanders in the Iceland–Faroe Ridge area. *Deep-Sea Res.,* **26A,** 1067–1082.

Hansen, M. D., 1978. Nahrung und Freßverhalten bei Sedimentfressern, dargestellt am Beispiel von Sipunculiden und Holothurien. *Helgoländer wiss. Meeresunters.,* **31,** 191–221.

Heezen, B. C. and C. D. Hollister, 1971. *The Face of the Deep.* Oxford University Press, New York, 659 pp.

Hemmingsen, A. M., 1960. Energy metabolism as related to body size and respiratory surfaces and its evolution. *Rep. Steno Hosp. (Kbh),* **9,** 1–110.

Hessler, R. R., 1971. Problems of meiobenthic sampling in the deep sea. In *Proceedings of the First International Conference on Meiofauna,* N. C. Hulings, ed. (Reprinted in *Smithsonian Contrib. Zool.,* **76,** 187–190.)

Hessler, R. R., 1974. The structure of deep benthic communities from central oceanic waters. In *The Biology of the Oceanic Pacific,* C. Miller, ed. Oregon State University Press, Corvallis, pp. 79–93.

Hessler, R. R. and P. A. Jumars, 1974. Abyssal community analysis from replicate box cores in the central North Pacific. *Deep-Sea Res.,* **21,** 185–209.

Hessler, R. R. and H. L. Sanders, 1967. Faunal diversity in the deep sea. *Deep-Sea Res.,* **14,** 65–78.

Hesthagen, I. H., 1970. On the near-bottom plankton and benthic invertebrate fauna of the Josephine Seamount and the Great Meteor Seamount. *"Meteor" Forsch. Ergebnisse D,* **8,** 61–70.

Höhnk, W., 1961. A further contribution to the oceanic mycology. *Rapp. P.-v. Réun. Cons. perm. int. Expl. Mer,* **149,** 202–208.

Holme, N. A. and A. D. McIntyre, 1971, eds. *Methods for the Study of Marine Benthos.* IBP Handbook, Vol. 16, pp. 1–344. Blackwell Scientifique, Oxford.

Horn, W., W. Hussels and J. Meincke, 1971. Schichtungs- und Strömungsmessungen im Bereich der Großen Meteor Bank. *"Meteor" Forsch. Ergebnisse A,* **9,** 31–46.

Hureau, J.-C., P. Geistdoerfer, and M. Rannou, 1979. The ecology of deep-sea benthic fishes. *Sarsia,* **64,** 103–108.

Isaacs, J. D., 1969. The nature of oceanic life. *Sci. Am.,* **221,** 146–162.

Jones, N. S., 1969. The systematics and distribution of Cumacea from depth exceeding 200 meters. *Galathea Rep.*, **10**, 99–180.

Jones, N. S. and H. L. Sanders, 1972. Distribution of Cumacea in the deep Atlantic. *Deep-Sea Res.*, **19**, 737–745.

Jumars, P. A., 1975a. Methods for measurement of community structure in deep-sea macrobenthos. *Mar. Biol.*, **30**, 245–252.

Jumars, P. A., 1975b. Environmental grain and polychaete species diversity in a bathyal benthic community. *Mar. Biol.*, **30**, 253–266.

Jumars, P. A., 1976. Deep-sea species diversity: Does it have a characteristic scale? *J. Mar. Res.*, **34**, 217–246.

Jumars, P. A., 1978. Spatial autocorrelation with RUM (Remote Underwater Manipulator): Vertical and horizontal structure of a bathyal benthic community. *Deep-Sea Res.*, **25**, 589–604.

Jumars, P. A. and R. R. Hessler, 1976. Hadal community structure: Implications from the Aleutian Trench. *J. Mar. Res.*, **34**, 547–560.

Kinzer, J., 1972. Makroplankton. In *Bericht über den Verlauf der Roßbreiten Expedition 1970*. G. Hempel and W. Nellen, eds., pp. 61–62. (Reprinted in *"Meteor" Forsch. Ergebnisse A*, **10**, 51–78.)

Knorr, G., 1969. Über die erforderliche Probenzahl bei quantitativen Untersuchungen des marinen Benthos. Diploma thesis, Math.-Nat. Fakultät, University of Hamburg, 71 pp.

Knudsen, J., 1970. The systematics and biology of abyssal and hadal Bivalvia. *Galathea Rep.*, **11**, 1–241.

Knudsen, J., 1979. Deep-sea bivalves. In *Pathways in Malacology*. S. v.d. Spoel, A.C. v. Bruggen and J. Lever, eds. Scheltema & Holkema, Utrecht, pp. 195–224.

Körte, F., 1966. Plankton- und Detritusuntersuchungen zwischen Island und den Färöer im Juni 1960. *Kieler Meeresforsch.*, **22**, 1–27.

Kohlmeyer, J., 1977. New genera and species of higher fungi from the deep-sea (1615–5315 m). *Rev. Mycologie*, **41**, 189–206.

Kullenberg, G., 1978. Light scattering observations in the northwest African upwelling region. *Deep-Sea Res.*, **25**, 525–542.

Lonsdale, P., 1977. Clustering of suspension-feeding macrobenthos near abyssal hydrothermal vents at oceanic spreading centers. *Deep-Sea Res.*, **24**, 857–863.

Lynn, R. J. and J. L. Reid, 1968. Characteristic and circulation of deep and abyssal waters. *Deep-Sea Res.*, **15**, 577–595.

Madsen, F. J., 1961. On the zoogeography and origin of the abyssal fauna in view of the knowledge of the Porcellanasterida. *Galathea Rep.*, **4**, 177–218.

Marcus, A., 1934. Tardigrada. In *Die Tierwelt Deutschlands*, **10**, 1–150.

Mare, M. F., 1942. A study of a marine benthic community with special reference to the micro-organisms. *J. Mar. Biol. Assoc. U. K.*, **25**, 517–554.

Marshall, N. B., 1979. *Developments in Deep-Sea Biology*. Blandford Press, Poole (England), 566 pp.

McIntyre, A. D., 1969. Ecology of marine meiobenthos. *Biol. Rev.*, **44**, 245–290.

Meincke, J., 1971a. Der Einfluß der Großen Meteorbank auf Schichtung und Zirkulation der ozeanischen Deckschicht. *"Meteor" Forsch. Ergebnisse A*, **9**, 67–94.

Meincke, J., 1971b. Observations of an anticyclonic vortex trapped above a seamount. *J. Geophys. Res.*, **76**, 7432–7440.

Meincke, J., 1978. On the distribution of low salinity intermediate waters around the Faroes. *Dt. hydrogr. Z.*, **31**, 50–64.

Meincke, J., E. Mittelstaedt, K. Huber, and K. P. Koltermann, 1975. W.F.S. "Planet" 12. 1.–30. 3. 1972, Seegebiet NW-Afrika. Strömung und Schichtung im Auftriebsgebiet von Nordwest-Afrika. *Dt. Hydrogr. Inst., Meereskundl. Beob. Ergebn.*, **41**, 1–117.

Menzies, R. J., R. Y. George, and G. T. Rowe, 1973. *Abyssal Environment and Ecology of the World Oceans*. Wiley, New York, 488 pp.

Menzies, R. J., and G. T. Rowe, 1968. The LUBS, a large undisturbed-bottom sampler. *Limnol. Oceanogr.*, **13**, 708–714.

Mittelstaedt, E., 1976. On the currents along the Northwest African coast south of 22°N. *Dt. hydrogr. Z.*, **29**, 97–117.

Monniot, C., 1979. Adaptations of benthic filtering animals to the scarcity of suspended particles in deep water. *Ambio, Spec. Rep.*, **6**, 73–74.

Morita, R. Y., 1979. Deep-sea microbial energetics. *Sarsia*, **64**, 9–12.

Müller, P. J. and E. Suess, 1979. Productivity, sedimentation rate and sedimentary organic matter in the oceans. I. Organic carbon preservation. *Deep-Sea Res.*, **26A**, 1347–1362.

Nakajima, K. and S. Nishizawa, 1972. Exponential decrease in particulate carbon concentration in a limited depth interval in the surface layer of the Bering Sea. In *Biological Oceanography of the Northern North Pacific Ocean*, A. Y. Takenouti, ed. Idemitsu Shoren, Tokyo, pp. 495–505.

Nellen, W., 1973. Untersuchungen zur Verteilung von Fischlarven und Plankton im Gebiet der Großen Meteorbank. *"Meteor" Forsch. Ergebnisse D*, **13**, 47–69.

Newton, J. G., O. H. Pilkey, and J. O. Blanton, 1971. *An Oceanographic Atlas of the Carolina Continental Margin*. North Carolina Dept. Com. Devel. (Division of Mineral Resources, NCD of C & D), 57 pp.

Nørvang, A., 1961. Schizamminidae, a new family of Foraminifera. *Atlantide Rep.*, **6**, 169–201.

Oliver, P. G., 1979. Adaptations of some deep-sea suspension feeding bivalves (Linopsis and Bathyarca). *Sarsia*, **64**, 33–36.

Pamatmat, M. M., 1973. Benthic community metabolism on the continental terrace and in the deep sea in the North Pacific. *Int. Revue Ges. Hydrobiol.*, **58**, 345–368.

Paul, A. Z. and R. J. Menzies, 1974. Benthic ecology of the high Arctic deep sea. *Mar. Biol.*, **27**, 251–262.

Peterson, W. H. and C. G. H. Rooth, 1976. Formation and exchange of deep water in the Greenland and Norwegian Seas. *Deep-Sea Res.*, **23**, 273–283.

Phleger, F. B. and A. Soutar, 1973. Production of benthic foraminifera in three east Pacific oxygen minima. *Micropaleontology*, **19**, 110–115.

Polloni, P., R. Haedrich, G. T. Rowe, and C. H. Clifford, 1979. The size–depth relationship in deep ocean animals. *Int. Revue Ges. Hydrobiol.*, **64**, 39–46.

Powell, E. N., M. A. Crenshaw, and R. M. Rieger, 1979. Adaptations to sulfide in the meiofauna of the sulfide system. I. ^{35}S-Sulfide accumulation and the presence of a sulfide detoxification system. *J. Exp. Mar. Biol. Ecol.*, **37**, 57–76.

Powell, E. N., M. A. Crenshaw, and R. M. Rieger, 1980. Adaptations to sulfide in sulfide-system meiofauna. Endproducts of sulfide detoxification in three turbellarians and a gastrotrich. *Mar. Ecol. Progr. Ser.*, **2**, 169–177.

Precht, H., J. Christopherson, H. Hensel, and W. Larcher, 1973. *Temperature and Life*. Springer-Verlag, Berlin, 779 pp.

Rachor, E., 1975. Quantitative Untersuchungen über das Meiobenthos der nordostatlantischen Tiefsee. *"Meteor" Forsch. Ergebnisse D*, **21**, 1–10.

Rad, U. v., 1974. Great Meteor and Josephine Seamounts (eastern North Atlantic). Composition and origin of bioclastic sands, carbonate and pyroclastic rocks. *"Meteor" Forsch. Ergebnisse C*, **19**, 1–61.

Reise, K. and P. Ax, 1979. A meiofaunal "thiobios" limited to the anaerobic sulfide system of marine sands does not exist. *Mar. Biol.*, **54**, 225–237.

Reise, K. and P. Ax, 1980. Statement on the thiobios-hypothesis. *Mar. Biol.*, **58**, 31–32.

Remane, A. and C. Schlieper, 1971. *Biology of Brackish Waters*, 2nd ed. Schweitzerbart'sche Verlagsbuchhandlung, Stuttgart, 372 pp.

Revsbeck, N. P., B. B. Jørgensen, and T. A. Blackburn, 1980. Oxygen in the sea bottom measured with a microelectrode. *Science*, **207**, 1355–1356.

Rex, M. A., 1973. Deep-sea species diversity: Decreased gastropod diversity at abyssal depths. *Science*, **181**, 1051–1053.

Rex, M. A., 1976. Biological accommodation in the deep-sea benthos: Comparative evidence on the importance of predation and productivity. *Deep-Sea Res.*, **23**, 975–987.

Rex, M. A., in press. Geographic patterns in species diversity in the deep-sea benthos. In *Deep-Sea Biology*, Vol. 8, *The Sea*, G. T. Rowe, ed. Wiley, New York, Chapter 11.

Rice, A. L., R. G. Aldred, D. S. M. Billett, and M. H. Thurston, 1979. The combined use of an epibenthic sledge and a deep-sea camera to give quantitative relevance to macro-benthos samples. *Ambio, Spec. Rep.*, **6**, 59–72.

RISE Project Group, 1980. East Pacific Rise. Hot springs and geophysical experiments. *Science*, **207**, 1421–1433.

Rowe, G. T., 1971a. Benthic biomass in the Pisco, Peru upwelling. *Inv. Pesq.*, **35**, 127–135.

Rowe, G. T., 1971b. Benthic biomass and surface productivity. In *Fertility of the Sea*, Vol. 2, J. D. Costlow, Jr., ed. Gordon and Breach, New York, pp. 441–454.

Rowe, G. T. and D. W. Menzel, 1971. Quantitative benthic samples from the deep Gulf of Mexico with some comments on the measurement of deep-sea biomass. *Bull. Mar. Sci.*, **21**, 556–566.

Rowe, G. T. and N. Staresinic, 1979. Sources of organic matter to the deep-sea benthos. *Ambio, Spec. Rep.*, **6**, 19–23.

Rowe, G. T., P. T. Polloni, and S. G. Hornor, 1974. Benthic biomass estimates from the northwestern Atlantic Ocean and the northern Gulf of Mexico. *Deep-Sea Res.*, **21**, 641–650.

Rowe, G. T., P. T. Polloni, and R. L. Haedrich, 1975. Quantitative biological assessment of the benthic fauna in deep basins of the Gulf of Maine. *J. Fish. Res. Board Can.*, **32**, 1805–1812.

Ryther, J. H., 1963. Geographic variations in productivity. In *The Sea*, M. N. Hill, ed. Interscience, London, pp. 347–380.

Saidova, K. M., 1967. The biomass and quantitative distribution of live Foraminifera in the Kurile–Kamchatka Trench area. *Doklady Akad. Nauk SSSR*, **174**, 207–209.

Saidova, K. M., 1970. Benthic foraminifers of the Kurile-Kamchatka Trench area. *Trudy Akad. Nauk SSSR*, **86**, 144–176.

Sanders, H. L., 1968. Marine benthic diversity: a comparative study. *Am. Nat.*, **102**, 243–282.

Sanders, H. L., 1969. Benthic marine diversity and the stability time hypothesis. *Brookhaven Symp. Biol.*, **22**, 77–81.

Sanders, H. L. and R. R. Hessler, 1969. Ecology of the deep-sea benthos. *Science*, **163**, 1419–1424.

Sanders, H. L., R. R. Hessler, and G. R. Hampson, 1965. An introduction to the study of deep-sea benthic faunal assemblages along the Gay Head–Bermuda Transect. *Deep-Sea Res.*, **12**, 845–867.

Schemainda, R. and D. Nehring, 1975. The annual cycle of the space-temporal dislocation of the North-West African upwelling region. *Communication 42, 3. Intern. Symp. Upwelling Ecosystems*, Kiel 1975.

Schoener, T. W., 1969. Models of optimal size for solitary predators. *Am. Nat.*, **163**, 277–313.

Scottish Marine Biological Association, 1977. *Annual Report 1976–1977*, pp. 26–27.

Scottish Marine Biological Association, 1979. *Annual Report 1978–1979*, pp. 28–29.

Sedberry, G. R. and J. A. Musick, 1978. Feeding strategies of some demersal fishes of the continental slope and rise off the mid-atlantic coast of the USA. *Mar. Biol.*, **44**, 357–376.

Sieburth, J. McN., V. Smetacek, and J. Lenz, 1978. Pelagic ecosystem structure: Heterotrophic compartments of the plankton and their relationship to plankton size fraction. *Limnol. Oceanogr.*, **23**, 1256–1263.

Smith, K. L., Jr., 1978. Benthic community respiration in the N. W. Atlantic Ocean: *In situ* measurements from 40 to 5200 m. *Mar. Biol.*, **47**, 337–347.

Sokolova, M. N., 1970. Weight characteristics of meiobenthos from different parts of the deep-sea trophic regions of the Pacific Ocean. (Engl. transl.). *Okeanologia*, **10**, 266–272.

Sokolova, M. N., 1972. Trophic structure of deep-sea macrobenthos. *Mar. Biol.*, **16**, 1–12.

Sournia, A., 1973. La production primaire planctonique en Méditerranée. Essai de mise à jour. *Bull. Étude en Commun Médit.*, **5**, 1–128.

Soyer, J., 1971. Bionomie benthique du plateau continental de la côte Catalane Française. V. Densités et biomasses du méiobenthos. *Vie Milieu*, **22** (Sér. B), 351–424.

Stackelberg, U. v., U. v. Rad, and B. Zobel, 1979. Asymmetric sedimentation around Great Meteor Seamount (North Atlantic). *Mar. Geol.*, **33**, 117–132.

Sushchenya, L. M., 1970. Food rations, metabolism and growth of crustaceans. In *Marine Food Chains*, J. H. Steele, ed. Oliver and Boyd, Edinburgh, pp. 127–141.

Taniguchi, A., 1973. Phytoplankton–zooplankton relationships in the western Pacific Ocean and adjacent seas. *Mar. Biol.*, **21**, 115–121.

Tendal, O. S., 1972. A monograph of the Xenophyophoria (Rhizopodea, Protozoa). *Galathea Rep.*, **12**, 7–103 and 17 plates.

Tendal, O. S., 1979. Aspects of the biology of Komokiacea and Xenophyophoria. *Sarsia*, **64**, 13–17.

Tendal, O. S. and R. R. Hessler, 1977. An introduction to the biology and systematics of Komokiacea (Textulariina, Foraminiferida). *Galathea Rep.*, **14**, 165–194 and 26 plates.

Thane, A. and T. Perry, 1975. Observations on the absence of cytochrome oxidase and mitochondria in some anaerobic marine ciliates. *J. Protozool.*, **22**, 45A. (Abstract.)

Thiel, H., 1966. Quantitative Untersuchungen über die Meiofauna des Tiefseebodens. *Veröff. Inst. Meeresforsch. Bremerh., Sonderbd.*, **2**, 131–147.

Thiel, H., 1970. Bericht über die Benthosuntersuchungen während der "Atlantischen Kuppenfahrten 1967" von F.S. "Meteor." *"Meteor" Forsch. Ergebnisse D*, **7**, 23–42.

Thiel, H., 1971. Häufigkeit und Verteilung der Meiofauna im Bereich des Island-Färöer-Rückens. *Ber. Dt. wiss. Komm. Meeresforsch.*, **22**, 99–128.

Thiel, H., 1972a. Die Bedeutung der Meiofauna in küstenfernen benthischen Lebensgemeinschaften verschiedener geographischer Regionen. *Verh. Dt. Zool. Ges., Helgoland*, **65**, 37–42.

Thiel, H., 1972b. Meiofauna und Struktur der benthischen Lebensgemeinschaft des Iberischen Tiefseebeckens. *"Meteor" Forsch. Ergebnisse D*, **12**, 36–51.

Thiel, H., 1975. The size structure of the deep-sea benthos. *Int. Revue Ges. Hydrobiol.*, **60**, 575–606.

Thiel, H., 1978a. Benthos in upwelling regions. In *Upwelling Ecosystems*, R. Boje and M. Tomczak, eds. Springer-Verlag, Berlin, pp. 124–138.

Thiel, H., 1978b. Zoobenthos of the CINECA area and other upwelling regions. Symp. The Canary Current: Studies of an Upwelling System; Las Palmas, 11–14 April 1978.

Thiel, H., 1979a. Structural aspects of the deep-sea benthos. *Ambio, Spec. Rep.*, **6**, 25–31.

Thiel, H., 1979b. First quantitative data on the deep Red Sea benthos. *Mar. Ecol. Progr. Ser.*, **1**, 347–350.

Thiel, H., 1980a. Benthic investigations of the deep Red Sea. Cruise Reports: R.V. "Sonne"—MESEDA I (1977) and R.V. "Valdivia"—MESEDA II (1979). *Courier Forsch. Inst. Senckenberg*, **40**, 35 pp.

Thiel, H., 1980b. Community structure and biomass of the benthos in the central deep Red Sea. *Proc. Symp. The Coastal and Marine Environment of the Red Sea, Gulf of Aden and Tropical Western Indian Ocean; Khartoum, 9–14 January 1980*, Vol. III, 127–134.

Thiel, H., 1981. Benthic investigations in the Northwest African upwelling area. Report on the cruises 26, 36, 44 and 53 of R.V. "Meteor." *"Meteor" Forsch. Ergebnisse D*, **33**, 1–15.

Thiel, H., 1982. Zoobenthos of the CINECA area and other upwelling regions. *Rapp. P.-v. Réun. Cons. perm. int. Expl. Mer*, **180**, 323–334.

Thiel, H. and R. R. Hessler, 1974. Ferngesteuertes Unterwasserfahrzeug erforscht Tiefseeboden. *Umschau*, **74**, 451–453.

Thiel, H. and H. Rumohr, 1979. Photostudio am Meeresboden. Neue Erkenntnisse für das Verhalten von Meerestieren. *Umschau*, **79**, 469–472.

Thiel, H., D. Thistle, and G. D. Wilson, 1975. Ultrasonic treatment of sediment samples for more efficient sorting of meiofauna. *Limnol. Oceanogr.*, **20**, 472–473.

Thistle, D., 1978. Harpacticoid dispersion patterns: Implications for deep-sea diversity maintenance. *J. Mar. Res.*, **36**, 377–397.

Thistle, D., 1979. Harpacticoid copepods and biogenic structures: Implications for deep-sea diversity maintenance. In *Ecological Processes in Coastal and Marine Systems*, R. J. Livingston, ed. Plenum Press, New York, pp. 217–231.

Thorson, G., 1957. Bottom communities. *Mem. Geol. Soc. Am.*, **67**, 461–534.

Tietjen, J. H., 1971. Ecology and distribution of deep-sea meiobenthos off North Carolina. *Deep-Sea Res.*, **18**, 941–957.

Tietjen, J. H., 1976. Distribution and species diversity of deep-sea nematodes off North Carolina. *Deep-Sea Res.*, **23**, 755–768.

Uhlig, G., 1970. Untersuchungen über die Bodenmikrofauna der Tiefsee. Jahresber. 1970. *Biol. Anst. Helgoland*, 23–25.

Uhlig, G., 1971. Untersuchungen über die Bodenmikrofauna der Tiefsee. Jahresber. 1971. *Biol. Anst. Helgoland*, 17–18.

Uhlig, G., H. Thiel, and J. S. Gray, 1973. The quantitative separation of meiofauna. A comparison of methods. *Helgoländer wiss. Meeresunters.*, **25**, 173–195.

Vernberg, W. B. and B. C. Coull, 1974. Respiration of an intertidal ciliate and benthic energy relationships. *Oecologia*, **16**, 259–264.

Vitiello, P. and M. H. Vivier, 1974. *Données Quantitatives sur la Méiofaune d'une Zone Profonde de Déversements Industriels*, Vol. IV, Bulletin 1. Union des Océanographes de France, pp. 13–16.

Vivier, M. H., 1975. Le méiobenthos du Canyon de Cassidaigne influencé des déversements du boue rouge d'alumine sur la nématofaune. Thèse, Université d'Aix-Marseille, 109 pp.

Vivier, M. H., 1978a. Conséquences d'un déversement de boue rouge d'alumine sur le méiobenthos profond (Canyon de Cassidaigne, Méditerranée). *Tethys*, **8**, 249–262.

Vivier, M. H., 1978b. Influence d'un déversement industriel profond sur la nématofaune (Canyon de Cassidaigne, Méditerranée). *Tethys*, **8**, 307–321.

Walch, C., 1978. Recent abyssal benthic Foraminifera from the eastern equatorial Pacific. Unpublished M.S. thesis, University of Southern California, Los Angeles, 117 pp.

Walker, D. A., A. E. Linton, and C. T. Schafer, 1974. Sudan Black B: A superior stain to rose bengal for distinguishing living from non-living Foraminifera. *J. Foram. Res.*, **4**, 205–215.

Walsh, J. J., 1981. Shelf-sea ecosystems. In *Analysis of Marine Ecosystems*, A. R. Longhurst, ed. Academic Press, London, pp. 159–196.

Wieser, W., 1956. Some free-living marine nematodes. *Galathea Rep.*, **2**, 243–253.

Wieser, W., 1960a. Benthic studies in Buzzard Bay. II. The meiofauna. *Limnol. Oceanogr.*, **5**, 121–137.

Wieser, W., 1960b. Populationsdichte und Vertikalverteilung der Meiofauna mariner Böden. *Int. Revue Ges. Hydrobiol.*, **45**, 487–492.

Wigley, R. L. and A. D. McIntyre, 1964. Some quantitative comparisons of offshore meiobenthos and macrobenthos south of Martha's Vineyard. *Limnol. Oceanogr.*, **9**, 485–493.

Wolff, T., 1960. The hadal community, an introduction. *Deep-Sea Res.*, **6**, 95–124.

Wolff, T., 1962. The systematics and biology of bathyal and abyssal Isopoda Asellota. *Galathea Rep.*, **6**, 1–320.

Wolff, T., 1970. The concept of the hadal or ultra-abyssal fauna. *Deep-Sea Res.*, **17**, 983–1003.

Wolff, T., 1977. Diversity and faunal composition of the deep-sea benthos. *Nature*, **267**, 780–785.

Zenk, W., 1971. Zur Schichtung des Mittelmeerwassers westlich von Gibraltar. *"Meteor" Forsch. Ergebnisse A*, **9**, 1–30.

Zeuthen, E., 1953. Oxygen uptake as related to body size in organisms. *Quart. Rev. Biol.*, **28**, 1–12.

Addenda

1. Vitiello, P. and A. Dinet, 1979. Définition et échantillon-nage du méiobenthos. *Rapp. Comm. int. Mer. Médit.*, **25/26**, 279–283.

Foraminifera are included in the microbenthos. Meiobenthos is restricted to metazoans, microbenthos covers all unicellular organisms, irrespective of size and processing methods.

2. Steedman, H. F., 1976. Miscellaneous preservation techniques. In *Zoo-*

plankton Fixation and Preservation, H. F. Steedman, ed. Monographs on Oceanographic Methodology. The UNESCO Press, 350 pp.

For sorting and storage the formalin-free preservative fluid containing 5% propylene glycol, 0.5% propylene phenoxetol, and 94.5% filtered seawater proved to be suitable, it does not dissolve carbonate structures and is harmless.

3. Dinet, A., 1980. Répartition quantitative et écologie du méiobenthos de la plaine abyssale atlantique. Thèse Université d'Aix-Marseille II, 180 pp. and annexes.

Meiofauna counts not published otherwise are given for the Norwegian Sea, the Eastern Atlantic, the Mediterranean, and the central Atlantic Ocean (Vema fracture zone). These data fall in the ranges of those presented in Figures 6–11, and would not change the conclusions.

4. The same. According to Khripounoff the calorific content on a weight specific basis differs considerably between taxa (Nematoda: 5617 cal/g; Copepoda: 3090 kcal/g). This may influence considerations on size structure if energy contents are compared.

5. Carey, Jr., A. G., 1981. A comparison of benthic infaunal abundance in two abyssal plains in the northeast Pacific Ocean. *Deep-Sea Res.*, **28,** 467–479.

Macro-infauna was found to be smaller on the average in samples from the more distant abyssal plain compared to those taken at the same depths but closer to the continents.

6. Grasshoff, M., 1981. Die Gorgonaria, Pennatularia und Antipatharia des Tiefwassers der Biskaya (Cnidaria, Anthozoa). Ergebnisse der französischen Expeditionen Biogas, Polygas, Geomanche, Incal, Noratlante und Fahrten der "Thalassa." II. Taxonomischer Teil. *Bull. Mus. Natl. Hist. Nat. Paris,* 4e sér. 3, Sec. A, No. 4, 941–978.

Umbellula thieli is synonymous to *U. monocephalus.*

7. Rhoads, D. C., A. Lutz, R. M. Cerrato, and E. C. Revelas, 1982. Growth and predation activity at the deep-sea hydrothermal vents along the Galápagos Rift. *J. Mar. Res.*, **40,** 503–516.

A new species of mussels recovered from the hydrothermal vents exhibits large size and high growth rates similar to those of shallow water mussels. This supports the hypothesis on organism size in the deep sea. Additionally, the authors cite several papers on the hydrothermal vent communities and species' mode of life under vent conditions.

6. MICROBIOLOGY OF THE DEEP SEA

Holger W. Jannasch and Carl O. Wirsen

1. Introduction

In the 1840s Edward Forbes, a leading British oceanographer, theorized from the observed decrease of the bottom fauna with increasing depth that life should be absent from the ocean floor below about 300 fathoms. It was this azoic-zone theory that instigated Thomas Henry Huxley, then pathologist at the London School of Mines, to use the first opportunity for a closer look. In 1857 the steamer H.M.S. *Cyclops* was sent out to conduct a series of soundings in the northern Atlantic in preparation for a cable laying. Huxley had asked the medical officer of the ship to preserve in alcohol whatever deposits might be brought back to the surface. In this material Huxley (1868) discovered the coccolithophores.

In addition, however, he believed that the coccolithophores were "produced by independent organisms." In the wake of early Darwinism and strongly influenced by Haeckel's protoplasm theory, he thought that the whitish, jellied material encasing the coccolithophores might represent Haeckel's "Urschleim" (primordial slime), which he named *Bathybius haeckeli*. During the *Challenger* expedition (1873 to 1876), commonly considered the beginning of deep-sea biology, it became obvious that Huxley's observation was based on an alcohol precipitate of organic debris. He admitted his error gallantly. This affair, which was historically analyzed by Rehbok (1975), occurred at a time when nonmedical microbiology was in its infancy. The conceptual and methodological basis for tracing microbial life in the deep ocean, other than by the use of the microscope, was not yet developed.

Certes (1884) found bacteria in water and sediment samples collected from depths to 5000 m during the *Travallier* and *Talisman* expeditions (1882 and 1883). In 1886, during a crossing of the Atlantic Ocean, Fischer plated water samples on solidified bacteriological media, a technique that had just then been described by Koch (1881). Fischer, again a ship's medical officer, found bacteria in every sample in varying numbers and contributed to the emerging perception of the microbe's ubiquity in the biosphere. He reported up to 30,000 bacteria/ml in surface waters, and he also described a decrease of their number with depth and the finding of "very few" bacteria in waters exceeding 1100 m (Fischer, 1894).

These beginnings of marine microbiology were described in detail in two monographs by Benecke (1933) and ZoBell (1946). From both these records of the early literature it is apparent that a definable deep-sea microbiology did not yet exist up to those years. Later ZoBell and Johnson (1949) started work on the effect of hydrostatic pressure on microbial activities. The former's participation in the

Danish *Galathea* expedition (1950 to 1952) and his subsequent work resulted in many publications that were later discussed comprehensively in a number of reviews (ZoBell, 1968, 1970; ZoBell and Kim, 1972; Morita, 1972, 1976). Some of these represent chapters in proceedings of symposia on barobiology edited by Zimmerman (1970), Sleigh and Macdonald (1972), and Brauer (1972) emphasizing temperature and hydrostatic pressure as the most important physical parameters in deep-sea biology. Recent treatises on pressure as an ecological, physiological, and molecular parameter in microbial metabolism are those by Marquis (1976), Marquis and Matsumura (1978), and Landau and Pope (1980).

In view of these rather extensive literature surveys, the present chapter concentrates on the more recent developments in the field. Furthermore, in accordance with the overall theme of this book, the major emphasis is on field work and the effect of deep-sea conditions on natural microbial populations. Strictly physiological, biochemical, and molecular effects of temperature and pressure on microbial metabolism are not part of this chapter. Its general purpose is to provide an overview of current research and trends in deep-sea microbiology with an emphasis on work done in the authors' laboratory.

2. Definitions

Any organism of microscopic dimensions may be defined as a microorganism. In present-day usage, however, microbiology deals largely with the study of procaryotic organisms, that is, those lacking a membrane-bound nucleus, and the viruses. The eucaryotic protists (algae, protozoa, and fungi) are included in a general treatise on marine microorganisms by Sieburth (1979) but are usually studied under their own specific label. Deep-sea microbiology as it appears today has concentrated on the study of bacteria with less emphasis on viruses and fungi. Occurrence and systematics of the latter with emphasis on the deep sea are treated, for instance, by Kohlmeyer and Kohlmeyer (1979) and are not covered in this chapter.

To a certain extent, deep-sea microbiology is high-pressure rather than low-temperature microbiology. The temperatures are confined to the very narrow range of 2 to 4°C. These are conditions that presumably select for low-temperature adapted or "psychrophilic" bacteria. They grow optimally at about 15°C or lower (by definition) and relatively fast within the range of −2 to 15°C as compared to the common "mesophilic" growth responses with optima at 30 to 40°C (Morita, 1976; Baross and Morita, 1978).

Hydrostatic pressure increases by about 1 atm (1.02 bar) per 10 m of water depth. Because of the compressibility of seawater, the deviation amounts to +9.0 atm at 5000-m and +27 atm at 10,000-m depth, neglecting additional effects of latitude and gravitational irregularities (Saunders and Fofonoff, 1976). Since the absence of a gas phase is self-evident in most of these considerations, any reference to pressure is always meant to be hydrostatic if not otherwise indicated.

ZoBell (1968) defines the deep sea as the area below 3800 m, the mean depth of the world's oceans. Because of the continual pressure increase with depth and the absence of (as far as observed) microbial responses to distinct ranges of physiologically effective pressures, no defined depth zoning is given here. Studies relating to deep-sea biology commonly deal with depths starting well below the light-affected surface waters and the nutrient and temperature discontinuities of the thermocline, that is, at depths below about 1000 m.

Deep-sea water and the upper layers of most sediments are oxygenated. This is largely due to the low rates of oxygen consumption in this oligotrophic environment. The term "oligotrophic" is used to describe waters with low nutrient concentrations. More recently it has also been applied to designate growth characteristics of organisms that are able to cope with low-nutrient conditions (Shilo, 1979; Poindexter, 1981a,b).

Growth of natural microbial populations or bacterial pure culture isolates have been found to be barotolerant to various degrees. In other words, they tolerate increased hydrostatic pressure more or less as determined by the increase of cell numbers or by metabolic activities as compared to controls incubated at atmospheric pressure. The term "barophilic" was used first by ZoBell and Johnson (1949) and is defined today by a requirement of increased pressure for growth or by increased growth at pressures higher than atmospheric pressure.

In the absence of light, the flow of nutrients and biochemical energy to the deep sea depends on the sedimentation of particulate food materials, originated by photosynthesis in surface waters, and on the transport of dissolved materials by currents and diffusion. In terms of particulate organic carbon, this flow has been estimated in a number of recent studies (Honjo, 1978; Hinga et al., 1979; Brewer et al., 1980) to be in the range of 20 to 800 $mg/m^2 \cdot yr$. Much higher quantities (2 to 6 $g/m^2 \cdot yr$) have been estimated for the continental slope (Rowe and Gardner, 1979). An exception to this surface-originated food source is the recently discovered geothermal supply of energy for chemosynthetic primary productivity at deep-sea spreading centers, which are discussed in a separate section of this chapter. Microorganisms depending on organic materials as sources for energy, carbon and electrons are described as chemoheterotrophic. Chemoautotrophic or chemosynthetic organisms use CO_2 as the primary source of carbon and reduced inorganic compounds, such as hydrogen sulfide or ammonia, as sources of energy.

3. Methodological Considerations

Deep-sea microbiology and microbial ecology in general attempt principally to answer three basic questions: (1) what there is (i.e., a description of the physiological and morphological types of microorganisms present); (2) how many are there (i.e., a quantification of the different types and their distribution, etc.); and (3) what they are doing (i.e., the *in situ* metabolic activities of the different types and their interaction with and significance for higher forms of life as well as for biogeochemical processes). The obvious difficulties in answering these questions are inherent in the characteristic properties of procaryotes that make them potentially ubiquitous: their small size and, therefore, high dispersibility combined with their ability to survive unfavorable growth conditions (including starvation, high pressure, and low temperature) for long periods of time. The metabolic diversity and specificity of microorganisms, specifically, their ability to grow at the expense of a wide variety of organic and inorganic substrates, is another characteristic of procaryotic organisms. For that reason, microbial ecologists are more concerned with describing the occurrence of metabolic types than species of microorganisms.

Because of this variable metabolic activity of the individual cell in the natural habitat, the significance of their standing crop or biomass is not the same as in higher organisms; that is, numbers of viable bacteria are not a measure of their *in situ* activity. In addition, errors in cell counts are high, often rendering such data statistically incompatible with those of physicochemical measurements. Many of the existing data on microbial populations have been obtained by the classical technique of streaking a small sample aliquot on the surface of an agar plate containing a certain growth medium and counting the grown colonies after a certain period of incubation at a certain temperature. As a result of the above-mentioned metabolic diversity and specificity, only a fraction of the actual microbial population can be expected to develop into visible colonies under the particular conditions provided. Furthermore, a colony-forming unit can be a single cell as well as a particle or conglomerate containing hundreds of cells. Agar plating techniques are of value, however, when the media's selective capacity is used, that is, for the differential counting of a specific metabolic type, but not for "total bacterial counts."

Estimation of total numbers can be approached by microscopic techniques, but difficulties in distinguishing microbial cells from organic debris often lead to overestimates (Jannasch and Jones, 1959). Differential staining with Rose Bengal and erythrosine has been used to distinguish between living and dead organic matter since the turn of the century (Winogradski, 1949). Strugger (1949) introduced the use of acridine orange in connection with fluorescence microscopy. This technique was considerably improved when used in combination with epi-illumination of samples passed through cellulose ester and polycarbonate filters (Zimmermann and Meyer-Reil, 1974; Jones and Simon, 1975; Daley and Hobbie, 1975). Yet, in the study of off-shore seawaters, where cell size varies greatly (e.g., Tabor et al., 1981b), even counts obtained with the epifluorescence technique should be considered only approximations because of the limited resolution of light microscopy. For this reason, attempts have been made by use of scanning electron microscopical spot tests (Daley, 1979; Krambeck et al., 1981). Such observations are useful for morphological studies but have obvious statistical limitations for quantitative assessments. Of more importance, however, are the above-mentioned inherent problems in relating total counts to metabolic activity.

In recent years, a number of sophisticated techniques have been developed to estimate microbial biomass in seawater by measuring characteristic cell constituents. The usefulness of these techniques depends basically on two conditions: the analytical procedure should be highly sensitive; and the material to be measured must be a constituent characteristic of the living microbial cell only. Starting with adenosine triphosphate (ATP), various nucleotides have been proposed and tested for this purpose. In a thorough survey of this work, Karl (1980) discusses the diverse applications of nucleotide determinations in biological oceanography and the interpretation of the data. Other materials partially fulfulling the above criteria have been the lipopolysaccharides (Watson et al., 1977) and muramic acid (King and White, 1977) as cell wall constituents and specific lipids (White et al., 1979) of bacteria.

For most marine microbiological studies, the determination of metabolic transformation activities appears to be more significant than that of biomass. However, in contrast to CO_2 as the substrate common to all photoautotrophic organisms, no single substrate can be used for measuring the activity of all chemoheterotrophic organisms at the same time. A multitude of assimilable and decomposable organic compounds may be available to and used simultaneously or in succession by the microbial population in natural seawater. The "heterotrophic CO_2 uptake" has been suggested as a comprehensive index of population activity (Romanenko, 1964), but its percentage of the total uptake of organic carbon ranges from 0.2 to about 10 and varies greatly with the type of organism and substrate used. This CO_2 uptake in the dark will also be affected by chemoautotrophic microorganisms if inorganic electron donors, such as reduced sulfur and nitrogen compounds, are available.

Many attempts to arrive at the most suitable activity measurement have been made, often in combination with biomass determinations. The following papers (references given in parentheses) may serve as guides to this literature: autoradiography in combination with epifluorescence microscopy (Meyer-Reil, 1978); metabolic reduction of tetrazolium salts (Christensen et al., 1980); and the same techniques in combination with the use of nalidixic acid as a specific inhibitor of deoxyribonucleic acid synthesis (Maki and Remsen, 1981). This chapter is not the place to discuss those various approaches in detail, especially as none of them have been used in deep-sea microbiology.

Two general approaches to activity determinations in oligotrophic waters are measurement of *potential* transformation rates of an added substrate and measurement of *unaltered in situ* activities. The former concerns the uptake or conversion of an added metabolizable compound, such as an amino acid or a carbohydrate, the concentration of which will be considerably above the natural level. Such data can be used for comparative studies describing effects of environmental factors, such as pressure or temperature, on the metabolic response of natural populations. The approach will also yield information on the potential *in situ* turnover of a given substrate, assuming that it becomes available through some normal event such as rapidly sinking solid food sources or pollutants. The consumption of a radiolabeled substrate can be expressed as incorporation or respiration (e.g., Wirsen and Jannasch, 1974), or it can be measured by autoradiography (e.g., Hoppe, 1976).

Measurement of unaltered in situ activities is more difficult. It should be understood that the *in situ* concentration of a particular substrate, that is, the "natural" level, corresponds theoretically to the "left-over" concentration in open systems. As in continuous culture, this steady-state concentration is independent of the actual input of substrate and is not indicative of *in situ* turnover rates (Herbert et al., 1956). Furthermore, at substrate concentrations below or near the saturation constant, growth responds exponentially to substrate increase. Thus the addition of an indicator substrate will induce or at least stimulate microbial activities more at low than at relatively high concentrations. This would be especially effective in oligotrophic waters, such as the deep sea, where a large part of the microbial populations is in an inactive or resting state.

One recent approach that might overcome this problem is the use of adenine as a non-ATP-generating substrate. As a radiolabeled tracer material it will function only as a precursor in the production of stable ribonucleic acid (RNA) without affecting the rate of incorporation (Karl, 1979, 1981). The ubiquity of the adenine uptake among the bacteria as well as the magnitude of error associated with measuring the nonbacterial adenine nucleotides in the particulate fraction remain to be established. Another approach is the use of sulfate assimilation into proteins as a universal indicator of microbial growth (Cuhel, 1981; Cuhel et al., 1982).

In deep-sea microbiology, measurements of microbial growth and activity may be carried out in samples incubated with the added indicator substrate at *in situ* temperature and pressure. Whereas the temperature may be maintained constant during sample retrieval from great depths by various means, the loss of pressure has largely been accepted as unavoidable. Samples are recompressed in the ship's laboratory for incubation in special pressure vessels. The equipment used for this purpose is, in principle, still the same as originally described by ZoBell and Oppenheimer (1950). In discussing large numbers of bacteria brought to the surface from great depths, ZoBell (1968) states: "Although many bacteria from the deep sea survived, this observation fails to prove that some bacteria, possibly the most sensitive ones, were not destroyed by decompression. Answering this question may require the examination of deep-sea bacteria at *in situ* pressures without subjecting them to decompression." The effect of decompression on microbial activity in samples from a depth of only 400 meters has been shown to be significant (Seki and Robinson, 1969). More recently, pressure- and temperature-retaining samplers have been developed which render work with undecompressed populations of deep-sea bacteria possible (Jannasch et al., 1976; Jannasch and Wirsen, 1977; Tabor et al., 1981a). Much of this work is done at sea, and sterile techniques are a requirement for the sampling procedures as well as for the handling and transferring of the samples aboard the ship.

4. *In Situ* Incubation Studies

A classical approach for studying microbial activities in terrestrial and aquatic environments is to measure the metabolic transformation of certain substrates during incubation of a soil or water sample directly in the environment. Except for the addition of the substrate, commonly at near-natural concentrations, and the containment of the sample for the quantitative recovery of the substrate, the conditions of incubation are meant to be unchanged. Types and rates of microbial transformation studied in this way are often of a "potential" nature but have yielded a tremendous amount of useful information.

Applied to the deep sea, the advantages of *in situ* studies are the complete avoidance of pressure and temperature changes that might affect the population of microorganisms prior to incubation. The obvious disadvantage is the inaccessibility of deep-sea sites for taking series of subsamples, that is, conducting time-course experiments leading to calculations of growth and transformation rates. Muraoka (1971) made qualitative tests on the microbial decomposition of solid materials in the deep sea by depositing cotton, hemp, jute, and wood at depths of 1600 to 2000 m. Clear indication of microbial attack was found after 6.3 months of incubation, but no comparison with samples incubated in surface waters was made.

We attempted to improve this approach by running a parallel l-atm control at *in situ* temperature of 3°C in the laboratory (Jannasch et al., 1971). Plastic sample containers were lowered to a depth of 5300 m and fastened to buoy cables about 10 m above the sea floor. These packages had numerous 25-mm-diameter holes to allow free passage of seawater to the inside. The test materials were held in plastic bottles with holes of 13-mm diameter (seaweeds, etc.) or mounted on racks (wood samples, etc.). In addition, materials were kept in 120-ml bottles inoculated with in-shore seawater and closed with a rubber stopper for pressure equilibration. This procedure was based on the assumption that most solid organic materials sinking to the sea floor are heavily contaminated with surface-originated microorganisms. In other bottles, initially containing only sterile substrates, pressure equilibrating serum caps were slit to allow filling by pressure during descent. The general result of comparing data from *in situ* incubations to those of 1-atm incubations indicated markedly reduced rates of microbial decomposition activities in the deep sea. A study by Sieburth and Dietz (1974) indicated that deep-sea scavengers consume organic foodstuffs quickly. They have to be protected by screening to allow an unobstructed observation of microbial activities (see Fig. 2).

Use of the research submersible *Alvin* offered new possibilities for deep-sea incubation studies (Jannasch and Wirsen, 1973). Negative effects of deep-sea conditions on surface-originated bacteria may be expected, but specifically adapted microorganisms may still exist. To avoid contamination of sterilized substrates by microorganisms of surface waters or the water column during descent, the glass bottles with slit serum caps were housed in pressure-tight aluminum cylinders (Fig. 1). Since the amount of air contained in the cylinders would result in a toxic concentration of oxygen after dissolution (ZoBell and Hittle, 1967), flushing with nitrogen gas was necessary. After transport by *Alvin* to depths of up

Fig. 1. Two "bottle racks" with twenty 120-ml bottles each deposited on the ocean floor (39°46′N, 70°41′W, 1800-m depth) by the research submersible *Alvin* after *in situ* inoculation (for explanation, see text). Two aluminum pressure housings are shown at lower right and the submersible's mechanical arm, at lower center and left. [From Jannasch and Wirsen (1973).]

to 4000 m, valves are operated by the mechanical arm, flooding the cylinders and thereby inoculating the sterile sample bottles. After equilibration of the pressure inside and outside of the cylinders, their lids were removed, the bottle racks lifted out and deposited on the sea floor for up to 1 year of incubation time. Parallel samples were returned for control incubations at 1 atm in the laboratory. This device, designated as "bottle racks," has been used for many years in studying the decomposition of solid materials, such as various kinds of seaweeds, wood, paper, chitin, agar, and so on (Fig. 2) and radiolabeled dissolved organic substrates (Jannasch and Wirsen, 1973).

An improved instrument for these studies, but limited to dissolved substrates,

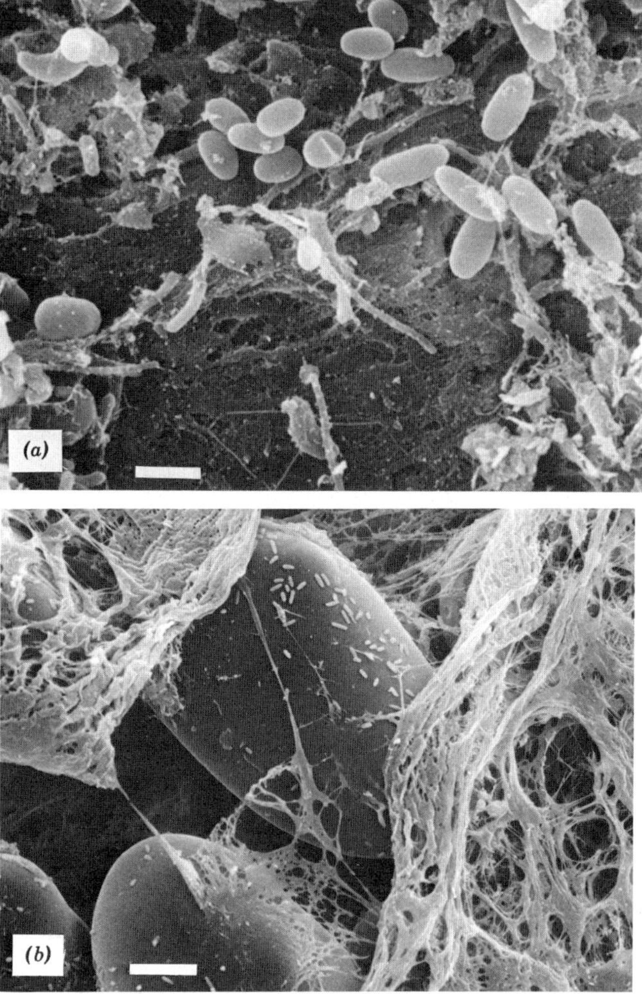

Fig. 2. Scanning electron micrographs of (*a*) pine wood contained behind 64-μm Nitex screening (scale bar 1 μm) and (*b*) a potato slice contained behind 28-μm Nitex screening (scale bar 10 μm) after exposure for 1 year on the northern Atlantic Ocean floor (32°19′N, 69°41′W) at a depth of 3640 m. All the potato slices, except for the ligneous peel, had disappeared behind screening with a mesh size of 64 μm and larger.

is the "syringe array." Six 200-ml plastic syringes containing radiolabeled substrate are filled *in situ* through a joint inlet nozzle by rotary movement of *Alvin*'s mechanical hand. After *in situ* incubation each syringe array is retrieved *in toto* and the parallel samples and controls fixed or processed in the ship's laboratory. Syringe arrays have been used extensively at depths of up to 4000 m at different geographic locations including the Galapagos Rift area (see below). Advantages over the bottle racks are that no pressure differential occurs during filling and that pressure housings are unnecessary.

The data obtained with bottle racks and syringe arrays demonstrate invariably that (1) the conversion of organic substrates is slower *in situ* than in the 1-atm controls, (2) *in situ* conversion rates were similar when the inoculation took place near the surface or in the deep sea, and (3) there was a definite substrate preference. These results become of particular interest considering the fact that pressure-adapted or barophilic bacteria do indeed exist (see below) but do not seem to become prevalent in these bottle or syringe experiments, which basically represent microbial enrichment cultures. These long-term (up to one year) incubations, however, are limited to single end-point measurements. Therefore, time-course experiments in pressure-retaining samplers (see below) became desirable.

Because experimentation with *Alvin* on the deep ocean floor is restricted by weather, expense, and a depth limitation of 4000 m at present, the possibilities of using *free vehicles* for microbiological experiments have been explored and found to be extremely useful. Since Isaacs's (1969) original work, free vehicles of various configuration and degree of technical sophistication have been used, with the more expensive ones controlled by acoustic means and the others by simple preset timers. Tripods equipped with flotation and releasable weights can be dropped over the side of the ship to descend to practically any depth (Fig. 3) (Jannasch and Wirsen, 1980; descriptive paper in preparation). Chapter 2 of this volume reviews other uses of free vehicles.

For microbiological work, five experimental approaches have been used with timer-operated tripods: (1) microbial growth in a peptone yeast extract medium (Seki et al., 1974) and microbial nitrate metabolism (Wada et al., 1975), (2) transformation of radiolabeled organic substrates injected into various levels of the upper sediment layers (Jannasch and Wirsen, 1980), (3) microbial growth on surfaces of various materials, (4) decomposition of substrates contained in compartmentalized tubes and inserted into the sediment (Wirsen and Jannasch, 1976), and (5) trapping and incubation of deep-sea amphipods in the presence of a radiolabeled solid food source for comparative studies on the metabolism of deep-sea scavengers and their intestinal microflora (Jannasch et al., 1980). The tripods used in these studies are intentionally of a relatively inexpensive and unsophisticated design in order to assure a statistically significant amount of data from a large number of vehicles and deployments.

An example of studying microbial activities in deep-sea sediments is given in Table I. It was found that the incorporation and respiration of acetate decreased with increasing sediment depth *in situ* as well as in the 1-atm controls whereas the overall rate of substrate conversion was substantially higher in the latter. When deep-sea amphipods were captured and incubated *in situ* with a ^{14}C-labeled protein food source (Jannasch et al., 1980), the conversion of carbon within various biochemical fractions (Table II) in the amphipod tissue differed from that in the

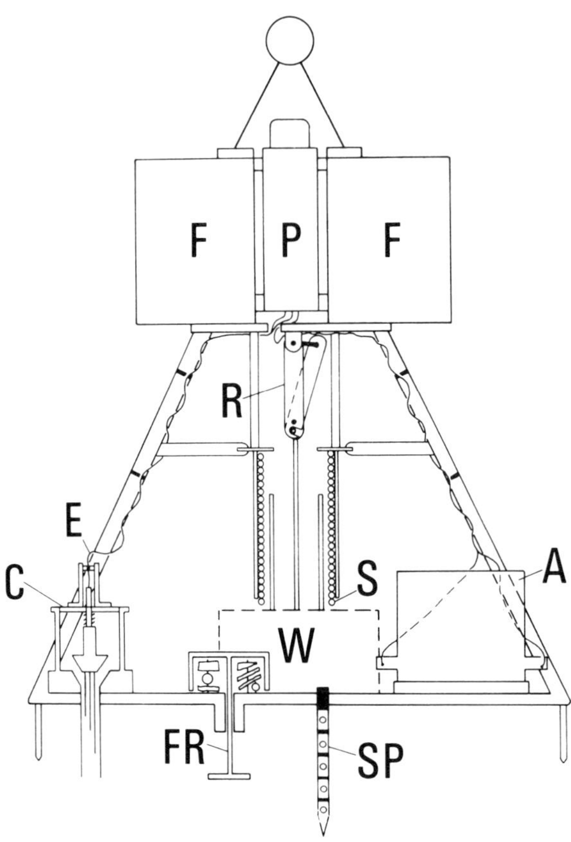

Fig. 3. Free vehicle for microbiological deep sea studies: F, syntactic foam flotation; P, pinger housing containing batteries and electronic timer; R, release hook; S, spring; W, weight; C, core unit; E, electrocorrosive release for automated sample injection (Jannasch and Wirsen, 1980); FR, fouling rack; SP, sediment poker (Wirsen and Jannasch, 1976); A, amphipod trap (Jannasch et al., 1980).

TABLE I

Free-Vehicle Measurements on Incorporation and Respiration of ^{14}C-Labeled Acetate in Upper Sediment Layers at a Northern Atlantic Site (39°10′N, 67°10′W) 4175 m Deep Compared to 1 atm Control[a]

	Points of Injection (Core Depth in cm)	Acetate Incorporated (μmol/cm^3 · day)	Acetate Respired (μmol/cm^3 · day)
In situ incubation	2	6.55×10^{-5}	2.80×10^{-5}
about 3°C, 6 days	4	3.19×10^{-5}	1.69×10^{-5}
	7	2.43×10^{-5}	1.45×10^{-5}
Control incubation at	2	6.61×10^{-4}	1.32×10^{-4}
1 atm, 3°C, 6 days	4	1.28×10^{-4}	4.58×10^{-5}
	7	1.91×10^{-4}	3.96×10^{-5}

[a] The amount of acetate injected was 2.68×10^{-1} μmol per 60 cm^3 core section. The sections 0 to 3, 3 to 6, and 6 to 9 cm from the sediment surface were analyzed.

TABLE II
Fractionation of ^{14}C from a Labeled Protein Food Source (A, control) by Intestinal Microflora of an Amphipod (B) and by Remaining Tissue and Carapace of Animal (C) after 3 Days' *in situ* Incubation at 4975-m Depth[a]

	Total Radioactivity (%)		
Fraction	A	B	C
LMW pools, sugars, free amino acids	0.6	18.7	95.0
Nucleic acids, polysaccharides	4.5	8.1	0.4
Lipids	5.4	6.5	1.1
Peptides, long-chain amino acids	8.5	4.8	0.5
Protein	81.0	62.0	3.0

[a]The conversion of the labeled carbon in B went into the pools of low-molecular-weight (LMW) compounds, sugar, and amino acids and to the nucleic acid fraction, indicating the increase of microbial biomass. In the amphipod tissue most of the label is found in the first fraction and biosynthesis appears to be minimal under the given conditions.

gut tract and intestinal microflora. This approach offers a high potential for studying metabolic strategies of procaryotes and eucaryotes in the deep-sea environment.

5. Studies with Undecompressed Microbial Populations

Although the *in situ* inoculation–incubation approach represents one way to overcome the decompression problem, it has the distinct disadvantage of being limited to mere end-point measurements. There is no assurance that the microbial activity is constant over an extended incubation period. Furthermore, radiolabeled carbon may be recycled, thereby indicating an unrealistic respiration:incorporation ratio. It was necessary, therefore, to construct pressure- and temperature-retaining deep-sea water samplers that permitted time-course experiments. This was accomplished (Jannasch et al., 1973) by a 1-liter sampling unit that, in connection with a transfer unit for the addition and withdrawal of 13-ml subsamples, can also be used as an incubation vessel aboard ship or in the laboratory (Figs. 4*a,b*) (Jannasch et al., 1976).

The samplers are lowered on a cable and triggered by a messenger to take in a 1-liter sample slowly and without pressure differential. There is sufficient insulation in the heavy steel housing to prevent temperature changes on retrieval. A nitrogen gas cushion (Fig. 4*a*,C) compensates for small volume changes that occur during retrieval due to vessel expansion or during the attachment of a pressure gauge to determine the exact depth of sampling. The hand-operated transfer unit (Figure 4*a*,e) permits introduction and removal of 13-ml subsamples without affecting the pressure within the sample. Thus the sampler, containing a magnetic stirring bar, is turned into a culture chamber.

In preparation for isolating pure cultures in the absence of decompression (see following text), a filter sampler was developed for the *in situ* 1:200 concentration of 3-liter water samples over Nuclepore filters (Fig. 5). This instrument made sea

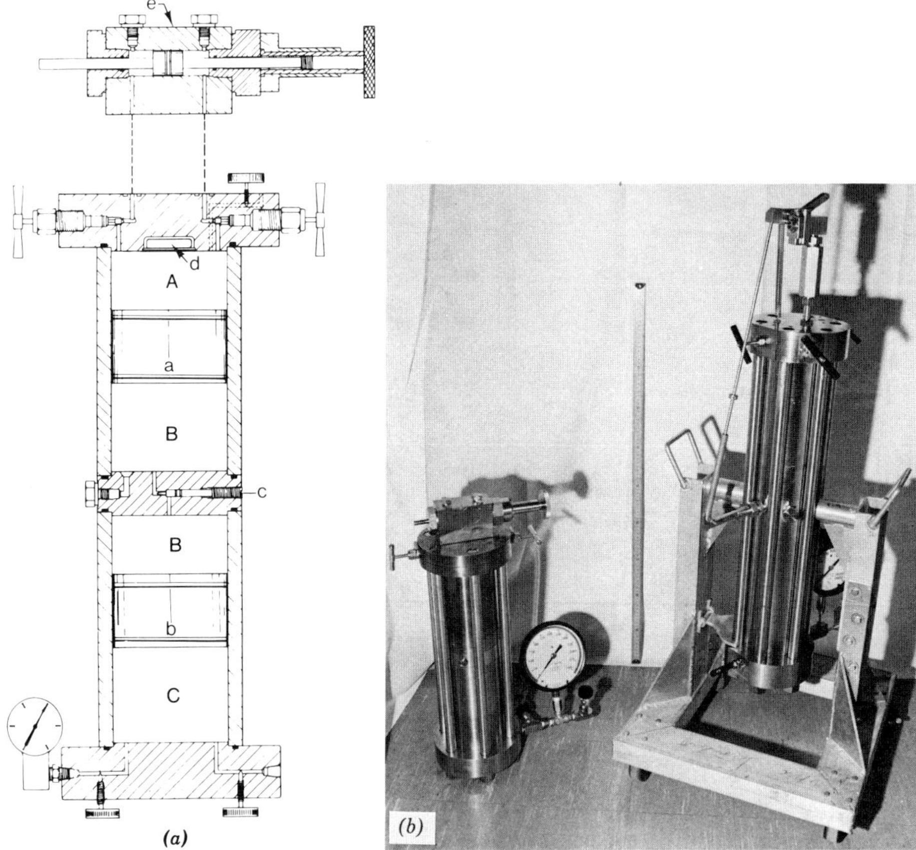

Fig. 4. Sampler–incubation vessels. (*a*) Scheme: (A) sample chamber (Teflon lined); (B) two sections of a freshwater-filled chamber separated by a pressure-snubbing device; (C) precharged gas (N_2) cushion; (a) and (b) free-floating pistons; (c) set screw for pressure-snubbing orifice; (d) stirring bar; (e) subsampling unit. Check valve, toggle valve, and intake cover are not shown. [From Jannasch et al. (1976).] (*b*) Pressure-retaining vessels. Left: for use at 200 atm, 316 stainless steel, intake mechanism replaced by subsampling unit. Right: for use at 600 atm (i.e., sampling depth, 6000 m), Nitronic 50 stainless steel, trigger device attached. [From Jannasch et al. (1976).]

operations more efficient because it can be sterilized and reused at sea. As many undecompressed samples can now be collected on a cruise as transfer–storage units are available. Scanning electron micrographs of cells collected at various depths by use of this sampler and fixed before (Fig. 6) and after decompression showed no significant differences in cell morphology. Similar results were obtained by Carlucci et al. (1976) by examining *in situ* fixed samples from 1200-m depth.

Metabolic uptake and turnover of a number of substrates by mixed microbial populations collected at various depths invariably showed an increase on decompression to 1 atm. The extent of pressure sensitivity appears to depend on the type of substrate used (Jannasch and Wirsen, 1982). The turnover rate as well as the total utilization of acetate were much less affected by pressure than are those of glutamate (Fig. 7). This can be interpreted in two ways: (a) the particular substrate

Fig. 5. Pressure-retaining "filter sampler" on retrieval from a depth of 5300 m; toggle valve tripped and protective intake cover removed by messenger. [For details, see Jannasch and Wirsen (1977).]

is utilized by only one or a limited number of organisms, which exhibit their own specific degree of barotolerance, or (b) the pathway and one or more enzymes involved in metabolizing the particular substrate express a specific barotolerance independent of the organisms carrying out the conversion. Only pure culture studies will be able to resolve this issue.

Tabor and Colwell (1976) and Tabor et al. (1981a,b) developed a somewhat different system for the sampling and cultivation of undecompressed microbial populations. They also found that the substrate utilization rates were generally stimulated by decompression. In an attempt to arrive at natural rather than potential rates, the substrate concentrations used in these studies were kept as close as possible to those estimated to occur naturally, specifically, 1 to 10 μg/liter (Lee and Bada, 1975) in the case of glutamate. When glutamate concentrations in the range of 31 to 96 μg/liter were added, the amount of total substrate utilization was not more than 2%.

6. Studies with Pure Cultures of Deep-Sea Bacteria

There is considerable literature [reviewed by Marquis and Matsumura (1978)] on growth studies at various pressures with bacterial pure cultures collected from

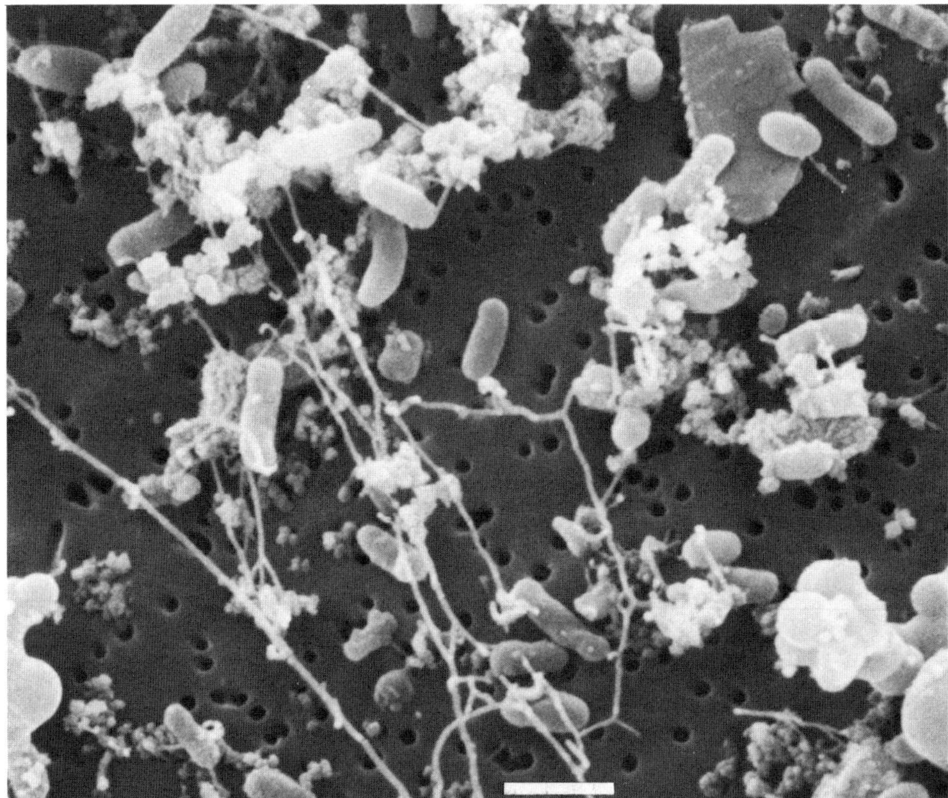

Fig. 6. Scanning electron micrograph of material collected with the filter sampler (Fig. 5) on a Nuclepore membrane (pore size 0.2 μm) at a depth of 4400 m and chemically fixed (glutaraldehyde) prior to decompression. Whereas the granular material has been shown to be an artifact of the fixation process, the slender filaments of 0.1-μm diameter and less are commonly observed next to typical bacterial cells collected at various depths (scale bar 1μm).

the deep sea and isolated at 1 atm. One set of data principally confirming results of the earlier work is given in Table III. The notion that psychrophilic (low temperature-adapted) bacteria might exhibit generally higher degrees of barotolerance was only partly confirmed in this study (Wirsen and Jannasch, 1975). The complexity of pressure effects is illustrated by the observations that small pressure increases may stimulate bacterial growth (Marquis and Matsumura, 1978) and that more highly barotolerant organisms could be isolated from garden soil than from the deep sea (Kriss, 1962, pp. 91–103).

In 1978, Marquis and Matsumura wrote "even today, we still have the question of whether there are bacteria specifically adapted for growth under deep-sea pressures." Shortly thereafter Yayanos et al. (1979) described pure cultures of truly barophilic bacteria that grow faster at pressures higher than 1 atm. These isolates were obtained from the guts of deep-sea amphipods by a technique described earlier (Dietz and Yayanos, 1978). A study by Schwarz et al. (1976) had already indicated that a whole gut population, if incubated with acetate, showed a near barophilic rate of uptake. More recently Deming et al. (1981) demonstrated that

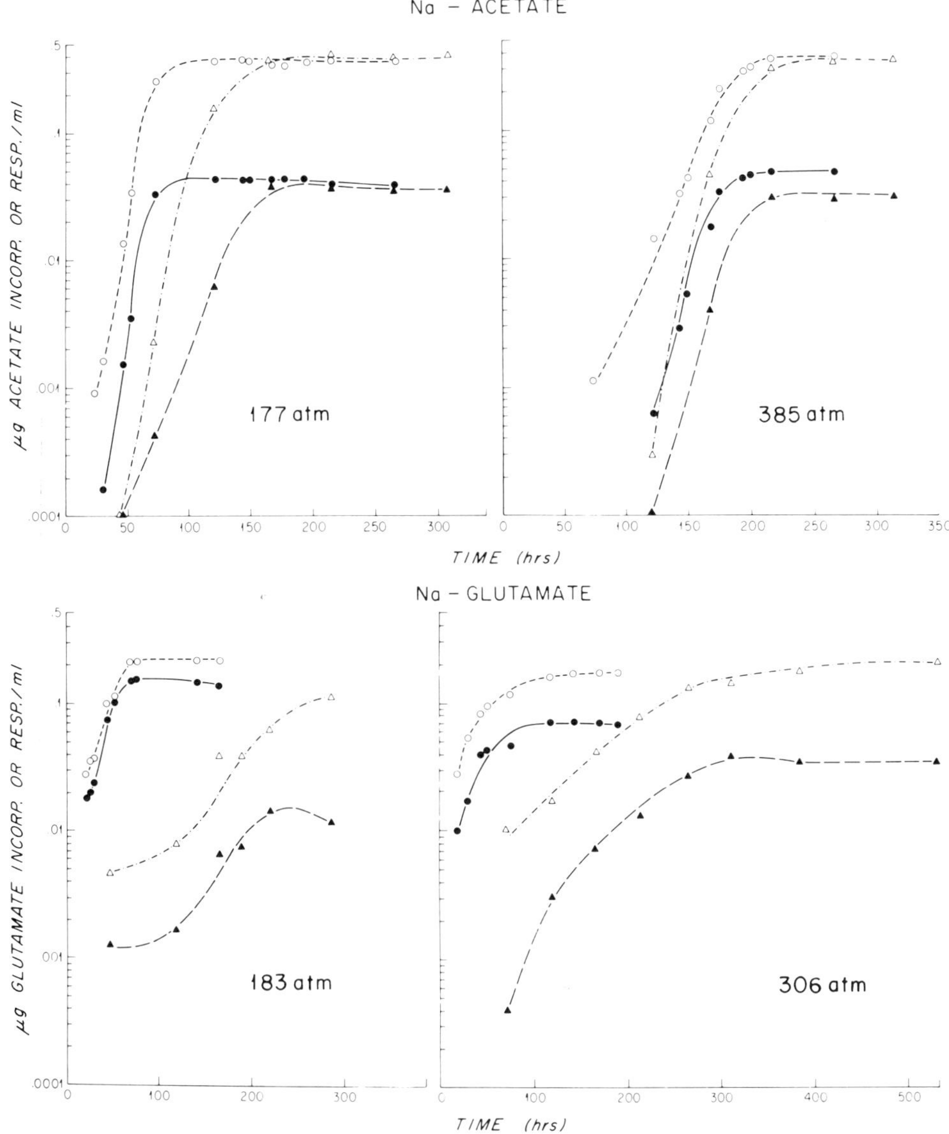

Fig. 7. Incorporation (filled symbols) and respiration (empty symbols) of ^{14}C-labeled acetate and glutamate by undecompressed microbial deep-sea populations incubated at *in situ* pressure (triangles) compared to decompressed 1-atm controls (circles). All incubations were done at 3°C, and the initial concentration of the substrates was approximately 0.5 μg/ml. The total utilization of acetate averaged 93% at 1 atm and 91% at *in situ* pressure, and that of glutamate averaged 69% at 1 atm and 37% at *in situ* pressure. Compared to the 1 atm controls the calculated average rates of acetate incorporation and respiration were lower by a factor of 0.59 at 177 atm and 0.99 at 385 atm. The corresponding values for the rates of glutamate turnover were 0.37 at 183 atm and 0.31 at 306 atm. [From Jannasch and Wirsen (1982).]

TABLE III
Incorporation of ^{14}C-Labeled Glutamate in μg/12 ml by Two Marine Bacterial Isolates (Psychrophile Ps-8 and Mesophile Ms-2) at 3°C and Various Pressures and Incubation Periods[a]

	4 hr		24 hr		7 days	
	Ps-8	Ms-2	Ps-8	Ms-2	Ps-8	Ms-2
1 atm	0.131	0.061	0.237	0.046	8.951	2.667
50 atm	0.097	0.037	0.091	0.041	5.073	0.042
400 atm	0.062	0.016	0.082	0.004	0.220	0.004

[a]The initial substrate concentration was 2 μg/ml. Data from Wirsen and Jannasch (1975).

growth of bacteria isolated from deep-sea amphipod and holothurian guts are enhanced at *in situ* pressure as compared to atmospheric pressure.

An explanation for this limited occurrence of barophilic bacteria seems to relate to the fact that their rate of growth, although relatively faster than that of the nonbarophilic bacteria, is still slower than the continuous input of surface-originated bacteria by sedimentation. Only where high nutrient concentrations prevail, such as in the guts of benthic animals, barophilic bacteria may establish the degree of predominance that facilitates their isolation.

The data in Fig. 7 indicate that the degree of barotolerance or barophilism can depend on the type of substrate used or on the metabolic pathway of a particular organism. As a consequence, an organism may be barophilic under natural conditions but not in the particular medium used for its cultivation and vice versa. This aspect of attributing the pressure response to the type of metabolism rather than to species of organisms must be investigated further.

While barotolerant bacteria are negatively affected by pressures higher than 1 atm, barophilic bacteria are also inhibited by pressures below a certain optimum. ZoBell and Morita (1957) termed organisms "obligately" barophilic if no growth occurred at 1 atm (Fig. 8). According to that definition, an organism growing at 1 atm at a lower rate than at elevated pressure may be called "nonobligately," rather than "facultatively," barophilic.

Referring to the early observations on negative effects of decompression on growth, ZoBell (1968) stated:

> Considerable difficulty has been experienced in trying to maintain barophilic bacteria in the laboratory. Most of the enrichment cultures from the deep-sea lose their viability after two or three transplants to new media, although some barophiles have survived in sediment samples for several months when stored at *in situ* pressures and temperatures. Even at refrigeration temperatures, the deep sea barophiles die off much more rapidly at 1 atm than when compressed to *in situ* pressures.

The probable physiological and molecular reasons for the loss of barophilism and barotolerance during decompression were discussed by Pope et al. (1976) and Landau and Pope (1980). In more recent studies, growth of a truly barophilic bacterium isolated from amphipods decomposing in a pressure-retaining trap (Yayanos et al., 1979) was reduced by about 90% when decompressed. Recently

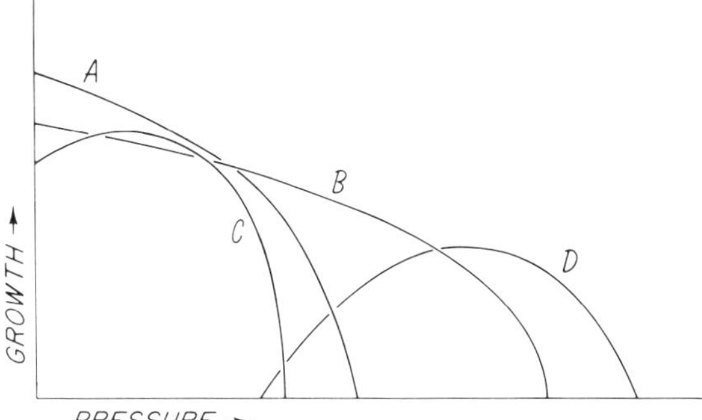

Fig. 8. Scheme of microbial growth responses to pressure based on data available at this time: *A* and *B* show barotolerance varying by degree and range; *C* and *D* show barophilic characteristics as defined by increased rates of growth at pressures higher than 1 atm. The obligate barophilic response *D* is defined by the inability to grow at 1 atm. The quantitative relationships between responses *A* and *D* as well as the shape of the curves are still most arbitrary in this scheme since, in the few studies done so far, the types of growth media and measurements used differed greatly.

described, highly pressure adapted, obligately barophilic bacteria do only grow above 350-atm pressure (Yayanos et al., 1981) and may be under strong stress during decompression.

It is always possible that some barophilic bacteria may not only be inhibited, but also killed during decompression, or that their growth behavior is irreversibly affected. Controlled decompression experiments can only be done with pure cultures of undecompressed bacteria.

It became a challenge, therefore, to construct a pressure chamber permitting the isolation of pure cultures of deep-sea bacteria in the absence of decompression. This was possible in connection with the sample-concentrating filter sampler (see above). To streak these concentrated seawater samples on a solidified medium, growth in a hyperbaric chamber, that is, one containing a gas phase, had to be ascertained. Taylor (1979) found that a certain helium/oxygen mixture did not inhibit microbial growth at high pressures. The isolation chamber was built and successfully used (Jannasch et al., 1982). Barotolerant as well as barophilic bacterial strains were isolated from deep-sea water, indicating that the latter also occurred outside the intestinal tracts of deep-sea animals, although at low densities.

7. Microbiology of Deep-Sea Thermal Vents

Below water layers of high photosynthetic productivity, the uptake of oxygen during decomposition of organic materials may lead to anoxic conditions and the subsequent bacterial reduction of sulfate to hydrogen sulfide. The latter may again be oxidized in oxic–anoxic interfaces by aerobic chemosynthetic bacteria with the concomitant reduction of carbon dioxide to organic carbon (Fig. 9). Because of the relatively high concentrations of sulfate in seawater, this auxiliary interaction of the sulfur cycle with the carbon cycle is particularly important in marine environments but it is largely limited to shallow water habitats such as lagoons and estuaries.

The water as well as the top sediment layers are normally oxygenated as a

AEROBIC

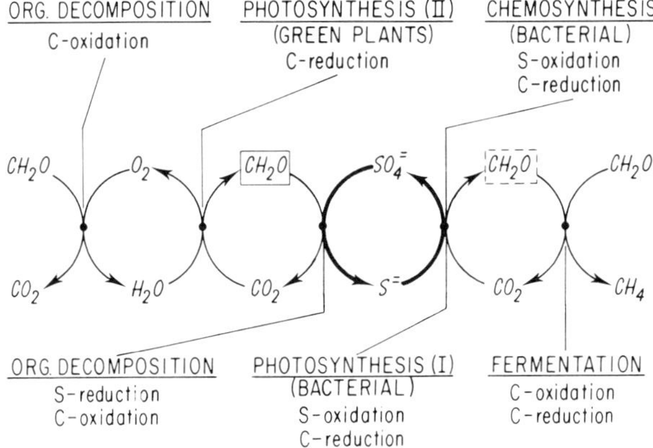

ANAEROBIC

Fig. 9. Scheme of the auxiliary nature of the sulfur cycle within the carbon cycle in the marine environment and the various microbial transformations involved. *Primary* photosynthetic production (photosystem II) of organic carbon by green plants is indicated by the solid rectangle. The dashed rectangle represents *secondary* chemosynthetic production, including some primary bacterial production by photosystem I. The chemosynthetic reduction of CO_2 at the deep-sea hydrothermal vent systems may be termed "primary" production if the definition is based on the geothermic origin of H_2S as the energy source, that is, independent of photosynthesis. The assimilatory reduction of sulfate to organic sulfur compounds and the release of hydrogen sulfide by deamination are quantitatively less important and are not included. [From Jannasch (in press).]

result of the low rate of oxygen uptake in the oligotrophic deep sea. According to a recent estimate by Jørgensen (in press), the oxygen uptake at the sea floor below a depth of 1000 m (87% of the total area) is about 8% of the total. By the same token, sulfate reduction is largely limited to the sediments of shallower water and is assumed to be negligible in the deep sea. Bottle rack experiments have shown, however, a potential thiosulfate oxidation and thereby the presence of thiosulfate-oxidizing bacteria (Tuttle and Jannasch, 1976). In only a few ocean areas do anoxic conditions reach the deep-sea floor. Examples are in the Black Sea [maximum depth ca. 2000 m (Degens and Ross, 1974)] and the Cariaco Trench [maximum depth ca. 1400 m (Richards, 1975)], where at the oxic–anoxic interface (Black Sea ca. 150 m, Cariaco Trench ca. 350 m) the reduced sulfur compounds become the source of energy for secondary production of organic matter (Sorokin, 1964, 1965).

In early 1977 unusually dense assemblages of benthic invertebrates were discovered in association with hydrothermal vents of the eastern Pacific Ocean floor spreading centers at depths of about 2550 m (Ballard, 1977; Corliss et al., 1979; Lonsdale, 1977). Photosynthetic production was not likely to be the major food source for these uncommonly active and copious deep-sea communities, and it was hypothesized that the hydrogen sulfide contained in the emitted vent water might represent the energy source for bacterial chemosynthesis providing the primary base of the food chain (Jannasch and Wirsen, 1979). This has largely been

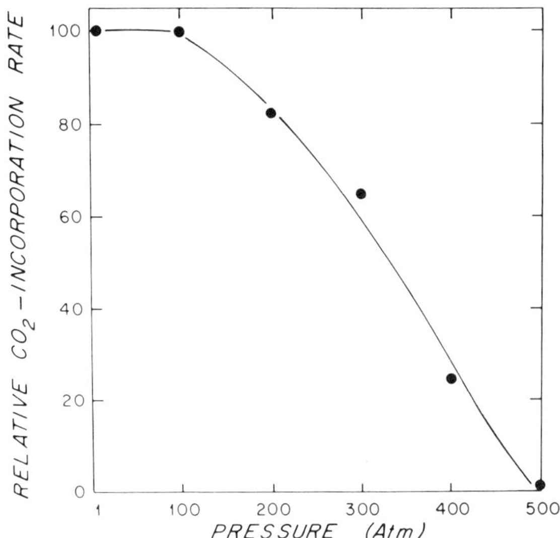

Fig. 10. Effect of hydrostatic pressure on the rate of $^{14}CO_2$ incorporation by *Thiomicrospira* sp. The organism was isolated from a hydrothermal vent at a depth of 2550 m. At the corresponding pressure of about 250 atm the activity was about 75% of its 1-atm level. [From Ruby and Jannasch (1982).]

confirmed (Karl et al., 1980), and several types of sulfur-oxidizing bacteria have been isolated (Ruby et al., 1981).

The reduction of sulfate occurs during the passage of seawater through hot parts of the Earth's crust (Edmond et al., 1979; Mottl et al., 1979) prior to emission. Thus the resulting hydrogen sulfide is produced by geothermal energy and is unrelated to photosynthesis, that is, solar energy. Therefore, the chemosynthetic conversion of inorganic into organic carbon based on hydrogen sulfide oxidation at the deep-sea vents represents a primary production. This definition of primary production is based on the origin of energy and disregards the fact that free oxygen as the electron sink required for chemosynthesis is also a product of photosynthesis. It is likely that other chemosynthetic organisms, such as hydrogen-, ammonium-, nitrite-, iron-, and possibly manganese-oxidizing bacteria contribute to this production. However, considering the relative quantity of reduced sulfur compounds (Mottl et al., 1979), it is reasonable to assume that by far the largest portion of chemosynthesis must be attributed to sulfur oxidation.

The aspect of pressure effects on the chemosynthetic activities at the vents might be discussed on a similar basis as above with respect to the preferential occurrence of pressure adaptation in enriched habitats. The first physiological study on a pure culture isolate, the obligately chemolithotrophic sulfur bacterium *Thiomicrospira* sp. (Ruby and Jannasch, 1982), showed a barotolerant rather than a barophilic response (Fig. 10). Since this organism might have grown deeper in the subsurface vent system at increased temperatures, we attempted to find thermophilic characteristics. We had no success with this particular organism; however, thermophilic bacteria have been isolated from a different vent system (J. Baross, personal communication).

At this stage of our knowledge, there appear to be three locations where the chemosynthesis takes place:

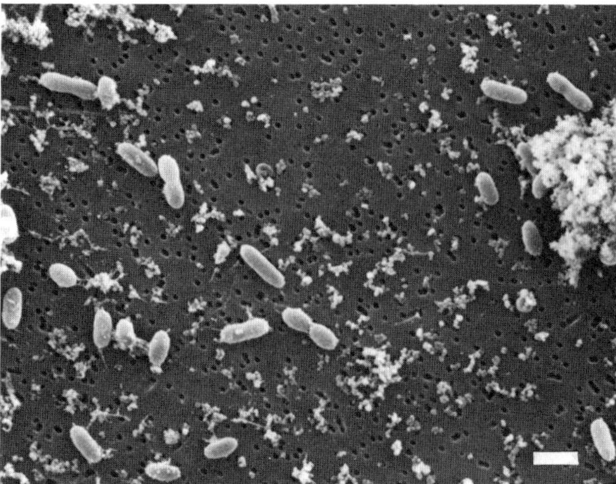

Fig. 11. Scanning electron micrograph of the particulate materials of turbid vent water from a depth of 2550 m collected on a Nuclepore filter with a pore size of 0.2 μm (scale bar 1 μm).

1. Chemical analyses of the vent water (Edmond et al., 1979) indicate that oxygenated seawater percolates downward through the porous lava layers, where it mixes with the highly reduced rising vent water. The high counts of bacteria as well as the ATP values measured in the emitted waters (Karl et al., 1980) indicate that subsurface growth occurs within the vent system. As filter samples indicate (Fig. 11), the turbidity in the vent water was largely due to microbial cells rather than particulate elemental sulfur or other amorphous material. The uniformity of cell shapes (rods and short spirals) was remarkable. Filter-feeding mussels are most commonly distributed very near vent openings where turbid water is emitted.

2. Multilayered microbial mats have covered all surfaces in the immediate vicinity of vents (Fig. 12), including rock, clam shells, and the outer layer of mussel shells (periostracum) (Jannasch and Wirsen, 1981). Scanning electron micrographs of this material showed a variety of filamentous forms (Fig. 12*a*) from *Beggiatoa* cells 5 to 40 μm wide, to *Thiothrix*-like structures 0.8 to 1.2 μm wide, and prosthecate bacteria with stalks 0.1 to 0.2 μm wide. Transmission electron micrographs revealed mostly procaryotic cells with walls typical for Gram-negative bacteria (Fig. 12*b*). An unexplained abundance of cyanobacterialike trichomes (Figs. 12*c* and 13*a*) were also found in these mats, but very few eucaryotic cells. These surfaces are intermittently exposed to hydrogen sulfide- and oxygen-containing waters. Limpets and other invertebrates seem to graze on these mats. Metal deposits within the microbial mats were heaviest in samples collected directly from the vents (Fig. 13*a*). Whereas the manganese: iron ratio in the microbial mats is close to 1.0 (Fig. 13*b*) here, the higher solubility of manganese in seawater causes an increase of this ratio with increasing distance from the vents. Active microbial precipitation of manganese and iron oxides and hydroxides can be expected but will be difficult to prove.

3. The newly described giant clam *Calyptogena magnifica* (Boss and Turner, 1980) and vestimentiferan tube worms *Riftia pachyptila* (Jones, 1980) are also found near the vents only but occur mostly in nonturbid waters. They appear to

Fig. 12. Electron micrographic views of microbial mats. (*a*) Scanning electron micrograph of a microbial mat from the surface of a mussel collected from the intermediate vicinity of a deep sea hydrothermal vent (Jannasch and Wirsen, 1981). The large filaments are likely to be *Beggiatoa* and the shorter septated forms *Leucothrix* or *Thiothrix*. The very thin filaments have been shown to be stalks of prosthecate bacteria (scale bar 10 μm). (*b*) Transmission electron micrograph of a section through a microbial mat showing procaryotic cells with Gram-negative walls and some metal deposits (scale bar 1 μm). (*c*) Transmission electron micrograph of another section showing large trichomes resembling the cyanobacterium *Calothrix* or *Homoeothrix* (scale bar 5 μm).

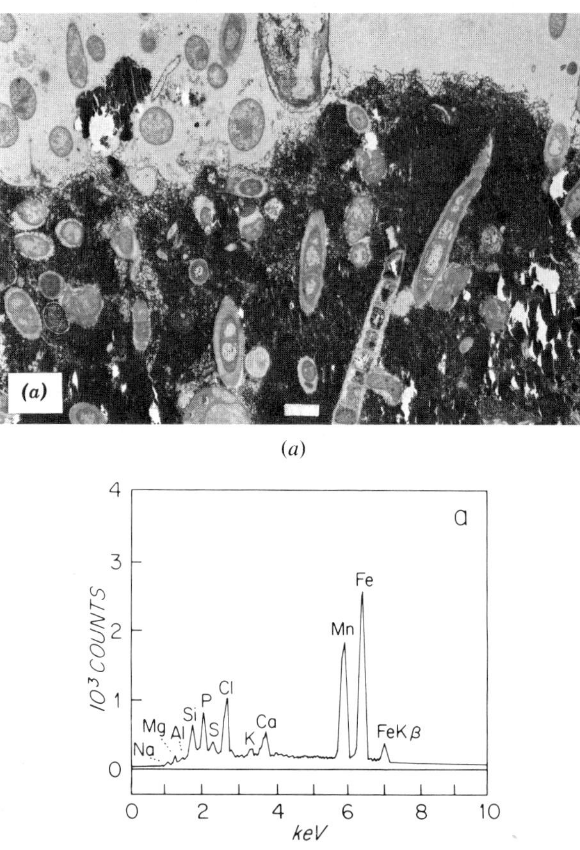

(*a*)

(*b*)

Fig. 13. (*a*) Transmission electron micrograph of a section through a microbial mat collected near the opening of a hydrothermal vent (scale bar 1 μm). (*b*) Line drawing from a KEVEX X-ray spectrum photo of a sample from the same area. The heavy deposition of metal oxides and hydroxides appears to cover layers of filamentous and unicellular microbial cells. The iron : manganese ratio decreases with increasing distance from the vents. [From Jannasch and Wirsen (1981).]

represent an early and a later evolutionary state of symbiosis between chemosynthetic microorganisms and invertebrates. As shown by electron microscopy (Cavanaugh et al., 1981) and enzymatic evidence (Felbeck, 1981), CO_2-assimilating procaryotic symbionts exist in the gills of the clam and appear to form the central tissue of *R. pachyptila,* the "trophosome." The latter is supplied with oxygen by a blood system containing an extracellular hemoglobin of a high oxygen affinity and storage capacity (Arp and Childress, 1981).

Ocean spreading centers as potential areas for the occurrence of hydrothermal vents are extensive. However, a speculation on the amount and rate of energy released in the form of reduced sulfur compounds for primary production appears premature. For an explanation of the manganese–magnesium anomaly in the ion-composition of seawater, it has been theorized (Edmond et al., 1979) that the total amount of seawater may percolate every 7 million years through the Earth's crust. The temptation to follow this conjecture up by estimating the amount of terrestrial chemical energy potentially available for chemosynthetic production is thwarted by the large number of unknown variables such as the proportion between chemical and biological sulfur oxidation; the rate of vent appearance and die-off; and the vent topography, including the presence or absence of subsurface seawater mixing. At this time, the interest lies in the particular phenomenon of this unique deep-sea ecosystem, the only one known to be independent of—with the exception of the supply of free oxygen—solar energy.

8. Conclusions

The progress of deep-sea microbiology has often depended on technological developments. Starting with the use of simple and practical laboratory and shipboard pressure equipment and deep-sea sampling techniques originally conceived by ZoBell, much information on bacterial deep-sea isolates has been gained since the late 1940s. During the last decade the use of a research submersible has stimulated *in situ* inoculation–incubation studies. The results of this work emphasize the ecological aspects of deep-sea microbiology and are of great significance in the interpretation of the physiologically and biochemically oriented pure culture studies.

Experimentally measured growth characteristics of deep-sea isolates at *in situ* pressures and temperatures have commonly been measured in rich bacteriological media and may not necessarily reflect the behavior of these organisms under natural conditions. Natural concentrations of metabolically utilizable organic carbon sources, on the other hand, may be reduced to low values at which growth ceases. The existence of such threshold concentrations has been verified in chemostat experiments and was found to depend on factors other than the growth-limiting substrate (Jannasch, 1979). For example, at low population densities—a consequence of low growth-limiting substrate concentrations—commonly microaerophilic marine bacteria are unable to cope with the near-saturation levels of free oxygen and stop growing in the presence of a certain substrate level. Reduction in the redox potential of the seawater medium or the partial pressure of free oxygen in the aeration gas enables the population to continue growing until a new and lower threshold value of the particular substrate is reached. Oxygen toxicity on aerobic organisms is well known (Kuenen, 1978), but its effect on natural populations of microaerophilic bacteria in the pelagic and deep ocean has received little attention so far. The amounts of free amino and fatty acids in seawater are more likely to represent threshold levels, as defined above, than true steady-state concentrations.

Of the two most important environmental factors of the deep sea, low temperature and increased hydrostatic pressure, effects of temperature on microbial metabolism have been well studied, as reviewed by Morita (1975) and Baross and Morita (1978). Psychrophilic bacteria might be lost if samples of natural microbial populations from deep-sea water or sediment were to reach temperatures over 20° during retrieval. However, it is technically not difficult to prevent the rise of temperature during sampling, and many psychrophilic deep-sea bacteria have been described.

On the other hand, studies of the effects of hydrostatic pressure on the metabolism of deep-sea bacteria have always been limited to organisms isolated from populations that were decompressed during sampling. The selective effects of this decompression on natural deep-sea populations are still unknown. The technique suggested by Dietz and Yayanos (1978), specifically, to isolate organisms in silica shake tubes under pressure, eliminates one of the major problems, namely, the successful growth competition by those organisms that grow most rapidly at 1 atm. It is a common observation, originally made by ZoBell and Morita (1959), that a large proportion of the bacteria contained in samples from deep-sea sediments grow readily on relatively rich bacterial media. By using the shake tube technique (Yayanos et al., 1979), it was possible to enrich for and to isolate barophilic bacteria that survived a brief decompression to 1 atm. This does not prove the absence of highly pressure adapted organisms that are indeed killed by decompression. Study of the possible existence of such forms is now facilitated by the above-mentioned pressurized isolation chamber, which can be used in connection with decompression-free sampling equipment.

The discovery of rich invertebrate populations at deep-sea hydrothermal vents and the available evidence for their primary subsistence on bacterial chemosynthesis proves that neither deep-sea temperatures of approximately 2°C nor pressure of about 250 atm prevent rapid microbial growth in a nonoligotrophic situation. Some of the microbial activity might occur at high temperature in subsurface lava cracks as indicated by the dense cell suspensions in the water emitted by some of the vents. However, at the present state of our knowledge, it appears that most of the organic matter used for growth by the major populations of vent invertebrates, namely, clams and vestimentiferan tube worms, occurs in symbiotic association between chemosynthetic bacteria and these animals and largely at the 2.1°C of the ambient seawater (Cavanaugh et al., 1981). At present, this phenomenon appears to be one of the most challenging problems of deep-sea microbiology.

Acknowledgments

We thank E. Seling (Harvard University), J. Waterbury, and S. M. Houghton (both of Woods Hole Oceanographic Institution) for their expert assistance with the scanning and transmission electron microscopy. We are indebted to C. C. Woo (U.S. Geological Survey) for the KEVEX analyses and their interpretations. We also thank D. C. Nelson, S. J. Molyneaux, and J. M. Peterson for their help in preparing the manuscript. The work was supported by the National Science Foundation grants OCE79-19178 and OCE80-24253, ONR contract 71.80 and a Mellon Senior Study Award by the Woods Hole Oceanographic Institution. This is Woods Hole Oceanographic Institution Contribution No. 4961.

References

Arp, A. G. and J. J. Childress, 1981. Blood function in the hydrothermal vent vestimentiferan tube worm. *Science*, **213**, 342–344.

Ballard, R. D., 1977. Notes on a major oceanographic find. *Oceanus*, **20**, 35–44.

Baross, J. A. and R. Y. Morita, 1978. Microbial life at low temperatures: Ecological aspects. In *Microbial Life in Extreme Environments*, D. J. Kushner, ed. Academic Press, New York, pp. 9–71.

Benecke, W., 1933. Bakteriologie des Meeres. *Abderh. Handb. d. Biol. Arbeitsmeth.*, **9** (5), 717–854.

Boss, K. J. and R. D. Turner, 1980. The giant white clam from the Galapagos Rift *Calyptogena magnifica* species novum. *Malacologia*, **20**, 161–194.

Brauer, F. W., ed., 1972. *Barobiology and the Experimental Biology of the Deep Sea*. North Carolina Sea Grant Program, University of North Carolina, Chapel Hill, 428 pp.

Brewer, P. G., Y. Nozaki, D. W. Spencer, and A. P. Fleer, 1980. Sediment trap experiments in the deep North Atlantic: Isotopic and elemental fluxes. *J. Mar. Res.*, **38**, 703–728.

Carlucci, A. F., S. L. Shimp, P. A. Jumars, and H. W. Pearl, 1976. In situ morphologies of deep sea and sediment bacteria. *Can. J. Microbiol.*, **22**, 1667–1671.

Cavanaugh, C. M., S. L. Gardiner, M. L. Jones, H. W. Jannasch, and J. B. Waterbury, 1981. Procaryotic cells in the hydrothermal vent tube worm *Riftia pachyptila* Jones: Possible chemoautotrophic symbionts. *Science*, **213**, 340–341.

Certes, A., 1884. Sur la culture, a l'abri des germes atmospheriques, des sediments rapportes par les expeditions du Travailleur et du Talisman. *Compt. Rend. Acad. Sci.* (Paris), 690–693.

Christensen, J. P., T. G. Owens, A. H. Devol, and T. T. Packard, 1980. Respiration and physiological state in marine bacteria. *Mar. Biol.*, **55**, 267–276.

Corliss, J. B., J. Dymond, L. I. Gordon, J. M. Edmond, R. P. von Herzen, R. D. Ballard, K. Green, D. Williams, A. Bainbridge, K. Crane, and T. H. van Andel, 1979. Submarine thermal springs on the Galapagos Rift. *Science*, **203**, 1073–1083.

Cuhel, R. L., 1981. Assimilatory sulfur metabolism in marine microorganisms. Ph.D. thesis. Massachusetts Institute of Technology, Woods Hole Oceanographic Institution, WHOI-81-29.

Cuhel, R. L., C. D. Taylor, and H. W. Jannasch, 1982. Assimilatory sulfur metabolism in marine microorganisms: Considerations for the application of sulfate incorporation into protein as a measurement of natural population protein synthesis. *Appl. Environ. Microbiol.*, **43**, 160–168.

Daley, R. J., 1979. Direct epifluorescence enumeration of native aquatic bacteria: Uses, limitations, and comparative accuracy. In *Native Aquatic Bacteria: Enumeration, Activity, and Ecology*, J. W. Costerton and R. R. Colwell, eds. ASTM Publication STP 695, pp. 29–45.

Daley, R. J. and J. E. Hobbie, 1975. Direct counts of aquatic bacteria by a modified epifluorescence technique. *Limnol. Oceanogr.*, **20**, 875–882.

Degens, E. T. and D. A. Ross, eds., 1974. *The Black Sea—Geology, Chemistry and Biology*. Am. Assoc. Petrol. Geol., Tulsa, OK.

Deming, J. W., P. S. Tabor, and R. R. Colwell, 1981. Barophilic growth of bacteria from intestinal tracts of deep-sea invertebrates. *Microb. Ecol.*, **7**, 85–94.

Dietz, A. S. and A. A. Yayanos, 1978. Silica gel media for isolating and studying bacteria under hydrostatic pressure. *Appl. Environ. Microbiol.*, **36**, 966–968.

Edmond, J. M., C. Measures, R. E. McDuff, L. H. Chan, R. Collier, B. Grant, L. I. Gordon, and J. B. Corliss, 1979. Ridge crest hydrothermal activity and the balances of the major and minor elements in the ocean: The Galapagos data. *Earth Planet. Sci. Letters*, **46**, 1–18.

Felbeck, H., 1981. Chemoautotrophic potentials of the hydrothermal vent tube worm, *Riftia pachyptila* (Vestimentifera). *Science*, **213**, 336–338.

Fischer, B., 1866. Bakteriologische Untersuchungen auf einer Reise nach Westindien. *Zeitschr. f. Hygiene*, **1**, 421–464.

Fischer, B., 1894. Die Bakterien des Meeres nach den Untersuchungen der Planktonexpedition unter gleichzeitiger Berücksichtigung einiger älterer und neurer Untersuchungen. *Centralbl. f. Bakteriol.*, **15**, 657–666.

Herbert, D., R. Elsworth, and R. C. Telling, 1956. The continuous culture of bacteria: A theoretical and experimental study. *J. Gen. Microbiol.*, **14**, 601–622.

Hinga, K. R., J. McN. Sieburth, and G. R. Heath, 1979. The supply and use of organic material at the deep sea floor. *J. Mar. Res.*, **37**, 557–579.

Honjo, S., 1978. Sedimentation of materials in the Sargasso Sea at a 5367 m deep station. *J. Mar. Res.*, **36**, 469–492.

Hoppe, H. G., 1976. Determination and properties of actively metabolizing bacteria in the sea investigated by means of micro-autoradiography. *Mar. Biol.*, **36**, 291–302.

Huxley, T. H., 1868. On some organisms living at great depths in the North Atlantic Ocean. *Quart. J. Microscop. Sci., N. S.*, **8,** 203–212.

Isaacs, J. D., 1969. The nature of oceanic life. *Sci. Am.*, **221,** 146–162.

Jannasch, H. W., 1979. Microbial ecology of aquatic low nutrient habitats. In *Strategies of Microbial Life in Extreme Environments*, M. Shilo, ed. Life Sci. Res. Rep. 13, Verlag Chemie, Weinheim/New York, pp. 243–260.

Jannasch, H. W., in press. Interactions between the carbon and sulfur cycles in the marine environment. In *SCOPE Workshop on Interactions of Biogeochemical Cycles*, B. Bolin and R. B. Cook, eds. Wiley, New York.

Jannasch, H. W. and G. E. Jones, 1959. Bacterial populations in seawater as determined by different methods of enumeration. *Limnol. Oceanogr.*, **4,** 128–139.

Jannasch, H. W. and C. O. Wirsen, 1973. Deep-sea microorganisms: In situ response to nutrient enrichment. *Science*, **180,** 641–643.

Jannasch, H. W. and C. O. Wirsen, 1977. Retrieval of concentrated and undecompressed microbial populations from the deep sea. *Appl. Environ. Microbiol.*, **33,** 642–646.

Jannasch, H. W. and C. O. Wirsen, 1979. Chemosynthetic primary production at East Pacific sea floor spreading centers. *BioScience*, **29,** 592–598.

Jannasch, H. W. and C. O. Wirsen, 1980. Studies on the microbial turnover of organic substrates in deep sea sediments. In *Biogeochimie de la Matiere Organique a l'Interface Eau-Sediment Marin*, R. Roumas, ed. (Reprinted in *Actes des Colloques du CNRS*, **293,** 285–290.)

Jannasch, H. W. and C. O. Wirsen, 1981. Morphological survey of microbial mats near deep-sea thermal vents. *Appl. Environ. Microbiol.*, **41,** 528–538.

Jannasch, H. W. and C. O. Wirsen, 1982. Microbial activities in undecompressed and decompressed deep sea water samples. *Appl. Environ. Microbiol.*, **43,** 1116–1124.

Jannasch, H. W., C. O. Wirsen, and C. L. Winget, 1973. A bacteriological, pressure-retaining, deep-sea sampler and culture vessel. *Deep-Sea Res.*, **20,** 661–664.

Jannasch, H. W., C. O. Wirsen, and C. D. Taylor, 1976. Undecompressed microbial populations from the deep sea. *Appl. Environ. Microbiol.*, **32,** 360–367.

Jannasch, H. W., C. O. Wirsen, and C. D. Taylor, 1982. Deep sea bacteria: Isolation in the absence of decompression. *Science*, **216,** 1315–1317.

Jannasch, H. W., K. Eimhjellen, C. O. Wirsen, and A. Farmanfarmaian, 1971. Microbial degradation of organic matter in the deep sea. *Science*, **171,** 672–675.

Jannasch, H. W., R. L. Cuhel, C. O. Wirsen, and C. D. Taylor, 1980. An approach for in situ studies of deep-sea amphipods and their microbial gut flora. *Deep-Sea Res.*, **27,** 867–872.

Jones, J. G. and B. M. Simon, 1975. An investigation of errors in direct counts of aquatic bacteria by epifluorescence microscopy, with reference to a new method for dying membrane filters. *J. Appl. Bacteriol.*, **39,** 317–329.

Jones, M. L., 1980. *Riftia pachyptila*, n. gen., n. sp., the vestimentiferan worm from the Galapagos Rift geothermal vents (Pogonophora). *Proc. Biol. Soc. Wash.*, **93,** 1295–1313.

Jørgensen, B. B., in press. Processes at the sediment-water interface. In *SCOPE Workshop on Interactions of Biogeochemical Cycles*, B. Bolin and R. B. Cook, eds., Wiley, New York.

Karl, D. M., 1979. Measurement of microbial activity and growth in the ocean by rates of stable ribonucleic acid synthesis. *Appl. Environ. Microbiol.*, **39,** 850–860.

Karl, D. M., 1980. Cellular nucleotide measurements and applications in microbial ecology. *Microbiol. Rev.*, **44,** 739–796.

Karl, D. M., 1981. Simultaneous measurements of rates of RNA and DNA syntheses for estimating growth and cell division of aquatic microbial populations. *Appl. Environ. Microbiol.*, **42,** 802–810.

Karl, D. M., C. O. Wirsen, and H. W. Jannasch, 1980. Deep-sea primary production at the Galapagos hydrothermal vents. *Science*, **207,** 1345–1347.

King, J. D. and D. C. White, 1977. Muramic acid as a measure of microbial biomass in estuarine and marine samples. *Appl. Environ. Microbiol.*, **33,** 777–783.

Koch, R., 1881. Zur Untersuchung von pathogenen Organismen. *Mittl. d. Kaiserl. Gesundh. Amt.*, **1**, 1–48.

Kohlmeyer, J. and E. Kohlmeyer, 1979. *Marine Mycology: The Higher Fungi*, Academic Press, New York.

Krambeck, C., H. J. Krambeck and J. Overbeck, 1981. Microcomputer assisted biomass determination of plankton bacteria on scanning electron micrographs. *Appl. Environ. Microbiol.*, **42**, 142–149.

Kriss, A. E., 1962. *Marine Microbiology (Deep Sea)*, Wiley-Interscience, New York.

Kuenen, J. G., 1978. Oxygen toxicity. In *Strategies of Microbial Life in Extreme Environments*. Life Sciences Research Report 13, Verlag Chemie, Weinheim, pp. 223–241.

Landau, J. V. and D. H. Pope, 1980. Recent advances in the area of barotolerant protein synthesis in bacteria and implications concerning barotolerant and barophilic growth. *Adv. Aquat. Microbiol.*, **2**, 49–76.

Lee, C. and L. Bada, 1975. Amino acids in equatorial ocean water. *Earth Planet. Sci. Letters*, **26**, 61–68.

Lonsdale, P., 1977. Clustering of suspension-feeding macrobenthos near abyssal hydrothermal vents at oceanic spreading centers. *Deep-Sea Res.*, **24**, 857–863.

Maki, J. S. and C. C. Remsen, 1981. Comparison of two direct count methods for determining metabolizing bacteria in freshwater. *Appl. Environ. Microbiol.*, **41**, 1132–1138.

Marquis, R. E., 1976. High-pressure microbial physiology. *Adv. Microbiol. Physiol.*, **14**, 158–241.

Marquis, R. E. and P. Matsumura, 1978. Microbial life under pressure. In *Microbial Life in Extreme Environments*. D. J. Kushner, ed., Academic Press, New York.

Meyer-Rèil, L.-A., 1978. Autoradiography and epifluorescence microscopy combined for the determination of number and spectrum of actively metabolizing bacteria in natural waters. *Appl. Environ. Microbiol.*, **36**, 506–512.

Morita, R. Y., 1972. Pressure. 1. Bacterial, fungi and blue-green algae. In *Marine Ecology*, Vol. 1, O. Kinne, ed. Wiley-Interscience, pp. 1361–1388.

Morita, R. Y., 1975. Psychrophilic bacteria. *Bacteriol. Rev.*, **39**, 144–167.

Morita, R. Y., 1976. Survival of bacteria in cold and moderate hydrostatic pressure environments with special reference to psychrophilic and barophilic bacteria. 26th Symp. Soc. Gen. Microbiol. In *The Survival of Vegetative Microbes*, T. Gray and J. R. Postgate, eds. Cambridge University Press, New York, pp. 279–298.

Mottl, M. J., H. D. Holland, and R. F. Corr, 1979. Chemical exchange during hydrothermal alteration of basalt by seawater. II. Experimental results for Fe, Mn, and sulfur species. *Geochim. Cosmochim. Acta*, **43**, 869–884.

Muraoka, J. S., 1971. Deep ocean biodeterioration of materials. *Ocean Industry* (February 1971), 21–23.

Poindexter, J. S., 1981a. The caulobacters: Ubiquitous unusual bacteria. *Microbiol. Rev.*, **45**, 123–179.

Poindexter, J. S., 1981b. Oligotrophy: feast and famine existence. *Adv. Microb. Ecol.*, **5**, 67–93.

Pope, D. H., W. P. Smith, M. A. Orgrinc, and J. V. Landau, 1976. Protein synthesis at 680 atm: Is it related to environmental origin, physiological type, or taxonomic group? *Appl. Environ. Microbiol.*, **31**, 1001–1002.

Rehbock, P. F., 1975. Huxley, Haeckel, and the oceanographers: The case of *Bathybius haeckeli*. *Isis*, **66**, 504–533.

Richards, F. A., 1975. The Cariaco Basin (Trench). *Oceanogr. Mar. Biol. Ann. Rev.*, **13**, 11–67.

Romanenko, W. I., 1964. Heterotrophic assimilation of CO_2 by the aquatic microflora. *Microbiologia*, **33**, 679–683.

Rowe, G. and W. Gardner, 1979. Sedimentation in the slope water of the northwest Atlantic Ocean measured directly with sediment traps. *J. Mar. Res.*, **37**, 581–600.

Ruby, E. G. and H. W. Jannasch, 1982. Physiological characteristics of *Thiomicrospira* sp. L-12 isolated from deep sea hydrothermal vents. *J. Bacteriol.*, **149**, 161–165.

Ruby, E. G., C. O. Wirsen, and H. W. Jannasch, 1981. Chemolithotrophic sulfur-oxidizing bacteria from the Galapagos Rift hydrothermal vents. *Appl. Environ. Microbiol.*, **42**, 317–324.

Saunders, P. M. and N. P. Fofonoff, 1976. Conversion of pressure to depth in the ocean. *Deep-Sea Res.*, **23,** 109–111.

Schwarz, J. R., A. A. Yayanos, and R. R. Colwell, 1976. Metabolic activities of the intestinal microflora of a deep sea invertebrate. *Appl. Environ. Microbiol.*, **31,** 46–48.

Seki, H. and D. G. Robinson, 1969. Effect of decompression on activity of microorganisms in seawater. *Int. Revue ges Hydrobiol.*, **54,** 201–205.

Seki, H., E. Wada, I. Koite, and A. Hattori, 1974. Evidence of high organotrophic potentiality of bacteria in the deep ocean. *Mar. Biol.*, **26,** 1–4.

Shilo, M., 1979. *Strategies of Microbial Life in Extreme Environments.* Life Sciences Research Report 13, Verlag Chemie, Weinheim.

Sieburth, J. McN., 1979, *Sea Microbes.* Oxford University Press, New York, 491 p.

Sieburth, J. McN. and A. A. Dietz, 1974. Biodeterioration in the sea and its inhibition. In *Effect of the Ocean Environment on Microbial Activities,* R. R. Colwell and R. Y. Morita, eds. University Park Press, Baltimore, pp. 318–326.

Sleigh, M. A. and A. G. MacDonald, eds., 1972. The effects of pressure on organisms. *Symposium of Society of Experimental Biology XXVI.* Academic Press, New York, 516 pp.

Sorokin, Y. I., 1964. On the primary production and bacterial activities in the Black Sea. *J. Cons. Int. Explor. Mer.*, **29,** 41–60.

Sorokin, Y. I., 1965. On the trophic role of chemosynthesis and bacterial biosynthesis in water bodies. *Mem. Ist. Ital. Idrobiol.*, **18** (Suppl.), 187–205.

Strugger, S., 1949. *Fluoreszenzmikroskopie and Mikrobiologie.* M. & H. Schaper, Hannover.

Tabor, P. S. and R. R. Colwell, 1976. Initial investigations with a deep ocean in situ sampler. *Proc. MTS/IEEE OCEANS '76,* Washington, DC, 13D-1-13D-4.

Tabor, P. S., J. W. Deming, K. Ohwada, H. Davis, M. Waxman, and R. R. Colwell, 1981a. A pressure-retaining deep ocean sampler and transfer system for measurement of microbial activity in the deep sea. *Microb. Ecol.*, **7,** 51–65.

Tabor, P. S., K. Ohwada, and R. R. Colwell, 1981b. Filterable marine bacteria found in the deep sea: Distribution, taxonomy, and response to starvation. *Microb. Ecol.*, **7,** 67–83.

Taylor, C. D., 1979. Growth of a bacterium under a high pressure oxyhelium atmosphere. *Appl. Environ. Microbiol.*, **37,** 42–49.

Tuttle, J. H. and H. W. Jannasch, 1976. Microbial utilization of thiosulfate in the deep sea. *Limnol. Oceanogr.*, **21,** 697–701.

Wada, E., I. Koike, and A. Hattori, 1975. Nitrate metabolism in abyssal waters. *Mar. Biol.*, **29,** 119–124.

Watson, S. W., T. J. Novitsky, H. L. Quinby, and F. W. Valois, 1977. Determination of bacterial number and biomass in the marine environment. *Appl. Environ. Microbiol.*, **33,** 940–946.

White, D. C., R. J. Bobbie, J. D. King, J. Nickels, and P. Amoe, 1979. Lipid analyses of sediments for microbial biomass and community structure. In *Methodology for Biomass Determinations and Microbial Activities in Sediments,* C. D. Litchfield and P. L. Seyfried, eds. ASTM Publication STP 673, pp. 87–103.

Winogradski, S., 1949. *Microbiologie du Sol.* Massonet Cie, Paris.

Wirsen, C. O. and H. W. Jannasch, 1974. Microbial transformations of some ^{14}C-labeled substrates in coastal water and sediment. *Microb. Ecol.*, **1,** 25–37.

Wirsen, C. O. and H. W. Jannasch, 1975. Activity of marine psychrophilic bacteria at elevated hydrostatic pressures and low temperatures. *Mar. Biol.*, **31,** 201–208.

Wirsen, C. O. and H. W. Jannasch, 1976. Decomposition of solid organic materials in the deep sea. *Environ. Sci. Technol.*, **10,** 880–886.

Yayanos, A. A., A. S. Dietz, and R. Van Boxtel, 1979. Isolation of a deep-sea barophilic bacterium and some of its growth characteristics. *Science,* **205,** 808–810.

Yayanos, A., A. S. Dietz, and R. Van Boxtel, 1981. Obligately barophilic bacterium from the Mariana Trench. *Proc. Natl. Acad. Sci. (USA),* **78,** 5212–5215.

Zimmerman, A. M., ed., 1970. *High Pressure Effects on Cellular Processes*, Academic Press, New York, 324 pp.

Zimmerman, R. and L.-A. Meyer-Reil, 1974. A new method for fluorescence staining of bacterial populations on membrane filters. *Kieler Meeresforschungen*, **30**, 24–27.

ZoBell, C. E., 1946. *Marine Microbiology*. Chronica Botanica Company, Waltham, MA.

ZoBell, C. E., 1968. Bacterial life in the deep sea. *Bull. Misaki Mar. Biol. Inst., Kyoto Univ.*, **12**, 77–96.

ZoBell, C. E., 1970. Pressure effects on morphology and life processes of bacteria. In *High Pressure Effects on Cellular Processes*, H. M. Zimmerman, ed. Academic Press, New York, pp. 85–130.

ZoBell, C. E. and L. L. Hittle, 1967. Effects of hyperbaric oxygenation on bacteria at increased hydrostatic pressure. *Can. J. Microbiol.*, **13**, 1311–1319.

ZoBell, C. E. and F. H. Johnson, 1949. The influence of hydrostatic pressure on the growth and viability of terrestrial and marine bacteria. *J. Bacteriol.*, **57**, 179–189.

ZoBell, C. E. and J. Kim, 1972. Effects of deep sea pressures on microbial enzyme systems. In *The Effects of Pressure on Organisms*, M. A. Sleigh and A. G. MacDonald, eds. Academic Press, New York, pp. 125–146.

ZoBell, C. E. and R. Y. Morita, 1957. Barophilic bacteria in some deep sea sediments. *J. Bacteriol.*, **73**, 563–568.

ZoBell, C. E. and R. Y. Morita, 1959. Deep sea bacteria. In *Galathea Report, I*, Copenhagen, pp. 139–154.

ZoBell, C. E. and C. H. Oppenheimer, 1950. Some effects of hydrostatic pressure on the multiplication and morphology of marine bacteria. *J. Bacteriol.*, **60**, 771–781.

7. BIOCHEMICAL AND PHYSIOLOGICAL ADAPTATIONS OF DEEP-SEA ANIMALS

George N. Somero, Joseph F. Siebenaller, and Peter W. Hochachka

1. Physical and Biological Influences on Evolution of Deep-Sea Animals

The object of this chapter is to relate the physiological and biochemical characteristics of deep-sea animals to the unique suite of environmental properties present in the deep sea. Historically, studies of deep-sea biology have a common theme: an interest in the potential effects of the physical environment on biological processes. Influenced by the impressive physical attributes of the deep-sea environment—namely, high hydrostatic pressures and low temperatures—coupled with the absence of primary productivity, early investigators believed that the deeper regions of the seas were likely to be azoic deserts (Forbes, 1859; Menzies et al., 1973; Mills, this volume, Chapter 1). The subsequent demonstrations of life at great depths and, indeed, the discovery of great species diversity in the deep sea (Hessler and Sanders, 1967) in no way preclude the importance of the environmental characteristics of the deep sea in affecting the biology of deep-sea organisms. Instead, the discovery of a diverse fauna at depth has raised interesting questions about the types of physiological and biochemical adaptation of deep-sea animals that permit their successful function at high pressures and low temperatures and in the apparent absence of a rich food supply. This chapter considers adaptations to the physical and biological features of the deep sea. Attention is given to those physiological and biochemical traits that appear to make life possible at great depths and to those traits that may be instrumental in establishing the rates of biological activity in the deep sea.

Our analysis begins with a consideration of the types of adaptation in enzymatic properties that permit the existence of a satisfactory level of closely regulated metabolic activity under conditions of low temperature (abyssal temperatures average only 1 to 3°C) and high pressure. [Pressure increases by 1 atm for each 10-m increment in depth; abyssal pressures range from approximately 400 to 600 atm, and hadal pressures reach approximately 1100 atm (Saunders and Fofonoff, 1976).] Certain enzymes of cold-adapted, shallow-living fish are not preadapted to high hydrostatic pressure, and successful colonization of the deep sea has necessitated major alterations in the pressure sensitivities of enzyme function. Apparently, a certain price must be paid to gain pressure insensitivity of enzyme function; namely, the high pressure-adapted enzymes of deep-sea fish have lower catalytic efficiencies (rate of function, as measured by substrate turnover rates) than do the homologous enzymes of cold-adapted, shallow-living fish.

We propose that the benefits resulting from enhanced pressure insensitivity more than outweigh losses of catalytic rate.

A second broad array of adaptations is linked to the dramatic differences between the biological characteristics of deep and shallow water habitats. For example, the apparently low food input into the deep sea, reflected in terms of both total biomass and caloric content per unit mass of deep-sea organisms, has had pervasive and often highly conspicuous influences on the "design" of deep-sea animals. Among these influences are reduced metabolic rates, reduced protein and lipid contents, elevated water contents, and reduced levels of enzymatic activity in muscle tissue. These latter biochemical adjustments may be crucial in reducing the metabolic rates of muscle tissue, which comprises the bulk of most animals. Feeding strategies of midwater animals also seem to reflect adaptation to a low food supply and, further, are reflected in the fundamentals of body form and locomotory mechanisms. Buoyancy mechanisms also appear to reflect adaptations to a low food supply. Thus examination of the physiological and biochemical properties of deep-sea animals provides an excellent means for discerning how both the physical and the biological characteristics of the environment affect organism "design."

2. Enzyme Kinetic Adaptations and Macromolecular Structural Adaptations to Low Temperatures and High Pressures

A. Temperature and Pressure Stresses on Enzyme Function and Structure

An obvious requirement for life at depth is a battery of enzymes capable of maintaining an adequate level of controlled metabolic function under conditions of low temperature and high, and often variable, hydrostatic pressure. Clearly, in the deep sea not only may enzymatic systems be stressed by the extremes of temperature and pressure, but for many species the large changes in pressure and generally smaller changes in temperature also pose problems. Deep-living species whose life histories include shallow-living, pelagic larval stages and bathyal species that maintain populations over a broad depth range must cope with considerable variations in both pressure and temperature. Vertically migrating animals may encounter similar types of stress on a diurnal basis. Thus, in discussing enzymatic adaptations to deep-sea conditions, we must consider mechanisms for coping with changes in pressure and temperature as well as adaptations that offset the stresses of extremes of pressure and temperature.

Both temperature and pressure may exert large-scale effects on (1) rates of enzymatic catalysis, (2) the precise regulation of catalysis, and (3) enzyme structure (Hochachka and Somero, 1973; Somero and Hochachka, 1976a,b). For a detailed discussion of the physical and chemical bases of temperature and pressure effects, the reader is directed to the volume by Johnson et al. (1974) and to the abbreviated summary of key points from this volume presented in Appendix A at the end of this chapter.

The effects of temperature changes on enzymatic reaction rates are well studied, and many of the principal adaptations that serve to compensate metabolic rates for changes in temperature are known (Hazel and Prosser, 1974). The maximal velocities ($V_{\max}$) of enzymatic reactions typically increase about twofold with each 10°C increase in measurement temperature. That is, Q_{10} values of enzymatic reactions, when measured under conditions of saturating substrate concentrations, are near 2. The implications of these temperature effects on enzymatic reaction rates for species migrating in the water column are clear. For an organism moving from surface water with a temperature near 15°C to colder bathyal and abyssal temperatures, the rate of metabolism would be expected to fall by at least 50%. Thus, whether we are considering long-term colonization of deeper waters by shallow-living species or the short-term processes of life-stage-related depth changes and diurnal vertical migration, movement to greater depths may lead to a substantial slowing down of metabolic function. Are these low-temperature depressions of metabolic rate offset by the types of compensatory changes in enzymatic systems, namely, increased enzyme concentrations (Hazel and Prosser, 1974; Shaklee et al., 1977) and the evolutionary development of more efficient enzymes (Somero, 1978) that have been discovered when shallow-water species are exposed to a colder environment? Or do the biological conditions of the deep sea render these temperature-compensatory modifications in metabolic activity unnecessary or even maladaptive? We show that the latter circumstance appears to pertain and that the critical problems facing enzyme systems in deep-sea animals relate more to the regulation of catalysis and the maintenance of correct protein structure than to the absolute rate at which catalysis occurs.

Although the selective advantage of maintaining the high metabolic rates typical of shallow-living animals may be lost in the deep sea, the necessity of maintaining a closely regulated metabolic apparatus remains. Whatever the overall metabolic rate of an organism, close coordination of the relative activities of different reactions and pathways is essential (Atkinson, 1977). The regulation of enzymatic catalysis in the sense of altering the activities of enzymes that are already present in the cells is achieved principally through control of enzyme–ligand (substrate, cofactor, and modulator) binding (Atkinson, 1969, 1977). Enzymes typically function in the presence of substrate concentrations that are lower than those necessary for saturation of the enzyme, that is, to yield V_{max} values. At subsaturating concentrations of ligands, changes in the binding affinities of enzymes for ligands thus can promote large increases or decreases in enzymatic activity. Increase in binding ability at subsaturating ligand concentrations can lead to rate increases, whereas decreased binding can reduce catalytic rates dramatically.

Another exceedingly important result of the subsaturating substrate concentrations present in cells is that enzymes maintain a vital *reserve capacity*. If substrate concentrations increase during elevated metabolic activity, such as during vigorous exercise, enzymes are capable of increasing their rates of catalysis since they are operating below V_{max}. If some enzymes normally functioned under saturating conditions of substrate concentration, this reserve capacity would not exist, and increased metabolic flux would be accompanied by rapid and potentially large increases in the concentrations of those metabolites that are the substrates of the already saturated enzymes. As Atkinson (1969) has argued, the high chemical reactivities of many substrates and the limited quantities of cellular water available to dissolve substrates and other intracellular solutes preclude large increases in substrate concentration in properly functioning metabolic systems. It follows from Atkinson's reasoning that selection will favor the retention of subsaturating relationships between enzymes and their substrates under most, if not all, conditions. In the present context it is important to realize that both temperature and pressure can affect this relationship and, as shown below, an essential event in the successful invasion of deep-water regions is the acquisition of the proper ability to bind substrates, cofactors, and modulators under conditions of low temperature and high pressure.

B. Binding Adaptations

Adaptations to Low Temperatures

The binding of substrates, cofactors, and modulators by enzymes generally is temperature dependent (Hochachka and Somero, 1973; Somero, 1978). Most commonly, decreases in temperature appear to facilitate binding (binding is exothermic), as estimated from studies of temperature effects on apparent Michaelis constant (K_m) values.[1] Thus K_m values for most enzyme–substrate combinations vary directly with temperature, although plots of K_m versus experimental temperature are rarely linear over the full range of measurement (Hochachka and Somero, 1973).

Let us consider the implications of temperature-dependent K_m values under circumstances in which the interspecific homologues of a particular type of enzyme, such as a common dehydrogenase, all display the same absolute K_m and the same temperature dependence of K_m. If we assume that substrate concentrations also are virtually the same in all species, an assumption that appears to be valid

[1]The relationship between the Michaelis constant (K_m) of an enzyme and the actual dissociation constant for the equilibrium between free enzyme and substrate and the enzyme substrate complex is as follows. Consider the simple, one-substrate reaction

$$E + S \underset{k_2}{\overset{k_1}{\rightleftharpoons}} ES \underset{k_4}{\overset{k_3}{\rightleftharpoons}} E + P$$

where E is free enzyme, S is free substrate, ES is the enzyme–substrate complex, and P is product. The rate constants for each individual reaction are given as k values (k_4 is neglected here since we are considering the instance where P is approximately zero and S is high). The K_m for this system is defined as

$$K_m = \frac{k_2 + k_3}{k_1}$$

Thus, for estimation of the K_m for the system, three separate rate constants must be determined. In practice, these individual rate constants are not determined; instead, an "apparent" K_m is determined by using one or more of the standard transformations of velocity versus substrate concentration (Wilkinson, 1961; Fersht, 1977). In this chapter we employ the term K_m throughout to refer to "apparent K_m."

If k_1 and k_2 are known, then the substrate dissociation constant K_s, defined as

$$K_s = \frac{k_2}{k_1}$$

can be determined. This dissociation constant provides an estimate of the reciprocal of enzyme–substrate "affinity." In much conventional usage, the apparent K_m of substrate is treated as an approximate estimate of enzyme–substrate affinity. However, this is correct only if K_m is approximately equal to K_s, that is, when $k_2 >> k_3$. Thus an enzyme with a low K_m is referred to as having a high affinity for substrate, and changes in K_m due, for example, to pH changes, are regarded as true affinity changes if V_{max} effects are excluded.

(Yancey and Somero, 1978b), potentially serious regulatory problems would arise. For low-temperature ectotherms, this hypothetical enzyme would function at or near V_{max} since the intracellular substrate concentration might be well above the K_m value of the enzyme at low temperatures. Thus the low-temperature species would lack the important reserve capacity discussed above. In organisms with high body temperatures, such as birds and mammals, the K_m of substrate might be vastly higher than physiological substrate concentrations. Under this circumstance, very high substrate concentrations would be necessary to provide a high rate of enzymatic activity. This, too, is maladaptive.

The scenario given above is not found in nature. Comparative studies of homologous forms of a given type of enzyme from species differing widely in body temperatures have shown that each species possesses a particular variation on a given enzyme theme that displays the correct kinetic properties at the normal body temperatures of the organism. Figure 1 illustrates this adaptive pattern for

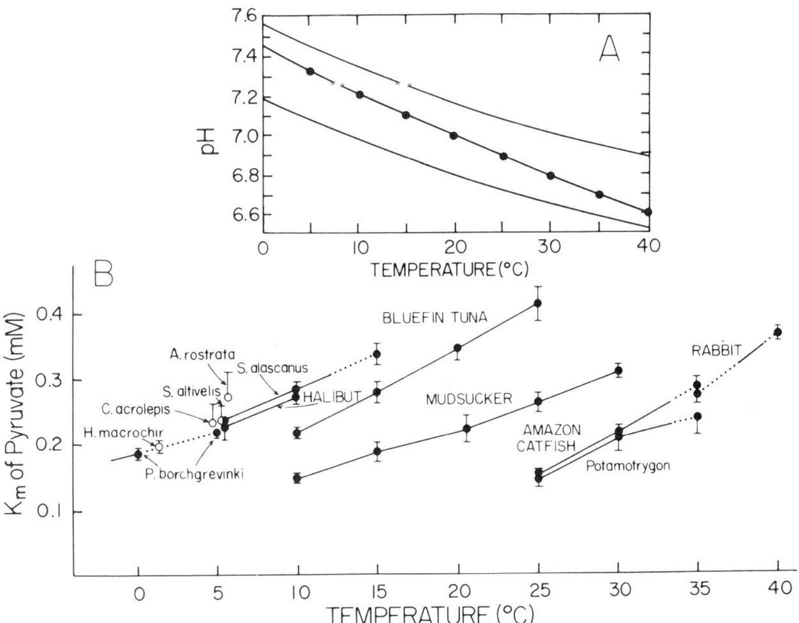

Fig. 1. (A) Relationship between temperature and pH for intracellular (cytosol) fluid of several vertebrates. The observed ranges are bracketed by thin lines. The points represent the effect of temperature on the pH of imidazole–HCl buffer. [From Yancey and Somero (1978a).] (B) Effect of temperature on apparent Michaelis constants K_m of pyruvate for purified muscle-type M_4 LDHs from several vertebrate species. *Pagothenia borchrevinki* is an Antarctic teleost; the mudsucker (*Gillichthys mirabilis*) is a temperate estuarine teleost; *Potamotrygon* is a freshwater Amazon ray, and *Hippoglossus stenolepis* is the Pacific halibut. The depth distributions of some of the teleosts are given in Fig. 3. Open symbols indicate deep-living (bathyal) species. [Data for all terrestrial and shallow water species are from Yancey and Somero (1978a). *Sebastolobus altivelis* data are from Siebenaller (1978b).] All assays were performed at 1-atm pressure in 80 mM imidazole–HCl buffer with 150 μM reduced nicotinamide adenine dinucleotide (NADH) and various concentrations of sodium pyruvate. At each temperature, activity was measured at 7 to 10 pyruvate concentrations between 0.08 and 4 mM. The pH of the imidazole–HCl buffer at the assay temperatures is shown in panel A. Apparent K_m values and the indicated 95% confidence intervals were calculated according to Wilkinson's (1961) method. The solid lines indicate the approximate temperature ranges of the species.

skeletal muscle type (M_4) lactate dehydrogenase (LDH) (EC 1.1.1.27; NAD^+ : lactate oxidoreductase). Even though each LDH variant exhibits a temperature-dependent K_m of substrate (pyruvate) at the respective physiological temperatures of the different species, an extremely high degree of K_m conservation is observed. Under these conditions, variation in K_m of pyruvate is only about twofold among all the species studied (a range of approximately 0.15 to 0.35 m*M*). Thus the capacity of M_4 LDHs to increase their activities as pyruvate concentrations rise, as during vigorous muscle activity, is maintained in all cases (Yancey and Somero, 1978a,b). The M_4 LDHs of bathyal fish (*Halosauropsis macrochir, Coryphaenoides acrolepis, Antimora rostrata,* and *Sebastolobus altivelis*) exhibit K_m of pyruvate values that are very close to those of the M_4 LDHs of cold-adapted, shallow-water fish, such as the Antarctic nototheniid teleost, *Pagothenia (Trematomus) borchgrevinki.* The M_4 LDHs of bathyal fish and cold-adapted shallow-living fish also display similar K_m values for coenzyme (NADH) at 5°C and 1-atm pressure (Fig. 2). Thus, in terms of the K_m values for both substrate and cofactor, the M_4 LDHs of deep-sea fishes have typically cold-adapted properties.

Adaptations to High and Variable Pressures

The maintenance of physiologically appropriate K_m values in the face of high and/or variable hydrostatic pressures appears to have necessitated important changes in the pressure sensitivities of at least some enzymes. The enzyme for which the most data are available is again the M_4 isozyme of lactate dehydrogenase (Fig. 2) (Baldwin et al., 1975; Hochachka, 1975a, Moon et al., 1971; Siebenaller and Somero, 1978, 1979). The M_4 LDHs of shallow-living cold-adapted teleosts display pressure-dependent values for the K_m values of pyruvate and NADH (Fig. 2). The pressure sensitivity of the K_m of NADH is especially large, and even the M_4 LDHs of deep-living species exhibit a slight increase in K_m of cofactor between 1- and 68-atm pressure. However, for the deep-living fish, the increase in K_m of NADH over this pressure range is only about 25%, whereas for the shallow-living species, the increase is about twofold. At pressures between 68 and 476 atm, the K_m of NADH for the deep-living species LDHs is essentially pressure insensitive, whereas the K_m values for the shallow species' LDHs continue to rise. For the M_4 LDHs of the latter species, the increase in K_m of NADH appears sufficient to impair LDH function at the depths at which the deep-living fish typically occur (Fig. 3). This conclusion is based on the observation that nucleotide binding sites normally are fully saturated, and the direction of enzyme function (either pyruvate reduction or lactate oxidation in the case of LDH) is determined by the ratio of oxidized to reduced nucleotides in the cytosol (Atkinson, 1977). Thus, unlike the situation pertaining for substrates, where concentrations may vary widely during changes in metabolic rate, the total concentration of oxidized and reduced pyridine nucleotides may remain stable, whereas the balance between NADH and NAD^+ may shift. For the M_4 LDHs of shallow-living fish, the large increases in K_m of NADH with rising pressure may prevent the enzymes from remaining cofactor saturated, with resulting perturbation of metabolic regulation and metabolic rate. We assume in this argument that concentrations of NAD^+ and NADH in fish tissues are similar to the concentrations present in mammalian tissues (Tischler et al., 1977). This assumption seems reasonable in view of the conservation in substrate concentrations found among species. As in the case of

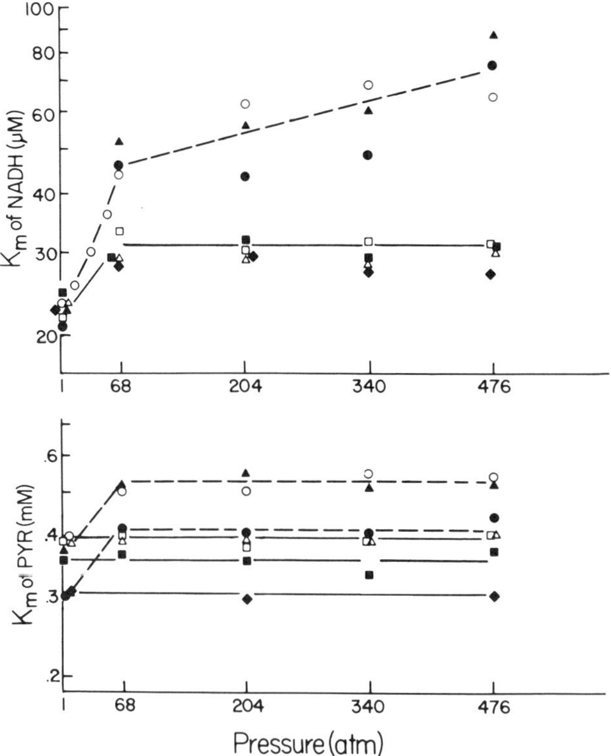

Fig. 2. Effects of hydrostatic pressure on apparent K_m of NADH (upper panel) and pyruvate (lower panel) for purified M_4 LDHs of several marine teleosts. Deep-living species: *Sebastolobus altivelis* (□); *Antimora rostrata* (△); *Coryphaenoides acrolepis* (■); and *Halosauropsis macrochir* (◆). Shallow-living species: *Pagothenia borchgrevinki* (●); *Scorpaena guttata* (▲); and *Sebastolobus alascanus* (○). The 95% confidence intervals around the K_m of pyruvate values are approximately ±10% of the K_m value, and the 95% confidence intervals around the K_m of NADH values are approximately ±14% of the K_m value. The 95% confidence limits for the increased K_m values between 1 and 68 atm (and, for the *S. alascanus* K_m of NADH, between 1 and 51 atm) do not overlap. The assay mixture contained 80 m*M* Tris–HCl, pH 7.5, at the assay temperature of 5°C and 100 m*M* KCl. Apparent K_m of NADH values were determined by using 8 to 10 NADH concentrations (10 to 300 μ*M*) and 5 m*M* sodium pyruvate. Apparent K_m of pyruvate values were determined by using 7 to 9 sodium pyruvate concentrations (0.1 to 4 m*M*) and 150 μ*M* NADH. The apparent K_m values and standard errors were calculated by use of a weighted linear regression (Wilkinson, 1961). [From Siebenaller and Somero (1979).] Although not shown in this figure, the purified M_4 LDHs of the rattails *Nezumia bairdii* and *Coryphaenoides leptolepis* (depth ranges in Fig. 3) display pressure insensitivity of apparent K_m of NADH and pyruvate identical to that of the LDHs of the deep-living species shown here. (Siebenaller and Hennessey, unpublished data.)

K_m of pyruvate values (Figs. 1 and 2), the observed K_m values of NADH at typical habitat temperatures and pressures for the different fish are markedly similar among species (Fig. 2).

The finding that the M_4 LDHs of deep-sea teleost fishes belonging to four different families are so similar in their pressure–K_m relationships offers a striking example of convergent evolution at the molecular level and argues strongly that K_m conservation at high and/or varying pressures is a critical feature of enzyme evolution in deep-sea species. These data further show that the M_4 LDHs of cold-

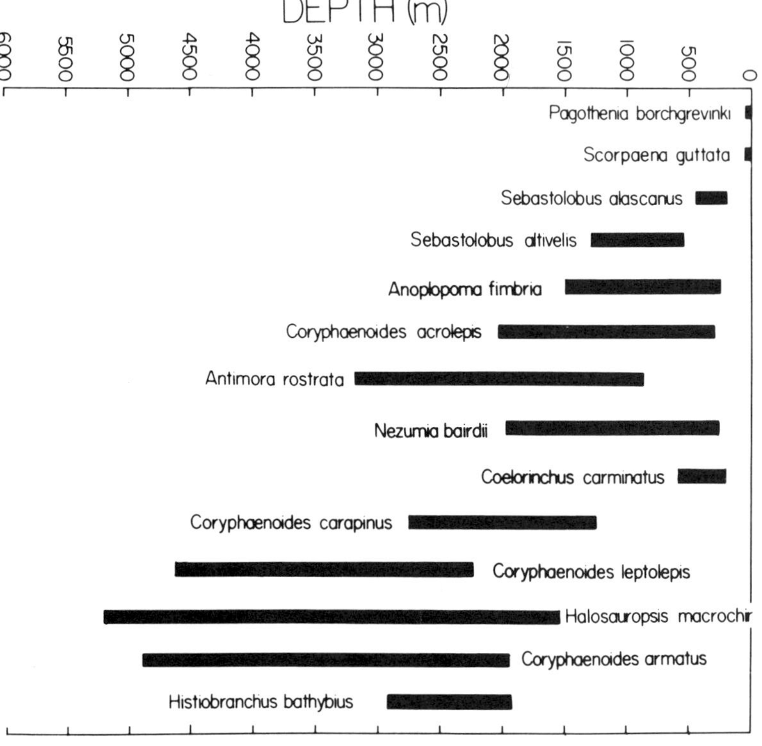

Fig. 3. Depths of maximal abundance for the seven teleost fish used in the M_4 LDH studies and some of the species used in the enzyme activity–depth studies. Sources: *Sebastolobus altivelis* and *S. alascanus* (Moser, 1974), *Antimora rostrata* (Wenner and Musick, 1977; Iwamoto, 1975), *Coryphaenoides acrolepis* (Okamura, 1970; Iwamoto and Stein, 1974; T. Matsui, personal communication), *Halosauropsis macrochir* (McDowell, 1973; Haedrich et al., 1980), *Scorpaena guttata* (Miller and Lea, 1972), *Pagothenia borchgrevinki* (Wohlschlag, 1964), *Anoplopoma fimbria* (Miller and Lea, 1972; K. M. Sullivan, personal communication), *Nezumia bairdii* (Marshall and Iwamoto, 1973a; Haedrich et al., 1975; Haedrich et al., 1980), *Coelorinchus carminatus* (Marshall and Iwamoto, 1973b; Haedrich et al., 1975; Haedrich et al., 1980), *Coryphaenoides (= Lionurus) carapinus* (Marshall, 1973; Haedrich and Polloni, 1976; Merret, 1978; Haedrich et al., 1980), *Coryphaenoides (= Chalinura) leptolepis* (Marshall, 1973; Haedrich et al., 1980), *Coryphaenoides (= Nematonurus) armatus* (Marshall, 1973; Haedrich and Henderson, 1974; Haedrich et al., 1980), and *Histiobranchus bathybius* (DeWitt, 1971; Haedrich et al., 1980).

adapted, shallow-water fish are not preadapted for function at depth and that even relatively modest pressures, of the order of 50 to 100 atms, are sufficiently perturbing of enzyme function to favor pressure-adaptive modifications of enzymes.

This latter conclusion is most dramatically supported by the differences noted between the M_4 LDHs of the two scorpaenid congeners, *Sebastolobus altivelis* and *S. alascanus* (Siebenaller and Somero, 1978). The depth distribution profiles for these two species are shown in Fig. 3 [see also Fig. 1 of Siebenaller and Somero (1978)]. Because these two congeners provide an important index of the role of moderate hydrostatic pressures in establishing pressure-adaptive differences in enzymic properties, a brief description of their biologies and life histories is merited. The two species occur in the northeastern Pacific. *Sebastolobus altivelis* is found from the southern tip of Baja California to the Aleutian Islands

(Moser, 1974). *Sebastolobus alascanus* is reported from Cedros Island, Baja California (Miller and Lea, 1976, addendum) to the Bering Sea and the Commander Islands off the Asian mainland (Moser, 1974). Although the bathymetric distributions of the species overlap to some degree (Moser, 1974) and the two species may occasionally be taken in the same trawls (Siebenaller, 1978b; Hubbs, 1926), their distributions of abundance are such that *S. altivelis* always occurs deeper than *S. alascanus*. Moser (1974) has documented their life histories on the basis of midwater trawl collections of the pelagic juvenile stages and benthic otter trawls of the adults. The adults of both species are benthic. Spawning occurs in a 4- to 5-month period, peaking in April (Moser, 1974). Gelatinous egg masses that float to the surface are produced (Pearcy, 1962). Hatching and larval development take place in surface waters. *Sebastolobus alascanus* spends approximately 14 to 15 months in the water column from spawning to settlement and *S. altivelis,* 20 months (Moser, 1974). The amount of genetic differentiation between these two species is similar to the differentiation observed among other scorpaenid fish congeners (Siebenaller, 1978b). Siebenaller (1978b) has also examined the distribution of electrophoretically detectable genetic variation among populations of *S. altivelis* in the Southern California Continental Borderland, as discussed later in this chapter (Section 2.D). Thus these two congeners provide an excellent study system for evaluating the importance of hydrostatic pressure in adaptation to the deep sea, since they experience similar thermal regimes and are close phylogenetically. Many past comparisons of pressure sensitivities of enzymes have involved studies of enzymes from species that are phylogenetically remote and have widely different body temperatures, rendering isolation of pressure influences difficult.

For the M_4-LDH comparisons shown in Fig. 2, it is clear that the LDH of the shallow-dwelling congener, *S. alascanus,* is kinetically indistinguishable from the LDH of another shallow-water scorpaenid teleost, *Scorpaena guttata*. In contrast, the M_4 LDHs of the deeper-living congener, *S. altivelis,* displays the pressure insensitivities characteristic of the LDHs of deep-living teleosts. It thus appears that pressures of less than 100 atm may, in effect, function as barriers to shallow-water species and that the colonization of deep water depends on the acquisition of enzymes having reduced pressure sensitivities. In the case of these two *Sebastolobus* species, we speculate that an ancestral, shallow-living population contained some individuals having pressure tolerant enzymes similar to those of *S. altivelis*. These individuals were capable of functioning efficiently at greater depths than were other members of the species (Siebenaller and Somero, 1979), and these pressure-tolerant individuals eventually differentiated into a second, deeper-occurring species. Below we present data that suggest that the acquisition of pressure-insensitivity by enzymes carries, as a price, a loss of competitive ability with shallow species' enzymes under conditions of low temperature and pressure (see Section 2.C). Thus both the upper and lower limits of species' depth distribution ranges may be determined, at least in some part, by enzymatic factors.

Up to this point we have focused entirely on a single type of enzyme, the M_4 isozyme of lactate dehydrogenase, which has received more study from the standpoints of temperature and pressure effects than any other enzyme. However, M_4 LDH is not unique in displaying both temperature- and pressure-adaptive differences among interspecific homologues. In the case of temperature adaptation, conservation of K_m values and catalytic function (see below) has been documented for several sets of enzyme homologues (Somero, 1978). Pressure studies have been more restricted, and, as mentioned earlier, many studies of pressure effects have used enzymes from distantly related species that also differed widely in body temperature. Nonetheless, the data from these comparative pressure studies tend to support the conclusions reached above concerning the importance of maintaining K_m values for substrates and cofactors within precise ranges in the face of high and/or varying pressures. Examples of enzymes for which adaptations of this type have been noted are acetylcholinesterase (Hochachka, 1974; Hochachka et al., 1975a), citrate synthase (Hochachka et al., 1975b), isocitrate dehydrogenase (Moon and Storey, 1975), and teleost gill Na-K ATPase (Pfeiler, 1978).

Not all enzymes display pressure-related differences between deep- and shallow-living species, however. An example is the skeletal muscle pyruvate kinase (PK) of the two *Sebastolobus* congeners (Fig. 4) (Siebenaller and Somero, 1982). Pressure effects on the K_m values of both substrates, phosphoenolpyruvate (PEP) and adenosine diphosphate (ADP), are the same for the PKs of the two species. This similarity contrasts with the situation found for the M_4 LDHs of these same two species (Fig. 2). It seems appropriate to conclude, therefore, that the need for pressure-adaptive changes in enzymes to permit colonization of deeper regions of the water column is not ubiquitous and that at least some enzymes may be preadapted for function at high and/or variable pressures. Further comparative studies of deep- and shallow-living species may reveal what fraction of the cellular enzymes are modified for function in the deep sea.

The mechanistic bases of pressure effects on K_m values have received some study, although no general conclusions can presently be drawn. Using pressure-jump fluorescence techniques, Coates et al. (1977) and Hardman et al. (1978) found that the binding of NADH to pig heart LDH was impeded by increased pressure. Morild (1977 a,b) observed a similar perturbation of cofactor (NAD^+)

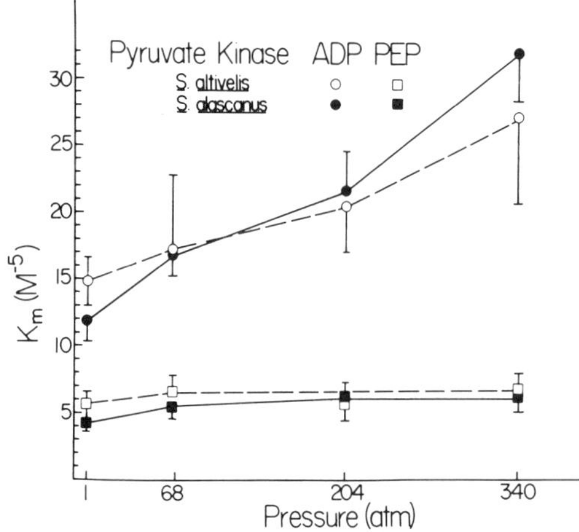

Fig. 4. Effects of pressure on K_m values of adenosine-5′-diphosphate (ADP) and phosphoenolpyruvate (PEP) for skeletal muscle pyruvate kinases isolated from *Sebastolobus altivelis* and *S. alascanus*. [Data from Siebenaller and Somero (1982).]

binding to alcohol dehydrogenase by increased pressure. Although rate constants other than the dissociation constant may contribute to the expression for K_m of NADH (or NAD^+) (Borgmann et al., 1975; Coates et al., 1977; Hardman et al., 1978; Greaney and Somero, 1980), the major contribution to perturbation of the K_m of cofactor by pressure may result from actual disruption of cofactor-binding site interactions. These disruptions may be the result of at least two distinct effects. First, when the strongly hydrated cofactor molecule binds to the active site, the densely organized water around the cofactor may be displaced, leading to an increase in system (enzyme + cofactor + solvent) volume. (The relationships of volume changes to pressure sensitivities are discussed in Appendix A.) In addition to pressure sensitivity arising from the removal of electrostricted water around the cofactor (and the charged residues with which the cofactor interacts), pressure sensitivity may arise from the change in enzyme conformation that usually accompanies ligand (substrate or cofactor) binding. For example, in the case of LDH, a large conformational change occurs during cofactor binding, in which the "loop" region of the active center folds down on top of the cofactor and remainder of the binding site (Eventoff et al., 1977). Water displacement would again be expected, with a concomitant increase in system volume, and the precise orientation of the loop with respect to the remainder of the LDH molecule might also be sensitive to pressure. Thus the observed pressure sensitivities of K_m values for NADH may be the net result of inherent binding volume effects due to water displacement and direct pressure effects on enzyme structure.

The types of amino acid replacement in primary structures of enzymes that are needed to offset these types of perturbation may include alterations in active site residues that interact with ligands. Studies by Hochachka and colleagues (Hochachka, 1974, 1975a; Hochachka et al., 1975c) of the effects of temperature and pressure on substrate–cofactor analogue binding have provided evidence that the pressure- and temperature-adaptive differences in ligand binding properties may derive in part from modifications in binding site amino acid compositions that have the effect of shifting the relative importance of different types of weak interactions (hydrogen bonds, electrostatic interactions, and hydrophobic interactions) in differently adapted species. Of particular significance is the possible role played by hydrophobic interactions, that are weakened by low temperatures and high pressures. Hochachka (1974, 1975a) obtained kinetic evidence that enzymes of deep-sea fish tend to rely less on hydrophobic interactions for stabilizing substrate (acetylcholine : acetylcholinesterase) and cofactor (NADH : LDH) complexes than the homologous enzymes of mammals. Amino acid sequencing of homologous active sites could resolve the importance of these types of evolutionary change.

In addition to amino acid replacements in the active site regions where ligand binding occurs, replacements at other regions of a protein could also influence the temperature and pressure sensitivities of ligand binding (and catalysis). If, for example, a conformational change during ligand binding involves the exposure to solvent of a water-constricting group formerly "buried" within the enzyme, a decrease in system volume could result that is independent of events occurring at the active site (Somero and Low, 1977). Modification of these non-active-site volume changes could be used to "titrate" the overall volume changes that accompany ligand binding. The involvement of water structure changes around amino acid side chains during the catalytic steps in an enzymatic reaction sequence (Greaney and Somero, 1979) argues that pressure effects on catalytic rates may also be due in large measure to hydration density effects and that modifications of the latter effects may provide "raw material" for the evolutionary modification of the pressure sensitivities of enzymatic reactions in deep-sea animals.

Adjustments in Enzyme-Modulator Binding

From the preceding discussion of substrate and cofactor binding, it is evident that a critical mechanism for the regulation of enzymatic activity is the maintenance of K_m (or $S_{0.5}$, the analogous kinetic parameter for enzymes that display sigmoidal binding properties) values close to *in vivo* substrate concentrations. Even if the full catalytic potential of enzymes is not realized as a result, this relationship between K_m values and substrate concentrations is advantageous because small changes in substrate availability or in enzyme–substrate binding ability can lead to relatively large changes in reaction velocity. This is why positive and negative modulators of metabolic pathways usually regulate enzymatic activity through changing K_m (or $S_{0.5}$) values, rather than through influencing the V_{max} values of the reactions. This key regulatory feature, calling for specific interactions between strategically placed regulatory enzymes and modulators, is displayed only by allosteric (or regulatory) enzymes and seldom applies to nonregulatory enzymes catalyzing reactions that are usually at, or near to, their thermodynamic equilibrium positions.

The binding of modulators to regulatory sites on enzymes should be as pressure- and temperature-sensitive as the binding of substrates. Both classes of enzyme–ligand interaction involve weak bonds (hydrogen bonds, hydrophobic interactions, electrostatic interactions) that are known to be highly sensitive to changes in pressure and temperature (Somero and Hochachka, 1976 a,b; Appendix A, this chapter). Also, binding ability for all classes of ligand will depend strongly on the conformation of the ligand binding sites, and both pressure and temperature can have marked effects on enzyme conformation. Thus the precise regulation of metabolism that is effected through enzyme–modulator interactions appears to be a highly sensitive site for selection in deep-sea organisms.

For the enzyme fructose bisphosphatase (FBPase), the enzyme for which the most data on temperature and pressure effects on regulation exist, the predicted sensitivity of regulation to temperature and pressure is supported (Hochachka et al., 1971a,b; Rosenmann et al., 1977). Fructose bisphosphatase catalyzes the hydrolysis of fructose bisphosphate (FBP), in a reaction that is usually considered an irreversible step in the process of gluconeogenesis (the synthesis of glucose from smaller precursors such as pyruvate):

$$\text{Fructose bisphosphate} \longrightarrow \text{fructose-6-phosphate} + P_i$$

Adenosine monophosphate (AMP) is a negative modulator of the reaction; that is, rising AMP concentrations slow down, or stop, the FBPase reaction. In contrast, high AMP concentrations lead to an activation of the oppositely poised reaction, that catalyzed by phosphofructokinase (PFK):

$$\text{Fructose-6-phosphate} + \text{ATP} \longrightarrow \text{FBP} + \text{ADP}$$

The key regulatory role of AMP can be understood on the basis of the fact that the AMP concentration in the cell is an excellent index of the extent to which the adenylate system, consisting of AMP, ADP, and ATP is "charged"; that is, the AMP concentration signals the need of the cell for generating ATP via reaction sequences such as glycolysis. The key regulatory site in this sequence of reactions is PFK.

In initial studies of FBPase from mammals, Taketa and Pogell (1963, 1965) found that FBPase–AMP interactions are strongly stabilized at low temperatures. This is the expected effect for an enzyme–ligand pair that is stabilized by charge–charge interactions (see Appendix A). In an ectothermic organism such a strong enhancement of modulator binding by reduced temperature would, in the absence of counteracting effects, lead to a blockage of this important reaction even under conditions where the AMP concentration was low enough to call for high levels of FBPase activity. An illustration of the importance of temperature perturbation of FBPase–AMP interactions has been given by van Tol (1975), who showed that in the absence of AMP the Q_{10} of the FBPase reaction is approximately 2, whereas in the presence of AMP the Q_{10} is near 20. For an enzyme displaying this degree of temperature sensitivity in modulator binding, a mere 3°C change in temperature would lead to a fivefold change in catalytic rate.

For organisms that can tolerate large temperature and/or pressure changes, the metabolic apparatus must either be able to tolerate such large alterations in regulatory abilities and reaction rates or, alternatively, adaptations must be effected that minimize or eliminate these environmentally induced changes in regulatory properties. The latter option seems the more likely. In the case of AMP regulation of FBPase, studies of homologues from species inhabiting environments with differing temperatures and pressures indicate that AMP binding is strongly compensated with respect to both parameters (Hochachka et al., 1971a,b; Rosenmann et al., 1977). Thus for FBPase from *Coryphaenoides* sp., the binding of AMP was found to be essentially pressure insensitive (Hochachka et al., 1971a,b) and the binding of AMP by a fish FBPase was found to be weaker than in the case of the mammalian FBPase, an effect that could lessen the perturbing effects of reduced temperatures on the fish enzyme (Rosenmann et al., 1977).

These data, plus information from studies of other enzymes such as pyruvate kinase (Moon et al., 1971) suggest that evolutionary modifications of modulator binding abilities—and probably of modulator binding sites—may play an important role in molecular adaptation to deep-sea conditions. Much more work is needed to reveal the extent and mechanisms of these regulatory adaptations.

C. *Rates of Catalysis by Enzymes of Deep-Sea Fishes*

Thermodynamic Efficiencies of M_4 LDHs of Shallow- and Deep-Living Fish

Data presented in the foregoing discussion indicate that the M_4 LDHs of shallow-water and deep-sea fishes have highly similar K_m values for substrate and cofactor at the appropriate habitat temperatures and pressures of the different species. What, then, of the actual rates at which a single M_4-LDH molecule can convert pyruvate to lactate and regenerate NAD^+ from NADH at biological temperatures and pressures? Is there a similar conservation of catalytic activity among the M_4 LDHs of fish adapted to different temperatures and pressures?

To approach this question, we must first consider the enzymatic properties that determine how rapidly cofactor and substrate, once they are bound to the enzyme, can be converted to products. We must also consider how these catalytic rate-determining properties vary among homologues of a particular type of enzyme from species adapted to different temperatures but similar hydrostatic pressures. These considerations will enable us to make some specific predictions about the ways in which enzymes of deep-sea animals might be adapted to overcome the effects of low temperature and high pressure on catalytic rates.

Under conditions of saturating substrate concentrations—that is, when the formation of enzyme–substrate–cofactor complexes does not contribute to the reaction rate—the velocity of an enzymatic reaction is determined by the activation free energy ($\Delta G\ddagger$) "barrier" to the reaction [see Johnson et al. (1974) and papers in the volume edited by Gandour and Schoner (1978)]. The absolute rate of an enzymatic reaction is related to the "height" of this free-energy barrier according to the equation

$$K = \kappa \frac{kT}{h} \exp \frac{-\Delta G\ddagger}{RT}$$

where K is the rate constant of the reaction, $\kappa \simeq 1$, k is the Boltzmann constant, h is the Planck constant, R is the universal gas constant, and T is the absolute temperature (°K). The activation free energy is a composite of the activation enthalpy ($\Delta H\ddagger$) and activation entropy ($\Delta S\ddagger$):

$$\Delta G\ddagger = \Delta H\ddagger - T\,\Delta S\ddagger$$

The activation enthalpy is related to the Arrhenius activation energy (E_a) by the equation

$$\Delta H\ddagger = E_a - RT$$

Arrhenius activation energies are obtained from plots of log K versus the reciprocal of absolute temperature (see Fig. 5).

It is thus seen that the activation free energy of an enzymatic reaction is a quantitative index of the catalytic efficiency of the enzyme catalyzing the reaction. The more an enzyme can reduce the $\Delta G\ddagger$ "barrier" to the reaction, the more rapidly substrates can be converted to products at any given temperature. Since body temperatures of different species vary so widely, biologists have asked whether selection has favored adaptive differences in $\Delta G\ddagger$ values, such that enzymes of cold-adapted species possess greater abilities for reducing $\Delta G\ddagger$ "barriers" than do the homologous enzymes of warm-adapted species. Reductions in $\Delta G\ddagger$ could readily compensate for the decreased thermal energy present in the cells of low-body-temperature species and serve as an important component of metabolic rate compensation to temperature (Hochachka and Somero, 1973; Somero, 1978).

Comparative studies of homologues of M_4 LDHs (Borgmann et al., 1975; Borgmann and Moon, 1975; Low et al., 1973), pyruvate kinases (Low and Somero, 1976), glyceraldehyde-3-phosphate dehydrogenases (Cowey, 1967), and myosin ATPases (Johnston and Walesby, 1977) have demonstrated that the heights of the

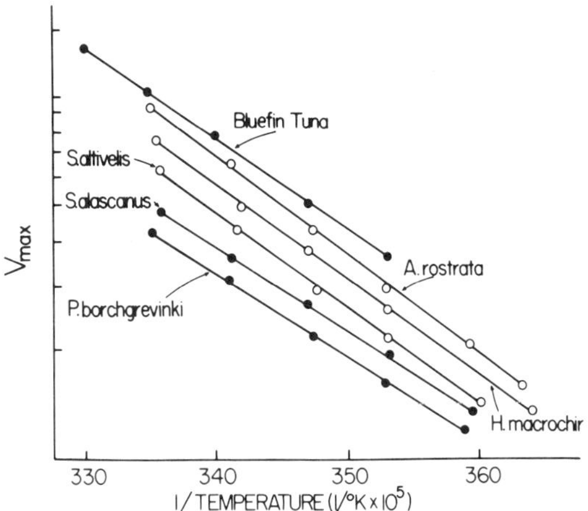

Fig. 5. Plots of log V_{max} versus the reciprocal of absolute temperature for purified M_4 LDHs of shallow- and deep-living teleost fish; V_{max} values are in arbitrary units and do not reflect the actual turnover numbers of the enzymes (see Table I). The slopes of the regression lines are proportional to the Arrhenius activation energies (E_a) of the reactions. The assays were conducted in 80 m*M* imidazole–HCl buffer with temperature-dependent pH values (see Figure 1*a*), 150 μ*M* NADH, and five to eight concentrations of sodium pyruvate; V_{max} was determined by the weighted linear regression technique of Wilkinson (1961). [Data from Somero and Siebenaller (1979).]

$\Delta G\ddagger$ "barriers" to enzymatic reactions are, in fact, lower for the enzymatic reactions of low-body-temperature ectotherms than those for high temperature ectotherms, birds, and mammals. Over evolutionary time periods, modifications of enzyme amino acid sequences have led to significant gains in the abilities of enzymes of cold-adapted species to compensate for the effects of low temperatures on metabolic reaction rates. (This analysis does not, however, explain why enzymes of species adapted to high temperatures, e.g., birds and mammals, should benefit from a loss of catalytic efficiency, i.e., high $\Delta G\ddagger$ values; we consider this apparent evolutionary paradox later.)

In view of the high catalytic efficiencies of enzymes from low-temperature-adapted fish, one would expect that deep-sea animals would also possess enzymes with relatively low $\Delta G\ddagger$ "barriers." In the case of M_4 LDH, the one type of enzyme that has been examined from this perspective, this expectation is not realized (see Table I and Figure 5). The Arrhenius activation energies ([E_a]—from which the other activation parameters are calculated; see Table I footnotes) of the enzymatic reactions of the deep-sea fish are homogeneous [analysis of variance (ANOVA), $F_{[3,19]} = 0.99$, $0.5 < p < .25$], as are the activation parameters of the shallow species' reactions (ANOVA, $F_{[1,19]} = 2.24$, $0.25 < p < .10$). However, the E_a values of the deep and shallow species groups differ (ANOVA, $F_{[1,19]} = 9.09$, $p < .01$). The activation enthalpies and free energies of the M_4 LDH reactions of deep-sea fish from four families lie above those of polar and temperate zone shallow-water fishes. In fact, the M_4 LDHs of the deep-sea fishes resemble mammalian M_4 LDHs in terms of activation thermodynamic parameters.

As in the case of pressure effects on K_m values, the M_4 LDHs of the *Sebastolobus* congeners differ in $\Delta G\ddagger$ and $\Delta H\ddagger$ values. The shallower species, *S. alascanus,* has an M_4 LDH that has a catalytic efficiency comparable to that of the M_4

TABLE I

Activation Energy Parameters and Relative Absolute Velocities for Purified M_4 LDHs from Differently Temperature- and Pressure-Adapted Fish and a Mammal[a]

Species[b] (Body Temperature)	E_a[c] (cal/mol)	$\Delta H\ddagger$[c] (cal/mol)	$\Delta S\ddagger$[d] (cal/mol · °K)	$\Delta G\ddagger$[d] (cal/mol)	Relative[d] Velocity
Pagothenia borchgrevinki (−2°C)	11,020 (197)	10,467	−12.7	14,000	1.00
Sebastolobus alascanus (4 to 12°C)	11,068 (196)	10,515	−12.6	14,009	0.98
Sebastolobus altivelis (4 to 12°C)	12,538 (216)	11,985	−8.1	14,249	0.64
Coryphaenoides acrolepis (2 to 10°C)	12,366 (220)	11,813	−8.7	14,222	0.67
Halosauropsis macrochir (2 to 5°C)	12,396 (259)	11,843	−8.6	14,227	0.66
Antimora rostrata (2 to 5°C)	13,110 (223)	12,557	−6.4	14,343	0.54
Thunnus thynnus (15 to 30°C)	11,937 (153)	11,384	−10.0	14,152	0.76
Rabbit (37°C)	13,100	12,550	−6.4	14,342	0.54

[a] *Notation:* E_a, Arrhenius activation energy; $\Delta G\ddagger$, activation free energy; $\Delta H\ddagger$, activation enthalpy; $\Delta S\ddagger$, activation entropy. Assay conditions are given in the legend of Fig. 5. All values have been standardized to 5°C. (Modified from Somero and Siebenaller, 1979.)

[b] The depth ranges for the first six species are given in Fig. 3. *Thunnus thynnus* (bluefin tuna) is abundant from the surface to less than 300 m (Suzuki et al., 1977). For the analysis of variance tests referred to in the text, *P. borchgrevinki* and *S. alascanus* were the shallow-living species, and *S. altivelis, H. macrochir, A. rostrata,* and *C. acrolepis* were the deep-living species.

[c] The Arrhenius activation energy (E_a) values were computed from the slopes, determined by linear regression, of plots such as those shown in Fig. 5. Standard errors for the E_a values are given in parentheses beneath the E_a values: $\Delta H\ddagger = E_a - RT$ (T = 278.2°K; R is the universal gas constant).

[d] Estimates of $\Delta S^{\ddagger}$ and $\Delta G^{\ddagger}$ were computed by use of the following rationale. For all sets of homologous enzymes that have been studied to determine activation parameters, a highly regular covariation between $\Delta H^{\ddagger}$ and $\Delta S^{\ddagger}$ has been found (Low et al., 1973; Borgmann et al., 1975; Borgmann and Moon, 1975; Johnston and Walesby, 1977). For LDHs, a $\Delta H^{\ddagger}/\Delta S^{\ddagger}$ plot gives a slope of approximately 333°K (Low and Somero, 1974; Low et al., 1973; Borgmann and Moon, 1975). This slope is called the *compensation temperature* (Lumry and Rajender, 1970) to emphasize that simultaneous increases in $\Delta H^{\ddagger}$ and $\Delta S^{\ddagger}$ have compensating and offsetting effects on $\Delta G^{\ddagger}$ ($\Delta G^{\ddagger} = \Delta H^{\ddagger} - T\,\Delta S^{\ddagger}$). All the M_4 LDHs used in this study were assumed to fit the compensation relationship noted for other vertebrate M_4 LDHs. On the basis of this assumption $\Delta S^{\ddagger}$ values were estimated, using experimentally determined differences in $\Delta H^{\ddagger}$, as well as the experimentally determined $\Delta S^{\ddagger}$ value for the M_4 LDH of flounder (i.e., $\Delta S^{\ddagger} = -12.6$ cal/mol · deg and $\Delta H^{\ddagger} = 10{,}500$ cal/mol) (Borgmann et al., 1975). For each 333-cal/mol increment in $\Delta H^{\ddagger}$, the compensation relationship predicts a 1 entropy unit (cal/mol · °K) increase in $\Delta S^{\ddagger}$. Using the flounder enzyme as a baseline, the other $\Delta S^{\ddagger}$ values were computed and using these, the $\Delta G^{\ddagger}$ values. The relative velocity values are the reaction rates computed from the differences in $\Delta G^{\ddagger}$ among the M_4 LDHs, using the reaction rate of the *Pagothenia* M_4-LDH reaction (the reaction having the lowest $\Delta H^{\ddagger}$ and $\Delta G^{\ddagger}$ values) as the reference (value = 1.0). The calculated relative rates agree extremely well with the experimentally determined rates for the M_4-LDH reactions of *Sebastolobus alascanus* and *S. altivelis*. The specific activities of these two enzymes, in micromoles of NADH oxidized to NAD^+ per mole of LDH per minute, are 4.66×10^{10} and 2.82×10^{10}, respectively at 5°C and 1 atm pressure. The values reported for *A. rostrata* in this study are in excellent agreement with those reported by Baldwin et al. (1975); they obtained values for E_a and $\Delta H^{\ddagger}$ of 13,200 and 12,648 cal/mol (at 5°C), respectively. Their experimentally determined substrate turnover number for the *A. rostrata* enzyme (2.48×10^{10} μmol NADH oxidized to NAD^+ per mole of LDH per minute) is also in excellent agreement with the relative rate calculated from the activation parameters. The $\Delta G^{\ddagger}$ values were estimated by using the compensation relationship rather than through direct experimental measurement of velocities and protein concentrations because a small amount of enzymatically inactive protein can greatly distort specific activity values, whereas a high degree of precision is possible in the measurement of $\Delta H^{\ddagger}$ values. The regression coefficients for all Arrhenius plots were greater than 0.996. The bluefin tuna data are from Yancey and Somero (1978b). The rabbit data (obtained by use of a phosphate buffer system) are from Low et al. (1973).

LDH of the cold-adapted Antarctic fish, *Pagothenia borchgrevinki,* whereas the M_4 LDH of the deeper-occurring species, *S. altivelis,* resembles the enzymes of other bathyal fish. As in the case of K_m adaptations, pressure increases in the range of only 50 to 100 atm may be adequate to promote significant changes in the catalytic properties of enzymes.

Table I also presents estimates of the relative LDH reaction velocities (in the direction of pyruvate reduction to lactate—the predominant direction of M_4 LDH function in fish skeletal muscle) for the M_4 LDHs of the different fish at 5°C and 1-atm pressure. The assumptions underlying these calculations are given in the footnotes to Table I. At pressures near 1 atm, the relative inefficiencies of the M_4 LDHs of the deep-sea species are apparent. The higher activation energy "barriers" to the LDH reactions of the deep-sea species translate into pyruvate reduction to lactate that is about 40% less at low temperatures, relative to the rates characteristic of the enzymes of shallow water, cold-adapted fish. Thus, whereas the pressure insensitivity of K_m values and catalytic rate (see below) make the M_4 LDHs of deep-living fish superior for function under high or variable pressure, these pressure-adapted enzymes are catalytically inferior at low pressures.

Is there a basis for interpreting or rationalizing this apparent "failure" of evolutionary processes to provide deep-sea fish with the types of highly efficient enzyme characteristic of other cold-adapted but shallow-occurring, fish? We speculate that one basis for this alleged failure is the relationship that exists between enzyme structural stability and catalytic efficiency [Borgmann et al. (1975); Borgmann and Moon (1975); Johnston and Walesby (1977); Low and Somero (1976); reviewed in Somero (1978)]. In the case of M_4 LDHs, Borgmann et al. (1975) have shown that the differences in $\Delta G\ddagger$ and $\Delta H\ddagger$ between a mammalian and flounder enzyme can be related to different numbers of weak bonds in the enzyme–substrate–cofactor complexes formed with these two LDH homologues. In the mammalian (beef muscle) system, additional weak bonds are formed to stabilize the complex. Two important consequences follow. First, the beef enzyme–substrate–cofactor complex is more thermally stable than the complex containing the flounder enzyme. The advantages of this situation for a high-body-temperature species seem obvious, and these advantages in heat stability appear to outweigh the second important effect of additional weak bond formations in the mammalian complex, an increase in the energy barrier to catalysis. The bonds that are formed when the enzyme–substrate–cofactor complex is assembled must be broken during the subsequent steps of the reaction so that product and oxidized cofactor can be released from the enzyme. Thus the cost of enhanced complex stability is a higher activation energy. This discovery by Borgmann et al. resolves the paradox mentioned above concerning the possible selective advantage of *reduced* catalytic efficiencies (higher $\Delta G\ddagger$ values) for enzymatic reactions of warm-adapted species. The resolution of this paradox is seen to involve more than just the rate of enzymatic activity under conditions of saturating substrate and cofactor concentrations. Borgmann et al. show that the benefits of a stable enzyme–substrate–cofactor complex outweigh the disadvantages of slightly higher $\Delta G\ddagger$ values for reactions of high-body-temperature organisms. Besides, the relatively large amounts of thermal energy in these latter species lead to a high acceleration of metabolic rates, especially since the enzymes of warm-adapted species are characterized by large $\Delta H\ddagger$ values (Fig. 5; Table I).

If we apply reasoning similar to that used in the temperature adaptation arguments given above to the case of pressure adaptation, a similar "compromise" between functional and structural attributes of enzymes may be discerned (Somero, 1979). The LDH reaction occurs with large changes in enzyme conformation (Eventoff et al., 1977) and in exposure of the enzyme surface to surrounding solvent (White et al., 1976), two types of event that may lead to large volume changes and, hence, large pressure effects (see Appendix A). If the conformational changes accompanying the LDH reaction are highly pressure sensitive, adaptation to deep-sea pressures may necessitate modifications in LDHs that strengthen enzyme structure. The analogy with high-temperature adaptation is clear. Thus we suggest that during evolutionary adaptation to high temperatures and to high pressures, selection favors an increase in enzyme structural rigidity that is accompanied by an increase in the energy barriers to catalysis.

Pressure Effects on Catalytic Rate

Although the M_4 LDHs of deep-sea fish do not possess the catalytic efficiencies of M_4 LDHs of shallow-water fish at 1-atm pressure (Table I), the relevant comparisons of these enzymes to obtain estimates of *in situ* activities must be made at physiologically appropriate temperatures and pressures. The data given in Table II show that, whereas the M_4 LDHs of deep-sea fish are less efficient in catalyzing the conversion of pyruvate to lactate, they are able to maintain their catalytic abilities better in the face of increases in pressure than are the more efficient LDHs of shallow-water fish. For all of the M_4 LDHs studied, log V_{opt} (optimal velocity, obtained under optimal cofactor and substrate concentrations) varied in a linear manner with pressure. From the slopes of plots of log V_{opt} versus pres-

TABLE II
Apparent Volume Changes Associated with Pressure Inhibition of Optimal Velocities of Deep- and Shallow-Living Fish M_4-LDH Reactions[a]

		Percent Inhibition of 1-atm Rate at		
Species	$\Delta V \pm$ SE (cm^3/mol)	68 atm	204 atm	340 atm
Deep-living species				
Sebastolobus altivelis	8.1 ± 0.22	2	7	11
Antimora rostrata	0.3 ± 1.45	0	0	0
Coryphaenoides acrolepis	2.6 ± 0.83	1	2	4
Halosauropsis macrochir	4.2 ± 0.40	1	3	5
Shallow-living species				
Sebastolobus alascanus	12.8 ± 0.53	4	11	17
Scorpaena guttata	12.8 ± 0.94	4	11	17
Pagothenia borchgrevinki	5.2 ± 1.53	2	5	7

[a] Computation of the volume changes is detailed in Appendix A. Experiments were conducted in 80 mM Tris–HCl, pH 7.5, at the reaction temperature of 10°C, 100 mM KCl, and 150 μM NADH and with a pyruvate concentration (4 or 5 mM) yielding optimal velocities. Data from Siebenaller and Somero (1979).

sure, the apparent volume changes associated with the rate-determining step(s) in the reactions ΔV was computed (Table II). In general, the M_4 LDHs of the deep-sea fish are characterized by the smallest ΔV's values; in other words, the inhibition of V_{opt} by pressure is less for the deep-sea species' enzymes. For the deepest-occurring species studied (see the depth distribution data given in Fig. 3), the V_{opt} of the M_4-LDH reaction is inhibited by no more than 5% as pressure is increased from 1 to 340 atm. For the shallow-water species, inhibition by 340-atm pressure is approximately 17%. Moreover, since the V_{opt} effects were obtained at high concentrations of pyruvate and NADH, the rate inhibition observed may belie the true extent of rate inhibition that occurs under physiological substrate and cofactor concentrations, where pressure effects on K_m values may also contribute to rate inhibition. If, for example, the large increases in the K_m of NADH with rising pressure found for the shallow-living species (Fig. 2) are adequate to prevent cofactor saturation from occurring, the actual pressure inhibition of the LDH reactions of the shallow-living species may be much larger than the observed V_{opt} inhibitions.

In conclusion, although the M_4 LDHs of deep-sea fish have lower inherent catalytic efficiencies than do the homologous LDHs of shallow-living species, when comparisons are made at 1-atm pressure, the former LDHs exhibit a much lower reduction in catalytic rate in the face of increased pressure. Much of the catalytic rate enhancement advantage seen for the shallow species' LDH^{-} at 1 atm would disappear at depth. Conversely, at pressures typical of those experienced by the shallow-water species studied (Fig. 3), the M_4 LDHs of the deep-sea fish would have substrate and cofactor binding abilities similar to those of shallow species, and would be less efficient catalytically. These considerations suggest that both the lower and upper distribution limits of marine species may be established, in part, by the kinetic properties of certain enzymes. Of particular significance is the finding that the M_4 LDH of *Sebastolobus altivelis* is catalytically inferior to the M_4 LDH of *S. alascanus* at low pressures. This discovery resolves an issue raised earlier as to why species possessing pressure-insensitive enzymes are generally restricted to the deeper regions of the water column. These species (e.g., *S. altivelis*) may compete poorly with similar species (e.g., *S. alascanus*) under shallow-water conditions where the reduced catalytic efficiencies of the enzymes of deep-sea species render these species less metabolically efficient than shallow-water species.

D. *Pressure Effects on Macromolecular Structure*

Protein Conformation and Subunit Aggregation

The preservation of correct functional (kinetic) properties by enzymes is dependent on the maintenance of the proper enzyme structure. The binding, catalytic, and regulatory properties of enzymes are particularly sensitive to changes in the conformations and the subunit aggregation states of enzymes. In the case of conformational changes, it is likely that even minor alterations in the three-dimensional structures of enzymes, especially near the active centers, will have disruptive effects on enzyme function. Disruption of subunit aggregation generally leads to a complete loss of enzymatic activity, and in the case of structures such as microtubules (Engelborghs et al., 1976; O'Conner et al., 1974; Salmon, 1975) and

F-actin (Estes, 1975; Ikkai and Ooi, 1966) and ribosomes (Pope et al., 1975), such disruption leads to the loss of important cellular components associated with processes such as cell division, motility, and protein biosynthesis (e.g., Walker and Wheatley, 1979).

Theoretical considerations suggest that high hydrostatic pressures will have considerable effect on protein conformation and, especially, on protein subunit aggregation (Brandts et al., 1970; Penniston, 1971; Zipp and Kauzmann, 1973). Indeed, a number of recent studies have demonstrated that protein conformations (Brandts et al., 1970; Carter et al., 1978; Hawley, 1971; Hawley and Mitchell, 1975; Schmid et al., 1978; Torgerson et al., 1979; Zipp and Kauzmann, 1973) and aggregation states (Engelborghs et al., 1976; Halvorson, 1979; Schmid et al., 1979; Schade et al., 1980; Paladini and Weber, 1981) are readily disrupted by the application of high pressures. A shortcoming of most of these studies, in the context of the present discussion of deep-sea species, is that the pressures that have been employed to disrupt these systems have almost invariably been higher than pressures found in the deepest regions of the ocean. This shortcoming, when paired with the fact that these studies have generally been conducted under nonphysiological conditions of pH, temperature, and ionic strength with the use of enzymes from terrestrial organisms, renders the available data on pressure effects on protein structure only marginally useful for an analysis of actual pressure stresses and adaptive responses in deep-sea organisms. At best these studies provide suggestions for the types of questions to ask in designing more biologically relevant investigations.

The first consideration in studies of the structural effects of elevated hydrostatic pressure on the protein systems of deep-sea animals is the use of appropriate pressures. As discussed earlier in the context of pressure effects on M_4 LDH function, the sharp changes in K_m values observed at modest pressures (of the order of several tens of atmospheres) for the enzymes of shallow-water teleosts may derive from slight pressure-induced changes in enzyme structure at these low pressures. Selectively important pressure perturbation of enzyme structures need not involve the gross denaturation effects noted in the studies cited above. Instead, only minor shifts in the orientations of amino acid residues in or near the active sites of enzymes may suffice to seriously impair enzyme function. These minor pressure-induced changes in structure may not eliminate the ability of an enzyme to function; nonetheless, the reduced binding and/or catalytic abilities that result from these structural alterations will render the enzyme suboptimal in its function. Amino acid substitutions that confer a high degree of structural stability, that is, pressure resistance, to the enzyme will thus be of high selective advantage.

The correct approach to the study of pressure perturbation of enzymes of marine species is thus seen to necessitate a refocusing of experimental outlook to search for minor conformational changes that may have important functional manifestations. To date, we are aware of no studies that meet this criterion, but with the advent of techniques such as high-pressure fluorescence measurements, slight yet biologically important changes in protein conformation should be detectable (e.g., Clegg et al., 1975). Changes in subunit aggregation state may also be detected by use of biologically realistic pressures (Halvorson, 1979). Ultracentrifugation is an especially simple and useful technique for this purpose (Harrington, 1975; Kegeles et al., 1967). In the one study that has used the high pressures generated in the ultracentrifuge to examine subunit aggregation for an enzyme of a deep-sea species, Hochachka (1975b) found that high pressure may be necessary for dissociation of enormously large subunit aggregations to yield the smaller aggregation state noted for the enzyme (citrate synthase) in terrestrial species. Thus, for citrate synthase of this deep-sea fish (*Antimora rostrata*), pressure seems less a stress to the enzyme than a requirement for the acquisition of the proper quaternary structure. This discovery is interesting in view of Penniston's (1971) conjecture that deep-sea organisms may be unable to maintain the same complex, multisubunit structures found for enzymes of shallow-water or terrestrial forms. It is possible that, for some enzymes, subunits of deep-sea organisms have very high inherent tendencies to aggregate and that elevated pressures are indeed necessary to reduce this aggregation tendency to biologically appropriate levels.

The conjecture that subunit aggregation processes will display strong pressure sensitivities is based on the fact that these processes are thought to depend on hydrophobic interactions. Formation of hydrophobic interactions occurs with an increase in system volume as densely organized water around the nonpolar groups on the subunit contact sites is displaced into the bulk water (Low and Somero, 1975). Hydrophobic interactions form with positive changes in enthalpy ΔH and entropy ΔS, and the entropy change is the dominant contributor to the overall free-energy change ΔG of the reaction. It is appropriate, therefore, to examine the thermodynamics of subunit polymerization to obtain at least circumstantial evidence concerning different reliances on hydrophobic interactions on subunit assembly events in differently pressure-adapted species. To this end Swezey and Somero (1982a) studied the polymerization of skeletal muscle actins from fourteen vertebrates, including deep-sea fish. The assembly of filamentous (F) actin from globular (G) actin was found to occur with the smallest ΔH and ΔS changes for the two very deep-living fishes *Coryphaenoides armatus* and *Halosauropsis macrochir*. Preliminary data also show that these actins assemble with a low volume increase relative to other actins. Thus these thermodynamic data are consistent with the prediction that deep-sea organisms may have protein subunit assembly processes that are much less dependent on hydrophobic interactions than are the homologous or analogous assembly processes in shallow-living organisms.

Pressure and Temperature Effects on Lipid-Based Systems

A second macromolecular system of cells that is extremely sensitive to changes in hydrostatic pressure and temperature are the cell membranes (Ceuterick et al., 1978; Cossins and Prosser, 1978). Membrane function, for example, enzyme-mediated active transport, depends critically on the physical state of the membrane phospholipids. If the phospholipids are either too flexible or too rigid, membrane function will be seriously impaired. As changes in temperature and pressure have marked influences on phospholipid organization (Ceuterick et al., 1978), it is to be expected that adaptation to different zones in the water column will entail modifications in this important component of cellular membranes.

In the case of temperature, the nature of these phospholipid adaptations is well understood (Cossins and Prosser, 1978). To preserve a proper balance between flexibility and rigidity, generally referred to as an optimal *liquid–crystalline* state, the fatty acid composition of the membrane phospholipids is altered. Adaptation to reduced temperatures is associated with increases in the quantities of unsaturated low-melting-point fatty acids in membrane phospholipids; high-temperature adaptation occurs with increases in the quantities of fatty acids that are highly saturated and thus have high melting temperatures (Cossins and Prosser, 1978). Sinensky (1974) has termed these fatty acid modifications of phospholipids a "homeoviscous" adaptation process to denote that membrane viscosity is preserved at or near some optimal value at all cell adaptation temperatures.

Hydrostatic pressure is also known to have large effects on the viscosity or physical state of lipids (Ceuterick et al., 1978; Griest et al., 1958; Macdonald, 1978; Weale, 1967; Yayanos et al., 1978). Increases in pressure generally raise the melting temperatures of lipids, indicating that high pressure stabilizes a rigid, crystalline lipid structure. This effect is understandable because highly ordered lipid crystalline systems have higher density (less volume) than disordered, fluid systems (see Appendix A).

These lipid effects have broad implications for deep-sea organisms. First, pressure-induced changes in phospholipid organization could affect the function of membrane-associated enzymes, especially those enzymes that are lipoproteins (enzymes requiring a lipid moiety for function). The recent studies by Ceuterick et al. (1978) have demonstrated that increases in hydrostatic pressure that shift the organization of phospholipids from a disordered state to a highly ordered crystalline state also increase the activation energy of an enzymatic reaction (the nitrogenase enzyme of the bacterium *Azotobacter*). Thus, for an organism traveling downward in the water column, a threshold pressure may be reached at which the membrane phospholipids undergo a phase transition, with the result that the activities of membrane lipoprotein enzymes are drastically reduced. Since reductions in temperature also favor a more rigid, crystalline phospholipid organization, sinking or downward migration in the water column is seen to pose two major problems for membrane enzyme function. Unfortunately, there exist few data on the phospholipid composition of membranes of deep-sea animals, so we are currently unable to determine how deep-sea species cope with high-pressure and low-temperature stresses on membrane systems. In the only analysis of phospholipid composition of a deep-sea species conducted to date, Patton (1975) found that no significant differences existed between the phospholipid compositions of a deep-sea fish (*Antimora rostrata*) and a shallow-water, cold-adapted fish (*Pagothenia borchgrevinki*). Patton's findings suggest that the reductions in phospholipid fatty acid saturation that occur during adaptation to low temperature may be sufficient to ensure that membrane phospholipids retain a correct viscosity even at elevated pressures. However, much more study of these phenomena is needed before any strong generalizations should be attempted. In particular, pressure effects on membrane-associated enzymes from deep- and shallow-living animals must be compared. Existing data on membrane-associated, ion-dependent ATPases (Pfeiler, 1978; Pequeux and Gilles, 1978) suggest that pressure may have pronounced effects on inorganic ion transport across membranes. Morita (1980) stresses the potential importance of pressure effects on membranes and transport processes for deep-sea organisms. Studies of pressure effects on membrane systems may also reveal the mechanisms by which pressure affects neural function. Both electrophysiological (Harper et al., 1977; Henderson and Gilbert, 1975;

Wann et al., 1979a,b) and behavioral (Macdonald and Teal, 1975; Menzies and George, 1972; Otter and Salmon, 1979) studies show that pressure effects on the electrophysiological properties of membranes may be exceedingly important in adaptation to depth. In an interesting study of pressure effects on the locomotory behavior of the protozoan *Paramecium caudatum,* Otter and Salmon (1979) showed that a pressure of only 68 atm was sufficient to disrupt the "avoiding" response of the swimming *Paramecium*. The relatively low pressures found to perturb LDH function (Fig. 2) are thus seen to affect a behavioral–electrophysiological process as well. The mechanism underlying the behavioral effects noted with *Paramecium* appears to involve a pressure-induced disruption of calcium transport by the ciliary epithelium. Increased pressure blocks the entry of calcium into the cilia, preventing the reversal of ciliary beat necessary for the "avoiding" response. The pressure responses noted for wild-type *P. caudatum* and *P. aurelia* were not found in a mutant, *P. aurelia* "pawn B," which has nonfunctional calcium channels in the ciliary epithelium. In the wild-type *P. caudatum,* release of pressure leads to a rapid influx of calcium, but this response was not observed in the mutant. Otter and Salmon (1979) speculate on several possible mechanisms for the pressure disruption of calcium transport. One mechanism involves changes in the conformation of the membrane components forming the calcium channel. If the transport of calcium involves postulated changes in channel protein conformation, pressure might favor a low-volume conformation of the protein(s) having low calcium transport ability. Alternatively, pressure might locally "freeze" a lipid component associated with the calcium channel, thereby blocking transport. Similar mechanisms could be responsible for pressure-induced changes in the electrical characteristics of *Helix pomatia* and *H. aspersa* neurons observed by Wann et al. (1979a,b). These workers also noted significant pressure effects at only a few tens of atmospheres. Thus, whereas the mechanistic bases of these pressure effects on behavioral and electrophysiological phenomena are not completely understood, these studies have demonstrated unequivocally that relatively modest pressures are capable of interfering with the normal electrophysiological properties of membranes of shallow-living species. Consequently, this suggests that electrophysiological adaptations play an important role in permitting life at depth. Hopefully, with the advent of more sophisticated capture–retrieval gear for the recovery of live deep-sea animals, analysis of membrane physiology and biochemistry can be extended to truly deep-living forms.

A second implication of the large effects of temperature and pressure on the physical state of lipid systems concerns the alterations in density and, hence, buoyancy properties, of the lipids of marine species. Yayanos et al. (1978) have shown that lipids of a common marine copepod (*Calanus plumchrus*), which consisted primarily of wax esters, exhibit a high degree of density change with pressure and, especially, with temperature. A copepod migrating vertically through the water column would encounter significantly different buoyancy problems at different depths. The expansion of lipids at high temperatures and low pressures would increase the organism's buoyancy near the surface. The biological significance of these temperature- and pressure-dependent density changes remains to be explored.

E. *The Roles of Genetic Polymorphism in Adaptation to Depth*

In view of the potentially strong effects of temperature and pressure changes on enzyme function and structure, it is reasonable to enquire as to whether a single form of a particular type of enzyme is adequate for species that undergo broad excursions, ontogenetically or diurnally, through the water column, or for conspecific populations living at different depths. Do single isozymic[2] forms of any given enzyme perform satisfactorily at all the temperature–pressure combinations that such species encounter during their lifetimes? Or, alternatively, must multiple isozyme or allozyme forms be employed to permit organisms to successfully cope with widely different pressures and/or temperatures?

The possible significance of genetic polymorphisms in adaptation to heterogeneous environments has received extensive discussion. [See Lewontin (1974); Hedrick et al. (1976); the papers in the volume edited by Ayala (1976); and Somero (1978). Somero's (1978) review considers the roles of both isozymes and allozymes in temperature adaptation.] Although there is a considerable literature demonstrating a correlation of allelic frequencies with environmental gradients in temperature (Schopf and Gooch, 1971; Johnson, 1971, 1977; Merrit, 1972; Mitton and Koehn, 1975; Powers and Place, 1978), and salinity (Lassen and Turano, 1978; Koehn et al., 1976), there are few data demonstrating a functional basis for such correlations [however, see Merrit (1972); Place and Powers (1979); Powers et al. (1979)]. The importance of multiple isozyme forms in adaptation to heterogeneous environments has also been demonstrated in very few cases (Somero, 1975, 1978; Moon, 1975; Shaklee et al., 1977). With the exception of hemoglobins (Hochachka and Somero, 1973), multiple-gene-locus isozymes do not appear to be broadly important in adaptation to heterogeneous environments.

In the case of marine organisms that encounter wide variation in both hydrostatic pressure and temperature, however, the imposition of two protein-perturbing stresses on enzyme function may lead to relatively major disruptions of metabolic activity. Thus species that encounter wide variations in pressure and temperature may employ multiple protein forms to adapt to the range of conditions that they experience.

There have been a number of investigations of the levels of variation at gene loci encoding proteins in deep-sea invertebrate species, encompassing a variety of taxa (Table III). The levels of genetic variation observed in these species have been as high or higher than those observed for marine invertebrates from other environments (Powell, 1975; Nevo, 1978; Siebenaller, 1978a). These results have contradicted the predictions of those who reasoned that species inhabiting physically stable environments, such as the deep sea, would display little genetic variation (Manwell and Baker, 1970, p. 311; Grassle, 1972; Grassle and Sanders, 1973).

However, most of the studies cited in Table III have been based on samples from single localities, which are potentially open to immigration from surrounding shallower and deeper populations. If populations employ different allozyme vari-

[2]Multiple protein forms performing a single function can broadly be categorized as "isozymes." Here we distinguish multiple forms of a single protein type that result from genetic variation at a single gene locus as *allozymes* and multiple protein forms resulting from polypeptides encoded at more than one gene locus as *isozymes*.

TABLE III
Levels of Electrophoretically Detectable Genetic Variability in Invertebrates from Bathyal Depths in Deep Sea

Species	N[a]	Number of Loci	Hetero-zygosity[c]	Percent Loci Polymorphic	Source
Frieleia halli	45	18	0.169	50	Valentine and Ayala, 1975
Ophiomusium lymani	257	15	0.170	53	Ayala and Valentine, 1974
4 asteroid spp.	31	24	0.164	28–62	Ayala et al., 1975
Benthopecten armatus	30	12	0.094	36.4	Murphy et al., 1976
Dytaster insignis	30	9	0.092	33.3	
Psilaster florae	18	4	0.111	50	
Zoroaster fulgens	30	5	0.140	40	
Ophiomusium lymani	125	7	0.301	85.7	
Ophiomusium planum	12	3	0.306	66.7	
Ophiura sarsi	42	10	0.005	0	
Ophiura signata	15	7	0.057	28.6	
Pandalopsis ampla	13	15	0.07	33	Gooch and Schopf, 1972
Munidopsis diomedeae	6	12	0.12	30	
Nuculana pontonia	13	12	0.23	45	
Malletia sp.	8	10	0.12	22	
Ophiomusium lymani	12	8	nd	33	
Echinus affinus	10	4	nd	0	
Psolus sp.	11	9	nd	25	
Sipunculid	10	4	nd	0	
Siboglinum atlanticum	nd[b]	18	nd	6	Manwell and Baker, 1968
Bathybembix bairdii	479	18	0.162	50	Siebenaller, 1978a
Buccinum sp.	22	29	0.092	35.7	Costa and Bisol, 1978
Ophioglypha bullata	25	24	0.137	56.5	
Munidopsis hameta	23	29	0.079	25	
Ophiomusium lymani	47	17	0.191	52.9	

[a] Number of individuals examined.
[b] No data.
[c] Heterozygosity, an index of genetic variability, is calculated as the mean proportion of gene loci at which an individual possesses two allelic forms.

ants to adapt to particular depth regimes, gene flow into centrally located populations may account for the high levels of genetic variation that have been observed in these bathyal populations (Gooch and Schopf, 1972; Gooch, 1975; Siebenaller, 1978a).

Doyle (1972) reported depth-related heterogeneity at a diallelic esterase locus in five bathyal populations of the ophiuran *Ophiomusium lymani* sampled over a 1000-m-depth range in the Atlantic. However, this heterogeneity was not clinal; that is, the changes in gene frequency did not appear to be directly correlated with increases in hydrostatic pressure. Ayala and Valentine (1974) reported that samples of *O. lymani* and the brachiopod *Frieleia halli* (Valentine and Ayala, 1975) taken at two stations 1 km apart in the San Diego Trough (Pacific) were not genetically different.

Siebenaller (1978a) examined five populations of the bathyal trochid gastropod *Bathybembix bairdii* over a depth range of 759 to 1156 m in the Southern California Continental Borderland at 18 presumptive gene loci. All the populations were highly similar genetically, and none of the populations were characterized by unique alleles. Four of five polymorphic loci showed heterogeneity of allelic frequencies among localities. One locus (LAP-1) with four alleles was homogeneous over all five localities. Three loci, typified by the phosphoglucomutase locus, with nine alleles, were "randomly" patterned with respect to depth and distance between stations. Only one locus, fumarase, showed a trend in allelic frequencies with depth, a pattern not strongly one with geographic distance between stations. However, without more knowledge than is currently available about the modes and magnitudes of dispersal of this species, and in the absence of any knowledge about functional differences (if any) between the different fumarase allozymes, it is difficult to ascribe a mechanism for maintaining this pattern. There was no statistical difference over all of the polymorphic loci in the number of effective alleles[3] among the stations (Friedman test, $.75 < p < .90$), indicating that all of the populations maintained similar levels of genetic variability. However, the broad-scale geographic distribution of genetic variability among populations of this widely distributed species is unknown, and thus the contribution of such populations to the genetic diversity of the populations examined cannot be evaluated.

There is only limited information available on the patterns of genetic variability in deep-sea fishes. Table IV lists the heterozygosities, the percentage of polymorphic loci, and the number of individuals examined in three deep-sea fish species. The levels of heterozygosity found in these species are somewhat below the average observed for fish generally (Somero and Soulé, 1974; Powell, 1975; Nevo, 1978), but the two *Sebastolobus* species have slightly higher levels of variation than that observed for three shallow-water scorpaenid species (Johnson et al., 1973).

The geographic and bathymetric distributions of genetic variability among populations of *Sebastolobus altivelis* taken at 10 stations in the Southern Califor-

[3]The effective number of alleles is a measure of the contribution of the observed alleles to heterozygosity. Since rare alleles contribute little to the genetic variance, and unless allelic frequencies are equal, the number of effective alleles will be less than the observed number. The effective number of alleles is calculated as the inverse of the sum of the squares of allelic frequencies ($1 / \Sigma p_i^2 = n_e$) (Crow and Kimura, 1970).

TABLE IV
Levels of Genetic Variability in Deep-Sea (Bathyal) Fish Species
[Data from Siebenaller (1978b)]

Species	N	Number of Loci	Heterozygosity	Percent Loci Polymorphic
Sebastolobus alascanus	63	20	0.049	20
Sebastolobus altivelis	352	20	0.045	30
Coryphaenoides acrolepis	27	25	0.033	16

nia Continental Borderland have been examined (Siebenaller, 1978b). These stations range from 463 to 1415 m, and encompass the entire depth range over which this species is common (see Fig. 3). Allelic and genotypic frequencies were homogeneous over the 10 samples, with the exception of the phosphoglucomutase locus at a single station (see Table V). The samples were pooled by depth into four groups and analyzed for homogeneity. Allelic frequencies were homogeneous with depth, and there were no trends of allelic or zygotic frequencies with depth. Specimens taken within a 3-month period were pooled by size to examine the

TABLE V
Geographic Distribution of Allelic Frequencies at Two Polymorphic Loci in *Sebastolobus altivelis*[a]

Station	Depth (m)	N	Allele 0.96	0.98	1.00	1.01	1.02
				PGM			
A15	1332	28	0	0	0.982	0	0.018
A24 and A33	1135	21	0	0.048	0.738	0.190	0.024
A27	759	72	0.007	0.035	0.812	0.146	0
A28 and A29	1140	85	0.006	0.059	0.829	0.100	0.006
A30 and A31	1378	68	0	0.029	0.868	0.088	0.015
E3	550	47	0	0.053	0.851	0.096	0
BLB	463	29	0	0.052	0.845	0.103	0
				PGI-1			
A15	1332	29			0.757		0.241
A24 and A33	1135	21			0.619		0.381
A27	759	73			0.658		0.342
A28 and A29	1140	81			0.679		0.321
A30 and A31	1378	63			0.596		0.405
E3	550	43			0.616		0.384
BLB	463	29			0.741		0.259

[a]The samples were taken in the Southern California Continental Borderland (PGM, phosphoglucomutase; PGI, phosphoglucose isomerase; N, number of individuals). Alleles are given numerical designations referring to the relative mobility of their protein products under the electrophoretic conditions used. [From Siebenaller (1978b).]

TABLE VI
Genotypic and Genic Variation with Body Length in *Sebastolobus altivelis*[a]

Size Class (mm)	PGI-1 Genotype			PGM Allele		
	1.02/1.02	1.02/1.00	1.00/1.00	1.00	1.01	Others
175.5 to 245.5	13 (7.69)	17 (24.39)	28 (25.92)	81 (90.16)	21 (10.15)	4 (5.69)
130.5 to 175.5	7 (9.15)	39 (29.01)	23 (30.84)	121 (119.08)	13 (13.41)	6 (7.51)
95.5 to 130.5	9 (9.15)	24 (29.01)	36 (30.84)	126 (119.08)	7 (13.41)	7 (7.51)
50.5 to 95.5	6 (9.02)	31 (28.59)	31 (30.39)	116 (115.68)	9 (13.03)	11 (7.30)
	G = 14.64; 6 DF; $p < .025$			G = 17.88; 6 DF; $p < .01$		

[a]Abbreviations: PGI, phosphoglucose isomerase; PGM, phosphoglucomutase. The number in parentheses is the number of individuals expected for a homogeneous distribution with length; G is the log likelihood ratio for the fit to a homogeneous distribution with length. Allelic and genotypic designations refer to the relative mobilities of the respective allozymes under the electrophoretic conditions used. [From Siebenaller (1978b).]

possible existence of trends in allelic or genotypic frequencies related to size classes. The phosphoglucose isomerase-1 zygotic classes were heterogeneous with length (Table VI), but allelic frequencies were not. The phosphoglucomutase allelic frequencies were heterogeneous over these size classes (Table VI). In the case of both of these enzymes, the heterogeneity stemmed mainly from the largest size classes. However, there was no directionality to this heterogeneity when all size classes are considered; thus there is no indication of strong directional selection acting on the benthic adults. Neither was there an increase in levels of heterozygosity with increasing size. The distribution of genetic variation among populations of *S. altivelis* seems to be determined by the wide-ranging dispersal of this species, rather than by strong short-term selection by depth-related factors acting on the benthic adults. The differentiation of the larger size classes may reflect differential contributions of geographically distant populations to the studied areas and/or temporal variation in selection acting on the pelagic dispersal stages. In spite of the hydrostatic pressure differences (approximately 88 atm) between the sampled populations, strong short-term selection adapting *S. altivelis* to different depths is not apparent. Neither are there differences in isozyme patterns among the smallest, newly settled individuals and larger adults. Thus we propose that eurybathal species such as *S. altivelis* possess enzymes that themselves are eurybathal in functional and structural properties. The M_4 LDH of *S. altivelis* represents a paradigm of this condition, since, at least above 68 atm, the binding properties of the enzyme are pressure insensitive up to at least 476 atm (Fig. 2). Pressure effects on rate are also relatively small for M_4 LDHs of deep-living fish. These data suggest, in fact, that the initial acquisition of the ability to function at relatively moderate pressures, specifically, in the 50- to 100-atm range, preadapts an enzyme for pressure-insensitive function to pressures of up to approximately 500 atm (Fig. 2). Thus the M_4 LDH of *Sebastolobus altivelis* seems as capable of binding substrate and coenzyme at 476 atm, as are the M_4 LDHs of deeper-occurring species such as *Coryphaenoides acrolepis, Halosauropsis macrochir,* and *Antimora rostrata*. The amino acid substitutions that provide an enzyme with the ability to function at increased pressures also appear to provide the ability to function over broad depth ranges. In this view, then, for at least some enzymes, there may be no "need" for multiple allelic or isozymic forms to permit species to function over wide depth ranges. The available genetic information discussed above is consistent with this view.

F. Summary

To summarize the conclusions reached in this section of the chapter, we briefly reiterate the putative roles assigned to adaptations of enzymes to temperature and pressure in establishing the depth optima and depth limits of species in the water column. Three questions within this general context have been considered: (1) the types of adaptation that appear necessary to permit controlled enzymatic function at high pressures and low temperatures; (2) whether the changes that permit successful enzymatic function at high pressures exact a price in terms of function at low pressures (i.e., whether adaptation to deep-sea pressures places an enzyme at a competitive disadvantage at low pressures); and (3) the possible roles of genetic variation in the deep sea, especially in the context of whether more than one allozyme or isozyme form of a given class of enzyme is needed to permit eurybathal species to perform well metabolically over the entire depth range they experience.

Because the study of biochemical adaptation by deep-sea organisms is still in its infancy, large amounts of data are not available for use in answering these questions. Nonetheless, we have attempted to phrase some tentative generalizations using what appear to be the most appropriate available data, especially recent work on M_4 LDHs from different marine teleost fish.

The conclusions reached in our analysis of the available enzyme data are as follows. First, the preservation of ligand binding abilities at pressure is a key factor in adapting metabolic function for deep-sea existence. Although the enzymes of shallow-living species may have appropriate ligand binding abilities at the low temperature characteristic of the deep sea, these enzymes are unable to maintain these abilities at deep-sea pressures (Fig. 2). The importance of acquiring a high degree of pressure insensitivity in ligand binding is most dramatically emphasized by the findings from the studies of the two *Sebastolobus* congeners. For these two species, the confounding effects of different adaptation temperatures and phylogenetic distance are absent, and the different kinetic properties of the M_4 LDHs of these two species seem solely attributable to the adaptation of these species to different depths.

A second conclusion reached in this discussion is that pressures of only 50 to 100 atm may be sufficiently stressful to enzyme function to favor selection for pressure insensitivity. Pressures of this magnitude have never been found to fully denature proteins, so it may seem surprising that these low pressures are associated with significant stresses to enzyme systems. Our point is that even minor pressure-induced structural changes may have functionally important consequences, for instance, greatly impaired ligand binding ability with consequent disruption of metabolic rates and regulation.

A third point reached from the comparative studies of M_4 LDHs is that the capacity to function in a pressure-insensitive manner may be attained only at the cost of a loss in inherent catalytic efficiency, as estimated by activation energy parameters. The implications of this finding for depth zonation patterns are clear. Even though the enzymes of shallow- and deep-living species may have identical binding abilities at low pressures and temperatures (Figs. 1 and 2), the high-pressure-adapted enzymes are at a competitive disadvantage at low pressures because of their lower catalytic rates (Table I). Conversely, the rates of catalytic activity of the shallow-adapted enzymes are significantly reduced at high pressures (Table II). Thus on the basis of these M_4 LDH data, we propose that enzymatic adaptations to pressure may be instrumental in establishing the upper and lower depth distribution limits of marine species, at least in the sense of setting the pressure limits defining the zone where metabolic function is optimal.

A fourth conclusion reached in this discussion is that enzymes that are adapted to function at high pressure are likely to be eurybaric. Once an enzyme acquires the ability to function properly at pressures of 50 to 100 atm, it appears capable of maintaining the correct kinetic properties to pressures of at least 476 atm (Fig. 2). Thus neither the kinetic data nor the genetic data (Table V) obtained in studies of enzymes of deep-sea species provide any suggestion that multiple enzyme forms are needed to enable species that occur over broad depth ranges to cope with different pressure regimes.

In conclusion, whereas we do not wish to imply that enzymatic factors alone are responsible for establishing depth zonation patterns in the water column—and these zonation patterns are certainly the result of a vast array of ecological interactions as well as specific biochemical adaptations to varying temperatures and pressures—the limited data available from studies of homologous enzymes of deep- and shallow-water fish do suggest that the interacting effects of low temperature and high pressure on enzyme function and, consequently, on enzyme evolution have led to the development of enzymes in deep-sea species that permit these organisms to conduct metabolism efficiently at depth but that lack the efficiencies of enzymes from cold-adapted, shallow-water species. These differences, insofar as they are reflected in the organisms' metabolic performances, may be important in zonation processes through the water column.

3. Metabolic and Compositional Changes with Depth

A. *Introduction: Biomass and Metabolism Decrease with Depth*

The enzymatic adaptations discussed in Section 2 of this chapter are, in a certain sense, "qualitative" differences between deep- and shallow-living animals, since these adaptations involve modifications in the functional and structural traits of macromolecules rather than differences in the quantities of these molecules in tissues. To only a limited extent, namely, in the context of $\Delta G\ddagger$ differences, do these "qualitative" adaptations indicate what the overall metabolic rates of deep-sea and shallow species can be. We now consider ecological, physiological, and biochemical factors that play key roles in establishing the metabolic levels—the quantity of metabolic activity—of deep-sea species.

Perhaps the most striking depth-related trend in marine organisms is a general decrease in "life" with depth. This trend is observed in biomass, measured both in terms of numbers of individuals and by total mass (Banse, 1964). There is also a decrease in the caloric content of individuals (Childress and Nygaard, 1973, 1974;

Smith, 1978), in respiratory metabolism (Childress, 1968, 1975, 1977a,b; Gordon et al., 1976; Meek and Childress, 1973; Quetin and Childress, 1976; Smith, 1978; Smith et al., 1979; Smith and Hessler, 1974; Torres et al., 1979), in community metabolism (Smith and Hinga, this volume, Chapter 8), in microbial growth rates (Jannasch et al., 1971, 1976; Jannasch and Wirsen, 1977; Wirsen and Jannasch, 1976), and in tissue enzyme content (Childress and Somero, 1979). Associated with these quantitative changes in physiological and biochemical parameters are major changes in the composition of taxa with depth (Menzies et al., 1973; Carney et al., this volume, Chapter 9).

The low biomass of the deep-sea fauna has been attributed to the low input of utilizable organics into this region, as a result of the great distances that separate the deep sea from the photic zone (Agassiz, 1888; Dayton and Hessler, 1972; Ekman, 1953; Marshall, 1954; Menzies, 1962; Sanders and Hessler, 1969; Rowe, this volume, Chapter 3). The paucity of food input into the deep sea is regarded as a key factor influencing the nature of adaptations essential for life in this environment (Agassiz, 1888; Dayton and Hessler, 1972; Hessler and Jumars, 1974; Hessler et al., 1978; Marshall, 1954; Menzies, 1962; Sokolova, 1972). Aspects of our analysis are predicated on a low food availability and its influence on the "design" of deep-sea animals.

B. *Biological Rates versus Depth*

The decrease in metabolic processes with depth spans all groups of organisms that have been examined and extends to the levels of tissues and enzymatic reactions. Jannasch and co-workers (e.g., Jannasch and Wirsen, 1973; Jannasch et al., 1976; Jannasch and Wirsen, this volume, Chapter 6) have demonstrated that microbial activity, in at least certain regions of the deep sea, may be exceedingly low. Packard et al. (1971) showed that the respiratory activity of plankton decreases exponentially with depth. Childress (1969, 1971a, 1975) has shown a similar relationship between metabolism and depth in studies of respiratory rates of individual midwater crustaceans (Fig. 6). Studies of fish respiration have yielded comparable results (Fig. 7) (Childress, 1971a; Smith, 1978; Smith and Hessler, 1974; Smith and Laver, 1981; Torres et al., 1979). Both "routine" and "active" respiratory rates decrease exponentially with depth (Figs. 6 and 7). Many deep-living animals thus have lower basal metabolic rates plus an apparently reduced ability to sustain high rates of respiration to support vigorous movement.

Metabolic rate differences between shallow- and deep-living animals may reach almost two orders of magnitude and raise a number of questions concerning the physiological and biochemical changes that underlie such low respiratory rates. Before examining these mechanisms, however, it is appropriate to consider how the limited food supply of the deep sea has influenced the feeding strategies and, thereby, the activity and respiratory levels of deep-living animals.

C. *Energy Supplies, Feeding Strategies, and Metabolic Demands in the Deep Sea*

Not only does the quantity of biomass decrease exponentially with depth through the water column (Banse, 1964; Rowe, 1971; Rowe and Menzel, 1971; Rowe, this volume, Chapter 3), but the calories available from 1 g of this biomass

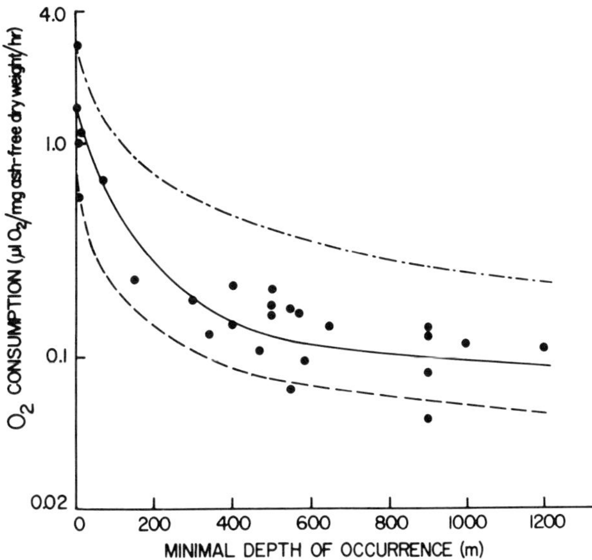

Fig. 6. Relationship between maximal respiration (—·—·—·), respiration at 30 to 70 mm Hg O_2 (———) and minimal respiration (---) and species' minimal depths of occurrence for a number of midwater marine crustaceans. [Modified after Childress (1975).]

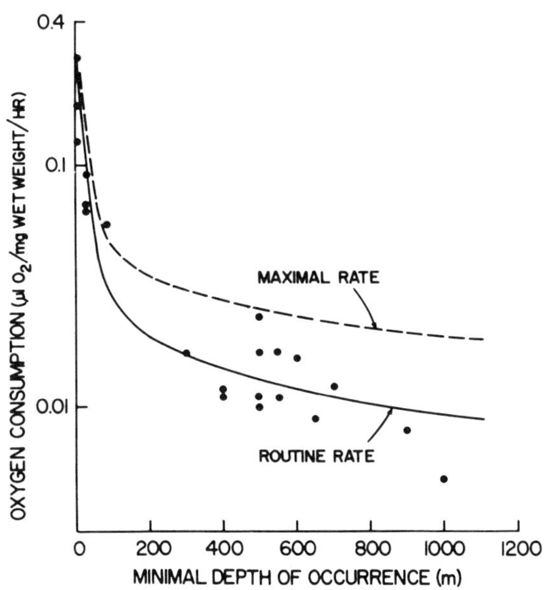

Fig. 7. Relationship between routine respiration (———), maximal respiration (---), and minimal depth of occurrence for a number of midwater marine teleost fish. For species occurring at depths of less than 100 m, measurements were made at 10°C; deeper-occurring species were measured at 5°C. [Modified after Torres et al. (1979).]

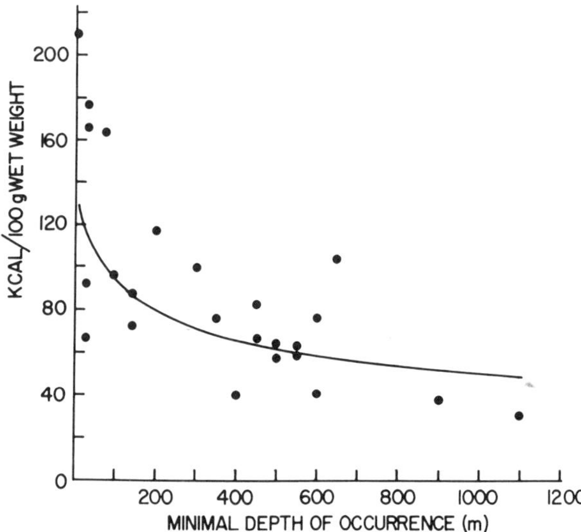

Fig. 8. Relationship between minimal depth of occurrence and energy content (kcal/100 g wet weight) for several species of midwater teleost fish. [Modified after Childress and Nygaard (1973).]

also decrease sharply with depth (Fig. 8). An animal living at great depths thus may be faced with a food regime in which relatively few food parcels are present and those that are present may have a very low energy content. It is not surprising, therefore, that the scarcity of food has major ramifications in the "design" of deep-living animals. Moreover, although it is not possible to consider all of these evolutionary outcomes of life in a food-limited, deep-water environment here, we feel that it is necessary to discuss certain of the behavioral and morphological characteristics of deep-sea animals as a basis for better interpreting the unique physiologies and biochemistries of these organisms.

Of particular significance in this regard is the feeding strategy of many of these species. Many deep-living midwater fish appear to be "sit and wait" predators (Denton and Marshall, 1958). Remaining relatively motionless in the water column, awaiting prey organisms, would be highly advantageous where food parcels are few and difficult or impossible to find by visual means. Also, each food parcel may have a low caloric content and represent a low payback for energy expended in searching behavior. Indeed, some species have amplified the "sit and wait" strategy with the use of lures to attract their prey (e.g., Denton and Marshall, 1958; Pietsch, 1974).

A passive feeding strategy such as this allows profound changes in an organism. Vast reductions in the propulsive system, including the muscular contractile apparatus and the enzymatic power supply that drives it (Childress and Somero, 1979; Somero and Childress, 1980), and the skeletal framework to which the muscles are attached, are possible. Although deep-sea fish can move rapidly (Cohen, 1977; Denton and Marshall, 1958), their propulsive systems are merely a shadow of those of actively foraging pelagic species such as the scombrids. Thus, for a great many deep-sea species, especially pelagic fishes, a reduced food supply has favored adoption of a feeding strategy that, in turn, has permitted a great reduction in the need for a propulsive system that demands a high metabolic rate.

Modest currents in the deep sea may also minimize demands for a powerful muscular-metabolic-skeletal system [see Childress and Nygaard (1973) for a discussion of body structure in these fish].

Reduced muscle (Blaxter et al., 1971; Childress and Nygaard, 1973; Denton and Marshall, 1958) and skeletal structure (Childress and Nygaard, 1973; Denton and Marshall, 1958) in deep-sea fish will not only facilitate a low basal or maintenance metabolic requirement, but will also reduce sinking by reducing heavy elements—protein and bone mineral. In an energy-poor environment like the deep sea, any reduction in energy expenditure has a high selective advantage. Maintaining neutral buoyancy may be especially important for "sit-and-wait" predators by enhancing invisibility through reducing or eliminating swimming motions.

Adaptations to the deep-sea environment by deep-sea invertebrates have received much less study. Chemical compositions of a number of deep-living, midwater crustaceans contrast somewhat with those of deep-living, midwater fish (Childress and Nygaard, 1974). In these crustaceans, protein content decreased slightly with depth, although a much larger decrease in the protein component's contribution to ash-free dry weight was found. Water content, which increased with depth of occurrence in midwater fishes (Childress and Nygaard, 1973) and in muscle of several deep-water teleosts studied by Sullivan and Somero (1980) (Table VII), was lowest in intermediate-depth species. In spite of these irregular patterns in chemical composition, midwater crustaceans' respiration decreased with increasing depth of occurrence, following a pattern similar to that found for midwater fishes (Figs. 6 and 7).

Even less is known about benthic invertebrates. Metabolic rates and composition may again be related to feeding strategies. Bottom-associated amphipods with a "mobile scavenger" feeding strategy (Dayton and Hessler, 1972; Hessler et al., 1978; Schulenberger and Hessler, 1974) studied by K. L. Smith *in situ* at 3650 m in the northwest Atlantic have extremely low rates of "resting" metabolism, but high metabolic rates during activity associated with scavenging for food (Smith, personal communication). On the other hand, Yayanos (1978), who recovered abyssal and hadal amphipods under *in situ* pressures, found that these deep-water amphipods display ventilation rates (pleopod beating frequencies) similar to those of shallow-water species.

A final aspect of the deep sea that may have important effects is the aseasonality in the chemical and physical factors and, probably, in the food supply as well. One effect of this aseasonality may be on the reproductive patterns in deep-sea species. Orton (1920) first suggested that breeding in the seasonless deep sea and tropics would be year-round. In the absence of seasonal changes in the environment that may affect reproductive success at different times of the year, continuous breeding may be selected for. Rokop (1974, 1977a,b, 1979) concluded, on the basis of a sampling program at 1200 m in the San Diego Trough specifically designed to examine reproductive seasonality, that the predominant reproductive pattern among deep-sea benthic invertebrates was aseasonality. Eleven species were examined. Of these, three bivalves, two ophiuroids, two isopods, one amphipod, and one polychaete breed year-round; a brachiopod and a scaphopod spawn seasonally. Time-series studies of a number of benthic invertebrates have been undertaken in the Rockall Trough (Lightfoot et al., 1979). At 2900 m the cosmopolitan ophiuroid, *Ophiomusium lymani,* displayed asynchronous year-

Table VII
Enzymatic Activities of White Skeletal Muscle and Brain Tissue, Muscle Water, Protein Content, and Buffering Capacity of Marine Fish Having Different Depths of Distribution, Feeding, and Locomotory Habits

	Enzyme Activity (Units per Gram Wet Weight)							
Species (*n*)	LDH	PK	MDH	CS	Water (%)	Protein (mg/g)	Buffering Capacity[a]	Depth Range
Shallow-living muscle								
Medialuna californiensis (2)	981	125	23	0.70	75.6	198	63	Surface to 40 m
Engraulis mordax (16)	540	60	61	1.52	80.8		71	Surface to 100 m
Phanerodon furcatus (1)	414	90	77	0.71	77.3	251	66	Surface to 45 m
Atherinops affinis (8)	412	107	20		77.2	222		Surface to 40 m
Paralabrax nebulifer (6)	397	71	21	0.52	75.6	211	62	Surface to 185 m
Paralabrax clathratus (7)	389	75	23	0.79	77.4	200	62	Surface to 50 m
Chromis puntipennis (1)	388	92	51	0.85	77.8	230	62	Surface to 50 m
Rhacochilus toxotes (1)	351	42	63	1.15	78.5	211		Shallow to 50 m
Gillichthys mirabilis (7)	321	28	26	0.90			60	Shallow to 10 m
Paralabrax maculatofaciatus (3)	287	41		0.50				Shallow to 60 m
Genyonemus lineatus (6)	267	88		0.67				Shallow to 100 m
Caulolatilus princeps (1)	209	32	32	0.50	80.8	213	59	Shallow to 95 m
Sebastes mystinus (1)	116	72		0.74				Shallow to 95 m
Scorpaena guttata (2)	77	15		0.25			52	Shallow to 95 m
$\bar{x}$	367	67	40	0.75	77.9	217	62	
SD	215	32	21	0.32	1.9	17	5	
Deep-living muscle								
Coryphaenoides acrolepis (2)	154	10	9	0.31	79.8	102	42	1460 to 1830 m
Anoplopoma fimbria (5)	107	16	16	0.48	71.5	112		200 to 1550 m
Paraliparis rosaceus (1)	66	10		0.10	93.9			1800 to 2500 m
Coryphaenoides armatus[b] (13)	53	7	19	0.79	83.7	177		1885 to 4815 m

Histiobranchus bathybius[b] (2)	53	8	12	0.61			42	1885 to 2830 m
Dichrolene intranegra[b] (1)	46	6	13	1.22	81.1	100		720 to 1960 m
Antimora rostrata (1)	36	10	7	0.37	82.4	102	44	825 to 2500 m
Coryphaenoides rupestris[b] (5)	16	5	10	0.58	84.6	142		550 to 1960 m
Bathysaurus agassizi[b] (2)	35	11	9	0.81	80.9	108		1500 to 2967 m
Halosauropsis macrochir[b] (3)	12	2	4	0.40	81.2	91	48	1500 to 5179 m
Nezumia bairdii[b] (8)	7	5	18	0.62	81.2	144	46	260 to 1965 m
Coryphaenoides carapinus[b] (11)	5	6	7	0.50	85.3	120		1250 to 2740 m
Coryphaenoides leptolepis[b] (7)	4	3	7	0.41	82.3	144	46	2288 to 4639 m
$\bar{x}$	46	8	11	0.55	82.3	122	47	
SD	44	4	5	0.28	5.0	26	2	
Brain tissue								
Medialuna californiensis (1)	35	21		2.88				
Paralabrax nebulifer (6)	35	22		1.75				
Paralabrax clathratus (5)	36	20		1.87				
Chromis puntipennis (1)	32	19		2.65				
Genyonemus lineatus (1)	18	14		1.60				
Caulolatilus princeps (1)	31	19		1.88				
Scorpaena guttata (2)	31	17		1.40				
Coryphaenoides acrolepis (2)	36	12		1.46				
Anoplopoma fimbria (2)	40	16		1.65				
Coryphaenoides rupestris[b] (4)	28	22	44	2.0				
Coryphaenoides armatus[b] (4)	22	13	51	1.4				
Coryphaenoides leptolepis[b] (4)	18	14	35	1.3				
$\bar{x}$	30	17	43	1.82				
SD	7	4	8	0.49				

[a] Buffering capacity as μmol of base needed to titrate the pH of a 1-g sample of muscle by one pH unit, between pH values of approximately 6 and 7. [Data from Castellini and Somero (1981).]
[b] Data from Siebenaller et al. (1982); all other data from Sullivan and Somero (1980). The enzyme activities were measured at 10°C under optimal substrate conditions.

round breeding (Tyler and Gage, 1979, 1980) as was observed in the study in the San Diego Trough at 1200 m (Rokop, 1974). At 2200 m in the Rockall Trough an ophiuroid (Tyler and Gage, 1979) and an asteroid (Tyler et al., 1982) both display reproductive cycles that are asynchronous among individuals of the populations. Two bivalves and an ophiuroid at the 2900-m study site appear to have seasonal reproduction (Lightfoot et al., 1979; Tyler and Gage, 1979, 1980; Gage and Tyler, 1981a,b).

A review by Gordon (1979) considers aspects of seasonality in deep-sea anacanthine fishes, particularly the macrourids and morids. Temperate-latitude species occurring on continental slopes to about 2000 m display seasonal characteristics.

D. *Physiological and Biochemical Correlates of Reduced Metabolic Rates*

Influences of Low Temperature and High Pressure on Respiratory Rates

To what extent are the low *in situ* metabolic rates of deep-sea species due to low-temperature and high-pressure inhibition of metabolism? Are the low metabolic rates necessitated by ecological conditions such as a reduced food supply largely the result of such inhibition, or have other biochemical adaptations come to play an important role in reducing the metabolic rates of deep-sea species?

In the case of pressure inhibition of metabolism, the available data suggest that increases in hydrostatic pressure are, at most, a minor factor in reducing the rates of respiratory metabolism in deep-sea animals (Childress, 1975; Gordon et al., 1976; Meek and Childress, 1973; Smith and Teal, 1973; Teal and Carey, 1967). Widely varying effects of increased pressure on rates of oxygen consumption have been reported, but in no case has pressure been found to have an order of magnitude influence on respiratory rate. Some species exhibit complete pressure insensitivity of respiration, at least up to pressures of 50 to 100 atm (Meek and Childress, 1973), whereas others show slight stimulation (Childress, 1977a; Teal and Carey, 1967). Slight increases in respiration rate with rising pressure may offset the effects of decreased deep-sea temperatures on metabolism. Certain other species show a slight pressure inhibition of metabolism (Teal and Carey, 1967). These respiration studies thus provide no basis for predicting a significant degree of pressure inhibition of metabolism in deep-sea species, a conclusion that is obviously consistent with studies of enzyme–pressure interactions (see Section 2 of this chapter).

Much more pronounced inhibitory effects on metabolism can be expected in the case of temperature. If an animal migrated from surface waters having a temperature between 10 and 15°C to deeper waters where the temperature was 2 to 3°C, the metabolic rate predictably would be decreased by at least 50%, assuming that a typical Q_{10} value near 2 characterized the organism's metabolism. Even this reduction, however, is minor compared to the differences between deep- and shallow-living animals (Figs. 6 and 7). The one- to two-order-of-magnitude differences in respiratory rate between deep- and shallow-living species appear, therefore, to be a reflection of profound biochemical, including enzymatic, differences in the cells of these different species.

Enzymatic Activities of Muscle and Brain and Muscle Buffering Capacity as a Function of Depth of Occurrence and Feeding and Locomotory Habits

This prediction has been verified in studies of enzymes of energy metabolism from a variety of fish that have different depths of occurrence (Fig. 9; Tables VII and VIII) (Childress and Somero, 1979; Somero and Childress, 1980; Sullivan and Somero, 1980; Siebenaller and Somero, 1982; Siebenaller et al., 1982). In white skeletal muscle, the tissue comprising the largest fraction of body mass [ca. 45 to 55%; see Somero and Childress (1980)], the amount of enzymatic activity (expressed as international units of activity measured at 10°C per gram wet weight of tissue) decreases exponentially with increasing minimal depth of occurrence. For midwater fish, this decrease in enzymatic activity per gram of muscle closely parallels the decrease in respiration rates with increasing depth (Figs. 7 and 10). This strong correlation between enzymatic activity of skeletal muscle and whole organism respiration rate (Childress and Somero, 1979) (Fig. 10) indicates that the low metabolic rates of deep-sea fish may result largely from the decreased enzymatic capacity in white muscle.

The dominant effect of skeletal muscle enzymatic activities on total respiration rate is further indicated by the data given in Fig. 9*a* (MDH) and Table VII on brain metabolism. Enzymatic activities of brain tissue are highly similar among all fish examined, a finding that has several implications for our discussion. First, this shows that the reductions in muscle enzymatic activities are a specific adaptation allowing reduction of metabolic rate under conditions where intense locomotory function is inconsistent with the physical and biological features of the environment. Thus, whereas reductions in metabolic and locomotory functions with increasing depth involve major alterations in muscle biochemistry, other tissues whose activities must be maintained at certain levels regardless of depth may exhibit little, if any, depth-related changes in enzymatic activity. Second, the similarities in brain enzymatic activities among different species suggests that the needs for, and the costs of providing, neural function do not change significantly with depth. This appears to be true at least in the case of activities per gram of brain tissue. Reductions in total brain mass have been suggested by the work of Childress and Nygaard (1973), Denton and Marshall (1958), and Marshall (1979). Marshall (1979) proposes that certain brain regions may be reduced in deep-living fish, as a consequence, for instance, of reduced amounts of locomotory muscle in certain species. Thus, for example, comparisons of two *Gonostoma* species having different depth distributions showed a large reduction in the size of the corpus cerebellum, the brain region whose size is proportional to the amount of lateral swimming musculature in the deeper-living species (Marshall, 1971). However, in spite of some reduction in certain brain regions, the results of the enzyme studies shown in Fig. 9*a* and Tables VII and VIII suggest that brain cells of fish living at different depths possess extremely similar metabolic capacities.

Certain compositional features of white skeletal muscle, namely, protein concentration, water content, and buffering capacity (Table VII), reflect the reductions in locomotory performance suggested by the decreases in enzymatic activity with increasing depth of occurrence. Muscle water content is slightly higher, on the average, in deep-sea fishes, and protein concentration is significantly lower in these species (Table VII). It should be noted, however, that the decreases in

MALATE DEHYDROGENASE (units · g wet weight⁻¹)
PROTEIN (% wet weight)
MINIMAL DEPTH OF OCCURRENCE (m)

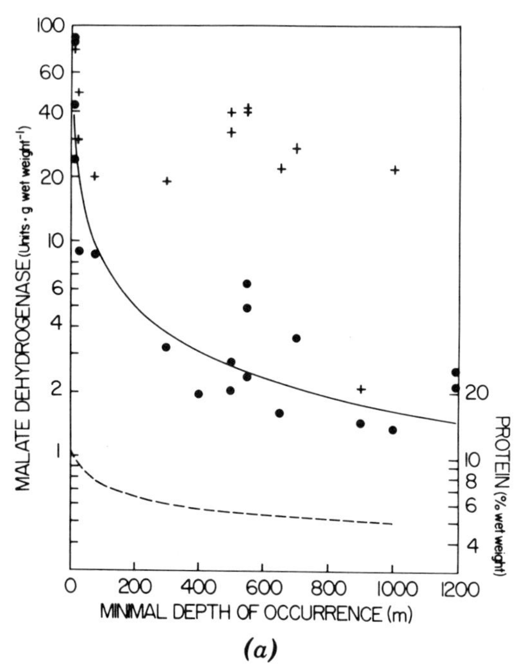

(a)

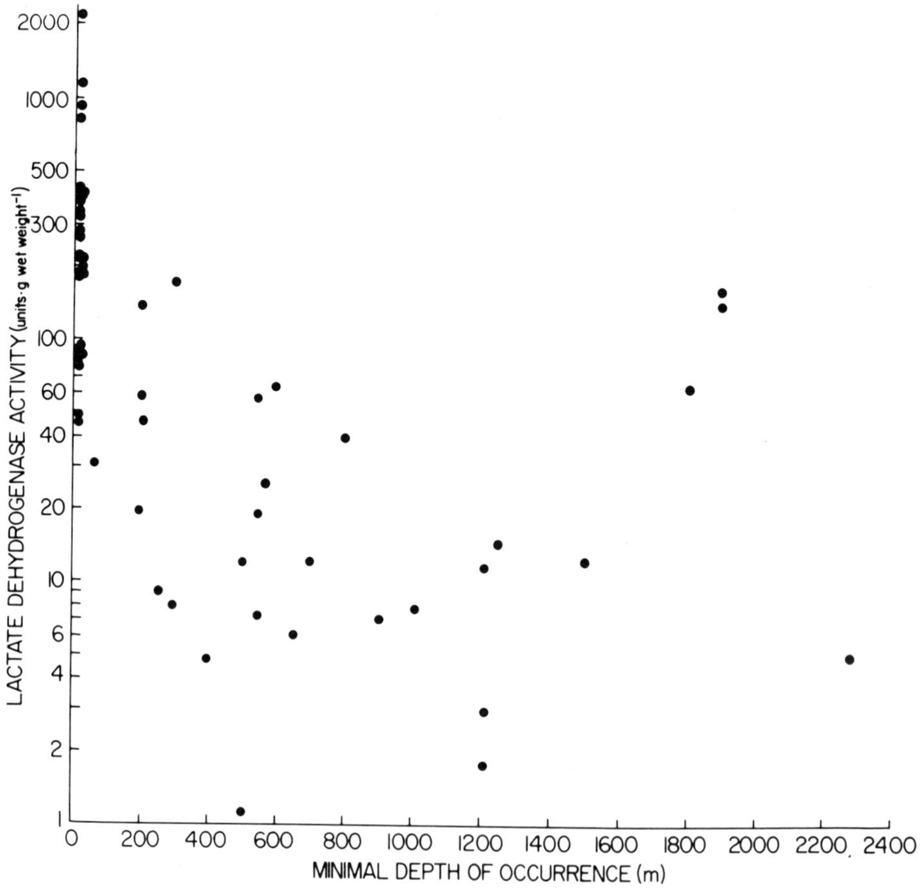

LACTATE DEHYDROGENASE ACTIVITY (units · g wet weight⁻¹)
MINIMAL DEPTH OF OCCURRENCE (m)

(b)

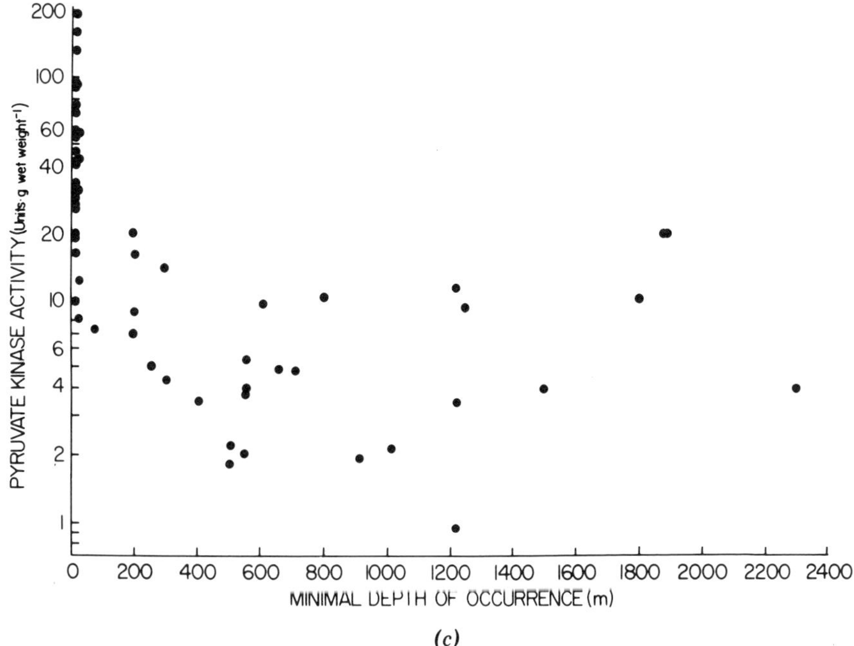

(c)

Fig. 9. Relationship between minimal depth of occurrence and enzymatic activity in a number of marine teleost fish. Units of enzymatic activity are expressed as micromoles of substrate converted to product per minute per gram wet weight of tissue. All assays were conducted at 10°C, using optimal concentrations of substrates and cofactors. The activity values thus reflect maximal activities of the enzymes in all cases. (*a*) Malate dehydrogenase (MDH): ● muscle activities; + brain activities. Protein contents of these species are shown in the lower portion of this figure by the dashed line. The protein data are from Childress and Nygaard (1973). [Modified after Childress and Somero (1979).] (*b*) Lactate dehydrogenase (LDH): muscle activities. [From Sullivan and Somero (1980).] (*c*) Pyruvate kinase (PK): muscle activities. [From Sullivan and Somero (1980).]

enzymatic activity with depth are much larger than the decreases in muscle protein concentration. For example, LDH activity differs by approximately three orders of magnitude among species, whereas protein concentration differs by only about 50%. Thus the decreases in enzymatic activity are not merely a reflection of a general dilution of muscle proteins, but instead are the consequence of specific reductions in certain enzymatic proteins. Part of the decrease in enzymatic activity with depth of occurrence is the result of lowered enzyme efficiency, at least in the case of LDH (see Section 2.C) (but probably not for PK; see below). However, the decrease in activity due to efficiency differences is small, on the order of 40 to 50%, compared to the total activity decreases shown in Fig. 9.

Because white muscle is primarily a glycolytic tissue, used mostly for short bouts of locomotion during which the catabolism of glycogen to lactic acid is primarily responsible for ATP generation, some provision must be made for coping with the production of protons (formed by the dissociation of hydrogen ions from lactic acid) during muscle function. Specifically, there is likely to be a correlation between the glycolytic capacity of a muscle and its buffering capacity. Comparisons of fish having different depth distributions and life-styles have shown a very strong correlation of this type (Castellini and Somero, 1981) (Table

Table VIII

Comparisons of Certain Biochemical Characteristics of White Skeletal Muscle and Brain Tissues of *Sebastolobus alascanus* and *S. altivelis* [Abbreviations as in Table VII; Data from Siebenaller and Somero (1982)]

	Enzymatic Activities[a]					Protein	Water	Buffering
	CPK[e]	MDH	LDH	PK	CS	Content[b]	Content[c]	Capacity[d]
Muscle								
S. alascanus	119	5	58	7	0.35	149	81.2	47.4
S. altivelis	75	3	25	4	0.19	152	81.6	43.0
S. altivelis/ *S. alascanus*	0.63	0.66	0.45	0.61	0.57			
Brain								
S. alascanus			35	23	1.6			
S. altivelis			31	29	1.6			

[a] Enzymatic activities are expressed in international units per gram wet weight of tissue at a measurement temperature of 10°C. All muscle activities are significantly different between the two species ($p < .001$). Brain activities do not differ between species ($p > .05$).
[b] Protein content is expressed as milligrams of protein per gram wet weight of tissue. The interspecific differences are not significant ($p > .05$).
[c] Water content is expressed as percent of wet weight. The interspecific difference is not significant ($p > .05$).
[d] Buffering capacity is expressed as milliequivalents of base required to titrate the pH of 1 g of muscle by one pH unit, between pH values of approximately 6 and 7. The interspecific difference is significant ($p < .01$).
[e] Creatine phosphokinase.

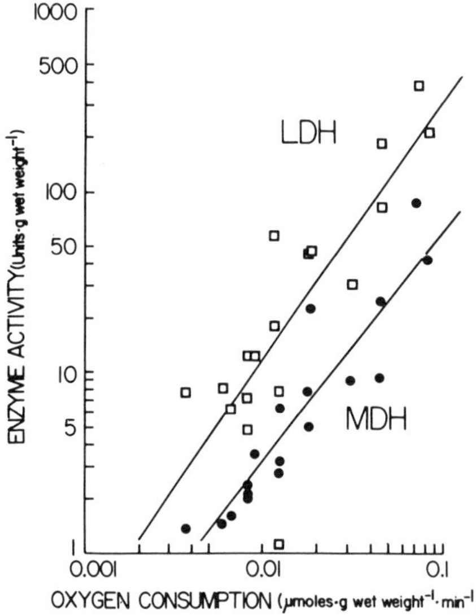

Fig. 10. Relationship between routine oxygen consumption rates of midwater fish and muscle enzymatic activities. Malate dehydrogenase (MDH) ●; lactate dehydrogenase □. Enzymatic activity is expressed as μmoles substrate converted to product per minute per gram wet weight of muscle tissue. [Modified after Childress and Somero (1979).]

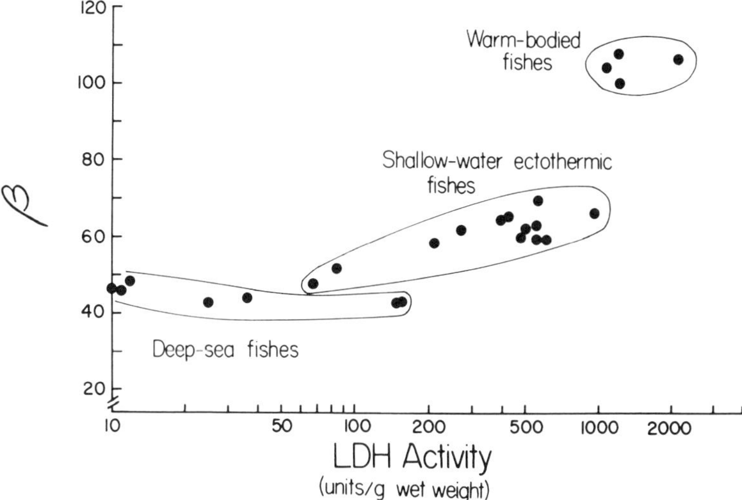

Fig. 11. Relationship between lactate dehydrogenase (LDH) activity and buffering capacity β [measured in slykes = milliequivalents of base needed to raise the pH of one gram (fresh weight) of muscle by 1 pH unit between pH values of approximately 6 and 7] for white skeletal muscle of different marine fish. [Data from Castellini and Somero (1981).]

VII). The highest muscle buffering capacities reported to date are those of the white skeletal muscles of warm-bodied fish (tunas and related species; see Fig. 11). The lowest buffering capacities are for the muscle of deep-sea fish. Shallow-living ectothermic fish are intermediate. The strong correlation between the capacity for lactate production, as measured by muscle LDH activity, and buffering ability is illustrated in Fig. 11. The striking trends shown in this figure indicate that warm-bodied fishes are truly a unique group in terms of muscle metabolism and that deep-sea fish likewise appear to form a more or less distinct class in terms of the capacity of muscle to generate and then buffer acidic end products such as lactic acid.

Within the groups of shallow-living ectothermic species and deep-sea species there is considerable interspecific variation. Within each of these two groups, approximately twenty- to thirtyfold differences in enzymatic activity are found. It is important to consider the bases of this within group variation. Some of this variation may be due to differences in body size among the species. Somero and Childress (1980), Siebenaller and Somero (1982), and Siebenaller et al. (1982) have reported increases in glycolytic enzyme activity in larger-sized individuals of a species (e.g., Fig. 12). This body-size-related scaling pattern contrasts with the "normal" scaling function between aerobic metabolic rate and aerobic enzyme activity (Somero and Childress, 1980), wherein larger individuals of a species, or larger species per se, show lower mass-specific rates. The rise in the capacity for anaerobic glycolysis in larger individuals of a species has been interpreted as a means for conserving a constant ability for burst locomotion, as measured on a relative velocity scale (body lengths per second) in different-sized individuals of a species (Somero and Childress, 1980). This anaerobic metabolism scaling pattern has been noted in a wide variety of shallow-living and deep-sea fish and thus appears to be largely independent of the *absolute* level of anaerobic metabolism

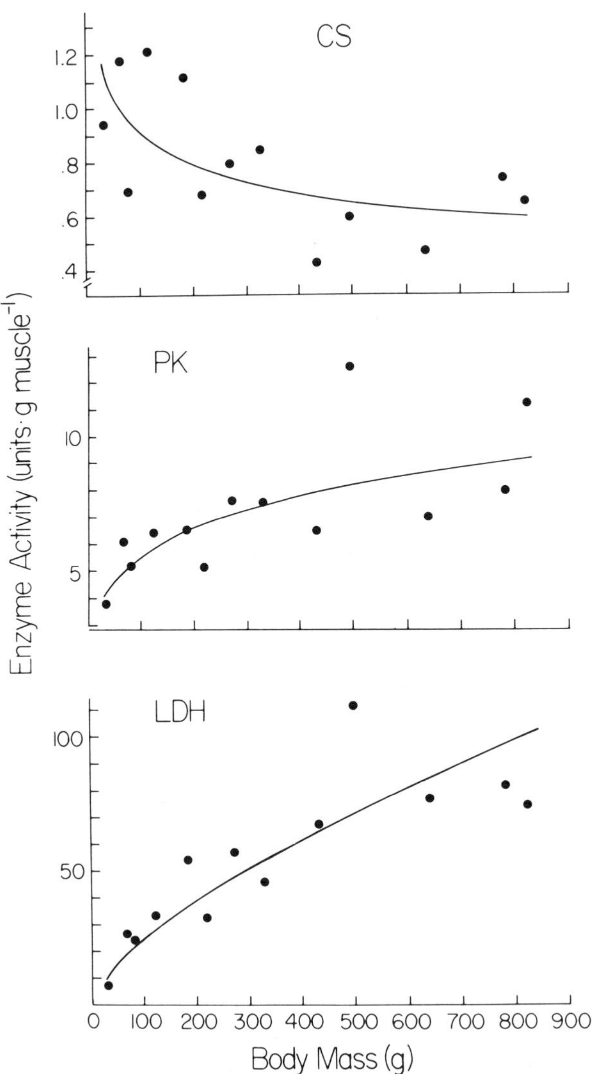

Fig. 12. Scaling of enzymatic activity in white skeletal muscle versus body mass for individuals of *Coryphaenoides armatus*. Citrate synthase (CS), pyruvate kinase (PK), and lactate dehydrogenase (LDH) displayed statistically significant scaling of activity versus body mass. The equations for these scaling relationships are as follow: $A = 1.0\ W^{-0.59 \pm 0.005}$ for CS; $A = 1.83\ W^{0.24 \pm 0.13}$ for PK; and $A = 1.16\ W^{0.66 \pm 0.20}$ for LDH. The 95% confidence intervals are given for the scaling exponents; A is the enzymatic activity, and W is the weight of the entire fish in grams. [Data from Siebenaller et al. (1982).]

the species is capable of. It is not yet clear how meaningful comparisons of this particular scaling relationship are. It is most likely that differences in enzymatic activity due to locomotory and feeding habits will more than offset size-related differences among species [e.g., the comparison of *Coryphaenoides armatus* and *C. leptolepis* given in Table VII and in Siebenaller et al. (1982)]. Thus differences in LDH activity as a function of body size within a species are of the order of three- to fourfold (Somero and Childress, 1980), whereas interspecific differences, which we attribute largely to selection for different locomotory capacities, reach almost three orders of magnitude (Fig. 9).

Thus we conclude that the major determinants of the within-group variabilities in enzymatic activities and buffering capacities are a consequence of varied feeding and locomotory habits among the species within each group. This conclusion is strongly supported by comparisons among the "warm-bodied" fish, shallow-living ectotherms, and deep-sea species. The extremely powerful swimmers such as tunas, which must swim continuously at high speeds to maintain adequate irrigation of the gills, have the highest enzymatic activities and buffering capacities of muscle tissue, whereas sluggish "sit-and-wait" or "float-and-wait" predators have the lowest values for these two parameters. Within each group, differences in feeding and locomotory habit probably correlate with the differences in enzymatic activity and buffering capacity.

Smith (1978) and Smith and Hessler (1974) have demonstrated that the *in situ* oxygen consumption rates of *Coryphaenoides armatus* and *C. acrolepis* are extremely low. If there is a relationship between LDH activity and oxygen consumption in benthopelagic rattails similar to that observed for midwater fish by Childress and Somero (1979), then *Nezumia bairdii, C. rupestris, C. carapinus,* and *C. leptolepis* will have extremely low rates of oxygen consumption (Siebenaller et al., 1982). Any correlation between LDH activity and oxygen consumption for macrourids does not fall on the same curve as the data for midwater fish.

The question of the role of depth per se in establishing the physiological and biochemical attributes of species inhabiting different regions of the water column is difficult to answer because of the plethora of physical and biological factors that select for locomotory capacities. It is appropriate, therefore, to seek a study system that involves species having highly similar foraging habits, morphologies, life histories, and so on, but that differ in their depth distributions. Such a set of species may allow an especially clear resolution of adaptations related strictly to depth of occurrence. Such a study system is offered by the two *Sebastolobus* species discussed earlier in this chapter. These two species are extremely similar but differ in their bathymetric distributions. Comparisons of the biochemical attributes of their muscle and brain tissues (Table VIII) thus provide important tests of many of the statements made earlier concerning adaptation to the deep sea. Moreover, the comparisons of *S. alascanus* and *S. altivelis* yield insights into the mechanisms by which muscle enzymatic activities are varied in species inhabiting different depths.

The data in Table VIII show that the shallow-living species, *S. alascanus,* has approximately twice the activity for each enzyme studied in muscle tissue compared to the deeper-living species, *S. altivelis*. Brain enzymatic activities are the same in both species. The buffering capacity is higher in *S. alascanus* muscle, although this difference is slight. These data are consistent with those obtained in the broader comparisons involving fish of different genera and families (Table VII), and, indeed, provide a strong confirmation of the conclusions made on the basis of the broader comparative study.

The comparisons of the two *Sebastolobus* species (Siebenaller and Somero, 1982) also included analysis of the mechanistic bases for the approximately twofold differences in muscle enzymatic activity. Two possible mechanisms are available for adjusting enzymatic activity: (1) changes in enzymatic efficiency (as measured by substrate turnover number or activation energy characteristics; see Section 2.C of this chapter) could lead to the twofold differences; or (2) different numbers of enzyme molecules could be maintained in the cell; that is, enzyme concentrations could be adjusted. In the *Sebastolobus* species, both activity-adjusting mechanisms are used. In the case of muscle LDH, the differences in activity per gram wet weight of muscle are due to the differences in substrate turnover number of the LDH homologues (see Table I of this chapter). For PK, however, the turnover numbers are identical (Siebenaller and Somero, 1982). Therefore, the differences in PK activity are probably the result of differences in PK concentration. The bases of the differences in activity for the other enzymes studied were not established. What these findings indicate is that two distinct mechanisms for controlling enzymatic activities, catalytic efficiency changes and concentration adjustments, are used to (1) maintain a consistent ratio of activities among enzymes of energy metabolism within the muscle of each species and (2) establish approximately twofold differences in activity of all of these enzymes between the two species. No doubt the depth-related differences shown in Fig. 9 are also based on both of these activity-adjusting mechanisms.

The discovery that enzymatic activity decreases more rapidly with increasing minimal depth of occurrence than does total muscle protein (see Fig. 9*a*) suggests that different classes of proteins have different depth-dependent concentration patterns. In fact, Swezey and Somero (1982b) have shown that the concentration of actin in skeletal muscles of a wide variety of marine teleost fishes from different depths exhibits no depth-related pattern. Despite an almost thousandfold variation in LDH activity per gram of white skeletal muscle among these species (see Fig. 9*b*), there was no significant depth-related trend in actin concentration. This finding suggests that the adaptations in deep-sea fish that are related to a reduced locomotory energy expenditure are achieved by reducing the concentrations of the ATP generating enzyme systems, and not by reducing the ATP utilizing systems of the contractile apparatus.

An important technical point arising from the differential rate of decrease in total protein content and enzymatic activity is that enzymatic activity measurements expressed in terms of units of activity per unit protein would not show the true activity versus depth pattern illustrated in Fig. 9 for these fish. Expression of enzymatic activity on a protein-specific basis would largely obscure the depth pattern found for these fishes, and the differential changes in protein content and enzymatic activity with depth would further obscure the important relationship between metabolic rate and enzymatic activity.

E. Metabolic Activities in Low Oxygen and Anoxic Waters

The preceding discussions of deep-sea animals have centered on organisms that have adequate oxygen to support aerobic metabolism, at least under resting or moderately active conditions. Plentiful oxygen is, of course, available in most deep-sea habitats. There are, however, certain habitats in deeper waters where oxygen concentrations are very low, and, in certain trenches, anoxic zones may exist. What is known about the metabolic functions of organisms living in these regions?

Childress and co-workers (Belman and Childress, 1976; Childress, 1968, 1971b;

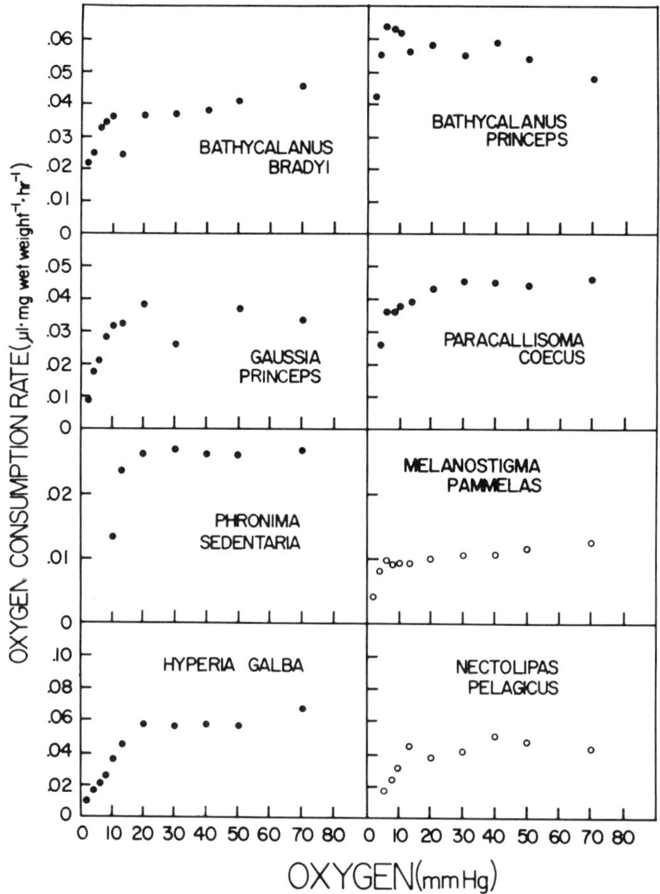

Fig. 13. Oxygen consumption rates as functions of oxygen concentration for some midwater crustaceans and fish, showing capacities for regulation of oxygen consumption at low oxygen concentrations. Species studied: crustaceans—*B. bradyi, G. princeps, P. sedentaria, H. galba, B. princeps*, and *P. coecus*; fish—*M. pammelas* and *N. pelagicus*. [From Childress (1975).]

1975, 1977a,b; Quetin and Childress, 1976) have studied the physiological and anatomical characteristics of several oxygen minimum layer species, principally crustaceans. Most appear to be capable of a fully aerobic metabolism, even at the low oxygen pressures characteristic of their habitats. This ability is reflected in the capacities of these species to regulate their rates of oxygen consumption down to extremely low oxygen partial pressures (Fig. 13). The critical oxygen pressures (P_c) of these species often are near 10 mm Hg O_2. An important adaptation permitting aerobic metabolism at these low oxygen pressures is respiratory pigments with extremely high oxygen affinities (Childress, personal communication). Studies of enzymatic activities in oxygen minimum layer fishes (Childress and Somero, 1979) have shown that these species possess similar levels of aerobically and anaerobically poised enzymes to those found in other fish living at comparable depths; no indication of enhanced activity of glycolytic enzymes was detected, suggesting that the oxygen minimum layer species are not adapted for high levels of anaerobic function. Like the oxygen minimum layer crustaceans, these fishes generally possess respiratory pigments (hemoglobins) with extremely high oxygen affinities (Childress, personal communication), although low affinity hemoglobins, but elevated hematocrits, have been found in certain oxygen minimum layer fishes (Douglas et al., 1976).

In the case of anoxic basins such as the Cariaco Trench, the complete absence of oxygen, and the presence of hydrogen sulfide, create vastly more stressful conditions than do those in oxygen minimum layers. It is astonishing, in fact, that animals are capable of spending significant periods of time in these anoxic waters. Baird et al. (1973) reported that the fish *Bregmaceros nectabanus* (Whitley), migrates diurnally into the anoxic waters of the Cariaco Trench for periods of 10 to 11 hours. Nothing is known about this fish's metabolic chemistry other than that adaptations must be present for tolerance of long periods of anoxia and exposure to hydrogen sulfide. The latter compound, HS^-, is, indeed, coming to be appreciated as an exceedingly important environmental factor not only in anoxic basins, but also in the waters of the deep-sea hydrothermal vents, a habitat that we now consider.

F. Hydrothermal Vent Communities: Sulfide-Driven Primary Productivity

The discovery of dense animal assemblages associated with the deep-sea hydrothermal vents [Lonsdale, 1977; Corliss et al., 1979; Spiess et al., 1980; *Oceanus* 1979, Vol. 22 (2)] presented biologists with a community of organisms that in many ways appears to break some of the key ''rules'' concerning life in the deep sea. In fact, as we argue below, the hydrothermal vent communities are really an ''exception'' to the ''rules'' of deep-sea biology that ''proves'' the rules elaborated in the preceding sections of this review.

The hydrothermal vent ecosystem differs from typical (nonvent) deep-sea habitats in physical, chemical, and biological characteristics. As the sketch of the vent habitat (Fig. 14) illustrates, waters issuing from the vents are much warmer than typical deep-sea waters. Extremes of temperature are found at the ''black-smoker''-type vents (Spiess et al., 1980). Black-smoker water represents the geothermally heated and modified vent water that is largely, if not entirely, undiluted with ambient seawater. In contrast, water issuing through the ''Galápagos-type'' vent is geothermally heated–modified water that has been diluted with ambient seawater. Thus, depending on the extent of dilution, the vent waters range in temperature from approximately 350°C to near-ambient (2 to 3°C) temperatures. The animals inhabiting the vents usually are no warmer than approximately 20°C and typically are regarded to have body temperatures near 2 to 10°C (R. R. Hessler, personal communication).

The chemical properties of the vent waters (Fig. 14) (Edmond et al., 1979; Spiess et al., 1980) differ strikingly from those of the ambient deep-sea water. Again, the black smoker vents represent the extreme case. Black-smoker water is anoxic, depleted of nitrate, but rich in HS^-. Concentrations of HS^- in the range of several millimoles per liter have been found (H. Craig, personal communication). In the Galápagos-type vents, the concentrations of oxygen, nitrate, and sulfide are variable, reflecting the varying degree to which the vent water per se is diluted with ambient seawater. Galápagos-type vent waters may contain sulfide at

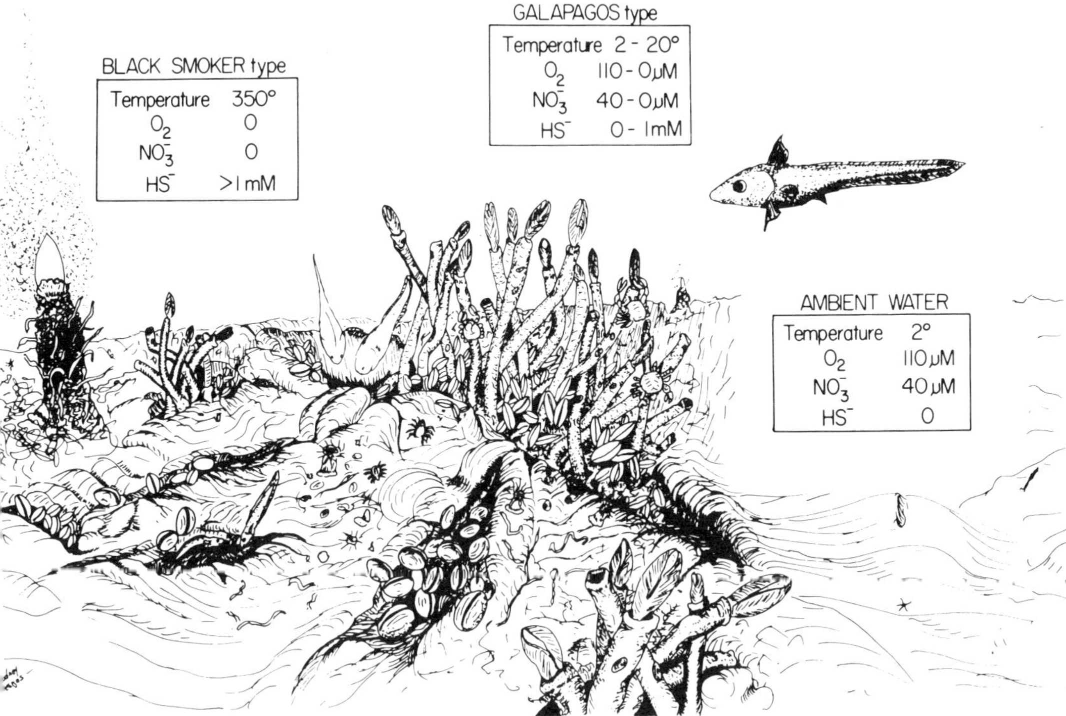

Fig. 14. Hydrothermal vent communities. A composite sketch of the different hydrothermal vents showing many of the animal species found at the Galápagos and 21°N sites and the water chemistry of ambient seawater, undiluted "black-smoker" water, and "Galápagos-type" vent water, which is black-smoker-type water highly diluted with ambient seawater. Note the dominant contribution made by the large tubeworms (*Riftia pachyptila* Jones) and the bivalve mollusks to the vent community biomass.

concentrations approaching millimolar (m*M*) levels (Edmond et al., 1979), as well as substantial concentrations of oxygen and nitrate. As discussed below, it is the simultaneous access to O_2, HS^- and NO_4^- that appears to provide the opportunity for a unique type of deep-sea energy metabolism and primary productivity in the vent communities.

The biological characteristics of the vent ecosystem possess two key attributes. First, the density of life is vastly higher than in other deep-sea regions. Childress (personal communication) has estimated that the large pogonophoran tube worm, *Riftia pachyptila* Jones (Jones, 1981) (Fig. 14) is clustered at densities of up to 10 to 15 kg/m^2 at the Galápagos-type vents. High densities of bivalves, such as *Calyptogena magnifica,* also are present at the vents (Fig. 14). This extremely high density of animal life suggests that the vent animals have access to greater amounts of food than is typical at deep-sea depths (the average depth of the Galápagos and 21°N vents is approximately 2400 m).

The second striking feature of the vent fauna is that it is not typical in species composition of the fauna found at similar depths at nonvent sites. For example, *Riftia pachyptila* is known only at the vents. One important distinction between the vent animals and typical deep-sea species appears to be the type of energy metabolism found in at least the biomass-dominating vent animals.

The key to success in the vent habitat is the capacity to metabolize sulfide. Two distinct metabolic tasks are, of course, involved in the processing of sulfide: (1) HS^- must be detoxified, for this compound is extremely poisonous; and (2) mechanisms must be found to exploit the bond energies of sulfide. The avoidance of poisoning of aerobic respiration by sulfide has been shown to involve sulfide binding proteins in the case of the tube worm, *Riftia pachyptila* (Arp and Childress, 1983; Powell and Somero, 1983). Proteins with high affinity for sulfide and a high sulfide transport capacity in the blood of *R. pachyptila* appear capable of preventing any significant amount of HS^- from entering the cells of the animals, where the sulfide-sensitive respiratory enzyme, cytochrome c oxidase, is located (Powell and Somero, 1983). Mechanisms used to prevent poisoning of respiration by HS^- in other vent species, or in marine species from other sulfide-rich habitats, may also involve sulfide binding proteins (Arp and Childress, personal communication) or high activities of sulfide-oxidizing systems (Powell and Somero, unpublished observations).

The pathways by which the bond energy of HS^- is exploited are now partially understood. These pathways almost certainly are critical components of the primary productivity of the vent community and help to account for both the high biomass and the taxonomic composition of the vent fauna.

Our understanding of the sulfide-driven energy metabolism of vent organisms is based primarily on studies of the classes of enzyme found in certain tissues of the dominant vent animals (Felbeck, 1981; Felbeck et al., 1981). Table IX lists the activities of three important classes of enzymes associated with (1) utilizing the bond energy of HS^- to drive the production of ATP and reducing power (NADPH), (2) the fixation of carbon dioxide (the Calvin–Benson cycle), and (3) the reduction of nitrate in select tissues of vent animals and other animals found in habitats offering access to both sulfide and oxygen. In the first of these sets of reactions, HS^- is oxidized (e.g., to sulfate), with the concomitant production of the ATP and NADPH needed to drive reductive biosynthetic reactions, such as those of the Calvin–Benson cycle and nitrate reduction. Thus the energy contained in HS^- eventually appears in the form of reduced carbon and nitrogen compounds. As shown in Fig. 14, all the key "ingredients" for a sulfide-driven production of reduced carbon and nitrogen compounds are present in the Galápagos-type vent water [black smoker water is apparently sterile (Spiess et al., 1980)]. A key feature of this water chemistry is the simultaneous availability of both sulfide and oxygen, for the efficient extraction of sulfide energy demands that the HS^- be "burned" to the level of SO_4^{2-}.

Figure 15 portrays the current model of sulfide-driven energy metabolism in *Riftia pachyptila*. Sulfide from the ambient seawater enters the animals by diffusion through the tentacles and is transported by the blood to the trophosome, the only tissue in which the enzymatic potentials listed in Table IX have been found. Through a series of reactions, HS^- is oxidized to SO_4^{2-}, which eventually is passed back to the ambient seawater. The reduction of CO_2 and NO_3^- also takes place exclusively in the trophosome. Carbon dioxide may be transported to the trophosome in the form of malate, for high (ca. 10 m*M*) concentrations of malate have been found in the blood of *Riftia,* and the enzymes for malate synthesis have been detected in the animal's tentacles (Felbeck and Somero, unpublished observations). The synthesis of malate as the transport form of CO_2 is common in green

TABLE IX

Enzymes of Calvin–Benson Cycle, Sulfur Metabolism, and Nitrate Reduction in Marine Animals from Habitats in Which Sulfide and Oxygen Are Present [Data from Felbeck et al. (1981)]

Species (Phylum) and Habitat	Enzyme Activities[a]					
	Calvin-Benson Cycle		Sulfur Metabolism			
	RuBPCase	Ru-5-P Kinase	ATP-Sulfurylase	APS-Reductase	Rhodanese	Nitrogen Metabolism Nitrate Reductase
Rift vents						
Riftia pachyptila (Pogonophora)	0.22	19.0	74.0	23.3	7.6	0.07
Riftia sp. (Guyamas basin)	1.13		133.0	30.1	5.2	0.34
Calyptogena magnifica (Mollusca)	0.4					
Vent mytillid (unnamed) (Mollusca)	0.05					
Fault Vent (San Diego Trough)						
Lamellibrachia barhami + unnamed species (Pogonophora)	0.4		30.0			
Calyptogena pacifica (Mollusca)	0.6		25.0			
Sewage Outfall						
Solemya reidi (Mollusca)	2.4	4.4	77.0	4.1	0.7	0.23
Parvilucina tenuisculpta (Mollusca)	0.01		3.8			0.08
Santa Barbara Basin						
Lucinoma annulata (Mollusca)	0.1	1.25	0.6		18.0	

[a] Enzymatic activities are expressed as micromoles substrate converted to product per minute per gram wet weight of tissue at 25°C. Abbreviations for enzymes are ribulose-1,5-bisphosphate carboxylase (RuBPCase), ribulose-5-phosphate kinase (Ru-5-P Kinase), and adenylylsulfate reductase (APS reductase). Where no enzymatic activity is indicated, the assay failed to show the presence of the enzyme in all cases. Negative results may, in most cases, reflect lability of the enzymes (Felbeck et al., 1981).

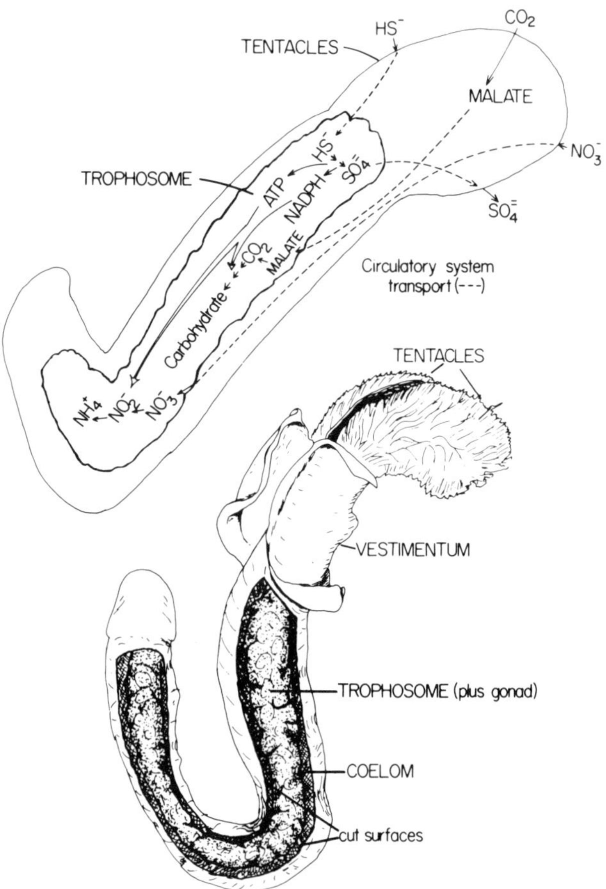

Fig. 15. Anatomic and biochemical characteristics of *Riftia pachyptila*. The lower illustration shows the worm, removed from its tube, and the major body regions. Note the large fraction of the body cavity filled by the trophosome, which is the site of the chemoautotrophic metabolic activities shown in the upper diagram. The upper diagram presents the sulfide-driven metabolic scheme proposed to exist in *Riftia pachyptila*. [Felbeck et al. (1981); see text.]

plants with a C-4 type of metabolism and represents an effective way to concentrate CO_2 at the site of carbohydrate synthesis.

The sulfide-driven metabolic schemes of the other tube worms studied (e.g., of the *Lamellibrachia* species found in the San Diego Trough at a depth of approximately 1200 m), as well as those of the bivalve mollusks found to possess these metabolic capacities (Table IX), probably have the same general organizations as the scheme portrayed for *Riftia pachyptila* (Fig. 15). One distinction between the tube worms and the bivalves is access to CO_2 by the cells responsible for CO_2 fixation into carbohydrate. In the bivalves, the enzymes associated with sulfide metabolism and CO_2 and NO_3^- reduction are found only in the gills (Table IX) and CO_2 transport may not require the malate-trap mechanism hypothesized to exist in *Riftia pachyptila* and the other tube worms.

The data available on the enzymes of sulfide-driven CO_2 and NO_3^- reduction do not, of course, allow one to make quantitative statements about the amount of primary productivity occurring in tissues of these vent animals. It is noteworthy, however, that the activity of the diagnostic enzyme of the Calvin–Benson cycle, ribulose-1,5-bisphosphate carboxylase, in trophosome tissue of the tube worms and gills of the bivalves reaches levels found in fresh spinach leaves (Felbeck, 1981; Felbeck et al., 1981). This evidence is suggestive of a high capacity for CO_2 fixation in these organisms and, in turn, of an important role for these species in the primary productivity of the vent ecosystem. These high activities of enzymes associated with chemoautotrophic metabolism also raise questions about the extent to which the tube worms and bivalves depend on external food sources. This important ecological question cannot be answered at present, but several lines of evidence do suggest that the tube worms and bivalves may be largely independent of reduced organic compounds originating in the euphotic zone and transported to the vent sites. First, in the case of *Riftia pachyptila* the absence of mouth and gut, as in other pogonophorans, makes the uptake of particulate matter problematic. Uptake of dissolved organic matter also appears unable to support the metabolism of pogonophorans, as discussed by Southward et al. (1981), who report the presence of CO_2-fixing enzyme systems in more typical, minute pogonophoran species. In addition, the ratio of carbon isotopes found in *Riftia pachyptila* (Rau, 1981) and other pogonophorans (Southward et al., 1981) suggests a source of reduced carbon different from photosynthetically produced compounds. The fact that the carbon isotope ratio of trophosome and muscle of *Riftia pachyptila* is the same indicates a common pool of reduced carbon in this organism; that is, reduced carbon compounds formed in the trophosome probably are transported throughout the organism.

In the case of the bivalves possessing the enzymatic activities listed in Table IX, the quantitative contribution of reduced carbon and nitrogen compounds synthesized within the gills to total energy budgets is less clear. The bivalves may possess abilities to filter particulate matter from the seawater, such as bacteria, which have been reported at high densities in the Galápagos-type vents (Karl et al., 1980). Williams et al. (1981) estimate that the vent mussels derive at most 25% of their carbon from photosynthetic sources. The carbon isotope ratios of the vent bivalves (Rau and Hedges, 1979) and of the sewage outfall clam, *Solemya reidi* (Rau, personal communication) are suggestive of a reduced carbon source(s) other than green plant photosynthesis. *Solemya reidi,* a member of a genus noted for reduced or, in some cases, no gut structures, may have a high dependence on reduced carbon and nitrogen compounds made within its gills. This is suggested by the unusual anatomies of these clams (large and fleshy gills in conjunction with reduced or absent guts) and by the observation of Mr. David Montagne [discussed by Felbeck et al. (1981)] that populations of *S. panamensis* disappear from sediments near sewage outfalls when changes in current patterns lead to the oxidation of the sediments and, concomitantly, the disappearance of sulfide. Clearly, the quantitative contribution of biosynthesis within animal tissues of reduced carbon and nitrogen compounds to the total energy budget of species living at the hydrothermal vents and in other habitats offering simultaneous access to HS^- and O_2 is an important question for future study.

The enzymatic activities listed in Table IX are unusual for animal tissues. Although the enzymes associated with sulfide oxidation may be present at low levels in certain animal tissues, as in mammalian liver, for functions of sulfide detoxification, enzymes of the Calvin–Benson cycle and nitrate reduction are thought to occur exclusively in green plants and bacteria. It is not surprising, therefore, that the tube-worm and bivalve tissues found to contain these activities also harbor high densities of bacteria that may function as symbionts. The trophosome tissue of *Riftia pachyptila* contains large procaryotic cells at densities of approximately 10^9 cells per gram wet weight (Cavanaugh et al., 1981). Gills of the vent bivalves *S. panamensis* and *Lucinoma annulata* also contain large numbers of bacteria (Cavanaugh, personal communication; Felbeck and Somero, unpublished data). It is not clear in any case whether these bacteria are intracellular or extracellular. The key point is that these bacteria may be functioning in an analogous fashion to the algal symbionts found in many corals and bivalves (Felbeck et al., in press). The details of this putative animal–bacterial symbiosis in the tube worms and certain bivalves await elucidation.

In conclusion, the dense aggregations of unusual animal species at the deep-sea hydrothermal vents provide an exception to the rules of deep-sea biology, as formulated from study of the typical deep sea, and, in effect, substantiate certain of these rules. We have argued throughout this chapter that the food characteristics of the deep-sea environment play a dominant role in establishing the physiological and biochemical attributes of deep-sea animals. In a sense, the hydrothermal vent communities provide us with a powerful test of the food-limitation hypothesis. Given an energy source to drive primary production (HS^- in the case of the vents), a dense bacterial and animal population can exist in the deep sea. This dense community shows that temperature and pressure are not, in and of themselves, deterrents to a high level of productivity and metabolism in the deep sea. The effects of high pressure and low temperature can be offset, if necessary, by compensatory biochemical adaptations, as by increases in enzyme concentrations.

G. *Summary*

The basic objective of this chapter has been to relate the physiological and biochemical "design" features of deep-sea animals to the physical, chemical, and biological features of the deep-sea environment. In Section 2 we discussed the types of adaptation in enzyme and lipid-based systems that appear to be necessary to permit successful metabolic function in the face of low temperatures and high and/or variable pressures. In Section 3 we have tried to relate the compositional and metabolic rate characteristics of deep-sea organisms to the unique suite of environmental features of the deep sea. Among these features, the apparently low food input into typical deep-sea regions stands out as a factor of paramount importance. Many of the traits of typical (nonvent) deep-sea animals are adaptations to life under food-poor conditions, that is, conditions where little food is available and where the locating of this food is difficult and, metabolically, potentially expensive. Reductions in body protein, enzymatic activity, and skeletal components reflect decreases in capacity for vigorous locomotion and, hence, in metabolic rate. Deep-sea fish having "float-and-wait" predatory strategies are especially low in enzymatic potential and, by extrapolation, in metabolic rate. Other deep-sea fish are more active predators. Rattails such as *Coryphaenoides acrolepis* and *C. armatus* exhibit higher muscle enzymatic activities in accordance with their apparently more active foraging behavior. Nonetheless, even these more active deep-sea fish possess drastically lower enzymatic activities and metabolic rates than those of most shallow-living, pelagic fish.

Certain groups of organisms, notably many deep-sea fish, have received sufficient study to permit generalizations about their adaptations to deep-sea conditions; however, other groups of organisms have received far less attention. Thus care must be taken in extending the statements made about some deep-sea animals to all deep-sea animals. For example, we know relatively little about the biochemical and physiological characteristics of deep-sea invertebrates. Fortunately, with the rapid evolution of capture devices capable of returning deep-sea invertebrates to the surface under *in situ* pressures and temperatures (Yayanos, 1978, 1981; Macdonald and Gilchrist, 1980), we may expect to learn a great deal more about the physiology and biochemistry of deep-sea invertebrates in the near future. Hopefully, studies of benthic fishes and invertebrates will progress rapidly, so that the applicability of conclusions drawn from studies of pelagic species to benthic species can be determined. Our prediction is that the adaptations discussed in Section 2, dealing with adaptations that appear essential for permitting controlled metabolic function at depth—regardless of the absolute rate of that metabolic function—will be found to pertain in most, if not all, deep-sea species. On the other hand, many of the metabolic-rate-determining adaptations discussed in Section 3 may be applicable only to those species that face a low energy and temporally uncertain food supply. In fact, the recent studies of hydrothermal vent organisms have provided initial evidence that the deep-sea environment is not inimical to a high metabolic rate; the limiting factor is the presence of *in situ* primary productivity.

In addition to outlining what is known about the physiological and biochemical adaptations of deep-sea animals, we have tried to emphasize questions that appear especially important for future examination. Pressure and temperature tolerance adaptations merit intense study at the level of membrane systems and protein assembly processes. Metabolic rate studies with the use of *in situ* respirometry deserve broader application, and, in particular, questions concerning the relationship between food supply and metabolic turnover will require a great deal more study, both in typical deep-sea habitats and at the hydrothermal vent ecosystems.

Appendix A. Physical and Chemical Bases of Pressure and Temperature Effects

This brief appendix is included to provide a short summary of the physical and chemical factors instrumental in establishing the pressure and temperature sensitivities of biochemical reaction rates and equilibria. This discussion is drawn largely from the volume by Johnson et al. (1974), to which the reader is directed for a fuller treatment of these topics. Johnson and Eyring (1970) provide a more encapsulated treatment of these subjects.

1. *Pressure Effects*

The basis of hydrostatic pressure effects on chemical reactions are the volume changes that may accompany the reactions. In the absence of any change in system volume—and it is important to emphasize that the system in question includes the solvent (generally water) as well as the compounds undergoing reaction—a reaction will exhibit no sensitivity to pressure. If a volume decrease occurs during the reaction, pressure will facilitate the process. Conversely, pressure impedes reactions that occur with a volume increase. These generalizations apply both to chemical equilibria, where the difference in system volume between the reactant- and product-containing systems is important, and to chemical reaction rates, where the change in system volume that occurs as the ground-state complex is excited to form the transition-state complex, is instrumental in establishing the pressure sensitivity of the rate of reaction. The following equations summarize these relationships:

$$\left(\frac{\partial \ln K_{eq}}{\partial P}\right)_T = \frac{-\Delta V}{RT}$$

where K_{eq} is the equilibrium constant of the reaction, P is pressure in atmospheres, R is the universal gas constant (equal to 82 $cm^3/atm \cdot {}^\circ K \cdot mol$), T is absolute temperature in Kelvin, and ΔV is the volume change accompanying the conversion of reactants to products, and

$$k_p = k_0 \exp \frac{-P \, \Delta V\ddagger}{RT}$$

where k_p and k_0 are the reaction rate constants at an elevated pressure and 1 atm, respectively, and $\Delta V\ddagger$ (the activation volume) is the change in system volume that occurs as the ground-state complex is excited to the transition state.

The sources of volume changes during biochemical reactions are generally thought to involve (1) alterations in the actual volumes of the reacting molecules, such as in the void volumes of enzymes or (2) changes in the density of solvent surrounding the reactants (Laidler and Bunting, 1973; Low and Somero, 1975). Since changes in enzyme structure generally accompany the binding and/or activation events of an enzymatic reaction, changes in enzyme void volume due to altered "packing efficiency" of amino acid residues may contribute to binding and activation volumes. However, the major source of volume changes during enzymatic reactions is thought to be solvent reorganization around the enzyme and the ligands bound to it (Low and Somero, 1975). For example, the binding of a negatively charged ligand to a positively charged site on an enzyme will probably be accompanied by the release of tightly electrostricted water from both the ligand and the binding site residues the ligand coordinates with. Thus the solvent reorganization contribution to the binding volume change will be positive, since "bulk" water is less dense than tightly electrostricted water. Analogous reasoning applies in the case of enzyme conformation changes, during which water-structure-modifying groups alter their exposure to solvent. For example, the coalescence of two oppositely charged groups during a conformational change could lead to a large positive volume change due to release of electrostricted water. Nonpolar groups such as the hydrocarbon side-chains of valine and isoleucine residues also have a pronounced effect on water structure, with a decrease in system volume accompanying the transfer of these nonpolar groups from the "interior" of a protein molecule to the solvent-protein interface (Low and Somero, 1975). In summary, many features of enzymatic reaction systems, including the charged nature of most metabolites and the alterations in enzyme conformation that accompany binding and catalysis, suggest that large volume changes and, therefore, high pressure sensitivities, are likely to be a nearly ubiquitous feature of metabolic reactions. It seems pertinent to note, however, that the possibility for modifying volume changes by means of adaptations involving appropriate conformational changes seems available (Somero and Low, 1977). For example, if the release of electrostricted water during ligand binding creates a large positive volume change (making the binding step difficult at high pressures), a compensatory negative volume change due to exposure of a polar side-chain to solvent could effectively "titrate" the increase in system volume.

Whereas in theory it is possible to distinguish clearly activation volumes from the volume changes that accompany binding or release of ligands, experimentally such distinctions may be difficult (Laidler, 1951; Laidler and Bunting, 1973). Only under conditions where the true theoretical V_{max} of a reaction is affected by pressure is it correct to refer to an activation volume. Under conditions where substrate or cofactor concentrations are not saturating, binding volumes may contribute significantly to the observed pressure effects on reaction rate. Even under substrate and cofactor concentrations yielding optimal reaction velocities, pressure effects may be due to additional volume changes besides the $\Delta V^{\ddagger}$ of the reaction. For example, elevated pressures could offset substrate inhibition, a phenomenon frequently noted at high substrate concentrations. These considerations argue for a conservative approach in assigning the activation symbol (‡) to an observed volume change unless the reaction is known to be occurring at its theoretical V_{max}. For this reason we have not assigned the activation symbol to the volume changes observed under optimal pyruvate and NADH concentrations for the M_4-LDH reactions studied (see Table II).

2. *Temperature Effects*

The effects of temperature changes on enzymatic reaction velocities were discussed briefly in Section 2. The absolute rate of a reaction was shown to depend on the activation free energy ($\Delta G^{\ddagger}$) of the reaction, and the temperature dependence of rate was found to result from the activation enthalpy ($\Delta H^{\ddagger}$) of the reaction.

We wish to consider here the effects of temperature on the equilibria that involve the formation or rupture of weak chemical bonds: hydrogen bonds; electrostatic interactions; hydrophobic interactions; and Van der Waals forces. These weak bonds are primarily responsible for maintaining protein subunit aggregation, the correct protein conformations (tertiary and secondary structure), the integrity of enzyme–ligand complexes, and the physical state of lipid-based structures. As their name implies, these noncovalent bonds form and break with small energy changes [enthalpies of formation range from ca. 1 to 7 kcal/mol; see Table 7-1 of Hochachka and Somero (1973)]. These weak bonds thus are readily disrupted by temperature changes, and the integrities of weak-bond-dependent structures are consequently a major focal point in temperature adaptation processes. We have discussed in Section 2 the increase in amounts of weak bonding that appear to occur in enzyme systems during adaptation to high temperatures. We also mentioned the adjustments in phospholipid fatty acid composition that offset temperature effects on lipid viscosity. Adaptations of these types are essentially adjustments in the total number of weak bonds formed or broken in the systems or reactions or, more specifically, in the net energy changes that accompany bond formation or rupture. In addition to this type of weak bond adaptation process, it is possible to envision adaptations that involve the types, rather than the numbers, of weak bond formed or broken. An important attribute of the different classes of weak bonds is that, whereas hydrogen bonds, electrostatic interactions, and Van der Waals interactions form with a decrease in enthalpy (decreases in temperature favor the formation of these three types of weak bond), the formation of hydrophobic interactions proceeds with an uptake of heat from the environment (i.e., increases in temperature favor the formation of hydrophobic interactions). The possibility thus exists that adjustments in, for example, binding enthalpies can be effected by altering (1) the relative contributions of hydrophobic interactions and (2) exothermic weak bonds (Hochachka, 1975a,b). In the case of substrates, cofactors, and modulators that bind to enzymes by means of both endothermic and exothermic bonding reactions, this type of adaptation seems accessible. For other classes of ligand that bind only through charge–charge interactions, for example, adjustments of binding energies may require modifications of the structural changes in the enzyme that accompany binding (Somero and Low, 1977). It is also important to note that weak bonds are pressure sensitive as a result of the changes in water organization that accompany bond formation or rupture (Low and Somero, 1975), and much of the solvent reorganization around enzymes and ligands that occurs during an enzymic reaction reflects the forming and breaking of weak bonds (see above).

Torgerson et al. (1979, 1980) have developed an interpretation of pressure effects on proteins in terms of whether the volumes of binding sites or domains can be reduced under pressure. A compressible ("soft") site, whose volume can be reduced by rotation around some of its covalent bonds, is stabilized by increased pressure. The volume of an incompressible or "hard" site cannot be reduced under pressure because of the molecular architecture of the site. Pressure increases will promote the dissociation of complexes at hard binding sites. These dissociations are thought to be largely unrelated to the nature of the forces that provide the driving free energy for complex formation at atmospheric pressure. An example of such a "hard" site is the binding site for flavin mononucleotide in the flavodoxins (Torgerson et al., 1979). This interpretation of pressure effects based on compressibilities has been extended to subunit aggregation by Paladini and Weber (1981).

Appendix B. Additional Readings in Deep-Sea Physiology

The topics discussed in the preceding sections of this chapter represent a relatively broad focus on what is known about the physiological and biochemical features of deep-sea animals. However, certain topics either have not been discussed at all or have been mentioned in only a cursory fashion. Therefore, we wish to supplement our treatment with a list of readings and also list a number of volumes devoted to deep-sea and/or high pressure phenomena.

1. Books and Articles Dealing with High-Pressure and Deep-Sea Topics

1. Pressure effects on biochemical systems of abyssal fishes: The 1970 *Alpha Helix* Expedition to the Galapagos Archipelago. *American Zoologist*, 1971, **11** (3), 399–576.
2. Brauer, R. W., ed., 1972. *Barobiology and the Experimental Biology of the Deep Sea*. North Carolina Sea Grant Program, Chapel Hill, North Carolina.
3. Pressure effects on biochemical systems of abyssal and midwater organisms: The 1973 Kona Expedition of the *Alpha Helix*. In P. W. Hochachka, ed., 1975, *Compar. Biochem. Physiol.*, Vol. 52B, pp. 1–99.
4. Heremans, K., 1982. High pressure effects on proteins and other biomolecules. *Ann. Rev. Biophys. Bioeng.*, **11**, 1–21.
5. Jaenicke, R., 1981. Enzymes under extremes of physical conditions. *Ann. Rev. Biophys. Bioeng.*, **10**, 1–67.
6. Kinne, O., ed., 1972. Pressure. In *Marine Ecology*, Vol. I, Part 3, O. Kinne, R. Y. Morita, W. Vidaver, and H. Flugel, contributors. Wiley-Interscience, London.
7. Macdonald, A. G., 1975. *Physiological Aspects of Deep Sea Biology*. Monographs of the Physiological Society, No. 31, Cambridge University Press.
8. Marshall, N. B., 1979. *Deep-Sea Biology. Developments and Perspectives*. Garland STPM Press, New York.
9. Marshall, N. B., 1954. *Aspects of Deep-Sea Biology*. Hutchinson, London.
10. Morild, E., 1981. The theory of pressure effects on enzymes. *Adv. Protein Chem.*, **34**, 93–166.
11. Sleigh, M. A. and A. G. Macdonald, eds., 1972. The effects of high pressure on living organisms. *Society of Experimental Biology Symposium 26*. Cambridge University Press, London.
12. Zimmerman, A. M., ed., 1970. *High Pressure Effects on Cellular Processes*. Academic Press, New York.

2. Buoyancy

1. Alexander, R. McN., 1972. The energetics of vertical migration by fishes. *Symp. Soc. Exp. Biol.*, **26**, 273–294.
2. Butler, J. L. and W. G. Pearcy, 1972. Swimbladder morphology and specific gravity of myctophids off Oregon. *J. Fish. Res. Board Can.*; **29**, 1145–1150.
3. Denton, E. J., 1961. The buoyancy of fish and cephalopods. *Prog. Biophys.*, **11**, 117–234.
4. Denton, E. J., 1963. Buoyancy mechanisms of sea creatures. *Endeavour*, **22**, 3–8.
5. Denton, E. J., 1971. Examples of the use of active transport of salts and water to give buoyancy in the sea. *Phil. Trans. Roy. Soc. Lond. B.*, **262**, 277–287.
6. Phleger, C. F., J. Patton, P. Grimes, and R. F. Lee, 1976. Fish-bone oil: Percent total body lipid and carbon-14 uptake following feeding of 1-^{14}C-palmitic acid. *Mar. Biol.*, **35**, 85–89.
7. Horn, M. H., P. W. Grimes, C. F. Phleger, and L. L. McClanahan, 1978. Buoyancy function of the enlarged fluid-filled cranium in the deep-sea ophidiid fish *Acanthonus armatus*. *Mar. Biol.*, **46**, 335–339.
8. Lee, R. F., C. F. Phleger, and M. W. Horn, 1975. Composition of oil in fish bones: Possible function in neutral buoyancy. *Comp. Biochem. Physiol.*, **50B**, 13–16.
9. Steen, J. B., 1970. The swimbladder as a hydrostatic organ. In *Fish Physiology*, Vol. IV, W. S. Hoar and D. J. Randall, eds. Academic Press, New York, pp. 413–443.
10. Yayanos, A. A., A. A. Benson, and J. C. Nevenzel, 1978. The pressure-volume-temperature (PVT) properties of a lipid mixture from a marine copepod, *Calanus plumchrus:* Implications for buoyancy and sound scattering. *Deep-Sea Res.*, **25**, 257–268.
11. Wittenberg, J. B., D. E. Copeland, R. L. Haedrich, and J. S. Child, 1980. The swimbladder of deep-sea fish: The swimbladder wall is a lipid-rich barrier to oxygen diffusion. *J. Mar. Biol. Assoc. U. K.*, **60**, 263–276.

References

Agassiz, A., 1888. Three cruises of the United States Coast and Geodetic Survey Steamer "Blake." *Bull. Harvard Mus. Comp. Zool.*, **14,** 1–314; **15,** 1–220.

Arp, A. J. and J. J. Childress, 1983. Sulfide binding by the blood of the hydrothermal vent tube worm *Riftia pachyptila. Science,* **219,** 295–297.

Atkinson, D. E., 1969. Limitation of metabolite concentrations and the conservation of solvent capacity in the living cell. *Cur. Top. Cell. Regul.*, **1,** 29–43.

Atkinson, D. E., 1977. *Cellular Energy Metabolism and Its Regulation.* Academic Press, New York.

Ayala, F. J., ed., 1976. *Molecular Evolution.* Sinauer, Sunderland, Massachusetts.

Ayala, F. J. and J. W. Valentine, 1974. Genetic variability in the cosmopolitan deep-water ophiuran *Ophiomusium lymani. Mar. Biol.*, **27,** 51–57.

Ayala, F. J., J. W. Valentine, D. Hedgecock, and L. G. Barr, 1975. Deep-sea asteroids: High genetic variability in a stable environment. *Evolution,* **29,** 203–212.

Baird, R. C., D. F. Wilson, and D. M. Milliken, 1973. Observations on *Bregmaceros nectabanus* Whitley in the anoxic, sulfurous water of the Cariaco Trench. *Deep-Sea Res.*, **20,** 503–504.

Baldwin, J., 1975. Selection for catalytic efficiency of lactate dehydrogenase M_4: Correlation with body temperature and levels of anaerobic glycolysis. *Compar. Biochem. Physiol.*, **52B,** 33–37.

Baldwin, J., K. B. Storey, and P. W. Hochachka, 1975. Lactate dehydrogenase M_4 of an abyssal fish: Strategies for function at low temperature and high pressure. *Compar. Biochem. Physiol.*, **52B,** 19–23.

Banse, K., 1964. On the vertical distribution of zooplankton in the sea. In *Progress in Oceanography,* Vol. II, M. Sears, ed. Pergamon Press, Oxford, pp. 53–125.

Belman, B. W., 1978. Respiration and the effects of pressure on the mesopelagic vertically migrating squid *Histioteuthis heteropsis. Limnol. Oceanogr.*, **23,** 735–739.

Belman, B. W. and J. J. Childress, 1976. Circulatory adaptations to the oxygen minimum layer in the bathypelagic mysid *Gnathophausia ingens. Biol. Bull.*, **150,** 15–37.

Blaxter, J. H. S., C. S. Wardle, and B. L. Roberts, 1971. Aspects of the circulatory physiology and muscle systems of deep-sea fish. *J. Mar. Biol. Assoc. U. K.*, **51,** 991–1006.

Borgmann, U. and T. W. Moon, 1975. A comparison of lactate dehydrogenases from an ectothermic and an endothermic animal. *Can. J. Biochem.*, **53,** 998–1004.

Borgmann, U., K. J. Laidler, and T. W. Moon, 1975. Kinetics and thermodynamics of lactate dehydrogenases from beef heart, beef muscle, and flounder muscle. *Can. J. Biochem.*, **53,** 1196–1206.

Brandts, J. F., R. J. Oliveira, and C. Westort, 1970. Thermodynamics of protein denaturation: Effect of pressure on the denaturation of Ribonuclease A. *Biochemistry,* **9,** 1038–1047.

Brauer, R. W., M. Y. Bekman, J. B. Keyser, D. L. Nesbitt, G. N. Sidelev, and S. L. Wright, 1980a. Adaptation to high hydrostatic pressures of abyssal gammarids from lake Baikal in eastern Siberia. *Comp. Biochem. Physiol.*, **65A,** 109–117.

Brauer, R. W., M. Y. Bekman, J. B. Keyser, D. L. Nesbitt, S. G. Shvetzov, G. N. Sidelev, and S. L. Wright, 1980b. Comparative studies of sodium transport and its relation to hydrostatic pressure in deep and shallow water gammarid crustaceans from lake Baikal. *Comp. Biochem. Physiol.*, **65A,** 119–127.

Brett, J. R. and T. D. D. Groves, 1979. Physiological Energetics, In *Fish Physiology,* Vol. VII. W. S. Hoar, D. J. Randall, and J. R. Brett, eds. Academic Press, New York, pp. 279–352.

Carter, J. V., D. G. Knox, and A. Rosenberg, 1978. Pressure effects on folded proteins in solution: Hydrogen exchange at elevated pressures. *J. Biol. Chem.*, **253,** 1947–1953.

Castellini, M. A. and G. N. Somero, 1981. Buffering capacity of vertebrate muscle: Correlations with potentials for anaerobic function. *J. Comp. Physiol.*, **143,** 191–198.

Cavanaugh, C. M., S. Gardiner, M. K. Jones, H. W. Jannasch, and J. B. Waterbury, 1981. Prokaryotic cells in the hydrothermal vent tube worm *Riftia pachyptila* Jones: Possible chemoautotrophic symbionts. *Science,* **213,** 340–342.

Ceuterick, F., J. Peeters, K. Heremans, H. DeSmedt, and H. Olbrechts, 1978. Effects of high pressure, detergents and phospholipase on the break in the Arrhenius plot of *Azobacter* nitrogenase. *Eur. J. Biochem.*, **87,** 401–407.

Childress, J. J., 1968. Oxygen minimum layer: Vertical distribution and respiration of the mysid *Gnathophausia ingens*. *Science*, **160**, 1242–1243.

Childress, J. J., 1969. The respiratory physiology of the oxygen minimum layer mysid *Gnathophausia ingens*. Ph.D. dissertation, Stanford University, 144 pp.

Childress, J. J., 1971a. Respiratory rate and depth of occurrence of midwater animals. *Limnol. Oceanogr.*, **16**, 104–106.

Childress, J. J., 1971b. Respiratory adaptations to the oxygen minimum layer in the bathypelagic mysid *Gnathophausia ingens*. *Biol. Bull.*, **141**, 109–121.

Childress, J. J., 1975. The respiratory rates of midwater crustaceans as a function of depth of occurrence and relation to the oxygen minimum layer off southern California. *Compar. Biochem. Physiol.*, **50A**, 787–799.

Childress, J. J., 1977a. Effects of pressure, temperature and oxygen on the oxygen consumption rate of the midwater copepod *Gaussia princeps*. *Mar. Biol.*, **39**, 19–24.

Childress, J. J., 1977b. Physiological approaches to the biology of midwater organisms. In *Oceanic Sound Scattering Prediction*, N. R. Andersen and B. J. Zahuranec, eds. Plenum Press, New York, pp. 301–324.

Childress, J. J. and M. H. Nygaard, 1973. The chemical composition of midwater fishes as a function of depth of occurrence off southern California. *Deep-Sea Res.*, **20**, 1093–1109.

Childress, J. J. and M. H. Nygaard, 1974. Chemical composition and buoyancy of midwater crustaceans as function of depth of occurrence off southern California. *Mar. Biol.*, **27**, 225–238.

Childress, J. J. and G. N. Somero, 1979. Depth-related enzymic activities in muscle, brain and heart of deep-living pelagic marine teleosts. *Mar. Biol.*, **52**, 273–283.

Chothia, C., 1975. Structural variants in protein folding. *Nature*, **254**, 304–308.

Clegg, R. M., E. L. Elson, and B. W. Maxfield, 1975. A new technique for optical observation of the kinetics of chemical reactions perturbed by small pressure changes. *Biopolymers*, **14**, 883–887.

Coates, J. H.. M. J. Hardman, and H. Gutfreund, 1977. Pressure relaxation of the equilibrium of the reaction catalysed by pig heart lactate dehydrogenase: A test of the kinetic mechanism. *Int. Symp. Pyridine Nucleotide-Dependent Dehydrogenases*, pp. 409–415.

Cohen, D. M., 1977. Swimming performance of the gadoid fish, *Antimora rostrata* at 2400 meters. *Deep-Sea Res.*, **24**, 275–277.

Corliss, J. B., J. Dymond, L. I. Gordon, J. M. Edmond, R. P. von Herzen, R. D. Ballard, K. Green, D. Williams, A. Bainbridge, K. Crane, and T. H. van Andel, 1979. Submarine thermal springs on the Galapagos Rift. *Science*, **203**, 1073–1083.

Cossins, A. R. and C. L. Prosser, 1978. Evolutionary adaptation of membranes to temperature. *Proc. Natl. Acad. Sci. (USA)*, **75**, 2040–2043.

Costa, R. and P. M. Bisol, 1978. Genetic variability in deep-sea organisms. *Biol. Bull.*, **155**, 125–133.

Cowey, C. B., 1967. Comparative studies on the activity of D-glyceraldehyde-3-phosphate dehydrogenase from cold- and warm-blooded animals with reference to temperature. *Compar. Biochem. Physiol.*, **23**, 969–976.

Crow, J. F. and M. Kimura, 1970. *An Introduction to Population Genetics Theory*, Harper and Row, New York.

Dayton, P. K. and R. R. Hessler, 1972. Role of biological disturbance in maintaining diversity in the deep sea. *Deep-Sea Res.*, **19**, 199–208.

Denton, E. J. and N. B. Marshall, 1958. The buoyancy of bathypelagic fishes without a gas-filled swimbladder. *J. Mar. Biol. Assoc. U. K.*, **37**, 753–767.

de Smedt, H., R. Borghgraef, F. Ceuterick, and K. Heremans, 1979. Pressure effects on lipid-protein interactions in ($Na^+ + K^+$)-ATPase. *Biochim. Biophys. Acta*, **556**, 479–489.

DeWitt, H. H., 1971. Coastal and deep-water benthic fishes of the Antarctic. *Am. Geogr. Soc., Antarctic Map Folio Series*, No. 15, 1–10.

Douglas, E. L., W. A. Friedl, and G. V. Pickwell, 1976. Fishes in oxygen minimum zones: Blood oxygenation characteristics. *Science*, **191**, 957–959.

Doyle, R. W., 1972. Genetic variation in *Ophiomusium lymani* (Echinodermata) populations in the deep sea. *Deep-Sea Res.*, **19**, 661–664.

Edmond, J. M., C. Measures, B. Mangum, B. Grant, F. R. Sclater, R. Collier, and A. Hudson, 1979. On the formation of metal-rich deposits at ridge crests. *Earth Planet. Sci. Letters*, **46**, 19–30.

Ekman, S., 1953. *Zoogeography of the Sea*. Sidgwick Jackson, London.

Engelborghs, Y., K. A. Hereman, and L. C. M. De Maeyer, 1976. Effect of temperature and pressure on polymerisation equilibrium of neuronal microtubules. *Nature*, **259**, 686–688.

Estes, J. E., 1975. A study of the polymerization of G-ADP-Actin. *Biophys. J.*, **15**, 34a.

Eventoff, W., M. G. Rossmann, S. S. Taylor, H.-J. Torff, H. Meyer, W. Keil, and H.-H. Kiltz, 1977. Structural adaptations of lactate dehydrogenase isozymes. *Proc. Natl. Acad. Sci. (USA)*, **74**, 2677–2681.

Felbeck, H., 1981. Chemoautotrophic potential of the hydrothermal vent tube worm, *Riftia pachyptila* Jones (Vestimentifera). *Science*, **213**, 336–338.

Felbeck, H. and G. N. Somero, 1982. Primary production in deep-sea hydrothermal vent organisms: Roles of sulfide-oxidizing bacteria. *TIBS* **7**, 201–204.

Felbeck, H., J. J. Childress, and G. N. Somero, 1981. Calvin–Benson cycle and sulphide oxidation enzymes in animals from sulphide-rich habitats. *Nature*, **293**, 291–293.

Felbeck, H., G. N. Somero, and J. J. Childress, in press. Biochemical interactions between molluscs and their symbionts. In *Biochemistry of Mollusca*, P. W. Hochachka, ed. Academic Press, New York.

Fersht, A., 1977. *Enzyme Structure and Mechanism*. Freeman, San Francisco.

Forbes, E., 1859. *The Natural History of the European Seas*. John Van Voorst, London, 306 pp.

Gage, J. D. and P. A. Tyler, 1981a. Non-viable seasonal settlement of larvae of the upper bathyal brittle star *Ophiocten gracilis* in the Rockall Trough Abyssal. *Mar. Biol.*, **64**, 153–161.

Gage, J. D. and P. A. Tyler, 1981b. Re-appraisal of age composition, growth and survivorship of the deep-sea brittle star *Ophirua ljungmani* from size structure in a sample time series from the Rockall Trough. *Mar. Biol.*, **64**, 163–172.

Gandour, R. D. and R. L. Schoner, eds., 1978. *Transition States of Biochemical Processes*. Plenum Press, New York.

Gooch, J. L., 1975. Mechanisms of evolution and population genetics. In *Marine Ecology*, Vol. II, Part 1, O. Kinne, ed. Wiley, New York, pp. 349–409.

Gooch, J. L. and T. J. M. Schopf, 1972. Genetic variability in the deep sea: Relation to environmental variability. *Evolution*, **26**, 545–552.

Gordon, J. D. M., 1979. Lifestyle and phenology in deep-sea anacanthine teleosts. *Symp. Zool. Soc. Lond.*, **44**, 327–359.

Gordon, M. S., B. W. Belman, and P. H. Chow, 1976. Comparative studies on the metabolism of shallow-water and deep-sea marine fishes. IV. Patterns of aerobic metabolism in the mesopelagic deep-sea fangtooth fish *Anoplogaster cornuta*. *Mar. Biol.*, **35**, 287–293.

Grassle, J. F., 1972. Species diversity, genetic variability and environmental uncertainty. *Proc. Eur. Mar. Biol. Symp.*, **5**, 19–26.

Grassle, J. F. and H. L. Sanders, 1973. Life histories and the role of disturbance. *Deep-Sea Res.*, **20**, 643–659.

Greaney, G. S. and G. N. Somero, 1979. Effects of anions on the activation thermodynamics and fluorescence emission spectrum of alkaline phosphatase: Evidence for enzyme hydration changes during catalysis. *Biochemistry*, **18**, 5322–5332.

Greaney, G. S. and G. N. Somero, 1980. Contributions of binding and catalytic rate constants to evolutionary modifications in K_m of NADH for muscle-type (M_4) lactate dehydrogenases. *J. Compar. Physiol.*, **137**, 115–121.

Griest, E. M., W. Webb, and R. W. Schiessler, 1958. Effect of pressure on viscosity of higher hydrocarbons and their mixtures. *J. Chem. Physics*, **29**, 711–720.

Haedrich, R. L. and N. R. Henderson, 1974. Pelagic food of *Coryphaenoides armatus*, a deep benthic rattail. *Deep-Sea Res.*, **21**, 739–744.

Haedrich, R. L. and P. T. Polloni, 1976. A contribution to the life history of a small rattail fish, *Coryphaenoides carapinus*. *Bull. So. Calif. Acad. Sci.*, **75**, 203–211.

Haedrich, R. L., G. T. Rowe, and P. T. Polloni, 1975. Zonation and faunal composition of epibenthic populations on the continental slope south of New England. *J. Mar. Res.,* **33,** 191–212.

Haedrich, R. L., G. T. Rowe, and P. T. Polloni, 1980. The megabenthic fauna in the deep sea south of New England, USA. *Mar. Biol.,* **57,** 165–179.

Halvorson, H. R., 1979. Relaxation kinetics of glutamate dehydrogenase self-association by pressure perturbation. *Biochemistry,* **18,** 2480–2487.

Hardman, M. J., J. H. Coates, and H. Gutfreund, 1978. Pressure relaxation of the equilibrium of the pig heart lactate dehydrogenase reaction. *Biochem. J.,* **171,** 215–233.

Harper, A. A., A. G. Macdonald, and K. T. Wann, 1977. The action of high hydrostatic pressure on voltage-clamped *Helix* neurons. *J. Physiol.,* **273,** 70–71.

Harrington, W. F., 1975. The effects of pressure in ultracentrifugation of interacting systems. *Fractions,* **1975** (1), 10–18.

Hawley, S. A., 1971. Reversible pressure–temperature denaturation of chymotrypsinogen. *Biochemistry,* **10,** 2436–2442.

Hawley, S. A. and R. M. Mitchell, 1975. An electrophoretic study of reversible protein denaturation: Chymotrypsinogen at high pressures. *Biochemistry,* **14,** 3257–3264.

Hazel, J. and C. L. Prosser, 1974. Molecular mechanisms of temperature compensation in poikilotherms. *Physiol. Rev.,* **54,** 620–677.

Hedrick, P. W., M. E. Ginevan, and E. P. Ewing, 1976. Genetic polymorphism in heterogeneous environments. *Ann. Rev. Ecol. Syst.,* **7,** 1–32.

Henderson, J. V., Jr. and D. L. Gilbert, 1975. Slowing of ionic currents in the voltage-clamped squid axon by helium pressure. *Nature,* **258,** 351–352.

Hessler, R. R., 1974. The structure of deep benthic communities from central oceanic waters. In *The Biology of the Oceanic Pacific,* C. B. Miller, ed. Oregon State University Press, Corvallis, pp. 79–93.

Hessler, R. R. and P. A. Jumars, 1974. Abyssal community analysis from replicate box cores in the central North Pacific. *Deep-Sea Res.,* **21,** 185–209.

Hessler, R. R. and H. L. Sanders, 1967. Faunal diversity in the deep sea. *Deep-Sea Res.,* **14,** 65–78.

Hessler, R. R., C. L. Ingram, A. A. Yayanos, and B. R. Burnett, 1978. Scavenging amphipods from the floor of the Philippine Trench. *Deep-Sea Res.,* **25,** 1029–1047.

Hochachka, P. W., 1974. Acetylcholinesterase: Temperature and pressure adaptation of the anionic binding site. *Biochem. J.,* **143,** 535–539.

Hochachka, P. W., 1975a. Fitness of enzyme binding sites for their physical environment: Coenzyme and substrate binding sites of M_4 lactate dehydrogenases. *Compar. Biochem. Physiol.,* **52B,** 25–31.

Hochachka, P. W., 1975b. How abyssal organisms maintain enzymes of the "right" size. *Compar. Biochem. Physiol.,* **52B,** 39–41.

Hochachka, P. W. and G. N. Somero, 1973. *Strategies of Biochemical Adaptation.* Saunders, Philadelphia.

Hochachka, P. W., D. E. Schneider, and A. Kuznetsov, 1970. Interacting pressure and temperature effects on enzymes of marine poikilotherms: Catalytic and regulatory properties of FDPase from deep and shallow-water fishes. *Mar. Biol.,* **7,** 285–293.

Hochachka, P. W., H. W. Behrisch, and F. Marcus, 1971a. Pressure effects on catalysis and control of catalysis by liver fructose diphosphatase from an off-shore benthic fish. *Am. Zool.,* **11,** 437–449.

Hochachka, P. W., D. E. Schneider, and T. W. Moon, 1971b. The adaptation of enzymes to pressure. I. A comparison of trout liver fructose diphosphatase with the homologous enzyme from an off-shore benthic fish. *Am. Zool.,* **11,** 479–490.

Hochachka, P. W., K. B. Storey, and J. Baldwin, 1975a. Design of acetylcholinesterase for its physical environment. *Compar. Biochem. Physiol.,* **52B,** 13–18.

Hochachka, P. W., K. B. Storey, and J. Baldwin, 1975b. Gill citrate synthase from an abyssal fish. *Compar. Biochem. Physiol.,* **52B,** 43–49.

Hochachka, P. W., C. Norberg, J. Baldwin, and J. H. A. Fields, 1975c. Enthalpy–entropy compensation of oxamate binding by homologous lactate dehydrogenases. *Nature,* **260,** 648–650.

Hubbs, C. L., 1926. The supposed intergradation of two species of *Sebastolobus* (a genus of scorpaenoid fishes) of western America. *Am. Mus. Novit.*, **216,** 1–9.

Ikkai, T. and T. Ooi, 1966. The effects of pressure on F-G transformation of actin. *Biochemistry*, **5,** 1551–1560.

Iwamoto, T., 1975. The abyssal fish *Antimora rostrata* (Gunther). *Compar. Biochem. Physiol.*, **52B,** 7–11.

Iwamoto, T. and D. L. Stein, 1974. A systematic review of rattail fishes (Macrouridae: Gadiformes) from Oregon and adjacent waters. *Occ. Pap. Calif. Acad. Sci.*, **111,** 1–79.

Jannasch, H. W. and C. O. Wirsen, 1973. Deep-Sea microorganisms: *In situ* response to nutrient enrichment. *Science*, **180,** 641–643.

Jannasch, H. W. and C. O. Wirsen, 1977. Microbial life in the deep sea. *Sci. Am.*, **236** (6), 42–52.

Jannasch, H. J., C. O. Wirsen, and C. D. Taylor, 1976. Undecompressed microbial populations from the deep sea. *Appl. Environ. Microbiol.*, **32,** 360–367.

Jannasch, H. W., K. Eimhjellen, C. O. Wirsen, and A. Farmanfarmaian, 1971. Microbial degradation of organic matter in the deep sea. *Science*, **171,** 672–675.

Johnson, A. G., F. M. Utter, and H. O. Hodgins, 1973. Estimate of genetic polymorphism and heterozygosity in three species of rockfish (genus *Sebastes*). *Compar. Biochem. Physiol.*, **44B,** 397–406.

Johnson, F. H. and H. Eyring, 1970. The kinetic basis of pressure effects in biology and chemistry. In *High Pressure Effects on Cellular Processes*, A. M. Zimmerman, ed. Academic Press, New York, pp. 1–44.

Johnson, F. H., H. Eyring and B. J. Stover, 1974. *The Theory of Rate Processes in Biology and Medicine*. Wiley, New York, 703 pp.

Johnson, M. S., 1971. Adaptive lactate dehydrogenase variation in the crested blenny, *Anoplarchus*. *Heredity*, **27,** 205–226.

Johnson, M. S., 1977. Association of allozymes and temperature in the crested blenny, *Anoplarchus purpurescens*. *Mar. Biol.*, **41,** 147–152.

Johnston, I. A. and N. J. Walesby, 1977. Molecular mechanisms of temperature adaptation in fish myofibrillar adenosine triphosphatases. *J. Compar. Physiol.*, **119,** 195–206.

Jones, M. L., 1981. *Riftia pachyptila* Jones: Observations on the vestimentiferan worm from the Galapagos Rift. *Science*, **213,** 333–336.

Jumars, P. A., 1976. Deep-sea species diversity: Does it have a characteristic scale? *J. Mar. Res.*, **34,** 217–246.

Karl, D. M., C. O. Wirsen, and H. W. Jannasch, 1980. Deep-sea primary production at the Galapagos hydrothermal vents. *Science*, **207,** 1345–1347.

Kegeles, G., L. Rhodes, and J. F. Bethune, 1967. Sedimentation behavior of chemically reacting systems. *Proc. Natl. Acad. Sci. (USA)*, **58,** 45–51.

King, F. D. and T. T. Packard, 1975. The effect of hydrostatic pressure on respiratory electron transport system activity in marine zooplankton. *Deep-Sea Res.*, **22,** 99–105.

Koehn, R. K., R. Milkman, and J. B. Mitton, 1976. Population genetics of marine pelecypods. IV. Selection, migration and genetic differentiation in the blue mussel *Mytilus edulis*. *Evolution*, **30,** 2–32.

Laidler, K. J., 1951. The influence of pressure on the rates of biological reactions. *Arch. Biochem. Biophys.*, **30,** 226–236.

Laidler, K. J. and P. S. Bunting, 1973. *The Chemical Kinetics of Enzyme Action*. Oxford University Press, London.

Lassen, H. H. and F. J. Turano, 1978. Clinal variation and heterozygote deficit of the Lap-locus in *Mytilus edulis*. *Mar. Biol.*, **49,** 245–254.

Lee, R. F., C. F. Phleger, and M. H. Horn, 1975. Composition of oil in fish bones: Possible function in neutral buoyancy. *Compar. Biochem. Physiol.*, **50B,** 13–16.

Lewontin, R. C., 1974. *The Genetic Basis of Evolutionary Change*. Columbia University Press, New York.

Lightfoot, R. H., P. A. Tyler, and J. D. Gage, 1979. Seasonal reproduction in deep-sea bivalves and brittlestars. *Deep-Sea Res.*, **26A**, 967–973.

Lonsdale, P., 1977. Clustering of suspension-feeding macrobenthos near abyssal hydrothermal vents at oceanic spreading centers. *Deep-Sea Res.*, **24**, 857–863.

Low, P. S. and G. N. Somero, 1974. Temperature adaptation of enzymes: A proposed molecular basis for the different catalytic efficiencies of enzymes from ectotherms and endotherms. *Compar. Biochem. Physiol.*, **49B**, 307–312.

Low, P. S. and G. N. Somero, 1975. Pressure effects on enzyme structure and function *in vitro* and under simulated *in vivo* conditions. *Compar. Biochem. Physiol.*, **52B**, 67–74.

Low, P. S. and G. N. Somero, 1976. Adaptation of muscle pyruvate kinases to environmental temperatures and pressures. *J. Exp. Zool.*, **198**, 1–12.

Low, P. S., J. L. Bada, and G. N. Somero, 1973. Temperature adaptation of enzymes: Roles of the free energy, the enthalpy and the entropy of activation. *Proc. Natl. Acad. Sci. (USA)*, **70**, 430–432.

Lumry, R. and S. Rajender, 1970. Enthalpy-entropy compensation phenomena in water solutions of proteins and small molecules: A ubiquitous property of water. *Biopolymers*, **9**, 1125–1227.

Macdonald, A. G., 1978. A dilatometric investigation of the effects of general anaesthetics, alcohols and hydrostatic pressure on the phase transition in smectic mesophases of dipalmitoyl phosphatidylcholine. *Biochimica et Biophysica Acta*, **507**, 26–37.

Macdonald, A. G. and I. Gilchrist, 1980. Effects of hydraulic decompression and compression on deep-sea amphipods. *Compar. Biochem. Physiol.*, **67A**, 149–153.

Macdonald, A. G. and J. M. Teal, 1975. Tolerance of oceanic and shallow water crustacea to high hydrostatic pressure. *Deep-Sea Res.*, **22**, 131–144.

Manwell, C. and C. M. A. Baker, 1968. Genetic variation of isocitrate, malate and 6-phosphogluconate dehydrogenases in snails of the genus *Cepaea*—introgressive hybridization, polymorphism and pollution? *Compar. Biochem. Physiol.*, **26**, 195–209.

Manwell, C. and C. M. A. Baker, 1970. *Molecular Biology and the Origin of Species*. University of Washington Press, Seattle, WA.

Marshall, N. B., 1954. *Aspects of Deep Sea Biology*. Hutchinson, London.

Marshall, N. B., 1971. *Explorations in the Life of Fishes*. Harvard University Press, Cambridge, MA, 204 pp.

Marshall, N. B., 1973. Genus *Macrourus* Bloch 1786. Fishes of the western North Atlantic. *Mem. Sears Found. Mar. Res.*, **1** (6), 581–600.

Marshall, N. B., 1979. *Deep-Sea Biology. Developments and Perspectives*. Garland STPM Press, New York.

Marshall, N. B. and T. Iwamoto, 1973a. Genus *Nezumia* Jordan 1904. Fishes of the western North Atlantic. *Mem. Sears Found. Mar. Res.*, **1** (6), 624–649.

Marshall, N. B. and T. Iwamoto, 1973b. Genus *Coelorhynchus* Giorna 1809. Fishes of the western North Atlantic. *Mem. Sears Found. Mar. Res.*, **1** (6), 538–563.

McDowell, S. B., 1973. Order Heteromi (Notochanthiformes). Family Halosauridae. Family Notocanthidae. Family Lipogenyidae. Fishes of the western North Altantic. *Mem. Sears Found. Mar. Res.*, **1** (6), 1–228.

Meek, R. P. and J. J. Childress, 1973. Respiration and the effect of pressure in the mesopelagic fish *Anoplogaster cornuta* (Beryciformes). *Deep-Sea Res.*, **20**, 1111–1118.

Menzies, R. J., 1962. On the food and feeding habits of abyssal organisms as exemplified by the Isopoda. *Int. Revue ges Hydrobiol.*, **47** (3), 339–358.

Menzies, R. J. and R. Y. George, 1972. Temperature effects on behavior and survival of marine invertebrates exposed to variations in hydrostatic pressure. *Mar. Biol.*, **13**, 155–159.

Menzies, R. J., R. Y. George, and G. T. Rowe, 1973. *Abyssal Environment and Ecology of the World Oceans*. Wiley-Interscience, New York.

Merret, N. R., 1978. On the identity and pelagic occurrence of larval and juvenile stages of rattail fishes (family Macrouridae) from 60°N, 20°W and 53°N, 20°W. *Deep-Sea Res.*, **25**, 147–160.

Merrit, R. B., 1972. Geographic distribution and enzymatic properties of lactate dehydrogenase allozymes in the fathead minnow, *Pimephales promelas*. *Am. Natur.*, **106**, 173–184.

Miller, D. J. and R. N. Lea, 1972. Guide to the coastal marine fishes of California. *Calif. Dept. Fish Game, Fish Bull.* No. 157, 235 pp. (1976 addendum, pp. 236–249).

Mitton, J. B. and R. K. Koehn, 1975. Genetic organization and adaptive response of allozymes to ecological variables in *Fundulus heteroclitus*. *Genetics,* **79,** 97–111.

Moon, T. W., 1975. Temperature adaptation: Isozymic function and the maintenance of heterogeneity. In *The Isozymes,* Vol. II. C. L. Markert, ed. Academic Press, New York, pp. 207–220.

Moon, T. W. and K. B. Storey, 1975. The effects of temperature and hydrostatic pressure on enzymes of an abyssal fish, *Antimora rostrata:* Liver NADP-linked isocitrate dehydrogenase. *Compar. Biochem. Physiol.,* **52B,** 51–57.

Moon, T. W., T. Mustafa, and P. W. Hochachka, 1971. Effects of hydrostatic pressure on catalysis by different lactate dehydrogenase isoenzymes from tissues of an abyssal fish. *Am. Zool.,* **11,** 473–478.

Morild, E., 1977a. Pressure neutralization of substrate inhibition in the alcohol dehydrogenase reaction. *J. Physical Chem.,* **81,** 1162–1166.

Morild, E., 1977b. Pressure variation of enzymatic reaction rates: Yeast and liver alcohol dehydrogenase. *Biophys. Chem.,* **6,** 351–362.

Morita, R. Y., 1980. Microbial life in the deep sea. *Can. J. Microbiol.,* **26,** 1375–1385.

Moser, H. G., 1974. Development and distribution of juveniles of *Sebastolobus* (Pisces; Family Scorpaenidae). *U.S. Natl. Mar. Fish. Ser. Fishery Bull.,* **72,** 865–884.

Murphy, L. S., G. T. Rowe, and R. L. Haedrich, 1976. Genetic variability in deep-sea echinoderms. *Deep-Sea Res.,* **23,** 339–348.

Musick, W. D. L. and M. G. Rossmann, 1979. The structure of mouse testicular lactate dehydrogenase isoenzyme C_4 at 2.9 Å resolution. *J. Biol. Chem.,* **254,** 7611–7620.

Nevo, E., 1978. Genetic variation in natural populations: Patterns and theory. *Theor. Pop. Biol.,* **13,** 121–177.

Oceanus, 22 (2), 1979. Initial findings of a deep-sea biologist quest. Galapagos Expedition Biology Participants.

O'Conner, T. M., L. L. Houston, and F. Samson, 1974. Stability of neuronal microtubules to high pressure *in vivo* and *in vitro*. *Proc. Natl. Acad. Sci.* (*USA*), **71,** 4198–4202.

Okamura, O., 1970. *Marourina* (*Pisces*). Academic Press of Japan, Tokyo.

Orton, J. H., 1920. Sea-temperature, breeding and distribution in marine animals. *J. Mar. Biol. Assoc. U. K.,* **12,** 339–366.

Otter, T. and E. D. Salmon, 1979. Hydrostatic pressure reversibly blocks membrane control of ciliary motility in *Paramecium*. *Science,* **206,** 358–361.

Packard, T. T., M. L. Healy, and F. A. Richards, 1971. Vertical distribution of the activity of the respiratory electron transport system in marine plankton. *Limnol. Oceanogr.,* **16,** 60–70.

Paladini, A. A., Jr. and G. Weber, 1981. Pressure-induced reversible dissociation of enolase. *Biochemistry,* **20,** 2587–2593.

Patton, J. S., 1975. The effect of pressure and temperature on phospholipid and triglyceride fatty acids of fish white muscle: A comparison of deepwater and surface marine species. *Comp. Biochem. Physiol.,* **52B,** 105–110.

Pearcy, W. G., 1962. Egg masses and early developmental stages of the scorpaenid fish, *Sebastolobus*. *J. Fish. Res. Board Can.,* **19,** 1169–1173.

Penniston, J. T., 1971. High hydrostatic pressure and enzymatic activity: Inhibition of multimeric enzymes by dissociation. *Arch. Biochem. Biophys.,* **142,** 322–332.

Pequeux, A. and R. Gilles, 1978. Effects of high hydrostatic pressures on the activity of the membrane ATPases of some organs implicated in hydromineral regulation. *Compar. Biochem. Physiol.,* **59B,** 207–212.

Pfeiler, E., 1978. Effects of hydrostatic pressure on ($Na^+ + K^+$)-ATPase and Mg^{2+}-ATPase in gills of marine teleost fish. *J. Exp. Zool.,* **205,** 393–402.

Phleger, C. F., J. Patton, P. Grimes, and R. F. Lee, 1976. Fish-bone oil: Percent total body lipid and carbon-14 uptake following feeding of 1-^{14}C-palmitic acid. *Mar. Biol.,* **35,** 85–89.

Pietsch, T. W., 1974. Osteology and relationships of ceratioid anglerfishes of the Family Oneirodidae, with a review of the genus *Oneirodes* Lukten. *Nat. Hist. Museum, Los Angeles Co. Sci. Bull.*, **18**, 1–113.

Place, A. R. and D. A. Powers, 1979. Genetic variation and relative catalytic efficiencies: Lactate dehydrogenase B allozymes of *Fundulus heteroclitus*. *Proc. Natl. Acad. Sci. (USA)*, **76**, 2354–2358.

Pope, D. H., W. P. Smith, R. W. Swartz, and J. V. Landau, 1975. Role of bacterial ribosomes in barotolerance. *J. of Bacteriology*, **121**, 664–669.

Powell, J. R., 1975. Protein variation in natural populations of animals. *Evol. Biol.*, **8**, 79–119.

Powell, M. A. and G. N. Somero, 1983. Blood components prevent sulfide poisoning of respiration of the hydrothermal vent tube worm *Riftia pachyptila*. *Science*, **219**, 297–299.

Powers, D. A., G. S. Greaney, and A. R. Place, 1979. Physiological correlation between lactate dehydrogenase genotype and haemoglobin function in killifish. *Nature*, **277**, 240–241.

Powers, D. A. and A. R. Place, 1978. Biochemical genetics of *Fundulus heteroclitus* (L.) I. Temporal and spatial variation in gene frequencies of LDH-B, MDH-A, GPI-B, and PGM-A. *Biochem. Genet.*, **16**, 593–607.

Quetin, L. B. and J. J. Childress, 1976. Respiratory adaptations of *Pleuroncodes planipes* to its environment off Baja California. *Mar. Biol.*, **38**, 327–334.

Rau, G. H., 1981. Hydrothermal vent clam and tube worm $^{13}C/^{12}C$: Further evidence of nonphotosynthetic food sources. *Science*, **213**, 338–340.

Rau, G. H. and J. I. Hedges, 1979. Carbon-13 depletion in a hydrothermal vent mussel: Suggestion of a chemosynthetic food source. *Science*, **203**, 648–649.

Rokop, F. J., 1974. Reproductive patterns in the deep-sea benthos. *Science*, **186**, 743–745.

Rokop, F. J., 1977a. Seasonal reproduction of the brachiopod *Frieleia halli* and the scaphopod *Cadulus californicus* at bathyal depths in the deep sea. *Mar. Biol.*, **43**, 237–246.

Rokop, F. J., 1977b. Patterns of reproduction in the deep-sea benthic crustaceans: A re-evaluation. *Deep-Sea Res.*, **24**, 683–691.

Rokop, F. J., 1979. Year-round reproduction in the deep-sea bivalve molluscs. In *Reproductive Ecology of Marine Invertebrates*, S. E. Stancyk, ed., University of South Carolina Press, pp. 189–198.

Rosenmann, E., A. M. Gonzalez, S. Hein, and F. Marcus, 1977. Carp (*Cyprinus carpio*) muscle fructose 1,5-bisphosphatase: Purification and some properties. *Compar. Biochem. Physiol.*, **58B**, 291–295.

Rowe, G. T., 1971. Benthic biomass and surface productivity. In *Fertility of the Sea*, J. D. Costlow, ed. Gordon and Breach, New York, pp. 441–454.

Rowe, G. T. and D. W. Menzel, 1971. Quantitative benthic samples from the deep Gulf of Mexico with some comments on the measurement of deep-sea biomass. *Bull. Mar. Sci.*, **21**, 556–566.

Salmon, E. D., 1975. Pressure-induced depolymerization of brain microtubules *in vitro*. *Science*, **189**, 884–886.

Sanders, H. L. and R. R. Hessler, 1969. Ecology of the deep-sea benthos. *Science*, **163**, 1419–1424.

Saunders, P. M. and N. P. Fofonoff, 1976. Conversion of pressure to depth in the ocean. *Deep-Sea Res.*, **23**, 109–111.

Schade, B. C., R. Rudolph, H.-D. Ludemann, and R. Jaenicke, 1980. Reversible high-pressure dissociation of lactic dehydrogenase from pig muscle. *Biochemistry*, **19**, 1121–1126.

Schmid, G., H.-D. Ludemann, and R. Jaenicke, 1978. Oxidation of sulfhydryl groups in lactate dehydrogenase under high hydrostatic pressure. *Eur. J. Biochem.*, **86**, 219–224.

Schmid, G., H.-D. Ludemann, and R. Jaenicke, 1979. Dissociation and aggregation of lactic dehydrogenase by high hydrostatic pressure. *Eur. J. Biochem.*, **97**, 407–413.

Schopf, T. J. M. and J. L. Gooch, 1971. Gene frequencies in a marine ectoproct: A cline in natural populations related to sea temperature. *Evolution*, **25**, 286–289.

Schulenberger, E. and R. R. Hessler, 1974. Scavenging abyssal benthic amphipods trapped under oligotrophic central North Pacific gyre waters. *Mar. Biol.*, **28**, 185–187.

Shaklee, J. B., J. A. Christiansen, B. D. Sidell, C. L. Prosser, and G. S. Whitt, 1977. Molecular aspects of temperature acclimation in fish: Contributions of changes in enzyme activities and isozyme patterns to metabolic reorganization in the green sunfish. *J. Exp. Zool.*, **201**, 1–20.

Siebenaller, J. F., 1978a. Genetic variation in deep-sea invertebrate populations: The bathyal gastropod *Bathybembix bairdii*. *Mar. Biol.*, **47**, 265–275.

Siebenaller, J. F., 1978b. Genetic variability in deep-sea fishes of the genus *Sebastolobus* (Scorpaenidae). In *Marine Ogranisms*, B. Battaglia and J. Beardmore, eds. Plenum Press, New York, pp. 95–122.

Siebenaller, J. F. and G. N. Somero, 1978. Pressure-adaptive differences in lactate dehydrogenases of congeneric fishes living at different depths. *Science*, **201**, 255–257.

Siebenaller, J. F. and G. N. Somero, 1979. Pressure-adaptive differences in the binding and catalytic properties of muscle-type (M_4) lactate dehydrogenases of shallow- and deep-living marine fishes. *J. Compar. Physiol.*, **129**, 295–300.

Siebenaller, J. F. and G. N. Somero, 1982. The maintenance of different enzyme activity levels in congeneric fishes living at different depths. *Physiol. Zool.*, **55**, 171–179.

Siebenaller, J. F., G. N. Somero, and R. L. Haedrich, 1982. Biochemical characteristics of macrourid fishes differing in their depths of distribution. *Biol. Bull.*, **163**, 240–249.

Sinensky, M., 1974. Homeoviscous adaptation—A homeostatic process that regulates the viscosity of membrane lipids in *Escherichia coli*. *Proc. Natl. Acad. Sci. (USA)*, **71**, 522–525.

Smith, K. L., Jr., 1978. Metabolism of the abyssopelagic rattail *Coryphaenoides armatus* measured *in situ*. *Nature*, **274**, 362–364.

Smith, K. L., Jr. and R. R. Hessler, 1974. Respiration of benthopelagic fishes: *In situ* meausurements in 1230 meters. *Science*, **184**, 72–73.

Smith, K. L. and M. B. Laver, 1981. Respiration of the bathypelagic fish, *Cyclothone acclinidens*. *Mar. Biol.*, **61**, 261–266.

Smith, K. L., Jr. and J. M. Teal, 1973. Temperature and pressure effects on respiration of thecosomatous pteropods. *Deep-Sea Res.*, **20**, 853–858.

Smith, K. L., Jr., G. A. White, and M. B. Laver, 1979. Oxygen uptake and nutrient exchange of sediments measured *in situ* using a free vehicle grab respirometer. *Deep-Sea Res.*, **26A**, 337–346.

Sokolova, M. N., 1972. Trophic structure of deep-sea macrobenthos. *Mar. Biol.*, **16**, 1–12.

Somero, G. N., 1975. The roles of isozymes in adaptation to varying temperatures. In *The Isozymes*, Vol. II, C. L. Markert, ed. Academic Press, New York, pp. 221–234.

Somero, G. N., 1978. Temperature adaptation of enzymes: Biological optimization through structure–function compromises. *Ann. Rev. Ecol. System.*, **9**, 1–29.

Somero, G. N., 1979. Interacting effects of temperature and pressure on enzyme function and evolution in marine organisms. In *Biochemical and Biophysical Perspectives in Marine Biology*, Vol. 4, D. C. Malins and J. R. Sargent, eds. Academic Press, New York, pp. 1–27.

Somero, G. N., 1982. Physiological and biochemical adaptations of deep-sea fishes: Adaptive responses to the physical and biological characteristics of the abyss. In *Ecosystem Processes in the Deep Oceans*, W. G. Ernst and J. Morin, eds. Prentice-Hall, Englewood Cliffs, NJ., pp. 257–278.

Somero, G. N. and J. J. Childress, 1980. A violation of the metabolism-size scaling paradigm: Activities of glycolytic enzymes muscle increase in larger size fishes. *Physiol. Zool.*, **53**, 322–337.

Somero, G. N. and P. W. Hochachka, 1976a. Biochemical adaptations to temperature. In *Adaptation to Environment: Essays on the Physiology of Marine Animals*, R. C. Newell, ed. Butterworths, London, pp. 125–190.

Somero, G. N. and P. W. Hochachka, 1976b. Biochemical adaptations to pressure. In *Adaptation to Environment: Essays on the Physiology of Marine Animals*, R. C. Newell, ed. Butterworths, London, pp. 480–510.

Somero, G. N. and P. S. Low, 1977. Eurytolerant proteins: Mechanisms for extending the environmental tolerance range of enzyme–ligand interactions. *Am. Natur.*, **111**, 527–538.

Somero, G. N. and J. F. Siebenaller, 1979. Inefficient lactate dehydrogenases of deep-sea fishes. *Nature*, **282**, 100–102.

Somero, G. N. and M. Soulé, 1974. Genetic variation in marine fishes as a test of the niche-variation hypothesis. *Nature*, **249**, 670–672.

Southward, A. J., E. C. Southward, P. R. Dando, G. H. Rau, H. Felbeck, and H. Flugel, 1981.

Occurrence of bacterial symbionts and a low ratio of ^{13}C to ^{12}C in the tissues of pogonophora point to unusual nutrition and metabolism. *Nature,* **293,** 616–620.

Spiess, F. N., K. C. Macdonald, T. Atwater, R. Ballard, A. Carranza, D. Cordoba, C. Cox, V. M. Diaz Garcia, J. Francheteau, J. Guerrero, J. Hawkins, R. Haymon, R. Hessler, T. Juteau, M. Kastner, R. Larson, B. Luyendyk, J. D. Macdougall, S. Miller, W. Normark, J. Orcutt, and C. Rangin, 1980. East Pacific rise: Hot springs and geophysical experiments. *Science,* **207,** 1421–1433.

Sullivan, K. M. and G. N. Somero, 1980. Enzyme activities of fish skeletal muscle and brain as influenced by depth of occurrence and habits of feeding and locomotion. *Mar. Biol.,* **60,** 91–99.

Suzuki, Z., Y. Warashina, and M. Kishida, 1977. The comparison of catches by regular and deep tuna longline gears in the Western and Central Equatorial Pacific. *Bull. Far. Seas Fish. Res. Lab.,* **15,** 51–72.

Swezey, R. R. and G. N. Somero, 1982a. Polymerization thermodynamics and structural stabilities of skeletal muscle actins from vertebrates adapted to different temperatures and hydrostatic pressures. *Biochemistry,* **21,** 4496–4503.

Swezey, R. R. and G. N. Somero, 1982b. Skeletal muscle actin content is strongly conserved in fishes having different depths of distribution and capacities of locomotion. *Mar. Biol. Lett.,* **3,** 307–315.

Taketa, K. and B. M. Pogell, 1963. Reversible inactivation and inhibition of liver fructose 1,6-diphosphatase by adenosine nucleotides. *Biochem. Biophys. Res. Commun.,* **12,** 229–235.

Taketa, K. and B. M. Pogell, 1965. Allosteric inhibition of rat liver fructose 1,6-diphosphatase by adenosine 5′-monophosphate. *J. Biol. Chem.,* **240,** 651–662.

Teal, J. M. and F. G. Carey, 1967. Effects of pressure and temperature on the respiration of euphausiids. *Deep-Sea Res.,* **14,** 725–733.

Tischler, M. E., D. Friedrichs, K. Coll, and J. R. Williamson, 1977. Pyridine nucleotide distributions and enzyme mass action ratios in hepatocytes from fed and starved rats. *Arch. Biochem. Biophys.,* **184,** 222–236.

Torgerson, P. M., H. G. Drickamer, and G. Weber, 1979. Inclusion complexes of poly-β-cyclodextrin: A model for pressure effects upon ligand–protein complexes. *Biochemistry,* **18,** 3079–3083.

Torgerson, P. M., H. G. Drickamer, and G. Weber, 1980. Effect of hydrostatic pressure upon ethidium bromide association with transfer ribonucleic acid. *Biochemistry,* **19,** 3957–3960.

Torres, J. J., B. W. Belman, and J. J. Childress, 1979. Oxygen consumption rates of midwater fishes off California. *Deep-Sea Res.,* **26A,** 185–197.

Tyler, P. A., 1982. The reproductive biology of *Ophiacantha bidentata* (Echinodermata: Ophiuroidea) from the Rockall Trough. *J. Mar. Biol. Assoc. U. K.,* **62,** 45–55.

Tyler, P. A. and J. D. Gage, 1979. Reproductive ecology of deep-sea ophiuroids from the Rockall Trough. In *Cyclic Phenomena in Marine Plants and Animals,* E. Naylor and R. G. Hartnoll, eds. Pergamon Press, Oxford, pp. 215–222.

Tyler, P. A. and J. D. Gage, 1980. An abyssal time-series II. Reproduction and growth of the deep-sea brittlestar *Ophiura ljungmani* (Lyman). *Oceanologica Acta,* **3,** 177–185.

Tyler, P. A., S. L. Pain, and J. D. Gage, 1982. The reproductive biology of the deep-sea asteroid *Bathybiaster vexillifer. J. Mar. Biol. Assoc. U. K.,* **62,** 57–69.

Valentine, J. W. and F. J. Ayala, 1975. Genetic variation in *Frieleia halli,* a deep-sea brachiopod. *Deep-Sea Res.,* **22,** 37–44.

van Tol, A., 1975. On the occurrence of a temperature coefficient (Q_{10}) of 18 and a discontinuous Arrhenius plot for homogeneous rabbit muscle fructose diphosphatase. *Biochem. Biophys. Res. Commun.,* **62,** 750–756.

Walker, E. and D. N. Wheatley, 1979. Effects of hydrostatic pressure in the range 100–300 atmospheres on cell division and protein synthesis in synchronized *Tetrahymena pyriformis:* A comparison with cycloheximide and emetine. *J. Cell. Physiol.,* **99,** 1–14.

Walsh, P. J. and G. N. Somero, 1982. Interactions among pyruvate concentration, pH and K_m of pyruvate in determining *in vivo* Q_{10} values of the lactate dehydrogenase reaction. *Can. J. Zool.,* **60,** 1293–1299.

Wann, K. T. and A. G. Macdonald, 1980. The effects of pressure on excitable cells. *Compar. Biochem. Physiol.,* **66A,** 1–12.

Wann, K. T., A. G. Macdonald, A. A. Harper, and S. E. Wilcock, 1979a. Electrophysiological measurements at high hydrostatic pressure: Methods for intracellular recording from isolated ganglia and for extracellular recording *in vivo*. *Compar. Biochem. Physiol.*, **64A,** 141–147.

Wann, K. T., A. G. Macdonald, and A. A. Harper, 1979b. The effects of high hydrostatic pressure on the electrical characteristics of *Helix* neurons. *Compar. Biochem. Physiol.*, **64A,** 149–159.

Weale, K. E., 1967. *Chemical Reactions at High Pressures*. Spon, Ltd., London, 349 pp.

Wenner, C. A. and J. A. Musick, 1977. Biology of the morid fish, *Antimora rostrata*, in the western North Atlantic. *J. Fish. Res. Board Can.*, **34,** 2362–2368.

White, J. L., M. L. Hackert, M. Buehner, M. J. Adams, G. C. Ford, P. J. Lentz, Jr., I. E. Smiley, S. J. Steindel, and M. G. Rossmann, 1976. A comparison of the structures of apo dogfish M_4 lactate dehydrogenase and its ternary complexes. *J. Mol. Biol.*, **102,** 759–779.

Wilkinson, G. N., 1961. Statistical estimations in enzyme kinetics. *Biochem. J.*, **80,** 324–334.

Williams, P. M., K. L. Smith, E. M. Druffel, and T. W. Linick, 1981. Dietary carbon sources of mussels and tubeworms from Galapagos hydrothermal vents determined from tissue ^{14}C activity. *Nature*, **292,** 448–449.

Wilson, D. F., 1972. Diel migration of sound scatterers into, and out of, the Cariaco Trench anoxic water. *J. Mar. Res.*, **30,** 168–176.

Wirsen, C. O. and H. W. Jannasch, 1976. Decomposition of solid organic materials in the deep sea. *Environ. Sci. Technol.*, **10,** 880–886.

Wohlschlag, D. E., 1964. Respiratory metabolism and ecological characteristics of some fishes in McMurdo Sound, Antarctica. In *Biology of the Antarctic Seas*. Vol. 1, *Antarctic Research Series*. M. O. Lee, ed. American Geophysical Union, Baltimore, pp. 33–62.

Yancey, P. H. and G. N. Somero, 1978a. Temperature dependence of intracellular pH: Its role in the conservation of pyruvate apparent K_m values of vertebrate lactate dehydrogenases. *J. Compar. Physiol.*, **125,** 129–134.

Yancey, P. H. and G. N. Somero, 1978b. Urea-requiring lactate dehydrogenases of marine elasmobranch fishes. *J. Compar. Physiol.*, **125,** 135–141.

Yayanos, A. A., 1978. Recovery and maintenance of live amphipods at a pressure of 580 bars from an ocean depth of 5700 meters. *Science*, **200,** 1056–1059.

Yayanos, A. A., 1981. Reversible inactivation of deep-sea amphipods (*Paralicella capresca*) by a decompression from 601 bars to atmospheric pressure. *Compar. Biochem. Physiol.*, **69A,** 563–565.

Yayanos, A. A., A. A. Benson, and J. C. Nevenzel, 1978. The pressure–volume–temperature (*PVT*) properties of a lipid mixture from a marine copepod, *Calanus plumchrus:* Implications for buoyancy and sound scattering. *Deep-Sea Res.*, **25,** 257–268.

Zipp, A. and W. Kauzmann, 1973. Pressure denaturation of metmyoglobin. *Biochemistry*, **12,** 4217–4227.

8. SEDIMENT COMMUNITY RESPIRATION IN THE DEEP SEA

KENNETH L. SMITH, JR. AND KENNETH R. HINGA

1. Introduction

The deep ocean is an open ecosystem with energy exchanges across both the air–sea interface and the sediment–water interface. In vertical profile this ecosystem can be divided into layers defined by depth, salinity, pressure, temperature, and surface illumination. Although a community of organisms is associated with each layer, vertical migrations that transcend boundaries add to the complexity of the system.

The solar energy that primarily drives this system is fixed into chemical energy by the phytoplankton in the euphotic zone. This energy is then disseminated throughout the water column by passive sinking or active dispersion mediated by organisms. Eventually this chemical energy passes downward through the mesopelagic and bathypelagic zones to the benthic boundary layer.

The benthic boundary layer in the deep ocean can be defined biologically as the sediment community and the assemblages of organisms in the overlying water column associated with the bottom (within about 100 m of the bottom). The primary difficulty with this definition is putting bounds on a continuum. Recent evidence shows that many animals that feed on the bottom also make vertical migrations to feed away from the bottom (Haedrich, 1974; Pearcy, 1976; Smith et al., 1979b). Conversely, some midwater species have been observed on the bottom.

The benthic boundary layer of the deep sea likely contains a number of interdependent biological components in the sediment and the overlying water column. The sediment has an assemblage of bacteria and invertebrates that inhabit generally the top 2 to 3 cm of the sediment surface. Above this layer, at the sediment–water interface, large animals ingest the sediments and their biota for nourishment. In the water column immediately above the sediment–water interface pelagic fish and invertebrates, including plankton, feed in both the water column and sediments. Sediment-dwelling animals also make excursions into the water column to feed, providing an active two-way avenue of food exchange. Many of the benthopelagic animals communicate directly with higher midwater animals above the benthic boundary layer, thus providing an active exchange of food that probably exists throughout the water column.

The continuum of communities in the deep sea makes it difficult to isolate and study the energy transformations and energy flow through any one component. Ideally the benthic boundary layer would serve as a defined and thus observable functional unit of the much larger open ocean ecosystem. The scope of this review, however, has been limited to one component of the benthic boundary layer, the sediment community, because it is the only functional unit of the system on which adequate information is available.

It is the intended purpose of this discourse to (1) describe the methodology involved in measuring sediment community respiration, (2) present all the available data on sediment community respiration in the deep sea, (3) formulate from that data a predictive equation for rates in the world ocean, and (4) compare the global sediment community respiration for the world ocean with the supply of organic carbon produced in the euphotic zone.

2. Methodology

The deep-sea sediment community can be defined as that assemblage of organisms inhabiting the soft bottom sediments at a water depth in excess of 200 m, exclusive of the large (megafauna) epibenthic animals that live at the sediment–water interface. The distinction is made primarily for convenience because the metabolism of each of the epibenthic fauna, whose numerical density is many orders of magnitude less than that of the corresponding sediment biota, must be measured individually using different techniques.

Oxygen consumption has been the primary measurement made to assess the energetic importance of these sediment communities. An area of the sediment and overlying water is generally enclosed either *in situ* or on board ship, and the oxygen depletion over the sediment is recorded with time. Indirect estimates of respiration have also been made from analyses of electron transport system activity and from oxygen gradients in sediments. Since growth rates of deep-sea organisms are largely unknown, and for most organisms the energy used in growth is considered small, it is assumed that the energy usage measured by oxygen consumption will be essentially equal to the total energy requirements of the organism.

Four techniques have been developed to measure the respiration of the deep-sea sediment community. One is a shipboard core incubation technique that returns sediment cores to the surface for oxygen consumption measurements. The second technique involves the *in situ* incubation of minimally disturbed enclosures of sediment with either continuous monitoring of dissolved oxygen by use of a polarographic sensor or end-point determinations of dissolved oxygen on withdrawn water samples. The third technique is a biochemical assay for indirectly determining oxygen consumption through respiratory electron-transport system activity. The fourth technique uses oxygen gradients in sediment pore waters to calculate flux of oxygen through those sediments.

A. In situ *Methods*

In situ techniques have been developed to minimize sampling and recovery problems, as well as eliminating effects of decompression on the biota. Sediment community respiration has been measured with submersible-manipulated and free-vehicle instruments such as the bell-jar respirometers (Smith and Teal, 1973; Smith et al., 1976; Hinga et al., 1979) and the grab respirometer (Smith, 1978; Smith et al., 1979a; Smith et al., in press).

A simple example of a bell-jar respirometer is the free-vehicle benthic respirometer described by Hinga et al. (1979). This respirometer was designed for attachment to the end of various instrument arrays. The unit consists of four cylindrical aluminum respiration chambers (24 cm diameter × 8 cm) mounted in an aluminum tubing frame. Suspended below the frame is a disposable anchor held by an acoustic or timed-release mechanism.

On the bottom, the respirometer chambers are initially held in an open position above the sediments by a dissolving link to assure total flushing and to prevent a bow-wave effect. On dissolution of this link, the chambers free fall on guide rods and penetrate 4 cm into the sediment. A Savonious rotor external to each chamber drives an internally mounted stirring bar that prevents stratification of dissolved oxygen and nutrients. Two reversing syringe withdrawal units are provided on each chamber to extract final water samples for dissolved oxygen and nutrient analyses just prior to release of the descent weights. A control water sample is taken with a Niskin bottle attached to the benthic respirometer. Respiration is then ascertained from end-point determinations of dissolved oxygen by use of the Winkler titration method in the laboratory.

A similar free-vehicle bell-jar respirometer is described by Smith et al. (1976), but with three fundamental differences: (1) dissolved oxygen tension within each respirometer chamber is monitored continuously with a polarographic sensor; (2) the bell-jar respirometer chambers are released into the sediment by acoustic command after the free vehicle reaches the bottom, providing the exact time of closure and commencement of the measurements with verification by photography; and (3) water samples can be withdrawn from the respiration chambers at any predetermined time by using syringes controlled by a precision electronic timer that activates a solenoid release.

The free-vehicle grab respirometer represents the latest development in the design of such instruments (Smith et al., 1979a; Smith and Baldwin, in press). It consists of two parts: an aluminum tripod that supports the instrumentation and a flotation array. The whole unit is controlled through a centrally located acoustic release that can be commanded from the surface. A grab respirometer assembly consisting of four replicate stainless steel grabs, each with a polarographic oxygen electrode, stirring motor and syringe withdrawal–injection system, is suspended from the acoustic release. After deployment and settlement to the bottom the grab respirometer assembly is released and guided to the sediment on four guide rods. Each grab penetrates 10 to 15 cm into the sediment. During the continuous measurement of the dissolved oxygen tension in the overlying water of each grab, the enclosed water is slowly circulated with a stirring motor to prevent stratification over the sediment–water interface and around the polarographic oxygen sensor. Control water samples are automatically taken externally to the grabs at the initiation and termination of each incubation (prior to the preset timed closure of the grab jaws). At the termination of the incubation, which generally lasts 1 to 5 days, an acoustic command from the surface allows the anchor weights of the free-vehicle grab respirometer to be dropped. The net positive buoyancy of the flotation pulls the sediment-laden grabs of the respirometer to the surface to be retrieved by a surface vessel. The grab jaws close automatically.

Sediment samples from each grab are sieved for macrofaunal biomass and abundance and organic carbon and nitrogen content. The control water samples are analyzed for dissolved oxygen.

One major criticism of *in situ* measurements of sediment community respiration is that enclosure of the sediments and overlying water creates an artificial system. Although stirring motors prevent oxygen stratification, the effects of decreased oxygen concentration and increased metabolite levels within the enclosure are unknown. Several arguments against the importance of such criticisms were presented by Smith (1978); however, definitive experiments are still needed.

Another free vehicle that will measure *in situ* benthic respiration is presently under development as part of the Manganese Nodule Project (MANOP) whose goal is to study the processes in the formation of manganese nodules.

B. *Shipboard Core Incubation Method*

The shipboard core incubation technique for deep waters is used largely by Pamatmat (1971, 1973, 1975, 1977). A multiple corer is employed to gently take four replicate cores of sediment ($\leq$15 cm) and return them along with overlying bottom water to the surface. The fiber-glass epoxy coring tubes (5.7-cm inner diameter) are removed from each core barrel at the surface and sealed with a polyvinyl chloride (PVC) stopper holding an oxygen sensor and magnetic stirring bar. Each core is then placed upright in a water bath at ambient bottom temperature, and an external motor-driven magnet centrally located in the water bath drives the stirring bar (Pamatmat, 1971). The dissolved oxygen tension is continuously monitored for periods up to several days.

C. *Electron-Transport System Method*

The third method used to calculate sediment community respiration is the biochemical assay of respiratory electron-transport system (ETS) activity (Christensen and Packard, 1977). Sediment cores are taken from shipboard bottom samplers and returned to the laboratory. Electron-transport system activity is determined for 1- to 2-cm sections of the core by use of the technique described by Owens and King (1975) and modified by Christensen and Packard (1977). Sediment samples are sonified to cause cellular disruption and enzyme dissolution. This homogenate is then assayed for ETS activity. The ETS activity is calculated from the observed activity corrected for ambient bottom temperature by using the Arrhenius equation and an activation energy of 15 kcal/mol · deg. *In vivo* rates of sediment oxygen consumption are calculated from these corrected ETS values by using the ratio of respiration to ETS activity (R/ETS) of 0.0078 that was determined from laboratory culture measurements on senescent, carbon-limited cultures of the marine bacterium *Vibrio anguillarum* (Christensen and Packard, 1977).

D. *Oxygen Gradient Method*

The fourth method used to calculate benthic oxygen consumption is based on the oxygen concentration gradient in sediment pore waters (Murray and Grundmanis, 1980). A few milliliters of pore water are withdrawn from the sediment *in situ* by using a winch-operated harpoon sampler (Sayles et al., 1973). The sampler draws the pore water through porous frits at given depths below the sediment–water interface and stores the samples until the sampler is returned to the surface where the dissolved oxygen in the samples is analyzed by use of a gas chromatograph. Samples are usually taken at intervals of a few centimeters. A curve is fitted to the data and the oxygen gradient calculated. The oxygen gradient times the diffusion constant for oxygen in the sediment gives an estimate of the flux of oxygen along that gradient.

E. Compartmentalization

Attempts have been made by several investigators to compartmentalize the total oxygen uptake in both shipboard core incubations and *in situ* into biological, chemical, and bacterial oxygen demands using various inhibitors and poisons. Formalin treatment (2.5 to 10%) has been widely used to poison the sediment organisms, leaving a residual oxygen consumption termed *chemical demand*. This has been equated to chemical oxidation of reduced substrates provided by anaerobic respiration where reduced sediments exist. The value of such a distinction is questionable since both are ultimately the result of biological processes, either aerobic or anaerobic (Pamatmat, 1971, 1977; Pamatmat and Bhagwat, 1973; Davies, 1975). In the majority of the deep-sea stations considered in this chapter, the sediments were oxidized and to date no one has detected oxygen consumption in *in situ* measurements after formalin treatment (Smith and Teal, 1973; Smith, 1978).

Antibiotic treatments have also been employed to inhibit the bacterial component of sediment community respiration (Smith, 1974). However, little is known of their effectiveness as either a bacteriostatic or bacteriocidal agent in deep-sea sediments. Interpretation of the formalin and antibiotic treatments is difficult, so we have opted to deal only with the total dissolved oxygen consumed within the enclosure.

F. Comparison of Techniques

Deep-sea benthic respiration has been measured at 54 locations by use of the four techniques (Fig. 1). The values shown in Fig. 1 for the *in situ* measurements represent the means of a number of replicate chambers used at each location. The *in situ* measurements typically vary by a factor of 2 between replicate chambers. The error within a single chamber measurement is largely a result of uncertainty in the volume in the chamber resulting from irregularities in the sediment surface or from incomplete or overpenetration of the chamber. This possible error is estimated to be as much as 25% for the 4-cm chamber height in the instrument used by Hinga et al. (1979) and a few percent for the 10- to 15-cm chamber height in the measurements by Smith and Teal (1973), Smith et al. (1976), Smith (1978), Smith et al. (1979a), and Smith et al. (in press).

There are two primary concerns with the shipboard core method for measurement of deep-sea metabolism: the effect of decompression and the effect of temperature increase on sediment cores during collection. In a preliminary comparison with *in situ* methods at a 3000-m station in the northwestern Atlantic (Smith, 1978), order-of-magnitude higher values were obtained with the coring method, suggesting that the sediment biota oxygen consumption is affected by decompression and/or temperature change. Deep-sea bacterial studies have shown that decompression and temperature increase during collection can have deleterious ef-

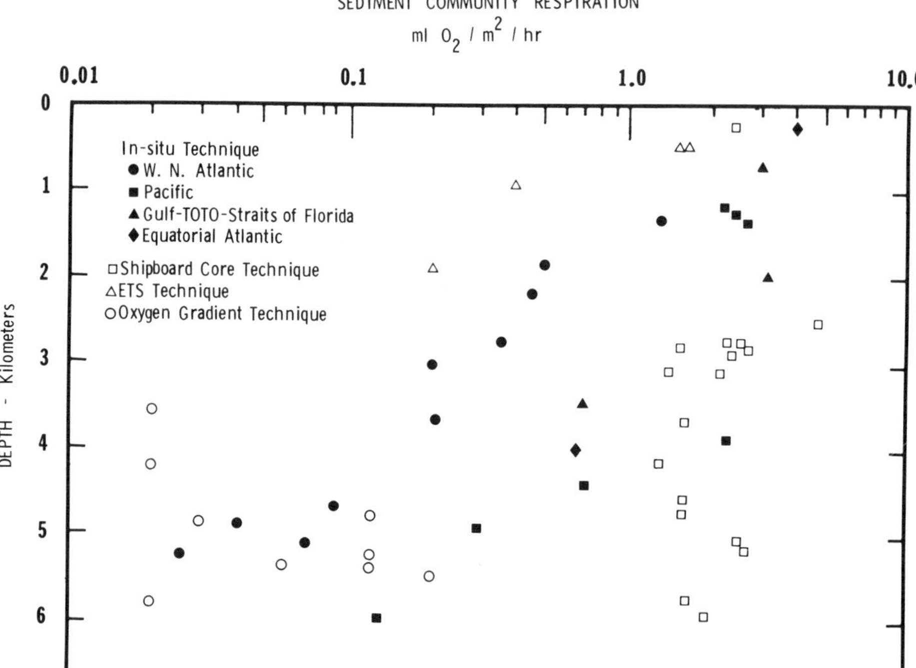

Fig. 1. Deep-sea sediment community respiration measurements by all four methods.

fects on growth (Seki and Robinson, 1969; ZoBell, 1970; Yayanos et al., 1979). Also, deep-sea macrofauna have rarely been recovered alive from depths greater than 1800 m without the aid of pressure-retaining or temperature-insulated traps (Menzies et al., 1974). Pressure and temperature changes have also been shown to exert significant effects on the respiration of some pelagic animals (Childress, 1977; Yayanos, 1981). The data generated by the shipboard core technique are suspiciously constant with water depth and consumption rates appear higher than the rest of the deep-water data (Fig. 1). The temperature change on recovery and decompression may have destroyed organisms that are decaying with subsequent elevated oxygen demand.

A possible criticism of the ETS method is the same as that for the shipboard core incubation method: the unknown effects of decompression and temperature change on samples during collection.

There are no published stations where sediment community respiration has been measured by both the ETS and *in situ* techniques. However, ETS measurements from four deep water stations [480 to 1820 m (Christensen and Packard, 1977)] located in the productive upwelling area off northeastern Africa are half to one quarter the *in situ* measurements at similar depths on the Gay Head–Bermuda transect and an order of magnitude lower than the *in situ* stations that underlie waters of similarly high primary production. Questions remain concerning the influence of decompression and temperature change during sample retrieval, the pressure dependence of sediment organism ETS activity (King and Packard,

1975), the applicability of the R/ETS factor, and the presence of enzyme inhibitors extracted from the sediment.

The oxygen gradient method has several limitations. The gradient must be well defined, which is difficult with a winch–wire-operated probe in the top few centimeters of sediment, and the effective oxygen diffusion coefficient must be accurately known. Also, the technique does not consider the effect of organisms that ventilate directly to the sediment surface. Figure 1 shows the oxygen gradient measurement from the equatorial Pacific to be generally low and highly variable compared to *in situ* measurements and one to two orders of magnitude lower than an *in situ* station in the equatorial Atlantic.

3. Sediment Community Respiration of Major Oceanic Regions

Sediment community respiration has been measured in a variety of deep-sea environments in the world ocean by employing the four techniques discussed above with an *in situ* technique used at 22 of these stations (Tables I, III, V, VI). The remainder of the discussion is restricted to treatment of the *in situ* data. The depths of these stations varied from 278 to 5900 m. Distance from shore spanned from 18 to 3200 km. Studies of the overlying water of these areas have reported annual primary productivities ranging from 50 to 307 g C/m^2 with annual particulate organic carbon flux to the bottom, available for a few of the stations, from 1 to 10 g C/m^2.

The benthic environments at these stations were highly variable. Water temperatures ranged from 2 to 11°C, whereas dissolved oxygen varied from 0.41 ml O_2/liter in a basin off southern California to 7.05 ml O_2/liter in the northwestern Atlantic. Sediments were generally pelagic oozes and clays, of which coarser constituents, such as silts and sands, were more prevalent at the shallower stations on the continental slope. Organic carbon and nitrogen content of these sediments was generally higher at stations with either a high particulate organic carbon flux rate or in close proximity to land. Under these circumstances their respective values ranged from 0.8 to 30.5 mg C/g dry sediment and from 0.1 to 3.8 mg N/g dry sediment.

Organisms inhabiting these sediments include bacteria, foraminiferans, nematodes, polychaetes, bivalves, and crustaceans. The most commonly measured component of the biota is the macrofauna, made up of animals retained by either a 420- or 297-μm-mesh sieve. Abundance and biomass generally reflect the food available to the bottom at each station. These values spanned three orders of magnitude, ranging from 19 to 22,988 individuals/m^2 for abundance and from 31 to 15,600 mg wet weight/m^2 for biomass.

In situ measurements of sediment community respiration ranged from 0.02 ml O_2/m$^2\cdot$hr in the oligotrophic Sargasso Sea at 5200 m to 3.93 ml O_2/m$^2\cdot$hr in the equatorial Atlantic at a depth of 278 m.

To analyze the large-scale variations in sediment community respiration and relate them to particular environmental parameters, we have arbitrarily divided these measurements and associated environmental descriptions into four geographic regions: the northwestern Atlantic, the Gulf of Mexico–Tongue of the Ocean, the eastern equatorial Atlantic, and the eastern and central northern Pacific. We describe each region with its unique characteristics below to provide

TABLE I

Sediment Community Respiration Measured *in*
of Northwestern Atlantic–Sargasso Sea

Station	Sediment Community Respiration (ml O_2/m^2·hr)	Number of Replicate Chambers	Depth (m)	Distance from Shore (km)	Annual Primary Productivity (g C/m^2·yr)	Bottom Water Temperature (°C)	Dissolved Oxygen Bottom (ml/liter)
DOS-1	0.50	10	1850	176	120	4	7.05
DWD	0.46	2	2200	172	100	3	6.34
ADS	0.35	4	2750	259	160	3	6.52
HH	0.20	6	3000	291	160	3	6.15
DOS-2	0.21	5	3650	352	100	3	6.54
JJ	0.09	3	4670	497	68	3	6.43
KK	0.04	1	4830	612	68	3	6.04
NN	0.07	2	5080	880	72	3	6.25
MM	0.02	4	5200	806	72	3	6.15
77DE	1.31[b]	4	1345[b]	148	85[c]	4[b]	5.65[b]

[a] Rowe and Gardner (1979).
[b] Hinga et al. (1979, with correction of organic carbon sedimentation rate values in Table 6 of that paper).
[c] Smith and Cowles (1975), Smith and Barber (1974) and Barber (1972).
[d] Menzies and Rowe (1968).
[e] Tietjen (1971).
[f] Deuser and Ross (1980).

a better framework from which to interpret the community respiration measurements.

A. *Northwestern Atlantic*

General Description

The North American Basin of the northern Atlantic is bounded on the west by the North American continent and on the east by the Mid-Atlantic Ridge. The main topographic features of the basin are the centrally located Bermuda Rise and the New England Seamount chain, which runs in a southeasterly direction off Cape Cod (Fig. 2). On the western boundary, the continental slope and rise conform to the classical topographic concept north of Cape Hatteras, which is heavily dissected by submarine canyons. South of the Cape the slope is complicated by the broad Blake Plateau (Emery and Uchupi, 1972).

Primary hydrographic features of the northwestern Atlantic are the Gulf Stream and the Western Boundary Undercurrent, which traverse or underlie the North Atlantic Central Water mass. The warm Gulf Stream flows out of the Straits of Florida in a northerly direction along the edge of the continental slope. This current thus separates the more eutrophic shelf–slope regime waters from the oligotrophic waters of the Sargasso Sea. At Cape Hatteras, the nutrient-poor Gulf Stream veers to the northeast toward the Grand Banks, leaving the boundary constraints of the slope, and meanders over abyssal depths and the New England Seamount chain (Stommel, 1966). The Gulf Stream current extends to the bottom

situ and Associated Environmental Characteristics
Stations [Adapted from Smith (1978)]

Benthic Abundance (No./m²)	Benthic Biomass (mg wet wt./m²)	Abundance: Biomass (No./mg wet wt.)	Sediment Organic Carbon (mg C/g dry wt.)	Sediment Organic Nitrogen (mg N/g dry wt.)	Organic Carbon: Organic Nitrogen	Particulate Organic Carbon Flux (g C/m²·yr)
3218	9450	0.3	10.0	1.1	9.1	
22,988	556	41.4	12.1	1.5	8.0	6.3[a]
8764	2143	4.1	13.3	1.6	8.3	2.3[a]
2146	653	3.3	9.1	1.1	8.3	2.3[a]
1632	771	2.1	13.0	0.9	14.4	4.2[a]
753	220	3.4	0.8	0.1	7.5	
285	180	1.6	6.9	0.7	9.8	
117	78	1.5	6.4	0.9	7.1	0.7[f]
259	142	1.8	6.4	0.9	7.1	0.7[f]
752[d]			15.6[e]			5.4[b]

along portions of its route. The Western Boundary Undercurrent flows southward at about the 3000-m contour and crosses the Gulf Stream in the vicinity of Cape Hatteras (Amos et al., 1971; Richardson and Knauss, 1971; Richardson, 1977). Surface mixed layer depths in the western North Atlantic Water mass vary seasonally from a few meters in summer to 150 m in winter in this predominantly temperate region. Generally this layer deepens with distance from shore and reaches maximum depth in the northern Sargasso Sea (Schroeder, 1965). The winter mixing enriches the surface waters with subsurface nutrients essential for phytoplankton growth (Ryther and Menzel, 1960). This winter mixing occurs as far south as the convergence that is identified between 27 and 30°N latitude (Katz, 1969; Backus et al., 1969). South of the convergence are the permanently stratified nutrient-poor surface waters of the southern Sargasso Sea.

Underlying the western North Atlantic Water mass with increasing depth is the North Atlantic Deep Water, the Norwegian Sea Overflow Water, and the Antarctic Bottom Water (Wright and Worthington, 1970; Emery and Uchupi, 1972). Some of the Antarctic Bottom Water is redirected in a southerly flow by the Western Boundary Undercurrent (Amos et al., 1971).

The deep-sea sediments of the northwestern Atlantic are spatially variable and reflect various origins and processes of transport. Continental slope and rise sediments are primarily clayey silts and silty clays with the finer-grained clays distributed on the abyssal plains (Sanders et al., 1965; Emery and Uchupi, 1972). Sediment is transported in the deep basin by turbidity currents and the Western Boundary Undercurrent. Both contribute to the 500-m-thick nepheloid or sus-

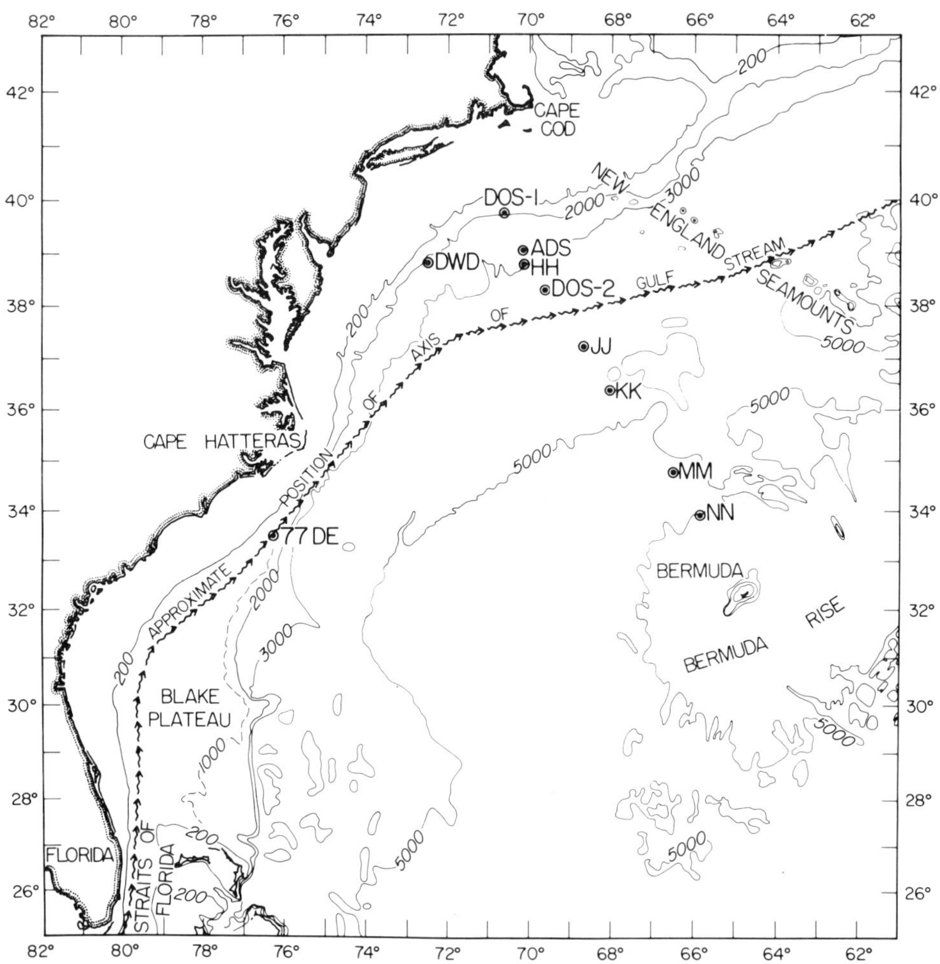

Fig. 2. Sediment community respiration stations in the northwestern Atlantic.

pended sediment layers typically found near the bottom below 3000-m depth (Eittreim et al., 1969).

Station Description

Sediment community respiration has been measured at 10 stations between depths of 1345 and 5200 m, which are representative of the primary topographic provinces from the continental slope to the abyssal plain (Fig. 2; Table I). Annual primary production of the surface waters overlying these stations generally decreases from the continental slope and rise stations toward the abyssal plain and Bermuda Rise (Table I). Available estimates of the amount of this surface productivity reaching the bottom range from about 1% at abyssal depths in the Sargasso Sea to as much as 6% on the slope south of Cape Hatteras (Table II). Radiolarians, diatoms, foraminiferans, and fecal pellets are dominant constituents of the sedimenting particulate matter at these stations (Rowe and Gardner, 1979; Hinga et al., 1979). This organic matter input to the benthic environment is probably aug-

TABLE II

Comparison of Annual Primary Productivity, Particulate Organic Carbon Flux, and Sediment Community Respiration Rates in Northwestern Atlantic (Respiration Rates Converted to g $C/m^2 \cdot yr$, Assuming a Respiratory Quotient of 0.85)

Station	Depth (m)	Annual Primary Productivity ($g\ C/m^2 \cdot yr$)	Annual Particulate Organic Carbon Flux ($g\ C/m^2 \cdot yr$)	Percent Primary Productivity in POC Flux to Sediment	Annual Sediment Community Respiration		Percent Particulate Organic Carbon Flux Utilized
					(liters $O_2/m^2 \cdot yr$)	($g\ C/m^2 \cdot yr$)	
77DE	1345	85	5.4[a]	6	11.48	5.25	97
DWD	2200	100	6.4[b]	6	4.03	1.84	29
ADS	2750	160	2.3[b]	1-3	3.07	1.40	61
			4.6[b]				30
HH	3000	160	2.3[b]	1-3	1.75	0.80	35
			4.6[b]				17
DOS-2	3650	100	4.2[b]	3-4	2.10	0.96	23
			3.8[a]				34
NN	5080	72	0.7[d]	0.5-1	0.61	0.28	40
			0.4[c]				78
MM	5200	72	0.7[d]	0.5-1	0.18	0.08	11
			0.4[c]				22

[a] Hinga et al. (1979).
[b] Rowe and Gardner (1979); there were two sequential deployments at the same site close to the depth of Stations ADS and HH.
[c] Honjo (1978).
[d] Deuser and Ross (1980).

mented at many of these stations by the sinking of larger parcels of food such as *Sargassum natans* and *S. fluitans* and the sea grass, *Thalassia testudinum* (Menzies et al., 1967; Menzies and Rowe, 1969; Schoener and Rowe, 1970). The high densities of deep-sea wood-boring mollusks in the vicinity of the Gay Head–Bermuda transect (Turner, 1973) also suggest the importance of large parcels as a food source to deep-sea benthic communities (Smith, 1978).

The bottom water at the abyssal stations of this region is uniformly low in temperature and high in dissolved oxygen, reflecting the general characteristics of the Antarctic Bottom Water and the Norwegian Sea Overflow Water (Table II). The Gulf Stream and slope water influence the temperature and dissolved oxygen concentrations for stations 77DE and DOS-1, respectively.

Sediments along the Gay Head–Bermuda transect grade from clayey silts at the shallower stations on the continental slope and rise to silty clays on the abyssal plain and Bermuda Rise (Sanders et al., 1965). North of the Blake Plateau (Station 77DE) the slope sediments are silts with a high calcareous content, attributable to coral and foraminiferans (Emery and Uchupi, 1972). Sediment organic carbon and nitrogen generally decrease with distance from shore with some anomalies along the continental rise probably caused by canyon funneling or turbidity currents. There is no evidence of reducing conditions in the surface sediments at any station along the Gay Head–Bermuda transect (Smith, 1978). High-sediment organic carbon is evident at the station north of the Blake Plateau. This may be due to its proximity to shore or the depositional character of fine sediments in an area influenced by the Gulf Stream and the Western Boundary Undercurrent.

Benthic infauna sampled at the stations along the Gay Head–Bermuda transect were dominated by polychaetes, nematodes, bivalves, and crustaceans. Generally, the polychaetes dominated larger size ranges and the nematodes the smaller fractions (Smith, 1978; Smith et al., 1978). Infaunal abundance and biomass generally decreased with increased depth and distance from shore (Table I). The anomalously high abundance and low biomass at Station DWD has been attributed to the funneling effect of the adjacent Hudson Canyon and the influence of waste dumping in the area (Smith, 1978). Infaunal abundance : biomass ratios generally decrease with depth, suggesting an increase in animal size with depth and distance from shore (Table I). Again, the anomalous DWD station exhibits an extremely high ratio, indicating that the infauna are small. The low infaunal abundance at Station 77DE, north of the Blake Plateau (Menzies and Rowe, 1968), when compared to the stations along the Gay Head–Bermuda transect, should be viewed with caution. This estimate of infaunal abundance at Station 77DE includes only the infauna retained by a 420-μm-mesh sieve. Smith (1978) found at stations DWD and ADS that a 420-μm sieve captured only 24% of the infauna by number caught by a 297-μm sieve. If this correction is applied to the 77DE station abundance values, an estimate of 3133 individuals/m^2 results, which is comparable to the infaunal abundance reported at the slope station (DOS-1) along the Gay Head–Bermuda transect (Table I). It must be kept in mind, however, that the above infaunal abundance and biomass do not include the foraminiferans and smaller biota that are also contributing members of the sediment community. An attempt to determine the biomass of the benthic foraminiferans, without tests, at Station DWD was made by Smith et al. (1978). They found that the foraminiferan biomass was greater than that of all the other taxa combined, suggesting the functional importance of this group in deep-sea sediment communities.

Sediment Community Respiration

Total oxygen uptake by the sediment community spans three orders of magnitude in the northwestern Atlantic. Along the Gay Head–Bermuda transect, respiration decreased with depth and distance from shore, reaching a low of 0.02 ml $O_2/m^2 \cdot hr$ at 5200 m in the Sargasso Sea northwest of Bermuda (Table II).

Smith (1978) developed a predictive equation for sediment community respiration along the Gay Head–Bermuda transect using available environmental parameters (Table I). The best fit equation was

$$\ln Y = 2.93 - 0.001D + 0.15(\mathrm{C:N}) + 0.11\frac{1}{\mathrm{N}} - 0.65T + 142.7\frac{1}{\mathrm{B}} \qquad (1)$$

where Y is the sediment community respiration, D is the depth in meters, C : N is the sediment organic carbon : nitrogen ratio, N is the sediment organic nitrogen in milligrams per gram dry weight of sediment, T is the bottom water temperature in degrees Celsius, and B is the biomass in milligrams wet weight per square meter. This equation accounts for 92.4% (r^2 = .924) of the variation in sediment community respiration along the Gay Head–Bermuda transect. Depth is the most important independent variable, contributing 83.1% of the variation. The importance of depth over the other parameters in predicting sediment community respiration must be kept in perspective because many of the other variables are interdependent.

Sediment community respiration at the continental slope station north of the Blake Plateau (77DE) was over twice as great as the shallowest slope station measurement (DOS-1) made on the Gay Head–Bermuda transect (Table I). The depth differential of 500 m could account for some of the variation. To test the predictive ability of the equation from the Gay Head–Bermuda transect, we have estimated the sediment nitrogen and biomass for Station 77DE based on the known sediment carbon and infaunal abundance values in Table II, using a C : N ratio of 9.1 and an abundance : biomass ratio of 0.3 with the previously calculated abundance of 3133 individuals/m^2. The predicted value is 1.53 and the observed value is 1.31 ml $O_2/m^2 \cdot hr$. The fit is reasonable, and it might be concluded, in the absence of much data, that the predictive value of the equation is good for that portion of the North American Basin examined to date north of the thermal front or convergence between 27 and 30°N. This convergence seems to be a physical and biological demarcation line. In terms of primary production (Ryther and Menzel, 1960) and mesopelagic fish distribution (Backus et al., 1969), the area south of this convergence is thought to be poorer. This impoverishment is probably reflected in the sediment community respiration as well, although this has not been examined.

Carbon Input and Utilization

An obvious question to ask is how much of the carbon fixed by primary producers at the surface reaches the bottom and how much of this is then utilized by the sediment community. The annual primary production at these stations in the northwestern Atlantic generally decreases with increased depth and distance from shore. As the water depth increases, the percent annual production reaching the bottom as particulate organic carbon generally decreases (Table II). This may be explained by either the increased efficiency of the water column food web in deeper, more depauperate water or just the increased complexity of the food web and the greater probability of food being ingested or broken down with increased exposure distance and travel time. Only a portion of the organic carbon that supposedly reaches the bottom is utilized by the sediment community (Smith, 1978). The accumulation rate of organic carbon in the sediments appears to be small when compared to the POC flux to the sediments and to the benthic respiration (Hinga et al., 1979). The proportion utilized by the sediment community generally decreases with depth and distance from shore and ranges from 97% at 1345 m (Station 77DE) down to 11% at 5200 m (Station MM) (Table II). It should be remembered that percentages listed in Table II must be considered approximate. Deuser and Ross (1980) and Deuser et al. (1981) have shown that the POC flux to the sediment varies by a factor of 2 to 3 during the year. In a discussion of the rather low utilization of the particulate organic carbon flux by the sediment community, Smith (1978) stresses that other integral members of the benthic boundary-layer community such as the epibenthic megafauna (holothurians, ophiuroids, asteroids, etc.) and benthopelagic animals are not considered in these measurements. That benthopelagic fishes and crustaceans congregate rapidly at bait drops (Hessler et al., 1972; Isaacs and Schwartzlose, 1975) and that they make considerable vertical excursions into the water column (Haedrich and Henderson, 1974; Pearcy and Ambler, 1974; Smith et al., 1979b) suggest that a substantial food requirement is necessary to sustain such activity. Although the respiration rate of animals such as the rattails *Coryphaenoides acrolepis* and *C. armatus* is significantly lower than that of comparable shallow-water species (Smith and Hessler, 1974, Smith, 1978), they could consume a significant proportion of the food energy reaching the bottom.

B. *Eastern Gulf of Mexico–Tongue of the Ocean–Straits of Florida*

General Description

This section combines three tropical–subtropical areas off the southeastern coast of the United States: the eastern Gulf of Mexico, the adjoining Straits of Florida, and the Tongue of the Ocean. For the purposes of this study, the eastern Gulf of Mexico is bounded on the north and east by western Florida and in the south by Cuba and the Yucatan Peninsula. The prominent topographic features of the area are the broad continental shelf and slope off western Florida and the Yucatan Peninsula (Emery and Uchupi, 1972) (Fig. 3). To the west of the Florida slope is the Mississippi Cone, which parallels the continental slope off western Florida and extends south to the Yucatan slope. To the west the Mississippi Cone gradually slopes to the Sigsbee Abyssal Plain. The Straits of Florida begin between Florida and Cuba, veer northeasterly along the east coast of Florida to the Blake Plateau, and are bordered on the east by the Bahama Banks. The straits are relatively shallow, having a maximum depth of 1000 m that slopes upward axially from west to north (Emery and Uchupi, 1972). Further east a northwesterly-tending deep steep-sided embayment, the Tongue of the Ocean, dissects the shallower Bahama Banks (Fig. 3).

Circulation of surface waters in the eastern Gulf of Mexico begins with a northerly influx through the Yucatan Strait between the Yucatan Peninsula and Cuba. Part of the water moves eastward into the Straits of Florida and becomes the Florida Current, the southern segment of the Gulf Stream system. Another

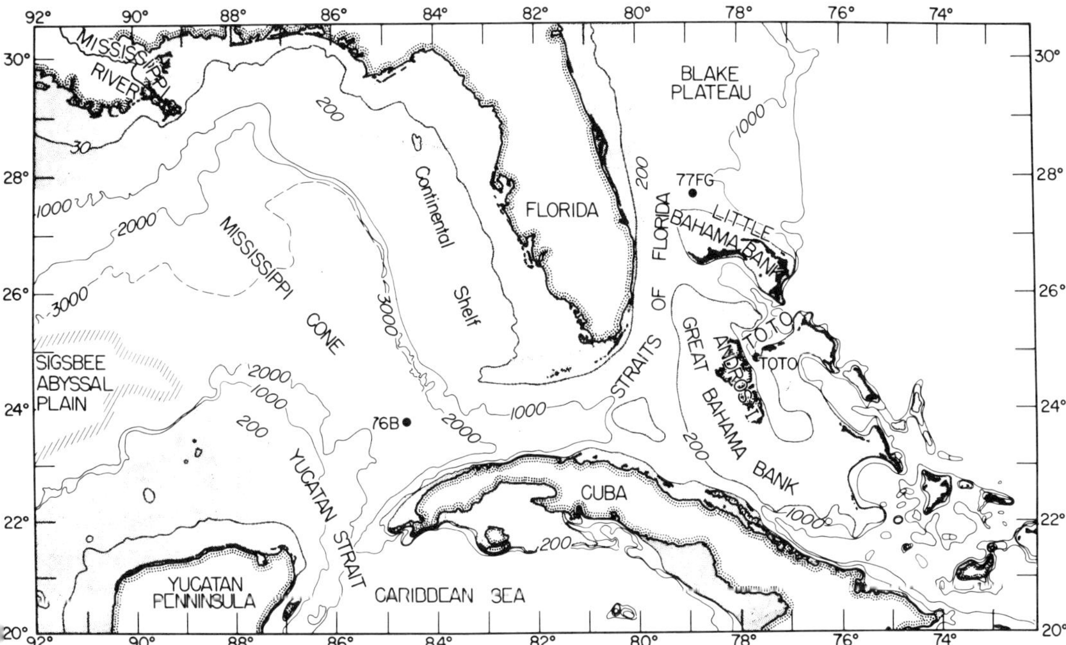

Fig. 3. Sediment community respiration stations in the eastern Gulf of Mexico–Straits of Florida–Tongue of the Ocean.

portion of the inflowing water moves northward across the Mississippi Cone and then southeasterly along the continental margin of Florida to join the Florida Current (Nowlin and McLellan, 1967). The sill depth of the Yucatan Strait (1500 to 1900 m) prevents the entry of all deeper water that is colder than 4°C (Nowlin, 1972). Hence all the deep water of the Gulf of Mexico originates as part of the North Atlantic Deep Water mass.

Sediments of the eastern Gulf of Mexico, Straits of Florida, and Tongue of the Ocean can generally be categorized as silty clays and generally have a high percentage of carbonate deposited from adjacent shelf reefs and platforms (Emery and Uchupi, 1972). A dominant source of the sediments in the eastern Gulf of Mexico is the Mississippi River, as evidenced by the large Mississippi Cone deposits.

A higher energy environment is found in the Straits of Florida. The sediments are coarse and of largely biogenic origin. Carbonate sands are found on the flanks of the straits, with patches of finer silts in the central axis. In the adjacent Tongue of the Ocean, sediments range from silty clays to clays with the finer sediments in the axial and cul-de-sac regions (Busby, 1962).

Station Description

We are concerned with three stations in this tropical region, one in the eastern Gulf of Mexico, one just outside the Straits of Florida north of the Bahamas, and one in the Tongue of the Ocean (Table III). Station 76B in eastern Gulf of Mexico is located at a depth of 3450 m at the southern extremity of the Mississippi Cone area due north of the Yucatan Strait (Fig. 3). Annual primary production in the

TABLE III

Sediment Community Respiration Measured *in situ* and Associated Environmental Characteristics of Eastern Gulf of Mexico–Tongue of the Ocean–Straits of Florida Stations

Station	Sediment Community Respiration (ml $O_2/m^2 \cdot hr$)	Number of Replicate Chambers	Depth (m)	Distance from Shore (km)	Annual Primary Productivity (g $C/m^2 \cdot yr$)	Bottom Water Temperature (°C)	Dissolved Oxygen Bottom (ml/liter)	Benthic Abundance (No./m^2)	Benthic Biomass (mg wet wt./m^2)	Abundance : Biomass (No./mg wet wt.)	Sediment Organic Carbon (mg C/g dry wt.)	Particulate Organic Carbon Flux (g $C/m^2 \cdot yr$)
77FG	2.95[b]	4	675	74	72[c]	10[b]	3.25[b]				10.8[b]	2.6[b]
TOTO	3.10[a]	6	2000	18	72[c]	4[a]	5.56[a]	102[e]			4.4[g]	2.1[a]
76B	0.69[b]	1	3450	220	50[d]	5[b]	4.16[b]	19[f]	31[f]	0.6	6.0[h]	

[a] Wiebe et al. (1976) and K. L. Smith (unpublished data).
[b] Hinga et al. (1979), with correction of organic carbon sedimentation rate values in Table 6 of that paper.
[c] Corcoran and Alexander (1963).
[d] Corwin (1969).
[e] J. F. Grassle (personal communication).
[f] Rowe et al. (1974).
[g] Busby (1962).
[h] Gromly and Sackett (1975) and Yemel'yanov (1975).

overlying surface water there is low, which is generally typical of the deeper regions of the Gulf of Mexico (Table III). Bottom water temperature and dissolved oxygen reflect the characteristics of the North Atlantic Deep Water mass that flows through the Yucatan Straits (Nowlin, 1972). The organic carbon content of the sediments and the low macrofauna abundance and biomass reflect the low surface productivity (Table III). The correlation of benthic biomass and surface productivity in the Gulf of Mexico was pointed out by Rowe et al. (1974), who analyzed data from a number of stations in both the eastern and western Gulf. They found that both abundance and biomass of macrobenthic invertebrates were lower in the Gulf of Mexico than in the northwestern Atlantic along the Gay Head–Bermuda transect.

The second station, just outside the Straits of Florida (Station 77FG), is located at the southern extremity of the Blake Plateau at a depth of 675 m (Fig. 3). Annual primary production in the vicinity of this station is low (Table III) and reflects the low nutrient concentration of the Florida Current water (Corcoran and Alexander, 1963) and nutrient-poor water of the equatorial current that flows along the eastern side of the Bahamas as the Antilles Current and joins the Florida Current in the vicinity of Station 77FG. The relatively high bottom water temperature and the lower dissolved oxygen concentration (Table III) are characteristic of the warm tropical water that borders the slope at this shallow depth. The particulate organic carbon flux to the bottom at Station 77FG was lower than that found at deeper slope depths in the northwestern Atlantic (Tables I and III). This organic sedimenting material contained diatoms, foraminiferans, and radiolarians. A significant input of pine pollen that probably originated from the Bahamas was also noted (Hinga et al., 1979).

Station TOTO was located in the axis of the Tongue of the Ocean about 18 km east of Andros Island. The annual primary productivity of the surface waters was estimated to be the same as at the nearby Straits of Florida station. However, at 2000-m depth, this station had a lower bottom water temperature and higher dissolved oxygen concentration (Table III). The organic carbon content of the sediments and benthic macroinfaunal abundance were low. This correlates with the low surface productivity and particulate organic carbon flux of the area (Table III). Identifiable organic material sedimenting to the bottom consisted of fecal pellets, some containing intact phytoplankton cells (Wiebe et al., 1976).

Sediment Community Respiration

Sediment community respiration in this tropical–subtropical region ranged from 0.69 ml $O_2/m^2 \cdot hr$ at the deep station in the eastern Gulf of Mexico to 3.10 ml $O_2/m^2 \cdot hr$ in the axis of the Tongue of the Ocean (Table III). The eastern Gulf of Mexico sediment community respiration was significantly higher than rates at a comparable depth (Station DOS-2) in the northwestern Atlantic (Table I). The influence of the diverging currents pouring through the Yucatan Straits and the ultimate convergence of the southeasterly current flowing along the Florida continental margin in close proximity to this station could partially explain this difference. This complex current pattern could serve as a concentrating mechanism for organic matter derived from river discharge and shallow-water sources along the western Florida continental shelf. However, other biological parameters such as the low primary productivity and benthic macrofaunal abundance and biomass do not support this premise. It must be kept in mind that such environmental parameters were not necessarily taken at the same station as the rate measurements but represent the best estimates available for the area. Hence, local phenomena reflected in sediment community respiration may not necessarily be reflected in the other parameters.

The Tongue of the Ocean station had a high rate of sediment community respiration that was significantly greater than even those stations beneath the eutrophic California Current in the northeastern Pacific (Table VI). A possible factor affecting this high rate was the close proximity to land, both to the east and west, and the probable supply of organic material derived from terrestrial and shallow-water sources. The channeling of such material into the axis of the canyon is an attractive explanation.

Carbon Input and Utilization

The annual particulate organic carbon flux at the Straits of Florida (Station 77FG) was similar to that measured in the Tongue of the Ocean. This seems reasonable since both stations have the same rate of annual primary productivity (Table III).

At both the Tongue of the Ocean and Straits of Florida stations, the particulate organic carbon flux was not sufficient to fulfill the food energy requirements of the sediment community as estimated from respiration. Sedimenting organic carbon supplied only 17% of that required by the sediment community in the Tongue of the Ocean and 22% of that required in the Straits of Florida (Table IV). This discrepancy was noted by Wiebe et al. (1976) and Hinga et al. (1979). Wiebe et al. (1976) suggested an additional organic carbon source such as turtle grass or other macroscopic plants. Submersible observations of turtle grass on the bottom in this vicinity (Rowe and Staresinic, 1979) verify that such material is present. Hinga et al. (1979) suggest that vertically migrating organisms may be responsible for carrying particulate material to the sediment communities, especially at Station 77FG. It is unlikely that macroscopic plants supply the entire magnitude of the discrepancy between POC flux and sediment community respiration at the TOTO station, especially if this station is representative of the entire Tongue of the Ocean.

C. Eastern Equatorial Atlantic Ocean

General Description

Two stations off the coast of northwestern Africa are of interest in this survey, both in the Gulf of Guinea (Fig. 4). At about 5°N, the coast of northwestern Africa turns sharply to the east. The major indentation of the continent at this point is known as the Gulf of Guinea and includes the body of water north of 5°S (Fig. 4). Along the northern boundary of the Gulf of Guinea (Ivory Coast) the continental shelf is 22 to 35 km wide with a shelf–slope break between 85- and 110-m depth (Martin, 1971). The continental slope in this area is dissected transversely by many small canyons. The upper slope has a gradient ranging from 2.5 to 14.5° with a break at 1400 to 1600 m; the lower slope declines at 2 to 3° (Martin, 1971). The Trou sans Fond Canyon off Abidjan tends in a southwesterly direction downslope and deeply carves the shelf and slope (Martin, 1971). Beyond the slope is the Guinea Continental Rise, which follows the general east–west contours between the Ivory Coast Rise and the Ivory Coast Escarpment (Emery et al., 1974b). The

Table IV

Comparison of Annual Primary Production, Particulate Organic Carbon Flux, and Sediment Community Respiration Rates in Eastern Gulf of Mexico–Tongue of the Ocean–Straits of Florida[a]

Station	Depth (m)	Annual Primary Production (g $C/m^2 \cdot yr$)	Annual Particulate Organic Carbon Flux (g $C/m^2 \cdot yr$)	Percent Primary Production in POC Flux to Sediment	Annual Sediment Community Respiration		Percent Respiration Supplied by Organic Carbon Flux
					(liters $O_2/m^2 \cdot yr$)	(g $C/m^2 \cdot yr$)	
77FG	675	72	2.6	4	25.84	11.81	22
TOTO	2000	72	2.1	3	27.16	12.41	17

[a]Respiration rates converted to g $C/m^2 \cdot yr$ assuming a respiration quotient of 0.85.

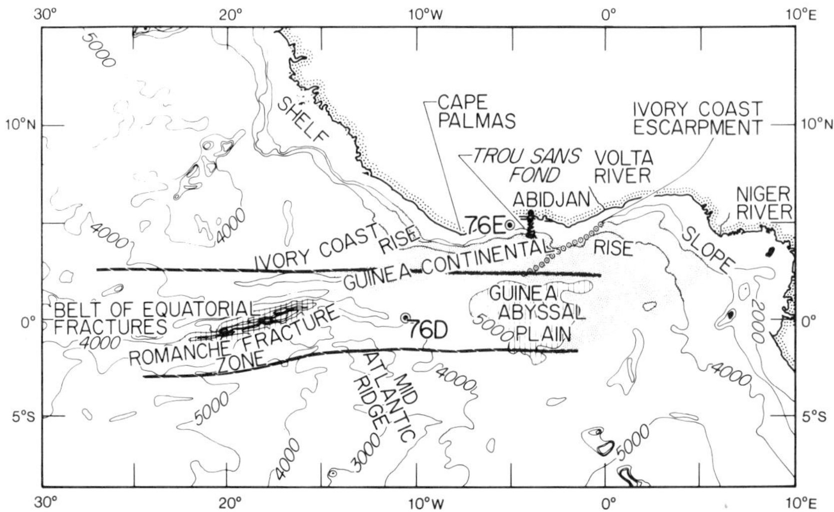

Fig. 4. Sediment community respiration stations in the eastern equatorial Atlantic.

rise is bounded on the south by a sequence of major fracture zones (Belt of Equatorial Fractures) that extend perpendicularly east from the Mid-Atlantic Ridge to the Guinea Abyssal Plain (Fig. 4). The Belt of Equatorial Fractures area has a highly irregular surface as a result of the underlying basement topography (Emery at al., 1974b).

The water masses and currents of the tropical eastern Atlantic are strongly influenced by southern Atlantic–formed water. A shallow surface layer, the Tropical Surface Water, lies between 10°S and 10°N across the Atlantic, creating a thermocline at 30 to 50 m in the eastern portion off Africa (Longhurst, 1962). Beneath this warm surface layer is the South Atlantic Central Water (SACW), distinguishable by its low salinity and low dissolved oxygen content. Surface water sinking at the Antarctic Convergence to the south forms the Antarctic Intermediate Water, which flows northward below the SACW at depths between 700 and 900 m (Lacombe, 1970). The North Atlantic Deep Water mass flows southward and in turn is underlain by the northerly Antarctic Bottom Water, which bathes the abyssal depths below 3500 m, where it gains access to the basins east of the Mid-Atlantic Ridge through the Romanche Fracture Zone area (Metcalf et al., 1964). In the Gulf of Guinea, Longhurst (1962) reports that local upwelling occurs along the Ivory Coast shelf in July and August, thus providing seasonal nutrient enrichment in this area also.

The continental margin of the Ivory Coast shows a complexity of sediment types. The shelf sediment ranges from coarse sands to silts (Martin, 1971). Easterly longshore currents deposit sediments in the Trou sans Fond Canyon, which, in turn, funnels them as turbidity currents downslope toward the continental rise. The sediments of the slope, continental rise, and abyssal plains are terrigenous silts and clays with 10 to 30% calcium carbonate content (Emery et al., 1974b). Sedimentation in the belt of equatorial fractures and the Mid-Atlantic Ridge is very patchy and is governed by local water currents.

Station Description

Two stations were occupied in the Gulf of Guinea (Table V). One was located at 278 m on the upper slope off the Ivory Coast west of the Trou sans Fond Canyon (Station 76E). The other was to the southwest in the belt of equatorial fractures at a depth of 4000 m (Station 76D) (Fig. 4). The upper slope station was relatively close to shore, situated below relatively high primary productivity surface waters enriched by both periodic upwelling and discharge from the Niger and Volta Rivers (Emery et al., 1974a). This enrichment is also reflected in high organic content of the terrigenous silt sediments at this station. The South Atlantic Central Water mass is over this station.

The second station in the Gulf of Guinea is 540 km from shore in a fracture zone of irregular relief with clay sediments. The surface productivity of the area is somewhat reduced in comparison to the inshore area but is higher than that found at stations in the northwestern Atlantic at comparable distances from shore. Emery et al. (1974a) refer to this area as a marine divergence with moderately abundant suspended material. Sediment organic carbon is low and comparable to that found in the northwestern Atlantic at abyssal depths (Tables I and V).

Sediment Community Respiration

As before, there is a decrease of sediment community respiration with increased depth (Table V). The influences of river discharge and the close proximity to land on sediment community respiration cannot be discounted at Station 76E. This station is probably enriched by the organic material of terrestrial and shallow-water origin that is transported by the longshore currents along the Ivory Coast shelf and slope. Station 76D, in the belt of equatorial fractures, has a relatively high sediment community respiration in comparison to that of stations of similar depth in the northwestern Atlantic (Table I). However, the rate is comparable to that found in the eastern Gulf of Mexico and northeastern Pacific (Tables III, V, and VI).

D. Eastern and Central Northern Pacific

General Description

The region of the northern Pacific of importance in this review is bordered on the west by the Emperor Seamount Chain, to the north by the Aleutian Island Chain, to the east by the North American continent, and to the south by the Hawaiian Islands and the Molokai Fracture Zone (Fig. 5). The general topography of the region within these boundaries is complex and marked by many irregularities such as fracture zones, troughs, and ridges. Depths of the abyssal areas not associated with these topographic features reach 6000 m (Fig. 5).

The Emperor Seamount Chain and its northerly extension, the Komandorskie Ridge, form the western boundary of our study area. Extending across the northern perimeter of the Pacific is the Aleutian Trench, which shoals in its eastern extremity in the Gulf of Alaska. The highly irregular coastline of western Canada is characterized by a broad continental shelf, slope, and rise. Further south, the North American continental margin narrows to the southwest toward the Mendocino Ridge (Fig. 5). The continental margin between this ridge south to Point Conception is highly irregular and narrows again in the vicinity of Monterey Bay, only to widen again along the southern California coast.

TABLE V

Sediment Community Respiration and Associated Characteristics of Eastern Equatorial Atlantic Stations

Station	Sediment Community Respiration (ml $O_2/m^2 \cdot hr$)	Number of Replicate Chambers	Depth (m)	Distance from Shore (km)	Annual Primary Productivity (g $C/m^2 \cdot yr$)	Bottom Water Temperature (°C)	Dissolved Oxygen Bottom (ml/liter)	Sediment Organic Carbon (mg C/g dry wt.)
76E	3.93[a]	2	278	40	159[b]	11[a]	1.01[a]	23.9[c]
76D	0.65[a]	4	4000	540	107[b]	2[a]	5.64[a]	4.8[c]

[a] Hinga et al. (1979).
[b] Mahnken (1969).
[c] Van Vleet and Quinn (1979).

The Patton Escarpment extends south from Point Conception to Baja California and delimits that broad portion of the continental margin known as the Southern California Continental Borderland. The topography of this region is made highly complex by a series of submarine basins with generally northwest–southeast-tending longitudinal axes, becoming progressively deeper to the southeast (Emery, 1960). These basins tend to have higher sedimentation rates shoreward as a result of their trapping of primarily terrestrial detritus. This process causes a gradual shallowing, broadening, and smoothing of the basin floors with decreased distance from shore. The Borderland extends west from shore 50 to 160 miles to the continental slope, which breaks at 1800 m (Emery, 1960), forming the Patton Escarpment. This precipitous escarpment, approximately 20 km wide, then drops to 3300 m, where it levels off into a narrow abyssal plain region. Further west the abyssal area becomes the Baja California Seamount Province, an east–west belt of irregular topography marked by seamounts, troughs, and ridges. This province lies between the Murray and Molokai Fracture Zones that intersect the continental margin at the Patton Escarpment. The irregular topography of this region when compared to the abyssal areas north of Point Conception is attributed to lower deposition rates of terrestrial and shallow-water material. This is due to decreased river discharge and to the trapping efficiency of the Borderland basins to the east, thus preventing turbidity current and slump transport seaward (Emery, 1960).

The Baja California Seamount Province extends westward toward the Hawaiian Island arc, showing less relief with increased distance from the continent (Fig. 5). The Molokai Fracture Zone, the southern boundary of this province, intersects the Hawaiian Islands at the western boundary of the study area. The Murray Fracture Zone forming the northern limit of this seamount province terminates in the north–south-trending Muscians Seamount region, which roughly parallels the Emperor Seamount Chain to the west. From the Hawaiian Island group, the Hawaiian ridge extends northwest, intersecting the Emperor Seamount Chain and thus closing the perimeter of our study area.

The central Pacific is characterized by a Central Water mass bounded on the north (above 40°N) by the Subarctic Water mass and to the south by the Equatorial Water mass, which occasionally reaches northward to Hawaii (Sverdrup et al., 1942). A permanent thermocline between 400 and 600 m separates this warm surface water mass from the underlying Intermediate Water formed in the subarctic. Northward the thermocline ascends to 100 m in the subarctic region (Masuzawa, 1972). The Intermediate Water has a pronounced oxygen minimum and phosphate maximum between 500 and 1200 m (Reid, 1965). Beneath the Intermediate Water is the Pacific Deep Water or Bottom Water, with salinity and temperature characteristics indicating formation in both the Norwegian–Greenland and Weddell Seas (Knauss, 1962; Reid and Lynn, 1971). The latter water mass is believed to upwell in the subarctic regions of the due northern Pacific (Wooster and Volkmann, 1960), eventually supplying nutrients to the photic zone.

The northern Pacific east of Hawaii consists of an Eastern Central Water mass bounded to the north by the Subarctic Water mass (Sverdrup et al., 1942). Subarctic water flows east as the Subarctic Current across the Aleutian and Tufts Abyssal Plains and then diverges to the north and south off Vancouver Island. The

TABLE VI
Sediment Community Respiration Measured *in situ* and Associated

Station	Sediment Community Respiration (ml O_2/m^2·hr)	Number of Replicate Chambers	Depth (m)	Distance from Shore (km)	Annual Primary Productivity (g C/m^2·yr)	Bottom Water Temperature (°C)	Dissolved Oxygen Bottom (ml/liter)
A-26	2.22[a]	4	1193	28	273[b]	4[a]	0.62[a]
SDT	2.40[b]	20	1230	26	273[b]	4[b]	0.71[b]
SCB	2.54	4	1300	40	273[b]	2	0.41
C-66	2.28[a]	2	3815	315	307[c]	2[a]	3.86[a]
F	0.70	4	4400	499	116	2	3.55
G	0.29	6	4900	1110	60	2	3.50
CNP	0.13	28	5900	3219	60	2	3.76

[a] Smith et al. (1979a).
[b] Smith (1974).
[c] Owen and Sanchez (1974).
[d] A. Soutar, K. Bruland, and W. T. Reed (unpublished data).

northerly component forms the Alaskan Current and the southerly component forms the California Current, which carries cold water along the North American continental margin to as far south as Baja California. There it is entrained in the westerly North Equatorial Current that can extend as far north as the Hawaiian Islands. Coastal upwelling of subsurface nutrient-rich water is associated with the California Current system as it is in other eastern boundary current areas. The upwelling is driven by north to northwest winds during the spring and summer; the strongest period of upwelling moves northward as the season progresses. The effects of this upwelling on surface water nutrients, phytoplankton, and zooplankton are manifest in waters up to 1000 km offshore (Reid et al., 1958). A deep countercurrent flows northwest along the southern California coastline below 200 m carrying Equatorial Pacific Water northward (Smith, R. L., 1968). The nutrient-rich upwelled water consists of subsurface subarctic water (Intermediate Water) and Equatorial Pacific Water (Reid et al., 1958).

The Pacific Deep or Bottom Water is believed to have a clockwise circulation from west to east entering the northern Pacific on the western side and flowing into the northern Pacific on the eastern side at depths below the Subarctic or Intermediate Water masses (Knauss, 1962). In contrast, Ewing and Connary (1970) hypothesized counterclockwise circulation of the bottom water, carrying a distinct suspended sediment load, a nepheloid layer, originating along the North American continental margin.

Station Description

Seven stations were occupied in a transect from southern California to the central northern Pacific (Fig. 5). The transect spans a variety of topographic provinces from the Southern California Continental Borderlands through the Baja California Seamount Province to the vicinity of the Muscians Seamount Chain. Near the coast the surface water primary productivity is high and is strongly

Environmental Characteristics of Northern Pacific Stations

Benthic Abundance (No./m^2)	Benthic Biomass (mg wet wt./m^2)	Abundance: Biomass (No./mg wet wt.)	Sediment Organic Carbon (mg C/g dry wt.)	Sediment Organic Nitrogen (mg N/g dry wt.)	Organic Carbon: Organic Nitrogen	Particulate Organic Carbon Flux (g C/m^2·yr)
7285[a]	8050[a]	0.9	30.5[a]	3.8[a]	8.0	9.8[d]
7285[a]	8050[a]	0.9	30.5[a]	3.8[a]	8.0	9.8[d]
9721	4014	2.4	58.2	4.2	13.8	
5274[a]	24190[a]	0.2	12.2[a]	1.7[a]	7.2	9.8[d]
1100	112.7	9.8	7.9	1.5	5.3	
213	57.3	3.7	4.2	0.5	8.3	
387	22.2	17.5	5.0	0.5	9.9	

influenced by the California Current (upwelling) system. In contrast, the primary productivity of the central Pacific is very low (Table VI). The nearshore annual particulate organic carbon flux, as determined from a 2-month collection by use of sedimentation traps suspended above the sediment–water interface, was estimated from values obtained from the San Nicholas Basin (Soutar, Bruland, and Reed, unpublished data), equidistant between the San Diego Trough and the Patton Escarpment (Table VI). These rates are higher than those obtained at similar depths but greater distance from shore in the northwestern Atlantic (Table I).

Three stations were located in the continental Borderland (SDT, A-26, SCB); two were in the axis of the San Diego Trough, and one was in the Santa Catalina Basin to the north (Fig. 5). Turbidity currents have a profound influence on the sediments of these basins, transporting predominantly stream-discharged terrestrial materials into them from submarine canyons (Emery, 1960; Shepard and Einsele, 1962). This source of sediments is far more important than hemipelagic deposition (Moore, 1969). The bottom water of the San Diego Trough and Santa Catalina Basin consists of Equatorial Pacific and Intermediate Water furnished from the south by the northerly flowing countercurrent. Bottom currents are insufficient to resuspend the fine sediments (Moore, 1969), which consist of silty clay of high organic carbon and nitrogen content (Table VI). The low dissolved oxygen concentration (Table VI) is due to the oxygen minimum of the Intermediate Water impinging on the Borderland at that depth and to local oxidation processes within the basins (Emery, 1960).

The abundance and biomass of the benthic macrofauna are high in the San Diego Trough and Santa Catalina Basin, as would be expected in a food-rich area. Biomass is comparable to that found at slope depths along the Gay Head–Bermuda transect (Table I; Station DOS-1).

Another station (C-66) off southern California is located on a relatively level

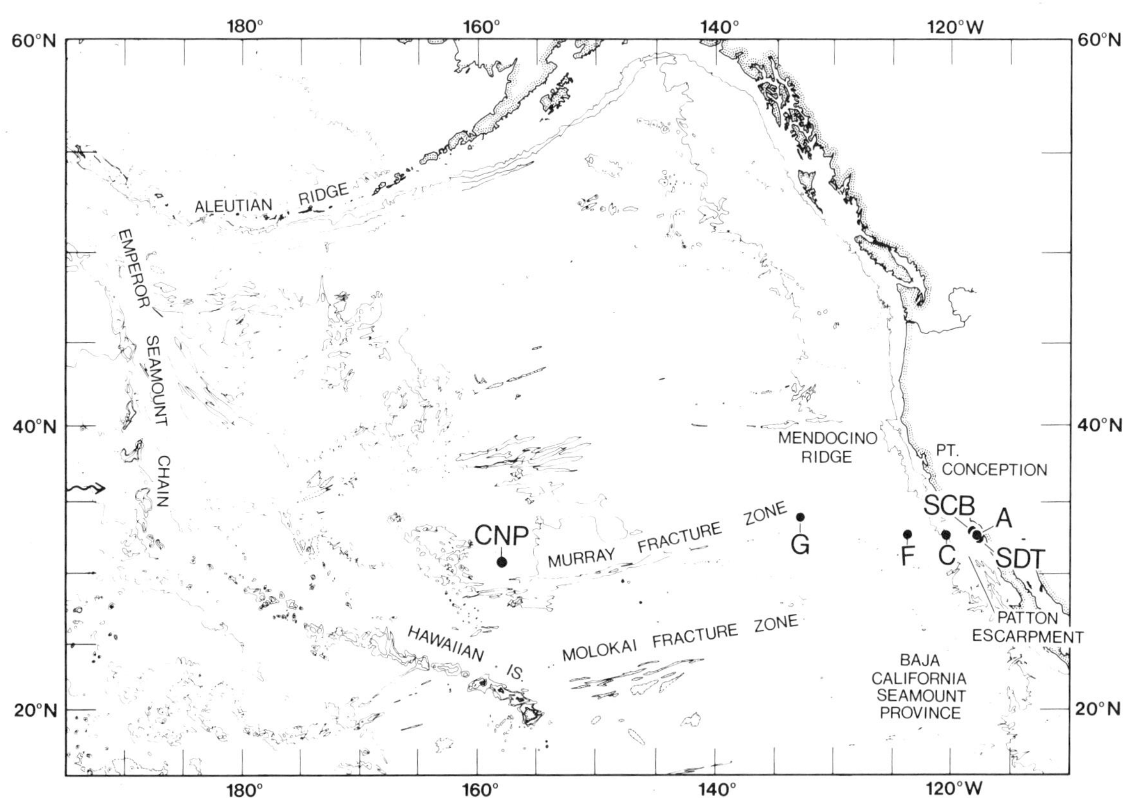

Fig. 5. Sediment community respiration stations in the northern Pacific.

abyssal area between the Patton Escarpment and the Baja California Seamount Province (Fig. 5). The overlying California Current (upwelling) has a primary productivity higher than that measured inshore over the Continental Borderland. Particulate organic carbon flux has been assumed to be equal offshore and over several basins in the Borderland (Table VI). Bottom water at Station C-66 had a dissolved oxygen content considerably higher than that found at the nearshore stations. The silty clay sediments have less than half the organic carbon and nitrogen of those in the San Diego Trough and Santa Catalina Basin.

The abundance and biomass of benthic macrofauna is lower at the abyssal station off the Patton Escarpment than in the San Diego Trough. Additionally, there is a decrease in the abundance to biomass ratio from Stations SDT and A-26 to Station C-66. This indicates an increase in animal size with increased depth.

Station F was located 499 km east of the southern California coast in the Baja California Seamount Province on a flat plain at a depth of 4400 m (Table VI, Fig. 5). The sediments were predominately clay. The sediment organic carbon, sediment nitrogen, biomass, and abundance were all of intermediate values between the nearshore stations and those stations further to the west. The sediments had the lowest carbon : nitrogen ratio of the Pacific stations, indicating a relative richness of nitrogen in the sediments (Table VI).

Station G was located east of the Murray Fracture Zone over 1000 km offshore in the northwestern portion of the Baja California Seamount Province at a depth of 4900 m. Station CNP was located north of the Murray Fracture Zone, 3200 km offshore, at 5900 m of depth just east of the Muscians Seamount Chain. Both these stations had very fine clay sediments with very low sediment organic carbon and sediment nitrogen concentrations. The primary production of the surface waters in the central Pacific is very low, as reflected in the very low abundance and biomass of the sediment organisms. The dissolved oxygen is typically low for the deep central Pacific (Table VI).

Sediment Community Respiration

Sediment community respiration values for the three borderland stations were all quite similar (Table VI). These values were nearly twice as high as the station (77DE) at similar depth in the northwestern Atlantic north of the Blake Plateau (Table I).

The Patton Escarpment station (C-66), even though 2600 m deeper and 270 km further from the coast than the three borderland stations, had a sediment community respiration equivalent to the three shallow stations (Table VI). This high respiration is likely a result of high primary productivity (307 g/cm^2·yr) associated with the California Current system.

The sediment community respiration decreased more than an order of magnitude as the distance offshore increased out to Stations F, G, and CNP. However, the three stations seaward of the California Current had respiration rates nearly an order of magnitude higher than comparable stations in the northwestern Atlantic (Tables I and VI). This difference cannot be explained by differences in primary productivity or macrofaunal biomass. The deep stations in the Pacific were under slightly lower primary productivity and had slightly lower macrofaunal biomass estimates than did the Atlantic stations. Since the sediment accumulation of organic carbon is small relative to the consumption of carbon (respiration) in both the Atlantic and Pacific, it would appear that an order of magnitude more organic carbon is being delivered to the deep Pacific than to the deep northwestern Atlantic, even though these regions have comparable primary productivity.

Smith et al. (in press) presented a predictive equation for sediment community respiration generated by use of a stepwise regression analysis with depth (Z) and macrofaunal abundance (MA) as independent variables. The equation

$$Y = 115.0 - 0.0200Z + 0.0043MA \quad (2)$$

accounts for 99.1% of the variability in sediment community respiration.

Carbon Input and Utilization

Flux estimates of the annual primary productivity reaching the bottom at stations off southern California (Stations SDT, A-26, C-66) ranged from 3% at the deeper station off the Patton Escarpment to 4% at the shallower San Diego Trough stations (Table VII). Sediment community respiration consumes 91 to 99% of this annual input. These estimates of sedimenting organic carbon utilization by the sediment community are comparable to that calculated for a station off the Blake Plateau (77DE) in the northwestern Atlantic (Table II) where 97% of sedimenting

TABLE VII

Comparison of Annual Primary Productivity, Particulate Organic Carbon Flux, and Sediment Community Respiration Rates in the Northeastern Pacific[a]

Station	Depth (m)	Annual Primary Productivity (g $C/m^2 \cdot yr$)	Annual Particulate Organic Carbon Flux (g $C/m^2 \cdot yr$)	Percent Primary Productivity in Sedimenting Carbon	Annual Sediment Community Respiration		Percent Particulate Organic Carbon Flux Utilized
					(liters $O_2/m^2 \cdot yr$)	(g $C/m^2 \cdot yr$)	
A-26	1193	273	9.8	4	19.45	8.95	91
SDT	1230	273	9.8	4	21.02	9.67	99
C-66	3815	307	9.8	3	19.97	9.19	94

[a] Respiration rates converted to g $C/m^2 \cdot yr$, assuming a respiratory quotient of 0.85.

carbon was utilized. With no other factors taken into consideration, these calculations would indicate that primary production is the main food source of the sediment community off southern California and that other organic carbon sources such as terrestrial and shallow-water material are not important. This reasoning is contrary to the evidence for the San Diego Trough and Santa Catalina Basin, where turbidity currents discharge terrestrial materials through the Trough (Emery, 1960; Shepard and Einsele, 1962). The importance of other consumers in the benthic boundary layer, such as the epibenthic megafauna and benthopelagic scavengers, must be considered in these estimates before a thorough evaluation of a carbon budget can be made.

4. Combined Analysis of Sediment Community Respiration

The next step in our analysis is to identify the environmental parameters that can be correlated with sediment community respiration and to develop a descriptive equation from the combined data. The subsequent step will then be to use the equation to predict sediment community respiration in the deep sea.

Attempts were made to utilize the data from the ETS and shipboard coring methods in conjunction with the *in situ* data to generate a predictive equation. Although various transformations and fitting procedures were employed to fit the sediment community data with the available environmental parameters, a satisfactory regression was not obtained. Further efforts were then limited to investigation of only the *in situ* data.

Sediment community respiration data were collected at 22 stations in the world ocean by use of *in situ* methods. However, the environmental parameters characterizing these sites were not available for all the sites. The variable most limiting was particulate organic carbon flux to the bottom, available for only 12 of the stations (based on sediment trap deployments at nine locations near sediment community respiration measurements). Deuser and Ross (1980) and Deuser et al. (1981) have shown that organic carbon flux to the bottom and surface primary productivity are directly correlated on a seasonal basis in the Sargasso Sea. Such a relationship has also been inferred by Eppley and Peterson (1979), who estimated new production from annual primary production. Other inferences of this relationship have been drawn from the sinking rates of herbivorous copepod fecal pellets containing phytoplankton carbon that has been identified in sedimentation trap samples taken at great depth (Honjo, 1978; Turner, 1977). On the basis of this rationale, simple regression and transformation techniques were utilized to create an equation to predict organic carbon flux from primary production. The resulting regression equation accounted for 70% (R^2 = .70) of the variation in organic carbon flux to the bottom. Conversion of annual primary production to new production following the equation proposed by Eppley and Peterson (1979) for use in the regression did not improve the fit. Considering the poor regression equation, the small size of the sedimentation trap data set, and the variability in the trap data, we decided not to use the sedimentation data in further analyses.

A correlation matrix for the 16 stations with all 11 other environmental variables showed considerable independence between variables (Table VIII).

Sediment community respiration and the 11 environmental parameters were regressed for the 16 stations with complete data. Stepwise regression analysis

TABLE VIII

Correlation Matrix for Variables Associated with *in situ* Measurements of Sediment Community Respiration Y for the 16 Stations with Complete Sets of Environmental Variables[a]

	Y	D	DFS	PP	T	DO	A	B	A : B	SOC	SON
D	−0.679[b]										
DFS	−0.421	0.707[b]									
PP	0.945[b]	−0.721[b]	−0.511[b]								
T	0.072	−0.529[b]	−0.456	0.141							
DO	−0.836[b]	0.425	0.060	−0.685[b]	0.101						
A	0.367	−0.644[b]	−0.407	0.367	0.156	−0.150					
B	0.683[b]	−0.362	−0.304	0.744[b]	0.065	−0.303	0.152				
A : B	−0.206	−0.002	0.214	−0.272	−0.175	0.153	0.681[b]	−0.278			
SOC	0.809[b]	−0.737	−0.415	−0.741[b]	0.077	−0.739[b]	0.425	0.246	−0.140		
SON	0.897	−0.806	−0.483	0.850[b]	0.250	−0.812[b]	0.468	0.371	−0.130	0.922[b]	
SOC : SON	0.106	−0.228	0.002	0.071	−0.067	−0.095	0.061	−0.126	−0.093	0.484	0.192

[a] Variables include depth (D), distance from shore (DFS), annual primary productivity (PP), bottom water temperature (T), dissolved oxygen (DO), macrofaunal abundance (A), macrofaunal biomass (B), abundance : biomass (A : B), sediment organic carbon (SOC), sediment organic nitrogen (SON), and sediment organic carbon : sediment organic nitrogen (SOC : SON).

[b] Significant correlations (≥.95).

(Statistical Analysis System, SAS Institute, Inc., Cary, NC) revealed an equation that accounted for 96% (R^2 = .9595) of the variation to be

$$Y = 0.3789 + 0.007577PP - 0.14692DO \quad (3)$$

where PP is annual primary productivity and DO is bottom water dissolved oxygen. The next step in the regression would further improve the fit to 97% of the variation by adding biomass to the equation. However, we chose equation 3 because of the questionable utility of a predictive equation with one more variable that is not readily available for most areas of the ocean. The residual values ($Y - \hat{Y}$) for the points entered in the regression (Table IX) show a good scatter and have a mean of absolute values of .13. Although this calculated regression line is a reasonable fit to the data, it must be remembered that there are significant (> 100%) differences between many of the sediment community respiration values predicted by equation 3 and the measured values. The predictions are

TABLE IX
Comparison of Observed Y and Predicted $\hat{Y}$ Sediment Community Respiration Measured *in situ* Using the Single Predictive Equation (2)[a]

Station	Y	$\hat{Y}$[c]	$Y - \hat{Y}$	PP	DO
77DE[b]	1.30	0.19	1.11	85	5.65
DOS-1	0.50	0.25	0.25	120	7.05
DWD	0.46	0.21	0.25	100	6.34
ADS	0.35	0.63	−0.28	160	6.52
HH	0.20	0.68	−0.48	160	6.15
DOS-2	0.21	0.18	0.03	100	6.54
JJ	0.09	−0.05	0.14	68	6.43
KK	0.04	0.01	0.03	68	6.04
NN	0.07	0.01	0.06	72	6.25
MM	0.02	0.02	0.0	72	6.15
SDT	2.40	2.34	0.06	273	0.71
A-26	2.22	2.36	−0.14	273	0.62
SCB	2.54	2.39	0.15	273	0.41
C-66	2.28	2.14	0.14	307	3.86
F	0.70	0.74	−0.04	116	3.55
G	0.29	0.32	−0.03	60	3.50
CNP	0.13	0.28	−0.15	60	3.76
TOTO[b]	3.10	0.10	3.00	72	5.56
76B[b]	0.69	0.15	0.54	50	4.16
76E[b]	3.93	1.43	2.50	159	1.01
76D[b]	0.65	0.36	0.29	107	5.64
77FG[b]	2.95	0.45	2.50	72	3.25

[a] Abbreviations: PP, primary productivity; DO, dissolved oxygen.
[b] Stations not used in the stepwise regression analysis.
[c] $\hat{Y} = 0.3789 + 0.007577\,PP - 0.14692DO$.

particularly poor for the six stations with incomplete environmental parameters that were not used in the regression.

Because of the previously noted discrepancy between sediment community respiration in the Atlantic and Pacific, we divided the data into two groups, the seven stations from the Pacific and the nine Atlantic stations (Gay Head–Bermuda transect), and a stepwise regression was run on each group. The resulting equation for the Pacific stations using the same regression analysis as for the overall equation (3) is

$$Y = 0.3508 - 0.0001142D + 0.007680PP \tag{4}$$

where D is the depth of the station and PP is the primary productivity. This equation accounts for 99% of the variance ($R^2 = .9965$). The equation for the stations on the Gay Head–Bermuda transect is

$$Y = 0.9421 - 0.0001621D - 0.001252PP \tag{5}$$

and accounts for 96% of the variance of these stations ($R^2 = .9646$).

Equations 4 and 5 provide much better prediction of sediment community respiration for each region than does the overall equation (3) (Table X). Equation 4 generated for the Pacific stations also provides an improvement in predictions for the stations not used in the stepwise regression analysis. The difference between the predictive equation (2) for the Pacific transect presented in Smith et al. (in press) and equation 4 are a result of slightly different data sets. Smith et al. used a random subsample of the individual measurements, whereas equation (2) is an averaging of all the chambers.

Thus, to predict the sediment community oxygen consumption for areas of the ocean, we could choose either an overall equation (3) with a modestly good fit or use two equations [(4) and (5)] that better fit the data. We have an insufficient data base on which to make more than an approximate estimation of sediment community respiration for areas of the ocean for which no measurements exist. We cannot at this time explain the differences between regions.

5. Global Estimate of Sediment Community Respiration

Equation 3 was used to estimate the sediment community respiration for the world ocean excluding shallow areas of less than 200-m depth. To generate such an estimate, it was necessary to have the global distribution of annual primary productivity and bottom-water dissolved oxygen. Distribution of annual primary productivity in the world ocean was determined from data given by Koblentz-Mishke et al. (1970) and Platt and Rao (1975) (Fig. 6). Dissolved oxygen of the bottom water was estimated primarily from data due to Kuo and Veronis (1973) and Emery and Uchupi (1972) (Fig. 7).

The global distribution of the two variables, annual primary productivity, and dissolved oxygen were produced on equal area projections, superimposed, and square area estimates were made with a planimeter for all the various overlapping combinations. The Arctic Ocean and other adjacent seas were not included in these initial estimates because of insufficient data. Sediment community respiration was then calculated by using a predictive equation (3). The mean dissolved oxygen and the lower primary productivity range estimates were used in these

TABLE X
Comparison of Observed Y and Predicted $\hat{Y}$ Sediment Community Respiration Measured *in situ* Using Two Predictive Equations (3 and 4)[a]

Station	Y	$\hat{Y}$	$Y - \hat{Y}$	PP	Depth
Northwestern Atlantic: $\hat{Y} = 0.9421 - 0.0001624\ Depth - 0.001252\ PP$					
77DE[b]	1.31	0.62	0.69	85	1345
DOS-1	0.50	0.50	0	120	1850
DWD	0.46	0.47	−0.01	100	2200
ADS	0.35	0.30	0.05	160	2750
HH	0.20	0.26	−0.06	160	3000
DOS-2	0.21	0.23	−0.02	100	3650
JJ	0.09	0.11	−0.02	68	4670
KK	0.04	0.08	−0.04	68	4830
NN	0.07	0.04	0.03	72	5080
MM	0.02	0.02	0	72	5200
Pacific and Other Areas: $\hat{Y} = 0.3508 - 0.0001142\ Depth + 0.007680\ PP$					
A-26	2.22	2.32	−0.10	273	1193
SDT	2.40	2.31	0.09	273	1230
SCB	2.54	2.30	0.14	273	1300
C-66	2.28	2.29	−0.01	307	3815
F	0.70	0.76	−0.06	116	4400
G	0.29	0.27	0.02	60	4900
CNP	0.13	0.16	−0.03	60	5900
TOTO[b]	3.10	0.68	2.42	72	2000
76B[b]	0.69	0.36	0.33	50	3450
76E[b]	3.93	1.54	2.39	159	278
76D[b]	0.65	0.73	−0.08	107	4000
77FG[b]	2.95	0.83	2.12	72	675

[a] Abbreviation: PP, primary productivity.
[b] Stations not used in the stepwise regression analysis.

calculations. The calculated sediment community respiration was 1.92×10^{15} liters O_2/yr as corrected for the surface area of adjacent seas in excess of 200-m depth for a total global square area of $334.9 \times 10^6\ km^2$ (Menard and Smith, 1966) (66% of the Earth's surface). This global sediment community respiration was not corrected for the surface area of the deep-sea bottom covered by hard substrate such as that found along mid-ocean ridges. Primary productivity calculated on a global scale, using the lower range values from Fig. 6, was 33.8×10^{15} g C/yr for the same area.

Global sediment community respiration of 1.92×10^{15} liters O_2/yr, when converted to carbon, assuming a respiratory quotient of 0.85, is 0.88×10^{15} g C/yr. Thus the benthic respiration accounts for 2.6% of the global primary production of carbon. This is within the range of values for the sediment trap measurements of the flux of organic carbon to the ocean floor (see Tables II, IV, and VII). This

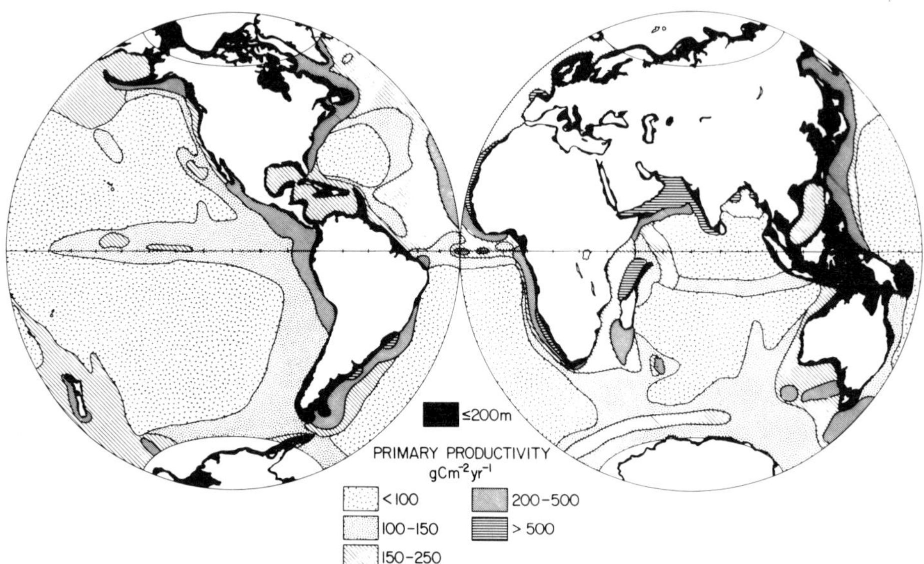

Fig. 6. Distribution of primary productivity over depths exceeding 200 m in the World Ocean.

suggests that most of the organic carbon settling to the deep-sea floor is utilized and does not accumulate.

In examining these calculations, two major questions come to mind: (1) what other consumers of sedimented organic matter exist in the benthic boundary layer and (2) what other possible food sources are available to the benthic boundary-layer organisms.

Some of the sedimenting particulate organic carbon, not including that small portion buried in the sediments, must be available to other biological components of the benthic boundary layer such as the epibenthic megafauna, benthopelagic animals, zooplankton, and water column microbiota. The food energy requirements of these populations are largely unknown and represent an important area for future study.

Sedimentation traps measure only the flux of small (i.e., fecal pellet size) particulate matter to the bottom. Other potential food sources include carcasses of dead pelagic animals (Clarke and Merrett, 1972), macrophyte detritus (Menzies et al., 1967; Menzies and Rowe, 1969; Schoener and Rowe, 1970), prey species occupying the benthic boundary layer, and dissolved organic carbon. The importance of these various food sources to the sediment community or to the overall benthic boundary layer is unknown. The rapid consumption of large parcels of bait by benthopelagic and epibenthic scavengers on the deep-sea floor (Hessler et al., 1972; Isaacs and Schwartzlose, 1975; Jannasch and Wirsen, 1977) suggests the importance of carrion as a food source. Macrophytes such as *Thalassia testudinum, Sargassum natans,* and *Sargassum fluitans* have been found on the bottom at abyssal depths in the northwestern Atlantic (Menzies et al., 1967; Menzies and Rowe, 1969). The pelagic phaeophyte alga, *Sargassum,* has been found in the gut contents of epibenthic ophiuroids along the Gay Head–Bermuda transect (Schoener and Rowe, 1970). Gut contents of benthopelagic fish and zoo-

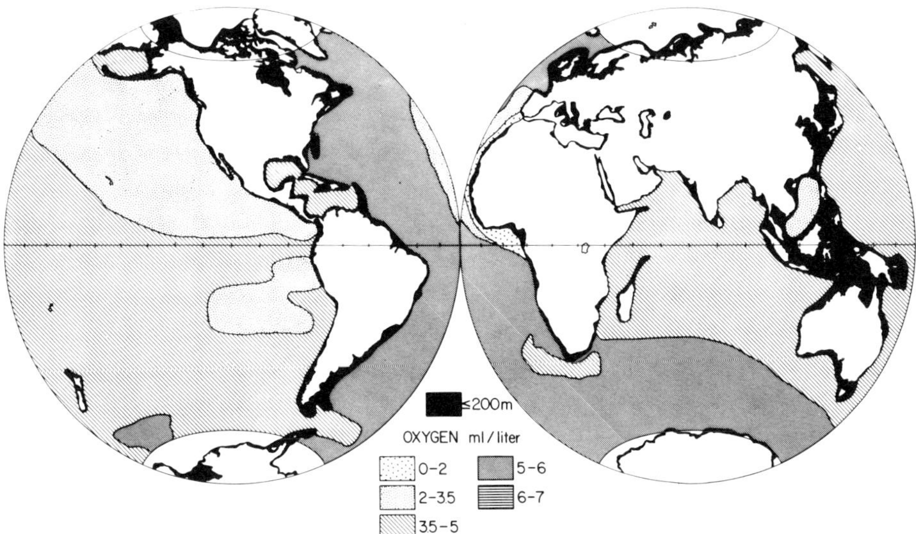

Fig. 7. Distribution of bottom water dissolved oxygen at depths exceeding 200 m in the World Ocean.

plankton suggest feeding on both pelagic and benthic prey (Haedrich and Henderson, 1974; Pearcy and Ambler, 1974; Barnes et al., 1976; McLellan, 1977; Smith et al., 1979b). The "swimming" ability of deep-sea infauna, such as the polychaete *Biremis blandi* (Polloni et al., 1973), also indicates trophic exchange between the sediment community and other component populations of the benthic boundary layer. Finally, heterotrophic bacteria are believed to be primary users of dissolved organic carbon in the deep sea (Williams and Carlucci, 1976).

The global estimate of sediment community respiration raises far more questions than are answered. Future studies concerning the functioning of the deep-sea benthic boundary-layer system within the overall open ocean ecosystem should include measuring activity rates (e.g., metabolism, mobility) of major biological components (benthopelagic animals, epibenthic megafauna, plankton, etc.) and quantitatively evaluating other food inputs to the system from the overlying water column.

Acknowledgments

Support for writing this review was provided by NSF (OCE-78-08640, OCE-81-17661) and Sandia National Laboratories. We thank many colleagues for critically reviewing the manuscript and Susan Hamilton, Roberta Baldwin, and Gil Rowe for editorial comments. We also thank Cindy Jones for her assistance.

References

Amos, A. F., A. L. Gordon, and E. D. Schneider, 1971. Water masses and circulation patterns in the region of the Blake–Bahama outer ridge. *Deep-Sea Res.*, **18**, 145–165.

Backus, R. H., J. E. Craddock, R. L. Haedrich, and D. L. Shores, 1969. Mesopelagic fishes and thermal fronts in the western Sargasso Sea. *Mar. Biol.*, **3**, 87–106.

Barber, R. T., 1972. *Data Report for R/V Eastward Cruise E-12-72, July 3–8, 1972*. Duke University Marine Laboratory Data Report No. 72-1.

Barnes, A. T., L. B. Quetin, J. J. Childress, and D. L. Pawson, 1976. Deep-sea macroplanktonic sea cucumbers: Suspended sediment feeders captured from deep submergence vehicle. *Science*, **194,** 1083–1085.

Busby, R. F., 1962. *Submarine Geology of the Tongue of the Ocean, Bahamas*. U.S. Navy Oceanography Office Technical Report No. TR-108, 84 pp.

Childress, J. J., 1977. Physiological approaches to the biology of midwater organisms. In *Oceanic Sound Scattering Prediction*, N. R. Andersen and B. J. Zahuranec, eds. Plenum Press, New York, pp. 301–324.

Christensen, J. P. and T. T. Packard, 1977. Sediment metabolism from the northwest African upwelling system. *Deep-Sea Res.*, **24,** 331–343.

Clarke, M. R. and N. R. Merrett, 1972. The significance of squid, whale and other remains from the stomachs of bottom-living deep sea fish. *J. Mar. Biol. Assoc. U. K.*, **52,** 589–603.

Corcoran, E. F. and J. E. Alexander, 1963. Nutrient, chlorophyll and primary production studies in the Florida Current. *Bull. Mar. Sci.*, **13,** 527–541.

Corwin, N., 1969. *Reduced Data Reports for Atlantis II: 31, 42 and 48*. Woods Hole Oceanographic Institution Technical Report No. 69-20 (unpublished manuscript).

Davies, J. M., 1975. Energy flow through the benthos in a Scottish sea loch. *Mar. Biol.*, **31,** 353–362.

Deuser, W. G. and E. H. Ross, 1980. Seasonal change in the flux of organic carbon to the deep Sargasso Sea. *Nature*, **283,** 364–365.

Deuser, W. G., E. H. Ross, and R. F. Anderson, 1981. Seasonality in the supply of sediment to the deep Sargasso Sea and implications for the rapid transfer of matter to the deep ocean. *Deep-Sea Res.*, **28A,** 495–505.

Eittreim, S., M. Ewing, and E. M. Thorndike, 1969. Suspended matter along the continental margin of the North American Basin. *Deep-Sea Res.*, **16,** 613–624.

Emery, K. O., 1960. *The Sea Off Southern California: A Modern Habitat of Petroleum*. Wiley, New York, 366 pp.

Emery, K. O. and E. Uchupi, 1972. *Western North Atlantic Ocean: Topography, Rocks, Structure, Water, Life and Sediments*. American Petroleum Geology Memoir No. 17, 532 pp.

Emery, K. O., E. Uchupi, J. Phillips, C. Bowin, and J. Mascle, 1974a. Suspended matter and other properties of the surface waters of the northeastern Atlantic Ocean. *J. Sed. Petrol.*, **44,** 1087–1110.

Emery, K. O., E. Uchupi, J. Phillips, C. Bowin, and J. Mascle, 1974b. *The continental margin off Western Africa: Angola to Sierra Leone*. Woods Hole Oceanographic Institution Technical Report No. 74-99, 152 pp.

Eppley, R. W. and B. J. Peterson, 1979. Particulate organic matter flux and planktonic new production in the deep ocean. *Nature*, **282,** 677–680.

Ewing, M. and S. D. Connary, 1970. Nepheloid layer in the North Pacific. In *Geological Investigations of the North Pacific*, J. D. Hays, ed., Geol. Soc. Am. Mem. No. 126, pp. 41–82.

Gromly, J. R. and W. M. Sackett, 1975. Carbon isotope evidence for the maturation of marine lipids. In *Advances in Organic Geochem.*, R. Campos and J. Goni, eds. Enadisma Servicio de Publicaciones, Madrid, pp. 321–339.

Haedrich, R. L., 1974. Pelagic capture of the epibenthic rattail, *Coryphaenoides rupestris*. *Deep-Sea Res.*, **21,** 977–979.

Haedrich, R. L. and N. R. Henderson, 1974. Pelagic food of *Coryphaenoides armatus* a deep benthic rattail. *Deep-Sea Res.*, **21,** 739–744.

Hessler, R. R., J. D. Isaacs, and E. L. Mills, 1972. Giant amphipod from the abyssal Pacific Ocean. *Science*, **175,** 636–637.

Hinga, K. R., J. McN. Sieburth, and G. R. Heath, 1979. The supply and use of organic material by the deep-sea benthos. *J. Mar. Res.*, **37,** 557–579.

Honjo, S., 1978. Sedimentation of materials in the Sargasso Sea at a 5367 m deep station. *J. Mar. Res.*, **36,** 469–492.

Honjo, S., 1980. Material fluxes and modes of sedimentation in the mesopelagic and bathypelagic zones. *J. Mar. Res.*, **38,** 53–97.

Isaacs, J. D. and R. A. Schwartzlose, 1975. Active animals of the deep-sea floor. *Sci. Am.*, **233,** 84–91.

Jannasch, H. W. and C. O. Wirsen, 1977. Microbial life in the deep sea. *Sci. Am.*, **236,** 42–52.

Katz, E. J., 1969. Further study of a front in the Sargasso Sea. *Tellus*, **21,** 259–269.

King, F. D. and T. T. Packard, 1975. The effect of hydrostatic pressure on respiratory electron transport system activity in marine zooplankton. *Deep-Sea Res.*, **22,** 99–105.

Knauss, J. A., 1962. On some aspects of the deep circulation of the Pacific. *J. Geophys. Res.*, **67,** 3943–3954.

Koblentz-Mishke, O. J., V. V. Vokovinsky, and J. G. Kabanova, 1970. Plankton primary production of the world ocean. In *Scientific Exploration of the South Pacific*, W. S. Wooster, ed. National Academy of Science, pp. 183–193.

Kuo, H. H. and G. Veronis, 1973. The use of oxygen as a test for an abyssal circulation model. *Deep-Sea Res.*, **20,** 871–888.

Lacombe, H., 1970. Physical oceanography of the eastern boundary current of the Atlantic Ocean. In *The Geology of the East Atlantic Continental Margin*, F. M. Delany, ed. Institute of Geology, ICSU/SCOR Working Party, 31st Symposium, Cambridge, 1970, pp. 47–65.

Longhurst, A. R., 1962. A review of the oceanography of the Gulf of Guinea. *Bull. de l'Ifan*, **24,** Series A.

Mahnken, C. V. W., 1969. Primary organic production and standing stock of zooplankton in the tropical Atlantic Ocean. Equalant I and II. *Bull. Mar. Sci.*, **19,** 550–556.

Martin, L., 1971. The continental margin from Cape Palmas to Lagos: Bottom sediments and submarine morphology. In *The Geology of the East Atlantic Continental Margin*, F. M. Delany, ed. Institute of Geology, ICSU/SCOR Working Party, 31st Symposium, Cambridge, 1970, pp. 79–95.

Masuzawa, J., 1972. Water characteristics of the north Pacific central region. In *Kuroshio: Physical Aspects of the Japan Current*, H. Stommel and K. Yoshida, eds. University of Washington Press, Seattle, pp. 95–127.

McLellan, T., 1977. Feeding strategies of the macrourids. *Deep-Sea Res.*, **24,** 1019–1036.

Menard, H. W. and S. M. Smith, 1966. Hypsometry of ocean basin provinces. *J. Geophys. Res.*, **71,** 4305–4325.

Menzies, R. J. and G. T. Rowe, 1968. The LUBS, a large undisturbed bottom sampler. *Limnol. Oceanogr.*, **13,** 708–714.

Menzies, R. J. and G. T. Rowe, 1969. The distribution and significance of detrital turtle grass, *Thalassia testudinum*, on the deep sea floor off North Carolina. *Int. Revue Ges. Hydrobiol.*, **54,** 217–222.

Menzies, R. J., R. Y. George, and A. Z. Paul, 1974. The effects of hydrostatic pressure on living aquatic organisms. 4. Recovery and pressure experimentation on deep-sea animals. *Int. Revue Ges. Hydrobiol.*, **59,** 187–197.

Menzies, R. J., J. A. Zanveld, and R. M. Pratt, 1967. Transported turtlegrass as a source of organic enrichment of abyssal sediments off North Carolina. *Deep-Sea Res.*, **14,** 111–115.

Metcalf, W. G., B. C. Heezen, and M. C. Stalcup, 1964. The sill depth of the mid-Atlantic ridge in the equatorial region. *Deep-Sea Res.*, **11,** 1–10.

Moore, D. G., 1969. Reflection profiling studies of the California Continental Borderland: Structure and quaternary turbidite basins. *Spec. Pap. Geol. Soc. Am.*, **107,** 1–42.

Murray, J. W. and V. Grundmanis, 1980. Oxygen consumption in pelagic marine sediments. *Science*, **209,** 1527–1530.

Nowlin, W. D., Jr., 1972. Winter circulation patterns and property distributions. In *Contributions on the Physical Oceanography of the Gulf of Mexico*, L. R. A. Capurro and J. S. Reid, eds. Texas A&M University Oceanographic Studies, 2. Gulf Publishing Company, Houston, TX, pp. 3–51.

Nowlin, W. D. and H. J. McLellan, 1967. A characterization of the waters of the Gulf of Mexico in winter. *J. Mar. Res.*, **25,** 29–59.

Owen, R. W., Jr. and C. K. Sanchez, 1974. *Phytoplankton Pigment and Production Measurements in the California Current Region, 1969–72.* National Marine Fisheries Service Data Report No. 91, 185 pp.

Owens, T. G. and F. D. King, 1975. The measurements of respiratory electron transport system activity in marine zooplankton. *Mar. Biol.*, **30**, 27–36.

Pamatmat, M. M., 1971. Oxygen consumption by the sea bed. IV. Shipboard and laboratory experiments. *Limnol. Oceanogr.*, **16**, 536–550.

Pamatmat, M. M., 1973. Benthic community metabolism on the continental terrace and in the deep-sea in the north Pacific. *Int. Revue Ges. Hydrobiol.*, **58**, 345–368.

Pamatmat, M. M., 1975. *In situ* metabolism of benthic communities. *Cah. Biol. Mar.*, **16**, 613–633.

Pamatmat, M. M., 1977. Benthic community metabolism: A review and assessment of present status and outlook. In *Ecology of Marine Benthos*, B. C. Coull, ed. University of South Carolina Press, Columbia, pp. 89–111.

Pamatmat, M. M. and A. M. Bhagwat, 1973. Anaerobic metabolism in Lake Washington sediments. *Limnol. Oceanogr.*, **18**, 611–627.

Pearcy, W. G., 1976. Pelagic capture of abyssobenthic macrourid fish. *Deep-Sea Res.*, **23**, 1065–1066.

Pearcy, W. G. and J. W. Ambler, 1974. Food habits of deep-sea macrourid fishes off the Oregon coast. *Deep-Sea Res.*, **21**, 745–759.

Platt, T. and D. V. Subba Rao, 1975. Primary production of marine microphytes. In *Photosynthesis and Productivity in Different Environments*. International Biology Program, Vol. 3, Cambridge University Press, New York, pp. 249–279.

Polloni, P. A., G. T. Rowe, and J. M. Teal, 1973. *Biremis blandi* (Polychaeta: Terebellidae), new genus, new species, caught by DSRV *Alvin* in the Tongue of the Ocean, New Providence. *Mar. Biol.*, **20**, 171–175.

Reid, J. L., Jr., 1965. Intermediate waters of the Pacific Ocean. *Johns Hopkins Oceanographic Studies*, Vol. 2. Johns Hopkins Press, Baltimore, 85 pp.

Reid, J. L. and R. J. Lynn, 1971. On the influence of the Norwegian–Greenland and Weddell seas upon the bottom waters of the Indian and Pacific Oceans. *Deep-Sea Res.*, **18**, 1063–1088.

Reid, J. L., Jr., G. I. Roden, and J. G. Wyllie, 1958. Studies of the California Current system. *Progr. Rep. Calif. Coop. Ocean. Fish. Invest.*, **6**, 28–57.

Richardson, P. L., 1977. On the crossover between the Gulf Stream and the Western Boundary Undercurrent. *Deep-Sea Res.*, **24**, 139–159.

Richardson, P. L. and J. A. Knauss, 1971. Gulf Stream and Western Boundary Undercurrent observations at Cape Hatteras. *Deep-Sea Res.*, **18**, 1089–1110.

Rowe, G. T. and W. Gardner, 1979. Sedimentation rates in the slope water of the northwest Atlantic Ocean measured directly with sediment traps. *J. Mar. Res.*, **37**, 581–600.

Rowe, G. T. and N. Staresinic, 1979. Sources of organic matter to the deep-sea benthos. *Ambio Spec. Rep.*, **6**, 19–27.

Rowe, G. T., P. T. Polloni, and S. G. Horner, 1974. Benthic biomass estimates from the northwestern Atlantic Ocean and the northern Gulf of Mexico. *Deep-Sea Res.*, **21**, 641–650.

Ryther, J. H. and D. W. Menzel, 1960. The seasonal and geographic range of primary production in the Western Sargasso Sea. *Deep-Sea Res.*, **6**, 235–238.

Sanders, H. L., R. R. Hessler, and G. R. Hampson, 1965. An introduction to the study of deep-sea benthic faunal assemblages along the Gay Head–Bermuda transect. *Deep-Sea Res.*, **12**, 845–867.

Sayles, F. L., T. R. S. Wilson, D. N. Hume and P. C. Mangelsdorf, Jr., 1973. *In situ* sampler for marine sedimentary pore waters: Evidence for potassium depletion and calcium enrichment. *Science*, **181**, 154–156.

Schoener, A. and G. T. Rowe, 1970. Pelagic *Sargassum* and its presence among the deep-sea benthos. *Deep-Sea Res.*, **17**, 923–925.

Schroeder, E. A., 1965. Average monthly temperatures in the North Atlantic Ocean. *Deep-Sea Res.*, **12**, 323–343.

Seki, H. and D. G. Robinson, 1969. Effect of decompression on activity of microorganisms in sea water. *Int. Revue Ges. Hydrobiol.*, **55**, 201–205.

Shepard, F. P. and G. Einsele, 1962. Sedimentation in San Diego Trough and contributing submarine canyons. *Sedimentology*, **1**, 81–133.

Smith, K. L., Jr., 1974. Oxygen demand of San Diego Trough sediments: An *in situ* study. *Limnol. Oceanogr.*, **19**, 939–944.

Smith, K. L., Jr., 1978. Benthic community respiration in the N. W. Atlantic: *In situ* measurements from 40 to 5200 meters. *Mar. Biol.*, **47**, 337–347.

Smith, K. L., Jr. and R. R. Hessler, 1974. Respiration of benthopelagic fishes: *In situ* measurements at 1230 meters. *Science*, **184**, 72–73.

Smith, K. L., Jr. and J. M. Teal, 1973. Deep-sea benthic community respiration: An *in situ* study at 1850 meters. *Science*, **179**, 282–283.

Smith, K. L., Jr., C. H. Clifford, A. H. Eliason, B. Walden, G. T. Rowe, and J. M. Teal, 1976. A free vehicle for measuring benthic community metabolism. *Limnol. Oceanogr.*, **21**, 164–170.

Smith, K. L., Jr., G. A. White, M. B. Laver, and J. A. Haugsness, 1978. Nutrient exchange and oxygen consumption by deep-sea communities: Preliminary *in situ* measurements. *Limnol. Oceanogr.*, **23**, 997–1005.

Smith, K. L., Jr., G. A. White, and M. B. Laver, 1979a. Oxygen uptake and nutrient exchange of sediments measured *in situ* using a free vehicle grab respirometer. *Deep-Sea Res.*, **26A**, 337–346.

Smith, K. L., Jr., G. A. White, M. B. Laver, R. R. McConnaughey, and J. P. Meador, 1979b. Free vehicle capture of abyssopelagic animals. *Deep-Sea Res.*, **26**, 57–64.

Smith, K. L., Jr. and R. J. Baldwin, in press. Deep-sea respirometry: *In situ* techniques. In *Handbook on Polarographic Oxygen Sensors: Aquatic and Physiological Applications*. H. Forstner and E. Gnaiger, eds. Springer-Verlag.

Smith, K. L., Jr., M. B. Laver, and N. O. Brown, in press. Sediment community oxygen consumption and nutrient exchange in the central and eastern North Pacific. *Limnol. Oceanogr.*

Smith, R. L., 1968. Upwelling. In *Oceanography and Marine Biology: An Annual Review*, H. Barnes, ed. Allen and Unwin, London, pp. 11–46.

Smith, W. O. and R. T. Barber, 1974. *An Investigation of the Chemical, Physical and Biological Properties of the Continental Shelf Between Cape Hatteras and Cape Lookout, August 5–10, 1974*. Duke University Marine Laboratory Data Report No. 74-3.

Smith, W. O. and T. J. Cowles, 1975. *An Investigation of the Chemical, Physical and Biological Properties of the North Carolina Continental Shelf, December 3–7, 1974*. Duke University Marine Laboratory Data Report No. 75-1.

Stommel, H., 1966. *The Gulf Stream: A Physical and Dynamical Description*. University of California Press, Berkeley, 248 pp.

Sverdrup, H. U., M. W. Johnson, and R. H. Fleming, 1942. *The Oceans: Their Physics, Chemistry and General Biology*. Prentice-Hall, Englewood Cliffs, NJ, 1087 pp.

Tietjen, J. H., 1971. Ecology and distribution of deep-sea meiobenthos off North Carolina. *Deep-Sea Res.*, **18**, 941–957.

Turner, J. T., 1977. Sinking rates of fecal pellets from the marine copepod *Pontella meadii*. *Mar. Biol.*, **40**, 249–259.

Turner, R. D., 1973. Wood-boring bivalves, opportunistic species in the deep sea. *Science*, **180**, 1377–1379.

Van Vleet, E. S. and J. G. Quinn, 1979. The diagenesis of marine lipids in ocean sediments. *Deep-Sea Res.*, **26**, 1225–1236.

Wiebe, P. H., S. H. Boyd, and C. Winget, 1976. Particulate matter sinking to the deep-sea floor at 2000 m in the Tongue of the Ocean, Bahamas, with a description of a new sedimentation trap. *J. Mar. Res.*, **34**, 341–354.

Williams, P. M. and A. F. Carlucci, 1976. Bacterial utilization of organic matter in the deep sea. *Nature*, **262**, 810–811.

Wooster, W. S. and G. H. Volkmann, 1960. Indications of deep Pacific circulation from the distribution of properties at five kilometers. *J. Geophys. Res.*, **65**, 1239–1249.

Wright, R. and L. V. Worthington, 1970. The water masses of the North Atlantic Ocean, a volumetric census of temperature and salinity. In *Serial Atlas of the Marine Environment*. American Geographic Society Folio No. 19.

Yayanos, A. A., 1981. Reversible inactivation of deep-sea amphipods (*Penelluld capenese*) by a decompression from 601 bars to atmospheric pressure. *Comp. Biochem. Physiol.*, **69A**, 562–565.

Yayanos, A. A., A. S. Dietz, and R. Van Boxtel, 1979. Isolation of a deep-sea barophilic bacterium and some of its growth characteristics. *Science*, **205**, 808–809.

Yemel'yanov, Y. M., 1975. Organic carbon in Atlantic sediments. *Doklady Akad. Nauk SSSR*, **220**, 220–223.

ZoBell, C. E., 1970. Pressure effects on morphology and life processes of bacteria. In *High Pressure Effects on Cellular Processes*, A. Zimmerman, ed. Academic Press, New York, pp. 85–130.

9. ZONATION OF FAUNA IN THE DEEP SEA

Robert S. Carney, Richard Lee Haedrich, and Gilbert T. Rowe

1. Introduction

In order of percentage of the Earth's surface, four great environmental gradients can be recognized. Over the entire Earth there are north–south (latitudinal) gradients with their associated climatic conditions. Second, there is depth and its related factors on all sloping parts of the ocean floor, from the intertidal across the abyssal plain to hadal trenches. The salinity gradients in estuaries, where freshwater and marine environments meet, and mountain "alpine" gradients throughout the world cover about the same area and rank third and fourth. The understanding of gradient-controlled distributions is of principal interest in the growing field of zoogeographic ecology (Pielou, 1980). The deep sea, as the second most expansive gradient, is an environment that can be studied to determine how large-scale gradients affect the structuring of communities and the shaping of evolution within taxa.

Throughout the oceans three depth-related faunal trends have been found so consistently that they can be treated as general phenomena. In brief form, they can be stated as follows:

1. The fauna of the ocean bottom changes in species composition with increasing depth. This change is so pronounced that taxa above the species level have restricted and circumscribed vertical ranges.
2. A diversity maximum occurs between 3000- and 4000-m depth (see Rex, this volume, Chapter 11 for a detailed discussion).
3. The standing stock of the meiofaunal to megafaunal size classes, as measured by biomass or counts, decreases rapidly with depth in a manner that resembles simple exponential decay. These are treated in Chapters 3 and 5 by Rowe and Thiel.

Elements of all three patterns were noted on the *Challenger* (Murray, 1895), but explanation of the causes of zonation in a seemingly homogeneous milieu has not attained the level of understanding as for biomass and diversity. This is due in large part to the ease with which the latter two can be parameterized, estimated, and explained. Both can be treated as simple variables in spite of underlying complex biological systems. As such, it is easy to map out diversity and biomass and to seek environmental relationships by inspection or more formal methods. Species zonation, however, unless limited to the tedious level of cataloging separate species' ranges, has no widely used definition or method of analysis. The patterns of biomass and diversity variation with depth are now sufficiently known that studies of processes can be initiated. Zonation, although well documented in many regions, is still in a more descriptive stage; experiments designed to explain its causes on an oceanic scale have not yet been devised.

In most of the great national expeditions that so typify the past century of ocean exploration, a major preoccupation was with questions of the vertical distribution of the deep ocean fauna, both pelagic and benthic. This interest was manifest in the publication of shelves of systematic monographs. Those from the *Challenger* alone totaled more than 40 volumes, and series based on the material from *Galathea, Dana, Vitjaz,* and others continue today. It was recognized early that the number of species in deep-sea assemblages was far greater than had been expected, and in this light the systematic monographs assumed a special importance.

Systematic work is time-consuming and, although cruise narratives could and often did draw general conclusions, it was not until decades after the original collecting that syntheses based on firm data bases began to appear. These tended to be divided along habitat and taxonomic lines (example, pelagic fish, benthic invertebrates, neritic copepods, etc.) establishing a practice still followed.

The classic synthesis is that by Ekman (1953), whose book is worldwide in scope. The emphasis is on the horizontal distribution of coastal and benthic invertebrates, but some data on fish and pelagic regions are included. LeDanois (1948) considered vertical distribution of the benthos off the coast of France. Zenkevitch (1963, 1969), Belyaev (1972), Vinogradova (1962), and Sokolova (1981) complete the general picture for the benthos through consideration of horizontal and vertical distribution in the deep sea and the trenches. Much of this has been summarized by Menzies et al. (1973). Ebeling (1967) and McGowan (1974) summarized much of the work on horizontal distribution in mesopelagic depths, and Parin (1968) did the same for epipelagic fish. Vinogradov (1968) deals with the vertical distribution of the oceanic zooplankton. A general synthesis for pelagic regions is provided by Van der Spoel and Pierrot-Bults (1979).

A number of themes are common to all these. Essentially descriptive, they stress zonal distributions aligned with the zonal climatic belts. Temperature is most often implied as a determinant, or at least the best correlate, for distribution. In the horizontal sense, therefore, oceanic distributions are shown to circle the globe in latitudinal bands that follow the isotherms. In the vertical sense, distributions are layered in accordance with the thermally layered structure of the ocean. The oceanic environment is thought to be very uniform and stable over extremely wide areas; in accordance with this, many species are horizontally very widespread. The notion of stability extends to time, too, and was especially popular in the early periods of deep-ocean exploration when it was thought that ancient species might be found in the depths. Because the gradient in physical parameters—particularly temperature—is quite steep vertically, species are not widespread in this sense and faunas replace one another successively with depth.

Basic to all these ideas, of course, is the notion of pattern. It is taken for granted that patterns both exist and persist in the communities and assemblages of deep-sea animals. This assumption becomes the point of departure for much scientific debate concerning the nature of deep-ocean life and provides the rationale for the application of statistical methodologies contrived to reveal pattern. Yet the data that underlie such an assumption are, for the deep sea, slender indeed and may apply at only gross levels. We attempt to address this difficult matter, although our major concern is with broad trends. Even considering the matter from quite different perspectives and on very different scales, we all agree that far more data are needed, particularly about the natural history of any deep-sea organism.

Concern with the distribution patterns assumed by animals in the ocean naturally must consider the nature of oceanic boundaries. This means, as expressed above, attention to the physical structure of the ocean. From the standpoint of the systematist, it also means a weighing of the strength of boundaries as barriers to gene flow and the nature of oceanic speciation. Thus studies of oceanic zoogeography frequently constitute an approach to the ever-green question of why the number of species is so large.

The stability of the ocean is perhaps an overworked assumption, and uncritical adherence to it can be misleading. The picture is dynamic, of course, as any reference to the paleontological literature will show (e.g., Berggren and Hollister, 1977; Cook and Taylor, 1975). Contemporary biogeographic studies in the ocean tend to focus on the immediate and often become mired in the details of particular patterns. The costs of oceanographic expeditions being what they are, it is extremely rare that a series of deep-ocean samples can be repeated. The situation is conducive to losing sight of the fact that faunal patterns are bound to change over any time scale considered; that this is not acknowledged we consider a failing of many discussions of the nature of deep-ocean life.

There is a sociologic aspect of deep-ocean biological studies that we cannot omit mentioning: ever since the *Challenger* expedition, relatively few people have participated in such work. For those that do, then, there is often an acute sense of personal involvement and some awareness of a personal place in the history of the science. Mills' chapter (this volume, Chapter 1) typifies this sense, as many practicing deep-sea biologists today can quite easily trace their scientific roots through short personal links to such names as Bigelow, Bruun, Agassiz, and Murray. Worthy of respect in many ways, this situation has led to one where there exists strongly held personal views and almost unconsciously an adherence to the professorial tradition. Schools of thought are not uncommon in deep-sea studies.

We can carry the argument one step further. Since deep-sea data are few and difficult to obtain, each school of thought stands on a considerable body of opinion. What is only opinion and what is known for certain are not easy to sort out in the literature. Even in the approach to what might be termed "hard facts," different sets of opinions lead to quite different interpretations. Such a situation, of course, is not unique to deep-ocean biology but can be identified in many branches of science.

We raise these perhaps obvious points for two reasons. On one hand, they interest us and we do consider them an important part of the framework of deep-sea biology; on the other hand, we do not consider ourselves aloof from the situation, and the reader is thereby forewarned.

2. Changing Emphasis in Explanations of Zonation

In general usage the concept of zonation is related to the manner in which some variable changes across space. When the change takes place at different rates, subregions or zones can be drawn that are internally homogeneous. With respect to the deep-sea fauna, there has never been any argument as to whether the fauna changes, but the existence of boundaries between zones has been debated. Some confusion has been introduced by the interchangeable use of zonation to denote the presence of homogeneous zones or uniform change without real subdivisions. In the latter case, a fauna that is zoned with depth changes continuously with depth. In the former case, a fauna that is zoned is one that does not change continuously with depth and whose boundaries are to some degree independent of or oblique to the depth profile (Mills, 1969).

In this chapter we employ the term "zonation" to mean that the fauna changes with depth in such a way that the general pattern of change is that of a nonrepeating sequence of species arrayed along the depth gradient. The more abundant species, at least, appear to have continuous ranges over some portion of the depth gradient. Bathymetrically discontinuous ranges in the same geographic region, which would produce repeating species sequences, are rare. Rex (1977) refers to this basic pattern as species replacement. As we discuss in Section 3, it is difficult in such a species sequence to detect faunal zones (homogeneous regions) analytically.

The interpretation and explanation of zonation has changed as more data became available and concepts of community ecology gradually replaced early dependence on physiological explanations for distribution. Prior to the synthesis of LeDanois (1948), bathymetric and hydrographic features of the ocean were far better known than the fauna distribution. Therefore, it is of no surprise that the early faunal boundary schemes of Murray and Hjort (1912) and Ekman (1953) coincided with bathymetric features. There was the Archibenthal (Bathyal) fauna corresponding to the continental slope and the Abyssal fauna corresponding with the abyssal plain. The work of LeDanois (1948) produced the first description of faunal change that was not heavily influenced by the presumed importance of the physical environment. Species composition changed, said LeDanois, with depth, but below 2500 m a relatively unchanging fauna was reached.

Invoking biological interactions to explain patterns is the mark of modern benthic studies (Paine, 1977), and aspects of this approach were extended to deep-sea zonation in studies of the Gay Head–Bermuda transect by Sanders and his colleagues (Sanders et al., 1965). It is largely from this point that our discussion begins. Although there are now enough data to make definitive statements about how fauna changes with depth, explanations as to the causes are highly conjectural, in that we are mostly restricted to inferring process from pattern. A major hindrance to the development of zonation concepts has been the difficulty of making the transition from the study of individual species to the study of an interacting system of animals (assemblage, community, etc.). Although some of the biological explanations may be just as spurious as the physical explanations of the past, they do encourage a trend toward the study of how systems operate along large-scale gradients.

3. Pelagic Zonation

The pelagic realm comprises all the waters of the deep ocean. It is a three-dimensional world wherein all life must float and is constantly in motion. The motions are on all scales, from those of microturbulence between local density surfaces to the ocean-wide sweep of great current systems. The physical structure is well described, and the relationship between the physical milieu and pelagic animal distribution is the subject of a recent review by Haedrich and Judkins (1979).

Summary charts and sections over broad regions impress the zoogeographer with the relative simplicity of the pelagic system. Temperatures are uniform over vast areas, currents are readily identifiable for thousands of kilometers, and the major water masses are no more than 10 in number. What such charts cannot hope to show is that considerable variability does occur. Although the system as a whole can be rather simply described, there are points quite basic to distributional questions that are unresolved. For example, in the northern Atlantic, probably the ocean area more studied than any other, it is still not clear whether the general circulation involves one or two major gyres (Veronis, 1973; Worthington, 1976).

Three parameters of special biological significance are nutrients, light, and oxygen. Ryther (1963) has considered the horizontal geographic variation in productivity and has shown (1969) that order-of-magnitude differences occur. The availability of both nutrients and light influences production. Light is attenuated so rapidly with depth that even in the clearest waters no photosynthesis appears to take place below about 200 m, and by about 1000 m light from the surface can no longer be distinguished. Therefore, a vertical zonation of the water column based on light is possible. In the euphotic zone, photosynthesis is supported. In the disphotic zone, surface light can have an influence but no photosynthesis occurs. In the aphotic zone, which comprises most of the pelagic volume, the only light is of local biological origin. Consideration of the temperature shows a similar picture. The euphotic zone is the region of near-surface and seasonal thermoclines, the disphotic zone occupies the region of the main thermocline, and the aphotic zone is the region of the uniformly cold (2 to 4°C) waters of the deep ocean. Inorganic plant nutrients enter into biological cycles in the euphotic zone, but below that they become conservative properties of the water until, transported by the subsurface circulation, they are upwelled into the light.

The important feature of the water column is its dynamism (Fig. 1). On the broad scale, large currents move warm water from tropical regions toward the poles on the western side of ocean basins, and cool water is returned either in eastern boundary currents, or, more generally, in the deep counterflow. The shelves have their own circulation pattern, with generally weak alongshore currents interrupted by freshwater inflows from rivers. The important site for interaction between the inshore and oceanic regimes lies over the transition from continental shelf to continental slope, the so-called shelf–slope break. Intrusions from offshore, such as the extensive rings or eddies associated with the western boundary currents, move in against the slope and may inject parcels of water onto the shelf. All along the shelf–slope break upwelling may occur either seasonally or throughout the year, making these regions important sites for primary production. Within the permanent thermocline offshore, there can be extensive internal waves fields. Where these intersect the bottom, along the upper slope and especially in submarine canyons, the waves break and in so doing move water a considerable distance vertically and resuspend fine sediments. On the shelf, seasonal cooling results in local pockets of very cold water that may settle in depressions or, if the

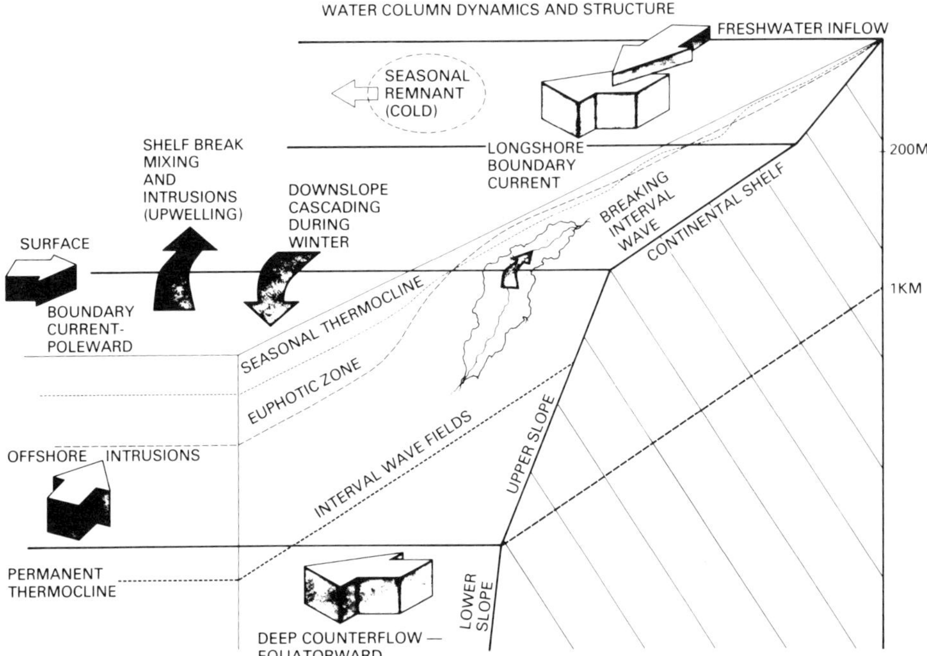

Fig. 1. Water column dynamics and structure that affect vertical zonation.

circulation carries them to the shelf–slope break, may cascade to the depths. Submarine canyons may play a special role in this respect.

As the waters move, they carry food and nutrients with them (Fig. 2). On the shelf, rivers import nutrients from the land, but an equally important source, if not the major one, is the erosion and resuspension of materials from the bottom. These contribute to the load of suspended matter typical of shelf waters and are transported along-, in-, and offshore by the shelf circulation. The nutrient load can be augmented by nutrients upwelled along the shelf-slope break, and the excess of organic matter locally produced is deposited again on the bottom in the quieter areas of that region (Walsh et al., 1981). Across this zone and in it, the periodic cascading of colder waters carries fine organic particles into the ocean depths. Offshore, fine particles settle from the region of production, the epipelagic zone, into the depths. To an extent, the diurnal vertical migration of pelagic organisms recycles this material in the surface layers. Below the mesopelagic zone, however, little vertical migration seems to take place, and the fine particles slowly and continuously rain to the bottom in great depths.

Because of the ocean-wide circulation patterns, pelagic species tend to be very widespread in the horizontal sense. The same species is often found in several oceans, and endemicity, when it occurs, is usually at the species level. Genera or families confined to a single ocean basin are very rare in the pelagial. It is often claimed that the tendency to be worldwide in distribution increases with depth and that the deepest-living pelagic animals are those with the widest distributions. In a general way, this does seem to be the case, but it is also true that the great depths

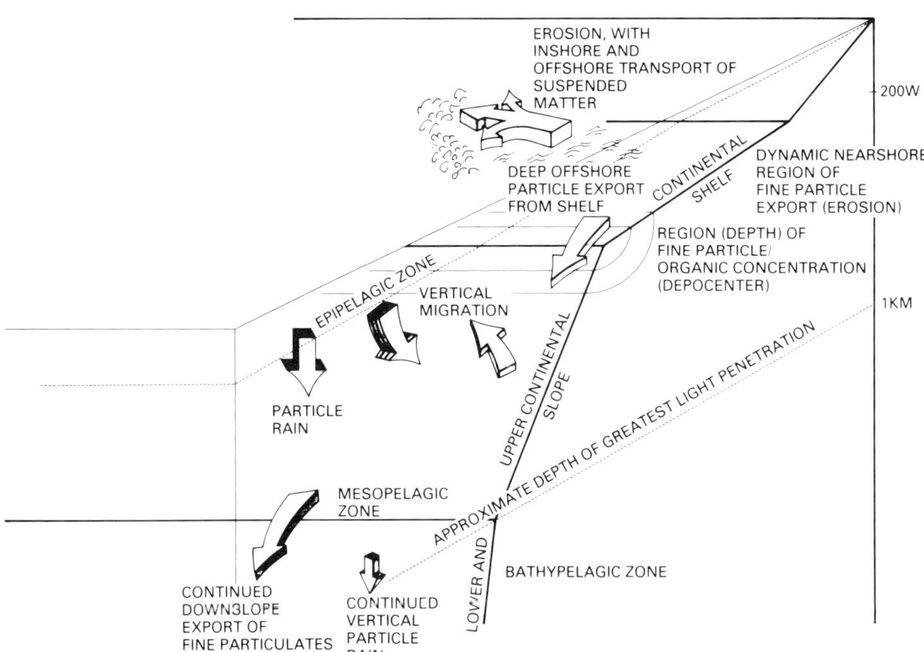

Fig. 2. Continental margin sources of sediment and organic detritus that affect zonation of the bottom and in the water column.

are poorly and spottily sampled and there are undoubtedly systematic problems that remain to be worked out.

Vertically, however, there is a pronounced replacement of species, genera, families, and even orders with depth. Such marked differences allowed even the earliest workers (Murray and Hjort, 1912) to distinguish several quite separate faunas. An epipelagic fauna occupied the uppermost layers, a mesopelagic fauna (including Hjort's "Lilliputian fishes") was found at middle depths, and a bathypelagic fauna was found below that. Quite recently, Marshall and Merrett (1977) have distinguished another category, a benthopelagic fauna that lives near the bottom in deep waters. In general, the epipelagic fauna lives in the euphotic zone, the mesopelagic fauna lives in the disphotic zone, and the bathy- and benthopelagic faunas live in the aphotic zone. Differing physical attributes can thus be associated with each zone that a fauna occupies (Fig. 3), but the zones themselves seem to have been defined first on faunal grounds.

Despite the fact that they are so neatly labeled, strict definitions of pelagic faunas by depth zone are not possible. A high proportion of the mesopelagic animals, for example, migrate at night up into epipelagic regions to feed, and most of the larvae of both mesopelagic and bathypelagic fish undergo their early development in the epipelagial. A similar blurring of clean distinctions occurs in the benthopelagic zone. Since all pelagic distributions are three-dimensional ones, there is added to this behavioral difficulty the vexatious problem of adequate sampling.

The three-dimensional aspect has of necessity been approached in pelagic studies through the use of opening–closing nets. These have had a long history of

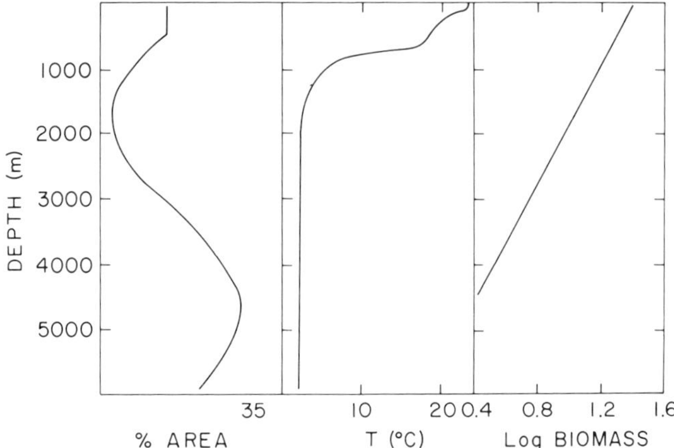

Fig. 3. The generalized relationships among depth of the ocean, the areal extent of the bottom for each depth, the hydrographic regime as typified by a temperate ocean temperature profile, and the exponential decrease in biomass with increasing depth (shown here as log gm/m^2 wet weight). It can be seen that the major faunal changes encountered between 1000 and 3000m occur over a small area between more extensive regions. They are below most simple hydrographic changes, but coincide with rapidly diminishing resources, if standing stock can be taken as indicative of the rate of resource supply to the bottom.

development and are still not perfect (Badcock, 1970; Angel and Fasham, 1975 and Pearcy et al., 1977). Considerable attention in such work has been to attempt to define what physical factors might determine the upper and lower limits of species occurrences. Temperature, salinity, and particularly light intensity have been favorite subjects. Results indicate that, although there are certainly ranges of physical parameters within which pelagic animals live, their responses are not the same everywhere (Fasham and Angel, 1975). In one part of its range, a species may appear to arrange itself according to light intensity but in another temperature may appear to play the dominant role. The question of biological interactions, for example, predation and competition, has received almost no attention in the pelagial, probably because the critical experiments seem logistically impossible. Nonetheless, such interactions have been shown to be quite significant in communities more amenable to experimentation, and there is no reason to believe that they are unimportant in pelagic systems.

Some hint that competition does influence pelagic distribution patterns can be found in the systematic literature. Lawson (1977) has shown that related species of candaciid copepods partition the physical environment so that there is little overlap in their centers of abundance. The deep-ocean fishes *Stomias affinis* and *S. nebulosus*, however, show only a slight suggestion of vertical depth separation in those regions where they occur (Gibbs, 1969). Community-based investigations (i.e., Ebeling et al., 1970) indicate that groups of mesopelagic species can be separated from one another in a single area according to physical and biological parameters, but the detail is insufficient to determine which parameters are predominant. Vertical separation is further complicated because the depth at which a species occurs is a function of age, size, sex, and reproductive state. In the few mesopelagic species for which detailed data are available (mostly fish), size is directly proportional to depth of occurrence, larvae are found at the shallowest depths, and gravid females occur the deepest.

4. Depth and Its Associated Gradients

Although the causes of vertical zonation remain unclear, it is traditional to seek correlations between distributions with physical factors such as temperature, sediment type, currents, and topography (Haedrich et al., 1980; Menzies et al., 1973, and references cited therein). If deep-sea faunal zonation is to be understood, it will be necessary to develop a theory of distribution along gradients that recognizes the possible effects of different types of gradient.

Even without the elaboration of a formal conceptual framework, three types of depth-related gradient can be identified. The first would be gradients of physiologically important factors such as temperature, salinity, and pressure. These factors are not modified by the organisms and cannot be directly partitioned (in the sense of resource allocation) by competition. They may, however, affect a life history parameter such as fecundity or otherwise affect competitive ability. The second group would be partitionable resources that change in type with depth, such as a change from granular sediments to fine sediments, or from a sediment dominated by vascular plants to one of pelagic detritus. The third would consist of those resources that change in availability with depth. A decrease in available food or increase in available space with depth are examples of the third type.

Physiologically important factors such as salinity, temperature, oxygen concentration, and hydrostatic pressure may impose known lethal limits or alter competitive abilities. Because the greatest vertical changes in temperature, salinity, and O_2 concentration (Fig. 3) all occur within the top 1000 m in the ocean, their effect on zonation should be largely restricted to above those depths. Any species that was adapted to conditions at 500 m would encounter little change in going to abyssal depths. An increasing understanding of pressure effects on metabolic reactions makes it clear that depth effects can be expected (Siebenaller and Somero, 1978; Somero et al., this volume, Chapter 7).

The direct determination of resource availability is a necessary step in the study of competition along gradients. Too often the physical environment is measured using techniques and scales of resolution selected for convenience rather than ecological relevance. Actual resource presence and quantity is more often inferred from the species composition of the consumer community than measured [see Peterson (1980) for a critique of competition studies in soft-bottom environments]. The deep-sea fauna is dominated by deposit feeding animals, but we do not know exactly what they are actually recognizing as food in the sediment and how coexisting species selectively partition available resources.

Considering the above caution, we suggest that depth-related changes in sediment facies represent a sequence or gradient of resource types that should regulate the distribution of specialized organisms (in the sense of having a marked competitive advantage on one type of resource). However, we suspect that the actual animal–sediment relationships may be complex. The importance of sediment-mediated distributions is quite obvious because sediments provide habitat and food and are an indicator of the hydrologic regime. In shallow water, distributions are often explained in terms of sediment interactions because some descriptive granulometric parameters of the sediment correlate well with animal distributions. Any or all of these could affect faunal zonation in the deep sea (Gray, 1974, and references cited therein). On the other hand, Rowe et al. (1982) found life in a submarine canyon much less affected by sediment parameters than might have been expected. In any soft-bottom environment with a detritus-based food web, the simple "sediment equals substrate" concept may be inadequate. Depth-related gradients in composition and structure of the inhabited sediment layer can be expected to affect the availability and utilization of resources. If this is the case, we must pursue an understanding of the processes that give the sediment its characteristics rather than its simple granulometric properties, organic content, and so on.

If evolutionary time scale phenomena are important in giving rise to the contemporary pattern of zonation, then it is important to consider that the areas over which gradients are found also change with depth. It has been suggested by Abele and Walters (1969) that the high diversity of the deep-sea is one of many examples of an unexplained area-diversity relationship. The bimodality of the depth–area curve, or hypsographic curve (Fig. 3), and the monotonic nature of the physiological and resource curves can form the basis of an explanation for zonation that invokes evolution and dispersion. It is apparent from the island area–species curves of living animals and the continent area–species curves for fossil forms that, all else being equal, area influences the outcome of immigrations, emigrations, speciation, and extinction in such a manner that the greater the area, the greater the number of species. Considering the empirical area–diversity relationship alone, we would expect a relatively large number of species to have arisen that were adapted to abyssal conditions. A smaller number would be expected on the continental shelf, and the fewest on the slope. In terms of faunal subdivisions, we should find a distinct shelf fauna above 1000 m, a distinct abyssal fauna below 2000 m, and an indistinct slope fauna that is partially obscured by immigration from the two larger areas shallower and deeper. Of course, the other gradients must modify this simple scenario. The vastness of the deeper mode might be counteracted by the decreased food supply, and mobility between the two modes would be restricted by the different physical conditions. Because of the climatically controlled variety of the continental shelf mode, exchange between shelf and slope would be more common at the poles. In spite of its intentional oversimplification, this area–evolution explanation for zonation is consistent with observed patterns. A test of this, unfortunately, would depend on determining the bathymetric sequence of the phylogenetic patterns of a large number of deep-sea taxa.

For our purposes, it is sufficient to accept the diversity–area relationship as a poorly understood empirical relationship. That leaves the area–evolution explanation, however, founded on an unexplained generality. Although we cannot offer a solution to the diversity–area relationship, it does seem that the popular MacArthur–Wilson island model (MacArthur, 1972) does not apply. The species composition on a MacArthur–Wilson island is determined by immigration and emigration over hostile barriers to a place where suitable habitats exist, but may be competed for. The deep-sea fauna must be controlled by immigration, emigration, and competition in a barrierless region that does not afford a habitat similar to the "mainland." Because of this apparent lack of applicability of the island model, we have not attempted to structure a gradient distribution model on it.

5. Analysis of Zonation

As with other topics in zoogeography, deep-sea zonation studies have recently made use of relatively objective methods of comparing different assemblages sampled along the depth gradient. In most cases, the techniques have been borrowed from the growing collection of classification or pattern recognition algorithms. Although extremely useful, the underlying strategies may lack any ecological interpretation. Their use beyond description carries the risk of confusing analytical artifact and meaningful pattern. Since there is no general theoretical definition of zonation, the risk of confusion is especially high.

The attempts to delimit vertical faunal zones have either explicitly or implicitly followed the suggestion by Ekman (1953) that boundaries be drawn to coincide with regions of greatest faunal change. Although simple in concept, this becomes more difficult in practice because of the lack of a widely accepted objective means of expressing the change in fauna among a series of samples. Four general approaches have been applied in the case of the deep sea: examination of all possible intersample similarities, simplification of intersample similarity by classification procedures, simplification by ordination procedures, and coincidence-of-range studies.

A square matrix (Fig. 4*a*), symmetrical about the diagonal, representing interlocation faunal similarity is common to the first three approaches. This can also be represented as values between all stations along the gradient (Fig. 4*b*). For each possible formulation of intersample similarity, at least one advocate can be found (Clifford and Stephenson, 1975). In ordination procedures the choice is restricted to expressions of faunal similarity that conform to the underlying geometric model. For classification, the choice may be restricted by the limitations of the data or the worker's opinions as to the ecological meaning of similarity (i.e., binary, proportional, or fully quantitative).

Measures of similarity that have a geometric analogue have the advantage of ease of interpretation and can be used in both classification and ordination. The Bray–Curtis similarity coefficient or a minor variant of it has been most widely used in depth zonation studies (Sanders, 1960; Sanders and Hessler, 1969; Pearcy et al., in press). More recently the probabilistic similarity coefficient, "expected number of species shared," has been used in a depth zonation study (Grassle, Sanders and Smith, 1979). Since the sampling distributions of such measures can be computed, it is possible to test the null hypothesis that the observed similarity is within the range of expected random variation. This is especially important when sample size varies greatly between samples.

Classification procedures are the most commonly used multivariate techniques in ecology today. The great appeal of these techniques is that they reduce an unmanageably large number of faunal samples into a smaller set of clusters based on similarity. The algorithms used are conceptually straightforward, and it is usually easy to produce ecologically acceptable explanations for the clusters produced. Several books give detailed explanations and instructions (Clifford and

		S1	S2	S3	S4
1000 m Sample	1	$D_{1,1}$	$D_{1,2}$	$D_{1,3}$	$D_{1,4}$
2000 m Sample	2	$D_{2,1}$	$D_{2,2}$	$D_{2,3}$	$D_{2,4}$
3000 m Sample	3	$D_{3,1}$	$D_{3,2}$	$D_{3,3}$	$D_{3,4}$
4000 m Sample	4	$D_{4,1}$	$D_{4,2}$	$D_{4,3}$	$D_{4,4}$

(a)

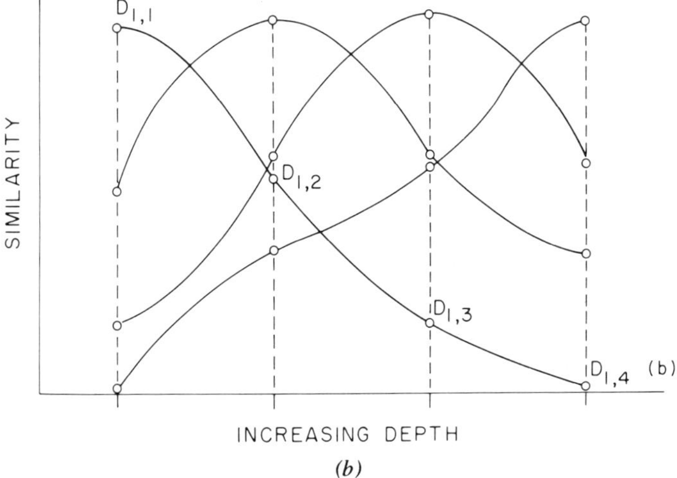

(b)

Fig. 4. (*a*) A typical matrix representation of any similarity values, $D_{i,j}$, between all possible pairs of samples, S. Most analyses which utilize this representation attempt to partition the similarity matrix through the application of sorting rules or formal mathematical procedures to produce clusters. (*b*) The similarity comparisons can also be plotted against depth. The similarities for each sample then form a curve symmetrical about the depth of each particular sample. In most instances of sampling deep-sea transects on a gross scale, the similarity curves are consistent with the simple statement that the greater the depth between samples, the less similar the fauna.

Stephenson, 1975; Anderberg, 1973; Sneath and Sokal, 1973), but critical evaluation sometimes suffers from an author's advocacy. A critical review, and one related to marine work, has been compiled by Boesch (1977). There are two central problems in applying cluster analysis to any gradient zonation: (1) the techniques are designed to form clusters, not to test their validity; and (2) many techniques are popular because they produce results that fit the preconceptions of the investigator. The best example of this is the "flexible linkage" strategy proposed by Lance and Williams (1967) (Fig. 5). By varying the flexible parameter, one produces clusters that could be interpreted as gradual faunal change or intensely separate zones.

When one is interested in the sequence and rate of change rather than classification, the limitations of cluster analysis can be partially overcome by comparing the results of classification when different similarities and sorting strategies are used. If the derivation and properties of the similarity measure and the sorting strategy are understood, an investigator can, in effect, change the definition of faunal similarity and faunal homogeneity. Regions of faunal homogeneity that persist through purposefully changed analyses are quite possi-

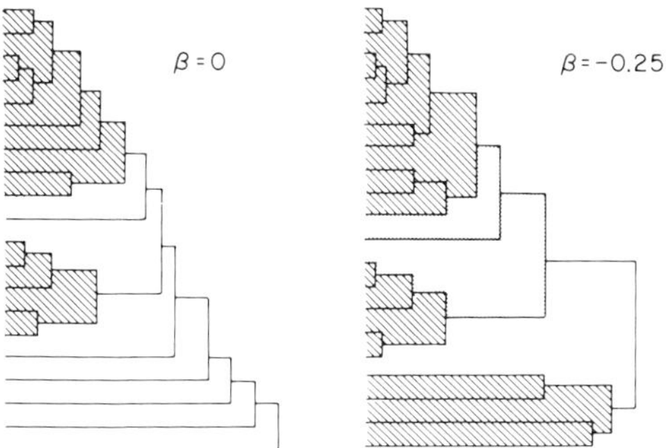

Fig. 5. The weakness of clustering techniques in zonation research is best shown by an example of "Flexible Linkage" strategy (Lance and Williams, 1967). When Beta was set at 0.0 and the analysis run on twenty samples, two distinct clusters were found (hatched portions of dendrogram) with six poorly related samples. However, when Beta was decreased to −0.25 there appeared to be two or three clusters, with no unassigned samples. The choice of Beta is purely arbitrary.

bly real. If the sole purpose of a cluster analysis is to delimit faunal zones, there is no best approach. However, for the purposes of comparisons the selected techniques should be identical to those reported in the pertinent literature.

Ordination procedures have two attractive features. They are based on exact vector models that sometimes allow for statistical testing of hypotheses. They do not require the assumption that distinct faunal clusters exist, but they are not especially useful for drawing definite boundaries. Unfortunately, they employ linear models that may have little ecological meaning, and interpretation of the results may be complicated by rotations that produce negative values. These techniques have been successfully applied to deep-sea zonation by a variety of workers (Carney and Carey, 1982; Haedrich and Kreft, 1978; Rex, 1977; Wigley and Theroux, 1981). In each case the greatest amount of faunal variation among samples has been consistently associated with depth. The relative merits of classification versus ordination have been debated in the phytogeography literature, and Anderson's (1965) conclusion agrees with the experience of benthic ecologists. When the sampled gradient is so long that faunal identity changes more than once along it, ordination techniques produce little more than a complex statement of that fact. In such cases, clustering produces a better initial subdivision. However, when a smaller portion of a vertical gradient is intensively sampled, ordination techniques afford a better means of relating Faunal change to environmental variables. This is basically the approach used by Whittaker (1973) in his extensive study of terrestrial faunal gradients.

Although careful examination of all pairs of intersample faunal similarity has been largely replaced by seemingly more sophisticated data processing techniques discussed above, such a humble approach can be very informative concerning the sequence and rate of faunal change. Ekman (1953), Vinogradova (1962), and Menzies et al. (1973) drew their initial bathymetric divisions on the basis of maximal changes between adjacent pairs of stations arrayed along the depth gradient. As crude as this seems, it actually works reasonably well in the deep sea. A consistent distance–similarity relationship allows one to compare the rate of faunal change between samples for different functional groups of organisms.

The fourth approach to zonation studies, coincidence of range, is the only one not to make direct use of Ekman's criteria. As applied to the deep-sea gradient (Gardiner and Haedrich, 1978), the important statistics are the number of upper depth limits or lower depth limits that occur within each subdivision of the bathymetric range. If there are regions of faunal homogeneity separated by regions of transition, the upper and/or lower limits should occur simultaneously within some subdivisions more often than would be predicted by chance. It was suggested by its originators that this procedure might be used as an initial test for zonation that could then be followed by more analytical approaches. Unlike simple presence or absence of a given animal from a sample, the presence of an upper or lower limit within a given part of the bathymetric range is dependent on the intensity of the sampling outside the observed range. Therefore, the coincidence-of-range limits approach will be very sensitive to differences in sampling intensity along the gradient. As such, it is best suited for heavily studied areas where individual species ranges have been firmly established.

All the above techniques share the common fault that they are very sensitive to the positioning of samples along a gradient and the relative size of the samples. When intersample Faunal similarity (any index) is plotted against intersample depth, there is an apparent, strong correlation. Because of this, gaps in the bathymetric series will be seen by the clustering algorithm as gaps in faunal similarity. A simple way to avoid selecting sampling discontinuities as faunal boundaries is to superimpose the dendrogram on the bathymetric sequence of the samples. If the clusters formed by the analysis coincide with sampling gaps, they can be accepted only with considerable trepidation. In classification analysis, the effect of different sample sizes is determined by the index selected, but it is quite possible for apparently meaningful dissimilarities to be due to sample size. The depression of the importance of the shelf-slope boundary in Hudson Canyon found by Rowe et al. (1982) may be an example of such an effect. In ordination techniques, if one portion of the faunal transition is more heavily sampled as a result of sample size or crowding of samples, it will dominate the analysis.

The effect of sampling on the coincidence-of-range approach is related to the probability of an upper or lower range falling within a specific depth zone. These probabilities are dependent on the actual distribution of the animal, the intensity of sampling within the specified zone, and the intensity of sampling in both shallower and deeper zones. Without going into detail, the simultaneous occurrence of upper and lower limits in a single zone can be affected by unequal sampling among zones. In the example given by Gardiner and Haedrich (1978), the depth intervals had been fairly evenly sampled. However, in the early compilation by Vinogradova (1962) the coincidence of first and last occurrences might be due to sampling.

6. Patterns Found and Points of Disagreement

The most conspicuous finding of all zonation studies regardless of region or taxa is agreement with the preliminary conclusions of the last century. The deep-sea fauna undergoes a nonrepeating sequential change with depth. The composition changes with depth, and most species have predictably restricted depth ranges. There is a distinct shallow-water fauna that corresponds well to the continental slope and an abyssal fauna that starts at about 1000 m. The nomenclature due to Menzies et al. (1973) is useful in making this distinction (i.e., the Shelf Province, Abyssal Province, and Archibenthal Zone of Transition). Between the two is a zone of transitions that is subdivided by some workers (Menzies et al., 1973; Mills, 1972; Haedrich et al., 1975, 1980). In light of our cautions about artifact boundaries imposed by the sampling and analysis, any subdivisions of this simple three-part scheme should be adopted with care.

When the sampling strategy allowed direct evaluation of the role of depth versus distance from land, depth was the dominant factor (Wigley and Theroux, in press; Carney, 1976; Carney and Carey, 1982; Pearcy et al., in press). This was strictly evident in the megafauna on Cascadia Abyssal Plain off the coast of Oregon and Washington.

With respect to delimiting relatively homogeneous zones, the northwestern Atlantic off the North American coast is the most thoroughly studied part of the deep sea. The various results are in good agreement, especially considering that different collecting techniques, faunal groups, and analytical methods were used. Detailed summaries can be found in Haedrich et al. (1980) and Rowe et al. (1982), with a comparison in Table I. Added to this should be the work on pogonophores (Southward, 1979) and forams (Douglas and Woodruff, 1981).

The megafauna of the New England coast to a depth of nearly 5000 m was found to be divisible into seven separate zones (Haedrich et al., 1980, and references cited therein). The analysis was based on approximately 460,000 specimens of 256 species collected in 105 trawl samples. The distribution of upper and lower limits was shown to be nonrandom. A classification of zones was constructed using percent similarity followed by group-average clustering. Faunal boundaries were not abrupt, and the authors noted that they were not immutable. This scheme was in good agreement with that due to Musick (1976) based on fish collected to the south and that of Menzies et al. (1973) based on isopods off North Carolina. Although it agrees with Rowe et al. (1982) in general, the lack of sampling at depth did not allow the latter to divide the deeper macrofauna into subzones.

The minor differences between results along the northwestern Atlantic may be due to local variations in the rate of species replacement with depth. This is evident at bathyal depths off North Carolina where the vertical range of megafauna was markedly depressed in a region impinged on by the Gulf Stream (Rowe and Menzies, 1969; Cutler, 1975). Local topographic features may also cause changes in the zonation pattern, as evidenced by a subdivision of the Archibenthal Zone of Transition within Hudson Canyon (Haedrich et al., 1975) compared to the lack of a similar break on the open slope (Rowe et al., 1982).

While Menzies et al. (1973) projected that similar zones extended around the entire world ocean moving shallower or deeper in response to physical factors, a detailed examination of patterns in other oceans to confirm this is hindered by a lack of reports that are comparable in scope and analytical approaches used in the Atlantic. The most extensive sampling suitable for looking at zonation over a wide depth interval has been conducted off the coasts of Oregon and Washington (Pearcy et al., 1982), but only a few faunal components have been reported. Patterns on the North American Pacific coast are complicated by the ridge and basin geomorphology. On the basis of vertical ranges for 90 species, two depth intervals were found where new species were encountered with an increased

TABLE I
Deep-Sea Benthos Classification of Zones (Units in Kilometers)

Hedgpeth (1957)	Depths	Pérès (1957)	Depths	Menzies et al. (1973)	North Carolina	Peru	Antarctic	Arctic	Haedrich et al. (1980)	Depths	Rowe et al. (1982)	Depths
Littoral	Continental shelf, <0.2	Littoral (with subdivisions)	Shelf	Shelf province	To 0.5	To 1.2	0.1	None		0.3		0.2
Bathyal	0.2 to 4 (>4°C)	Epibathyal, mesobathyal	Slope, rise	AZT[a]	0.5 to 0.9	1.9 to 3.3	0.1 to 0.9	0.001 to 0.4	AZT Upper Lower	0.3 to 1.3 0.3 to 0.7 0.7 to 1.3	AZT Upper Lower	0.26 to 1.7 0.26 to 1.0 1.0 to 1.7
Abyssal	4 to 6 (<4°C)	Infrabathyal	Abyssal plains	Abyssal province	0.9 to 5.4	3.3 to 6.3	0.9 to 5.5	0.4 to 2.8	Abyssal	1.4 to 5.0	Abyssal	>1.7
				Upper	0.9 to 2.6	3.3 to 4.5	0.9 to 3.5	0.4 to 0.6	Upper subzone,	1.4 to 2.0 2.1 to 2.5	Upper	1.7 to 2.2
				Meso	2.6 to 3.3	4.5 to 4.9	3.5 to 3.8	0.6 to 0.9	Meso subzone	2.5 to 3.1 3.2 to 3.7	Meso	>2.3
				Lower active	3.3 to 4.4	4.9 to 6.3	3.8 to 5.4	1.0 to 2.6				
				Lower tranquil	4.4 to 5.1	—	—	—	Lower	3.9 to 5.0		
				Lower red clay	5.1 to 5.3							
Hadal	>6	Hadal	Trenches		—	—	—	—		—		

frequency. The first of these coincided with the lower boundary of the very marked (0.5-ml/liter) oxygen minimum in the Pacific and lay between 400 and 700 m. Within that 300-m interval, 19 species were first encountered. At approximately midslope (1900 to 2200 m), 19 additional species were first encountered. When depths of last occurrence were tabulated, slightly different patterns were found. At the deep end, 24 species disappeared across the 2800- to 3100-m interval. This depth range was the most intensively sampled and coincides with the floor of the Cascadia Plain. At the shallow end, 34 species disappeared in the 500-m interval between 500 and 900 m.

The pattern off Oregon for fish seems quite simple. There is a distinctive group of species dominant on the continental shelf. The transition from shallow water to deep-sea fauna takes place just beyond the slope break between 400 and 900 m. Here the shallower forms drop out and the slope forms are first encountered. Beyond that point there may be just simple replacement with depth, although unevenness in sampling gives the impression of a major faunal change near the juncture of the Cascadia Basin and the continental slope. It is not possible to determine whether the deeper zone is subdivisible in the way reported for the northwestern Atlantic.

For the epibenthic holothuroids, a similar shelf–edge transition was found followed by continuous replacement (Carney and Carey, 1977). Using ordination techniques to study faunal change between 2000 and 4000 m, they found no discernible boundaries that could not be explained by irregularities in the sampling strategy and bottom topography. The fauna changed gradually, undergoing a complete replacement every 1000 m in depth.

7. The Vertical Pattern of Adaptations

Compilations of range and intersample similarity comparisons are informative but do not provide the ready information on life history adaptation that is required to develop an ecological explanation that can be tested. Toward such a goal it is necessary to proceed beyond a treatment of organisms as taxonomic entities to a study of the ways in which various taxa have adapted in parallel to the deep sea. There have been a number of general trends suggested concerning the vertical change in certain adaptive traits, and, whereas some of the proposed gross generalizations are only of historical interest, other patterns warrant closer examination. We restrict our discussion to size, feeding, mobility, and evolutionary age. Vertical patterns with plausible sensory–physiological explanations, such as reduction of eyes and coloration, have been previously discussed by Menzies et al. (1973).

Depth–size relationships have been frequently cited in the literature and are of general ecological interest because size tends to correlate with fecundity, age, mobility, and similar variables. There have been two different types of size–depth relationship mentioned in the literature: (1) that concerning the adult size of species in a genus or higher taxa that spans a wide depth range and (2) that concerning the size of the average individual in the fauna sampled at progressively greater depths. Unless the distribution and dominance of all the species within a taxon or the phylogenetic affinities of all individuals in a sample are known, the two types of observation cannot be compared.

For some deep-sea taxa, the largest sized species tend to be found in the deeper part of the taxa's range. The best documented cases are based on pericarid Crustacea. Restricting himself to 37 asellote isopod genera of at least four species, Wolff (1962) concluded (by inspection of plots) that the smallest genera tended to be restricted to less than 250 m whereas the largest species were below 1000 m. There were several exceptions discussed, but the relationship seemed well founded. It is important to note that the depth–size increases within taxa were shifts within broad size ranges, not distinct linear or curvilinear changes. Additional cases with varying degrees of analysis have been given for the wormlike holothuroid *Myriotrochus* (Belyaev, 1970), pericard Crustacea (Zenkevitch and Birstein, 1956), tanaidaceans (Gardiner, 1975), Gastropoda (Clarke, 1960), and Foraminifera (Bernstein et al., 1978).

Unlike the within-taxa pattern, there is a decrease in average body size with depth reported for most faunal components across the full meio–megafauna size range (Thiel, 1975; Polloni et al., 1979; Gage, 1978; Jumars and Gallagher, in press). This change appears to be caused by a replacement of larger species by smaller (Gage, 1978; Jumars and Gallagher, in press). When examined in the northwestern Atlantic, the greatest size decrease seemed to be restricted to shelf depths (<400 m) (Polloni et al., 1979). An exception has been reported for average size in deep-sea fish fauna. In the northwestern Atlantic, there is a progressive size increase with depth (Polloni et al., 1979; Haedrich and Rowe, 1978), but no such pattern was found in the northeastern Pacific (Pearcy et al., in press). Musick (personal communication) has suggested that the apparent increase in size is related to the general decrease in both the diversity and abundance of fish with depth.

Both the deeper–bigger within-taxa and the deeper–smaller within-fauna patterns may reflect different adaptations to a food decrease with depth. A foraging animal may be accommodated to low food levels by increasing its foraging area, decreasing its maintenance costs, or both. The first may result in an increase in size (Wolff, 1962; Belyaev, 1972), whereas the latter may result in a decrease (Thiel, 1975; Gage, 1978). The relatively rare, large species within a taxon represent the lineage that followed the increased foraging area course. The more common solution and the one exhibited by the dominant fauna is size decrease. In the case of fish, which are highly mobile foragers, increased size was the dominant adaptation.

There is a major reduction in the number of suspension feeders and increased dominance of deposit feeders in the deep sea that is readily explainable in trophic terms on the basis of the absence of a productive phytoplankton, a rich zooplankton, predictable transport of organic rich resuspended material, lack of substrate, or similar. However, as stressed by Jumars and Fauchald (1977), there is still a great deal of vertical change in composition at depths well below the elimination of suspension feeders. The older, simple feeding type classifications are not adequate for study of such deeper zonation.

The classification of deposit feeders into feeding–mobility guilds (Root, 1975) is a useful extension of the feeding-type analysis of communities introduced into shallow-water benthic ecology by Peterson (1914). As applied to the polychaetous annelids by Fauchald and Jumars (1979), it represents a refinement in that it provides a partitioning of the deposit feeders on the basis of mobility and feeding horizon. In analyses of depth series in the northeastern Pacific (Jumars and Fauchald, 1977) and northwestern Atlantic (Rowe et al., 1982), there is a general replacement of sedentary deposit feeders by more mobile forms. This same approach should be extended to other taxonomic groups when knowledge of the functional morphology is sufficiently advanced. Morphological classifications based on feeding type and locomotion are especially useful in that they allow application of maximal foraging theory (Jumars and Gallagher, in press).

Although suspension feeders are rare in the deep sea, they are not absent, and a comparative study of deep-sea versus related shallow-water forms might reveal common patterns of adaptation to the deep-sea environment. Some of these animals show very considerable depth ranges, such as the abyssal stalked crinoids and Coelenterata (Belyaev, 1972) (500 to 10,000 m) with no indication of intraspecies morphological change. Such forms may be taking advantage of local conditions that maintain high levels of resuspended sediments (Lonsdale, 1977). This might explain the repetitive pattern of suspension feeding (sestonophagous fauna) found by Sokolova (1959) at shelf, trench wall, and trench floor depths in the western Pacific. In addition, specialized suspension feeders have been found near deep hydrothermal activity (Corliss et al., 1979). Another group that might be expected to occur sporadically with depth are herbivores that utilize vascular plant food falls (Wolff, 1979).

The position held by deep-sea organisms within the evolutionary history of marine higher taxa was a topic of debate up until about two decades ago. Early claims were made and argued before continental drift was widely accepted and with virtually no information as to how the deep-sea fauna or environment had changed during geologic time. As a result of deep-sea drilling and micropaleontological studies, it is well documented that there have been major changes in composition and faunal zone depths. In the Miocene a 10° cooling of deep water resulted in a major faunal change. Much closer to the present, it has been found that the benthic foram assemblage of the deep northwestern Atlantic changed in dominance and depth from the last glacial period to the present (Streeter, 1973). In light of the growing evidence that the deep sea has not been a stable environment, it is necessary to consider that the current zonation pattern may have been strongly affected by historical events.

Prior to the *Challenger* expedition an extrapolation was made from early finds of some more archaic fauna in dredge samples. It was predicted that if stalked crinoids and other older forms could be collected at relatively shallow depths, an even older fauna must be found at greater depths. This first generalization about deep-sea faunal zones proved to be in error. The gross generalization that the deep sea is archaic does not hold because more examples of truly ancient taxa can be found at shelf and slope depths (Menzies and Imbrie, 1955). However, as pointed out by Boss (1971), some deep-sea taxa, especially the bivalves, are more closely related to older taxa than to common contemporary shallow-water forms. During the evolution of the bivalves the relatively primitive protobranches were apparently displaced in shallow water by the eulamellibranches. Similarly, among the fish, brotulids were once a more common littoral faunal component, whereas they are now most common in the deep sea, a relationship first pointed out by Andriashev (1953).

Clarke (1962) and Knudson (1979) have favored the explanations that the

Eulamellibranchiata are suspension feeders and poorly suited to the deep-sea soft sediment environment. Speciation that leads to feeding specializations is possible in shallow water because of high levels of productivity. However, a degree of feeding generalization would seem to have some preadaptive value with respect to limiting initial colonization of the food-poor deep sea. Therefore, in any lineage where there has been feeding specialization with time, the deep-sea representatives would probably resemble more primitive generalists. The time of colonization does not have to coincide with the shallow-water evolutionary origin of the primitive forms. On the basis of the holothurians and starfish, Hansen (1976) and Madsen (1961) suggested that it is actually the bathyal forms that are the most primitive, with abyssal and hadal species being more highly evolved.

The concept of separate evolutionary history for the deep sea is subject to refined examination with an application of techniques advocated by "cladist" zoogeographers [see Pielou (1980) for a brief summary and the necessary references]. If a probable evolutionary sequence can be assigned to deep-sea species within restricted taxa, an examination of distribution and phylogenetic sequence may suggest the history of that taxa. The best example in deep-sea biology is for the isopod family Ilyarachnidae (Hessler and Thistle, 1975) wherein these shallow-water Antarctic species were advanced relative to the diverse deep-sea fauna. The family appears to have arisen and undergone radiation in the deep sea with successful emergence and diversification in the Antarctic as a result of the lack of physiological barriers and competing decapod crustaceans. This finding was contrary to the earlier views that the deep-sea forms were derived from the shallow-water Antarctic fauna (Bruun, 1957; Wolff, 1960; Menzies et al., 1973).

8. Progressing from Description to Prediction

Whereas there is still great value to descriptive work in the deep sea, many of the earlier explanations, generalities, and hypotheses can now be tested by new field sampling that specifically examines their predictive value. Two approaches can be taken: those that test the predictive value of a particular zonation scheme and those that examine proposed mechanisms. Testing the validity and usefulness of proposed faunal boundaries could be as straightforward as designing a sampling program that would test predictions about faunal similarity over small segments of a bathymetric range. For example, both Musick (1976) and Haedrich et al. (1980) subdivide the fauna on the continental slope into two zones. From these findings it can be proposed that additional slope sampling should show significant within-zone homogeneity and between-zone differences. Such a validation of zones would have a practical benefit. If the existence of relatively homogeneous zones is tested and proved, basic and applied research in the deep sea can be greatly simplified by focusing on a subset of depths that include the major zones.

It is substantially more difficult to determine a process by testing predictions because of the large number of proposed ecological processes that give rise to similar predictions. If an unambiguous experiment could be designed, the needed data may be impractical to obtain [for a review of the application of ecological theory to the deep sea, see Jumars and Gallagher (in press)]. Nevertheless, it is extremely appealing to invoke predation and competition as factors that might play an important role in the zonation of the deep-sea fauna. This appeal stems from an inability to relate the patterns occurring below 1000 m to physiological limitations and the established importance of ecological phenomena in maintaining narrow zones in the rocky intertidal. If we accept the reasoning of Menge and Sutherland (1976), competition might be expected to play an even greater role in the deep sea than in shallow environments where physical disturbance is more severe.

The analytical problems of testing the significance of the similarity curves can be overcome simply by restricting the predictions and tests to the underlying species ranges. Rather than focusing on the upper and lower boundaries, which may be difficult to establish with confidence, population levels along the range should be examined to determined whether they are symmetrical, skewed, or discontinuous. A skewed distribution with depth, for example, was observed by Rowe and Menzies (1969) for the dominant ophiuroid *Ophiomusium lymani* in the northwestern Atlantic. Asymmetry may be due to an asymmetrical environmental gradient or to competitive compression.

We lack a definitive analytical technique for proving the existence of competitive partitioning along a gradient from distributional data; however, the ad hoc approach due to Terborg (1971) would seem to hold potential. Starting with species distributions along an alpine altitude gradient, Terborg computed the family of similarity curves that might be expected if one of three situations held. If species were randomly distributed about some optimum on a resource or physiological gradient and there were no interference between adjacent populations, ranges would overlap and similarity curves would be wide (Fig. 6*a*). If species were serially partitioning some resource along the gradient, with competition at the boundaries, boundaries would be sharp with little overlap, and the similarity curves would be peaked (Fig. 6*b*). If there were a major faunal discontinuity along the gradient, the similarity curves would show a marked change.

Rex (1977) applied this theory to deep-sea zonation with interesting results. The rate of faunal change between samples was dependent on the faunal group analyzed. Megafauna changed most rapidly, infaunal polychaetes changed the slowest, and predatory gastropods changed at an intermediate rate (Fig. 7). This was interpreted as supportive of Menge and Sutherland's (1976) findings in that higher trophic forms would lack a predatory reduction in population and would experience more crowding on the gradient. The actual relationship seems highly dependent on the actual fauna being examined. When holothuroids were compared to infaunal polychaetes and amphipods (Carney, 1976; Carney and Carey, 1982) on an eurybathal basin, the megafaunal holothuroids showed the slower faunal change. The comparison of replacement rate for selected functional or size groups is interesting but could easily be examined in a rigorous way without dependence on similarity indices whose properties remain poorly explored. The relationship between depth of capture and vertical range for different groups could be examined and tested between groups. Similarly, the frequency of adjacent ranges could be compared and tested between groups.

Although routine sampling techniques may not provide the estimates needed for some ecological testing, it does produce specimens. Simple studies of comparative morphology that stress feeding and locomotor structures can serve as a guide toward recognition and ultimate measurement of levels of resource availability. If specialization for different types or levels of resource does occur along

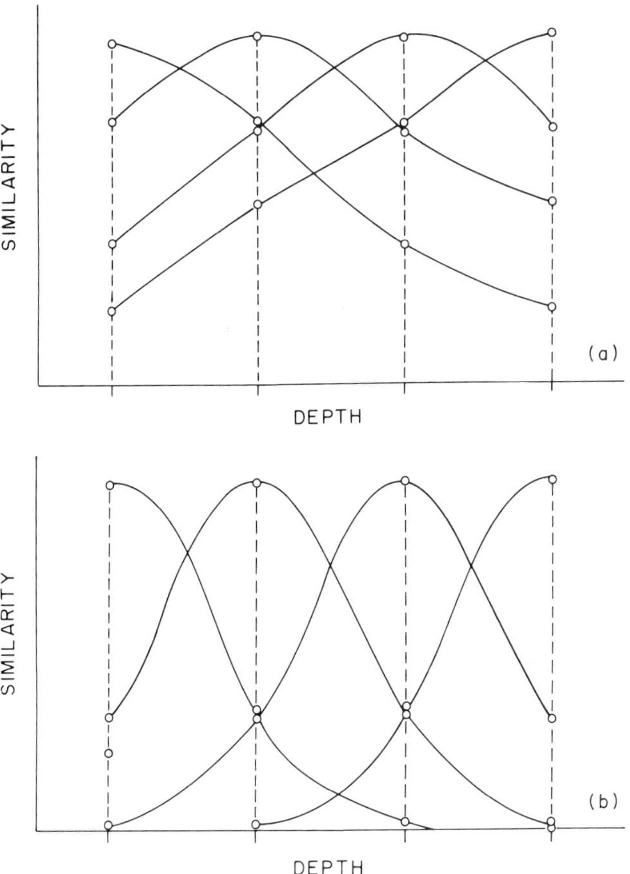

Fig. 6. (*a*) Wide overlap of hypothetical similarity curves resulting from random distribution of species about physiological or resource optima, with little interspecific interaction; (*b*) hypothetical similarity curves with competition at boundaries that results in little overlap and sharp peaks to the curves.

the depth gradient, there may be morphological adaptation of mouth and locomotor structures within wide-ranging taxa whose species have relatively narrow depth ranges. This type of inquiry could be descriptive or assume the form of testing formal predictions drawn from independent observations.

The area–evolution explanation leads to predictions about phylogenies that are potentially testable and are based on comparative morphology. If species have invaded the small slope region after speciation on the larger shelf and abyssal areas, this should be reflected in the phylogenetic pattern of the deep-sea fauna. The slope depth species should show closest affinities to shelf and abyssal congeners rather than slope congeners. It must be noted, however, that the testability of phylogenetic hypotheses is not universally agreed to. Hidden circularity is also a strong possibility in that depth of occurrence has often been used as a trait of taxonomic importance in order to subdivide species with broad depth or geographic ranges.

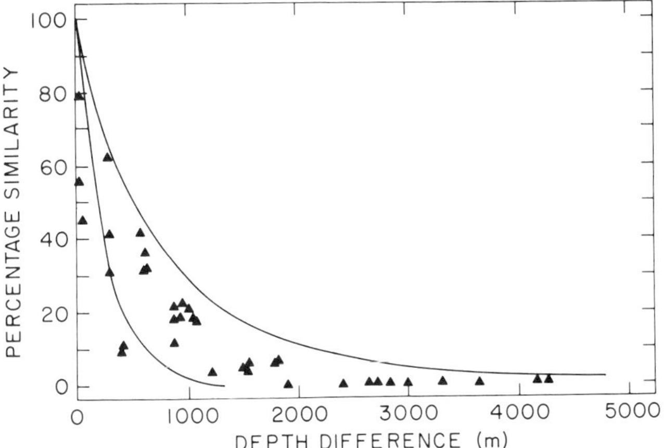

Fig. 7. When the depth between samples versus similarity between samples is considered for different faunal groups, different rates of change with depth are found. Off the coast of New England megafaunal invertebrates changed rapidly with depth (lower curve). Fishes, which might have a different mobility or be in a different trophic group, changed more slowly (upper curve). Predatory gastropods (triangles) fell between these two extremes. (Modified from Rex, 1981)

Although logistically complex, manipulation experiments have the potential to alter the resource–depth gradients and to move species beyond their realized ranges. Implantation of artificial substrates (Turner, 1973; Grassle, 1977; Desbruyérès et al., 1980) is an example of resource manipulation in the deep sea. Dayton et al. (1982) successfully transplanted deep fauna to shallow depths. Although the depth involved was not large, this is evidence from transplantation studies on a bathyal barnacle that at least one species is restricted to a zone smaller than it could potentially fill. Two hundred and fifty specimens of *Bathylasma corolliforme* (Hoek) were transplanted from 400 m in the Ross Sea to depths of 25 and 40 m. It was found that this species, which normally lives at 100 to 1500 m, survived at least 2 years, and that viable larvae were actually in the plankton, settling only when an adult was present. The authors eliminated several hypotheses about the restriction and favored one that related distribution to water movement. The principal point is that the distribution of an Archibenthal Transition Zone species appeared to be intimately related to quaternary events, suggesting that reasonably recent historical factors might have had an effect on the current zonation pattern.

In conclusion we anticipate future work. The pattern of zonation in the deep sea must be viewed as the result of several processes operating on evolutionary and ecological time scales. Although the zonal classification and total faunal similarity approach has been instructive, it adds a level of obscurity to a complex pattern that must be seen in detail. Our basic level of inquiry should be that of the individual species' range and the individual species' adaptations. Rather than add to the growing catalog of ecological explanations, there is now sufficient background against which to begin eliminating a few choices through experimental design and testing.

Acknowledgments

The authors wish to express appreciation to the National Science Foundation, the Office of Naval Research, the National Oceanographic and Atmospheric Administration, and the Department of Environment (Can.) for support of the work on which this review is based. Colleagues who aided us include Pam Polloni, C. H. Clifford, and David Judkins, primarily, and many others over the decade, too many to name. We trace the lineage of our interests in zonation to Robert J. Menzies.

References

Abele, L. and K. Walters, 1979. The stability-time hypothesis: reevaluation of the data. *American Naturalist* **114,** 559–568.

Anderberg, M. R., 1973. *Cluster Analysis for Applications*. Academic Press, New York, 359 pp.

Anderson, D. J., 1965. Classification and ordination in vegetation science: Controversy over a non-existent problem? *J. Ecol.,* **53,** 521–526.

Andriashev, A. P., 1953. Primary and secondary forms of deep-sea fishes and their significance in zoogeographic analysis. In *Ocherki po Voprosam Obschhei Ikhtiologii,* Izdatel'stvo AN USSR (in Russian), pp. 58–64.

Angel, M. V., 1976. Windows into a sea of confusion: Sampling limitations to the measurement of ecological parameters in oceanic mid-water environments. In *Oceanic Sound Scattering Prediction: Marine Science,* Vol. 5, N. Anderson and B. Zahuranec, eds. Plenum Press, New York, pp. 217–248.

Angel, M. V. and M. J. R. Fasham, 1975. Analysis of the vertical and geographic distribution of the abundant species of planktonic ostracods in the Northeast Atlantic. *J. Mar. Biol. Assoc. U. K.,* **55,** 709–737.

Badcock, J., 1970. The vertical distribution of mesopelagic fishes collected on the SOND cruise. *J. Mar. Biol. Assoc. U. K.,* **50,** 1001–1044.

Belyaev, G., 1970. Faunal of the Kurile–Kamchatka trench and living conditions thereof: Ultra abyssal holothuroids of the genus *Myriotrochus* (in Russian). *Trudy Inst. Okean., Akad. Nauk SSSR,* **86,** 458–486.

Belyaev, G., 1972. Hadal bottom fauna of the world ocean. Israel Program for Scientific Translation Ltd., Cat. No. 600107 (199 pp., Russian edition published in 1966). *Trudy Inst. Okean., Akad. Nauk SSSR,* **59,** 1–248.

Berggren, W. A. and C. D. Hollister, 1977. Plate tectonics and paleocirculation–commotion in the ocean. *Tectonophysics,* **38,** 11–48.

Bernstein, R., R. Hessler, R. Smith, and P. Jumars, 1978. Spatial dispersion of benthic Foraminifera in the abyssal central Pacific. *Limnol. Oceanogr.,* **23,** 401–406.

Boesch, D., 1977. Applications of numerical classification in ecological investigations of water pollution. U.S. Government Environmental Protection Agency Ecological Research Series EPA-600/3-77-033, pp. 155.

Boss, K. J., 1971. Review of the systematics and biology of abyssal and hadal Bivalvia by Knudsen. *Malacol. Rev.,* **4,** 219–220.

Bruun, A., 1957. Deep-sea and abyssal depths. In *Treatise on Marine Ecology and Paleocology,* Vol. 1, J. Hedgepeth, ed. Memoirs Geological Society America No. 67, pp. 641–672.

Carney, R. S. and A. G. Carey, Jr., 1977. Distribution pattern of holothurians on the Northeastern Pacific (Oregon, U.S.A.) continental shelf, slope and abyssal plain. *Thalassia Yugoslavica,* **12,** 67–74.

Carney, R. S. and A. G. Carey, Jr., 1982. Distribution and diversity of holothuroids (Echinodermata) on Cascadia Basin and Tufts Abyssal Plain. *Deep-Sea Res.,* **29,** 597–607.

Carney, R. S., 1976. Patterns of abundance and relative abundance of enthic holothuroids (Echinodermata: Holothurioidea) on Cascadia Basin and Tufts Abyssal Plain in the N. E. Pacific Ocean. Ph.D. thesis, Oregon State University, Corvallis, 185 pp.

Clarke, A. H., 1960. A giant ultraabyssal *Cocculina* (*C. superba*, n. sp.) from the Argentina Basin. *Nat. Hist., Nat. Museum, Can.*, **7**, 1–4.

Clarke, A. H., 1962. On the composition, zoogeography, origin and age of the deep-Sea molluscan fauna. *Deep-Sea Res.*, **9**, 291–306.

Clifford, H. T. and W. Stephenson, 1975. *An Introduction to Numerical Classification*, Academic Press, New York, 229 pp.

Cook, H. E. and M. E. Taylor, 1975. Early Paleozoic continental margin sedimentation, trilobite biofacies, and the thermocline, western United States. *Geology*, **3**, 559–562.

Corliss, J., J. Dymond, L. Gordon, J. Edmonds, R. R. von Herzen, R. D. Ballard, K. Green, D. Williams, A. Bainbridge, K. Crane, and T. H. van Andel, 1979. Explorations of submarine thermal springs on the Galapagos rift. *Science*, **203**, 1073–1083.

Cutler, E., 1975. Zoogeographic barrier on the continental slope off Cape Lookout, N.C. *Deep-Sea Res.*, **22**, 893–901.

Dayton, P., W. Newman, and J. Oliver, 1982. The vertical zonation of the deep-sea Antarctic acorn barnacle *Bathylasma corolliforme* (Hoek): Experimental transplants from the shelf into shallow water. *J. Biogeogr.*, **9**, 95–110.

Desbruyérès, D., J. Bervas, and A. Khripounoff, 1980. Un cas de colonization rapide d'un sediment profond. *Oceanologica Acta*, **3**, 285–291.

Douglas, R. and F. Woodruff, 1981. Deep-sea benthic foraminifera. In C. Emiliani, ed. *The Sea* Vol. 7. *The Oceanic Lithosphere*. Chapter 29, 1233–1328.

Ebeling, A. W., 1967. Zoogeography of tropical deep-sea animals. *Stud. Trop. Oceanogr.*, **5**, 593–613.

Ebeling, A. W., R. M. Ibara, R. J. Lavenberg, and F. Rohlf, 1970. Ecological groups of deep-sea animals off Southern California. *Bull. Los Angeles County Museum Nat. Hist.*, **6**, 1–43.

Ekman, S., 1953. *Zoogeography of the Sea*. Sidgwick and Jackson, London, 417 p.

Fasham, M. J. R. and M. V. Angel, 1975. The relationship of the zoogeographic distributions of the planktonic ostracods in the Northeast Atlantic to the water masses. *J. Mar. Biol. Assoc. U. K.*, **55**, 739–757.

Fauchald, C. and P. Jumars, 1979. The diet of worms: A study of polychaete feeding guilds. *Oceanogr. Mar. Biol. Ann. Rev.*, **17**, 193–284.

Gage, J. D., 1978. Animals in deep-sea sediments. *Proc. Roy. Soc. Edinburgh*, **76B**, 77–93.

Gardiner, F. and R. Haedrich, 1978. Zonation in the deep-sea benthic megafauna. *Oecologia*, **31**, 311–317.

Gardiner, L. F., 1975. The systematics, postmarsupial development and ecology of the deep-sea family Neotanaidae (Crustacea: Tanaidacea). *Smithsonian Contrib. Zool.*, **170**, 265 pp.

Gibbs, R. H., 1969. Taxonomy, sexual dimorphism, vertical distribution and evolutionary zoogeography of the bathypelagic fish genus *Stomias* (Stomiatidae). *Smithsonian Contrib. Zool.*, **31**, 1–25.

Grassle, J. F., 1977. Slow recolonization of deep-sea sediments. *Nature*, **265**, 618–619.

Grassle, F., H. Sanders, and W. Smith, 1979. Faunal changes with depth in the deep-sea benthos. *Deep-sea ecology and exploitation. Ambio Special Report*. No. 6, 47–50.

Gray, J. S., 1974. Animal-sediment relationships. In H. Barnes, ed. *Oceanography and Marine Biology, an Annual Review*. **12**, 223–262.

Haedrich, R. and G. Kreft, 1978. Distribution of bottom fishers in the Denmark Straits and Irminger Sea. *Deep-Sea Res.*, **25**, 707–720.

Haedrich, R., G. Rowe, and P. Polloni, 1975. Zonation and faunal composition of epibenthic populations on the continental slope south of New England. *J. Mar. Res.*, **33**, 191–212.

Haedrich, R., G. Rowe, and P. Polloni, 1980. The megabenthic fauna in the deep-sea south of New England. *J. Mar. Res.*, **57**, 165–179.

Haedrich, R. L. and D. C. Judkins, 1979. Macrozooplankton and its environment. In *Zoogeography and Diversity in Plankton*, S. van der Spoel and A. C. Pierrot-Bults, eds. Bimge, Utrecht, pp. 4–28.

Hansen, B., 1976. Systematics and biology of deep-sea holothurians. Part 1, Elasipoda. *Galathea Rep.*, **13,** 1–262.

Hedgpeth, J. W., 1957. Classification of marine environments. In *Treatise on Marine Ecology and Paleoecology*, Vol. 1, *Ecology*, J. Hedgpeth, ed. *Memoirs Geology Society of America* No. 67, pp. 17–27.

Hessler, R. and D. Thistle, 1975. On the place of origin of deep-sea isopods. *Mar. Biol.*, **32,** 155–165.

Jumars, P. and C. Fauchald, 1977. Between community contrasts in successful polychaete feeding strategies. In *Ecology of Marine Benthos*, B. Coull, ed. University of South Carolina Press, Columbia, pp. 1–20.

Jumars, P. and E. Gallagher, in press. Deep-sea community structure: Three plays on the benthic proscenium. In *Ecological Processes in the Deep-Sea*, W. G. Ernst and J. Morin, eds. Prentice-Hall, Englewood Cliffs, N.J., Chapter 10.

Knudson, 1979. The systematics and biology of abyssal and hadal bivalves. *Galathea Rep.*, **11,** 3–236.

Lance, G. N. and W. T. Williams, 1967. A general theory of classificatory sorting strategies. I. Hierarchical systems. *Computer J.*, **9,** 373–380.

Lawson, T. J., 1977. Community interactions and zoogeography of the Indian Ocean Candaciidae (Copepoda: Calanoida). *Mar. Biol.*, **43,** 71–92.

LeDanois, E., 1948. *Les Profondeurs de la Mer. Trente Ans de Recherches sur la Fauna Sous-Marine au Large des Cotes de France*. Payot, Paris, 303 pp.

Lonsdale, P., 1977. Clustering of suspension feeding macrobenthos near abyssal hydrothermal vents at ocean spreading center. *Deep-Sea Res.*, **24,** 857–863.

MacArthur, R. H., 1972. *Geographical Ecology*, Harper and Row, New York, 269 pp.

MacDonald, A. G., 1975. *Physiological Aspects of Deep-Sea Biology*. Cambridge University Press, Cambridge, 450 pp.

Madsen, F. J., 1961. On the zoogeography and origin of the abyssal fauna in view of the knowledge of the Porcellanasteridae. *Galathea Rep.*, **4,** 177–218.

Marshall, N. B., 1954. *Aspects of Deep-Sea Biology*. Hutchinsons, London. 380 pp.

Marshall, N. B. and N. R. Merrett, 1977. The existence of a benthopelagic fauna in the deep-sea. In *A Voyage of Discovery. George Deacon 70th Anniversary Volume*, M. Angel, ed. [Reprinted in *Deep-Sea Res.*, **24** (Suppl.), 483–497.]

McGowan, J. A., 1974. The nature of oceanic ecosystems. In *The Biology of the Oceanic Pacific*, C. B. Miller, ed. Oregon State University Press, Corvallis, pp. 7–28.

Menge, B. and J. Sutherland, 1976. Species diversity gradients: Synthesis of the roles of predation, competition, and temporal heterogeneity. *Am. Natur.*, **110,** 351–369.

Menzies, R. and J. Imbrie, 1958. On the antiquity of the deep-sea bottom fauna. *Oikos*, **9** (2), 192–201.

Menzies, R., R. George, and G. Rowe, 1973. *Abyssal Environment and Ecology of the World Ocean*, Wiley-Interscience, New York. 488 pp.

Mills, E. L., 1969. The community concept in marine zoology, with comments on continuation and instability in some marine communities: A review. *J. Fish. Res. Board Can.*, **26,** 1415–1428.

Mills, E. L., 1972. T. R. R. Stebbing, the *Challenger* and knowledge of deep-sea Amphipoda. *Proc. Roy. Soc. Edinburgh*, **72B,** 67–87.

Murray, J., 1895. A summary of the scientific results obtained at the sounding, dredging and trawling stations of the HMS *Challenger*. *Challenger Rep., Summary Res.*, **2,** 797–1608.

Murray, J. and J. Hjort, 1912. *The Depths of the Ocean*. MacMillan, London, 821 pp.

Musick, J., 1976. *Community structure of fishes on the continental slope and rise off the middle Atlantic Coast*. U.S. Joint Oceanographic Assembly, Edinburgh.

Paine, R., 1977. Controlled manipulations in the marine intertidal zone and their contributions to ecological theory. In *The Changing Scene in Natural Sciences 1776–1976*. Philadelphia Academy of Natural Sciences Special Publication, Vol. 12, pp. 245–270.

Parin, N. V., 1968. Ichthyofauna of the oceanic epipelagia. Acad. Sci. U.S.S.R. Moscow, 186 pp. (in Russian).

Pearcy, W., D. Stein, and R. Carney, 1982. The deep-sea benthic fish fauna of the Northeastern Pacific Ocean on Cascadia and Tufts abyssal plains and adjoining continental slopes. *Biological Oceanography*, **1**, 375–428.

Pearcy, W. G., E. Kuygier, R. Mesecar, and F. Ramsey, 1977. Vertical distribution and migration of oceanic micronekton off Oregon. *Deep-Sea Res.*, **24**, 223–245.

Pérès, J., 1957. Le probleme de l'etegement des formations benthiques. *Rec. Trav. Sta. Mar. d'Endoume*, **21**, 4–21.

Peterson, C., 1980. Approaches to the study of competition in benthic communities in soft sediments. *Estuarine Perspect.*, 291–302.

Peterson, C. G., 1914. On the distribution of the animal communities of the sea bottom. *Rep. Danish Biol. Sta.*, **22**, 1–7.

Pielou, E., 1980. *Biogeography*, Wiley, New York, 351 pp.

Polloni, P., R. Haedrich, G. Rowe, and C. Clifford, 1979. The size–depth relationship in deep-sea animals. *Int. Rev. Ges. Hydrobiol.*, **64**, 39–46.

Rex, M. A., 1977. Zonation in deep-sea gastropods: The importance of biological interactions to rates of zonation. In *Biology of Benthic Organisms*, B. F. Keegan, P. O. Ceidigh, and P. Boaden, eds. Pergamon Press, New York, pp. 521–530.

Rex, M. A., 1981. Community structure in the deep-sea benthos. *Ann. Rev. Ecol. Syst.*, **12**, 331–353.

Roe, H. S. J., 1974. Observations on the diurnal migrations of an oceanic animal community. *Mar. Biol.*, **28**, 99–113.

Root, R. B., 1975. Some consequences of ecosystem texture. In *Ecosystem Analysis and Prediction*, S. A. Levin, ed. Society for Industrial and Applied Mathematics, Philadelphia, pp. 83–97.

Rowe, G. and R. Menzies, 1969. Zonation of large benthic invertebrates in the deep-sea off the Carolinas. *Deep-Sea Res.*, **16**, 531–581.

Rowe, G., P. Polloni, and R. Haedrich, 1982. The deep-sea macrobenthos on the continental margin of the northwest Atlantic Ocean. *Deep-Sea Res.*, **29**, 257–278.

Ryther, J. H., 1963. Geographic variations in productivity. In *The Sea*, Vol. 2, M. N. Hill, ed. Wiley, New York, pp. 347–380.

Ryther, J. H., 1969. Photosynthesis and fish production in the sea. *Science*, **166**, 72–76.

Sanders, H. L., 1960. Benthic studies in Buzzards Bay. III. The structure of the soft-bottom community. *Limnol. Oceanogr.*, **5**, 138–153.

Sanders, H. L. and R. Hessler, 1969. Ecology of the deep-sea benthos. *Science*, **163**, 1419–1424.

Sanders, H., R. Hessler, and G. Hampson, 1965. An introduction to the study of the deep-sea benthic faunal assemblages along the Gay Head–Bermuda transect. *Deep-Sea Res.*, **12**, 845–867.

Siebenaller, J. and G. Somero, 1978. Pressure-adaptive differences in lactate dehydrogenases of congeneric fishes at different depths. *Science*, **201**, 255–257.

Sneath, P. and F. J. Sokal, 1973. *Numerical Taxonomy. The Principles and Practice of Numerical Classification*. Freeman, San Francisco, 573 pp.

Sokolova, M. N., 1959. On the distribution of deep water bottom animals in relation to their feeding habits and the character of sedimentation. *Deep-Sea Res.*, **6**, 1–4.

Sokolova, M. N., 1981. On characteristic features of the deep-sea benthic eutrophic regions of the world ocean. *Trudy Inst. Ocean., Acad. Sci. U.S.S.R.* **115**, 5–12.

Southward, E., 1979. Horizontal and vertical distribution of Pogonophora in the Atlantic Ocean. *Sarsia*, **64**, 51–55.

Streeter, S. S., 1973. Bottom water and benthonic Foraminifera in the North Atlantic glacial and interglacial contrasts. *Quaternary Res.*, **3**, 131–141.

Terborg, J., 1971. Distribution on environmental gradients: Theory and a preliminary interpretation of distributional patterns in the avifauna of the cordillera Vilacabamba, Peru. *Ecology*, **52**, 23–40.

Thiel, H., 1975. The size structure of the deep-sea benthos. *Int. Rev. Ges. Hydrobiol.*, **60**, 575–606.

Turner, R. D., 1973. Wood-boring bivalves, opportunistic species in the deep-sea. *Science*, **180**, 1377–1379.

Van der Spoel, S. and A. C. Pierrot-Bults, 1979. *Zoogeography and Diversity of Plankton*. Bunge, Utrecht, 410 pp.

Veronis, G., 1973. Model of world ocean circulation. I. Wind driven, two layer. *J. Mar. Res.*, **31**, 228–288.

Vinogradov, M. E., 1968. *Vertical Distribution of the Oceanic Zooplankton*. Acad. Sci. U.S.S.R., Institute of Oceanography, Moscow, 339 pp.

Vinogradova, N. G., 1962. The vertical distribution of the deep-sea bottom fauna in the abyssal zone of the ocean. *Deep-Sea Res.*, **8**, 245–250.

Walsh, J. W., G. T. Rowe, C. P. McRoy, and R. Iverson, 1981. Biological export of shelf carbon is a sink of the global CO_2 cycle. *Nature*, **291**, 196–201.

Whittaker, R. H., 1973. Approaches to classifying vegetation. In *Ordination and Classification of Animal Communities*, R. W. Whittaker, ed. *Handb. Veg. Sci.*, Vol. 5, Junk, The Hague, pp. 325–354.

Wigley, R. L. and R. B. Theroux, 1981. Macrobenthic invertebrate fauna of the middle Atlantic bight region: Part II. Faunal composition and quantitative distribution. U.S. Geological Survey Professional Papers. No. 529-N, 198 pp.

Wolff, T., 1960. The hadal community, an introduction. *Deep-Sea Res.*, **6**, 95–124.

Wolff, T., 1962. The systematics and biology of bathyal and abyssal Isopoda, Asellota. *Galathea Rep.*, **6**, 1–358.

Wolff, T., 1979. Macrofaunal utilization of plant remains in the deep-sea. *Sarsia*, **64**, 117–136.

Worthington, L. V., 1976. On the North Atlantic circulation. *Johns Hopkins Oceanogr. Stud.*, **6**, 1–110.

Zenkevitch, L. A., 1963. *Biology of the Seas of the U.S.S.R.* Wiley, New York, 955 pp.

Zenkevitch, L. A., ed., 1969. *Biology of the Pacific Ocean*, Vol. 2, *The Deep-Sea Bottom Fauna*. Acad. Sci. U.S.S.R., Inst. Ocean., Moscow, 353 pp.

Zenkevitch, L. A. and Y. A. Birstein, 1956. Studies of the deep-water fauna and related problems. *Deep-Sea Res.*, **4**, 54–64.

10. SPATIAL STRUCTURE WITHIN DEEP-SEA BENTHIC COMMUNITIES[1,2]

PETER A. JUMARS AND JAMES E. ECKMAN

1. Introduction

The late E. W. Fager taught that "pattern [of organisms in space] is one of the easiest obtainable clues to interaction in a non-uniform world." The following is an attempt to apply and extend the methods he taught for deciphering these clues to data from the deep sea, where spatial patterns have been among the few clues accessible to oceanographers. We limit our discussion to within-community patterns. Changing population abundances and species compositions from one community to the next are explored in detail by Carney et al. in Chapter 9 of this volume.

E. W. Fager also taught the statistics of sampling. He made it abundantly clear that whether one failed or succeeded in associating patterns with their causes, these patterns set the levels of precision attainable in estimates of population parameters derived from field sampling programs. So basic an ecological task as estimating population density thus requires some description of the pattern of variation observed. Perhaps less obviously, studies of parameters influenced by organisms also require at least a minimal knowledge of the pattern of organisms in space. For example, the locations and activities of animals in the sediments are critical to stratigraphic interpretations (Piper and Marshall, 1969; Guinasso and Schink, 1975; Risk et al., 1978), just as biogenic oxygen fluxes are pivotal in determining rates and directions of geochemical reactions (Goldhaber et al., 1977). Stratigraphic or geochemical models assuming spatial uniformity of biological activity are thus limited in accuracy and precision by the actual pattern of organisms in space.

Pielou (1969) suggested the practice of referring to these patterns of organism location in space as "dispersion patterns." She used this term to maintain the distinction between patterns of location and the statistical distributions used to describe them. We concur that this distinction should be maintained, but the dynamic sense in which "dispersion" frequently is used in oceanographic contexts unfortunately may again lead to confusion of concepts. Despite this potential difficulty, we refer to the pattern of organism location in space as the dispersion pattern of the individuals. More formally, by "dispersion pattern" we denote the

[1]Dedicated to E. W. Fager.
[2]Received 29 January 1979.

set of cartesian coordinates occupied at any one time by the centers of mass of the individuals in the (statistical) population studied.

We first explore the potential causes of deep-sea dispersion patterns. Next we briefly review statistical methods for describing and analyzing these patterns, as a prelude to discussing the dispersion patterns observed to date in the deep sea. After treating the categories of patterns observed, we examine the problem of relating these observations with their generating mechanisms. Then we address the kinds of pattern that are likely to have been missed. Finally, we briefly explore the consequences of organism dispersion patterns with regard to biological and nonbiological parameters of current interest in studies of the deep-sea benthic boundary layer.

We discuss (one-dimensional) vertical and (two-dimensional) horizontal dispersion patterns separately. No one would propose a uniform or random depth–frequency distribution of number of individuals per depth stratum in the benthic boundary layer; the task is thus to describe changes in abundance with depth. In the horizontal domain, on the other hand, a random arrangement of points on a plane provides a reasonable model to use as a null hypothesis in investigating dispersion patterns of deep-sea populations; the task is thus to document and describe departures from this null hypothesis or to discredit its alternatives. An obvious danger in treating vertical and horizontal dimensions separately is the inability to detect interactions. Individuals of potentially competing species, for example, may segregate vertically only when their horizontal coordinates coincide. Because of the lack of suitable data, we are unable to treat such potential interactions.

We similarly are restricted in our ability to study the evolution of patterns. Both theory (Levin and Paine, 1974; Connell and Slatyer, 1977) and deep-sea results (Carney et al., this volume, Chapter 9) argue strongly that time-series data on dispersion patterns aid greatly in identifying the causes of patterns. If patterns are easy clues, then time series of patterns are far easier clues. The added dimension of time, however, has rarely been achieved in the study of deep-sea dispersion patterns, due both to logistics problems (Rowe and Sibuet, this volume, Chapter 2) and to the statistical difficulties of treating time and space in the same analysis (Cliff et al., 1975).

2. Causes of Dispersion Patterns in the Deep Sea

The causes of dispersion patterns have been classified by Hutchinson (1953). *Vectorial* processes are chemical or physical causes that produce patterns in local abundances. *Reproductive* processes such as pair-bond formation or multiple births lead to patterns. *Social* behaviors such as schooling or herding can also alter or maintain single-species' dispersion patterns. Two or more species can interact in a *coactive* fashion (e.g., as predator and prey or as commensals) to modify or produce patterns in one or both species' populations. Finally, patterns also result in nondeterministic ways from *stochastic* processes.

Causes may be of second (or higher) order, adding to the difficulty of identifying them. For example, one species' burrowing may produce depressions in the sediment surface, and a second species may avoid these features. Whereas the first-order cause of the second species' pattern is the vectorial one due to local bottom topography, the second-order cause is coactive. Multiple causation is also likely, making clues derived from pattern much more challenging to interpret. E. W. Fager (unpublished) has thus suggested the term "historical" be applied to patterns that are likely due to undecipherably complex combinations of past processes. Causes of subsequent changes in patterns initially labeled as historical may then be more readily determined without addressing the manifold causes that led to the pattern observed at the first sampling. Despite these complexities, Hutchinson's list of five causes of pattern provides a convenient outline for examining the characteristics expected of deep-sea, pattern-generating processes.

A. Vectorial Causes

A naive first impression of the deep-sea floor is that, apart from major transform faults, spreading centers, and trenches, it is flat and featureless. Heezen's physiographic diagrams, coupled with photographs and other graphic evidence of active deep-sea geologic processes (Heezen and Hollister, 1971), have done much to dispel this popular notion on larger scales. On the smaller intracommunity scales of concern in the present context, there likewise is topographic relief that can be expected to correlate with spatial patterns of erosion (Johnson, 1972), with locally varying sedimentation rates (Lister, 1976; Hargrave and Nielsen, 1977), and with spatial patterns of secondary circulation (Hollister et al., 1976). On scales from meters to millimeters, bottom photographs (Heezen and Hollister, 1971; Grassle et al., 1975; Rhoads, 1974) (Figs. 1 through 4 in this chapter) give a vivid impression of physical heterogeneity. In regions of low current activity, this small-scale physical structure is predominantly biogenic (with the possible exception of manganese nodules, which may or may not be produced largely with the help of organisms), effecting what Grassle et al. (1975, p. 462) have called a "patina of biological activity"—a microrelief of fecal pellets, feeding depressions and mounds, tubes, tests, and tracks. Likewise in the vertical domain of the surface decimeter of sediments, organisms contribute substantially to variations in physical parameters (Rhoads, 1974).

Superimposed on, and partially controlled by, this background of physical variability in space, is chemical variability. Strong vertical gradients in chemical parameters are well documented in nearly all varieties of marine sediments (Berner, 1976). Horizontal variability has not been measured as frequently but must also be expected (Aller and Yingst, 1978).

Physical and chemical heterogeneities certainly are produced in shallower bottoms as well. They are, however, more rapidly homogenized by more active physical processes and by more abundant animals. Hence they frequently may not exist long enough to provide the requisite environmental heterogeneity for microhabitat or temporal partitioning by other species. This argument holds best for low-energy (in the physical sense), deep-sea environments; however, it may also extend to high-energy abyssal bottoms. Here animal structures provide the bottom roughness components that interact with the flow to produce characteristic crag-and-tail features (Gage, 1977). This argument for the importance of biogenic structure in affecting dispersion patterns is complicated by the paucity of knowledge of generation and other response times of deep-sea animals (Turekian et al., 1975). If food resources are in short supply, though, both ecological theory

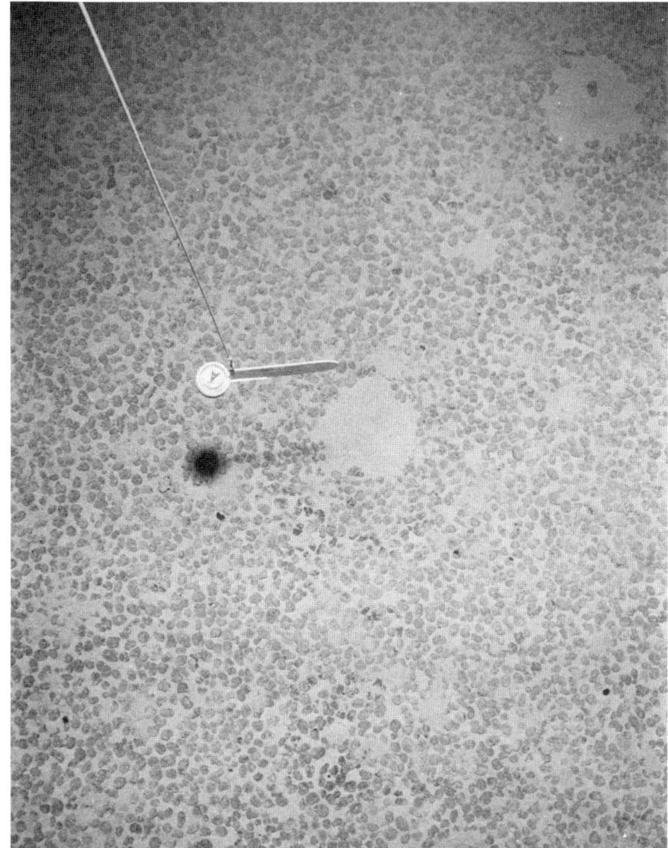

Fig. 1. Photograph of the bottom at 5200-m depth at Deep Ocean Mining Environmental Survey (DOMES) Site A (see Fig. 19). Note the multiple scales of environmental heterogeneity provided by the pattern of nodule cover. Bare spots may be the result of animal burrowing activity locally burying nodules. The compass is 7.6 cm in diameter. (DOMES cruise photograph courtesy of Larry Parsons, chief photographer.)

(MacArthur, 1972) and empirical results (Schoener, 1974) further suggest that selection will produce animals capable of partitioning resources by segregating among microhabitats. Analogous reasoning has been applied to other systems where physical disturbances are weak or infrequent, such as to particular plant communities (Westman, 1975).

B. Reproductive Causes

We are unaware of any theoretical studies that specifically address the question of reproductively caused dispersion patterns in the deep sea, although they are mentioned tangentially in several treatments. Clutch sizes of deep-sea species tend to be small (Grassle and Sanders, 1973), and planktonic dispersal is rare (Thorson, 1950). These two factors may act in opposition with respect to the probability of detecting reproductively produced aggregations. In contrast to the situation where dispersal of siblings to increasingly larger areas provides substan-

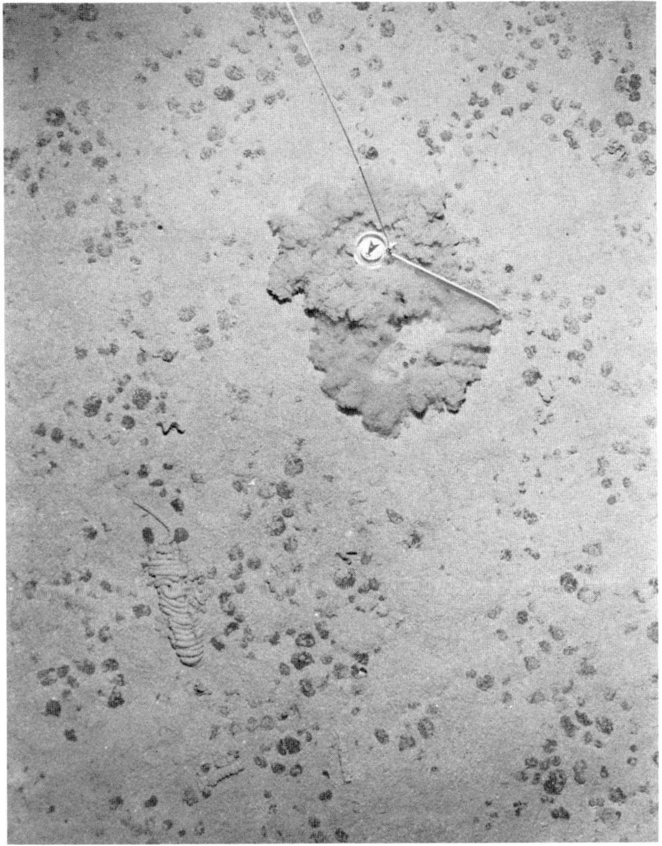

Fig. 2. Photograph of the bottom at 5200-m depth at DOMES Site A (see Fig. 19). The three neatly folded holothurian fecal mounds are almost certain to influence local meiofaunal and macrofaunal populations (scale and credit as in Fig. 1).

tially better odds of repeatedly finding favorable spots for production of the next generation (Strathmann, 1974), in the deep-sea benthos, where small-scale biogenic spatial heterogeneity may predominate, little may be gained (and considerable energy lost) by dispersing over more than a few square meters. Dispersal mechanisms and rates, reproductive rates, and areas over which siblings disperse are too poorly known, however, to allow accurate or precise prediction of expected patterns.

C. *Social Causes*

Intraspecific interactions may also act in divergent directions. Competition for space or for food supplies closely associated with available space classically is treated as a mechanism that will set a lower limit on interindividual distances (Moore, 1954). The consequent spacing might be expected among individuals of many deep-sea species if food reaches population-limiting levels for them. Alternatively, aggregative social behavior may set an upper limit (that of reciprocal detection by individuals) on interindividual distances. Ultimate causes of aggrega-

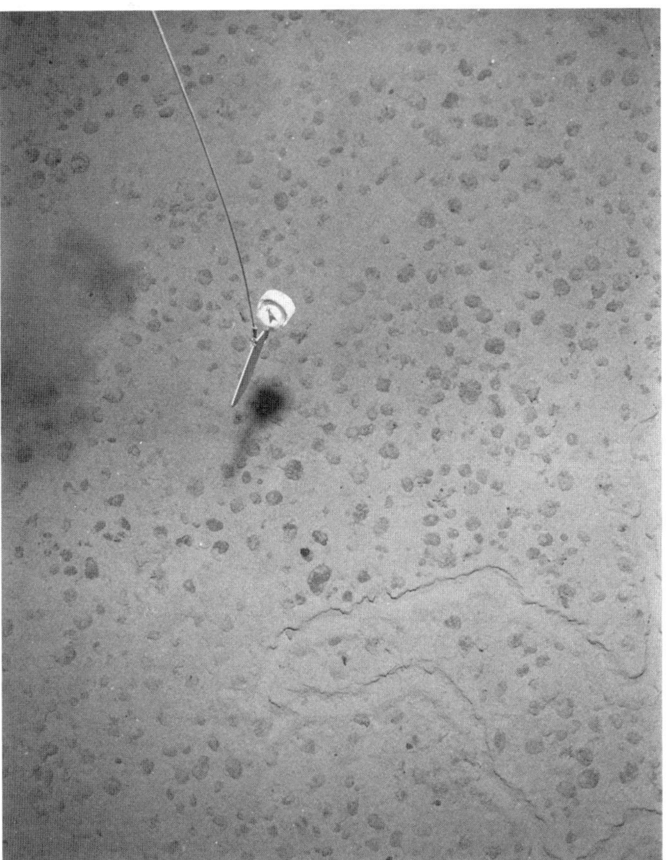

Fig. 3. Photograph of the bottom at 5200-m depth at DOMES Sites A (see Fig. 19). Note the irregular shape of the feeding trough apparently produced by a holothurian (scale and credit as in Fig. 1).

tive social behavior may come from one of the other categories of causes, such as from reproductive or coactive interactions. Aggregation may aid in fertilization of those species that exist at low average densities but shed gametes freely into the water column (Rokop, 1974). It may also be effective as a mechanism for reducing mortality due to predation (Taylor, 1976). Again, however, we are unaware of any predictions of social patterns peculiar to the deep sea.

D. *Coactive Causes*

Intraspecific interactions affecting deep-sea benthic community structure have been discussed at length (Slobodkin and Sanders, 1969; Dayton and Hessler, 1972; Grassle and Sanders, 1973; Menge and Sutherland, 1976; Rex, 1977). Predation is acknowledged (Grassle and Sanders, 1973; Jumars, 1975a) to be a likely driving force for local successional sequences (Connell and Slatyer, 1977), but no agreement has been reached on the dispersion patterns that would be expected to result. Again, different coactive processes are likely to drive patterns in disparate directions. Interspecific competition for space might be expected to lead toward regular spacing of individuals—whereas predation and other, perhaps amensa-

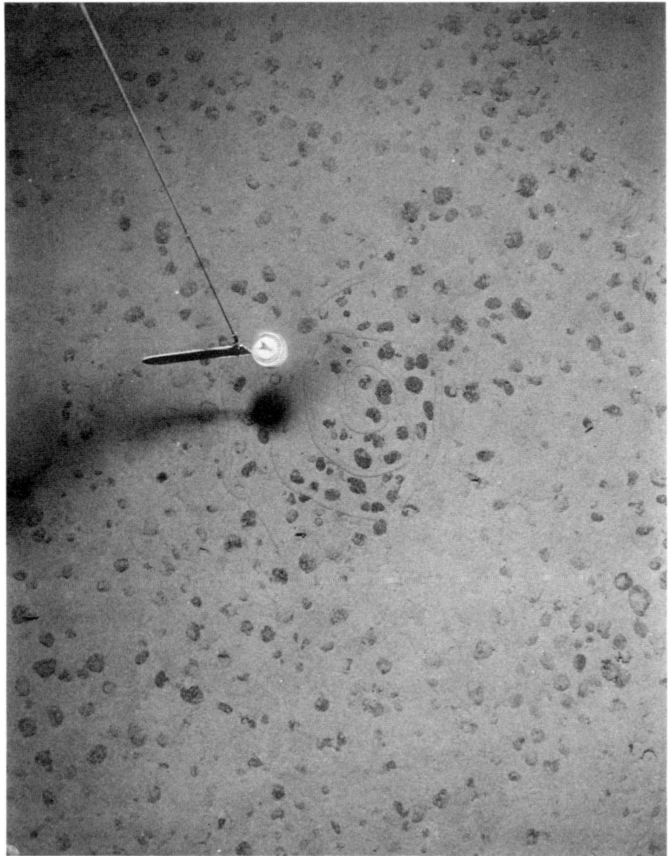

Fig. 4. Photograph of the bottom at 5200-m depth at DOMES Site A (see Fig. 19). In spite of the surprising regularity of this spirally arranged enteropneust feeding trace, consider how difficult its effects on the underlying fauna would be to document (scale and credit as in Fig. 1).

listic, disturbances can lead to a wide variety of patterns (Levin and Paine, 1974). Competition has been argued to be most likely at the highest trophic levels (Menge and Sutherland, 1976), but animals at these trophic levels are highly mobile, again making predictions (and measurements) of dispersion patterns difficult.

E. Stochastic Causes

Dispersion patterns produced by stochastic processes likewise can be diverse. Whereas a continual random walk (Feller, 1968) in a homogeneous medium might be expected to lead to truly random dispersion patterns, we have already established (Section 2.A) that deep-sea benthic habitats are far from homogeneous. As every commuter can intuit, if the random walks of population members are locally slowed for any reason whatsoever, a local aggregation is likely to result. Given the feeding trajectories of various deep-sea animals (as surmised from feeding traces), the stochastic portions of the feeding and moving processes can lead to a wide diversity of dispersion patterns (Papentin, 1973). Although not constructed for examining dispersion patterns per se, the bioturbation models of Piper and Marshall (1969) and Risk et al. (1978) illustrate some of this potential diversity. Nor is stochastic pattern limited to cases of animal movement. Pielou (1960), in a computer model, dropped "seeds" at random points on a plane and allowed the resultant "trees" to "grow" at a specified rate until their root diameter either attained a specified maximum size or impinged on a neighbor. Depending only on the specified maximum root diameter, patterns of aggregation, random location, or even spacing of the trees resulted. Connell (1963) showed a nearly analogous sequence of events in a tube-building amphipod, illustrating the potential applicability of similar stochastic models to the deep-sea benthos.

3. Statistics of Pattern Description and Analysis

A. *Horizontal Patterns*

Given our formal definition of dispersion pattern (Section 1), an exact description of the dispersion pattern of a population would simply be a complete listing of the set of coordinates occupied by each individual. Such a listing, however, is rarely either possible or desirable. Useful summary descriptions of dispersion patterns must allow comparisons from population to population and thus must be independent of the arbitrary origin of the coordinate system. Implicitly, by restricting consideration to the within-community scale, we have already assumed a system in which only relative spatial location of the individuals is of concern. Explicitly, we assume stationarity, in other words, that over the horizontal scales of concern, unidirectional or long-wavelength periodic changes in abundance are quantitatively unimportant. We also assume isotropy—that is, that on the average, the pattern of change is the same in all compass directions. Given known exceptions to these assumptions in such physical variables as deep-sea topography (mud waves) and current velocity (Hollister et al., 1976), both assumptions merit close scrutiny when data on animal dispersion patterns become available to test them. In the absence of contradictory data, however, the added complications of descriptors not assuming stationarity and isotropy are unwarranted.

We further assume that the ideal of knowing each individual's relative spatial coordinates cannot, in general, be achieved for benthic infauna. The closest approximation to this ideal comes from using a sample size that, on the average, collects no more than one individual of the population under consideration, thereby focusing on scales where interindividual interactions are possible. Since most of the sampling programs with which we deal used samples of fixed dimensions, we also limit statistical discussions to descriptions of such samples. The raw data available are thus in terms of numbers of individuals per quadrat (per core) and of geographic coordinates of those quadrats.

Single-Species Patterns

The most commonly employed descriptors and measures of dispersion pattern use only one of these two kinds of information, that is, the numbers per quadrat. A plethora of summary statistics of the observed frequency distribution of numbers per quadrat have been suggested; however, nearly all are based in one way or another on the first and second moments (mean and variance) of the sample distribution (Patil and Stiteler, 1974). We use the variance : mean ratio of the per-quadrat counts, which now appears to have good theoretical (Taylor and Taylor,

1977) and empirical (Taylor et al., 1978) support. Specifically, the variance and mean are usually related (Taylor et al., 1978) as

$$s^2 = a\bar{x}^b \tag{1}$$

making a log–log plot of s^2 versus $\bar{x}$ highly informative, because

$$\log(s^2) = \log a + b \log \bar{x} \tag{2}$$

We instead plot $\log(s^2/\bar{x})$ versus $\log(\bar{x})$. The relationship derived from equation 1 above remains linear:

$$\log\left(\frac{s^2}{\bar{x}}\right) = \log a + (b - 1) \log \bar{x}. \tag{3}$$

The Poisson distribution, for which the variance: mean ratio in theory equals 1, provides a special case, reducing equation 3 to

$$\log a + (b - 1) \log \bar{x} = 0, \tag{4}$$

requiring both a and b to equal 1. When the data are not Poisson distributed, and $b \neq 1$, it is apparent that $s^2/\bar{x}$ is correlated with $\bar{x}$.

To utilize effectively the information contained in the geographic coordinates of the quadrats, we again use a ratio, that of the spatial autocovariance to the sample variance. This ratio, suitably weighted as a function of intersample distance, is a measure of spatial autocorrelation. If samples at a particular distance from each other tend to be similar in per-quadrat abundance, positive spatial autocorrelation is said to exist at that intersample distance or "spatial lag." Conversely, if samples at a different spatial lag from each other tend to be dissimilar, negative spatial autocorrelation is said to exist at that intersample distance. Negative autocorrelation at one spatial lag does not preclude positive autocorrelation or a lack of autocorrelation in per-quadrat abundances at another. Hence a useful summary of the observed pattern is a plot of the sign and magnitude of autocorrelation versus spatial lag. Such a diagram is called a *correlogram*.

Various autocorrelation statistics have been developed, but we employ two that have recently been refined by Cliff and Ord (1973), Moran's I and Geary's c, whose formulations follow:

$$I = \left(\frac{n}{W}\right) \sum_{i=1}^{n} \sum_{\substack{j=1 \\ i \neq j}}^{n} w_{ij} z_i z_j \bigg/ \sum_{i=1}^{n} z_i^2, \tag{5}$$

$$c = \left(\frac{n-1}{2W}\right) \sum_{i=1}^{n} \sum_{\substack{j \neq 1 \\ i \neq j}}^{n} w_{ij}(x_i - x_j)^2 \bigg/ \sum_{i=1}^{n} z_i^2, \tag{6}$$

where x_i is the variate value (species' abundance) in sample i, n is the number of samples, $z_i = x_i - \bar{x}$, $w_{ij} = f$ (distance between samples i and j), and

$$W = \sum_{i=1}^{n} \sum_{\substack{j=1 \\ i \neq j}}^{n} w_{ij}.$$

Considerations in the selection of weights (w_{ij}) are detailed by Jumars et al. (1977). We will limit ourselves to two alternatives. For calculation of a single value of I or c from a given collection of data, we will set w_{ij} = (distance between samples i and $j)^{-2}$. In constructing correlograms, on the other hand, $w_{ij} = 1$ if the inter-sample distance falls within a given interval, while $w_{ij} = 0$ for all other spatial lags (Jumars et al., 1977; Henley, 1977). As can be seen by inspecting equations 5 and 6, I is sensitive to extreme values, whereas c is sensitive to similar or dissimilar values at the selected spatial lags. For the case of w_{ij} = (intersample distance)$^{-2}$, the smaller spatial lags are overridingly important in controlling the values of I and c.

We have found the nonparametric versions of these statistics most useful (I and c under the assumption of randomization, *sensu* Cliff and Ord (1973), as applied to ecological abundance data by Jumars et al. (1977) and Jumars (1978). The assumptions of the parametric form are clearly violated by equation 1 above. The details of alternative versions may be pursued in the excellent introduction to spatial autocorrelation given by Sokal and Oden (1978a,b).

Figure 5 illustrates the basic difference among the behaviors of I, c, and $s^2/\bar{x}$. Note that the assignment of different coordinates to abundances has no effect on $s^2/\bar{x}$, whereas I and c are "scale invariant" in the statistical sense. That is, one can multiply all the quadrat counts by any constant without changing the value of I or c (Cliff and Ord, 1973) (Fig. 5*a–c*). It is again semantically unfortunate that in pattern study the word "scale" is frequently taken to mean a function of horizontal distance across "patches," as distinct from "intensity," or difference in organism density between patches and areas outside patches. In this sense, the correlograms address the scale of pattern, whereas the variance:mean ratio addresses its intensity. By separating the pattern components to which I, c, and $s^2/\bar{x}$ respond in Fig. 5, we hope to avoid excessive semantic confusion. Whereas I and c take spatial coordinates into account, $s^2/\bar{x}$ does not. Figure 5 also shows the emptiness of saying that one population is more or less patchy than another without at the same time stipulating the parameters used in the pattern comparison.

As Fisher (1970) has suggested, one can use a χ^2 distribution with $(n - 1)$ degrees of freedom (where n is the number of quadrats tallied) to assess significant departure from the theoretically expected variance, σ^2. Specifically, the ratio of the observed to the theoretical variance, s^2/σ^2, should approximate the degrees of freedom and do so more closely with increasing n. This ratio is often referred to as the *index of dispersion*.

For a truly random dispersion pattern at typical deep-sea animal densities, it is reasonable to assume that, in a hypothetical sample of the same diameter as an individual, the probability of capturing an individual is small. Hence the Poisson approximation to a normal distribution should describe adequately the distribution of individual densities among quadrats. The theoretical variance of a Poisson distribution equals its observed mean, so that the appropriate index of dispersion is $s^2/\bar{x}$. In some respects, the dependence of $s^2/\bar{x}$ on n and, for non-Poisson distributions, on $\bar{x}$ (viz., equations 3 and 4 above) makes this ratio a better indicator of departure from Poisson assumptions than a descriptor of dispersion pattern over varying sample sizes and mean abundances. For this reason, we prefer to present $s^2/\bar{x}$ graphically (plotted against $\bar{x}$) and to emphasize the degree (probability level

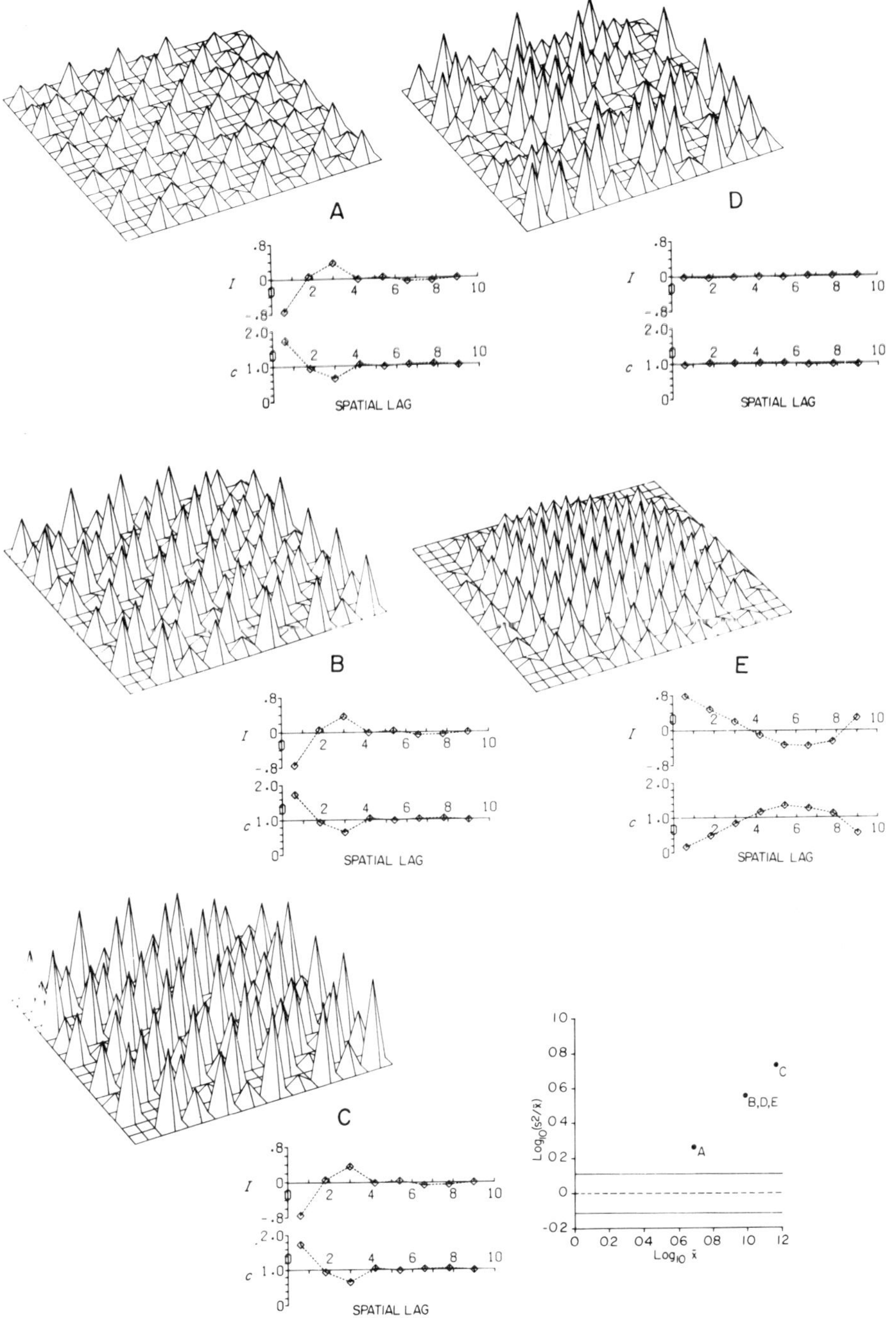

Fig. 5. Comparison of the behaviors of I, c, and $s^2/\bar{x}$ in analysis of spatial pattern of hypothetical faunal abundances in 10 × 10 arrays of quadrats (peak height proportional to per-quadrat abundance; one unit of distance equal to quadrat width). Patterns B and C were produced by taking the checkerboard pattern of A (its highest per-quadrat abundance equaling 9) and multiplying each per-quadrat abundance by 2 and 3, respectively. Pattern D was formed by randomly permuting the abundances in B, and E was formed by clustering all the high abundances in B. Note that the correlograms cannot distinguish patterns A, B, and C and that the plot of log $(s^2/\bar{x})$ versus log $\bar{x}$ cannot distinguish patterns B, D, and E. The dashed line in the latter graph is Poisson expectation, and the solid lines are its 95% confidence limits.

with the Poisson as null hypothesis) of departure of $s^2/\bar{x}$ from its expected value, rather than to assign some meaning to the absolute value of the isolated ratio. Although there are measures of pattern that can overcome some of the apparent difficulties with $s^2/\bar{x}$, other problems are often created (Taylor et al., 1978). Rather than introduce one or more of several potentially confusing alternative measures of pattern, we instead quote or illustrate the number of samples, sample dimensions, and mean density with $s^2/\bar{x}$, so that the interested reader can apply other such measures at will.

It is commonly said that $s^2/\bar{x} > 1$ is indicative of aggregation among individuals. Figure 5, however, should serve to temper such direct extrapolations from summary statistics of the distribution of numbers of individuals among quadrats to dispersion patterns of animals. A wide diversity of dispersion patterns (Fig. 5*b,d,e*) produce identical variance:mean ratios. If the observed quadrat counts are distributed randomly among the quadrat coordinates, then I should approximate the quantity $-(n - 1)^{-1}$, which approaches zero as n becomes large, and c should approximately equal 1 for all values of n. The methodology for detecting significant departures from these expectations is given by Cliff and Ord (1973) and is summarized by Jumars et al. (1977), Sokal and Oden (1978a,b), and Jumars (1978). Sokal and Oden (1978a) and Jumars (1978) give some useful corrections to conceptually minor but nonetheless troublesome typographic errors in formulae provided by Cliff and Ord (1973).

Multispecies Patterns

An additional question asked of pairs of species is whether their per-quadrat abundances are independent. The answer is an important though not definitive clue to coactive generating processes. Correlation of per-quadrat counts is an obvious test of independence, but it is quite weak at the abundance levels normally encountered in deep-sea samples. For this reason, Jumars (1975b) suggested testing the concordance in abundance among more than two species in providing a stronger answer to the less specific question of whether, on the average, deep-sea species are independently Poisson distributed in abundance among samples. This method takes advantage of the fact that independent χ^2 distributions are additive (Fisher, 1970) to draw its conclusions (see Fig. 6 and Table I). The total χ^2 determines whether species *on the average* depart from Poisson distributions in per-core abundances; the pooled term determines whether the per-core abundances of the individual species when summed by core produce a Poisson distribution (as they should if the species' abundances are independent and individually Poisson distributed), and the heterogeneity term determines whether species tend to be concordant (homogeneous) or discordant (heterogeneous) in their per-core abundances. When a large group of species is heterogeneous in per-quadrat abundances, it often can be divided into a small number of subgroups that internally are homogeneous or independent in per-species, per-quadrat abundances (Jumars, 1976).

An alternative means for grouping species when they are not independent in per-quadrat abundance is the "number of expected species shared (NESS)" community classification procedure developed by Grassle and Smith (1976). It groups species that occur simultaneously more frequently than would be expected by chance if their per-sample abundances were independent. The NESS classifica-

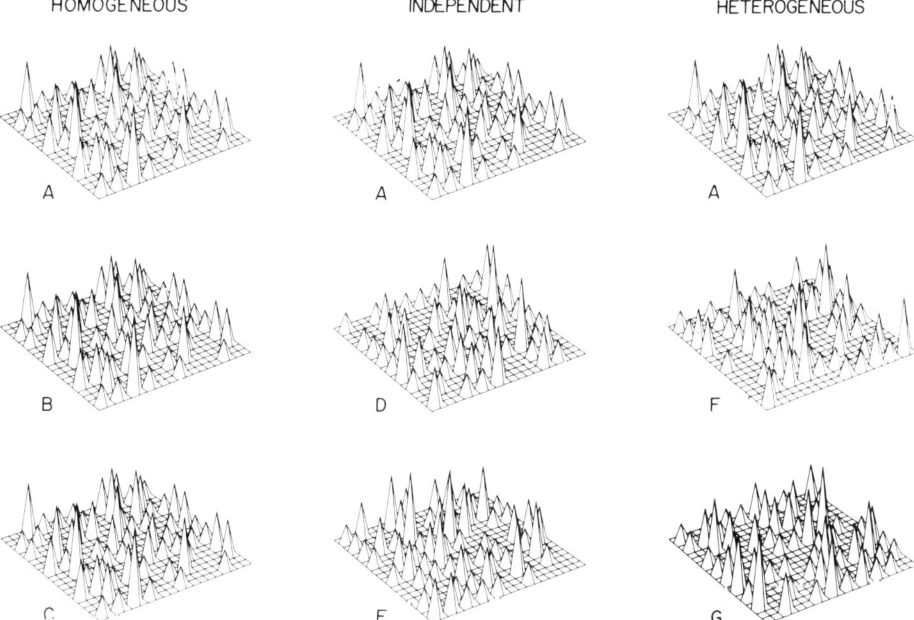

Fig. 6. Illustration of the heterogeneity χ^2 procedure of Table I in hypothetical 10 × 10 arrays of quadrats. Peak height is proportional to per-quadrat abundance, which is Poisson distributed, with $\mu = \sigma^2 = 1$, for A through G (maximum peak height = 4). Patterns A, B, and C are identical and thus perfectly concordant in per-quadrat abundances; D and E are independent random permutations of A; and F and G are permuted such that neither low nor high abundances in A, F, and G correspond in spatial location, making the three hypothetical species discordant in their local abundances.

tion is derived from the Hurlbert (1971) expected number of species diversity estimator, which had earlier been used in a less direct way as an indicator of discordance among species' abundances (Hessler and Jumars, 1974; Jumars, 1975a; Heck et al., 1975). Other clustering techniques and multivariate statistics can also aid in identifying species groups and the environmental variables to which they respond. We introduce such methods at the time of their application to particular examples.

It is unfortunate that no standardized, efficient, and generally applicable procedure exists for cross-correlating species' abundances, over a range of spatial lags, from irregularly spaced quadrats in two dimensions. One might expect, for example, that potential competitors might segregate spatially on some characteristic scale, or that prey and predators might have spatially displaced maxima in abundance. Available procedures for map comparisons (Cliff and Ord, 1973; Switzer 1975) tend to be cumbersome and error prone unless samples are abundant and regularly spaced.

B. Vertical Pattern

No sampler exists that will reliably retain organisms in the depth strata that they normally occupy. Predators frequent the sediment surface, and rapid withdrawal into subsurface tubes or burrows is a common means of prey escape.

TABLE I
Heterogeneity χ^2 Summary for Data in Fig. 6

Species	$\frac{s^2}{\bar{x}}(n-1) \cong \chi^2$	Degrees of Freedom	Two-Tailed Probability
	Homogeneous Case		
A	97.97	99	>.800
B	97.97	99	>.800
C	97.97	99	>.800
A + *B* + *C*	293.91	99	<.001
Total χ^2	293.91	297	>.800
Pooled χ^2	293.91	99	<.001
Heterogeneity χ^2	0.00	198	<.001
	Independent Case		
A	97.97	99	>.800
D	97.97	99	>.800
E	97.97	99	>.800
A + *D* + *E*	82.46	99	≅.200
Total χ^2	293.91	297	>.800
Pooled χ^2	82.46	99	≅.200
Heterogeneity χ^2	211.45	198	>.600
	Heterogeneous Case		
A	97.97	99	>.800
F	97.97	99	>.800
G	97.97	99	>.800
A + *F* + *G*	5.02	99	<.001
Total χ^2	293.91	297	>.800
Pooled χ^2	5.02	99	<.001
Heterogeneity χ^2	288.89	198	<.001

Deep-sea samplers further require considerable time in transit to the ocean surface, with little, if any, restraint on vertical excursions by animals within the sediments. Extant vertical depth–frequency data thus do not merit extensive quantification and analysis. We present such data simply as histograms and limit our analyses to nonparametric approaches at most.

4. Single-Species Patterns Observed

Random dispersion patterns are rare in nature (Taylor et al., 1978). They are rarer still in samples from nature because most sampling errors are likely to increase estimates of population variance but not of mean population density. Of the following list of likely errors, only those susceptible to human subjectivity (i.e., sorting and counting errors) are likely to reduce estimates of the population variance:

1. Active avoidance of, or attraction to, the sampler.
2. Bow-wave effects.
3. Imprecision of area taken by the sampler.
4. Escape or winnowing from the sampler during sample retrieval.
5. Loss during sample removal from the sampler.
6. Variation in retention efficiency during sample washing.
7. Variation in quality of fixation and preservation.
8. Variation in efficiency of animal removal from residual sediments.
9. Errors in identification.
10. Errors in counting, recording, or calculation.

Hence most sampling errors will artifactually inflate the variance:mean ratio. Marginally significant departures from Poisson expectation in the direction of apparent aggregation $(s^2/\bar{x}) > 1$ must thus be regarded with suspicion, whereas departures in the direction of apparent regularity $(s^2/\bar{x}) < 1$ are all the more surprising.

We summarize the available data in order of decreasing body size and motility. Our general arrangement is thus one of decreasing ambit (Lloyd, 1967) or volume of space in which the majority of day-to-day activities are carried out. If dispersion patterns are to be used to infer interindividual (either intra- or interspecific) interactions, ambits provide a natural scale for individual sampling units or quadrats. In sessile animals, the ambit is easy to describe; in more motile species, it is a statistical idea that may be conceptualized as a cloud whose local density is proportional to the relative length of the individual's life-span spent in that vicinity. In addition to the more obvious effects of animal size and motility, the topology of this cloud depends on whether the species utilizes homogeneously dispersed, fine-grained resources (Levins, 1968) or moves from patch to patch of coarse-grained resources.

Sufficient data are available for analyses of megafaunal, macrofaunal, and meiofaunal dispersion patterns. Although some counts of microbiota per unit area or per unit volume of sediments have been made (Burnett, 1977), the techniques are in the early stages of development and identifications to species level are essentially nonexistent. Furthermore, the complex statistics of subsampling are critically involved in assessing field dispersion patterns from these data. Consequently, we exclude microbiota from our review.

A. *Megafauna*

We choose to define the megafaunal category operationally rather than by some strict and arbitrary size scale. We include the taxa containing large individuals that are usually poorly sampled by cores and grabs. Their prevalent means of quantitative "sampling" has been photographic. Although photography eliminates some sources of apparent variance in abundance (e.g., winnowing, sample screening), it introduces others. Besides the obvious difficulties in taxonomic identification of specimens in arbitrary poses, there are more insidious problems of temporary burial of a portion of the population and frequently variable resolution among frames or between different photographic methodologies (Barham et al., 1967). It is difficult to specify precisely, for example, the size range of animals reliably "captured" on film. Nonetheless, the relative rapidity with which such data can be gathered and analyzed brings the statistical law of large numbers into play, greatly alleviating the burdens of imprecision in the analysis of single photographs.

Horizontal Patterns

The dispersion pattern study by Grassle et al. (1975) of continental slope megafauna off New England stands out in its thoroughness. They used transects of photographs, each covering roughly 4 m^2, to obtain information on linear scales up to a few hundred meters. Five species were sufficiently abundant for detailed treatment (listed in order of decreasing maximum density per transect): the brittle star *Ophiomusium lymani* at 1800- and 1500-m depths, a quill worm *Hyalinoecia* sp. at 500 m, the urchin *Echinus affinis* at 1800 m, the urchin *Phormosoma placenta* at 1300 m and a cerianthid anemone at 1800 m. Transects of contiguous or nearly contiguous (Grassle et al., 1975; Gage and Coghill, 1977) samples can be analyzed in a number of different ways (Hill, 1973) to address the two aspects of dispersion pattern that we have highlighted above. Insofar as possible from the published data analysis, we discuss the results in terms of both the variance: mean ratios observed and spatial autocorrelative trends expected. No critique of the original authors' alternative methods is implied. Rather, we attempt to introduce as few different and potentially confusing statistical methodologies as possible, dictating an approach that is not limited strictly to transects of quadrats.

In addition to being unusual in mean abundance, the surface deposit feeding–omnivorous *O. lymani* was unique in its dispersion pattern. It was the only species of the five that showed significant departure from Poisson expectation in the direction of low variance: mean ratio. It did so faithfully among abundances in 0.04- and 0.16-m^2 quadrats at the 1800-m depths, where the species is most abundant, but deviated in the opposite direction at the 1300-m locality, where the population was sparser. The variance: mean ratio of *O. lymani* in per-quadrat abundances in fact decreased consistently with increasing population density (Fig. 7). The explanation is not elusive. Individuals apparently avoid touching arms (maintain a minimum interindividual distance), which can be expected to produce a low variance: mean ratio in per-quadrat abundance for small quadrats (Holme, 1950) and negative spatial autocorrelation at small spatial lags but not large ones (similar to the pattern seen in the correlograms in Fig. 5*a–c*). At low densities, on the other hand, as found at the 1500-m locality, *O. lymani* avoided local depressions (apparently resulting from the burrowing of other species), giving an unexpectedly ($p < .001$) high variance: mean ratio (2.460, $n = 255$, $\bar{x} = 2.553$) at the smallest quadrat size (4 m^2 in this transect). The magnitude of these departures [as measured in several ways by Grassle et al. (1975)] in both directions from Poisson expectation decreased as the sampling scale increased. Vectorially mediated habitat selection at low densities apparently was overridden by social spacing at higher densities, but neither pattern-generating mechanism operated at scales above tens of meters.

The two urchin species both tended to show aggregation, but for widely diver-

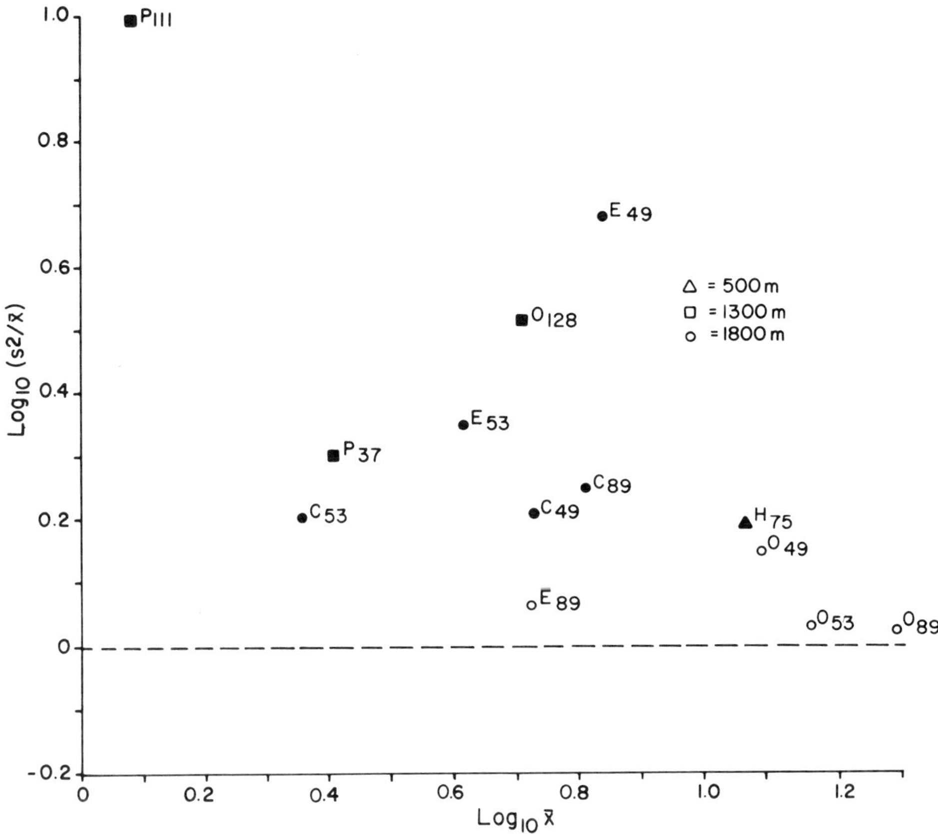

Fig. 7. Log $(s^2/\bar{x})$ versus log $\bar{x}$ of megafaunal abundances per 8 m^2 as determined from photographic transects (*Phormosoma placenta* data are for 10.25 m^2). Solid symbols: exceed 95% confidence limits for Poisson. Data from Grassle et al. (1975).

gent reasons. *Echinus affinis* showed apparently short-lived aggregations around pulsed inputs of its coarse-grained food resource of drift macroalgae, explaining the highly variable variance:mean ratios observed (Fig. 7). Spatial correlograms drawn for this species would thus be expected to vary between those in Fig. 5*d* and *e* in shape. On the other hand, *P. placenta*, feeds on the fine-grained resources within surface deposits but apparently moves in herds. Although the average density in an environment proved variable (Fig. 7), at the higher of the two densities encountered the linear extent of individual herds was approximately 40 to 50 m. The spatial correlogram of *P. placenta*'s per-quadrat abundance would likely resemble that in Fig. 5*e*, with positive spatial autocorrelation at lags up to roughly the patch radius of 20 to 25 m. In both urchin species the autocorrelation coefficients would approach their expected values at high spatial lags; no large-scale pattern (in excess of the herd size) was apparent.

Although the variance:mean ratios of their per-quadrat abundances do show some statistically significant departures from Poisson expectation at some quadrat sizes (Fig. 7), consistent patterns over the range of quadrat sizes from 4 to 128 m^2 were not evident in either the cerianthid or *Hyalinoecia* sp. The result is not surprising for the burrowing anemone; in one of the most thorough pattern studies ever accomplished in shallow water, a cerianthid showed no detectable departure from a random dispersion pattern (Fager, 1968). The scavenging quill worm *Hyalinoecia* sp., however, is known to congregate around meat bait (e.g., Dayton and Hessler, 1972). Just as in *E. affinis*, however, these aggregations are apparently short lived.

Grassle et al. (1975) thoroughly reviewed previous within-community studies of deep-sea megafaunal dispersion patterns. These earlier analyses do not add appreciably to the diversity of patterns discussed above, in part because the analyses were generally less thorough and based on fewer data. Aggregation, spacing, and apparently random dispersion patterns have all been observed. What is unique about these diverse patterns is their high degree of attributability to specific causes. The reason is not difficult to find; it is the relative ease with which natural history information can be gained with species that allow their activities to be watched. In addition, deep-sea megafaunal species have long been studied from qualitative trawls and dredges.

In fact, there are megafaunal groups whose natural histories are far better known than are their natural dispersion patterns. An obvious example is the suite of species that are drawn to baited monster camera deployments (Isaacs and Schwartzlose, 1975; Hessler et al., 1978). There are essentially no data on natural dispersion patterns of most of these species, thus indicating the difficulty of quantitatively sampling presumably rare and highly mobile species. They can escape nets and either avoid or congregate around cameras. Demersal fishes, for example, may be active at levels in the water column above those where cameras are normally set for bottom photography.

Vertical Pattern

The demersal fish sampling problem is thus also a glaring example of the general lack of data on depth–frequency distributions of megafauna. Many of these species may make excursions entirely out of the benthic boundary layer (Pearcy, 1976), and their patterns of vertical abundance and activity within the boundary layer are likewise poorly known. A less obvious problem is the quantitative sampling of deeply burrowing and relatively large megafauna. We have sampled large thalassinid shrimp (one specimen is 10 cm long) and their burrows at 200-m depth in Puget Sound, but many of their burrows continue through the bottom of 45-cm deep box cores. There is little reason to think they and other deeply burrowing taxa would be absent from deep-sea sediments of relatively high organic content. In a core taken at 7300-m depth in the Aleutian Trench (Jumars and Hessler, 1976), for example, a pair of well-maintained tubes of the sort often built by maldanid polychaetes extended from the surface of the sediments through the bottom of a 55-cm-deep core. Knowledge of megafaunal species is thus generally limited to relatively abundant species that frequent the sediment surface.

B. Macrofauna

Horizontal Pattern

Macrofaunal dispersion patterns have been studied far more frequently. Most of these studies have addressed only per-sample means and variances in abundance and have not utilized the kind of information contained in the geographic

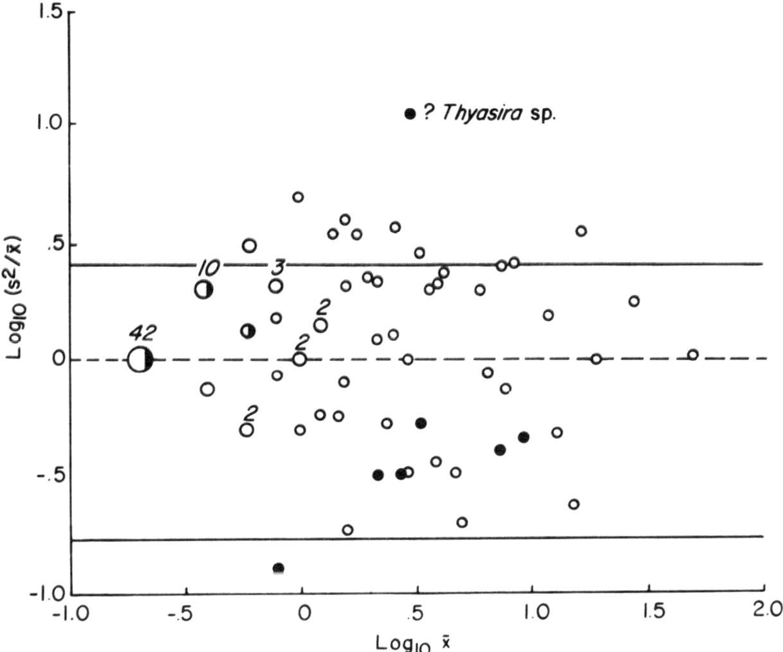

Fig. 8. Log ($s^2/\bar{x}$) versus log $\bar{x}$ for per-0.25-m^2-core abundances of bivalve (filled circles) and polychaete (open circles) species at 2875 m in the Rockall Trough. Numerals and larger circles indicate more than one species with the same value. Dashed lined shows Poisson expectation, and solid lines show its 95% confidence limits. (Reprinted with permission of Pergamon Press, Inc. from Gage, J. D., 1977. Structure of the abyssal macrobenthic community in the Rockall Trough. In *Biology of Benthic Organisms*, B. F. Keegan, P.O. Ceidigh, and P. J. S. Boaden, eds. Pergamon Press, Oxford, p. 258.)

coordinates of the samples. Nor would such analysis have been profitable. In the early, exploratory phases of quantitative deep-sea sampling it has not been cost-effective to take and process numerous samples from a limited geographic area, and the I and c measures, for example, often can behave erratically with fewer than ten samples (Cliff and Ord, 1973). Valuable information on per-sample variation in abundance has been obtained, however, for several hundred species. For the sake of quantitation and comparability, we restrict consideration to samples taken as replicate 0.25-m^2 box cores [as described by Hessler and Jumars (1974)]. Again, results from this subset of deep-sea observations encompass the range of variability observed as a whole.

The methods and results of Gage (1977) are typical of this limited set of data. He succeeded in obtaining six cores within an area of 1 × 2 nautical miles (1.9 × 3.7 km^2, centered on 55°0.7′N, 12°04.0′W) at a depth of 2875 m in the Rockall Trough off the western coast of Scotland. The bottom is relatively flat but shows local crag-and-tail features around projecting animal tubes, indicative of current activity (Gage, 1977). Figure 8 summarizes the variance : mean ratios in the abundances of the 105 polychaete species and 18 bivalve species encountered. It shows that few species exceed the 95% confidence limits for a variance : mean ratio derived from a Poisson distribution.

With minor changes of scale on the abscissa, this figure could have come from either the central northern Pacific at 5600-m depth (Hessler and Jumars, 1974) or from 1200-m depth in the San Diego Trough in the Southern California Continental Borderland (Jumars, 1975a, 1976). Hessler and Jumars (1974) studied 10 cores in two clusters 20 km apart (centered on 28°29′N, 155°23′W) and 2 cores 170 km north of these 10. Jumars (1975a, 1976) analyzed 5 cores within 2.20 km of 32°28.1′N, 117°29.9′W. The number of species exceeding the maximum expected variance : mean ratio for all three studies tends to be slightly greater than the one-in-forty level expected by chance, and the number of species showing a ratio under the minimum expected tends to be lower than the one-in-forty level expected by chance with two-tailed 95% confidence limits. The existence of known, variance-inflating bias in sampling (Jumars, 1975a,b, 1976), however, might easily lead one to conclude that the general dispersion pattern for deep-sea populations within communities is a random one. Occasional exceptions, such as the finding of 16 limpets (? *Cocculina* sp.) still attached to a blade of decaying eel grass (Jumars, 1976), because of their ready attributability, do not seem to weaken this generalization.

By equipping the corer with 10 × 10-cm^2 internal partitions, Jumars (1975a, 1976) also examined macrofaunal dispersion patterns in the San Diego Trough and the Santa Catalina Basin at spatial scales within the 50 × 50-cm^2 cores. On the basis of their variance : mean ratios, per-core abundances of single species in general again could not be distinguished from Poisson expectation (e.g., see Fig. 9).

Five 0.25-m^2 cores within 0.74 km of 32°28.1′N, 117°29.9′W in the Santa Catalina Basin, however, showed a somewhat different set of single-species variance : mean ratios of abundance (Jumars, 1974), best illustrated by the polychaete data (see Fig. 10 and Table II). Twelve polychaete species out of 57 significantly exceeded the two-sided Poisson confidence limits ($p \leq .05$) by having a larger variance in abundance than expected. Nine of these 12 were most abundant in a single core. This example serves two purposes. It helps to remove the temptation to generalize that random dispersion patterns are the overwhelming deep-sea rule, and it serves to point out, by inspection of Table II, the general lack of attributability of nonrandom dispersion patterns in deep-sea organisms of macrofaunal size range or smaller. Deep-sea macrofaunal species generally have not been observed *in situ*, and what meager natural history information is available comes through hazardous functional morphological extrapolations from better-known, shallow-water relatives (Jumars and Fauchald, 1977).

It may still be tempting to suggest, especially for the purposes of modeling deep-sea populations or their effects, that a random dispersion pattern holds in the overwhelming majority of cases. Similarly, contrasts are sometimes drawn between deep-sea population dispersion patterns, as summarized in Fig. 10, and analogous plots for shallow-water samples in which half or more of the macrofaunal species may significantly exceed the largest variance : mean ratio expected under the Poisson assumptions (Gage and Geekie, 1975). Table III should remove this temptation. More generally, under the prevailing falsificationist methodology of science (Lakatos, 1970), failure in discrediting a null hypothesis (Poisson-distributed abundances in our case) cannot provide strong support for that null hypothesis. In statistical terms, the degree to which the null hypothesis can safely be believed depends on the power (1 − β error probability) of the test against

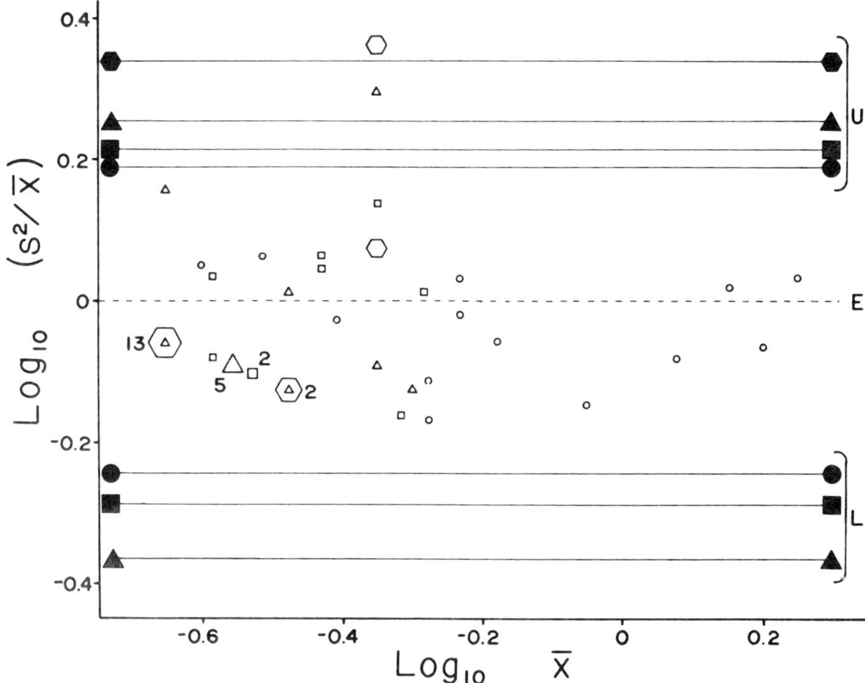

Fig. 9. Log $(s^2/\bar{x})$ versus log $\bar{x}$ for per-0.01-m^2-subcore abundances of polychaete species from 1230-m depth in the San Diego Trough. Shapes indicate the number of subcores from which data were drawn for that species: 9 (hexagon), 18 (triangle), 27 (square), or 36 (circle). Open figures indicate observed values, whereas closed figures and solid lines indicate upper (U) and lower (L) 95% confidence limits around Poisson expectation (dashed line). Confidence limits are approximate as a result of systematic sampling. [Reprinted with permission of Springer-Verlag New York, Inc. (*Marine Biology*) from Jumars, P. A., 1975. Environmental grain and polychaete species diversity in a bathyal benthic community. *Mar. Biol.*, **30**, 257.]

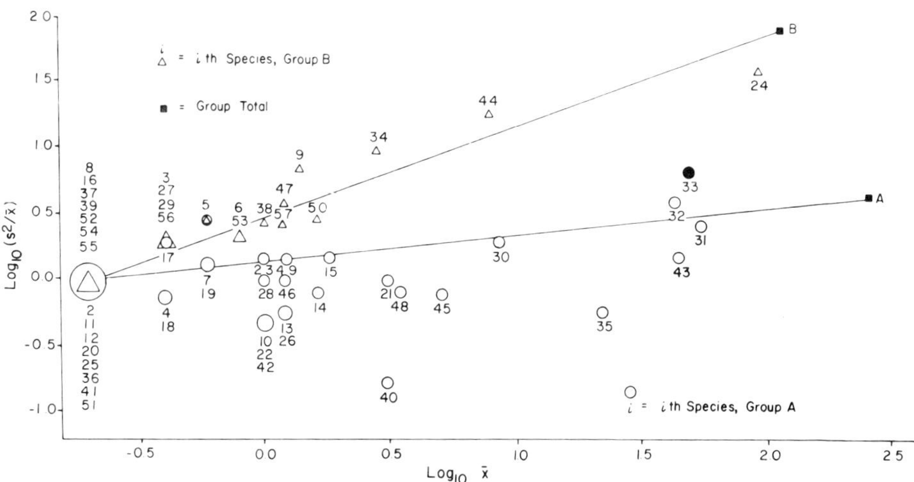

Fig. 10. Log $(s^2/\bar{x})$ versus log $\bar{x}$ for per-0.25-m^2-core abundances of 57 polychaete species from the Santa Catalina Basin at 1130 m (coded as in Table II). The solid lines summarize patterns of two major species groups as discussed in Section 5.B, subsection on horizontal patterns.

TABLE II
Numbers of Individuals by Species and Core Encountered at Santa Catalina Basin Locality[a]

Species Code and Group	Species	J9	J10	J11	J12	J13	Summed N_i	$\bar{x}$	$s^2/\bar{x}$
		Core Designation							
1 A	*Harmothoe forcipata* (Marenzeller)	0	0	3	0	0	3	0.60	3.00
2 A	*Macellicephala aciculata* (Moore)	0	0	1	0	0	1	0.20	1.00
3 B	Polynoid sp. C	0	2	0	0	0	2	0.40	2.00
4 A	Polynoid sp. D	0	1	1	0	0	2	0.40	0.75
5 B	*Euphrosine paucibranchiata* Hartman	0	3	0	0	0	3	0.60	3.00
6 B	Phyllodocid sp. A	0	3	0	0	1	4	0.80	2.13
7 A	Hesionid sp. A	0	1	2	0	0	3	0.60	1.33
8 B?	*Ancystrosyllis* cf. *groenlandica* McIntosh	0	1	0	0	0	1	0.20	1.00
9 B	*Exogone* sp.	0	7	0	0	0	7	1.40	7.00
10 A	*Ceratocephale pacifica* Hartman	1	2	1	0	1	5	1.00	0.50
11 A	*Clavodorum* sp. A	0	0	0	1	0	1	0.20	1.00
12 A	*Ephesiella brevicapitis* (Moore)	0	0	1	0	0	1	0.20	1.00
13 A	*Glycera capitata* Oersted	2	1	2	1	0	6	1.20	0.58
14 A	*Ninoe gemmea* Moore	3	2	0	1	2	8	1.60	0.81
15 A	*Ninoe* sp. B	4	3	1	1	0	9	1.80	1.50
16 B?	*Drilonereis falcata* Moore	0	1	0	0	0	1	0.20	1.00
17 A	*Dorvillea batia* Jumars	0	0	0	0	2	2	0.40	2.00
18 A	Dorvilleid sp. B	0	1	1	0	0	2	0.40	0.75
19 A	*Aricidea antennata* Annenkova	0	2	0	0	1	3	0.60	1.33
20 A	*Aricidea lopezi rubra* Hartman	0	0	0	1	0	1	0.20	1.00
21 A	*Aricidea ramosa* Annenkova	2	6	3	2	2	15	3.00	1.00
22 A	*Aricidea* cf. *suecica* Eliason	0	2	1	1	1	5	1.00	0.50
23 A	*Aricidea* sp. D	0	1	1	0	3	5	1.00	1.50
24 B	*Paraonis gracilis oculata* Hartman	85	191	57	49	55	437	87.40	40.58
25 A	*Paraonis* sp. B	1	0	0	0	0	1	0.20	1.00
26 A	*Prionospio* cf. *annuncata* Fauchald	0	2	1	1	2	6	1.20	0.58
27 B	*Prionospio* cf. *cirrifera* Wirén	0	2	0	0	0	2	0.40	2.00

28 A *Spiophanes* cf. *bombyx* Claparède	1	2	0	0	2	5	1.00	1.00
29 B *Spiophanes* cf. *kroyeri* Grube	0	2	0	0	0	2	0.40	2.00
30 A *Phyllochaetopterus limicolus* Hartman	6	9	15	4	8	42	8.40	2.06
31 A *Chaetozone* cf. *setosa* Malmgren	58	41	72	46	52	269	53.80	2.68
32 A *Tharyx* sp. A	27	33	61	46	45	212	42.40	4.08
33 A *Tharyx* cf. *monilaris* Hartman	39	77	51	30	41	238	47.60	6.84
34 B Cirratulid D	1	12	0	0	1	14	2.80	9.54
35 A *Cossura* cf. *pygodactyla* Jones	24	25	16	23	21	109	21.80	0.58
36 A *Cossura* sp. B	0	0	0	0	1	1	0.20	1.00
37 B? Flabelligerid sp. A	0	1	0	0	0	1	0.20	1.00
38 B *Fauveliopsis glabra* (Hartman)	1	4	0	0	0	5	1.00	3.00
39 B? *Ophelina* sp. A	0	1	0	0	0	1	0.20	1.00
40 A *Sternaspis fossor* Stimpson	4	2	3	3	3	15	3.00	0.17
41 A Capitellid sp. A	0	0	0	0	1	1	0.20	1.00
42 A *Neomediomastus glabrus* (Hartman)	1	1	0	1	2	5	1.00	0.50
43 A *Euclymene reticulata* Moore	39	41	57	47	36	220	44.00	1.57
44 B *Maldane* cf. *sarsi* Malmgren	4	29	4	0	1	38	7.60	19.25
45 A Maldanid sp. C	4	6	2	7	6	25	5.00	0.80
46 A *Lumbriclymene lineus* Hartman	0	1	1	3	1	6	1.20	1.00
47 B *Myriochele gracilis* Hartman	0	5	0	0	1	6	1.20	3.92
48 A Ampharetid sp. A	4	6	2	2	3	17	3.40	0.82
49 A Ampharetid sp. B	1	3	0	0	2	6	1.20	1.42
50 B *Anobothrus?* sp. A	4	4	0	0	0	8	1.60	3.00
51 A *Lysippe annectens* Moore	0	0	0	1	0	1	0.20	1.00
52 B? *Artacamella hancocki* Hartman	0	1	0	0	0	1	0.20	1.00
53 B *Leana caeca* Hartman	1	3	0	0	0	4	0.80	2.13
54 B? *Scionella japonica* Moore	0	1	0	0	0	1	0.20	1.00
55 B? Terebellid sp. D	0	1	0	0	0	1	0.20	1.00
56 B *Chone* sp. A	0	2	0	0	0	2	0.40	2.00
57 B *Oriopsis* sp. A	0	4	2	0	0	6	1.20	2.67
Total number of individuals observed	317	551	362	271	297	1798	359.60	34.92
Total number of species observed	24	47	26	21	28	57		

[a] For an explanation of species groups, see the beginning of Section 5.B.

TABLE III
Rejection Rate of a Range of Null Hypotheses Concerning Data from Fig. 10 and Table II, Using Index of Dispersion s^2/σ^2 as Rejection Criterion

Hypothetical Variance	Appropriate Index of Dispersion	Number of Species Exceeding 95% Confidence Limits: Too High	Too Low
$\bar{x}$ (Poisson)	$s^2/\bar{x}$	12	0
$2\bar{x}$	$s^2/2\bar{x}$	5	1
$3\bar{x}$	$s^2/3\bar{x}$	3	1
$4\bar{x}$	$s^2/4\bar{x}$	2	7
$5\bar{x}$	$s^2/5\bar{x}$	2	12
$6\bar{x}$	$s^2/6\bar{x}$	2	12
$7\bar{x}$	$s^2/7\bar{x}$	1	30
$8\bar{x}$	$s^2/8\bar{x}$	1	30
$9\bar{x}$	$s^2/9\bar{x}$	1	33
$10\bar{x}$	$s^2/10\bar{x}$	1	36

alternative hypotheses. Table III illustrates the distressingly low power of the index of dispersion in detecting departures from Poisson dispersion patterns at deep-sea animal densities and achieved sample sizes. In fact, Table III suggests the data to be more compatible with the alternative hypothesis that the variance: mean ratio exceeds Poisson expectation by a factor of 3!

Few studies of deep-sea macrofaunal dispersion patterns have utilized sample coordinates explicitly. Jumars (1975a) used the "joins" method of Pielou (1969) to examine spatial autocorrelation of polychaete species' per-subsample (0.01-m^2) abundances. Only the central, least biased 0.09 m^2 of the 0.25-m^2 cores from the San Diego Trough (discussed above) were utilized. The joins method uses only the nominal (versus ordinal, interval, or ratio) level of measurement, allowing data to be relatively easily pooled across species and cores (Jumars, 1975b). Although single species were too rarely encountered to allow firm conclusions to be drawn, pooling of data for all polychaete species showed a strong tendency ($p \cong 0.01$) toward negative spatial autocorrelation in abundance within species at small spatial lags (10 to 14.2 cm). In fact, it demonstrated that per-species, per-subcore abundances were, on the average, more similar in nonadjacent subcores (20 to 42.5 cm apart) than in adjacent subcores (10 to 14.2 cm apart). The simplest explanation of these observations invokes a spacing mechanism that maintains some average distance among individuals within species. To avoid the temptation to generalize that intraspecific spacing on small scales is the rule in the deep sea, it should be pointed out that the corresponding analysis revealed no departure from random dispersion patterns within species either in any taxon at the Santa Catalina locality (also discussed above) or in taxa other than polychaetes at the San Diego Trough site (Jumars, 1974, 1976).

A deep-sea sampling program designed from the outset to allow spatial autocorrelation over a broad range of spatial lags was Expedition Quagmire (Thiel and

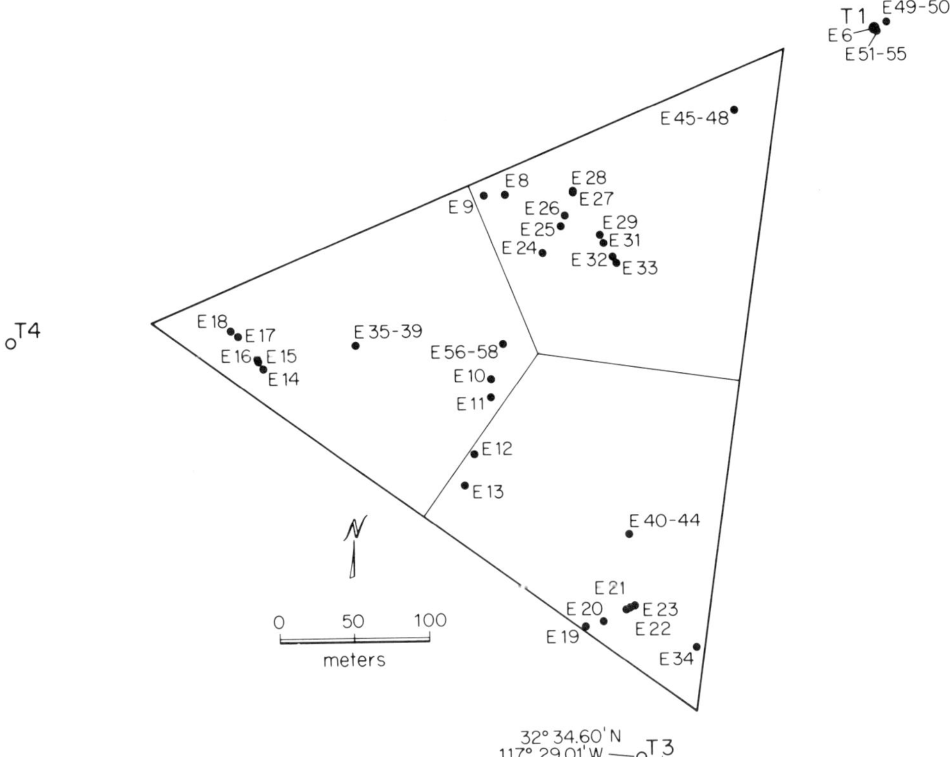

Fig. 11. Chart of study triangle from Expedition Quagmire showing locations of Ekman cores (E) and acoustic transponders (T, used in navigation). [Reprinted with permission of Pergamon Press, Inc. (*Deep-Sea Research*) from Jumars, P. A., 1978. Spatial autocorrelation with RUM (remote underwater manipulator): Vertical and horizontal structure of a bathyal benthic community. *Deep-Sea Res.*, **25,** 591. Copyright 1978, Pergamon Press, Ltd.]

Hessler, 1974; Jumars, 1978). One hundred twenty-five 0.01-m^2 subcores were taken for macrofaunal analysis very deliberately, through use of a transponder-navigated, unmanned submersible, remote underwater manipulator (RUM) (Marine Physical Laboratory, Scripps Institution of Oceanography). Each subcore came from an Ekman corer (20 $\times$ 20 cm^2) in which four such subcorers were nested. The core locations are mapped in Fig. 11. One to four subcores per core, sampled to a sediment depth of 10 cm, were used in spatial autocorrelation, giving a range of distances or spatial lags between subcore centers of 0.1 m to nearly 500 m.

Spatial autocorrelation data from two species will suffice to reveal the potential diversity of pattern hidden in summaries (e.g., Figs. 8 through 10) not taking spatial coordinates of sample values into account. *Ceratocephale pacifica* (Polychaeta, Nereidae) is a burrow-dwelling, generalist predator on meiofauna (Jumars, 1978), whereas *Polyophthalmus* sp. (Polychaeta, Opheliidae) is a motile, burrowing, shallow subsurface deposit feeder (Fauchald and Jumars, 1979). The variance : mean ratio for both species fails to show statistically significant departure from expected Poisson per-core or per-subcore abundances, either in the partitioned box core study of Jumars (1975a) or in the partitioned Ekman core sampling (Jumars, 1978).

Calculation of I for per-subcore abundances of *Polyophthalmus* sp., however, produced a value much larger [$p < 0.00001$, with w_{ij} set at (intersample distance)$^{-2}$] than expected; extreme high abundances were spatially segregated from extreme low abundances. Similar densities occurred in relatively close proximity, justifying (Jumars, 1978) spatial interpolation and hence mapping of local *Polyophthalmus* sp. abundance (Fig. 12). Causes of the pattern remained obscure, although a correlation of high density with large body size was suggested (Fig. 12). A correlogram for *Polyophthalmus* sp. confirmed the (albeit imperfect) trend toward decreasing similarity in per-subsample abundance with increasing spatial lag (Fig. 13).

Ceratocephale pacifica, on the other hand, did not show this general tendency (Fig. 13). In fact, its correlogram much more closely resembled that of Fig. 5*a–c*. Given *C. pacifica*'s mean per-subcore abundance of 1.35 individuals, the alternative pattern of positive and negative spatial autocorrelation observed at small spatial lags is highly suggestive of an interindividual spacing process, as modeled by Jumars (1978). Again, however, too little natural history of *C. pacifica* is known to view its observed dispersion pattern from an informed biological perspective.

Spatial autocorrelation in abundance has thus far been examined in 13 species encountered in Expedition Quagmire (Jumars, 1978). Significant departures from random patterns were detected only in *C. pacifica* and *Polyophthalmus*. Again, one might be tempted to conclude that random dispersion patterns are the deep-sea rule. Again an examination of the power of the test (the probability of rejecting the null hypothesis when it is indeed wrong) suggests caution. How likely is the achieved sampling pattern to have missed extant nonrandom patterns, especially given the low average standing stocks of most deep-sea species?

In the spirit of the Monte-Carlo approach often taken by E. W. Fager (e.g., Fager, 1972), we devised a very simple computer model to address this question of the power of our tests. We used the actual spatial coordinates of the Ekman samples as shown in Fig. 11, excluding those samples outside the triangle. Then we selected spatial coordinates within the triangle at random on which to center hypothetical, perfectly circular, nonoverlapping patches of uniform diameter. Whereas patch centers were restricted to the region within the triangle, other portions of the patches were allowed to extend over the edges of the triangle. The population density was made so uniform that every core within a patch would collect exactly one individual. (Recall that multiplication of all the sample values by any constant, say 10^6, *will not* affect the values of I or c. Also note that our patches cannot be detected by analyzing the variance: mean ratio of per-sample abundances provided no more than one individual occurs per sample.) No individuals were located outside patches. One would expect such clearly demarcated patches of uniform density to be relatively easy to detect. The results of this exercise (Table IV) quantitatively drive home the point, however, that even such "perfect" patches are difficult to recognize. Irregularly shaped patches of variable dimensions, with internally varying population densities and with some nonzero background density outside the patches, would be both more realistic and far more difficult to document. Even in the Quagmire locality, which has been more

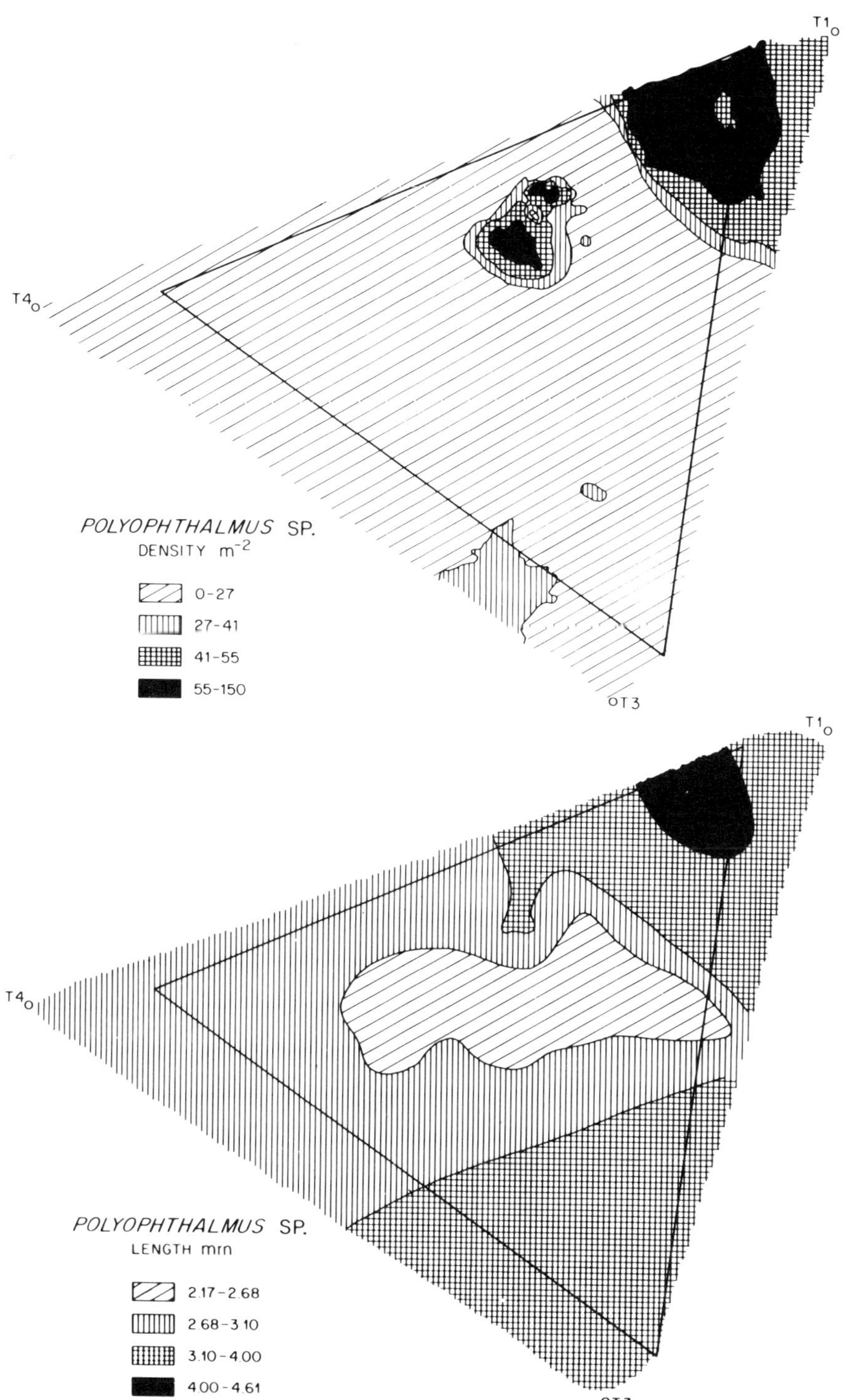

Fig. 12. Charts of estimated *Polyophthalmus* sp. density and individual size at 1220-m depth in the San Diego Trough (see Fig. 11 for scale). [Reprinted with permission of Pergamon Press, Inc. (*Deep-Sea Research*) from Jumars, P. A., 1978. Spatial autocorrelation with RUM (remote underwater manipulator): Vertical and horizontal structure of a bathyal benthic community. *Deep-Sea Res.*, **25,** 595. Copyright 1978, Pergamon Press, Ltd.]

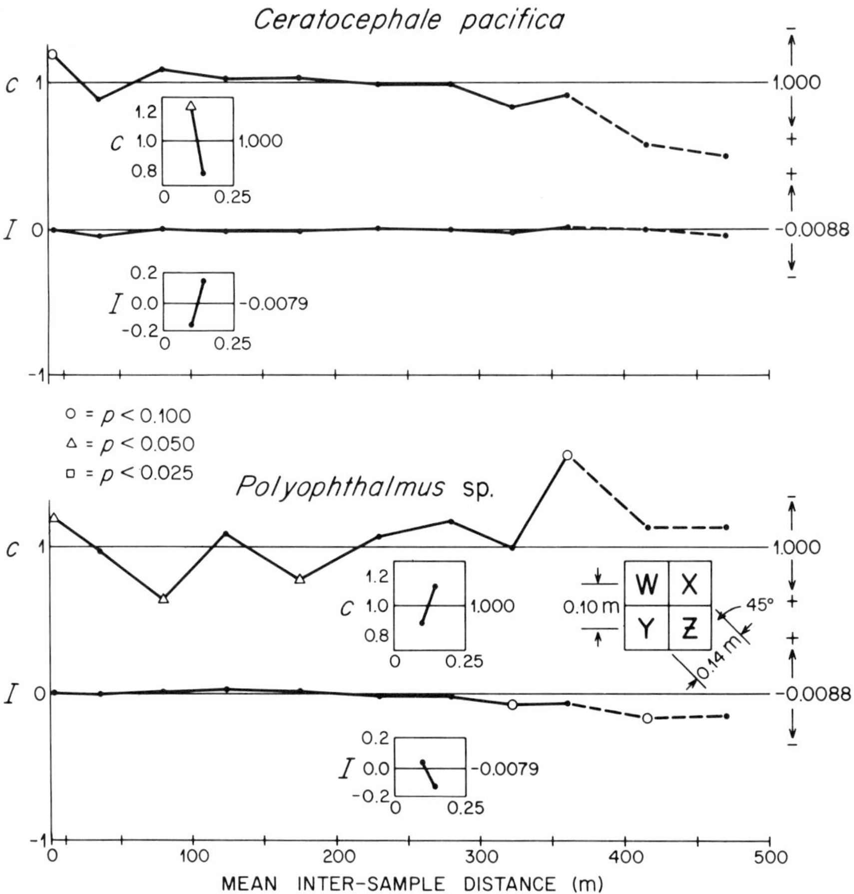

Fig. 13. Correlograms of I and c versus mean intersample distance for data for two polychaete species from Expedition Quagmire at 1220-m depth in the San Diego Trough. Intervals used in calculating I and c are shown by tick marks on the abscissa. Expected values are indicated by horizontal lines; deviations showing positive and negative spatial autocorrelation are given in the right-hand margin. Dashed lines are based on few intersample distances; the magnitudes are unreliable. Insets show results on within-core scales and arrangement of subcores (W, X, Y, Z). [Reprinted with permission of Pergamon Press, Inc. (*Deep-Sea Research*) from Jumars, P. A., 1978. Spatial autocorrelation with RUM (remote underwater manipulator): Vertical and horizontal structure of a bathyal benthic community. *Deep-Sea Res.*, **25**, 596. Copyright 1978, Pergamon Press, Ltd.]

intensively sampled with regard to spatial autocorrelative measures than has any other deep-sea region, one cannot reject the hypothesis that the majority of species exist in aggregations.

Vertical Patterns

Depth–frequency distributions for individual deep-sea macrofaunal species rarely have been documented. The reason is apparent from an inspection of Figs. 8 and 10 and the number of individuals per species (N_j) in Table II; many species are encountered only once in several box cores.

Differences among species clearly do exist, however. All species in Fig. 14 are polychaetes collected in vertically sectioned subcores from Expedition Quagmire

TABLE IV

Results of Monte Carlo Simulations with "Perfect" Patches as Described in Text; w_{ij} = (Intercore Distance)$^{-2}$ for I and c

Patch Diameter (m)	Number of Patches	Proportion of Simulations in Which no Patches Were Cored	Proportion of Values Significant ($p < .05$, One Tailed) I	c
100	1	10/40	1/30	0/30
100	2	2/40	1/38	0/38
100	3	2/40	0/38	0/38
100	4	0/40	0/40	0/40
50	4	5/40	2/35	0/35
50	8	0/40	0/40	0/40
50	12	0/40	0/40	0/40
50	16	0/40	0/40	0/40
10	100	5/40	0/35	0/35
10	200	0/40	0/40	0/40
10	300	0/40	0/40	0/40

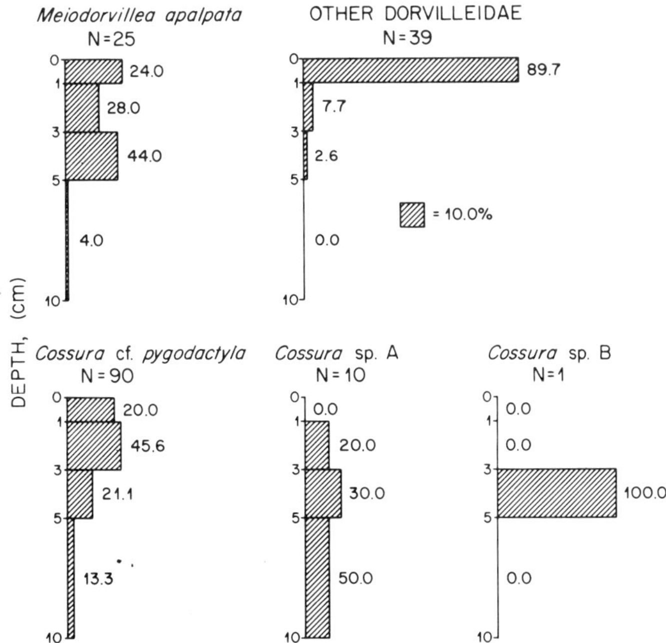

Fig. 14. Depth–frequency distributions of Dorvilleidae and Cossuridae (Polychaeta) at 1220-m depth in the San Diego Trough (N, total number of individuals). [Reprinted with permission of Pergamon Press, Inc. (*Deep-Sea Research*) from Jumars, P. A., 1978. Spatial autocorrelation with RUM (remote underwater manipulator): Vertical and horizontal structure of a bathyal benthic community. *Deep-Sea Res.*, **25,** 597. Copyright 1978, Pergamon Press, Ltd.]

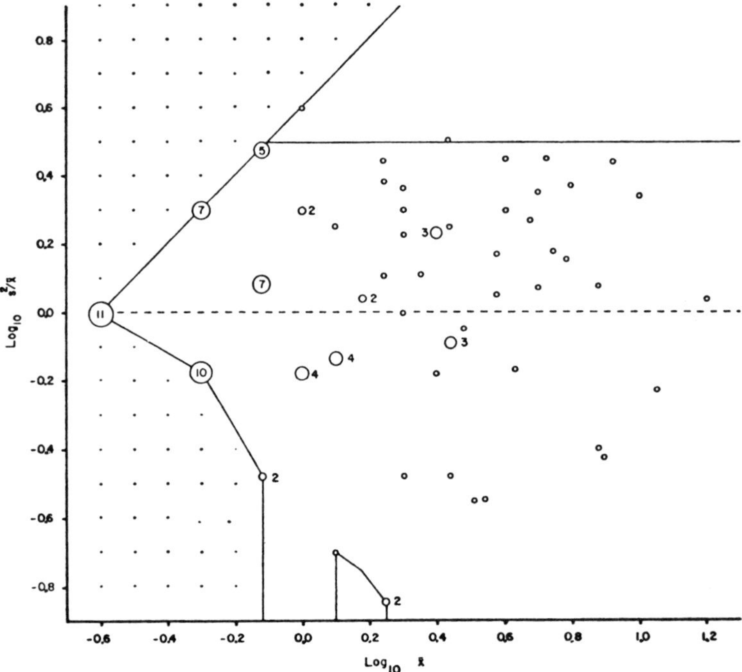

Fig. 15. Log $(s^2/\bar{x})$ versus log $\bar{x}$ for harpacticoids of Expedition Quagmire at the 100-m scale. Numerals and larger circles denote more than one species with the same value. Poisson expectation (dashed line) and the upper limit (solid line) of its 95% confidence interval are drawn (lower limit falls below abscissa). Stippled region indicates mathematical impossibility. (Reprinted with permission of *Journal of Marine Research* from Thistle, D., 1978. Harpacticoid dispersion patterns: Implications for deep-sea diversity maintenance. *J. Mar. Res.*, **36**, 386.)

(Jumars, 1978). *Meiodorvillea apalpata* (Dorvilleidae), whose morphology suggests a burrowing habit, showed a significantly different depth-frequency distribution from those of other dorvilleids, as did *Cossura* cf. *pygodactyla* (Cossuridae) from *Cossura* sp. A ($p < 0.05$, Komogoroff–Smirnov two-sample test). These depth–frequency distributions and all others as yet calculated from deep-sea samples must be interpreted with caution for reasons discussed in Section 3.B and Section 5.B, subsection on vertical patterns.

The lack of data on rare species is obvious. Perhaps less obvious is the glaring lack of information on deeply burrowing species, which may be of considerable importance in stratigraphy.

C. *Meiofauna*

Published dispersion pattern data for individual deep-sea meiofaunal species are extremely rare. Aside from the work of Thistle (1978) in association with the Quagmire Expedition, we know of no deep-sea studies directed toward this problem. He dealt exclusively with adult harpacticoid copepods retained on a 0.062-mm-aperture sieve after washing the surface 1-cm layer of sediment through it. Thistle succeeded in identifying all the harpacticoids in 14 (0.01-m^2) subcores from 8 Ekman cores (E10, 11, 12, 14, 45, 46, 47, and 48 in Fig. 11). Figures 15 through 17 summarize his findings. Without figure captions, one would be hard pressed to

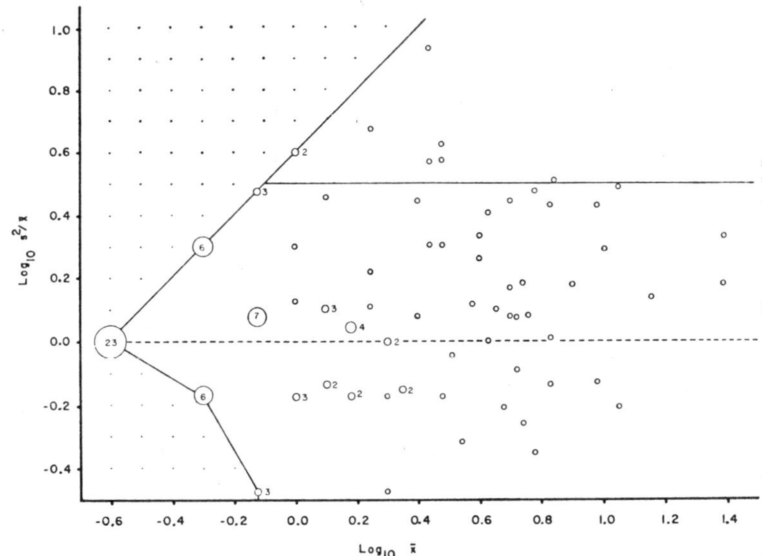

Fig. 16. Log ($s^2/\bar{x}$) versus log $\bar{x}$ for harpacticoids of Expedition Quagmire at the 1-m scale (see Fig. 15 for symbols). (Reprinted with permission of *Journal of Marine Research* from Thistle, D., 1978. Harpacticoid dispersion patterns: Implications for deep-sea diversity maintenance. *J. Mar. Res.*, **36**, 387.)

distinguish these results for harpacticoids from the polychaete and bivalve data in Figs. 8 and 9. The conclusion that most species individually are randomly dispersed must again be tempered according to the power of the index of dispersion. The test is weak at these abundance levels (cf. Table III), so that nonrandom patterns may again be prevalent.

Thistle also calculated I (equation 5) for 15 of the species he encountered. Instead of w_{ij} = (intersubcore distance)$^{-2}$, however, he used w_{ij} = (intersubcore distance)$^{-1}$. Although small spatial lags thus still played a relatively large role in determining the value of I, they did so to a far lesser degree than with the (distance)$^{-2}$ weighting. Of the 15 species only one showed strong ($p < 0.05$, no correction for multiple testing) autocorrelation. Extremes in a given direction from its mean abundance were observed in relatively close proximity. In view of the documented weakness (Table IV) of the method in detecting patches within the Quagmire triangle (Fig. 11), even when all the Ekman cores were utilized, one should be *extremely* reluctant to generalize about the presence or absence of nonrandom spatial structure in deep-sea meiofaunal populations. Single-species meiofaunal data are in fact so rare that we were unable to find other than anecdotal documentation of vertical dispersion patterns of deep-sea meiofaunal species.

5. Multispecies Patterns Observed

A. *Megafauna*

Some multispecies megafaunal patterns are more obvious from direct observation than they would be from extensive statistical analyses of samples. Every gorgonian Grassle et al. (1975, p. 464) encountered, for example, contained the

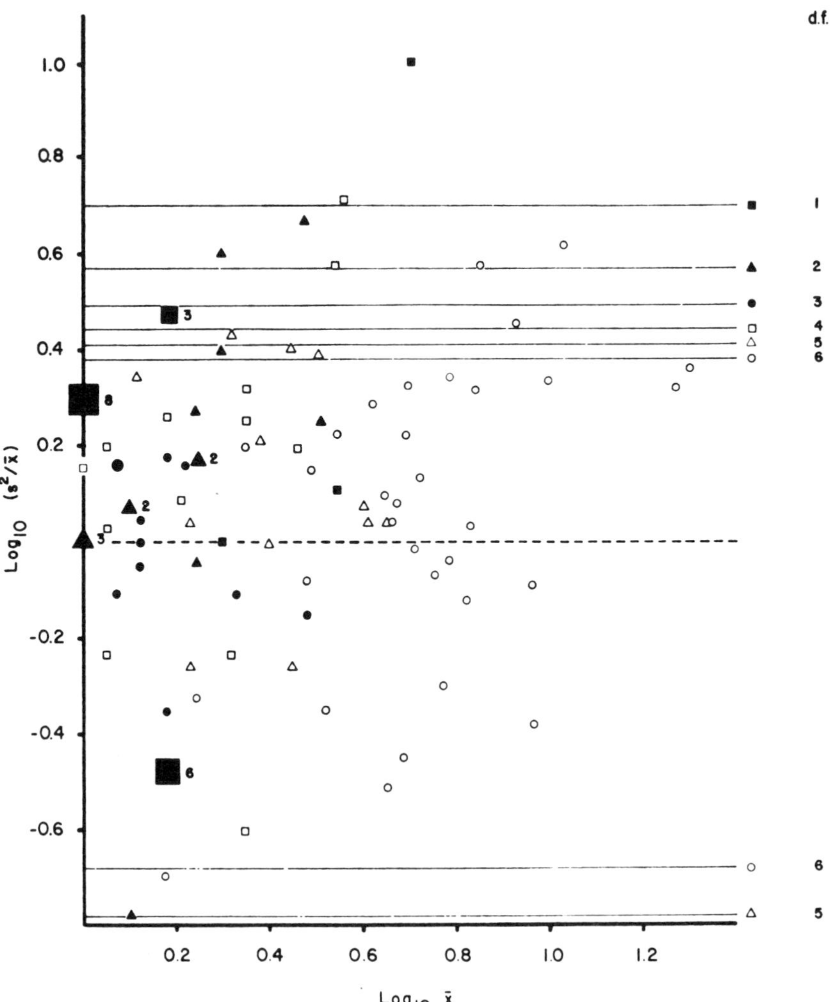

Fig. 17. Log $(s^2/\bar{x})$ versus log $\bar{x}$ for harpacticoids of Expedition Quagmire at the 0.1-m scale. Shapes of symbols indicate the number of degrees of freedom used in calculating the index of dispersion (see Fig. 15 for other symbols). (Reprinted with permission of *Journal of Marine Research* from Thistle, D., 1978. Harpacticoid dispersion patterns: Implications for deep-sea diversity maintenance. *J. Mar. Res.*, **36**, 388.)

ophiuroid *Asteronyx loveni* in its branches, whereas that ophiuroid was never seen elsewhere. The crab *Lithodes agassizii* was seen to be associated with *Phormosoma* aggregations. More subtle spatial associations in the megafaunal communities studied by Grassle et al. (1975) might, however, be suggested by cross-correlations in local abundances, which apparently have not been carried out.

Monster-camera photographs (Rowe and Sibuet, this volume, Chapter 2) provide one of the few examples of deep-sea time series and show stereotyped and integrated patterns of multispecies aggregation around meat baits, as documented by Isaacs and Schwartzlose (1975). At any one deep-sea locality, the sequence and timing of arrivals at such bait can be predicted with reasonable accuracy and precision. A typical sequence begins with highly mobile fish, such as rattails

(Macrouridae), and progresses to less mobile fish, such as hagfish (Myxinidae), which have means of sequestering bait from other fish depredation. Various invertebrates, including quill worms (*Hyalinoecia*), brittle stars, and decapod shrimps and crabs, arrive in as yet less well known sequences, but typically remain about the periphery of the bait, feeding on scraps loosed by fish activity. Often not so obvious in the photographs, with the exception of some truly megafaunal species (Hessler et al., 1972), is the presence of scavenging lysianassid amphipods, which begin arriving near the start of the sequence. In some deep-sea regions, notably in trenches, these amphipods are so aggressive and abundant that they seem to exclude other scavengers (Hessler et al., 1978). The predictability of the sequence and the stereotyped behaviors observed—especially the apparent sequestering strategies—argue strongly that these multispecies aggregations are a relatively frequent, if short-lived, natural phenomenon.

Equally spectacular are the recently discovered (Lonsdale, 1977; Ballard, 1977) multispecies associations in the vicinity of deep-sea hydrothermal vents. These multispecies aggregations appear to take the form of variably overlapping, concentric rings of vent-dependent species. The existence of such unique scavenging and vent-associated faunas tends to draw attention away from more general megafaunal activities and patterns. What impact do wandering, deposit-feeding herds of sea urchins (Grassle et al., 1975) and sea cucumbers (Barham et al., 1967; Stanley and Kelling, 1968) have on their megafaunal neighbors or on the background of smaller species in which they wander? What of the seemingly more selective and local depredations of fish (Barham et al., 1967) or the incessant rummaging (LaFond, 1967) of the cosmopolitan, omnivorous ophiuroid *Ophiomusium lymani*? There can be little doubt that such activities, in providing sediment structures similar to those of Figs. 1 through 4, influence dispersion patterns and local successional series of macrofauna, meiofauna, and microbiota. Once again, however, patterns in the vertical habit dimension are essentially unknown.

B. Macrofauna

Horizontal Patterns

Of all multispecies associations, macrofaunal patterns have been examined most frequently. Not surprisingly, these patterns are even more diverse than those found among individual species. Using the lesson on statistical power learned from inspecting single-species dispersion patterns, we focus on the nonrandom (i.e., nonindependent) patterns seen; an apparent lack of association among species may be real but is more likely a result of insufficient sampling on the spatial and temporal scales needed to reveal complex patterns of interaction.

Using nonparametric measures of correlation and concordance in their study of an abyssal, central northern Pacific Community, Hessler and Jumars (1974) found the four most abundant species identified in their study, namely, three polychaetes (*Chaetozone* sp., a capitellid species, and *Flabelligella* sp.) and a harpacticoid, to be concordant in per-0.25 m^2 abundance. The total numbers of Polychaeta, Harpacticoida, and Nematoda per core varied in a similar fashion; these taxa "agreed" on where to be abundant and where to be rare. The net result of this concordance (a "homogeneous" pattern such as that indicated in Fig. 6 and Table I) was a decidedly nonrandom $(s^2/\bar{x}) > 1$ pattern in the total number of macrofaunal individuals and in the total number of individuals (macrofaunal plus meiofaunal) per core. The causes of this pattern remain obscure, but sampling bias (bow-wave effects) and local variation in organic sedimentation rate (Lister, 1976; Hargrave and Nielsen, 1977) are likely candidates.

Using the dispersion χ^2 and methodology and separating the information into among-0.25-m^2-core and within-0.25-m^2-core (among 0.01-m^2 subcores) categories, Jumars (1976, Tables 4 and 6) summarized multispecies patterns within major taxa (Annelida, Crustacea, Mollusca, Echinodermata, and the remaining species as a group) of two bathyal basins in the Southern California Continental Borderland. In all taxa, the total number of individuals per taxon per core tended to be more variable than expected under the Poisson distribution [$(s^2/\bar{x}) > 1$, pooled chi-square value exceeds its degrees of freedom], just as in the abyssal sampling of Hessler and Jumars (1974). Apparently contrary to the observed concordance among species seen in the cores from the central northern Pacific (Hessler and Jumars, 1974), however, heterogeneity χ^2 always exceeded its degrees of freedom (see Fig. 6 and Table I, this chapter), indicating discordance among species (within taxa) among cores. In other words, where one species was abundant, others in the same major taxon tended to be rare.

For the most part, reasons for these among-core patterns again remain obscure. As Jumars (1976) discusses in more detail, they are compatible both with coactive and relatively static habitat partitioning and with more dynamic asynchrony of local successional series, coactive or otherwise—and with some combination of the two. We explore one such potential combination by further analysis of the data in Table II and Fig. 10. Examination of these data from 1130-m depth in the Santa Catalina Basin suggested that the polychaete species fell into two groups. The majority of species (i.e., the 37 species in group A) tended to be independently Poisson distributed (e.g., Fig. 6*a,d,e*), whereas the remaining 20 (group B) were relatively homogeneous in per-core abundance and were most abundant in a single core (designated J10). These group patterns are summarized by the solid lines in Fig. 10, using a method detailed by Jumars (1976, Appendix I). For the present purposes, it is sufficient to note in Fig. 5*a–c* that species showing homogeneous patterns are collinear on a plot of $\log(s^2/\bar{x})$ versus $\log \bar{x}$. The converse, however, generally is not true (Fig. 5*a–e*).

Although correlative information on the basis of five cores is tenuous at best, a suggestion can be made as to why core J10 contained so many individuals and species. Many feeding guilds (Fauchald and Jumars, 1979) are represented in each group; the clustering of group B species in core J10 does not appear to have been a short-lived response to temporarily available resources, as seen in some megafaunal species (Sections 4.A on horizontal patterns and Section 5.A). During sorting of the Santa Catalina Basin samples, a relatively high abundance of nonliving hexactinellid sponge fragments was noted in the sediment (retained on a 0.42-mm-aperture sieve) of core J10. When the above analyses revealed the unusual number of polychaetes in this sample, the sediment remaining from all cores was reexamined. Sponge fragments having any two normal dimensions greater than or equal to 2 mm were recovered by sieving through a 1-mm-aperture screen. The approximate area of these nearly planar fragments was estimated very crudely by calculating the area of an ellipse having major and minor axes of the lengths measured for the fragment to the nearest millimeter. Greater precision was not

TABLE V
Incidence of Sponge Fragments and Species Groups in Five Core Samples from Santa Catalina Basin (Jumars, 1974)

Parameter	J9	J10	J11	J12	J13
Encrusted sponge fragments (mm^2)	179	2746	1964	148	167
Number of subcores containing encrusted sponge fragments	3	22	14	1	3
Total sponge fragments (mm^2)	569	2914	2418	607	167
Number of subcores containing sponge fragments	8	22	17	3	3
Number of individuals belonging to group A species	221	273	299	222	238
Number of individuals belonging to group B species	96	278	63	49	59
Total number of polychaetes per core	317	551	362	271	291

warranted because these fragile structures had already been subjected to the relatively harsh procedures of washing and sorting. In addition, fragments were classified as to whether they were encrusted with Foraminifera. Those fragments not encrusted were usually filled with consolidated sediment, indicating burial well below the sediment surface. Because care was not originally taken to prevent breakage, the number of fragments found is probably not representative of *in situ* conditions. Therefore, the data are reported (Table V) as total areas of the fragments (probably a more consistent underestimate) and as frequencies of occupied subcores (perhaps more accurate).

Rank difference correlation (Tate and Clelland, 1957) of total number of polychaetes per core with area of encrusted sponge per core is perfect (one-tailed $p = 0.008$) but should be interpreted cautiously because it is based on *a posteriori* testing. Although it is difficult to imagine that a cover of only 1% of the total area of core J10 by encrusted pieces of sponge could explain its apparent favorableness, several modes of influence should be considered. First, the detected patch of sponge fragments was no doubt larger than the area of the core, and only robust fragments were measured; group B species observed in J10 were not necessarily supported by or attracted to the amount of sponge fragments displayed in Table V. Second, although small, the sponge fragments may have substantial effects; Lonsdale and Southard (1974) have shown that, even in the absence of currents of erosive magnitudes, small projections above the bottom may affect local particulate fluxes (and hence rates of food supply). Some species may browse the *aufwuchs* of the sponge; this possibility seems likely, for example, for *Euphrosine*, whose species are often associated with sponges in better-known environments (Fauchald and Jumars, 1979). Other species may use the sponge as a substrate for attachment; two individuals of *Myriochele* were found still anchored to sponge fragments during sorting. These fragments may also provide protection from predation; individuals of *Paraonis gracilis oculata* were found very frequently with several loops of their bodies thrown over single, large sponge spicules. The presence of the fragments may also alter larval settlement patterns; larvae respond to current velocity, surface texture, surface contour, angle of surfaces, and particle size (Meadows and Campbell, 1972). The latitude for plausible speculation again accents a lack both of basic natural history information and of time-series observations on the natural community.

Whereas we have focused on the polychaetes, the most numerous macrofaunal taxon in these five cores, parallel effects of the sponge fragments or their correlates were seen in the other contained taxa as well (Jumars, 1976, Table 5). The detection of this decidedly nonrandom multiple species pattern, however, hinged critically on the contents of one core. If this single sample had happened to have been drawn from a locality more similar in community composition to that of the other four sampled areas, independence of local (among-core) species abundances might have been claimed. Rare but significant ecological occurrences in space are as difficult to detect and document as are rare but significant ecological events in time.

On the within-core, among-0.01-m^2-subcore scale of sampling, still different patterns emerged among the major taxa of these two bathyal basins considered by Jumars (1976). The trend toward a higher-than-expected (under the independent Poisson assumption depicted in Fig. 6*a,d,e*) pooled χ^2 value persisted but was neither as strong nor as consistent as in the among-core analysis. Within major taxa, heterogeneity χ^2 values in general failed to suggest any overriding pattern of species concordance or discordance in abundance per subcore.

Besides the by now familiar problem of the low power of the test, coupled with the potential importance of spatially rare (unobserved) but significant occurrences, at least three other factors suggest restraint in concluding that species are independently dispersed on small scales—that is, the potential for diluting important effects by using summary statistics, the possibility of effects disproportionate to local abundances of individuals, and the arbitrary choice of sampling scale. Again we attempt to illustrate the problem with a real example. Jumars (1975a) calculated the heterogeneity χ^2 statistics for all polychaete species within cores among subcores at the San Diego Trough locality. Heterogeneity χ^2 closely approximated its degrees of freedom, giving no reason to suspect lack of independence among species' per-subcore abundances. Species of Paraonidae taken alone, however, showed strong concordance in abundance (Jumars, 1975a). This concordance among species within the family was obscured by the other species included in the heterogeneity χ^2 analysis for polychaetes as a whole.

Paraonids are shallow-burrowing deposit feeders or predators on Foraminifera and have no feeding appendages (Fauchald and Jumars, 1979). Jumars (1975a) thus reasoned that they might physically be excluded from small areas by the tube-building cirratulid *Tharyx luticastellus* (Jumars, 1975c). The tube or "mudball" of this species occupies a circular surface area of up to 3 cm in diameter and extends to depths of over 1 cm into the sediments. Figure 18*a* shows the relationship between number of *T. luticastellus* mudballs and number of paraonids per subcore. The corner test (Tate and Clelland, 1957) was used to test the relationship because the extreme values were of the highest interest *a priori;* given the hypothesis of physical exclusion of paraonids by these mudballs, a high density of mudballs would preclude a high density of paraonids, whereas an intermediate density of mudballs would allow, but would not guarantee, an intermediate density of paraonids (which might be rare or absent for other reasons). The null hypothesis of independence in per-subcore abundance was rejected ($p < 0.025$).

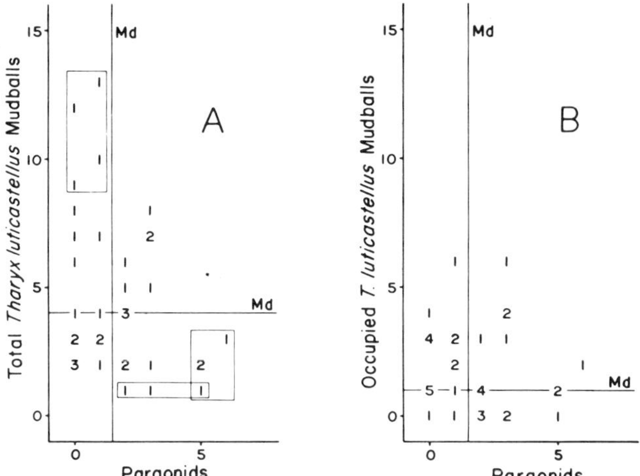

Fig. 18. Number of *Tharyx luticastellus* mudballs versus number of paraonid individuals per subcore for 0.01-m^2 subcores from 1230-m depth in the San Diego Trough. Numbers, frequencies of indicated coordinates; Md, median; A, total number of mudballs remaining after sample sieving, enclosed values of particular interest in the corner test; B, only those mudballs still occupied by *T. luticastellus*. [Reprinted with permission of Springer-Verlag New York, Inc. (*Marine Biology*) from Jumars, P. A., 1975. Environmental grain and polychaete species diversity in a bathyal benthic community. *Mar. Biol.*, **30**, 260.]

The finding of no apparent correlation when only the mudballs still occupied by *T. luticastellus* were considered supported the interpretation of physical exclusion of paraonids by habitat structures, whose local density did not correspond closely with numbers of living *T. luticastellus*. The first-order cause of the pattern may thus have been vectorial, whereas the second-order cause was coactive.

For similar reasons, the results of the heterogeneity χ^2 analysis would be expected to change, perhaps radically, with a slight change in the size of the subcores employed. Interindividual interactions may become most evident when the sampling scale approaches the scale at which these interactions occur, and the 0.01-m^2 subcores may greatly exceed such a scale insofar as sedentary or sessile macrofaunal species are concerned. Two macrofaunal species found in the same 0.01-m^2 sample may never have interacted. Excluding the possibility of vertical habitat partitioning, past or ongoing interactions are much more likely for individuals drawn from the same square centimeter of sediment.

Just as some patterns may become more evident with decreasing sample scales, others may be seen more clearly as sampling scales are increased. Perhaps the most ambitious quantitative sampling as yet attempted on these larger scales was the abyssal sampling program of Hecker and Paul (1979) in the equatorial Pacific. They gathered baseline information prior to commercial manganese nodule mining. Hecker and Paul collected a total of 80 0.25-m^2 box cores from three $2° \times 2°$ "squares" (from four to five replicate cores per numbered station shown in Fig. 19). All macrofaunal individuals recovered on a 0.30-mm sieve were sorted to the species level, and, for the analysis described below, all replicate cores per station were combined.

The NESS procedure (Grassle and Smith, 1976) was applied to allow clustering of stations according to the species composition of the composite samples repre-

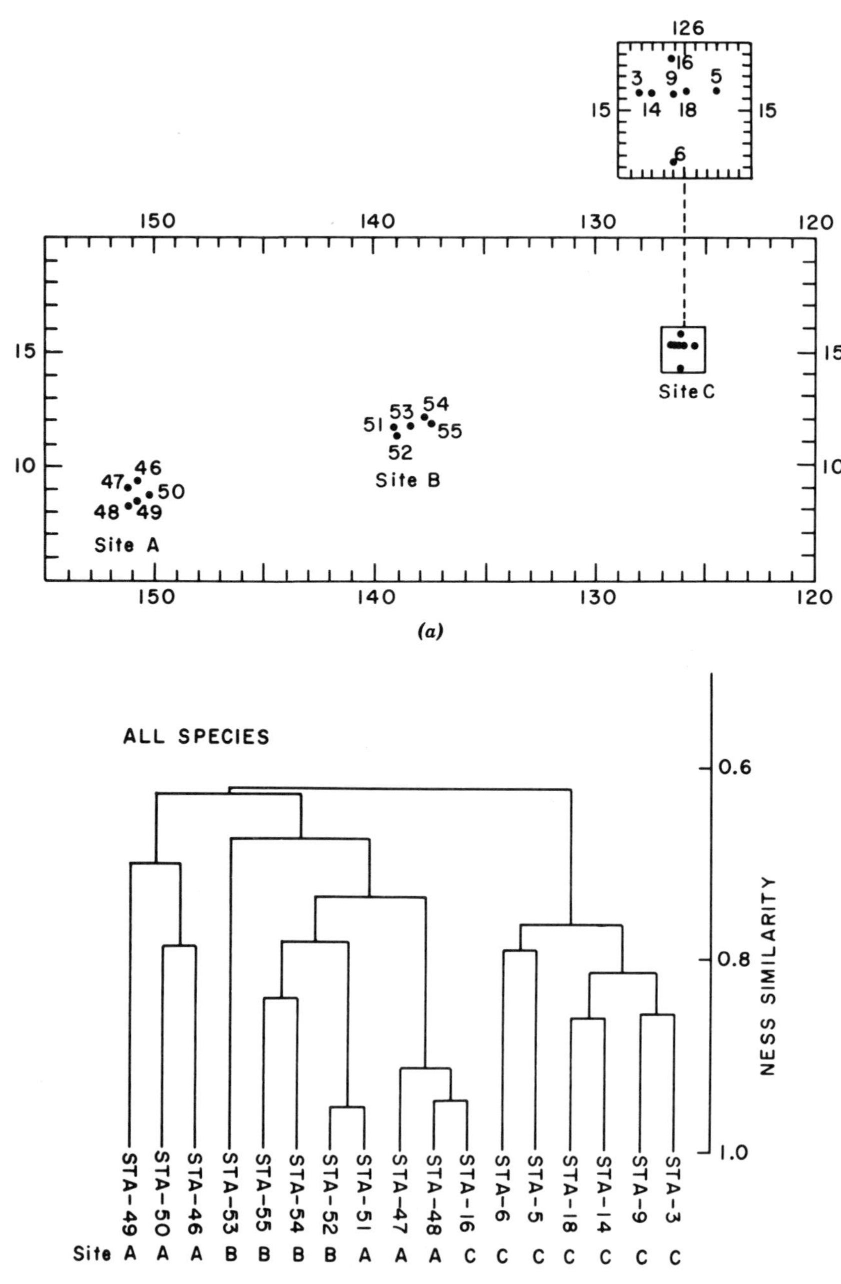

Fig. 19. (*a*) Chart and (*b*) dendrogram of site groups formed on the basis of all macrofauna retained from four to five replicate 0.25-m^2 cores per station (site). Number of expected species shared procedure (52) used for clustering. (Reprinted with permission of DOMES/PMEL Project NOAA from Hecker, B. and A. Z. Paul, 1978. *Benthic Baseline Survey*. DOMES Final Report, Contract 03-6-022-35141, pp. 7 and 85.)

senting individual stations. For the purpose of this clustering, each real composite sample was reduced statistically to a hypothetical sample size of 25 individuals. In terms of a loose analogy with the methods we have used above, the NESS method links (in the dendrogrammatic summary) those stations that are frequented by species belonging to homogeneous groups (*sensu* Fig. 6). Although some spatial coherence of community composition was evident in the dendrogram (Fig. 19), perhaps suggesting ordination as an alternative method of community description, the separation of Stations 49, 50, and 46 from the other Site A stations, and of Station 16 from other Site C stations, indicates a superimposed mosaic structure. We leave open the questions as to whether and where community boundaries should be drawn. This study suffices to show, however, that an arbitrarily selected 2° × 2° "square" of abyssal benthos may be relatively uniform (Site B) or decidedly more variable (Site A) in species composition. Hecker and Paul (1979) very plausibly speculated that the apparent mosaic structure of these benthic assemblages may be due vectorially to underlying patterns of sedimentary variation.

No simple, "typical," deep-sea, multispecies patterns thus emerge from quantitative studies to date, and future sampling can be expected to add to the diversity of patterns observed. The studies outlined above have, explicitly or implicitly, avoided major local topographic features, such as submarine canyons. Also excluded, for lack of quantitative sampling ability, were rock outcrops. Since decidedly nonrandom multispecies patterns have been documented, future sampling programs will be more obliged to collect correlative physiographic, geotechnical, and chemical data to begin to place some bounds on the latitude for speculation concerning vectorial causative factors. Identifying site-specific species groups is only half the numerical community analyst's task (Boesch, 1977).

Vertical Patterns

The cores collected on Expedition Quagmire (Thiel and Hessler, 1974; Jumars, 1978) were taken very deliberately with thorough television and motion picture documentation of the lack of a bow wave during sampling. One might, therefore, be tempted to believe the vertical depth–frequency data resulting from these samples, which were carefully sliced into vertical layers prior to sieving. The internal consistency of the data already presented in Fig. 14 adds to their credibility and suggests that vertical habitat segregation of species may be a mechanism of avoiding competition of species within some families of polychaetes. The deeper-living species within these families tend to be less abundant than the shallower members, perhaps supporting the idea of a competitive displacement further from the source of gradually sedimenting food.

Although vertical segregation among confamilial species is likely to exist, the data in Fig. 20 cast serious doubt on all depth–frequency distributions calculated for deep-sea samples. The "uncorrected" versions were calculated from the depth–frequency distributions observed. Species judged to have surface deposit feeding and suspension feeding habits were found, however, below the surface layer of sediments. The "corrected" version replaced surface deposit and suspension feeders, determined according to the functional classification by Fauchald and Jumars (1979) in the surface layer where they feed. This large discrepancy

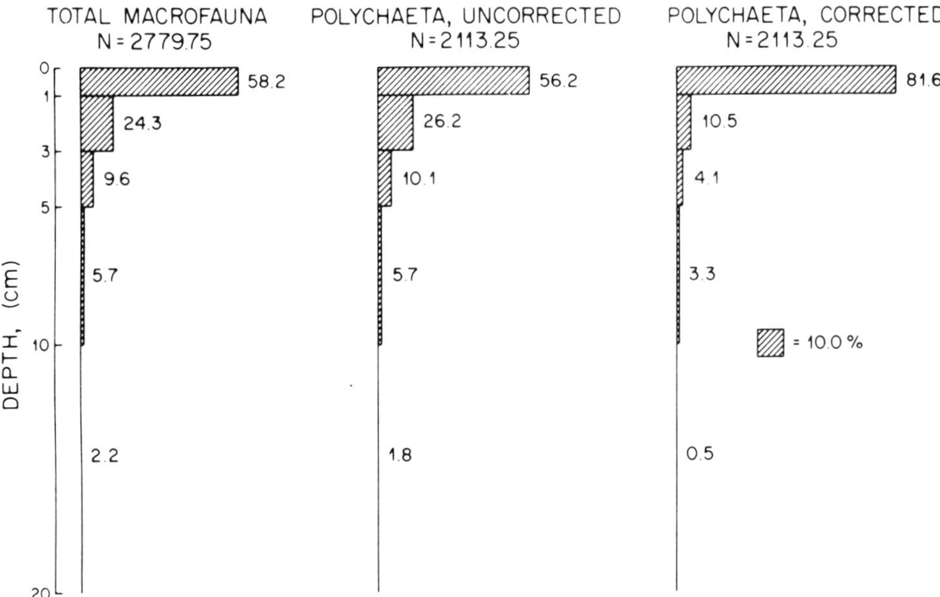

Fig. 20. Depth–frequency distributions of all macrofauna and polychaetes alone as collected at 1220 m in the San Diego Trough during Expedition Quagmire. Correction applied as described in text (*N*, total number of individuals). [Reprinted with permission of Pergamon Press, Inc. (*Deep-Sea Research*) from Jumars, P. A., 1978. Spatial autocorrelation with RUM (remote underwater manipulator): Vertical and horizontal structure of a bathyal benthic community. *Deep-Sea Res.*, **25**, 600. Copyright 1978, Pergamon Press, Ltd.]

between the two versions dissuaded us from extracting further depth–frequency data from the literature for quantitative comparisons.

Qualitatively, two general classes of patterns are seen. A steeper-than-linear decrease in abundance with depth in the sediment, with few macrofauna found below 10-cm depths, is characteristic of deep-sea regions having gradual sedimentation (e.g., Fig. 20). Paleontological data on the density of burrows (Piper and Marshall, 1969; Griggs et al., 1969) suggest more complex, time-varying patterns in areas of episodic sedimentation. Organic-rich sediments buried by turbidities may repay the energetic costs of mining by deeply burrowing benthos, such as thalassinid shrimp and maldanid or large opheliid polychaetes, but we are unaware of published faunal depth–frequency data for deep-sea turbidite deposits.

Better samplers alone are not likely to provide more useful data on animal depth–frequency distributions. We doubt that even *in situ* vertical subsampling could be done quickly enough to avoid the withdrawal into tubes and burrows evolved as an escape from epibenthic predators. Many polychaetes, for example, have evolved a so-called giant axon response, an extremely rapid withdrawal stimulated by one or a few large-diameter (and hence quickly transmitting) nerves (Dales, 1963). Improved sampling methods and continued correlative work are no substitutes for the insights, into both ecological relationships and their stratigraphic consequences, which could be gained by increased *in situ* observations and manipulative abilities.

C. *Meiofauna*

Horizontal Patterns

In the majority of published multispecies meiofaunal data (see Thiel, this volume, Chapter 5), species have not been discriminated. When only totals for major taxa (e.g., Nematoda, Harpacticoida) or for meiofauna as a whole are given, interspecific patterns of covariation remain unknown. In the terms of heterogeneity χ^2 analysis (Fig. 6, Table I), the same pooled χ^2 value can result from wildly different combinations of total and heterogeneity χ^2. A large variance in total numbers of meiofauna per sample, for example, could easily result from either independent but large variances in abundances of individual species or from concordant but much smaller individual variances.

Meiofaunal taxonomy is difficult, and meiofauna are comparatively abundant in the deep sea (Chapter 5), partly explaining the paucity of species-level data. To provide more manageable numbers of individuals from samplers that were designed for macrofaunal collection, meiofaunal researchers have resorted to subsampling on sample retrieval. Although the statistics of these subsampling procedures have been well worked out (Elmgren, 1973), their routine application to deep-sea samples seems unwarranted. Specifically, there is no guarantee against active or passive movement of meiofauna after sampling but before subsampling. The problem is especially severe with deep-sea samples: temperature gradients arise during the ascent, samples may take hours to retrieve, and bringing samples aboard a vessel on the open sea is hardly a delicate process.

Unfortunately, this problem is most severe where the analytical results are most sensitive. In Table VI we reproduce results from the recent review and analysis of quantitative meiofaunal sampling methods by Coull et al. (1977). The

TABLE VI
Two-Level, Nested ANOVA Calculated on $\log_{10}$–Transformed Total Faunal Densities[a]

Source of Variation	DF	SS	MS	F_s
Among depths	2	5.6413	2.82070	41.0932[b]
Among-box cores (within depths)	5	0.3629	0.07258	1.7811 NS
Within-box cores (meiostechers)	28	1.1941	0.04075	

[a]Highest level is among depths. Subgroups consist of box cores within a depth, and the basic sampling unit is the meiostecher, of which several were taken in each box core. Abbreviations: DF, degrees of freedom; SS, sums of squares; MS, mean squares; NS, not significant ($F_{0.05(5,28)} = 2.56$; $F_{0.001(2,5)} = 37.1$).
[b]Calculated from procedure of Sokal and Rohlf (1969, Box 10.4); significant at $p = .001$.
Source: Reprinted with permission of Springer-Verlag Inc., New York, from Coull, B. C., R. L. Ellison, J. W. Fleeger, R. F. Higgins, W. D. Hope, W. D. Hummon, R. M. Rieger, W. E. Sterrer, H. Thiel, and J. H. Tietjen, 1977. Quantitative estimates of the meiofauna from the deep sea off North Carolina, USA. *Mar. Biol.*, **39,** 237.

data were derived from a total of 36 (10-cm^2) meiostecher subsamples taken from eight 200-cm^2 box cores distributed among three sampling depths (400, 800, and 4000 m) off the coast of North Carolina. Any artifact due to alteration of natural patterns during retrieval would occur in the within-stations sum of squares, which unfortunately is the basis of the entire nested analysis of variance. Keeping in mind this problem, the generally low power of omnibus tests for nonrandomness, the lack of sufficient data for use of spatial coordinates of the samples explicitly (e.g., by means of spatial autocorrelation), let alone questions of homoscedasticity, we find tenuous the conclusion that most field variation is on the scale of the 10-cm^2 subsamples (Coull et al., 1977). The disparity of opinion noted by Coull et al. (1977) among meiofaunal workers with regard to the relative magnitudes of within-sample (among-subsample) versus among-sample variation may well reflect one or more of the problems we note above or may indicate real differences among the geographic regions studied.

Thistle (1978) circumvented many of these potential difficulties by (horizontal) subsampling *in situ*. As discussed in Section 4.C, he counted and determined to species all adult copepods found in the surface centimeter of sediments of 14 precisely positioned (0.01-m^2) subcores. With all species included, Thistle found the same qualitative result at all three sampling scales he used—10^3, 1, and 10^{-1} m (Figs. 15 through 17). Although most species taken individually showed no significant departures from Poisson expectations in per-subcore abundances, heterogeneity χ^2 analysis showed all three χ^2 quantities (total, pooled, and heterogeneity) to exceed their expected degrees of freedom ($p < 0.025$). Hence the total number of adult harpacticoids per subcore was not Poisson distributed, and the species tended to be discordant in abundance. These nonrandom patterns, then, did not appear to be limited to any single spatial scale.

Thistle went further to investigate the possible causes of the observed patterns by examining correlations of harpacticoid species' abundances with biogenic structures in the environment (Thistle, 1978) and (separately) with polychaete abundances (Thistle, 1979). Each structure that exceeded 0.50 mm in minimum dimension and was retained on a 1-mm sieve was classified as belonging to one of seven categories. The total volume of each structural class was then estimated, and the respective volumes were correlated with abundances of 124 individual harpacticoid species. We reproduce Thistle's results in Table VII. Biogenic structure appeared to account for part of the pattern, and the most abundant (in cubic millimeters) structures made the largest contributions (greatest number of significant correlations).

To investigate potential influences of polychaetes on the multispecies pattern observed in harpacticoids, Thistle (1979) grouped polychaete species found in the upper 10 cm of the subcores by Jumars (1978) by feeding guilds according to the classification of Fauchald and Jumars (1979). He then correlated the abundances of the 124 different species of harpacticoids with the six polychaete guilds encountered (Table VIII). Correlation relations need not be perfectly transitive (i.e., the fact that species A is positively correlated in abundance with the volume of polychaete tubes and that the abundance of species B is negatively correlated with that volume not necessarily implying that species A and B are negatively correlated in abundance). The fact that Tables VII and VIII show heterogeneity in apparent responses of different harpacticoid species to, respectively, biogenic

TABLE VII
Summary of Significant (Individual $p < 0.05$, Kendall's τ) Correlations Between Harpacticoid Species' Abundances and Volume (mm^3) of Biogenic Structures Found in San Diego Trough

Structural Class	Mean Volume per Subcore (mm^3)	Number of Positive Correlations	Number of Negative Correlations
Tharyx luticastellus (polychaete) mudballs	6725	17	4
Tharyx monilaris (polychaete) mudballs	73	6	5
Polychaete tubes	342	8	2
Orictoderma rotunda (foraminiferan) tests	238	7	7
Tube-shaped foraminiferan tests	1550	9	7
Dendritic foraminiferan tests	121	1	6
Tanaid (crustacean) tubes	16	3	6

Source: Reprinted with permission of Plenum Publishing Corporation from Thistle, D., 1979. Harpacticoid copepods and biogenic structures: Implications for deep-sea diversity maintenance. In *Ecological Processes in Coastal and Marine Systems,* R. J. Livingston, ed. Plenum Press, New York.

TABLE VIII
Summary of Significant (Individual $p < 0.05$, Kendall's τ) Correlations Between Harpacticoid Species' Abundances and Abundances of Members of Polychaete Feeding Guilds

Polychaete Feeding Guild	Number of Positive Correlations	Number of Negative Correlations
Motile, subsurface deposit feeders	6	4
Motile, surface deposit feeders	4	6
Carnivores	5	3
Discretely motile, subsurface deposit feeders	5	3
Discretely motile, surface deposit feeders	3	3
Sessile, surface deposit feeders	3	9

Source: Reprinted with permission of Springer-Verlag, Inc. New York, from Thistle, D., 1979. Deep-sea harpacticoid diversity maintenance: The role of polychaetes. *Mar. Biol.,* **52,** 371–376.

structures and polychaete feeding guilds suggests, however, that the high heterogeneity–χ^2 values reported by Thistle (1978) may have been due in part to divergent harpacticoid responses to these two features of their environment. If so, we would expect the correlations to become even more apparent at smaller sampling scales and the system to be amenable to experimental manipulation for judging cause and effect. As Thistle (1978, 1979) has pointed out, some, but not all, of the correlations may spuriously be due to the degree of multiple testing involved.

Bernstein et al. (1978) similarly examined all Foraminifera from several 0.25-m^2 cores from the central northern Pacific but used multivariate techniques to circumvent the problem of multiple testing. One of the five cores was partitioned [with the "vegematic" corer of Jumars (1975b)] *in situ* into 0.01-m^2 subsamples. Although a 0.30-mm mesh was used in this study and the Foraminifera were thus in the size class of macrofaunal taxa (Hessler and Jumars, 1974) from this locality, we, according to the usual taxon-based convention, include this study with the meiofauna.

Bernstein et al. (1978) used the indirect method of Hessler and Jumars (1974) to demonstrate heterogeneity in dispersion patterns both among the cores and among the subcores of the vegematic core. Using the Bray–Curtis index (Bray and Curtis, 1957) with a flexible sorting strategy (Lance and Williams, 1967), they proceeded to cluster the species (by subcores) and the subcores (by species) found within the vegematic core. Twelve species groups and five site groups (*A* through *E* in Fig. 21) were formed. To generate some hypotheses about the factors that may have been responsible for the heterogeneity represented by the site groups in Fig. 21, Bernstein et al. (1978) then performed multiple discriminant analysis. This analysis was used to identify the linear combinations of several variables that best distinguished among site groups A through E. The variables used (on a per-subcore basis) were as follows: number of suspension feeding metazoa; number of surface deposit feeding metazoa; number of subsurface deposit feeding metazoa; number of carnivorous metazoa; number of small ($\leq$5 mm) manganese nodules; number of medium-sized (6- to 10-mm) nodules, number of large (>10-mm) nodules, surface area (mm^2) of biogenic structures, and volume (mm^3) of biogenic structures. We reproduce their results for the first three discriminant axes in Fig. 21. As shown in Fig. 21, eight of the nine measured variables may have contributed to the formation of these site groups. Alternatively, the foraminiferal species that comprised the site groups may have accounted for the correlation by influencing the locations of the metazoan functional groups. Once again, basic natural history data and manipulative experiments, both of which might help differentiate cause and effect, are lacking.

The innovative approach of Bernstein and Meador (1979) uses dispersion patterns of living versus dead foraminiferans to give the only indication yet available of the evolution of horizontal dispersion patterns of deep-sea meiofauna. They found in the examination of the surface 2 cm of a partitioned 0.25-m^2 core from the central North Pacific that both living and dead forms showed horizontal segregation among 0.01-m^2 subcores (high heterogeneity chi square). Furthermore, they showed that in 12 of the 22 species represented by both living and dead individuals, abundances of living and dead individuals were significantly positively correlated (versus one significant negative correlation). Assuming the dead tests represented one or more previous generations, these data suggests the temporal persistence of multispecies patch structure. It would be extremely interesting to carry such an analysis deeper into the geological record.

Although the above discussion certainly implicates spatial heterogeneity—partly of biological origin—on areal scales smaller than 0.01 m^2 as being an important determinant of meiofaunal dispersion patterns, this small-scale pattern in no way precludes the existence of larger-scale patterns. What a large variability within samples (e.g., within-box cores variation revealed in Table VI) unquestionably does is to make variation between samples more difficult to establish. Furthermore, from the small-scale results there is good reason to expect that larger-scale variation does have effects. The discriminant analysis suggests, for example, that regional or topographical changes in the size-frequency distribution and abundance of nodules may influence regional foraminiferal community structure, as may regional variation in metazoan feeding-guild composition.

Vertical Patterns

Multispecies vertical patterns among the meiofauna are subject to all the qualifications noted for macrofauna (Section 5.B, on vertical patterns). We are unaware of published attempts to document vertical habitat segregation among species; most studies of vertical pattern (Thiel, this volume, Chapter 5) have identified specimens to major taxa (e.g., Nematoda, Harpacticoida) or only as "meiofauna."

Just as with the macrofauna, the typical vertical pattern (Fig. 22) is a (more rapid than linear) decrease in meiofaunal abundance with depth. The exception in the data from 4000 m, which show a subsurface maximum in meiofaunal abundance, are suspected by Coull et al. (1977) to be a sampling artifact due to core disturbance during sample retrieval. If they are correct, our distrust of shipboard sampling (this section, above, on horizontal patterns) is further justified.

We do not intend to imply, however, that all the depth-frequency distributions taken from deep-sea cores are artifactual. An excellent piece of evidence that suggests reliability of some of these abundance profiles was presented recently by Vivier (1978). She found subsurface maxima in meiofaunal abundances under red, unconsolidated mud (mostly iron and aluminum oxides) dumped as wastes from aluminum processing. These wastes were dumped at the head of a submarine canyon and were traced out to its bathyal fan. In natural sediments, however, there is no generally recognized means of determining whether the sample has been disturbed sufficiently, or whether the animals have migrated enough, to preclude valid subsampling. Consequently, the degree of confidence that can be placed in the patterns shown in Fig. 22 is unknown.

6. Summary and Conclusions

This chapter brings us to the unhappy conclusion that our actual knowledge of deep-sea faunal dispersion patterns allows fewer generalizations than one might have expected. It has been naive to expect sweeping generalizations to hold over the faunas that cover 60% of the Earth's surface—including bathyal, abyssal, and hadal communities encompassing multiple scales of environmental variation in the form of sediment type, physiography, boundary-layer dynamics, and biogenic structure. The paucity of well-documented, strongly nonrandom, horizontal patterns has been shown (Tables III and IV) to be a likely consequence of the sampling intensities and patterns achieved—even if strongly nonrandom patterns are the deep-sea rule. That nonrandom patterns are more prevalent than has been documented is supported by extant megafaunal data. Essentially continuous photographic coverage and comparatively abundant natural history information have revealed a wide diversity of patterns among the fauna of this size category. Spatial coverage achieved to date for deep-sea macrofauna or meiofauna has always been "gappy," presumably allowing many nonrandom patterns to escape or to be poorly characterized (Baer and Tribbia, 1976). Consequently, we suggest that attention be shifted from the bulk of patterns, in which nonrandomness cannot yet

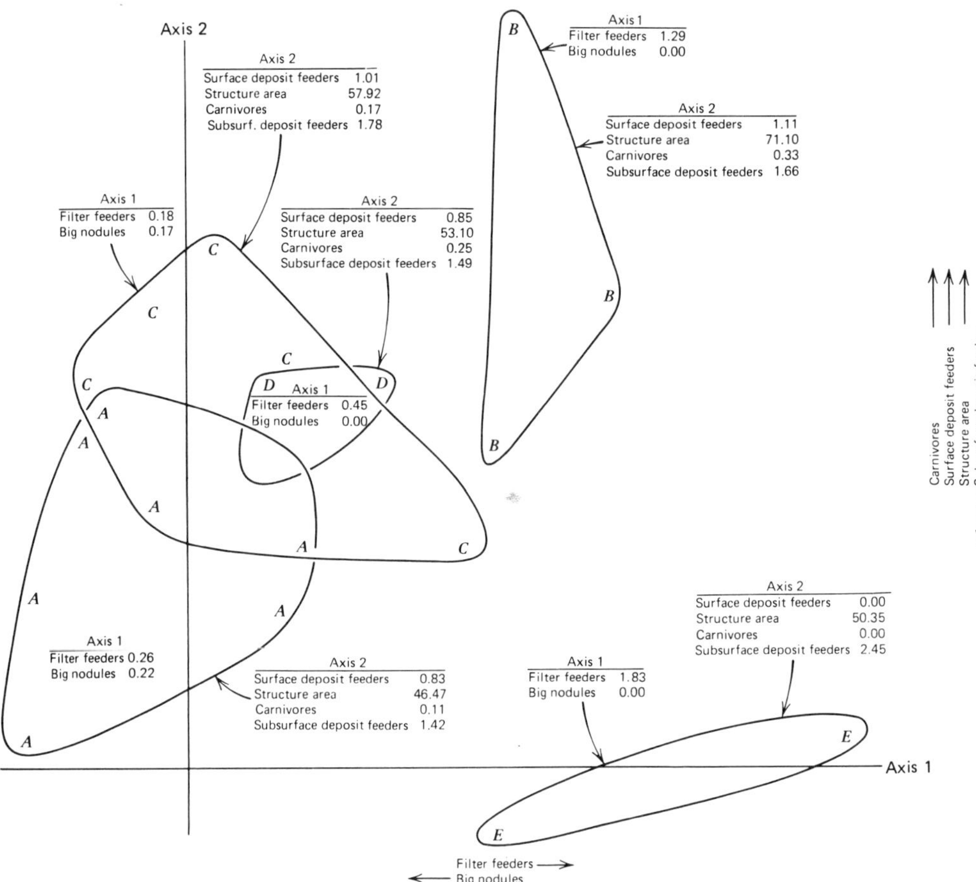

Fig. 21. Foraminiferan site groups and their correlates in twenty-four 0.01-m^2 subcores from a partitioned box core taken in the central northern Pacific (28°N, 155°W). Total numbers of fragments (upper-left corner), numbers of individuals (upper-right corner), numbers of species (lower-right corner), and site group letter codes are shown within the diagram of the corer, and the locations of these site groups along the first three discriminant axes are indicated. External variables important in constructing axes are inserted at the edges of the figure, the arrows indicating the directions of increase of these external variables. (Reprinted with permission of *Limnology and Oceanography* from Bernstein, B. B., R. R. Hessler, R. Smith, and P. A. Jumars, 1978. Spatial dispersion of benthic foraminifera in the abyssal Central Pacific. *Limnol. Oceanogr.*, **23**, 406, 410, 411.)

211 44 A 31	613 77 C 50	358 48 C 34	327 82 B 39	359 55 C 35
601 91 B 35	457 37 C 34	701 83 C 47	260 49 D 34	472 57 D 38
519 58 A 38	216 54 A 28	692 85 A 39	263 35 A 22	814 67 A 32
341 50 A 32	701 52 D 28	NO DATA	644 59 C 29	612 79 B 41
180 28 E 22	339 66 A 35	221 37 E 31	323 59 A 35	178 30 D 21

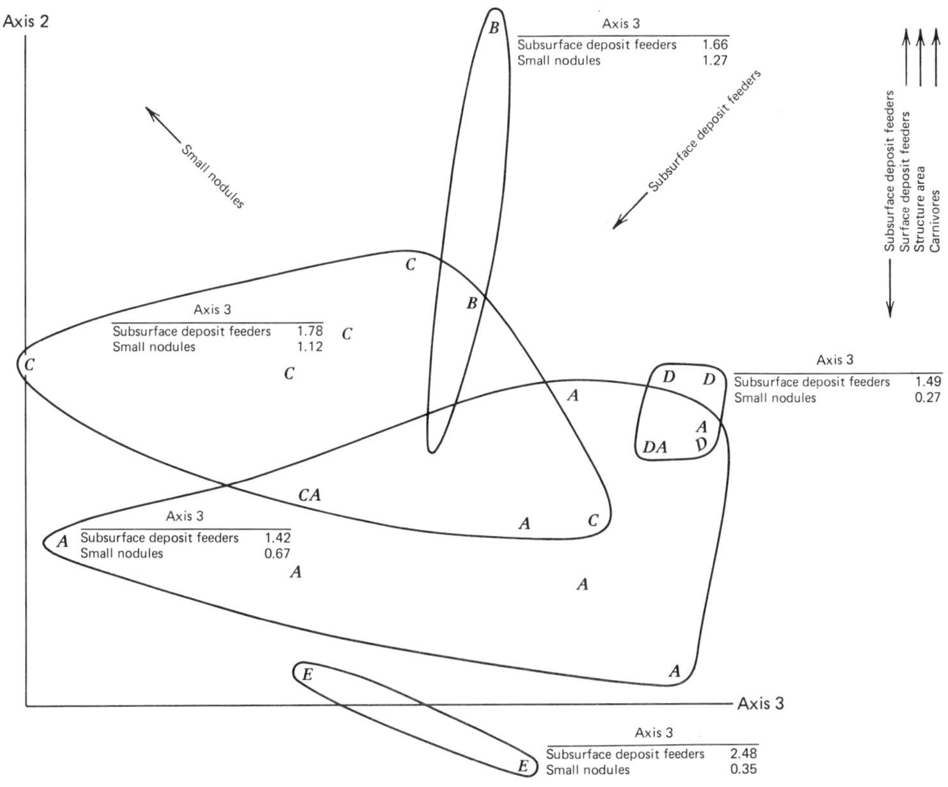

Fig. 21. (*Continued*)

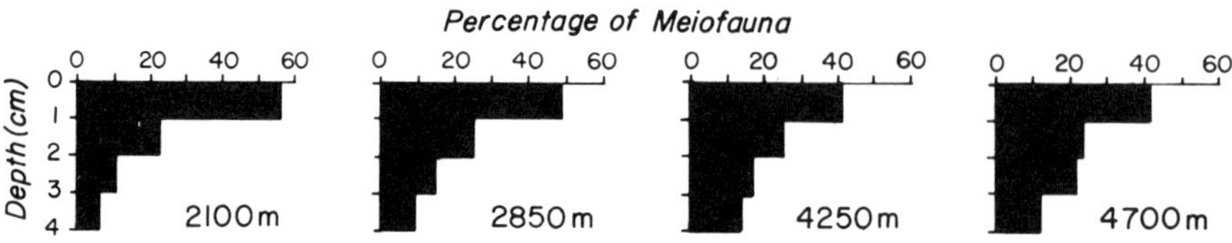

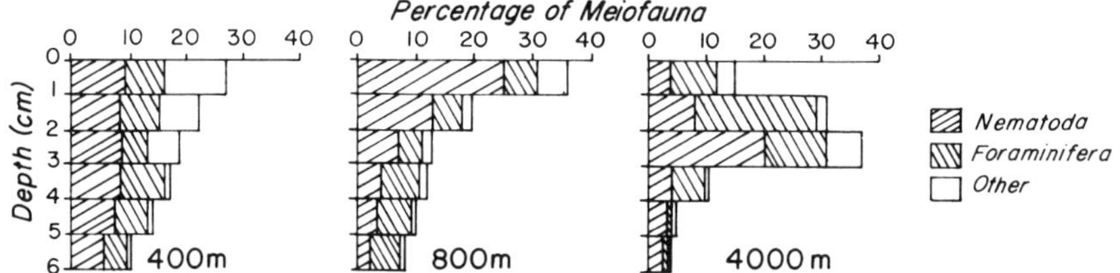

Fig. 22. Observed depth–frequency distributions of meiofaunal individuals in subsamples from box cores taken from the Gulf of Gascogne (Dinet and Vivier, 1977) and the deep sea off North Carolina (Coull et al., 1977) [Reprinted with permission of Springer-Verlag New York, Inc. (*Marine Biology*) and Cahiers de Biologie Marine from Coull, B. C., R. L. Ellison, J. W. Fleeger, R. F. Higgins, W. D. Hope, W. D. Humman, R. M. Rieger, W. E. Sterrer, H. Thiel, and J. H. Tietjen, 1977. Quantitative estimates of the meiofauna from the deep sea off North Carolina, USA. *Mar. Biol.*, **39**, 237 and Dinet, A. and M.-H. Vivier, 1977. Le meiobenthos abyssal du Golfe de Gascogne. I. Considerations sur les donneés quantitatives. *Cah. Biol. Mar.*, **18**, 93.]

be documented, to those instances in which nonrandomness has fortuitously been established.

Apparent spacing among individuals has been seen, as would be expected, when the sampling scale approached the scale of individual species' ambits. If the population in question is near its carrying capacity, this areal scale can be estimated as the reciprocal of population density. Two examples of apparent spacing among conspecific individuals on these scales are the patterns seen in *Ophiomusium lymani* (Section 4.A, on horizontal patterns) and *Ceratocephale pacifica* (Section 4.B, on horizontal patterns). The lack of similar examples among meiofauna probably stems from two sources: (1) sampling of abundant meiofaunal species on ambit scales has not been achieved, and documentation of nonrandom patterns in any rare species is difficult; and (2) meiofaunal species, because of their small size, are particularly likely to experience a coarse-grained or heterogeneous environment. If spacing occurs, it may not be detectable in a heterogeneous environment unless the preferred habitat of the species can be identified.

Aggregation, again, is best understood in megafaunal species, as feeding aggregations of several varieties now having been documented (Section 4.A, on horizontal patterns). Aggregation has been shown in macrofaunal (Section 4.B, on horizontal patterns) and meiofaunal (Section 4.C) species as well, but the generating processes are essentially unknown. Greater attributability of macro- and meiofaunal patterns cannot be expected until more natural history data become available. Until more is known of the life habits of these species, study of their spatial patterns is more or less a statistical exercise. Although these data might be considered useful for testing alternate models of community structure maintenance, their suitability is limited to the refutation of hypotheses suggesting random and independent dispersion patterns as deductive consequences. More realistic community structural models applied to the deep sea to date do not make explicitly different predictions concerning static dispersion patterns to be expected.

Specifically, there is no clear correspondence between the taxonomies of patterns and of their causes. Multispecies data, for example, frequently show discordance in local abundances among species (Sections 5.B and 5.C, on horizontal patterns). This pattern would be expected as a consequence of either habitat partitioning, with stable patterns relative to the generation times of populations, or of much more dynamic but asynchronous local successional series. Distinction between these two alternatives requires as yet unavailable information on the dynamics of the patterns. Available data do suffice to implicate biogenic and biological habitat heterogeneity, respectively, in the forms of identifiable physical features such as tubes, tests, "mudballs," and hexactinellid sponge fragments (e.g., Sections 5.B and 5.C, on horizontal patterns) and of sympatric species themselves (Section 5.C, on horizontal patterns) as correlates of discordant patterns. Manipulative experiments and time-series observations will be required, however, to clearly distinguish cause from effect and more static habitat partitioning from successional series.

The relationships between the dynamics of both horizontal and vertical dispersion patterns and sedimentation rate are critical to the assumption of one-dimensionality in stratigraphic analysis. It is unfortunate, therefore, that measured vertical depth–frequency distributions of organisms are of dubious validity (Sections 3.B; 4.A, on horizontal patterns; 4.B, on vertical patterns; 4.C). Even if the observed depth–frequency distributions were reliable, however, a lack of critical natural history information on the qualitative and quantitative effects of individual deep-sea species on sediments would inhibit extrapolations from these profiles of organism abundances to stratigraphic models of sediment mixing or layering.

Although measurements of dispersion patterns thus provide clues to deep-sea processes, they alone are not sufficient to test the manifold hypotheses that can be generated from these clues. In particular, studies of the evolution of these patterns and the biological causes effecting them are lacking. Without time-series observations of both natural and perturbed communities, without the basic biological information (e.g., identities of potential competitors and their shared resources or of predator–prey pairs) required to interpret these observations, and without hypotheses explicitly predicting spatial patterns *a priori*, the study of deep-sea dispersion patterns is fated to become a futile exercise in the statistics of sampling—a stagnant taxonomy of pattern.

Acknowledgments

We would like to thank Ms. Laura Lewis for help with the illustrations and Dr. Dan Wartenberg for rewriting the spatial autocorrelation program in an intelligible form. Support for much of the work came from NOAA contract 03-78-B01-17 and ONR contract #N00014-75-C-0502.

References

Aller, R. C. and J. Y. Yingst, 1978. Biogeochemistry of tube-dwellings: A study of the sedentary polychaete *Amphitrite ornata* (Leidy). *J. Mar. Res.*, **36,** 201–254.

Baer, F. and J. J. Tribbia, 1976. Spectral fidelity of gappy data. *Tellus*, **28,** 215–227.

Ballard, R. D., 1977. Notes on a major oceanographic find. *Oceanus*, **20** (3), 35–44.

Barham, E. G., N. J. Ayer, Jr., and R. E. Boyce, 1967. Megabenthic fauna of the San Diego Trough: Photographic census and observations from the bathyscaphe *Trieste*. *Deep-Sea Res.*, **14,** 773–784.

Berner, R. A., 1976. The benthic boundary layer from the viewpoint of a geochemist. In *The Benthic Boundary Layer,* I. N. McCave, ed., Plenum Press, New York, pp. 33–55.

Bernstein, B. B. and J. P. Meador, 1979. Temporal persistence of biological patch structure in an abyssal benthic community. *Mar. Biol.* **51,** 179–183.

Bernstein, B. B., R. R. Hessler, R. Smith, and P. A. Jumars, 1978. Spatial dispersion of benthic Foraminifera in the abyssal Central Pacific. *Limnol. Oceanogr.*, **23,** 401–416.

Boesch, D. F., 1977. *Application of Numerical Classification in Ecological Investigations*. Environmental Research Laboratory, Office of Research and Development, U.S. Environmental Protection Agency, Corvallis, OR, 115 pp. (EPA-600/3-77-033).

Bray, J. R. and J. T. Curtis, 1957. An ordination of the upland forest communities of southern Wisconsin. *Ecol. Monogr.*, **27,** 325–349.

Burnett, B. R., 1977. Quantitative sampling of microbiota of the deep-sea benthos. I. Sampling techniques and some data from the abyssal central North Pacific. *Deep-Sea Res.*, **24,** 781–789.

Cliff, A. D. and J. K. Ord, 1973. *Spatial Autocorrelation*. Pion, London, 178 pp.

Cliff, A. D., P. Haggett, J. K. Ord, K. A. Bassett, and R. B. Davies, 1975. *Elements of Spatial Structure*. Cambridge University Press, Cambridge, England, 258 pp.

Connell, J. H., 1963. Territorial behavior and dispersion in some marine invertebrates. *Res. Pop. Ecol.*, **5,** 87–101.

Connell, J. H. and R. O. Slatyer, 1977. Mechanisms of succession in natural communities and their role in community stability and organization. *Am. Natur.*, **111,** 1119–1144.

Cooley, W. W. and P. R. Lohnes, 1971. *Multivariate Data Analysis*. Wiley, New York, 364 pp.

Coull, B. C., R. L. Ellison, J. W. Fleeger, R. P. Higgins, W. D. Hope, W. D. Hummon, R. M. Rieger, W. E. Sterrer, H. Thiel, and J. H. Tietjen, 1977. Quantitative estimates of the meiofauna from the deep sea off North Carolina, USA. *Mar. Biol.*, **39,** 233–240.

Dales, R. P., 1963. *Annelids*. Hutchinson University Library, London, 200 pp.

Dayton, P. K. and R. R. Hessler, 1972. The role of disturbance in the maintenance of deep-sea diversity. *Deep-Sea Res.*, **19,** 199–208.

Dinet, A. and M.-H. Vivier, 1977. Le meiobenthos abyssal du Golfe de Gascogne. I. Considerations sur les donneés quantitatives. *Cah. Biol. Mar.*, **18,** 85–97.

Eckman, J. E. 1979. Small-scale patterns and processes in a soft-substratum, intertidal community. *J. Mar. Res.*, **37,** 437–457.

Elmgren, R., 1973. Methods of sampling sublittoral soft bottom meiofauna. *Oikos* (Suppl.), **15,** 112–120.

Fager, E. W., 1968. A sand-bottom epifaunal community of invertebrates in shallow water. *Limnol. Oceanogr.*, **13,** 448–464.

Fager, E. W., 1972. Diversity: A sampling study. *Am. Natur.*, **106,** 293–310.

Fager, E. W., Marine Ecology lectures at Scripps Institution of Oceanography, La Jolla, California. Unpublished.

Fauchald, K. and P. A. Jumars, 1979. The diet of worms: a study of polychaete feeding guilds. *Oceanogr. Mar. Biol. Ann. Rev.*, **17,** 193–284.

Feller, W., 1968. *An Introduction to Probability Theory and Its Applications*, 3rd ed. Wiley, New York, 359 pp.

Fisher, R. A., 1970. *Statistical Methods for Research Workers*, 14th ed. Hafner, Darien, CT, pp. 57–61.

Gage, J. D., 1977. Structure of the abyssal macrobenthic community in the Rockall Trough. In *Biology of Benthic Organisms*, B. F. Keegan, P. O. Ceidigh, and P. J. S. Boaden, eds. Pergamon Press, Oxford, 247 pp.

Gage, J. D. and C. G. Coghill, 1977. Studies on the dispersion patterns of Scottish sea loch benthos from contiguous core transects. In *Ecology of Marine Benthos*, B. C. Coull, ed., University of South Carolina Press, Columbia, 319 pp.

Gage, J. D. and A. D. Geekie, 1973. Community structure of the benthos in Scottish Sea-Lochs. III. Further studies on patchiness. *Mar. Biol.*, **20,** 89–100.

Goldhaber, M. B., R. C. Aller, J. K. Cochran, J. K. Rosenfeld, C. S. Martens, and R. A. Berner, 1977. Sulphate reduction, diffusion, and bioturbation in Long Island Sound Sediments: Report of the FOAM group. *Am. J. Sci.*, **277,** 193–237.

Grassle, J. F. and H. L. Sanders, 1973. Life histories and the role of disturbance. *Deep-Sea Res.*, **20,** 643–649.

Grassle, J. F. and W. Smith, 1976. A similarity measure sensitive to the contribution of rare species and its use in the investigation of variation in marine communities. *Oecologia,* **25,** 13–22.

Grassle, J. F., H. L. Sanders, R. R. Hessler, G. T. Rowe, and T. McLellan, 1975. Pattern and zonation: A study of the bathyal megafauna using the research submersible ALVIN. *Deep-Sea Res.*, **22,** 457–481.

Griggs, G. B., A. G. Carey, Jr., and D. L. Kulm, 1969. Deep-sea sedimentation and sediment-fauna interaction in Cascadia Channel and on Cascadia Abyssal Plain. *Deep-Sea Res.*, **16,** 157–170.

Guinasso, N. L., Jr. and D. R. Schink, 1975. Quantitative estimates of biological mixing rates in abyssal sediments. *J. Geophys. Res.,* **80,** 3032–3043.

Hargrave, B. T. and L. K. Nielsen, 1977. Accumulation of sedimentary organic matter at the base of steep bottom gradients. In H. L. Golterman, ed., *Interactions Between Sediments and Fresh Water.* Junk, The Hague, Netherlands, 168 pp.

Heck, K. L., Jr., G. van Belle, and D. Simberloff, 1975. Explicit calculation of the rarefaction diversity measurement and the determination of sufficient sample size. *Ecology*, **56,** 1459–1461.

Hecker, B. and A. Z. Paul, 1979. Abyssal community structure of the benthic infauna at the eastern equatorial Pacific: DOMES sites A, B, and C. In *Marine Geology and Oceanography of the Pacific Nodule Province,* J. L. Bischoff and D. Z. Piper, eds. Plenum Press, New York, pp. 297–308.

Heezen, B. C. and C. D. Hollister, 1971. *The Face of the Deep*, Oxford University Press, New York, 659 pp.

Henley, S., 1977. Autocorrelation coefficients from irregularly spaced areal data. *Computers Geosci.*, **2,** 437–438.

Hessler, R. R. and P. A. Jumars, 1974. Abyssal community analysis from replicate box cores in the central North Pacific. *Deep-Sea Res.*, **21,** 185–209.

Hessler, R. R., J. D. Isaacs, and E. L. Mills, 1972. Giant amphipod from the abyssal Pacific Ocean. *Science*, **175,** 636–637.

Hessler, R. R., C. L. Ingram, A. A. Yayanos, and B. R. Burnett, 1978. Scavenging amphipods from the floor of the Philippine Trench. *Deep-Sea Res.*, **25,** 1029–1047.

Hill, M. O., 1973. The intensity of spatial pattern in plant communities. *J. Ecol.*, **61,** 225–236.

Hollister, C. D., J. B. Southard, R. D. Flood, and P. F. Lonsdale, 1976. Flow phenomena in the benthic Boundary layer and bed forms beneath deep-current systems. In *The Benthic Boundary Layer,* I. N. McCave, ed. Plenum Press, New York, pp. 183–204.

Holme, N. A., 1950. Population dispersion in *Tellina tenuis* da Costa. *J. Mar. Biol. Assoc. U. K.*, **29,** 267–280.

Hurlbert, S. H., 1971. The nonconcept of species diversity: A critique and alternative parameters. *Ecology*, **52,** 577–586.

Hutchinson, G. E., 1953. The concept of pattern in ecology. *Proc. Acad. Nat. Sci.* (Philadelphia), **105,** 1–12.

Isaacs, J. D. and R. A. Schwartzlose, 1975. Active animals of the deep-sea floor. *Sci. Am.*, **233,** 84–91.

Johnson, D. A., 1972. Ocean floor erosion in the equatorial Pacific. *Bull. Am. Geol. Soc.*, **83,** 3121–3144.

Jumars, P. A., 1974. Dispersion patterns and species diversity of macrobenthos in two bathyal communities. Ph.D. dissertation, University of California, San Diego, 204 pp.

Jumars, P. A., 1975a. Environmental grain and polychaete species diversity in a bathyal benthic community. *Mar. Biol.*, **30,** 253–266.

Jumars, P. A., 1975b. Methods for measurement of community structure in deep-sea benthos. *Mar. Biol.*, **30,** 245–252.

Jumars, P. A., 1975c. Target species for deep-sea studies in ecology, genetics, and physiology. *Zool. J. Linn. Soc.*, **57,** 341–348.

Jumars, P. A., 1976. Deep-sea species diversity: Does it have a characteristic scale? *J. Mar. Res.*, **34,** 217–246.

Jumars, P. A., 1978. Spatial autocorrelation with RUM (remote underwater manipulator): Vertical and horizontal structure of a bathyal benthic community. *Deep-Sea Res.*, **25,** 589–604.

Jumars, P. A. and K. Fauchald, 1977. Between-community contrasts in successful polychaete feeding strategies. In *Ecology of Marine Benthos,* B. C. Coull, ed. University of South Carolina, Columbia, pp. 1–20.

Jumars, P. A. and R. R. Hessler, 1976. Hadal community structure: Implications from the Aleutian Trench. *J. Mar. Res.*, **34,** 547–560.

Jumars, P. A., D. Thistle, and M. L. Jones, 1977. Detecting two-dimensional spatial structure in biological data. *Oecologia*, **28,** 109–123.

LaFond, E. C., 1967. Movements of benthonic animals and bottom currents as measured from the bathyscaphe *Trieste*. In *Deep-Sea Photography,* J. B. Hersey, ed. Johns Hopkins University Press, Baltimore, pp. 295–302.

Lakatos, I., 1970. Falsification and the methodology of scientific research programmes. In *Criticism and the Growth of Knowledge,* I. Lakatos and A. Musgrave, eds. Cambridge University Press, London, 91 pp.

Lance, G. N. and W. T. Williams, 1967. A general theory of classificatory sorting strategies. I. Hierarchical systems. *Computer J.*, **9,** 373–380.

Levin, S. A. and R. T. Paine, 1974. Disturbance, patch formation, and community structure. *Proc. Natl. Acad. Sci. (USA)*, **71,** 2744–2747.

Levins, R., 1968. *Evolution in Changing Environments*, Princeton University Press, Princeton, NJ, 120 pp.

Lister, C. R. B., 1976. Control of pelagic sediment distributions by internal waves of tidal periods: Possible interpretation of data from the southern Pacific Rise. *Mar. Geol.*, **20,** 297–313.

Lloyd, M., 1967. Mean crowding. *J. Anim. Ecol.*, **36,** 1–30.

Lonsdale, P., 1977. Clustering of suspension-feeding macrobenthos near abyssal hydrothermal vents at oceanic spreading centers. *Deep-Sea Res.*, **24,** 857–863.

Lonsdale, P. and J. B. Southard, 1974. Experimental erosion of North Pacific red clay. *Mar. Geol.*, **17,** M51–M60.

MacArthur, R. H., 1972. *Geographical Ecology*. Harper and Row, New York, 59 pp.

Meadows, P. S. and J. I. Campbell, 1972. Habitat selection by aquatic invertebrates. *Adv. Mar. Biol.*, **10,** 271–382.

Menge, B. A. and J. P. Sutherland, 1976. Species diversity gradients: Synthesis of the roles of predation, competition, and temporal heterogeneity. *Am. Natur.*, **110,** 351–369.

Moore, P. G., 1954. Spacing in plant communities. *Ecology*, **35,** 222–227.

Papentin, F., 1973. A Darwinian evolutionary system. III. Experiments on the evolution of feeding patterns. *J. Theor. Biol.*, **39,** 431–445.

Patil, G. P. and W. M. Stiteler, 1974. Concepts of aggregation and their quantification: A critical review with some new results and applications. *Resp. Pop. Ecol.*, **15,** 238–254.

Pearcy, W. G., 1976. Pelagic capture of abyssobenthic macrourid fish. *Deep-Sea Res.*, **23,** 1065–1066.

Pielou, E. C., 1960. A single mechanism to account for regular, random and aggregated populations. *J. Ecol.*, **48,** 575–584.

Pielou, E. C., 1969. *An Introduction to Mathematical Ecology*. Wiley, New York, 286 pp.

Piper, D. J. W. and N. F. Marshall, 1969. Bioturbation of holocene sediments on La Jolla Deep Sea Fan, California. *J. Sed. Petrol.*, **39,** 601–606.

Rex, M. A., 1977. Zonation in deep-sea gastropods: The importance of biological interactions to rates of zonation. In *Biology of Benthic Organisms*, B. F. Keegan, P. O. Ceidigh, and P. J. S. Boaden, eds. Pergamon Press, Oxford, 521 pp.

Rhoads, D. C., 1974. Organism-sediment relations on the muddy sea floor. *Oceanogr. Mar. Biol. Ann. Rev.*, **12,** 263–300.

Risk, M. J., R. D. Venter, S. G. Pemberton, and D. E. Buckley, 1978. Computer simulation and sedimentological implications at burrowing by *Axius serratis*. *Can. J. Earth Sci.*, **15,** 1370–1382.

Rokop, F. J., 1974. Reproductive patterns in the deep-sea benthos. *Science*, **186,** 743–745.

Schoener, T. W., 1974. Resource partitioning in ecological communities. *Science*, **185,** 27–39.

Slobodkin, L. B. and H. L. Sanders, 1969. On the contribution of environmental predictability to species diversity. *Brookhaven Symp. Biol.*, **22,** 82–92.

Sokal, R. R. and N. L. Oden, 1978a. Spatial autocorrelation in biology. 1. Methodology. *Biol. J. Linn. Soc.*, **10,** 199–228.

Sokal, R. R. and N. L. Oden, 1978b. Spatial autocorrelation in biology. 2. Some biological applications of evolutionary and ecological interest. *Biol. J. Linn. Soc.*, **10,** 229–249.

Sokal, R. R. and F. J. Rohlf, 1969. *Biometry*. Freeman, London, 776 pp.

Stanley, D. J. and G. Kelling, 1968. Photographic investigation of sediment texture, bottom current activity, and benthonic organisms in the Wilmington Submarine Canyon. *USCG Oceanogr. Rep.*, **22,** 1–95.

Strathmann, R., 1974. The spread of sibling larvae of sedentary marine invertebrates. *Am. Natur.*, **108,** 29–44.

Switzer, P., 1975. Estimation of the accuracy of qualitative maps. In *Display and Analysis of Spatial Data*, J. C. Davis and M. J. McCullagh, eds. Wiley, London.

Tate, M. W. and R. C. Clelland, 1957. *Nonparametric and Shortcut Statistics*, Interstate Publishers and Printers, Danville, IL, 171 pp.

Taylor, L. R. and R. A. J. Taylor, 1977. Aggregation, migration and population mechanics. *Nature*, **265,** 415–421.

Taylor, L. R., I. P. Woiwod, and J. N. Perry, 1978. *J. Anim. Ecol.*, **47,** 383.

Taylor, R. J., 1976. Value of clumping to prey and the evolutionary response of ambush predators. *Am. Natur.*, **110,** 13–29.

Theil, T. and R. R. Hessler, 1974. Ferngesteurtes Unterwasserfahrzeug erforscht Tiefseeboden. *UMSCHAU in Wissenschaft und Technik*, **74,** 451–453.

Thistle, D., 1978. Harpacticoid dispersion patterns: Implications for deep-sea diversity maintenance. *J. Mar. Res.*, **36,** 377–397.

Thistle, D., 1979. Deep-sea harpacticoid diversity maintenance: The role of polychaetes. *Mar. Biol.*, **52,** 371–376.

Thistle, D., 1979. Harpacticoid copepods and biogenic structures: Implications for deep-sea diversity maintenance. In *Ecological Processes in Coastal and Marine Systems*, R. J. Livingston, ed. Plenum Press, New York, pp. 217–231.

Thorson, G., 1950. Reproductive and larval ecology of marine bottom invertebrates. *Biol. Rev.*, **25,** 1–45.

Turekian, K. K., J. K. Cochran, D. P. Kharkar, R. M. Cerrato, J. R. Vaisnys, H. L. Sanders, J. F. Grassle, and J. A. Allen, 1975. The slow growth rate of a deep-sea clam determined by 228 Ra chronology. *Proc. Natl. Acad. Sci. (USA)*, **72,** 2829–2832.

Vivier, M.-H., 1978. Consequences d'un deversement de boue rouge d'alumine sur le meiobenthos profond (Canyon de Cassidaigne, Mediterranee). *Tethys*, **8,** 249–262.

Westman, W. E., 1975. Pattern and diversity in swamp and dune vegetation, North Stradbroke Island. *Austral. J. Bot.*, **23,** 339–354.

11. GEOGRAPHIC PATTERNS OF SPECIES DIVERSITY IN THE DEEP-SEA BENTHOS

Michael A. Rex

1. Introduction

The high degree of species diversity found in the deep-sea benthos (Hessler and Sanders, 1967) is one of the most interesting and unexpected of recent ecological discoveries. There has been considerable controversy surrounding the problem of how so many species can coexist in the deep-sea environment. Alternative hypotheses have attributed it to competitive niche partitioning (Sanders, 1968), predation (Dayton and Hessler, 1972), both predation and competition mediated by productivity (Rex 1976a), contemporaneous disequilibrium (Grassle and Sanders, 1973; Jumars, 1975b, 1976) and a dynamic balance between rates of competitive displacement and the frequency of population reduction (Huston, 1979). A major difficulty in evaluating these theories is that the actual geographic patterns of species diversity that they attempt to explain are poorly known. The few published examples represent taxonomic subdivisions of the community or are confined to limited bathymetric ranges or both (e.g., Sanders, 1969; Coull, 1972; Rex, 1973; Jumars, 1975b). Comparisons of the known patterns are seldom possible because different indices of diversity have been used.

In this chapter I try to clarify the problem of species diversity along depth gradients in the deep sea. I concentrate on the northwestern Atlantic, where there is more known about large-scale patterns of community structure than for any other deep region of the world ocean. Using published and unpublished records, I have documented diversity from the continental shelf or upper slope to the abyss for six groups: the bivalves, gastropods, polychaetes, cumaceans, the invertebrate megafauna, and the fish. Diversity was calculated using one measure, Hurlbert's (1971) expected number of species. Although these data still have many shortcomings, they afford the most complete picture available of the general form of the relationship between species diversity and depth. All six groups appear to have parabolic patterns of diversity with maxima at intermediate depths (~2000 to 3000 m) and lesser values on the upper slope (<1000 m) and abyss (>4000 m). I then review major theories of species diversity in the deep sea in light of this predominant trend. Huston's (1979) dynamic equilibrium model provides the most consistent general explanation. However, our understanding of the specific events and interactions that shape community structure and the scales on which they act remains very fragmentary.

I begin by describing some features of the deep-sea ecosystem that are essential to the consideration of species diversity, particularly food availability and environmental predictability.

2. The Deep-Sea Environment

A. *Food Availability*

The deep sea is a nutrient-poor environment, and energy constraints become more and more severe with increasing depth and distance from land. Since there is no primary production by photosynthetic plants, the typical foundation of the trophic structure in terrestrial and shallow-water systems is entirely lacking. Instead, nutrients come from extrinsic sources and arrive in the deep sea in several forms [see reviews by Rowe and Staresinic (1979), Hinga et al. (1979), Stockton and DeLaca (1982), and Honjo et al. (1982)]. One is composed of small particulate organic debris that originates as either surface production or terrestrial runoff and settles in a relatively slow continuous fallout to the ocean floor. Another major form consists of large rapidly sinking parcels such as water-soaked plant material and carcasses. An additional source is the fecal material of large mobile benthopelagic species that either scavenge deadfalls or forage periodically on midwater populations. The quality, rate, and pattern of food input probably exert some of the most powerful selective forces acting on the deep-sea fauna and shaping the pattern of species diversity.

Our best indication of food availability, whatever its origin and nature, is the actual standing crop of various components of the benthos (see Rowe, this volume, Chapter 3). The pattern of standing crop with depth is well known in the general region of the Gay Head–Bermuda transect south of New England. Density values for the macrofauna collected by anchor dredge, grab (Sanders et al., 1965; Rowe et al., 1974) and box corer (Rowe et al., 1982) are shown in Fig. 1. Density declines exponentially with depth and reaches levels of 10 to 10^2 individuals/m^2 on the abyssal plain. All the sampling devices yielded roughly comparable density estimates, with the exception of values from box cores taken at lower bathyal depths, which appear to be slightly higher. Meiofaunal elements, which the macrofauna probably crops to some extent for food, have not been sampled as extensively along a depth gradient. They do become less abundant at abyssal depths but appear to be still several times more abundant than the macrofauna (Hessler and Jumars, 1974; Thiel, 1975, 1979, this volume, Chapter 5; Coull et al., 1977). The megafauna is also much less abundant at abyssal depths than at intermediate and shallow depths (Rowe and Menzies, 1969; Haedrich et al., 1980). Actual density of the megafauna is difficult to assess from trawl samples. However, catch rates (average number per hour) are considerably less at abyssal depths. For example, four trawls taken recently (Haedrich et al., 1980) between 4000 and 5000 m south of New England yielded only eight fish of three species.

The biomass of the macrofauna, estimated as wet weight per square meter, also declines exponentially with depth (Rowe et al., 1974; Haedrich and Rowe, 1977; Rowe et al., 1982) (Fig. 1). Smith (1978b) has shown that benthic community respiration measured *in situ* also drops three orders of magnitude from 21.5 ml O_2/$m^2 \cdot hr$ at 40 m to 0.02 ml $O_2/m^2 \cdot hr$ at 5200 m. There also appears to be a depth-correlated exponential decline in the biomass of benthopelagic plankton in the deep sea (Wishner, 1980). However, the pattern of biomass with depth is not as

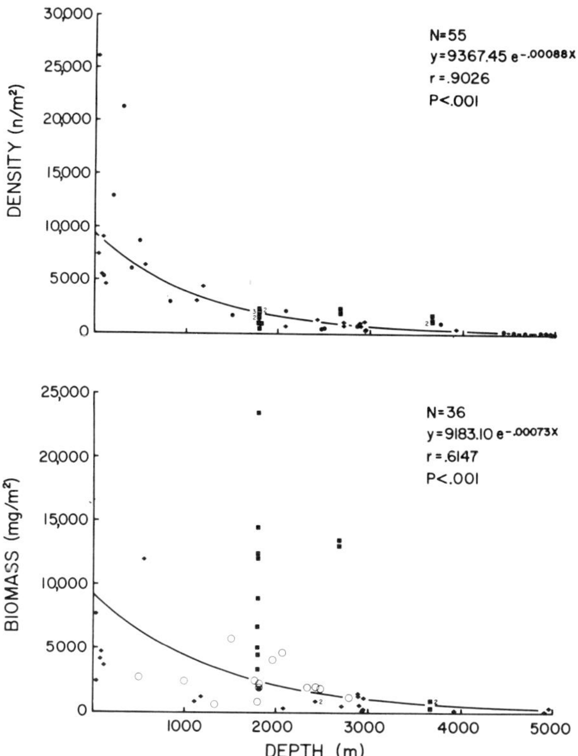

Fig. 1. Upper graph: animal density (number/m^2) plotted against depth (m) in the northwestern Atlantic. Solid circles, diamonds, and squares represent estimates based on anchor dredge samples (Sanders et al., 1965), grabs or anchor dredge (Rowe et al., 1974), and box cores (Haedrich and Rowe, 1977), respectively. All values represent the macrofaunal fraction, those animals retained on a 420 μm sieve. Lower graph: animal biomass (wet weight in milligrams per square meter) plotted against depth in the northwestern Atlantic. Solid symbols same as above, except that no biomass estimates were available from Sanders et al. (1965). Open circles are biomass estimates for demersal fish of the same region (Haedrich and Rowe, 1977).

straightforward as density. Although the values for biomass in Fig. 1 still fit an exponential regression better than a parabolic one, a secondary peak appears in biomass at the base of the continental slope. This is especially apparent in box core samples which seem to be more efficient in capturing the few large individuals that account for the higher biomass values. Biomass for the megabenthos also peaks at the slope base (Haedrich et al., 1980). This may be related to downslope transport of detritus and the accumulation of rich sediments at the slope base (Haedrich et al., 1980).

Interestingly, Haedrich and Rowe (1977) have shown that the biomass of the megafauna is roughly comparable to that of the macrofauna. In Fig. 1, I have included biomass values for demersal fish, and it can be seen that they correspond fairly well to the overall pattern in the macrofauna. Also, in contrast to shallow water, where the macrofaunal biomass exceeds that of the meiofauna, in the deep sea their biomasses seem to be roughly comparable (Thiel, 1975). The combined biomass of the macro- and meiobenthos in deep abyssal regions is only on the order of 1 g/m^2 and in some regions such as under the central oceanic gyres may be less (Thiel, 1975). Although much more information is needed, especially on the megafauna, present data suggest that the deep-sea community does not conform to a typical Eltonian food pyramid (Haedrich and Rowe, 1977), presumably because the megafauna and macrofauna exploit sources of food other than just smaller benthic organisms.

This trend of rapid reduction in density and biomass of the macrofauna with depth seems to be very general in the world ocean (Filatova, 1969; Rowe, this volume, Chapter 3). The rate of decline and average biomass at particular depths do vary geographically and depend *inter alia* on the level and seaward extent of local surface production and how much food it ultimately contributes to the bottom community (Rowe, 1971; Rowe et al., 1974). Average biomass is a positive function of surface productivity. The rate of decline is less in regions where high surface production extends out farther seaward over areas of great depth.

B. Environmental Predictability

Most theories of deep-sea species diversity rest, in part, on the assumption that the deep-sea environment is more stable and predictable than shallow-water environments. Seasonal temperature variation decreases with depth, and a permanent thermocline is reached at upper to midslope depths (Sanders, 1968). Similarly, other water-quality parameters vary so little that Sanders (1977) equated the deep sea to a chemostat. Variation in light and weather at the surface have an indirect and diminished influence on the deep milieu. Weak currents and sediment stability (Heezen and Hollister, 1971; Robb et al., 1981) over much of the deep-ocean floor have permitted a high degree of microhabitat specialization (Jumars, 1976) and the evolution of very delicate morphological structures (Tendal and Hessler, 1977). Unlike their shallow-water counterparts living in highly seasonal environments, most deep-sea organisms appear to have continuous reproduction (Rokop, 1974, 1977, 1979). Broad-scale distributional patterns (Rex, 1977, 1981; Haedrich et al. 1980) and the relative degree of phenotypic differentiation in individual species (Rex, 1976b, 1979; Rex and Etter, manuscript in preparation) suggest that the environment becomes more spatially uniform with increasing depth as well.

Although there does seem to be a general relationship between depth and environmental stability, it is becoming clearer that the deep sea is more dynamic and variable than previously supposed. Most deep currents are slow enough to give the sediment a highly stable structure, but currents sufficiently strong to cause lineation, scouring, and rippling have been observed at all depths (Heezen and Hollister, 1971). Richardson et al. (1981) have shown that strong near-bottom flows (4000 to 5000 m) in the northwestern Atlantic vary in both speed and direction. Fluctuations in deep currents (Dickson et al., 1982) and resuspension in the nepheloid layer (Gardner and Sullivan, 1981) appear to be related to variation in atmospheric wind conditions. In the northwestern Atlantic sporadic warm core rings impinging on the bottom may cause significant temperature variation at mid- to upper-slope depths (Wiebe, 1982). There can be little doubt that periodic massive turbidity currents, many of which have broken submarine cables, have a drastic effect on both the geomorphology and fauna, even at abyssal depths (Heezen and Hollister, 1971). Some submarine canyons of the slope may harbor persistently unstable sedimentary regimes (Rowe, 1972). Sediment slumps exert an important effect on community structure in deep-sea trenches (Jumars and Hessler, 1976) and at intermediate depths (Aldred et al., 1979) in some regions.

Nutrient input also appears to be more variable than earlier supposed. The drizzle of small particulate organic material, although probably a continuous source of food for the deep-sea macrofauna, shows seasonal variation in both amount and composition (Wiebe et al., 1974; Deuser and Ross, 1980; Honjo, 1982). In the Rockall Trough, trophic input is evidently seasonal enough to cause marked seasonality in the reproduction, growth, and recruitment of benthic invertebrates (Lightfoot et al., 1979; Gage and Tyler, 1981). There is considerable geographic variation in the amount and kind of fallout reaching the bottom, and thus its potential contribution to the food requirements of the benthic community (Smith, 1978b; Hinga et al., 1979; Rowe and Gardner, 1979). Sinking plant remains (Turner, 1973; Wolff, 1979) and animal carcasses (Isaacs, 1969; Dayton and Hessler, 1972) are important sources of enrichment that probably cause significant transient changes in local community structure (Stockton and DeLaca, 1982). Deadfalls are rare and unpredictable events compared to the background flux of fine particulates. However, the primary consumers of carcasses are cosmopolitan and highly adapted for scavenging (Smith, 1978a; Dahl, 1979), suggesting that falls are sufficiently frequent to impose an important form of natural selection. Wood is sufficiently abundant and predictable to support an entire endemic subfamily of obligate wood-boring bivalves (Turner, 1973, 1977). Rex et al. (1979) and Rex and Warén (1982) proposed that feeding strategies, physiology, and life histories of deep-sea organisms stabilize their populations against scarce and unpredictable food resources.

C. Summary

Patterns of animal density and biomass indicate that the rate of nutrient input to the deep-sea benthic community declines with increasing depth. In general terms, the deep-sea appears to be physically and biologically more stable than shallower realms. However, seasonal influences and sources of stochastic variation are clearly more prevalent than once assumed. There is also considerable geographic variation in both biotic and physical features. At this stage, our poor understanding of these recently discovered sources of variation and of how deep-sea organisms adapt to them makes it difficult to assess the contribution of stability to species diversity.

3. Species Diversity

A. Patterns of Diversity

I obtained data to calculate species diversity for four macrofaunal groups: the bivalves (H. L. Sanders, unpublished data), gastropods (Rex, 1973, 1976a), polychaetes (Hartman, 1965), and cumaceans (Jones and Sanders, 1972). Patterns of species diversity with depth are plotted in Fig. 2, using Hurlbert's (1971) expected number of species as a diversity measure. It is clear that in every case diversity increases with depth to a maximum at intermediate depths and then decreases at abyssal depths. Although the data show considerable scatter, parabolic regressions give the best fit and are all statistically significant (see figure

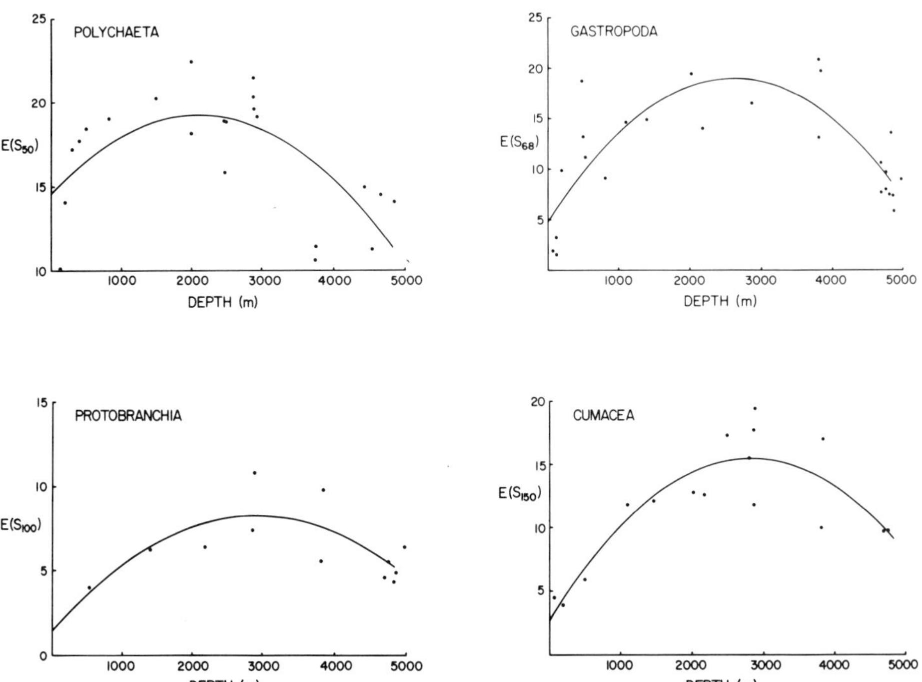

Fig. 2. Patterns of species diversity, using Hurlbert's (1971) expected number of species $E(S_n)$, with depth for four macrofaunal taxa in the northwestern Atlantic. All groups show a maximum diversity at intermediate depths. Subscripts n on $E(S_n)$ indicate the number of individuals at which the samples are compared within each taxon. The polychaete data (Hartman, 1965) are from anchor dredge samples, and all other groups represent epibenthic sled samples. Protobranch data were provided by H. L. Sanders, gastropod data are from Rex (1973, 1976a), and cumacean data are from Jones and Sanders (1972). Regression equations are: Polychaeta, $Y = 14.57 + 0.0044X - 0.000001X^2$, $r = 0.68$, $p < .01$; Gastropoda, $Y = 4.78 + 0.0109X - 0.000002X^2$, $r = 0.78$, $p < 0.01$; Protobranchia, $Y = 1.41 + .0047X - 0.000001X^2$, $r = 0.73$, $p < 0.05$; Cumacea, $Y = 2.70 + 0.0090X - 0.000002X^2$, $r = 0.88$, $p < 0.01$.

legends for regression equations and significance levels). Since polychaetes and bivalves are major elements of the deep-sea macrofauna (Sanders and Hessler, 1969, Rowe et al., 1982) and the gastropods (Sanders et al., 1965) and cumaceans (Jones and Sanders, 1972) are two important but less abundant groups, it seems likely that the macrofauna as a whole is characterized by a parabolic pattern of diversity with depth in the northwestern Atlantic. A recent analysis of the macrofaunal assemblage collected by small box cores from depths of 32 to 3659 m in this region revealed an increase in diversity with depth to about 3000 m followed by a decline at 3659 m (Rowe et al., 1982).

R. L. Haedrich, G. T. Rowe, and P. T. Polloni kindly made available to me the raw data from their studies of zonation in the deep-sea megafauna of the northwestern Atlantic (Haedrich et al., 1975, 1980). Both megafaunal invertebrates and fish show a trend similar to that of the macrofauna, a peak in diversity at intermediate depths, and lower diversity at shallower and abyssal depths (Fig. 3). The decline in diversity of megafaunal invertebrates at abyssal depths was previously pointed out by Vinogradova (1962) and Rowe and Menzies (1969). Recently,

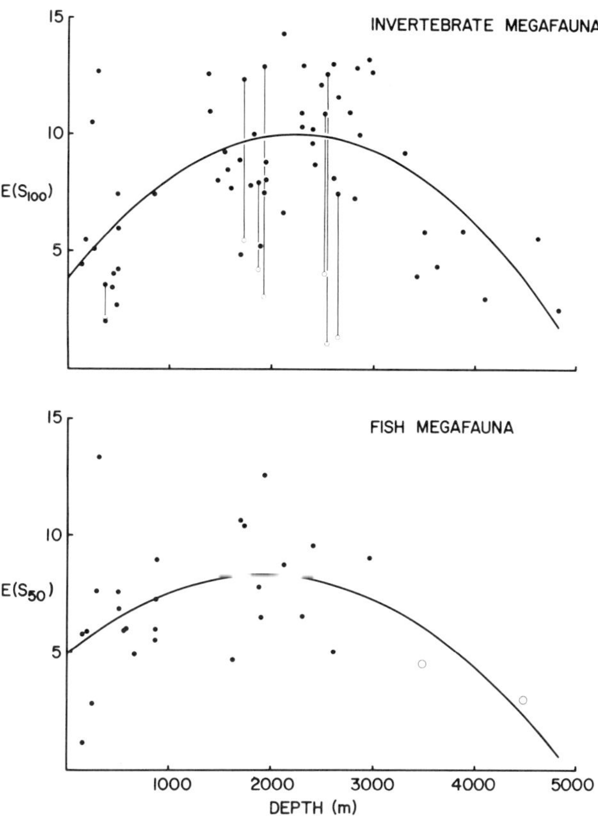

Fig. 3. Patterns of species diversity, using Hurlbert's (1971) expected number of species $E(S_n)$, with depth for the invertebrate megafauna and fish megafauna of the northwestern Atlantic. Both groups show a maximum diversity at intermediate depths. Subscripts n on $E(S_n)$ indicate the number of individuals at which the samples are compared within each group. Seven of the invertebrate samples (open circles) included unusually high abundances (≥89%) of patchily distributed ophiuroids. Vertical lines connect these values to estimates for the same samples with the dominant ophiuroids removed. For the fishes, the two deepest points (open circles) represent averages for samples from 3113 to 3879 m (10 samples, 98 individuals, 6 species) and 4099 to 4815m (4 samples, 8 individuals, 3 species). The last of these composite samples was too small to estimate diversity at $n = 50$, so $s = 3$ was used. All samples were trawls (Haedrich et al., 1975), and $E(S_n)$ estimates were calculated on data provided by R. L. Haedrich, G. T. Rowe, and P. T. Pollini (unpublished communication). Regression equations are: invertebrates, $Y = 3.83 + 0.0055X - 0.000001X^2$, $r = 0.61$, $p < 0.01$; fishes, $Y = 5.01 + 0.0034X - 0.000001X^2$, $r = 0.48$, $p < 0.05$.

Haedrich et al. (1980) have shown that when the megafauna is separated into depth regions by cluster analysis, the highest regional diversity is found at intermediate depths.

In the case of megafaunal invertebrates, one group in particular, the ophiuroids (*Ophiura ljungmani, O. sarsi,* and *Ophiomusium lymani*), has a very patchy distribution (Haedrich et al., 1980). Occasionally, when encountering a dense patch, the dredge is completely clogged with as many as 10^5 individuals. Seven such samples are included in Fig. 3, where one or two species of brittle star made up 89 to 99% of the sample. When the ophiuroids are removed, the remaining species reflect diversities much more representative of most samples in the regions from which they were dredged (Fig. 3). The diversity of the megafauna is obviously lower than that of the macrofauna as a whole even though the trawls used to collect the megafauna cover a larger area than those used for the macrofauna.

There are several potential sources of error for the diversity trends shown in Figs. 2 and 3. One is that the diversity values of the different taxa are estimated for different sample sizes, which represent a compromise between using the largest samples possible and including stations from all major depth regions, especially the abyss, where samples are generally smallest. Thus species numbers are not directly comparable among groups. Comparisons at equal sample sizes are made elsewhere (Rex, 1981) and show the same general trend. A more important difficulty is that all the groups in Figs. 2 and 3 were collected with dredges that sample an extensive and variable area often exceeding a kilometer in length. The megafauna was sampled with large trawls (Haedrich et al., 1980) and the macrofaunal taxa with either an epibenthic sled (Hessler and Sanders, 1967) or an anchor dredge (Sanders et al., 1965) in the case of the polychaetes. Jumars (1975b, 1976) has pointed out that if dredge hauls are long enough to traverse several distinctive faunal patches and if the spatial scales of patches vary with depth, patterns of diversity can be misrepresented. That the peak in diversity at intermediate depths (Figs. 2 and 3) is a sampling artifact is an obvious question. As we have seen (Fig. 3), some megafaunal elements can be patchy at intermediate depths between trawl samples. But the same organisms are known to exhibit mostly random dispersion patterns at scales of 1 to 10^2 m at the same depths (Grassle et al., 1975). Gray (1974), using data from Sanders et al. (1965) and Sanders (1968), attributed the high macrofaunal diversity at intermediate depths to the higher spatial heterogeneity of sediments found there. However, since the sediment samples were taken from mud recovered in an anchor dredge, we do not know the spatial distribution of sediments on the ocean floor. Also, neither the mean nor standard deviation of sediment grain size correlate with species diversity in the data that Gray used as evidence (Spearman's rank correlation $r_s = -0.70, p > 0.05, N = 5$, and $r_s = -0.80, p > 0.05, N = 5$, respectively: his Table VI, p. 252). Rowe et al. (1982) were unable to distinguish the faunas collected from quite different sediments at equivalent depths on the upper continental rise south of New England. Since we know so little about the spatiotemporal scales of patches or even what constitutes a patch in the deep sea of the northwestern Atlantic, it is difficult to assess how patchiness might bias sampling. The fact that such different organisms and sampling methods reveal parabolic patterns of diversity suggests that this is the predominant trend, but more quantitative sampling on smaller scales is needed to confirm this.

B. *Theories of Species Diversity in the Deep Sea*

The Stability–Time Hypothesis

A general positive relationship between species diversity and environmental stability on geologic time scales in a variety of marine and aquatic habitats led Sanders (1968, 1969, 1977, 1979) to formulate the stability–time hypothesis. Essentially, the theory proposed that physical environmental predictability permits biological interactions to stabilize, and this enables diversification by specialization on available resources. In evolutionary time, continued stability leads to the development of highly diverse and "biologically accommodated" communities. In contrast, where physical parameters are unpredictable, the primary selective agent was thought to be physiological stress per se. This favors adaptation to a broad range of physical conditions and prevents the refined competitive interactions that were believed to culminate in high diversity. Sanders (1968) argued that the long geologic history and unusual stability of the deep sea were ultimately responsible for the high diversity found there. Since the deep sea is so poor in nutrients, productivity was assumed to have only a subordinate role in the evolution of diversity except in extreme cases where enormously high production caused anoxic bottom conditions and mass mortality.

Biological Disturbance

The stability–time hypothesis emphasized the role of competitive niche diversification as the predominant biological mechanism acting to maintain high diversity in the deep sea. An alternative hypothesis was advanced by Dayton and Hessler (1972). Although concurring that stability is the factor that ultimately permits high diversity to evolve, they suggested that the apparent temporal and spatial homogeneity of the deep-sea environment would seriously limit the potential for competitive niche partitioning. They proposed instead that high diversity may be maintained by widespread biological disturbance in the form of "cropping" by large epibenthic invertebrates and fish that could alleviate competition among macrofaunal prey species. The macrofauna, in turn, might crop smaller meiofauna that crop the microfauna and thus foster diversity in much the same way envisioned by Paine (1966) in his predation hypothesis. Rather than having the highly specialized niches predicted by Sanders (1968), Dayton and Hessler (1972) suggested that deep-sea organisms in general have broadly overlapping diets. Generalization would be favored in large mobile croppers such as fish and decapods because of the scarcity of worthwhile prey items such as dead falls and sufficiently large patches of particular prey species. More slow moving large epibenthic invertebrates such as holothurians presumably ingest not only sediment, but the smaller inhabitants as well. Among the macrofauna, generalization results from the reduced competition associated with heavy predation by the megabenthos and the premium on consuming whatever food of appropriate size is encountered in a severely food-limited environment. Some degree of specialization in food types no doubt exists, but the food web was predicted to be very complex with numerous cross-linkages. Deep-sea fish do, in fact, have remarkably varied diets (Haedrich and Henderson, 1974).

Contemporaneous Disequilibrium

Grassle and Sanders (1973), in a rejoinder to Dayton and Hessler (1972), pointed out that the demographic parameters and life history tactics of many deep-sea species (low fecundity, slow growth and recruitment rates, and low density) would not enable them to sustain much mortality from predation, especially of the indiscriminate kind proposed by Dayton and Hessler (1972). Grassle and Sanders reasserted competition as the dominant mechanism regulating population size and maintaining diversity but emphasized that niche differentiation can be multidimensional, involving numerous biotic, physical, and temporal differences, and need not be based solely on the parameters of diet stressed by Dayton and Hessler (1972). The unusual physical stability of the environment may permit microhabitat specialization on an extremely small scale. Local disturbances resulting, for example, from occasional deadfalls may result in a mosaic of successional stages that are temporally out of phase with one another. Niche partitioning could involve adaptation to a particular stage of succession so that diversity could also be maintained, in part, by contemporaneous disequilibrium (Richerson et al., 1970).

The recent development and use of the box corer (Hessler and Jumars, 1974; Jumars, 1975a,b, 1976, 1978) for measuring species dispersion patterns has greatly increased our knowledge of how significant habitat partitioning could be in maintaining species diversity in the deep sea. Jumars has reviewed these advances in this volume, so they are discussed only briefly here. A great variety of dispersion patterns exists in the meiofauna, macrofauna, and megafauna (e.g. Grassle et al., 1975; Jumars, 1976; Thistle, 1978; Bernstein et al., 1978). Significant patchiness has been detected in members of some taxa on spatial scales ranging from centimeters (between 0.01-m^2 subcores within a single partitioned box core) to between box cores separated by distances of meters and kilometers up to 100 km (Jumars, 1976). In general, however, the degree of intraspecific spatial aggregation, and presumably the contribution of patchiness to species diversity among the small sedentary forms that comprise much of the deep-sea benthos are less than in shallow water situations (Jumars, 1976; Reys, 1971; Gage and Geekie, 1973). Like Grassle and Sanders (1973), Jumars (1975b, 1976) feels that the stability of deep-sea sediments may permit exploitation of microhabitats by either grain specialization, contemporaneous disequilibrium, or both. Dispersion patterns and anecdotal natural history evidence suggest that the relevant biological interactions and events take place on a very small scale, less than the 0.01-m^2 area sampled by subcores. To relate evidence on small-scale spatial dispersion to gradients of species diversity, it will be necessary to obtain complete depth series of partitioned box core samples. So far, most sampling with box cores has centered on small localities and restricted depths.

Predation, Competition, and Productivity

Little direct evidence is available to demonstrate the roles of competition and predation in structuring deep benthic communities. Using comparative data, Rex (1976a) suggested that both mechanisms were important but that their relative contribution varied with depth and depended on the rate and stability of production. Gastropod species diversity, which parallels the diversity of the other macrobenthos shown in Fig. 2, is lowest on the shelf, increases to a maximum at intermediate depths, and then drops to a lower level again in the abyss. Both the diversity and relative abundance of gastropod predators are positively correlated with overall gastropod diversity, supporting the corollary of Paine's (1966) predation hypothesis that predation should increase disproportionately along gradients of increasing community diversity. As we have already seen, standing crop and presumably the rate at which food reaches the bottom decrease exponentially with depth. It is also likely that the annual bursts of production on the shelf impose a greater degree of seasonality on the upper reaches of the slope than on deeper regions. The comparatively low diversity of gastropods and megabenthos on the upper slope could be related to seasonal variability in production. This might cause population fluctuations in deposit feeding species and in turn limit the ability of predators to diversify by specialization in diet (MacArthur and Levins, 1964; Menge, 1972; Menge and Sutherland, 1976). The high variance in diversity values found on the upper slope in groups with an important predatory role such as gastropods (Fig. 2) and the megabenthos (Fig. 3) may reflect the instability of prey resources. The diversity of gastropod predator species (Rex, 1976a) and the megafauna (Fig. 3) and the diversity of the community as a whole (Figs. 2, 3) reach their highest levels at intermediate depths. The influx of organic resources is lower here, but still adequate to support a diverse upper trophic level. If resources are also more stable, this would permit population stabilization and more efficient energy flow to upper trophic levels (Margalef, 1968). This may be a region where predators are permitted a higher degree of specialization and exert more diversifying feedback on the community by alleviating competition at lower trophic levels. At abyssal depths, diversity declines again (Figs. 2 and 3). The macrobenthic community appears to be quite monotonous here (Sanders and Hessler, 1969; Rex, 1977), both spatially and temporally; hence instability of food resources is not likely to be the cause of decreased diversity. Rather, the amount of production may become so low that there is simply insufficient energy available to support a diverse upper trophic level. Fish show a drop in diversity at these depths (Fig. 3). Apparently, their benthic food resources are too meager and midwater pelagic sources are too far above for efficient exploitation. The megafaunal invertebrate "croppers" also exhibit lower diversity (Fig. 3) and gastropod predators suffer a decline in both diversity and relative abundance (Rex, 1976a). These patterns suggest that the abyss is a region where predation is less significant as a structuring agent than at intermediate depths. The marked numerical dominance of gastropod deposit feeders found on the abyssal plain (Rex, 1976a) may indicate that it is a region of intensified competition among prey species exposed to only a low level of predation. The variation in values of species diversity and rapid transition in faunal makeup on the upper slope suggest that here also may be a region of intensified and unstable competitive interactions existing where predator diversity is comparatively low.

Preliminary work on gastropods (Rex, personal observation) and bivalves (Sanders, personal communication) shows that the decline in diversity at abyssal depths does not occur in areas where surface production is high and is conveyed to deep regions such as near the Antarctic Convergence. However, far from continents under the great central oceanic gyres, a decline in macrofaunal diversity may be universal. Hessler and Jumars (1974) claimed, to the contrary, that high benthic diversity found in oligotrophic regions in the central northern Pacific refuted the notion that productivity is related to diversity. However, when comparable data were collected from the California Continental Borderland, it became apparent that both the density and diversity of polychaetes were higher in this more productive area (Jumars, 1975b).

Interestingly, the foraminiferans (Buzas and Gibson, 1969; Bernstein et al., 1978) and possibly other meiobenthos (Hessler and Jumars, 1974) may reach their highest diversity at abyssal depths [cf. Lagoe (1976), however]. Rex (1973) suggested that the diversity of macrofaunal elements declined on the abyss because they had approached the limits of adaptation in reducing their size and structural complexity to further compensate for the reduction in available food. Perhaps the small size and/or simple structure of meiofaunal organisms, especially foraminiferans and nematodes, enables them to maintain high densities and diversity at abyssal depths.

Taxonomic Diversity and Competition

Another indication of the degree of ecological opportunity and intensity of biological interactions with depth is the pattern of taxonomic diversity, the number of species per genus *S/G*. If we assume that congeneric species are ecologically more similar to one another, on the average, than to members of other genera (e.g. Terborgh, 1971, 1973; Terborgh and Weske, 1975; Diamond, 1975; Mayr and Diamond, 1976), then *S/G* ratios should reflect the relative intensity of competition [see review in Simberloff (1978)]. Low values of *S/G* (fewer species per genus) should result where competition is intense, leading to frequent displacement; conversely, *S/G* values should be higher where competition is less severe.

Rex and Warén (1981) have calculated *S/G* ratios for gastropods collected along the Gay Head–Bermuda transect. Values of *S/G* [means of 1000 computerized random samples of $E(S_{68})$ species] for individual stations are plotted in Fig. 4. The two stations from the continental shelf (<200 m) have *S/G* values of 1; all species belong to different genera. Two large anchor dredge (Sanders et al., 1965) samples (total $N = 1997$) available from the shelf also yield *S/G* values of 1. Anchor dredge samples below the shelf seldom contained gastropods in any abundance (Sanders et al., 1965) and thus, were not used in our analysis. As with species diversity (Fig. 2), values of *S/G* were higher at continental slope and rise depths and then dropped again on the abyssal plain. The same parabolic patterns of taxonomic diversity with depth obtained whether we estimate *S/G* based on a random selection of $E(S_{68})$ species ($F_{2,20} = 5.11$, $p < 0.025$) or on the $E(S_{68})$ most abundant species in the samples ($F_{2,20} = 4.51$, $p < 0.025$).

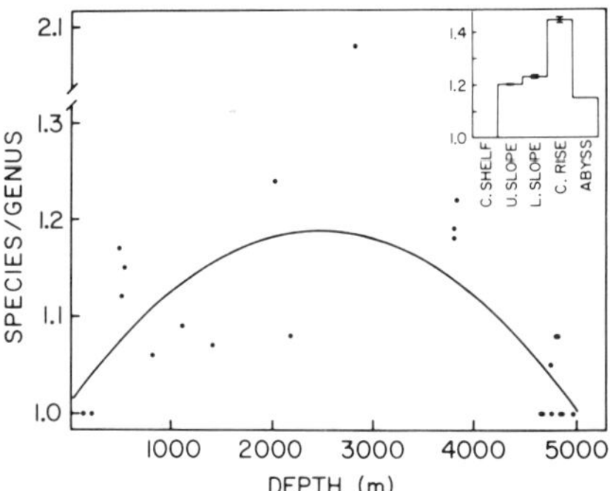

Fig. 4. Species per genus (*S/G*) ratios estimated as averages of 1000 computerized random samples of $E(S_{68})$ species from whole samples, plotted against depth in the northwestern Atlantic. The curve is calculated without the unusually high value at 2862 m ($Y = 1.012 + 1.427 \times 10^{-4}X - 2.878 \times 10^{-8} X^2$, $F_{(2,19)} = 9.08$, $p < 0.005$). When the sample from 2862 m is included, the relationship is also significant ($Y = 0.918 + 3.408 \times 10^{-4}X - 6.617 \times 10^{-8}X^2$, $F_{(2,20)} = 5.12$, $p < 0.025$). In the upper right, expected *S/G* ratios are shown for the upper slope, lower slope, and continental rise when their faunal sizes are reduced to that of the abyss (23 species). The five bathymetric regions indicated were separated by a factor analysis of gastropod composition (Rex, 1977). The regional faunas represent the combined lists of the $E(S_{68})$ most abundant species found in stations occurring within the regions. Five percent confidence limits are shown.

Williams (1964) and Simberloff (1970, 1978) have pointed out that raw values of S/G cannot be used as evidence of the relative intensity of competition, however, since S/G is largely just a function of S, the number of species. To make valid comparisons of S/G ratios that can be used in ecological arguments, it is necessary to normalize them to a common S. This can be accomplished by calculating the mean expected S/G from a large number of computerized random samples of S species. Using this approach, Rex and Warén (1981) estimated S/G values for combined species lists of samples falling into each of five bathymetric regions that were separated by a factor analysis of gastropod species composition along the transect (Rex, 1977). Figure 4 shows the expected S/G ratios for slope and rise regions when their faunal sizes (S) are equated to that of the abyss. After eliminating the influence of faunal size, S/G values still appear to be lowest on the shelf, increase to a maximum on the continental rise, and then decline again on the abyssal plain (Fig. 4).

This trend supports the above discussion of species diversity in suggesting that competition is more intense (lower S/G) in abyssal, upper slope, and shelf communities, and less so at intermediate depths. Of course, the S/G value is only a very indirect indication of competition (Simberloff, 1970, 1978), and it conveys no information about the importance of predation. It will be interesting to see if the other groups shown in Figs. 2 and 3, which have similar patterns of species diversity, also have similar patterns of taxonomic diversity.

The Dynamic Equilibrium Model

Recently, Huston (1979) proposed a general hypothesis of species diversity that reconciles the often contradictory predictions of theories centered on either competition, predation, or productivity by examining the effects of these mechanisms on population growth and reduction. The model assumes that natural communities exist in a state of competitive nonequilibrium because of fluctuating environmental circumstances and the constantly changing nature of biological interactions (Osman, 1977; Caswell, 1978; Connell, 1978; Sousa, 1979). Species diversity in a nonequilibrium system represents a dynamic balance between rates of competitive displacement and the frequency of population reduction caused by predation and environmental perturbations. As communities approach competitive equilibrium, diversity is expected to decline because of increased displacement by dominant competitors. In general, high population growth rates accelerate the approach to equilibrium and lower rates permit coexistence for a longer period. Diversity can be maintained at a high level by biotic or physical disturbances of moderate intensity and frequency that interrupt the approach to competitive equilibrium by periodically reducing population sizes. At a low frequency of population reduction, competitive equilibrium can be achieved, resulting in lowered diversity. Severe and frequent disturbances lower diversity by preventing populations from recovering.

The predicted relationships between diversity, rate of displacement, and frequency of reduction are shown in Fig. 5. Near the origin of the graph, diversity is depressed since, although population growth rates and consequently rates of displacement are very low, long periods of time between disturbances can still permit the community to closely approach competitive equilibrium. Low to intermediate rates of displacement and frequencies of reduction will result in maximum diver-

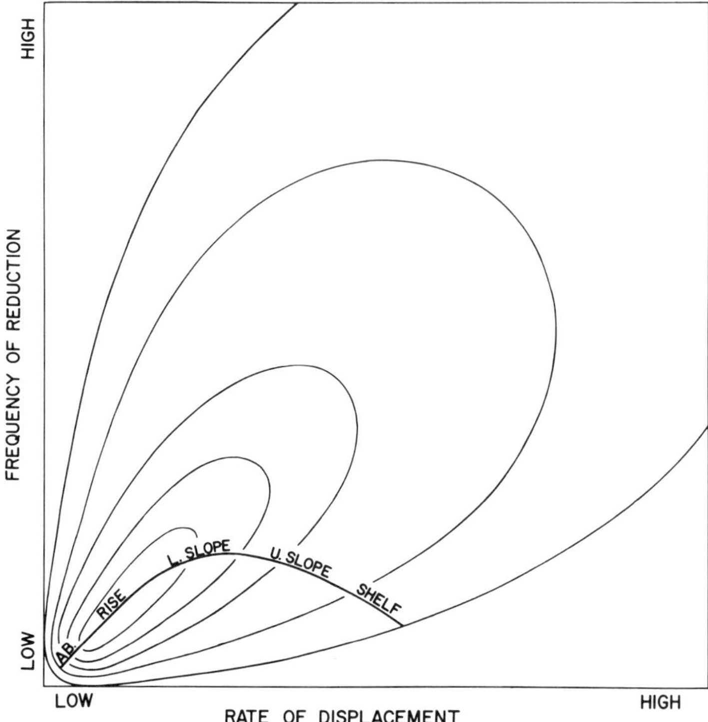

Fig. 5. General predictions of the dynamic equilibrium model of species diversity for various combinations of the frequency (or magnitude) of population reduction and rate of competitive displacement [after Huston (1979)]. Diversity is represented by contour lines with the highest value in the inner ellipsoid at the lower left. The combinations of diversity, frequency of reduction, and rate of displacement representing communities living in different bathymetric regions of the deep sea are indicated along the curved trajectory, and *AB* signifies the abyssal plain. (Basic format of the diagram has been reproduced with the permission of M. Huston and the University of Chicago Press. The diversity curve for the deep sea is added here.)

sity by prolonging coexistence of competitors. Higher values result in lower diversity by either preventing recovery of competitors in the case of reduction factors or permitting rapid approach to competitive equilibrium in the case of displacement.

Huston (1979) suggested that the parabolic pattern of species diversity with depth in the deep sea (Rex, 1973, 1976a) (Figs. 2, 3) is adequately explained by the dynamic equilibrium model and may not require the special explanations for different bathymetric regions that I (Rex, 1976a) proposed. In Fig. 5, I have indicated what combinations of reduction and displacement might represent different regions of the deep sea. Huston assumed that the decrease in density with depth (e.g., Fig. 1) is directly related to population growth rates and consequently to rates of competitive displacement. The frequency of population reduction resulting from predation is probably correlated with the patterns of predator diversity and abundance with depth.

The shelf and upper slope might support low diversity because the high rates of production lead to rapid population growth and high rates of competitive displacement, and there is a relatively low level of predation to alleviate the displacement.

At lower slope and rise depths, diversity might be enhanced by both moderate rates of displacement and predation pressure. Considering the high diversity of specialized predatory snails maintained at these depths, it seems very unlikely that predation has reached levels of sufficient intensity to actually depress diversity by not permitting prey populations to recover. At abyssal plain depths, predator diversity and abundance are much lower. The frequency of population reduction from predation may be too low to prevent the abyssal community from closely approaching competitive equilibrium even though rates of growth and competitive displacement here are probably quite low.

Huston (1979) suggested that diversity on the abyss is reduced because energy constraints are so extreme that population growth rates closely approach zero and consequently diversity cannot develop (his "breakpoint" situation). This may be the case for much of the megafauna. It is less likely for the macrofauna since its diversity on the abyss is comparable to or even higher than that on the shelf (Rex, 1973) (Fig. 2), although it could apply to some macrofaunal elements. The high degree of diversity of the meiofauna indicates that the explanation probably is not relevant for these organisms.

Rex (1976a) showed that overall diversity of deep-sea snails can be significantly predicted by using a multiple-regression analysis with predator diversity and total animal density as independent variables. I suggested that this outcome supported Paine's (1966) predation hypothesis, but it could also be interpreted as supporting the dynamic equilibrium model if predator diversity and density of organisms can be considered as correlates of the frequency of population reduction and the rate of displacement, respectively, as Huston (1979) maintained they are. Huston's model does provide a more parsimonious and unified explanation for geographic patterns of diversity in the deep sea. As with the previous theories, it should be cautioned that the evidence is entirely comparative and that the variables "reduction" and "displacement" subsume complex dynamic interactions that are still poorly understood in the deep-sea environment. Also, the model is concerned primarily with the maintenance of diversity in ecological time and does not attempt to incorporate evolutionary effects. Adaptive considerations, such as how predictability of food resources affects specialization in diet and how this, in turn, might limit diversity, can still be important.

4. Summary and Conclusions

Patterns of nutrient input and nutrient cycling in the deep sea are still poorly understood. Such information is crucial to understanding population dynamics and its relationship to species diversity. The best evidence for the amount of food available to the benthic community is the standing crop, which decreases exponentially with depth. It is likely that the rate of nutrient input is not only higher, but also more variable on the upper reaches of the slope. Recently discovered seasonal and stochastic phenomena may have important effects on community structure and require a reevaluation of the concept of stability in the deep sea.

The predominant pattern of species diversity in both the macrobenthos and megabenthos appears to be parabolic, with a peak at intermediate depths and lower values on the upper slope and abyss. This trend may, in part, reflect a dynamic balance between rates of competitive displacement and the frequency of population reduction by predation. The pattern of standing crop suggests that the rates of population growth and competitive displacement decrease with depth. The diversity of predatory forms, and presumably their impact on reducing population size, are highest at intermediate depths and lower on the upper slope and abyss. At upper slope depths, low overall species diversity may result from high rates of competitive displacement, counteracted by only a low level of predation. The high diversity at intermediate depths may be maintained by moderate rates of displacement and frequencies of reduction. At abyssal depths, rates of displacement are probably low, but infrequent reduction may allow sufficient time for the community to more closely approach competitive equilibrium, resulting in decreased diversity. The low species evenness and species/genus ratios on the abyss and upper slope suggest that competition is relatively more intense in these regions.

There is a tendency in the recent ecological literature (e.g., Abele and Walters, 1979a,b; Osman and Whitlatch, 1978) to refer to the "deep sea" as a single habitat with a characteristic fauna. This misconception has been engendered partly by the highly polarized controversy about the major cause of species diversity in the deep-sea benthos (Dayton and Hessler, 1972 vis-à-vis Grassle and Sanders, 1973). However, it is now clear that species diversity varies both regionally and bathymetrically within the deep sea. The causes of these diversity gradients are still incompletely known; however, it seems likely that food availability, predation, competition, and spatiotemporal heterogeneity are all significant and that the relative importance of these factors and the spatial scales on which they act vary geographically.

Acknowledgments

I thank J. Ebersole, J. F. Grassle, R. L. Haedrich, J. Hatch, M. Huston, and A. Rex for reading all or parts of the manuscript and for many helpful discussions. Unpublished data on the megafauna were provided by R. L. Haedrich, G. T. Rowe, and P. T. Polloni and on the Protobranchia by H. L. Sanders. This material is based on research supported by the U.S. National Science Foundation under grant No. OCE 77-05700.

References

Abele, L. G. and K. Walters, 1979a. Marine benthic diversity: A critique and alternative explanation. *J. Biogeogr.*, **6**, 115–126.

Abele, L. G. and K. Walters, 1979b. The stability–time hypothesis: Reevaluation of the data. *Am. Natur.*, **114**, 559–568.

Aldred, R. G., K. Riemann-Zürneck, H. Thiel, and A. L. Rice, 1979. Ecological observations on the deep-sea anemone *Actinoscyphia aurelia*. *Oceanol. Acta*, **2**, 389–395.

Bernstein, B. B., R. R. Hessler, R. Smith, and P. A. Jumars, 1978. Spatial dispersion of benthic Foraminifera in the abyssal North Pacific. *Limnol. Oceanogr.*, **23**, 401–416.

Buzas, M. A. and T. G. Gibson, 1969. Species diversity: Benthic Foraminifera in western North Atlantic. *Science*, **163**, 72–75.

Caswell, H., 1978. Predator-mediated coexistence: A nonequilibrium model. *Am. Natur.*, **112**, 127–154.

Connell, J. H., 1978. Diversity in tropical rainforests and coral reefs. *Science*, **199**, 1302–1310.

Coull, B. C., 1972. Species diversity and faunal affinities of meiobenthic Copepoda in the deep sea. *Mar. Biol.*, **14**, 48–51.

Coull, B. C., R. L. Ellison, J. W. Fleeger, R. P. Higgins, W. D. Hope, W. D. Hummon, R. M. Rieger, W. E. Sterrer, H. Thiel, and J. H. Tietjen, 1977. Quantitative estimates of the meiofauna from the deep sea off North Carolina, USA. *Mar. Biol.*, **39**, 233–240.

Dahl, E., 1979. Deep-sea carrion feeding amphipods: Evolutionary patterns in niche adaptation. *Oikos*, **33**, 167–175.

Dayton, P. K. and R. R. Hessler, 1972. Role of biological disturbance in maintaining diversity in the deep sea. *Deep-Sea Res.*, **19**, 199–208.

Deuser, W. G. and E. H. Ross, 1980. Seasonal change in the flux of organic carbon to the deep Sargasso Sea. *Nature*, **283**, 364–365.

Diamond, J. M., 1975. Assembly of species communities. In *Ecology and Evolution of Communities*, M. L. Cody and J. M. Diamond, eds. Belknap Press (Harvard University Press), Cambridge, MA, pp. 342–444.

Dickson, R. R., W. J. Gould, P. A. Gurbutt, and P. D. Killworth, 1982. A seasonal signal in ocean currents to abyssal depths. *Nature*, **295**, 193–198.

Filatova, Z. A., 1969. Quantitative distribution of the deep-sea benthic fauna. In *Deep-Sea Bottom Fauna, Pleuston*, L. A. Zenkevitch, ed. (in Russian). Reprinted in *Biol. Tikh. Okean. Tikii Okeana* (V. G. Kort, Isdated Nauka, Moskva), **7** (2), pp. 234–252. (Translation published by U.S. Naval Hydrographic Office, 1970.)

Gage, J., and A. D. Geekie, 1973. Community structure of the benthos in Scottish Sea-Lochs. II. Spatial pattern. *Mar. Biol.*, **19**, 41–53.

Gage, J. D. and P. A. Tyler, 1981. Re-appraisal of age composition, growth and survivorship of the deep-sea brittle star *Ophiura ljungmani* from size structure in a sample time series from the Rockall Trough. *Mar. Biol.*, **64**, 163–172.

Gardner, W. D. and L. G. Sullivan, 1981. Benthic storms: Temporal variability in a deep-ocean nepheloid layer. *Science*, **213**, 329–331.

Grassle, J. F., and H. L. Sanders, 1973. Life histories and the role of disturbance. *Deep-Sea Res.*, **20**, 643–659.

Grassle, J. F., H. L. Sanders, R. R. Hessler, G. T. Rowe, and T. McLellan, 1975. Pattern and zonation: A study of the bathyal megafauna using the research submersible *Alvin*. *Deep-Sea Res.*, **22**, 457–481.

Gray, J. S., 1974. Animal–sediment relationships. *Oceanogr. Mar. Biol. Ann. Rev.*, **12**, 223–261.

Haedrich, R. L. and N. R. Henderson, 1974. Pelagic food of *Coryphaenoides armatus*, a deep benthic rattail. *Deep-Sea Res.*, **21**, 739–744.

Haedrich, R. L. and G. T. Rowe, 1977. Megafaunal biomass in the deep sea. *Nature*, **269**, 141–142.

Haedrich, R. L., G. T. Rowe, and P. T. Polloni, 1975. Zonation and faunal composition of epibenthic populations on the continental slope south of New England. *J. Mar. Res.*, **33**, 191–212.

Haedrich, R. L., G. T. Rowe, and P. T. Polloni, 1980. The megabenthic fauna in the deep sea south of New England, USA. *Mar. Biol.*, **57**, 165–179.

Hartman, O., 1965. Deep-water benthic polychaetous annelids off New England to Bermuda and other North Atlantic Areas. *Occas. Pap. Allan Hancock Found.*, **28**, 1–378.

Heezen, B. C. and C. D. Hollister, 1971. *The Face of the Deep*. Oxford University Press, New York, 659 pp.

Hessler, R. R. and P. A. Jumars, 1974. Abyssal community analysis from replicate box cores in the central North Pacific. *Deep-Sea Res.*, **21**, 185–209.

Hessler, R. R. and H. L. Sanders, 1967. Faunal diversity in the deep-sea. *Deep-Sea Res.*, **14**, 65–78.

Hinga, K. R., J. M. Sieburth, and G. R. Heath, 1979. The supply and use of organic material at the deep-sea floor. *J. Mar. Res.*, **37**, 557–579.

Honjo, S., 1982. Seasonality and interaction of biogenic and lithogenic particle flux at the Panama Basin. *Science*, **218**, 883–884.

Honjo, S., S. J. Manganini, and J. J. Cole, 1982. Sedimentation of biogenic matter in the deep ocean. *Deep-Sea Res.*, **29**, 609–625.

Hurlbert, S. H., 1971. The nonconcept of species diversity: A critique and alternative parameters. *Ecology,* **52,** 577–586.

Huston, M., 1979. A general hypothesis of species diversity. *Am. Natur.,* **113,** 81–101.

Isaacs, J. D., 1969. The nature of oceanic life. *Sci. Am.,* **221,** 146–162.

Jones, N. S. and H. L. Sanders, 1972. Distribution of Cumacea in the deep Atlantic. *Deep-Sea Res.,* **19,** 737–745.

Jumars, P. A., 1975a. Methods for measurement of community structure in deep-sea macrobenthos. *Mar. Biol.,* **30,** 245–252.

Jumars, P. A., 1975b. Environmental grain and polychaete species' diversity in a bathyal benthic community. *Mar. Biol.,* **30,** 253–266.

Jumars, P. A., 1976. Deep-sea species diversity: Does it have a characteristic scale? *J. Mar. Res.,* **34,** 217–246.

Jumars, P. A., 1978. Spatial autocorrelation with RUM (remote underwater manipulator): Vertical and horizontal structure of a bathyal benthic community. *Deep-Sea Res.,* **25,** 589–604.

Jumars, P. A. and R. R. Hessler, 1976. Hadal community structure: Implications from the Aleutian Trench. *J. Mar. Res.,* **34,** 547–560.

Lagoe, M. B., 1976. Species diversity of deep-sea benthic Foraminifera from the central Arctic Ocean. *Bull. Geol. Soc. Am.,* **87,** 1678–1683.

Lightfoot, R. H., P. A. Tyler, and J. D. Gage, 1979. Seasonal reproduction in deep-sea bivalves and brittlestars. *Deep-Sea Res.,* **26,** 967–973.

MacArthur, R. H. and R. Levins, 1964. Competition, habitat selection, and character displacement in a patchy environment. *Proc. Natl. Acad. Sci. (USA),* **51,** 1207–1210.

Margalef, R., 1968. *Perspectives in Ecological Theory.* University of Chicago Press, Chicago, 111 pp.

Mayr, E. and J. M. Diamond, 1976. Birds on islands in the sky: Origin of the montane avifauna of Northern Melanesia. *Proc. Natl. Acad. Sci. (USA),* **73,** 1765–1769.

Menge, B. A., 1972. Foraging strategy of a starfish in relation to actual prey availability and environmental predictability. *Ecol. Monogr.,* **42,** 25–50.

Menge, B. A. and J. P. Sutherland, 1976. Species diversity gradients: Synthesis of the roles of predation, competition, and environmental stability. *Am. Natur.,* **110,** 351–369.

Osman, R. W., 1977. The establishment and development of a marine epifaunal community. *Ecol. Monogr.,* **47,** 37–63.

Osman, R. W. and R. B. Whitlatch, 1978. Patterns of species diversity: Fact or artifact? *Paleobiology,* **4,** 41–54.

Paine, R. T., 1966. Food web complexity and species diversity. *Am. Natur.,* **100,** 65–75.

Pearcy, W. G. and J. W. Ambler, 1974. Food habits of deep-sea macrourid fishes off the Oregon coast. *Deep-Sea Res.,* **21,** 745–759.

Rex, M. A., 1973. Deep-sea species diversity: Decreased gastropod diversity at abyssal depths. *Science,* **181,** 1051–1053.

Rex, M. A., 1976a. Biological accommodation in the deep-sea benthos: Comparative evidence on the importance of predation and productivity. *Deep-Sea Res.,* **23,** 975–987.

Rex, M. A., 1976b. Geographical variation in the deep-sea gastropod *Cithna tenella. Abstracts of the Joint Oceanographic Assembly,* Edinburgh, 1976, 151.

Rex, M. A., 1977. Zonation in deep-sea gastropods: The importance of biological interactions to rates of zonation. *Eur. Symp. Mar. Biol.,* **11,** 521–530.

Rex, M. A., 1979. *r*- and *K*-selection in a deep-sea gastropod. *Sarsia,* **64,** 29–32.

Rex, M. A., 1981. Community structure in the deep-sea benthos. *Ann. Rev. Ecol. Syst.,* **12,** 331–353.

Rex, M. A. and A. Warén, 1981. Evolution in the deep-sea: Taxonomic diversity of gastropod assemblages. In *Biology of the Pacific Ocean Depths,* N. G. Vinogradova, ed. Vladivostok: Far East Science Center, Academy of Sciences of the USSR, pp. 44–49.

Rex, M. A. and A. Warén, 1982. Planktotrophic development in deep-sea prosobranch snails from the western North Atlantic. *Deep-Sea Res.,* **29,** 171–184.

Rex, M. A., C. A. Van Ummersen, and R. D. Turner, 1979. Reproductive pattern in the abyssal snail

Benthonella tenella (Jeffreys). In *Reproductive Ecology of Marine Invertebrates*, S. E. Stancyk, ed. University of South Carolina Press, Columbia, pp. 173–188.

Reys, J. P., 1971. Analyses statistique de la microdistribution des especes benthique de la region de Marseille. *Tethys*, **3**, 381–403.

Richardson, M. J., M. Wimbush, and L. Mayer, 1981. Exceptionally strong near-bottom flows on the continental rise of Nova Scotia. *Science*, **213**, 887–888.

Richerson, P., R. Armstrong, and C. R. Goldman, 1970. Contemporaneous disequilibrium, a new hypothesis to explain the "paradox of the plankton." *Proc. Natl. Acad. Sci. (USA)*, **67**, 1710–1714.

Robb, J. M., J. C. Hampson, Jr., and D. C. Twichell, 1981. Geomorphology and sediment stability of a segment of the U.S. Continental Slope off New Jersey. *Science*, **211**, 935–937.

Rokop, F. J., 1974. Reproductive patterns in the deep-sea benthos. *Science*, **186**, 743–745.

Rokop, F. J., 1977. Patterns of reproduction in the deep-sea benthic crustaceans: A re-evaluation. *Deep-Sea Res.*, **24**, 683–691.

Rokop, F. J., 1979. Year-round reproduction in the deep-sea bivalve molluscs. In *Reproductive Ecology of Marine Invertebrates*, S. E. Stancyk, ed. University of South Carolina Press, Columbia, pp. 189–198.

Rowe, G. T., 1971. Benthic biomass and surface productivity. In *Fertility of the Sea*. J. D. Costlow, Jr., ed. Gordon and Breach, New York, pp. 441–454.

Rowe, G. T., 1972. The exploration of submarine canyons and their benthic faunal assemblages. *Proc. Roy. Soc. Edinburgh*, **73**, 159–169.

Rowe, G. T. and W. D. Gardner, 1979. Sedimentation rates in the slope water of the northwest Atlantic Ocean measured directly with sediment traps. *J. Mar. Res.*, **37**, 581–600.

Rowe, G. T. and R. J. Menzies, 1969. Zonation of large benthic invertebrates in the deep-sea off the Carolinas. *Deep-Sea Res.*, **16**, 531–537.

Rowe, G. T. and N. Staresinic, 1979. Sources of organic matter to the deep-sea benthos. *Ambio Spec. Rep.*, **6**, 19–23.

Rowe, G. T., P. T. Polloni and S. G. Hornor, 1974. Benthic biomass estimates from the northwest Atlantic Ocean and the northern Gulf of Mexico. *Deep-Sea Res.*, **21**, 641–650.

Rowe, G. T., P. T. Polloni, and R. L. Haedrich, 1982. The deep-sea macrobenthos on the continental margin of the northwest Atlantic Ocean. *Deep-Sea Res.*, **29**, 257–278.

Sanders, H. L., 1968. Marine benthic diversity: A comparative study. *Am. Natur.*, **102**, 243–282.

Sanders, H. L., 1969. Benthic marine diversity and the stability–time hypothesis. *Brookhaven Symp. Biol.*, **22**, 71–81.

Sanders, H. L., 1977. Evolutionary ecology and the deep-sea benthos. In *The Changing Scenes in Natural Sciences 1776–1976*, C. E. Goulden, ed. Academy of Natural Sciences Special Publ., Philadelphia, pp. 223–243.

Sanders, H. L., 1979. Evolutionary ecology and life-history patterns in the deep sea. *Sarsia*, **64**, 1–7.

Sanders, H. L. and R. R. Hessler, 1969. Ecology of the deep-sea benthos. *Science*, **163**, 1419–1424.

Sanders, H. L., R. R. Hessler, and G. R. Hampson, 1965. An introduction to the study of deep-sea benthic faunal assemblages along the Gay Head–Bermuda transect. *Deep-Sea Res.*, **12**, 845–867.

Simberloff, D. S., 1970. Taxonomic diversity of island biotas. *Evolution*, **24**, 23–47.

Simberloff, D. S., 1978. Use of rarefaction and related methods in ecology. In *Biological Data in Water Pollution Assessment: Quantitative and Statistical Analyses*, K. L. Dickson, J. Cairns, Jr., and R. J. Livingston, eds. American Society for Testing Materials, Special Technical Report, No. 652, pp. 150–165.

Smith, K. L., Jr., 1978a. Metabolism of the abyssopelagic rattail *Coryphaenoides armatus* measured *in situ*. *Nature*, **274**, 362–364.

Smith, K. L., Jr., 1978b. Benthic community respiration in the N. W. Atlantic Ocean: *In situ* measurements from 40–5200 m. *Mar. Biol.*, **47**, 337–347.

Sousa, W. P., 1979. Disturbance in marine intertidal boulder fields: The nonequilibrium maintenance of species diversity. *Ecology*, **60**, 1225–1239.

Stockton, W. L. and T. E. DeLaca, 1982. Food falls in the deep sea: Occurrence, quality and significance. *Deep-Sea Res.*, **29**, 157–169.

Tendal, O. S. and R. R. Hessler, 1977. An introduction to the biology and systematics of the Komokiacea (Texulariina, Foraminiferida). *Galathea Rep.*, **14,** 165–194.

Terborgh, J., 1971. Distribution on environmental gradients: Theory and a preliminary interpretation of distributional patterns in the avifauna of the Cordillera Vilacabamba, Peru. *Ecology,* **52,** 23–40.

Terborgh, J., 1973. Chance, habitat, and dispersal in the distribution of birds in the West Indies. *Evolution,* **27,** 338–349.

Terborgh, J. and J. S. Weske, 1975. The role of competition in the distribution of Andean birds. *Ecology,* **56,** 562–576.

Thiel, H., 1975. The size structure of the deep-sea benthos. *Int. Rev. ges. Hydrobiol.,* **60,** 575–606.

Thiel, H., 1979. Structural aspects of the deep-sea benthos. *Ambio Spec. Rep.,* **6,** 25–31.

Thistle, D., 1978. Harpacticoid dispersion patterns: Implications for deep-sea diversity maintenance. *J. Mar. Res.,* **36,** 377–397.

Turner, R. D., 1973. Wood-boring bivalves, opportunistic species in the deep sea. *Science,* **180,** 1377–1379.

Turner, R. D., 1977. Wood, mollusks, and deep-sea food chains. *Bull. Am. Malacol. Union,* **1977,** 13–19.

Vinogradova, N. G., 1962. Vertical zonation in the distribution of the deep-sea benthic fauna in the ocean. *Deep-Sea Res.,* **8,** 245–250.

Wiebe, P. H., 1982. Rings of the Gulf Stream. *Sci. Am.,* **246,** 60–70.

Wiebe, P. H., C. C. Remsen, and R. F. Vaccaro, 1974. *Halosphaera viridis* in the Mediterranean Sea: Size range, vertical distribution, and potential energy source for deep-sea benthos. *Deep-Sea Res.,* **21,** 657–667.

Williams, C. B., 1964. *Patterns in the Balance of Nature and Related Problems in Quantitative Ecology.* Academic Press, New York, 324 pp.

Wishner, K. F., 1980. The biomass of the deep-sea benthopelagic plankton. *Deep-Sea Res.,* **27,** 203–216.

Wolff, T., 1979. Macrofauna utilization of plant remains in the deep sea. *Sarsia,* **64,** 117–136.

12. PARASITISM IN THE DEEP SEA

RONALD A. CAMPBELL

The host is an island invaded by strangers with different needs, different food requirements, different localities in which to raise their progeny.

W. H. TALIAFERRO

Any comprehensive understanding of marine biology must include knowledge of parasites because they outnumber their hosts, and because they play a profound role in the biological economy of the sea.

E. R. NOBLE

1. Introduction

During the past decade parasite studies have proven of benefit to the comprehensive understanding of marine community dynamics by providing information that is cohesive to observations of free-living populations. Studies in fish parasitology, like other disciplines studying the marine environment, have been handicapped by the difficulties of working in an alien milieu. Because parasites are typically small, usually inconspicuous, and outnumber their hosts, how pervasive they are in ocean communities remains largely in the descriptive stage. Most early reports of fish parasites consist only of isolated descriptions of new species, often of insufficient detail, and include little data beyond locality, host, and intensity of infestation. Some studies attempted to list the helminth parasites of fishes as part of faunal surveys involving larger numbers of hosts but were more often directed at only a specific group. Despite the thousands of parasites described, the vast majority are believed still unknown. Manter (1969) speculated that not more than 15% of the estimated 10,000 species of digenea of marine fishes have been reported. His estimate was based on a prospective 50% occurrence among teleosts in future studies from an observed 93% average among 11 large surveys.

Species descriptions are a necessary foundation for other studies but in themselves are insufficient to provide an understanding of broad ecological relationships among oceanic habitats. The concept that parasites might be used as biological indicators of relationships between free-living organisms was realized as early as 1902 by Von Ihering in stating that parasites of genetically related hosts are also genetically similar. This idea received support from Manter (1925, 1931, 1934, 1947, 1954) in his surveys of digenetic trematodes of marine fishes. He confirmed Von Ihering's observations and expanded the concept of host–parasite coevolution by showing that the genetic divergence of the parasites proceeds more slowly than that of their hosts because of their more protected, endoparasitic, location and specialized body form. Both Metcalf (1929) and Manter (1963) used zoogeographic studies of parasites as evidence supporting continental drift. Manter (1955, 1967) presented an excellent discussion of the significance of trematode distributions to the fields of paleontology, geology, ichthyology, and oceanography. Numerous surveys have been published on shallow-dwelling fishes wherein more than 80 species and approximately 150 to 3400 hosts were examined (Linton, 1940; Manter, 1940, 1947, 1954; Osmanov, 1940; Shulman and Shulman-Albova, 1953; Hargis, 1957; Tripathi, 1957; Zhukov, 1960; Young, 1967, 1970; Sogandares-Bernal, 1959; Strelkov, 1960; Polyanski, 1955; Yamaguti, 1968, 1970; Unnithan, 1957, 1971; Rohde, 1977; Brinkmann, 1952; Fischthal, 1977). However, in the majority of such surveys attention was devoted to one particular group of helminths. Other surveys have been limited to fewer hosts and larger sample sizes (Polyanski, 1955; Overstreet and Martin, 1974, etc.). Various other studies have included helminth parasites of deep-sea fishes as part of faunal surveys (Bell, 1887; von Linstow, 1888; Linton, 1898; Price, 1934; Gallien, 1937; Guiart, 1935, 1938; Manter, 1934, 1947, 1960; Yamaguti, 1938, 1941, 1951; Dillon and Hargis, 1968; Bray, 1973; Prudhoe and Bray, 1973; Campbell, 1975a–c, 1977a–c, 1979; Campbell and Munroe, 1977; Zubchenko, 1981). Numerous species of parasites have been described from deep-sea fishes in taxonomic works in the past 10 years. Some are mentioned herein, but no attempt has been made to include all of them.

Present knowledge of the nature of deep-sea parasitism is based on only a few studies, most of which do not stress community and ecological interactions. Reviews by Manter (1934) and Noble (1973) indicate that little has been learned in the past 50 years. Nonetheless, sweeping predictive statements about the nature of parasitism in the deep-sea have been made, and these in almost total absence of comprehensive data where communities of true deep-sea fishes are concerned. Data have been recently obtained (Armstrong, 1974; Collard, 1968; Munroe, 1976; Campbell et al., 1980; R. Bray, D. Gibson, and M. H. Pritchard, personal communication), however, and it is largely on such data bases that present conclusions are drawn. Manter (1934) had concluded from his study of 710 deep-water fishes, of which only a few were deep-sea types, that (1) there was no attenuation of trematode species with increasing depth (to 1100 m), (2) temperature and depth were important in determining the distribution of the worms, and (3) the trematodes of deep-sea fishes have a wider geographic distribution than their inshore or pelagic counterparts. Records of parasites from deep-sea benthic fishes indicate that they are as heavily parasitized as most fishes in shallow-water habitats. Bathypelagic fishes (Noble and Orias, 1975; Orias et al., 1978) and mesopelagic fishes (Collard, 1970; Noble and Collard, 1970; M. H. Pritchard, personal communication) are parasite poor and ecologically quite different from benthic species. Collard and Noble discussed the broad ecological relationships among mesopelagic and bathypelagic fishes collected from several geographic localities. It has been difficult to accurately hypothesize about deep-sea parasitism because of lack of material and general ecological information from the deep ocean. In a realm where so much remains to be learned, this is an ambitious task and—as in all general or specific ecological interpretations involving parasites—is confronted with the historical problems of knowing both host and parasite life histories.

Seasonal fluctuations in the parasite fauna of marine fishes have been studied less extensively than those of freshwater fishes and seasonal dynamics of the invertebrate hosts, even less. Seasonal changes in the parasite fauna are complex phenomena involving biotic and abiotic factors. Climatic factors, the significance of temperature in particular, is one of the most important to be considered. Associated changes in the behavior of intermediate and definitive hosts manifest additions or deletions in their parasite fauna. Greatest fluctuations in the environment are associated with inshore waters and pelagic waters of the open sea above the permanent thermocline. In these waters the community fluctuates in abundance and diversity with the seasons; has considerable aquatic–terrestrial interaction; and lives in depth zones that are small, rather poorly defined, and subjected to changes in temperature, current flow, tides, salinity extremes, light, and pollution. Unlike most freshwater fishes, the parasite fauna of many young marine fishes is not like that of the adults because of separation of the young fishes from the adults, often in different habitats and with different food preferences. This is especially apparent in the delayed acquisition of parasites having direct life cycles.

Proceeding into the oceanic depths, the environment presents special problems for organisms living there. There is a marked tendency for physical stability in larger and more defined ecological zones. Stasis of temperature, salinity, water movement, and lighting are especially noticeable. Pressure increases with depth. The wide separation of communities created by depth, absence of light, and location of abundant food supplies in the euphotic and benthic zones is reflected in the morphological and behavioral adaptations of the fishes. Meals are few and far between at middepths, and many fish are modified to eat large organisms, even larger than themselves. Diel migrations into or near surface waters are made to follow the plankton community. Evidence for reverse migrations has also been found for some species. The variety of body forms seen among deep-dwelling teleosts tempts explanation of their evolution because of the variety of habitats offered in oceanic waters. Morphological variation seems almost endless among teleosts for some underlying evolutionary reasons, but selachians, the more ancient fishes, have only two basic body forms regardless of habitat. Few species of elasmobranchs are known from depths beyond 2000 m but baited camera and parasitological evidence indicate that large fish-eating types do occur there. The modifications of teleosts (sensory receptors, luminescence, coloration, presence or absence of swim bladder, elongate body, reduced metabolic rate, etc.) at great depths are directed at a predatory existence in a habitat where they must eat and keep from being eaten.

A popular assumption among biologists is that all parts of the deep ocean are essentially the same and that detailed study of part of it provides a model for all of it. Some basic generalizations about deep-sea fish populations can be made, but too little is known to say that specific groups of fishes everywhere would show very similar faunas without changes in diversity and incidence subject to environmental gradients, community structure, and host–parasite interaction. Parasitological studies in the Atlantic Ocean and recent data concerning seasonal transitions in the deep-sea fauna of the northern Atlantic gyre indicate that the concept of a monotonous continuum must be questioned in terms of community structure and interaction. Relatively few families of helminths predominate in deep-sea fishes, and many common genera are shared. This is indicative of the prolonged stability of the environment, isolation in the depths, and predominant groups of invertebrates available as intermediate hosts. Gradients in diversity of helminth faunas with changing latitude are not apparent thus far for deep-sea fishes as they are for shallow-water fishes. Decreased parasite diversity, incidence, and infestation rate among benthic fishes are indicated with increasing depth and distance from the continental slope.

The benefits of studying parasites of fishes have only recently received serious attention as indicators of community interaction, a concept promoted by numerous authors, including Noble (1973), who stated "Any comprehensive understanding of marine biology must include knowledge of parasites because they outnumber their hosts, and because they play a profound role in the biological economy of the sea." Host specificity of parasites and geographical distribution contribute through use as indicators of feeding habits, host distribution, and behavior. Dogiel (1962) pointed out that "the parasitological indicators of diet are among those clues which allow us to make deductions from the type of parasite fauna about various aspects of the ecology of the host." The fish's behavior, community diversity, and population density are of primary importance in determining the parasite load.

The idea of using parasites as natural tags has received more extensive use in the past 20 years or so but is attributed to Von Ihering (1891, 1902), who used the principle in studying the past zoogeography of South American vertebrates. His general conclusions have been fully corroborated, and subsequent workers using this principle (Metcalf, 1923, 1929, 1940; Harrison, 1928; Manter, 1960, 1963) began to refer to it as Von Ihering's method. Among uses as indicators variously realized for fishes are zoogeographic coherence in habitat, evolutionary relationships, feeding, and reproductive behavior. Knowledge about life styles of fish are implicit from such studies. Subclinal distributions, identification of syntrophic species, and supportive evidence for sex and habitat segregation of fishes may also be obtained, as well as revelation of environmental and faunal gradients. The requirements of a biological tag (Kabata, 1963; Templeman and Fleming, 1963) are (1) the parasite must be common in one host population but rare in another (of the same species), (2) a high degree of host specificity is preferable, (3) required intermediate hosts should not be dispersed over the entire area of study, (4) the parasite should live at least 1 year, (5) the incidence of the parasite should remain relatively stable, and (6) environmental conditions throughout the study area should be conducive to parasite survival. Examples from the literature are numerous, and a few are presented below.

One should not infer that biological tags are a panacea for solving identification problems but that they have proved very useful in a variety of ways. Parasites may also be used as biological indicators to characterize the life histories and community interactions of deep-sea fishes. Among deep-sea fishes, the family Macrouridae has more species than any other. Despite their abundance and diversity, very much remains to be learned about them. Species of *Coryphaenoides* and *Nezumia* are readily available that can be used to demonstrate subpopulations, latitudinal gradients, feeding strategies, feeding and reproductive behavior, migrations, and association with other species and their own young. Rattail fishes of current interest that could be used in such studies are *Coryphaenoides acrolepis,*

C. armatus, C. carapinus, C. filifer, C. leptolepis, C. pectoralis, Nezumia bairdi, and *N. aequalis*. For example, *C. acrolepis* shows habitat segregation of male and female fish with spawning in the spring; *C. pectoralis* and adult *C. armatus* vertically migrate, reflecting a change in food habits; and several combinations of these fishes occur together, thus increasing the likelihood of interspecific competition.

Recently, there has been increasing interest in studying parasitism at the population and ecosystem levels. Most parasites do not have a direct lethal effect on their host; however, an abundance of the more pathogenic parasites can serve as a regulatory mechanism of control (density dependent or density independent) on host population size. Kennedy (1975) discussed various examples of such host–parasite interactions, such as the density-dependent influence of larval *Schistocephalus solidus* (C) on the population size of sticklebacks (*Gasterosteus*). Both Kennedy (1975) and Esch et al. (1977) pointed out that the suprapopulations of parasites are characterized by overdispersion in a clumped distribution among the host population and that most parasite populations show annual fluctuations in response to density-independent factors. Extremes in temperature are easily correlated to these population changes, but altered host social and feeding behavior and population fluxes of the hosts due to mortality are also important. The larval stages of helminths are the greatest factor in the latter case. Density-dependent factors affecting the survival of the parasite infrapopulations depends on inter- and intraspecific competition among the parasites, effectiveness of the host's immune response, and the death of the host resulting from the pathology caused by the parasites.

Many of the difficulties inherent in understanding the regulation of free-living populations are also critical in the consideration of parasite populations. As discussed by Kennedy (1974), the growth of a fish parasite population does not differ in any fundamental way from the growth of any free-living animal population. The populations of all stages of the parasite will grow exponentially until density-dependent or density-independent constraints take effect. Persistence of the parasite population among fish hosts does not indicate stability. A stable population is obtained through regulation by density-dependent factors. An unstable population is constrained only by density-independent factors and is, therefore, unregulated and in constant peril for continued existence. Furthermore, May (1973) has shown that complexity, as occurs in helminth life cycles, does not ensure stability but decreases it instead. Anderson (1974) contends that only a single density-dependent control need operate in a complex life cycle to control the entire suprapopulation of parasites, and life cycle complexity creates time lags that destabilize the system (May et al., 1974); that is, larval survival is age dependent and decreases exponentially with time. Density-dependent regulation by parasite-induced host mortality, immune response, inter- and intraspecific competition, or unexplained parasite-mediated mechanisms have also been documented. Of the many studies on changes of infrapopulations in fish, few have been conducted over a sufficient length of time on all hosts involved in the life cycle to identify the density-dependent factors. Only density-independent factors, particularly temperature and host biology, have been shown to affect the parasite population. On the basis of present evidence, most fish parasite populations are considered unregulated and unstable (Kennedy, 1975). Unusual changes in climate are considered the typical threat to the further existence of the parasite population. In the deep ocean physical density-dependent factors are seemingly removed, especially the temperature change associated with shallow water, which is instead remarkably constant year-round.

Esch et al. (1977) compared the selection strategies of parasites and free-living populations in terms of the concepts of *r* and *K* selection. Such an approach contributed to the development of current views of parasite population regulation. An excellent discussion was given of host–parasite systems paralleling Pianka's (1970) summary of the concepts of *r* and *K* selection. Pianka (1970) believes that there is an *r–K* continuum with the *r* or *K* end points representing optimum strategies of reproductive ability (*r*) or competitive ability to utilize environmental resources (*K*). Host-associated variables influencing the success of parasite infrapopulations are the host's susceptibility (genetic), age, sex and maturation, and social and dietary behavior. Most researchers find that fish seem unable to mount an effective immune response. Few antibody classes have been specifically identified (IgG and IgM), and the production of those is temperature dependent (Harris, 1973). Other serum factors and mucoantibodies have been found, but there is no conclusive evidence that internal parasite infrapopulations of fish are controlled by "typical" immune responses. Ectoparasites are the only group to show regulation by fish immune response. Host age is one of the interacting variables that affects the density and diversity of parasite infrapopulations. The patterns of parasite faunal change in long-lived hosts can be expected to be quite variable in comparison to those of short-lived hosts. Far too little information is available for mesopelagic and bathypelagic parasite faunas over time, but there is general evidence that the hosts are short-lived, probably lasting no more than a year or two at most. Bottom-dwelling deep-sea fishes, however, are believed to have a long life-span, and the pattern of parasite recruitment seems to be the result of dietary preferences, changes in foraging behavior and distribution of hosts as they mature, and interaction among sympatric host species. Patterns of parasite recruitment, well known for shallow-living fishes, recognized by Campbell et al. (1980) among benthic deep-sea hosts, are (1) parasite recruited while host is young and then declines as host ages, (2) parasite absent in young fish and appearing with greater frequency as the host ages, and (3) recruitment begins with juvenile host and persists throughout its life [increasing infrapopulation density with host age may be attributed to intensity of feeding with increasing host size (endoparasites) or schooling behavior (ectoparasites); see Munroe et al. (1981)].

2. Parasite Life Cycles

A. Protozoa

The coccidians, myxosporida, and microsporidans are very numerous and, as a group, are the most important protozoan parasites of fishes. Among coccidians only *Eimeria* is known from marine fishes. Species are commonly found in the digestive tract, swim bladder, and testes. Records indicate that most microsporidans in marine fish are host specific and are species of the prolific xenoma-producing genus *Nosema*. Myxosporida are typically coelozoic and cause dysfunction of the liver and swim bladder, but others in this group invade and destroy somatic tissues. Discovery is uncertain for species that induce little or no inflammatory response and discontinue reproduction as soon as mature spores are formed. The size of the initial exposure is important in determining their presence unless "cysts" (xenomas) are formed or spores are abundant in a hollow organ. For this reason they are most easily found in fresh fish. In microsporidans (*Nosema*), and occasionally in coccidians (*Eimeria*) and myxosporidans (*Kudoa*), the size of the initial exposure is less important because asexual reproduction and growth cycles are continued or autoinfection occurs to create massive invasions. Life cycles among the myxosporida, which are commonly found in deep-sea fishes, are presumed to be direct. Spore viability and parasite host specificity vary. Infection duration may be permanent and multiple infections cumulative with age, making them especially useful as indicators of fish population interactions. Morphometric characters are few and focus on the size and shape of the spores and polar capsules. Variability in spore shape rarely causes them to resemble other genera. The reason for this occasional variability has been postulated to be polymorphism instead of abnormality or mixed infections. Variation in spore morphology has been reported to be more dependent on the host than the fixative. Meglitsch (1960) suggested that spore variability of myxosporidans would increase within a single host with higher rates of cross infection in the host population. If this hypothesis were true, it would be useful because a greater variety of spore morphologies and size divergence would allow estimation of infestation rates from small samples of the fish population. However, Moser (1977a) found no consistent relationship between spore size variance, infection rate, or season in three species of macrourids from eight different localities.

Moser (1977b) experimented and presented field data supporting Shulman (1966) and Donets (1969) suggesting that natural selection favors myxosporidan species whose spore shape results in the most effective transmission to physiologically suitable hosts. Spore morphology determines sedimentation rate (180 m/day for *Myxidium coryphaenoidium;* 132 m/day for *Ceratomyxa hokarari*) and the probability of being consumed by the fish host in a particular portion of the water column. Selection pressure is postulated not to be the aquatic environment itself, as long as it remains constant, but the presence of physiologically and behaviorally suitable hosts. Moser (1977b) and others found remarkable constancy in spore size and shape among numerous species of deep-sea macrourids from distant geographic regions. The constancy in spore size and shape was thought to reflect the unique and stable deep-sea environment but later was also found true of myxosporidans parasitizing shallow-water teleosts. There is evidence (Laird 1953; Moser 1977b) for host-induced variation in spore size possibly caused by autogamous reproduction or "random drift," but spore size and shape do not appear to depend on geographic location, depth of host habitat, or season of the year. He concluded that once a "successful" spore size and shape were attained, there was little selection pressure to change as long as critical environmental factors remained constant. Thus in time they were dispersed through suitable hosts and became geographically widespread.

In deep-sea fishes Noble (1973) found the gall bladder, urinary bladder, and kidneys of macrourids commonly invaded by myxosporidans. Moser and Noble (1976) and Moser et al. (1976) have reported several new species of *Ceratomyxa* and *Myxidium* (myxosporidia) from the biliary system of macrourids from various oceans. Five species of *Myxidium* were found in 225 (7 genera, 23 species), macrourids, and 8 species of *Ceratomyxa* in 161 (7 genera, 12 species). Twenty-six samples involved only one to three fish but five samples of 30 to 107 fish had infection rates of 30 to 73%. Most species showed a low degree of host-specificity and a wide geographic distribution. Marked consistency in spore size was noted for *M. coryphaenoideum* found in 15 species (five genera) of macrourids from Japan to New Zealand and from British Columbia to Chile and Norway. Noble and Collard (1970) found more (132 of 305; 18 of 32 species) bathy- and abyssopelagic fishes infected with more kinds of protozoa than mesopelagic fishes (29 of 307; 8 of 22 species). The parasites included an amoeba (*Entamoeba* sp.), a hemoflagellate (*Crytobia*), a ciliate (*Trichodina* sp.), coccidians (*Eimeria* spp.), and myxosporidan and microsporidan trophozoites. Despite the low population density of bottom-living deep-sea fishes, the common occurrence of these protozoa indicates that invertebrates may be responsible for maintenance of the parasites in the fishes by ingesting (concentrating) the spores. Some of the macrourids, among other teleosts, are noted for their predation on the detritus-feeding invertebrates inhabiting the ocean floor. Lom (1970) found oocysts of *Eimeria* from multiple sites in *Macrourus berglax,* and Noble (1973) listed six species of macrourids (five genera) infected with *Eimeria*. Noble et al. (1972) found that *Macrourus berglax* from 235 to 400 m off Newfoundland had more parasites of more kinds than three species of macrourids from deeper water in other geographic localities. The most commonly encountered parasites in the four species of macrourids (*Coryphaenoides rupestris, C. acrolepis, Coelorinchus scaphosis, M. berglax*) were juvenile nematodes and myxosporidan protozoa. Although he indicated that flagellates, coccidians, and microsporidans were also present, only three were named (*Myxidium* spp., *Contracaecum aduncum,* pseudophyllidean plerocercoids) from the nine major taxa reported. Zubchenko and Krasin (1980) have surveyed macrourids of the Northern Hemisphere for myxosporidans of the genus *Myxidium*. We found elongate intramuscular cysts containing spores of the microsporidan *Pleistophora* beneath the parietal peritoneum of about 10% of 337 *Nezumia bairdi* (Fig. 1*A,B*) from the New York Bight. Because these parasites are very host-specific, the species is probably new.

There are many factors known to affect metazoan parasites in fishes, but we still do not know to what extent they apply to protozoa. Not only are life cycles unknown, but data are so incomplete on their biology (Myxosporida, Microsporida, Haplosporida, Eucoccida) and development that identification of species can be difficult. Developmental stages as well as spores, especially of microsporidans, must be studied to identify them properly (Lom et al., 1980). Electron microscopy provides greater detail. Much work is needed on all aspects of the biology and development of species found in readily available inshore fishes before they can be used to full advantage in studying deep-sea hosts.

B. Helminths

The life histories of digenetic trematodes, cestodes, nematodes, and acanthocephalans are considered complex because they employ one or more intermediate hosts in addition to the definitive host. Larvae occur in intermediate hosts and adult worms in the definitive host. Monogenetic trematodes are mainly ectoparasitic flukes, having simple life histories (no intermediate hosts) and direct development. These flukes tend to be *k* strategists (complex adult, few progeny),

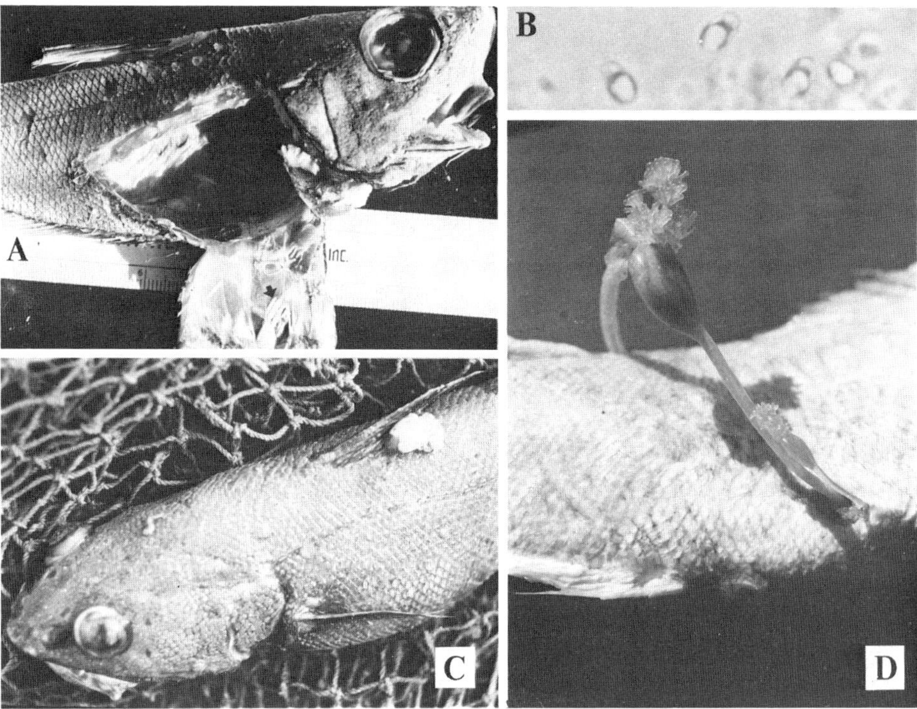

Fig. 1. Parasites of deep-sea fishes from Hudson Canyon: (A) intramuscular microsporidan cysts (*Pleistophora* sp.) beneath the parietal peritoneum of *Nezumia bairdi*; (B) section of cysts showing spores; (C) *Lophoura* sp. embedded in dorsum of *Coryphaenoides armatus;* (D) *Sphyrion lumpi* implanted in liver of *Antimora rostrata*.

resulting in a remarkable degree of host specificity (Bychowsky, 1957; Esch et al., 1977; Rohde, 1978). Host population density and schooling behavior are important factors in providing the opportunity for larval monogenea (oncomiracidia) to gain access to new hosts. The major families of monogenea in deep-sea fishes are Diclidophoridae, Choricotylidae, Discocotylidae, Gastrocotylidae, and Microcotylidae.

Digenetic trematodes are typically monoecious endoparasitic worms having indirect development and life cycles involving two intermediate hosts. Metagenesis occurs with sexually reproducing adults residing in the definitive host (fish) and asexual generations (polyembryony) in the first intermediate host. Digenea, therefore, are usually *r* strategists (simplified adult, many progeny) adapted to many hosts. Asexual multiplication in the first intermediate host culminates in the production of cercariae, usually free swimming, which invade the tissues of other invertebrates or vertebrate (second intermediate) hosts to become infective juveniles (metacercariae). Sexually reproducing adults develop when a suitable definitive host ingests the metacercariae (Figs. 2 and 3). A second, or alternative, cycle is known (hemiurids, opecoelids) in which progenetic adults develop in invertebrates. This abbreviated cycle would be a distinct advantage in waters of low population density. A high degree of host specificity for the first intermediate host, typically a mollusk but occasionally other invertebrates, is characteristic.

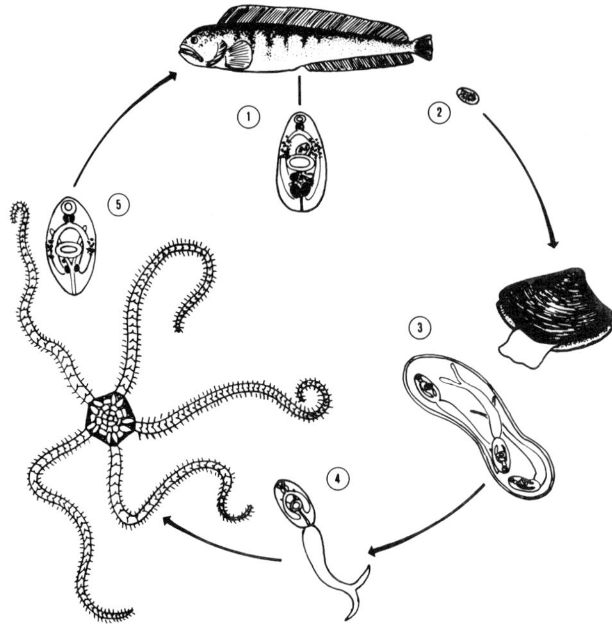

Fig. 2. Life cycle of a digenetic trematode, *Fellodistomum fellis:* (1) ovigerous adult in the fish *Anarhichas lupus;* (2) egg in fish feces; (3) sporocyst producing cercariae in the clam *Nucula tenuis;* (4) free-swimming cercaria; (5) metacercaria infective for *Anarhichas* in the brittle star *Ophiura sarsi* [based on Chubrik (1952, 1966)].

The degree of host specificity generally decreases as development proceeds. Margolis (1971) points out three instances in which tubiculous polychaetes are first intermediate hosts for digenea of fishes and many examples of their involvement as second intermediate hosts. Because fish are the primary predators on polychaetes, it is not surprising that digenea have exploited them, particularly when one considers their abundance. Representatives of five trematode families (Fellodistomidae, Hemiuridae, Lepocreadiidae, Monorchiidae, Zoogonidae) listed by Margolis have been found in deep-living fishes. Our observations indicate that hemiurids, fellodistomes (especially the genus *Steringophorus*), lepocreadiids (especially *Lepidapedon* and *Neolepidapedon*), and certain genera of opecoelids are common. However, with increasing depth we have found that abundance of all helminths decreases but that hemiurids and lepocreadiids are the most persistent. This, no doubt, is related to their type of life cycle and distribution of intermediate hosts. Host specificity of the adult worms seems to be related more to host ecology through prey selection than to host physiology in many cases. Dawes (1958), Overstreet and Hochberg (1975), and Yamaguti (1975) list many larval digenea from various intermediate hosts and provide some information on life cycles. Many life cycles are unknown, but some examples include the hemiurids using pelagic intermediates (copepods), fellodistomes using starfish and jelly animals, lepocreadiids (*Lepidapedon, Lepocreadium*), and zoogonids using polychaetes. Zoogonids are noted for their tailless cercariae (cercariaeum), which must crawl over the bottom to reach the intermediate hosts.

No detailed studies of the parasites of deep-sea plankton have been made.

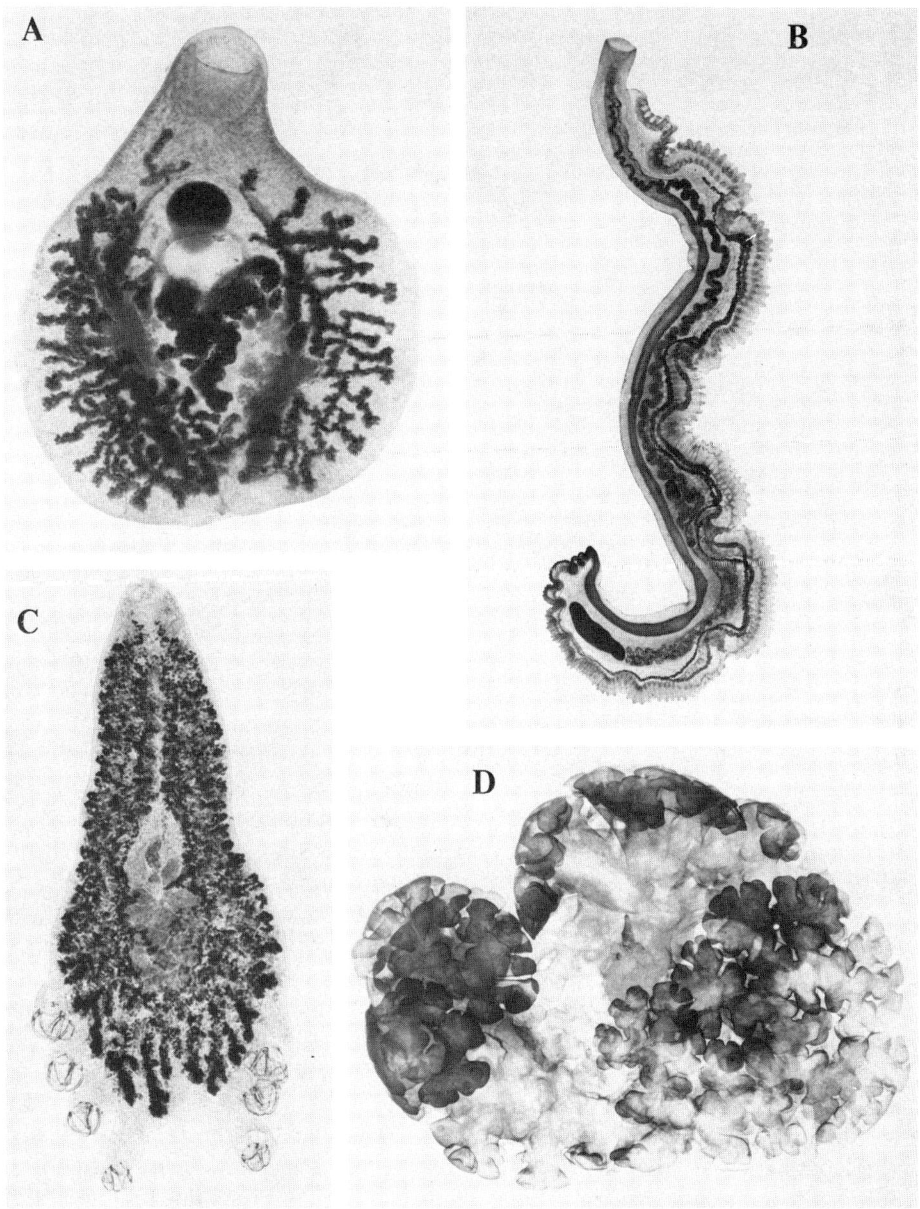

Fig. 3. Trematodes from marine fishes. A, *Degeneria halosauri,* urinary bladder fluke of the deep-sea halosaur, *Halosauropsis macrochir*. B, *Multicalyx cristata*, an aspidogastrean from gall bladder of cownose ray, *Rhinoptera bonasus*. C, *Diclidophora nezumiae*, monogenetic trematode from gills of macrourid, *Nezumia bairdi*. D, *Makairatrema* sp., didymozoid fluke from skin capsule of Atlantic Sailfish.

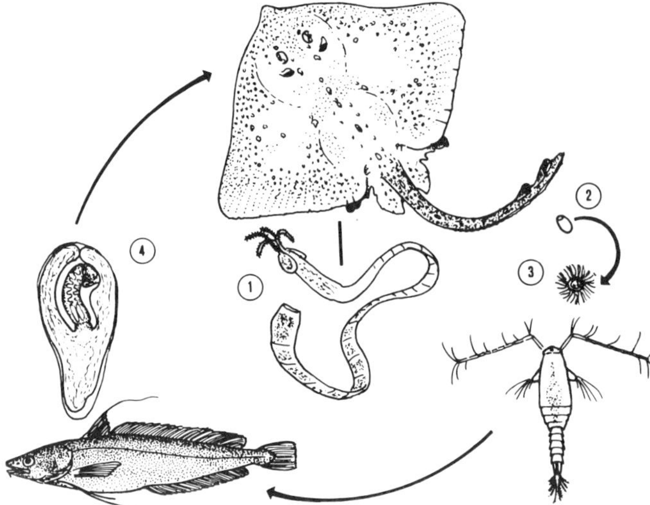

Fig. 4. Life cycle of a trypanorhynch cestode, *Grillotia erinaceus:* (1) ovigerous adult in skate *Raja clavata;* (2) egg in fish feces; (3) free-swimming coracidium ingested and forming procercoid in the copecod *Acartia longiremis;* (4) plerocercus infective for *Raja clavata* and living in various teleosts [based on Ruszkowski (1932, 1934)].

Attempts to study parasite larvae in shallow or euphotic waters have proved this to be a tedious and frustrating task because incidence is frequently low and large quantities of fresh material are needed for accurate identification. Stomach content analyses of the fishes and the most abundant components of the meiofauna and macrofauna suggest that those invertebrates should be examined. In abyssal communities important potential intermediate hosts (polychaetes, crustaceans, and bivalves) comprise 86 to 92% of the macrofauna; the relative number of individuals differs markedly between continental margins and oceanic gyres (Wolff, 1977). In zooplankton copepods are the most abundant at all depths, followed by chaetognaths, euphausiids, and amphipods (Vinogradov, 1968).

The life cycles of cestodes, nematodes, and acanthocephalans of fishes are similar to those of digenea in the sense that intermediate hosts are utilized but there is no polyembryony (Figs. 4 and 5). Three orders of cestodes predominate in marine fishes, the Pseudophyllidea, Tetraphyllidea, and Trypanorhyncha. In general, life histories are quite similar. Copepods are typical first intermediate hosts, and teleost fishes or other invertebrates may serve as second intermediate hosts. Adult pseudophyllideans parasitize teleosts, but tetraphyllideans and trypanorhynchs have exploited elasmobranchs. Juvenile cestodes tend to be less host specific than do the adults, and it is not unusual to find them in the viscera, mesenteries, or flesh of many teleosts. Passage of cestode larvae (plerocercoids, plerocerci) among paratenic (nonobligatory) second intermediate hosts commonly occurs between invertebrates, like squid, as prey of various teleosts or between teleosts themselves. This is particularly true of the tetraphyllideans and trypanorhynchs whose final hosts are fish-eating elasmobranchs. Unlike pseudophyllidean larvae, these cestodes may be identified to genus or even species and can

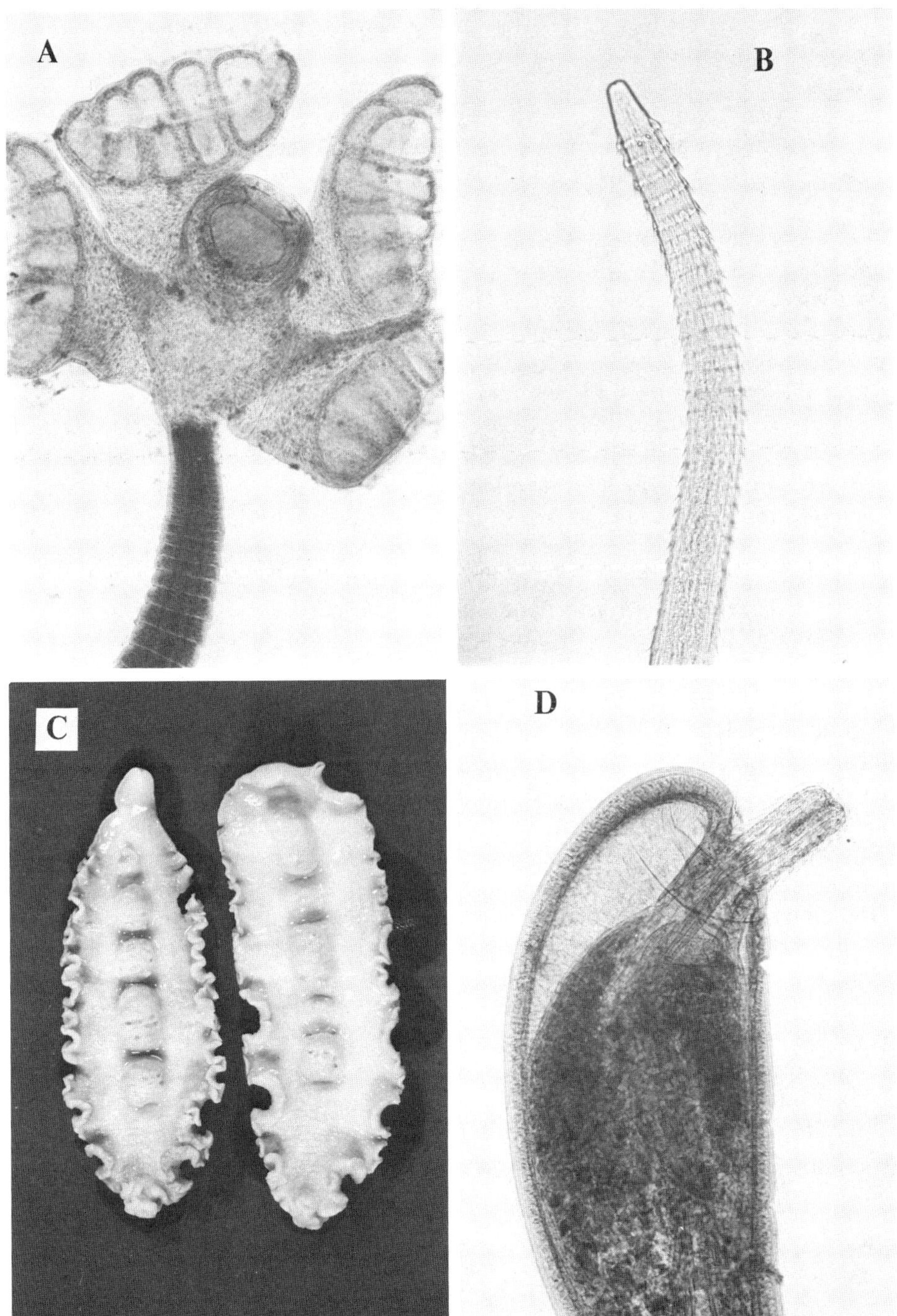

Fig. 5. Helminths from deep-water fishes: (A) scolex of *Echeneibothrium,* a tetraphyllidean cestode genus common in skates; (B) *Spinitectus oviflagellis,* nematode parasite of *Coryphaenoides leptolepis;* (C) specimens of *Gyrocotyle,* cestodarians from the spiral intestine of the chimaeroid, *Callorhynchus callorhynchus;* (D) Acanthocephalan, *Echinorhynchus* sp., from *Coryphaenoides carapinus.*

serve as indicators of trophic relationships involving the large predators that are difficult to capture in trawls.

Parasitic nematodes may utilize fish as both final and intermediate hosts. Eggs or larvae of adult worms are passed into the water with feces of their hosts. Fish become infected when they eat invertebrates that have ingested the eggs or larvae. When the final host is a piscivorous predator, another intermediate host, usually also a fish, may be incorporated into the life cycle. When the cycle involves only one intermediate host, it is typically a crustacean, particularly gammarid amphipods and cyclopoid copepods. Cycles involving a true second intermediate host involve plankton-feeding fishes. Juvenile nematodes from infected invertebrates (first intermediate hosts) molt and burrow into the viscera and body musculature to encyst and await ingestion by the final host. Probably the best known cycle of this type is that of *Thynnascaris aduncum,* in which predatory fishes are final hosts, clupeids and flatfishes are typical second intermediate hosts, and copepods are the first intermediates. Paratenesis, the use of nonessential hosts in which growth but no further development takes place, is common to many nematode life cycles and is believed important as an alternative means of dispersal.

Acanthocephalans are the most rarely encountered helminth parasites of fishes. The life cycle involves an arthropod intermediate host, most frequently benthic crustaceans such as amphipods, and occasionally a vertebrate paratenic host.

Complex helminth life cycles are of value in ecological interpretation because they utilize food chains to reach the final host wherein the sexually reproducing adults develop. By accurately defining the helminth fauna of a particular host, an indication is obtained of the involvement of the host within a community food web. Greater or lesser diversity of the endoparasitic helminth fauna provides a parallel indicator of host feeding behavior that can be verified by stomach contents. Involvement of the fish host as prey in other food chains is also indicated by its parasite fauna when juvenile parasites use the fish as an intermediate host. Cestode larvae and juvenile nematodes are of special significance in this regard. Both general and specific indications of community interaction can be obtained by thoroughly comparing the parasite faunas of as many community members as possible. Generalization or specialization in feeding behavior, benthic versus pelagic feeding, changes in diet, and similarities or differences between congeners are generally indicated by the parasite faunas of respective hosts. Specific predator–prey relationships can be predicted with absolute certainty only when the life history of the parasite is known. However, accurate identification and systematic allocation of the parasite is valuable even when life cycles are not known, for many closely related parasites have similar life histories.

C. *Copepoda*

The most commonly encountered parasitic copepods of deep-sea fishes are members of the families Lernaeopodidae, Sphyriidae, Naobranchiidae, and Chondracanthidae. The adult females characteristically lack copepod features, have a cephalothorax and enlarged trunk, and lack swimming legs. Only a few larval stages are known for marine species, and complete life histories of the numerous lernaeopodids are based on freshwater species in which larvae, copepodid, chalimus, and adult stages are parasitic on the same host (Kabata and Cousens, 1973). A common marine lernaeopodid, *Lernaeocera branchialis*, uses two hosts, flounder and cod, to complete its life cycle. Fertilization of the female occurs on the flounder prior to transformation and loss of swimming legs. They attach to their host by either implanting the bulla, at the tip of the maxillae, into the skin (*Brachiella*) or boring the longate cephalothorax into the viscera or somatic musculature (*Sphyrion*). Gills and fins are damaged by the feeding females, and burrowing species (*Sphyrion, Lophoura*) anchor themselves and cause considerable damage with the lateral outgrowths of their modified cephalothorax. The small males may be found attached to the genital segment of the female. Sphyriids were commonly encountered among the 1763 fishes (52 species) from deep waters off southern New England (discussed below). Sphyriids showed preferences for the liver (*Sphyrion lumpi; Lophoura* sp., Fig. 1*c,d*) of *Antimora rostrata* and the region of the dorsal aorta in the deep-sea fishes *Coryphaenoides armatus* and *Antimora*. The yellow atrophied livers of freshly caught fish were severely damaged by permeation of the enlarged spherical outgrowths of the copepod neck. *Lophoura gracilis* was also found on *Synaphobranchus kaupi*. A *Sarcotaces* species was found in the subcutaneous connective tissue of *Antimora rostrata* and *Chondracanthodes deflexus* from *Coryphaenoides brevibarbis, C. armatus*, and *C. carapinus*. *Acanthochondria phycidis* was common on the gills of *Urophycis chuss* and *Chondracanthodes deflexus* on *Nezumia bairdi; C. deflexus* is known from macrourids off New England, California, and Galapagos Islands. Ho (1975) reported chondracanthids, lernaeopodids, and naobranchids from Galápagos macrourids and brotulids noting that four of the five species he found also occur off Japan, Aleutian deep, California and Atlantic coast of North America. Ho (1970) transferred *Chondracanthus macrurus* Brady, 1883, taken from a macrourid during the *Challenger* Expedition, to *Lateracanthus*. Noble (1973) found *L. quadripedis* common on *Coryphaenoides acrolepis* off California and listed species of *Brachiella, Clavella, Chondracanthodes*, and *Lernaeenicus* (?) from macrourids. We have found *Clavella adunca*, and two new species of *Lateracanthus* from *Macrourus berglax*, and *Brachiella annulata* from *Antimora* in Newfoundland waters (verified by Cressey, personal communication). The works of Wilson (see Yamaguti, 1963), Ho (1970, 1971, 1975, 1977, 1978), and Kabata (1959, 1965, 1969, 1970, 1973) should be consulted for further information on this group.

Our knowledge of all aspects of the biology of this group is very poor, and many species, especially those from the northern Atlantic, are in need of redescription. There have been few studies of copepod parasites of deep-sea fishes, and detailed drawings and descriptions, including dissections, are necessary regardless of the scarcity of the material.

3. Methodology and Limitations

Assuming the opportunity exists to examine large numbers of fishes from a specific geographic locality in all seasons, the approach chosen by parasitologists to obtain information of ecological value is not universally agreed on. Some think it more important to choose a few specific fishes and examine several thousand of them and compare the parasite fauna with the usual parameters of sex, size, and season. Such an approach theoretically would allow assessment of regulatory factors influencing a population's stability if of sufficient duration. However, this approach requires years of sampling, is very costly, and provides a very narrow view of the interaction indicated by the total parasite fauna of the community. On the basis of my experience in working with deep-sea parasites, both at sea and in the laboratory, over the past 8 years, it would be blind optimism to contemplate samples of 1000 or more individuals for most deep-sea species in terms of current fishing methods, depth distribution of fishes, and relative numbers. It could take years to obtain 150 specimens of many fishes below 2500 m, and to catch 10 of certain species in a lifetime would be unusual. The second approach involves surveying as large a sample as feasible from as many members of the fish community as possible. This "cover everything approach," as it has been called, is ambitious with large samples but more profitable when coupled with studies of benthic invertebrates and fish diets to answer general questions of community ecology. It has a distinct advantage in characterizing the habitat because it gives wholeness and resolution to the study. Large sample sizes are, of course, necessary foı determining relationships of fish stocks, geographic comparisons, sex, season, size, recruitment, changes in feeding behavior, and longevity. More specific questions to be investigated are revealed once a general picture of the community has been resolved. Bias in sampling with small nets moving at slow speed makes collection of large predators a continuing problem. Scrupulous examination technique is critical to studies of parasite dynamics. Jose Amato (personal communication) has found that rinsing of organs under a faucet into 150- and 37-μm mesh sieves is extremely effective and practical for obtaining small and numerous helminths that could be overlooked and require considerable time to collect by conventional postmortem examination methods. Fagerholm (1982) has found a most satisfactory method for collecting nematodes from fish livers at sea. The livers are placed in separate plastic bags containing a pepsin digest solution for 30 min. After the nematodes have fallen into the sediment they are maneuvered into a corner of the bag that is then snipped off into a container of fixative. Large samples can be easily processed by this method, much time is saved and small juveniles that might be overlooked by dissection are discovered.

Fresh material is usually at a premium because of the small sample size, expense of shipping time, and limited space available on deep-sca sampling expeditions. The most likely reason parasites have been so inadequately studied from deep-sea fishes is the problem of obtaining sufficient samples of good material. Sampling of a sufficient number of hosts is difficult aboard ship, even with skilled assistants. On the average we have found that, depending on host size and parasite load, the number of fish that can be properly examined for helminths ranges from 6 to 10 every 8 hours. Large and heavily infected hosts greatly reduce this rate. Even then, on examination of moribund fish, it is difficult to find all parasites because the skin of some species is often abraded or denuded of scales. Blood parasites in dead fish, whether helminths or protozoans, are difficult to assess because only a small amount of blood can be obtained despite efforts to remove it from the heart or caudal vein. In more northerly latitudes this problem may be minimal because surface waters are cold and the fish may be brought to the surface in much better condition or even alive. Fluid should be kept to a minimum during microscopic examination. Once fixation of specimens is complete, they should be transferred to 70% ethanol for transport and storage. Aldehyde fixatives are considered dangerous chemicals by commercial airlines, and specimens should be transported in ethanol in "carry-on" luggage if possible instead of exposing the specimen containers to the temperature and pressure extremes of the aircraft's luggage compartment. Any containers placed in the luggage compartment should be tightly closed, sealed, and of sturdy construction to prevent bursting or loss of fluid and specimens.

Shipboard examinations are not extensive enough for ecological interpretation, but the fresh and often living parasites obtained are invaluable as a synoptic set for taxonomic purposes and can be used for identification of additional specimens from preserved hosts. Living nematodes and acanthocephalans are often obtained from fish that have been frozen for 24 hr prior to examination. However, frozen fish must be examined quickly once thawed because they rapidly deteriorate. Preservation of freshly caught fishes has proved very satisfactory in our studies of benthic fishes, provided the 10% buffered formalin gained access to the viscera. Good or, occasionally, excellent specimens were typical of preserved fish whose body cavities were slit and promptly immersed in 10-gallon containers of fixative. Eagle and McCauley (1965) reported that injected formalin also gave good results, and the hot-water immersion technique suggested by Sinclair and John (1973) deserves consideration for rapid fixation of organ systems or entire fish. Improved and applicable staining techniques have recently been described for locating protozoan spores in fixed tissue sections (Palmieri and Sullivan, 1977), helminth larvae in preserved tissues (Rau and Gordon, 1977), and viral inclusion bodies in smear preparations (Palmieri et al., 1978). Loss of ectoparasites was not a problem after preservation. Less than a half dozen of about 800 ectoparasites (636 monogenea; 137 copepods) recovered from 1763 deep benthic fishes examined were found at the bottom of the storage containers. Limited transport, filled containers, gentle handling, and brief storage are responsible for this degree of success. Separation of individual fish in plastic bags has been used on occasions where prolonged transport was involved. Comparison of the results of shipboard examinations with those of preserved hosts indicates that the accuracy of sampling is certainly very great. Identification of larvae is difficult when obtained from fresh material aboard ship and often impossible once preserved *in situ*. The low incidence levels of larval helminths in the abundant invertebrate populations makes surveys for intermediate hosts a frustrating task. Many helminth larvae have been found in zooplankton (see below), but no satisfactory mass quantification techniques have been developed. Invertebrate sampling provides clues to a major problem confronting specific interpretations of deep-sea parasitism—the parasite life cycle. Experimental determination of life cycles will remain impossible unless potential hosts can be removed from the depths and maintained alive in the laboratory. To date only indirect observations and systematic association temper our observations. Temperature change seems to be a primary factor in preventing retrieval of live animals from great depths. The effects of decreasing pressure are also involved, but both living fish and invertebrates have been recovered from depths down to 2500 m when surface waters were cold.

Deep-ocean invertebrates have been maintained in the laboratory for varying lengths of time with success. Culture of live helminths would provide interesting physiological data and clues regarding life cycles as well. Hatching stimuli and observation of helminth larvae would provide insight into behavior, longevity, and

TABLE I
Composition of Ringer's Solution for the Rattail, *Coryphaenoides armatus*[a]

Salt	Grams/Liter	m*M*/Liter
NaCl	12.82	219.0
KCl	0.14	1.9
$CaCl_2$	0.46	4.1
$MgSO_4 \cdot 7H_2O$	0.46	1.9
Na_2HPO_4	0.48	3.4
K_2PO_4	0.08	0.5
$NaHCO_3$	1.00	11.9
Glucose	0.99	5.5

[a]Osmolarity approximately 450 mosm/liter.

abbreviation of life histories and might allow opportunity to experiment with their dependence on chemotaxis. Miracidia of *Gonocerca haedrichi* from the macrourid *C. armatus* were observed hatching *in utero,* and live gravid worms were obtained from many fish. Helminths from fish should be washed in saline prior to being placed in culture solutions.

Robert Griffith (1981) has analyzed the blood of several species of deep-sea benthic fishes. Ranges of osmolarity and a suggested saline solution based on them is presented in Table I. According to Griffith, an impressive body of data exists on the chemistry of the blood and other fluids of fishes from freshwater, inshore marine and surface oceanic habitats [see the comprehensive review by Holmes and Donaldson (1969)], but there is a paucity of information on the blood chemistry of midwater or benthic deep-sea fishes. This gap in our knowledge is unquestionably due to a combination of difficult access to specimens and the generally poor physiological condition of fishes subjected to the stresses of decompression, abrasion in nets, and exposure to great changes in temperature. Fange et al. (1972) reported the composition of the blood of the deep-sea rattail, *Coryphaenoides rupestris,* and found it to be rather similar to that of inshore or oceanic surface marine teleosts. Griffith (1981) analyzed a variety of midwater and benthic deep-sea teleosts for osmolarity, chloride, urea, and protein. In general, midwater fishes in good physical condition show levels of osmolarity and chloride similar to or only slightly higher than those characteristic of inshore marine teleosts and have low blood urea levels, and most have extremely low serum protein levels and hematocrits. The low protein and hematocrit levels may be involved as adaptations to achieve neutral buoyancy as suggested by Blaxter et al. (1971). Benthic fishes such as the rattail *Coryphaenoides armatus* are usually characterized by moderately high osmolarities (ca. 450 mosm/liter) and chloride. Protein, urea, and hematocrits are not markedly different from those of inshore teleosts.

Culture media for macrourid helminths can be prepared by using several variations of components. The above rattail Ringer's solution can be used for repeated rinses of the worms to free them from bacteria and debris prior to incubating them in Ringer's enriched with a nutrient medium such as Parker 199. The worms might also be placed in this Ringer's solution on an autoclaved substrate of agar containing fluid filtered from a minced preparation of the host's somatic musculature. The worms can be held in simple but sterile containers in these media plus antibiotics. Control over temperature fluctuation may prove to be a factor critical to the survival of the worms. These simple media may serve as a suitable starting point for further experimentation.

Culturing of helminths has provided evidence of their probable longevity. Evidence of a long life span has been found for some deep-sea organisms and is indicated by the size of anisakid nematodes parasitizing macrourids and other deep-sea hosts that I have studied. Davey (1969) showed that *Contracaecum osculatum,* an anisakid nematode parasite of seals that uses fish as intermediate hosts, has larvae that hatch in 13 days at 16°C but require 6 months to hatch at 2°C. McClelland and Ronald (1974) found that this nematode needed 8 months at 15°C to attain a length of 6 mm, which was the minimum size necessary to become infective (molt) when the incubation temperature was raised to 35.5°C (body temperature of thc seal definitive host). If one extrapolated from the hatching time at 2°C, ambient temperature near the deep-sea floor, the larvae would reach the minimum infective size after about 7 years. Gary McClelland (personal communication) found that the size of anisakine nematode larvae in cod varied directly with the size (age) of the fish. Ichthyologists contend that fishes such as macrourids live a long time, but currently there is no method for accurate determination of their age. The large size (≥30 mm) of *Anisakis simplex* larvae (Beverley-Burton et al., 1977) in these fishes is consistent with ideas of fish longevity and may indicate a long relationship with the fish host. Extrapolation from McClelland's *in vitro* studies implies that worms of this size could be an almost unbelievable 35 years old. Greater longevity would offer obvious advantages to parasite larvae in an environment where host population densities are low. Among digenetic trematodes from deep-sea hosts, species of lepocreadiids (*Lepidapedon*) and opecoelids (*Podocotyle, Plagioporus*) are typically large in comparison with species from shallow-water hosts. Current evidence indicates that most adult helminths live no more than a year in cold-blooded vertebrates but larvae may survive for longer periods.

An intriguing but little studied phenomenon is host-induced variability. Works by Stunkard (1957), Watertor (1967), Blankespoor (1971), Palmieri (1976, 1977a–c), and myself (Campbell, 1972, 1973) have shown that most characters are subject to variation when trematodes develop in different hosts. These variations, sometimes extreme and of generic importance, have been shown experimentally for trematodes parasitizing all vertebrate classes except fishes. Although the experimental hosts are abnormal in the sense of true ecological relationships, it is important to know to what extent this variability applies to development in fish hosts. Most of the trematode species tested show great variability and capability of infecting many hosts. The similarity of many species of fellodistomes, lepocreadiids, and opecoelids so common in deep-sea fishes makes them suspect.

Varying abundance of host species and differential behavior of juveniles and adults may make sampling of the full size range of hosts difficult. Small and large fish are necessary for evaluation of changes in trophic behavior. Controlled closing nets are most desirable for quantitative studies of deep-living fishes, especially those from the mesopelagic and bathypelagic zones. Accurate environmental data characterizing the region of capture are relevant, but until recently little detailed ecological information has been available for any deep-sea benthic community that would help to clarify the role of parasitism. Refinement of sampling techniques has led to reassessment of basic ecological interpretations such as the characterization and causes of deep-sea diversity (Wolff, 1977; Rex, this volume, Chapter 11), magnitude of megafaunal and macrofaunal biomass with depth, importance of small organisms in the energetics of the deep benthos, general complexity of community interaction, environmental "stability," and life strategies. Finally, interpretation of data from limited sampling of the vast environment of the deep ocean where capture efforts are biased leaves the investigator in a tenuous position to advance hypotheses. However, the comments of critics are even more so if they are not based on their published accounts from freshly caught fishes. Variability in the distribution and kinds of intermediate hosts at different depths and in different geographic localities makes it necessary to intensively study hosts from more than one population before generalizations can be advanced for all deep-sea fishes with any certainty.

4. Habitats and Hosts

Studies of marine fish helminths began about 200 years ago in inshore European waters and have involved numerous well-known investigators. These and the great majority of parasitological studies in other parts of the world have been concentrated over the continental shelves. Total shelf area represents only 18% of the Earth's total land area and underlies only 7.5% of the ocean (Emery, 1969). A continuous gradient from freshwater to marine conditions exists in fish species that inhabit estuaries or spend varying portions of their lives in low salinities to a fully saline environment. Characterization of parasitism of fish populations in different habitats also involves a gradient in the open or closed nature of the ecosystem and trophic habits (both herbivorous and carnivorous) of the fishes living there. In the next sections, specific examples of fishes or fish groups were chosen that illustrate some of the aspects of parasitism in fishes that have different specializations from inshore to oceanic environments.

Characterization of parasitism is based not only on species lists from the general literature, but comprehensive population studies (both published and unpublished) and personal research. Given the volume of literature today many different citations could have been used, but the extensive faunal lists compiled provide ample evidence for our purposes here. Specific examples have been selected to illustrate some of the variations and transitions between inshore and oceanic habitats. Because categorization of habitat based on the fish life history is not easily defined in many cases, the designation given by the American Fisheries Society (Robins et al., 1980), Bigelow and Schroeder (1953), or current literature was used. Publication of complete host–parasite lists here for all fishes researched would be both prohibitive and of no general interest to the majority of readers. However, some detail is given for certain examples. Considerable effort has been devoted to this end in the case of deep-sea fishes.

A. *Marine Birds and Mammals*

Birds prey on fishes and invertebrates from the estuaries to the high seas. Fishes in shallow water commonly serve as intermediate hosts of bird parasites such as heterophyid, strigeoid, and echinostome trematodes as well as nematodes of the genus *Contracaecum*. The parasite fauna of gulls, for example, is listed in many papers (Lapage, 1961; I. C. Williams, 1961; Threlfall, 1964, 1965, 1966) or approached both qualitatively and quantitatively (Pemberton, 1963; Leonov, 1960; Guildal, 1964; Bakke, 1975). Dogiel (1962) used marine birds as examples in his categorization of the types of life cycle of avian parasites and considered migration of the definitive host to exert the greatest influence on the parasite fauna.

Marine mammals, like birds, are subject to a degree to aquatic–terrestrial interaction. Comprehensive checklists of parasites from pinnipeds and cetaceans include marine helminth genera common to marine fishes and invertebrates (Dailey and Brownell, 1972; Delyamure, 1968). Particularly noticeable in their reoccurrence are the records of nematodes (*Anisakis, Phocanema, Terranova, Contracaecum, Porrocaecum, Phocascaris*), adult pseudophyllidean and larval tetraphyllidean cestodes (*Phyllobothrium, Monorygma*), and acanthocephalans (*Bolbosoma, Corynosoma*). The genera are readily recognized from the faunas of marine fishes and invertebrates where they occur as larvae and, in some cases, adults. Paratenic and true intermediate hosts from these groups are well documented. Marine birds also share some genera occurring in marine mammals (*Cryptocotyle,* D; *Contracaecum,* N; *Tetrabothrius,* C; *Corynosoma,* A). The relationships of codworm infections to seal populations are described by Rae (1960), Wiles (1968), and Young (1972). Marine mammals also interact with deep-sea fishes. Mead et al. (1964) and Paxton (1967) gave examples of midwater oceanic fishes as prey for marine mammals, and it is known that some whales can dive to depths of 3000 m.

Strandings of marine mammals have been an enigma since the days of Aristotle. Recently, parasitic disease has been a popular explanation for solitary and mass strandings, especially of cetaceans. Worm parasites, usually obtained from their prey, are the cause of brain lesions, middle-ear infections, respiratory disorders, pneumonia, cirrhosis, mastitis, and a stress syndrome. Certain helminth species have been suspect as the cause of stranding because of their involvement with the central nervous system. The resulting damage is postulated to result in disorientation or a weakened condition that culminates in stranding. Ridgeway and Dailey (1972) described the extensive necrosis in the brains of dolphins caused by digenetic trematodes (*Campula*) that commonly inhabit the sinuses, invade the brain, and produce ova. Nematodes are also commonly found in the sinuses, middle ears, brains, and eustachian tubes of these animals. The resulting inflammation and abscesses could certainly lead to altered acoustic behavior and are believed to be a significant cause of natural mortality. The nematode *Stenurus globicephalae,* so common in the middle ear of porpoises, may interfere with echolocation by toxic effects on the inner ear. Crimean dolphin fishermen have long called harbor porpoises "deaf Azovka" that would not respond to their attempts to drive them. Geraci (1978) reviewed the mystery of marine mammal strandings, parasitism, and a new hypothesis (limbic response) to explain the phenomenon. Although the alteration of host behavior due to parasite damage is certainly significant, the presence of parasites is only one of several factors involved in strandings. Jon Lien of the Whale Research Group, Memorial University of Newfoundland, has found that the numerous strandings he has observed consistently occurred in areas with gradual slopes that are unsatisfactory for good echolocation. Animals drowned in offshore nets are also parasitized, and he feels that the significance of parasitism to strandings has been overemphasized because it is only one of a combination of factors that can lead to stranding (Lien, personal communication). A detailed comparison of the parasite of faunas of seals, sea lions, and whales is not given here, but it can be said that evidence of their eating habits prevails in the universality of some of the helminths shared. Shared parasites are also indicative of zoogeography, as in the case of sea lions, for example, who share some helminths according to their overlapping distributions (Dailey, 1975). Seals and sea lions also have helminths with terrestrial cycles shared with canids. The whales share species and genera of pseudophyllidean cestodes, nematodes (particularly *Crassicauda* and *Anisakis*), and acanthocephalans of the genus *Bolbosoma*. The Sirenia (sea cow, manatees) are faunally distinct.

B. Shallow-Living Teleosts

Among the better known parasitological studies on fishes from the northwestern Atlantic are those by Linton, Levinson, Stafford, MacCallum, Cooper, Manter, Miller, Heller, Price, Nigrelli, Brinkmann, Sindermann, Cable, Sogandares-Bernal, Hargis, Threlfall, Margolis, Appy, Ronald, Scott, Bray, Gibson, and Zubchenko. Stafford (1904) made an extensive survey of fish trematodes in Canadian waters. Many of these authors are noted for studies of particular fish and parasite groups. Linton, in studies spanning 50 years (1889 to 1940), made the first surveys of helminth parasites of marine fishes in the Woods Hole area, at Tortugas, Florida (Linton, 1909, 1910), and at Beaufort, North Carolina (Linton, 1905). Manter redescribed some of Linton's species and added many new species of trematodes from shallow-living fishes from Maine (Manter, 1925), North Carolina (Manter, 1931), and Florida (Manter, 1947). These and many other works provide ample proof that inshore fishes harbor a very abundant and diverse parasite fauna, especially in warm-water latitudes. Detailed studies of the parasite faunas based on large sample sizes are relatively few. Russian workers have directed their attention to inshore fishes for many years in the interest of commercial fishing but more recently have increased their interest in deep-sea hosts. Examples from the ecological studies of Polyanski (1955) and others investigating fishes on the continental shelves provide interesting contrasts with fishes from oceanic habitats. The type of parasite is indicated by a letter in parentheses as Acanthocephala (A), Cestoda (C), Digenea (D), Monogenea (M), Nematoda (N), Branchiura (B), Copepoda (Co), cysts (Cy), Protozoa (P), Mollusca (F), and Fungi (Fu).

Salmons (Salmoniformes)

Salmon are anadromous and show a parasite fauna indicative of their migrations between freshwater and marine environments. Their parasites have been used to identify the freshwater origins of fish caught at sea (Nyman and Pippy, 1972; Pippy, 1969a,b; Hare and Burt, 1976). Much work has been devoted to the study of the Atlantic salmon, *Salmo salar*. Populations are found in coastal waters on both sides of the Atlantic, and the European side extends well into the Arctic. It is known that the stocks intermix off Greenland, but much more remains to be learned of the salmon's life at sea. Adults spawn in freshwater, where they eat very little. At sea their diet includes a variety of smaller fish (herring, lances, mummichogs, blennies, small mackerel, haddock, small sculpins, flatfish, myctophids) and invertebrates (euphausiids, pelagic amphipods, gammarids, crabs). As mature fish, their marine enemies are harbor seals, large tunas, swordfish, sharks, and humans.

Dogiel and Petrushevski (1933, 1935) succeeded in obtaining the first complete picture of the changes in the parasite fauna of salmon (*Salmo salar*) throughout an entire life cycle. Their samples included fish 3 to 4 months old up to 10 years of age. The freshwater parasite fauna of young salmon was not specific to them but included parasites of related fishes from the same locality. Twelve species of parasites were accumulated gradually with age similar to the dynamics of the parasite fauna of freshwater fishes (i.e., gradual increase in parasite number and species with age). Typical freshwater parasites were *Crepidostomum farionis* (D), *Diplostomum spathaceum* (metacercariae, D), *Phyllodistomum conostomum* (D), *Triaenophorus nodulosus* (C), *Proteocephalus* sp. (plerocercoids, C), *Neoechinorhynchus rutili* (A), *Spiroptera tenuissima* (N), *Rhabdochona* and *Rhaphidascaris acus* (N). The average number of parasite species per host was used as an index of infection of each age group.

Subsequent investigations have shown that most freshwater parasites are lost when the fish go to sea. Only three of 15 parasite species of adult fishes returning from the sea were freshwater types [*Camallanus lacustris* (N) and *Capillaria* sp. (N), one occasion each; *D. spathaceum,* commonly], all acquired in the course of travel upstream. The few parasites acquired on return to freshwater were the result of altered trophic behavior. Anadromous salmon were commonly infected only by *Diplostomum* metacercariae derived from the active penetration of the fish by free-swimming cercariae. Unlike their other parasites, they do not gain access to the fish through a food chain. The remaining parasites were strictly marine. Among marine faunal components were hemiurid trematodes (*Brachyphallus, Hemiuris* spp., *Lecithaster, Derogenes varicus*), cestode larvae (*Tetrabothrius, Eubothrium*), the acanthocephalan *Echinorhynchus gadi,* and the larvae of the nematode *Porrocaecum capsularia*. *Derogenes, Echinorhynchus,* and *Porrocaecum* are all known from a great variety of marine teleosts.

As the salmon moved upstream, Dogiel and Petrushevski (1933, 1935) found that the marine parasites were gradually reduced. The ectoparasitic *Lepeophtheirus* (Co), once exposed to freshwater, were the first to be lost. Reduction of the number of intestinal marine species and intensity of infection was directly related to the amount of time spent in freshwater. Therefore, it is possible to distinguish spring-run spawners from those arriving in the fall. For example, *Derogenes varicus* infected 93.7% of autumn fish at 120 worms per fish as compared to 53.3% infection of spring-run fish, averaging 12 worms per host. Similar observations have been made by other investigators.

In recent years helminths have been used to distinguish between the origin of members of salmon populations caught at sea. Pippy (1969a,b) used *Crepidostomum farionis* (D), *Metabronema salvelini* (N), and *Pomphorhynchus laevis* (A) to separate smolts from Canadian and Irish river systems. Species of *Diplostomulum* were used to separate smolts originating in Greenland, Europe, and North America. Later, Nyman and Pippy (1972) were able to separate adults originating in Europe or North America by the incidence and intensity of *Eubothrium cras-*

sum (C) and *Anisakis simplex* (N). Margolis (1982) has used parasites of Pacific salmon to determine (1) freshwater areas of origin, (2) tributaries of origin in mixing areas, (3) ecological juvenile groupings during freshwater residence, (4) lake or river of fish origin taken by inshore fisheries, (5) migratory (anadromous) fish from resident charr populations, (6) oceanic feeding areas, and (7) illegal catch (forensic use). Similar examples involving different fishes can be found in the literature.

Herrings (Clupeiformes)

The parasite faunas of herrings also include species acquired during their migrations as well as those obtained with metamorphic changes in food preferences. Sea herring, *Clupea harengus,* are plankton feeders found in open water over the continental shelves of cool regions. Their dependence on phytoplankton gradually

TABLE II
Comparison of Parasite Faunas of *Alosa* spp. from the Black and Caspian Seas [After Petrushevski (1957)]

Parasite	(Taxon)	Black Sea	Caspian Sea
Mitraspora capsalosae	(P)	+	+
Mazocraes alosae	(M)	+	+
Hemiuris appendiculatus	(D)	+	–
Lecithaster confusus	(D)	+	–
Bacciger bacciger	(D)	+	–
Diplostomum spathaceum	(D)	+	+
Proteocephalus sp.	(C)	+	+
Acanthocephaloides incrassatus	(A)	+	–
Contracaecum aduncum	(N)	+	–
Ergasilus nanus	(Co)	+	–
Clavellisa emarginata	(Co)	+	–
Cymothoe punctata	(Co)	+	–
Trichodina domerguei capsalosae	(P)	–	+
Bunocotyle cingulata	(D)		+
Ascocotyle coleostoma	(D)	–	+
Cysticercus gen. sp.	(C)	–	+
Pseudophyllidea larvae	(C)	–	+
Corynosoma strumosum	(A)	–	+
Contracaecum squali	(N)	–	+
Porrocaecum sp.	(N)	–	+
Camallanus truncatus	(N)	–	+
Desmidocercella sp.	(N)	–	+
Philometra ovata	(N)	–	+
Eustrongylides sp.	(N)	–	+
Agamonema sp.	(N)	–	+
Anisakis sp.	(N)	–	+
Ergasilus sieboldi	(Co)	–	+
Caligus lacustris	(Co)	–	+
Argulus foliaceus	(B)	–	+
Glochidia of Unionidae	(F)	–	+

switches more to zooplankton (copepods, amphipods, euphausiid shrimps, decapods) as the fish grow older. In the absence of an abundant supply of crustacea, they will take molluscan larvae, fish eggs, *Sagitta,* pteropods, and annelids but rarely small fishes (lances, silversides, sardines, capelin). Adult fish eat young sand lances that they competed with for food as young. Records from the literature indicate an endemic marine fauna of 18 species of parasites.

The relationships between parasite faunas of related hosts, like clupeids, become more apparent as host feeding behavior is more clearly resolved. Acquisition of parasites from varied habitats were well illustrated by zoogeographic studies of *Alosa* (Table II) in the Caspian and Black Seas by Petrushevski (1957). The majority of parasites recovered from the Caspian Sea were endemic to freshwater or brackish water habitats and parasitized fishes of other families beside clupeids. In all, two species, *Mazocraes alosae* (M) and *Mitraspora caspialosae* (P), were specific to clupeids in both seas. Parasites of *Alosa* from the Black Sea were endemic marine species with low (trematodes, *Hemiuris, Lecithaster,* and nematode, *Contracaecum*) or high host specificity (copepods, *Ergasilus nanus, Clavellisa emarginata, Cymothoe punctata*). The fauna of *Alosa* from the Caspian Sea differed in being more diverse but was lacking in marine species and those specific to clupeids. Acanthocephalans were absent, and only a few digenetic trematodes were present, indicating the poorness of the invertebrate fauna, especially amphipods, isopods, and mollusks.

Flatfishes (Pleuronectiformes)

Numerous species of these carnivorous bottom fishes are found over the continental shelves and upper slope regions. Their sedentary habits, trophic preferences, and distribution from the estuarine zone to marine depths contribute to their typically diverse parasite faunas. A checklist of their parasites has been given by Ronald (1959). Helminth cycles to birds (digenetic trematodes and cestodes) and other fishes (helminths, copepods) are indicated by their larval parasites. Ecological studies of flounder (*Platichthys flesus* L.) and plaice (*Pleuronectes platessa* L.) populations from Scotland appear in several publications summarized by MacKenzie and Gibson (1970). They found that the entire parasite fauna of the plaice included 22 species of parasites and the flounders 27. Of 34 different parasites, at least 25 had been reported as common to both hosts. Fifteen species, including seven species of digenea, were common to their populations from Loch Ewe and Aberdeen on opposite coasts of Scotland. Gibson (1972) found variations in the parasite fauna of *Platichthys flesus* populations in estuaries of Aberdeen, Scotland. Species of *Podocotyle* (D) and *Zoogonoides* (D) proved very effective biological labels in distinguishing between the flounder populations. Crustacean and protozoan ectoparasites also varied with the habitat. *Trichodina* (P) was almost ubiquitous on flounders in the Ythan estuary but rare at sea, and *Lernaeocera branchialis* (C), so common at sea, was usually lost after the fish entered low-salinity waters. Parasite faunas of nine species of *Paralichthys* surveyed from the literature include 12 to 46 species of protozoans, helminths, and crustaceans.

Zhukov (1953) found that large-mouthed flatfishes had a richer helminth fauna than did small-mouthed species from deeper water. Thirteen species of helminths were common to both and included worms obtained from ingestion of planktonic

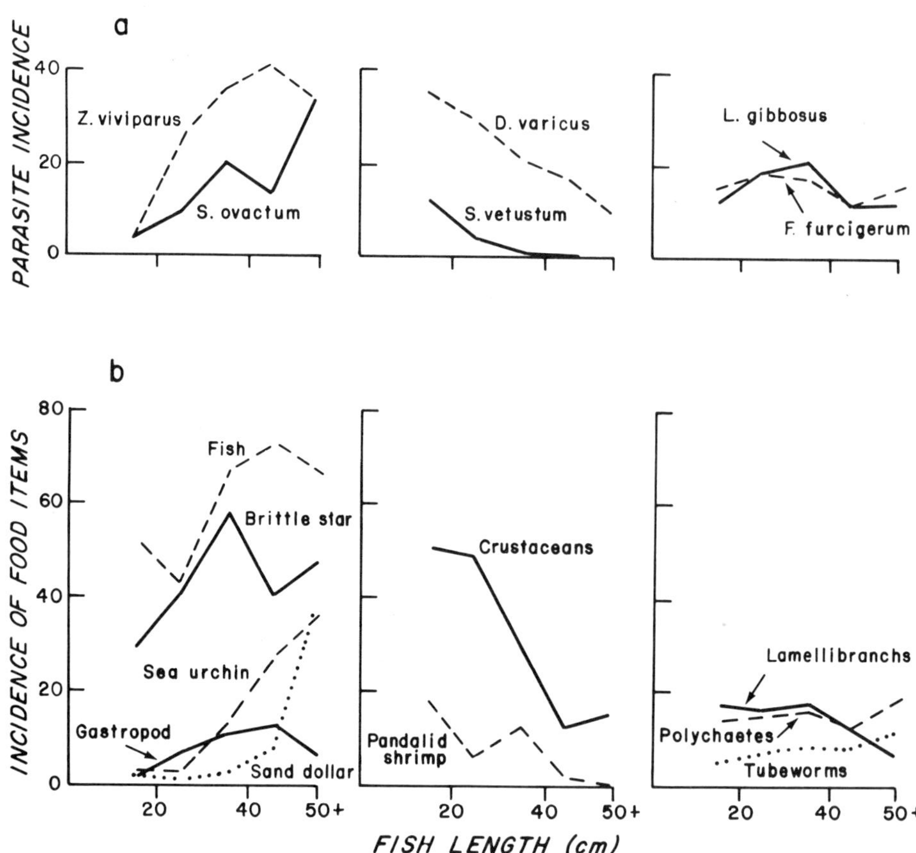

Fig. 6. Comparison of trematode incidence and changing diet of the American plaice (D, *Derogenes;* F, *Fellodistomum;* L, *Lecithaster; S, Stenakron vetustum, Steringotrema ovacutum;* Z, *Zoogonoides*). [Reprinted with permission of *Journal of the Fisheries Board of Canada* from Scott, J. S., 1975. Incidence of trematode parasites of American plaice (*Hippoglossoides platessoides*) of the Scotian Shelf and Gulf of St. Lawrence in relation to fish length and food. *J. Fish. Board Can.*, **32,** 482.]

and nektonic intermediate hosts. Such studies indicate that similar food and habitat preferences may result in similar helminth faunas of phylogenetically divergent hosts. Polyanski (1955) found that catfishes and flatfishes of the Barents Sea, both demersal and benthophagous, shared nine species of parasites. The converse is true of phylogenetically closely related hosts having widely divergent diets and habitats. Rysavy (1966) made similar conclusions after studying the cestode faunas and ecological and phylogenetic relationships among different orders of birds.

Scott (1975) used the incidence of digenetic trematodes of *Hippoglossoides platessoides* (American plaice) as indicators of changing diet associated with growth of the host (Fig. 6). Correlations between parasite incidence of four trematode species and the frequency of occurrence of small crustaceans, sea urchins, and brittle stars indicated the probable intermediate hosts involved in transferring the worms to the plaice. Scott (1969) and Shotter (1973) found that intestinal trematodes generally have a short life span, indicating that Scott's (1975) observations more likely reflect recent dietary changes by the host rather than accumulation of parasites over a long period of time. The relative food volume ingested by small versus large fish influences the incidence and number of parasites acquired, but the relative importance of prey species in transporting the parasites to fish of a particular size is clearly indicated.

The American plaice is a sluggish bottom fish of stationary habits whose distribution includes both sides of the northern Atlantic to about latitude 72°N in the Arctic Circle. They occur in a wide range of salinities and depths from the tide line to about 700 m. Planktonic diatoms and copepods are fed on by larval fish until they become demersal and begin to eat shrimp and other crustaceans. As they grow older, they progressively include a greater proportion of echinoids and ophiuroids in their diet. Shrimps, crabs, worms, mollusks, tunicates, and an occasional small fish are also eaten. Large cod, spiny dogfish, and Greenland sharks prey on them.

Mullets (Mugilidae)

A well-known generalization relating feeding behavior and parasitism in marine fishes is that phytoplankton feeders have fewer kinds, numbers, and frequency of infection than do carnivorous species. This observation is based on differences in food preferences or the quantities consumed that would contain infective larval parasites. Lom (1970) was also of the opinion that incidence and parasite abundance is very dependent on the host's mode of life. Studies on parasites and the biology of *Mugil cephalus* (Rawson, 1973; Skinner, 1975; Odum, 1966) and *M. auratus* [Reshetnikova, 1955, cited in Polyanski (1958)] have shown that these fishes have two distinct life history phases. Young fish lived a pelagic life feeding on plankton and acquired parasites cycling through planktonic crustacea. As the fish grew older, their inshore food preferences turned to bottom detritus and their "childhood" parasites were gradually exchanged for new ones. Parasites with simple life cycles were readily acquired by schooling with older fish. Skinner (1975) found that adult fish had more species of haplosplanchnid digenea of all families represented. Cable (1954) showed that cercariae of haplosplanchnids encyst on vegetation that is then ingested by the adult fish. Adult fish also became intermediate hosts in the life cycles of strigeoid and heterophyid trematodes that use shallow-water invertebrates as first intermediate hosts. Of the 35 species of parasites that Skinner found in Biscayne Bay, Florida, about one-third were also found in Georgia waters (Rawson, 1973), and the remainder were apparently more closely related to the fauna of the Gulf of Mexico, the Caribbean, and Brazil. Two species showed zoogeographic affinities with the Pacific Ocean.

Codfishes (Gadiformes)

Most adults of the some 70 cod species are found in the colder waters of the Northern Hemisphere. Atlantic cod, *Gadus morhua,* are ground fish that seldom frequent depths beyond about 250 m. Larvae and small cod are found in shoal waters until they are old enough to seek deeper waters of the continental shelf. The literature on cod, including the monograph by Dollfus (1953), could provide a long list of its parasites (36 species of trematodes, 31 cestodes, 22 nematodes, 9 acanthocephalans), but Polyanski (1955) was the first to show the host age dynamics of the parasite fauna. Moller (1975) examined 926 cod from the Kiel Fjord in

Germany and found infection rates of digenea, *Cryptocotyle* sp. (98%), *Podocotyle atomon* (20.8%), *Hemiuris communis* (7.9%), *Derogenes varicus* (1.1%), and *Stephanostomum pristis* (0.5%); cestode, *Bothriocephalus scorpii* (0.2%); nematodes, *Thynnascaris* (= *Contracaecum*) *aduncum* (39.4%) and *Ascarophis* sp. (0.9%); and the acanthocephalan *Echinorhynchus gadi* (97.8%). Platt (1975, 1976) found *Phocanema* (= *Terranova*) *decipiens* (N) useful as a biological tag to identify cod stocks off Iceland and Greenland. Scott and Martin (1957, 1959) and Scott and Black (1960) examined from 1500 to 73,000 cod fillets and found the nematode *P.* (= *Porrocaecum*) *decipiens* more often in the flesh of young cod caught closer to shore but in greater numbers (hundreds) in older migratory fish. Scott (1954) experimentally proved the passage of larval nematodes from smelt to cod. This is an alternative means of accumulating these larvae because the geographic concentrations of the worms in cod varied with the distribution of the seal (*Phoca, Halichoerus*) hosts.

Several genera of anisakine nematodes occur as larvae in the flesh of cod that are of economic or public health importance (*Phocanema, Thynnascaris, Anisakis*). The presence of these larvae affects marketability of the fillets and renders some unusable. Cod fillets are typically passed over a light table prior to packaging for consumers (Margolis, 1977), but freezing of the fish (48 hr at −20°C) will kill the worms (*Phocanema, Anisakis*), as will thorough cooking to an internal temperature over 60°C (Bier, 1976). However, prolonged exposure to freezing (−20°C) is necessary to kill *Thynnascaris* (Bier, 1976). The additional hazard of visceral larva migrans makes it necessary to take precautions in processing the fish (Oshima, 1972; Jackson, 1975; Myers, 1975; Cheng, 1976). Immediate evisceration of fish is required by some countries to prevent migration of larvae (*Anisakis*) into the flesh. Other commercially important fishes involved besides cod are herring, Alaskan pollack, haddock, bonito, and mackerel. More than 1200 cases of anisakiasis and about 50 cases of phocanemiasis are known (Marcial-Rojas, 1975) in addition to scattered reports implicating other genera. Of the two most common, *Anisakis* infection is the more serious because of the migrating activities of the larvae. *Phocanema* is short-lived (10 to 14 days) and usually causes localized granulomas in the intestinal mucosa of man.

The parasite fauna accumulated by cod illustrates well the changing ecology of a species moving from shoal to deeper water during its life cycle. Larvae spend their first months of life drifting in the upper water layers where they subsist on copepods and other minute crustacea. When the young fish first seek the bottom, they add amphipods, barnacle larvae, small worms, and other small crustacea to their diet. Polyanski (1955) and Polyanski and Kulemina (1963) found that larval cod in shoaler water were free of parasites but began to acquire them gradually as they moved toward the open sea (Table III). Fish 4 to 5 months old were rapidly infected almost exclusively by helminths having intermediate hosts in the cod's food web. For example, *Podocotyle atomon* (D) was obtained by eating littoral gammarids, whereas the hemiurids *Derogenes varicus* (D); *Hemiuris levinseni* (D); and *Lecithaster* sp. (D), *Scolex polymorphus* (C) larvae, and *Contracaecum* (N) were all derived from pelagic copepods. Twenty-seven species of parasites were eventually acquired, but adult cod were free of some parasites that were present during the first year of life. Examples are *Podocotyle atomon* (D) and *Lepidapedon gadi* (D), the last obtained by eating *Nereis pelagica*. Few parasites

TABLE III

Dynamics of Parasite Fauna of *Gadus morhua* from Barents Sea [After Polyanski (1955)] (All Numbers in Table Represent Percent Infection)

Parasites	Location	Fish Age and Number Examined		
		4 to 5 months (39)	10 to 11 months (29)	1+ years (50)
(P) *Trichodina* sp.	Gills, fins	—	10.3	—
(P) *Myxidium bergense*	Gall bladder	2.6	—	—
(P) *Pleistophora* sp.	Muscles	2.6	—	—
(M) *Gyrodactylus marinus*	Gills	—	3.4	—
(D) *Podocotyle atomon*	Intestine	23.6	34.8	40
(D) *Podocotyle reflexa*	Intestine	—	20.7	—
(D) *Lepidapedon gadi*	Intestine	—	62.1	16
(D) *Hemiuris levinseni*	Stomach	7.7	3.4	16
(D) *Derogenes varicus*	Stomach	18	3.4	10
(D) *Lecithaster* sp.	Intestine	2.6	—	—
(C) *Scolex polymorphus* (l.)	Intestine	28.2	20	46
(C) *Abothrium gadi*	Intestine	—	—	2
(C) Pseudophyllidean gen. sp. (l.)	Mesentery	2.6	—	2
(N) *Contracaecum aduncum* (l.)[a]	Intestine, liver, mesentery	28.2	79.3	92
(N) *Anisakis* sp. (l.)	Mesentery, liver	2.6	6.8	10
(A) *Echinorhynchus gadi*	Intestine	—	10.3	42
(Co) *Caligus curtus*	Body surface	2.6	—	—
(Co) *Lernaeocera branchialis*	Gills	—	3.4	—
(Co) *Clavella uncinata*	Gills	—	3.4	10

[a] *Contracaecum aduncum* is a member of *Thynnascaris*.

with direct development were found, and these had a rather low host specificity [*Trichodina* sp. (P), *Myxidium bergense* (P), and *Caligus curtis* (Co)].

Cod are relatively long-lived, 15 years or more, and adults usually lie within a fathom or so off the bottom where they consume a great variety of invertebrates, especially bivalve mollusks, large crustaceans, pelagic shrimp, ophiuroids, echinoids, holothurians, tunicates, ctenophores, squid, and various small fishes from upper waters or the bottom. The presence of indigestible objects in their stomachs indicates very generalized feeding habits similar to certain macrourids and the blue hake, *Antimora rostrata* (see below). Generalized feeding and longevity explain the progressive growth of their parasite fauna (Fig. 7), which can include very large numbers of worms. Ralph Appy (personal communication) has found more than 10,000 *Lepidapedon elongatum* (D) in a single host. A fact not generally appreciated about the trophic behavior of cod is their cessation of feeding during part of the year (John Anderson, personal communication), which may be related to fluxes in prey populations and their movements through the subzero

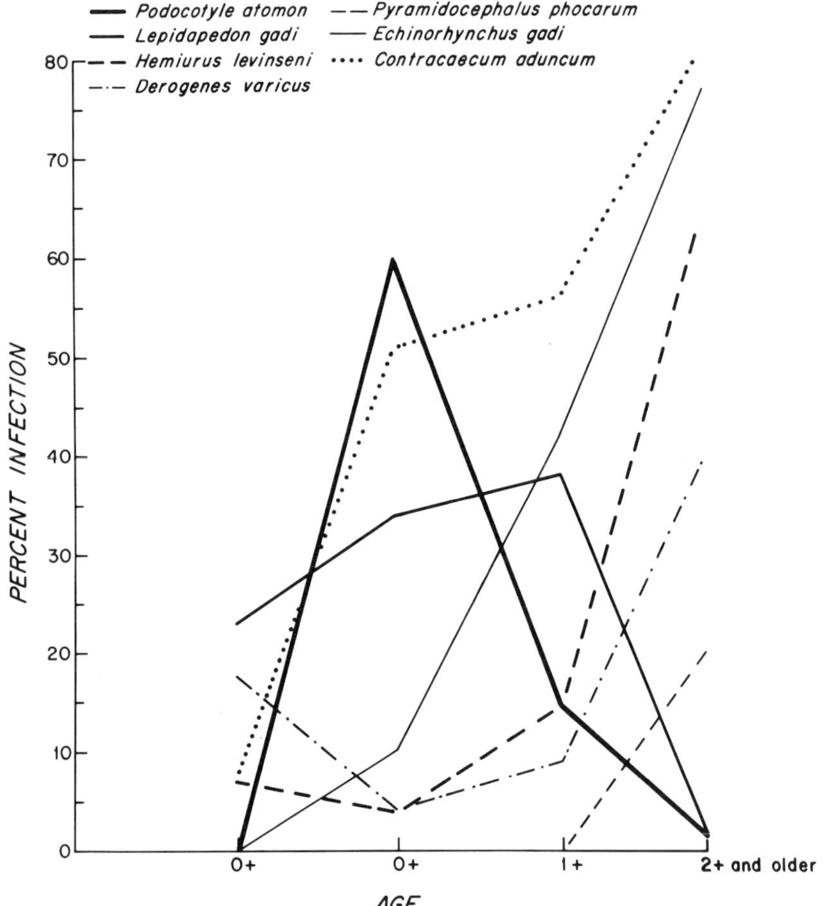

Fig. 7. Changes in the parasite fauna of the cod, *Gadus morhua*, with age [after Polyanski (1955)].

temperatures of the Labrador Current along the Grand Bank of Newfoundland. Such phenomena cause the seasonal loss of parasites.

Ralph Appy (personal communication), after a prolonged study, has found that he can easily identify the geographic origin of cod from eight localities in Canadian waters by studying the helminth parasites from samples of 30 hosts. It is even more significant that he was able to determine whether the fish had migrated after correlating the length–frequency curves of otoliths in the fish with seasonal cycles of the parasites (intensity and prevalence) and with parasite maturation cycles (unpublished dissertation). Chubb (1982) has shown the significance of this approach.

Silver hake, *Merluccius bilinearis,* are swift swimmers over the continental shelf with voracious appetites for small schoolfishes. Like deeper living benthic fishes, they will eat large invertebrates (squid and crabs) given the opportunity and will rise well off the bottom in pursuit of prey. In southern New England waters they are caught on the continental slope at depths of 275 to 750 m and show preference for cooler water. At least 26 species of helminths appear in the literature for this species from Atlantic coastal waters of North America. All helminth taxa except acanthocephala are common. Jorotaeva and Leont'eva (1972) examined 139 *Macruronus novae-zealandiae* taken in the Indian and Pacific Oceans. They found seven (only three species named) species of trematodes, one acanthocephalan, and five of nematodes. Comparison with reported species from macrourids reveals a similarity in the sharing of *Gonocerca phycidis* (D), *Pseudopecoelus* (D), *Derogenes varicus* (D), *Contracaecum tasmaniense* (N), *Capillaria tasmanica* (N), *Anisakis* and *Ascarophis* spp. (N), and *Echinorhynchus gadi.* However, *D. varicus, G. phycidis, Anisakis* sp., *Ascarophis* spp., and *E. gadi* are widespread among marine fishes.

C. Sharks, Skates, Rays, and Chimaeroids (Selachii)

Some 800 different species of sharks, skates, rays, and chimaeroids are known. The batoids, especially, are common and more abundant over the continental shelf and slope areas than in deeper water. Sharks roam freely as cosmopolitan inhabitants of coastal and epipelagic oceanic waters. The most distinctive feature of their parasite faunas is the disproportionate numbers of cestodes to their other helminths. More than 400 species of adult cestodes assigned to five orders have been found in spiral valves of about 150 species of elasmobranchs examined thus far. Juvenile cestodes are infrequently found in their tissues. Investigation of the parasites of these fishes and their food preferences in Atlantic coastal waters of the United States provided by the National Marine Fisheries Service at Woods Hole (Maurer and Bowman, 1975; Bowman et al., 1976; Bowman, 1977) indicates that parasite faunal diversity and abundance may be the result of variation observed in the diets of the species and even among congeners from different ecological regions. A variety of blood protozoa and ectoparasitic copepods have been described from them. Crustacean ectoparasitic diversity rivals that of cestodes in sharks. Monogenea have been reported from (skin, respiratory surfaces, cloaca, rectal gland) these fishes, but digenea (stomach, blood, body cavity), aspidogastrea (biliary system), nematodes (stomach, pancreas, spiral valve), and leeches (skin, cloaca) are less common. Acanthocephalans are extremely rare in elasmobranchs, and it has been suggested they cannot tolerate the high levels of urea in these fishes. Of interest to parasite evolution is the presence of the ectoparasitic planarian, *Micropharynx parasitica,* on the skin of *Raja radiata* (Ball and Khan, 1976) because the classes of parasitic platyhelminths are believed to have evolved from the turbellarians, first as ectoparasites that later became endoparasitic (Cameron, 1950; Hyman, 1951; Llewellyn, 1965). Piscicolid leeches, typically ectoparasitic but occasionally invading external orifices, are known vectors of trypanosomes for cartilaginous (Burreson, 1975) and bony fishes (Khan, 1974). Monopisthocotylean monogenea (Monocotylidae, Microbothriidae, Acanthocotylidae, Capsalidae) are often reported from the smooth skin of batoids but not from the rough-skinned sharks. The respiratory surfaces of selachians are sites of attachment for polyopisthocotylean monogenea of the families Hexabothriidae and Discocotylidae. Rare examples of endoparasitic monogenea from elasmobranchs are a curiosity among fishes in general. *Discocotyle* (body cavity) and *Calicotyle* (rectal gland, urogenital system) are common in some species but so far are unknown from most deep-sea hosts. Williams (1964) examined over 500 *R. radiata* and found gravid females uninfected with *Dictyocotyle coeliaca,* providing a rare example of differential parasitism between sexes. Encounters with this trematode seemed to be restricted to depths in excess of 100 m. Digenea are rarely found in the digestive system of most selachians; the one common exception is the genus *Otodistomum.* Stomach or intestinal infections by *Derogenes, Plagioporus, Ptychogonimus, Torticaecum* (larvae), *Paravitellotrema,* and *Plectognathotrema* are known. Gorgoderids (Annaporrhutinae, Problitrematinae), a family of digenea known from the urinary bladders of teleosts and amphibians, frequent the body cavity. Sanguinicolids (D) are restricted to blood vessels. Aspidogastrean trematodes (Fig. 3B; viz., *Multicalyx, Stichocotyle*) are found in the gall bladder and bile ducts of a few hosts. Habitation of the spiral intestine seems to have become physiologically improbable for all but the cestodes.

In general it would appear that sharks host more cestodes and batoids have the more diverse faunas but that decreases in parasite incidence, abundance, and diversity are indicated for sharks and batoids with higher latitude and increasing depth of habitat. Parasite diversity is greatest in warm waters and long lists of 30 or more metazoan parasites have been tabulated from the more common hosts. Tetraphyllidean cestodes, especially the phyllobothriids and oncobothriids, are extremely common. Trypanorhynchs are a widely distributed and diverse order of cestodes but tend to occur less frequently as adults and in few numbers. The lecanicephalans and systematically disputed *Echinobothrium* are common to batoids. Parasitological evaluations of elasmobranch populations are practically nonexistent. McVicar (1977) recently assessed the intestinal helminth fauna of 263 *Raja naevus* in British waters. Higher intensities of infection of *R. naevus* were found off Plymouth, England than near Aberdeen, Scotland. Although cestode diversity was greater in 206 *R. naevus* from Aberdeen (7 species) than Plymouth (5 species, 57 hosts), the average number of worms per fish was considerably less in the higher latitude (142 vs. 553). There was evidence of seasonality in the cold Scottish waters, with fewer worms from November to May, but not at Plymouth. Two distinct patterns of infection were noted in relation to host size: (1) increase in intensity with increasing size (*Acanthobothrium quadripartitum* and *Phyllobo-*

thrium pirei) and (2) increase in intensity with size to some intermediate length followed by a decrease in larger hosts (*Echeneibothrium* sp., *Grillotia erinaceus*). We have found that intensity of infections in *R. erinacea* and *R. ocellata* decreased in winter months, but diversity increased (Table IV) in studies in New England waters.

The variation in cestode faunal abundance and diversity for some of the species that my students (B. Hayden and A. Williams) and I have examined from the northwestern Atlantic coastal waters is presented in Table IV. Most hosts were taken during the months of May to August. A greater abundance and diversity per infestation was obvious for batoids from more southerly waters despite the fact that some northern hosts, like *Dasyatis centroura* and *Raja radiata,* have diverse faunas reported for them. We know of 23 different cestodes from *D. centroura* and four from *R. radiata* in New England waters. At least 16 cestodes have been specifically identified from *R. radiata*. Change in the diet of *R. radiata* from crustaceans, primarily amphipods (Table IV, footnote *d*), to onc including fish prey (Table IV, footnote *e*) is reflected by the numerous *Echeneibothrium* (C) in young hosts and addition of the trypanorhynch, *Grillotia* (C), in older hosts. Note also that the older *R. radiata* harbored fewer cestodes.

Differences in abundance and diversity are also apparent for other skates. A common genus of cestodes in New England waters is *Echeneibothrium* (*R. erinacea, R. ocellata, R. radiata*), but *Acanthobothrium* (two species) were more common and abundant in 37 *R. eglanteria* from Chesapeake Bay. Almost 2800 cestodes representing 3 orders, 5 families, and 10 species were recovered from a single *Dasyatis americana* taken in Chesapeake Bay, Virginia. Over 2000 of the worms belonged to a single species, *Acanthobothrium lineatum*. Williams (1968b) reported 13,000 *A. quadripartitum* in a single *Raja naevus* from the North Sea. Such large infrapopulations of cestodes, on the average, are more common among large stingrays from warm waters than populations obtained from the miscellaneous charcharinids, isurids, alopiids, scyliorhinids, triakids, requiem sharks, sphyrnids, and squalids that I have examined from New England waters to Bimini in the West Indies, Chilean Pacific, and Mediterranean.

The results of intraspecific crowding on tapeworm infrapopulations have been well established from laboratory experiments using mammalian hosts. The most striking effects were the marked reduction of size and fecundity of the worms. Even the casual observer of natural infections of elasmobranch cestodes would note that the worms are typically small, slender, and numerous. The phenomenon of hyperapolysis is common to a great many species (premature detachment of segments from the worm before they are gravid) of several orders. Detached segments wander within the spiral valve, grow several times their initial size, become gravid, and finally release eggs. The small size and hyperapolysis of so many species of elasmobranch cestodes may be the result of selection to minimize the effects of inter- and intraspecific competition. The greatly modified and specialized scoleces of these worms apparently are significant in radial and longitudinal site selection as well as host recognition (see below). Large cestodes ($\geq$20 cm) are usually found in small numbers, although they are widely distributed in the faunas of a great many of the larger batoids and sharks. Exceptions can be found, of course, such as the large (5-m) tiger sharks from Hawaiian waters, that harbor

Table IV
Cestode Faunal Abundance and Diversity of 434 Elasmobranchs (12 Spp.) from Chesapeake Bay (C) and Waters off Southern New England (NE)

Species	Locality		Inf/Ex[g]		Number of Worms Average (Range)	Number of Cestode spp.	Number of Cestode Families (Orders)
	C	NE					
Rhinoptera bonasus	+		20	20	202 (67 to 737)	11	6 (4)
		+	8	8	74 (41 to 169)	7[h]	4 (4)
Dasyatis americana	+		7	7	1472 (842 to 2719)	11	5 (3)
Raja eglanteria	+		37	37	42.3 (21 to 177)	3	2 (2)
Dasyatis centroura		+	19	19	232 (38 to 1610)	15	4 (3)
Raja erinacea[a]		+	21	45	3.9 (1 to 11)	2	1 (1)
—[b]		+	13	19	3.7 (1 to 9)	3	2 (2)
—[c]		+	45	84	2.0 (1 to 5)	3	2 (2)
Raja ocellata		+	39	51	10.2 (1 to 48)	2	2 (2)
—[d]		+	21	27	8.2 (1 to 23)	3	2 (2)
Raja radiata[e]		+	10	11	46.8 (16 to 159)	1	1 (1)
—[f]		+	9	11	3.4 (1 to 14)	2	2 (2)
Raja senta		+	29	35	6.8 (1 to 14)	2	1 (1)
Torpedo nobiliana		+	10	13	27.8 (10 to 90)	2	1 (1)
Prionace glauca		+	11	11	2083 (310 to 6376)	4	3 (1)
Mustelus canis		+	27	27	38.2 (4 to 77)	3	2 (2)
Rhizoprionodon terraenovae	+		6	8	54.7 (1 to 400)	3	2 (1)

[a] Mature fish, summer months.
[b] Immature fish (less than 20 cm TL), summer months.
[c] Mature fish (31 to 54.5 cm TL), winter months.
[d] Mature fish (40 to 108 cm TL), winter months.
[e] Immature fish (14 to 25 cm TL), summer months.
[f] Mature fish (48 to 96 cm TL), summer months.
[g] Indicates number of fish infected versus number examined (also in Table V). Infection averages and ranges pertain to hosts.
[h] Only two specimens of *Duplicibothrium* and one *Glyphobothrium* were recovered.

several hundred large phyllobothriids in a single host. Intuitively, one might suspect that surface area of the spiral valve, tapeworm abundance and faunal diversity, host diet and longevity, and temperature regime (internal and external) influence tapeworm size.

Very few elasmobranchs, about 15 species, are known to occur below about 2000-m depth. Few helminths have been reported from deep-water elasmobranchs, but my observations, based on study of a few mature fishes, indicate that both abundance and incidence of helminths may be considerably lower than that of shoaler dwelling species (Table V). Current evidence indicates that parasite diversity also decreases with depth for elasmobranchs. I have found few adult tetraphyllideans (*Onchobothrium, Phyllobothrium, Echeneibothrium*), trypanorhynchs (*Grillotia*), and diphyliideans (*Echinobothrium*) in examination of a shark (*Centroscymnus*) and five species of skates (*Raja, Bathyraja*) taken at mean trawl depths of 500 to 2400 m between Hudson Canyon and Newfoundland. The presence of larval trypanorhynchs in deep benthic teleosts indicates the presence of large elasmobranchs. At shallower depths trypanorhynch larvae of cestodes common to batoids inhabiting the continental shelf are common in vertically migrating teleosts that prefer zooplankton. Dailey and Vogelbein (1982) have recently reported a new and unusual family of trypanorhynchs (C) from an undescribed deep-sea shark, "megamouth," in the Pacific. *Ceratobothrium xanthocephalum* has long been known from *Centroscymnus coelolepis*. The limitation of the majority of cestodes to shallower depths apparently indicates their dependence on an abundance of pelagic intermediate hosts.

Halvorsen and Williams (1968) examined 90 *Chimaera montrosa* from Oslo Fjord, Norway and found that incidence of infection with *Gyrocotyle,* a cestodarian, increased with host age (size). About 30 species of parasites, some 10 of which are claimed to be different species of *Gyrocotyle,* have been described from this host. Most of the remaining species are ectoparasitic monogenea, a digenean, an aspidogastrean, leech, copepods, and an isopod. Numerous studies on parasites of chimaeras have focused on the monozoic cestodarians, which tend to occur in pairs, are protandric, and are believed to act as a reproductive unit. Dienske (1968) found eight species of adults and two juvenile helminths in 215 *C. montrosa* from Norwegian waters. There was a positive correlation between host weight and weight of specimens of *Gyrocotyle urna*. Incidence of infection increased rapidly to 100% in fish 35 to 50 cm long and then decreased to near 70% in fish up to 58 cm in length. To Dienske this observation indicated a long adult life span for *G. urna*. A similar fauna is reported for the chimaeroid fish *Callorhynchus* and *Hydrolagus*. Aspidogastreans are more common among chimaeroids than other selachians. Schell (1973) found another, *Rugogaster,* in *H. colliei* and Carvajal, and I (Campbell, unpublished data) recovered *Multicalyx elegans* from *C. callorhynchus* at Puerto Montt, Chile. Chimaeroids occur in relatively shallow waters, but species of *Hydrolagus* have been taken from depths of 2250 m. The chimaeroids have long been considered faunally distinct as hosts for cestodarians, but Williams and Bray (unpublished data) have recently found the first true tapeworm, a new family of tetraphyllideans, from a chimaeroid fish that is of evolutionary and taxonomic significance (5th International Congress of Parasitology, Toronto, Canada, 1982).

Table V
Cestode Faunas of Elasmobranchs Taken at Depths of 500 m or More from North Atlantic Ocean, Gulf of Mexico, and Caribbean Sea

Species	Inf/Ex	Number of Worms Total (Range)	Number of Cestode spp.	Number of Families (Orders)
Acanthobatis longirostris	0 1	0	0	0
Bathyraja richardsoni	1 1	30	4	3 (2)
Benthobatis marcida	0 1	0	0	0
Breviraja spinosa	0 1	0	0	0
Centroscyllium fabricii	1 2	1	1	1 (1)
Cruriraja poyei	0 2	0	0	0
Pseudoraja atlanta	1 4	2	1	1 (1)
Raja bathyphila	3 3	15 (1 to 9)	2	2 (2)
Raja jenseni	0 1	0	0	0
Raja radiata	1 1	38	3	2 (2)
Raja spinicauda	1 1	72	1	1 (1)
Springeria filirostris	1 3	2	1	1 (1)

Life cycles of elasmobranch cestodes have not been worked out completely in the laboratory, but experimentation and host surveys indicate that a variety of crustaceans, particularly copepods, are first intermediate hosts and mollusks, larger crustaceans, or fish serve as second intermediate hosts. Selachians are considered to be the only definitive hosts, but at least one exception involving a teleost (*Myoxocephalus scorpius*) final host is known (Threlfall, 1969). Mudrey and Dailey (1971) gave evidence of an alternative life cycle involving development through the infective plerocercoid stage in a single intermediate host (copepod). Tom E. Mattis (personal communication) has experimental evidence that three intermediate hosts may be typical of trypanorhynch life cycles to elasmobranchs. First intermediate hosts are small crustaceans. Second and third intermediates are teleost fishes or crustaceans of successively greater size that must be considered necessary to the life cycle, not paratenic, in ascending the food chain to the final (elasmobranch) host. Cake (1976, 1977a), and Cake and Menzel (1980) listed cestode larvae from a variety of experimental and natural molluscan infections. Cheng (1967) and Cake (1977b) discussed those found in commercially important mollusks. Cake believes that pelecypods may serve as primary intermediate hosts and predatory gastropods are paratenic hosts serving to collect many fully developed larvae.

Rex (1976), in a study of mollusks along the Gay Head–Bermuda transect at depths of 478 to 4862 m, found a significant positive correlation between numbers of gastropod predators and the diversity of their polychaete and protobranch bivalve prey. Gastropod predator diversity is greatest on the lower continental slope and rise but is lowest on the abyssal plain. If paratenesis in gastropods is important to the life cycles of tetraphyllidean cestodes of elasmobranchs (*Echeneibothrium, Acanthobothrium,* etc.), the paucity of these mollusks on the abyssal plain indicates that these cestodes would be unsuccessful there. Stunkard (1977) reviewed records from squids. Extensive lists of cestode larvae appear in a series of papers on parasites of marine plankton by Dollfus (1923, 1931, 1942, 1953, 1963, 1970, 1971). Teleosts may serve as true second intermediate or as paratenic hosts for cestodes. Many well-known sources contain records of larval cestodes from marine teleosts whose adults occur in elasmobranchs [Linton (1898, 1900, 1901, 1905, 1909, 1910, 1924, 1940), Dollfus (1942), etc.]. Tetraphyllidean plerocercoids and postlarvae of trypanorhynchs commonly are found within the intestine or viscera of teleosts. Plerocerci of trypanorhynchs, distinguished by a blastocyst, are common in extraintestinal locations. They are often numerous in the body musculature, and very young fish may be killed once vital organs are invaded; however, analysis of large numbers of hosts indicates that there probably is no serious detrimental effect on most fish. Overstreet (1977) made such a conclusion on analysis of plerocerci of *Poeciloancistrum caryophyllum* from more than 3000 sciaenids (10 species) from the Gulf of Mexico. As the fish grew larger, the incidence of infection increased, but the number of worms did not for all host species. Lubieniecki (1976) reported continual accumulation of plerocerci of *Grillotia erinaceus* in haddock, but Overstreet (1977) found evidence of immunity to challenge doses in adult sciaenids.

The striking spatial preferences of cestodes within the spiral intestine of their host has been noted by several authors. Williams (1960) believed that mucosal topography and the form of the cestode scolex were major factors in determining host-specificity and site recognition by the parasites. Williams (1963) subsequently described two distinct biological groups of species of *Echeneibothrium* (C) from shallow-end deep-water skates near England and revised the genera *Echeneibothrium* (Williams, 1966), *Phyllobothrium* (Williams, 1968a), and *Acanthobothrium* (Williams, 1969). Allen Williams and Campbell (unpublished data) found a very different longitudinal distribution of the same species of cestode in both *Dasyatis centroura* and *Rhinoptera bonasus*. Dramatic changes in mucosal topography and widespread distribution of *Polypocephalus medusiae* (C) occur in the valve of the roughtail stingray, *D. centroura,* but only subtle, although statistically different (*a posteriori,* nonparametric analysis), changes in mucosal regions are evident with a very limited distribution of the same cestode in the valve of the cow-nose stingray, *R. bonasus*. In addition, preferences were noted for transverse locations on the spiral valves by different species of cestodes. Repetitive samples are being obtained to consider other variables, but we suspect that physiological factors are of primary importance in site location of the worms and that scolex morphology is a secondary adaptation.

D. Epipelagic Fishes

Inhabitants of the sunlit surface waters of the open ocean to depths of about 100 m include a variety of far-ranging fishes, sharks, whales, and seals. Sportfishes such as sailfish, marlins, swordfish, dolphinfish, and commercially important mackerels and tunas are infected by numerous and varied helminths, mostly didymozoid digenetic and monogenetic trematodes and crustaceans. These fishes are rapid swimmers of the upper waters in tropical and occasionally temperate seas. All of them are carnivorous and have many parasites associated with their warm habitat and varied diet, but much remains to be learned about them in terms of their populations and general biology. Some of them share the same parasites, and many have closely related parasites. Skin parasites are especially common. Large monopisthocotyleans and bizarre didymozoids (D; see Fig. 3D) infest the gills and skin. A variety of trypanorhynch larvae (C), also found in smaller epipelagic teleosts, develop to adults in large sharks found in this zone. Species of *Tentacularia, Nybelinia, Dasyrhynchus, Callitetrarhynchus, Otobothrium, Hepatoxylon,* and so on are commonly encountered, and trematodes of the families Capsalidae, Didymozoidae, Accacoeliidae, Hemiuridae, and Bucephalidae are well known among epipelagic fishes. Life histories of the predominant digenea involve planktonic invertebrates. The slow-moving molids are particularly good examples of hosts harboring a variety of parasites from this habitat, so much so in fact that they have been referred to as the "parasitologist's dream."

On *Mola mola,* a high-ocean circum-global fish of warm waters (Ekman, 1953), ectoparasites are particularly common despite the fish's thick, rough, protective skin. Monogenea (3 species), copepods (9 species), accacoeliids (10 species) and didymozoids (2 species), digenea, larval *Floriceps* (C), and *Ancistrocephalus microcephalus* (C) have been reported from Atlantic and Pacific waters. Eight of the 10 species of accacoeliids from the North Atlantic *Mola mola* have recently been redescribed in detail by Bray and Gibson (1977). *Mola mola* prefers a diet of coelenterates, ctenophores, and salps. The last two commonly contain hyperiid amphipods. Reports of accacoeliid metacercariae from chaetognaths, ctenophores, scyphomedusae, and siphonophores have been summarized by Dollfus (1963) and Rebecq (1965). Pelagic teleosts may act as paratenic (obligatory?) hosts (Nikolaeva, 1968) by feeding on medusae and cycling into carnivorous teleosts.

Little information has been available to date on parasitism of billfishes beyond descriptions of parasites obtained from a few fish. Opportunity to assess the parasite fauna of Atlantic sailfish, *Istiophorus platypterus,* in waters off southeastern Florida was made possible for this volume through the donation of unpublished data and specimens collected by Mr. Robert C. Richardson, working in cooperation with the Florida Department of Natural Resources. All four species of billfishes taken in the area bounded by Miami to the Florida Keys east to the Bahamas were examined between February 1973 and December 1974. The study included 404 sailfish (183 males, 221 females), 8 *Tetrapterus albidus* (white marlin), 3 *Makaira nigricans* (blue marlin), and 1 *Xiphias gladius* (swordfish). Sample size from the sailfish population during the 2-year study is sufficient to characterize parasitism of these epipelagial hosts and to be statistically relevant. A summary is presented in Table VI. Unfortunately, intensity of infection was estimated by Mr. Richardson, so exact numbers of parasites are not available. As many as 11 species of parasites were found in a single fish. There appears to be no biologically significant qualitative difference between infection of males or females. The most common parasites were didymozoid trematodes (*Makairatrema, Metadidymozoon, Neonematobothrioides*-like, unidentified), monogenea (*Tristomella* spp.), the cestode *Bothriocephalus manubriformis,* and the copepod *Gloiopotes costatus.* The complete list of symbionts includes 14 species of helminths, 2 identified copepods, *Pseudomonas* bacterium, a remora, lernaeopodid copepods, an isopod, and a goose-necked barnacle (Table VI). Multiple locations of the parasites included most of the exterior and interior of the body. Most common in order of frequency of observation were *Pseudomonas* sp. (46) isolated from blood and abrasions; *T. laevis* (M), venter of bill (131), mouth (31), and gill arches (14); *T. pricei* (*provis*) primarily in groove for dorsal fin (173), branchiostegals (35), dorsal fin (9), or on body (4); *H. ventricosa* (D) from the stomach (52) and dislodged into the mouth (14) (also common in the stomach of prey species, *Euthynnus alletteratus*); *Makairatrema* within red–orange subdermal cysts up to 4-cm diameter embedded along the lateral line (122), body surface (15), dorsal fin (1), dorsal and pelvic grooves (2), gill cavity, and body cavity (2); unidentified didymozoids encysted in 20 locations over the body, in the gill cavity (105), and in the viscera; *Metadidymozoon* in the gill lamellae (264); *Neonematobothrioides*-like didymozoid up to 2 m long from the body cavity (91), mesenteries (9), viscera (6) and subcutaneous locations (50); *Bothriocephalus manubriformis* from the intestine (315); trypanorhynch larvae *Tentacularia, Otobothrium,* and *Hepatoxylon* from the mesenteries (31) and viscera (15); acanthocephalan cystacanths of orange or white color in the viscera and also found ingested *E. alletteratus; T. histiophori* from intestine (60) and stomach (20); *Gloiopotes* on body (331); *Caligus,* mouth (50) and gills (7); Remora commonly on the body and small individuals (±10 cm) in gill cavity (36); and incidental symbionts (lernaeopodid copepods, an isopod, and a goose-necked barnacle).

Seasonal increases in prevalence (Table VII), generally April through September, were noted for *T. laevis* (M), didymozoids (D), *T. histiophori* (N), and *Caligus quadratus* (Co). The average number of parasite species per fish was also

TABLE VI
Summary of Symbionts Collected from 404 Atlantic Sailfish from Atlantic Waters off Miami, Florida[a]

Parasite	Number Infected (Males)	Number Infected (Females)	Total Infected	%
Pseudomonas sp.	20	26	46	11.4
Tristomella laevis	84	106	190	47
Tristomella pricei (?)	104	123	227	56.1
Hirudinella ventricosa	23	43	66	16.3
Makairatrema sp.	63	79	142	35.1
Didymozoidae	136	121	257	63.4
Metadidymozoon (?)	119	145	264	65.3
Neonematobothrioides (?)	80	83	163	40.3
Fistulicola sp. *Bothriocephalus manubriformis*	157	171	328	81.1
Otobothrium dipsacum (?) *Tentacularia coryphaenae*	41	43	84	20.7
Hepatoxylon megacephalum	4	5	9	2.2
Acanthocephala (cystacanths)	3	4	7	1.7
Thynnascaris histiophori	49	46	95	23.5
Gloiopotes costatus	153	179	332	82.1
Caligus quadratus	28	33	61	15
Lernaeopodidae Isopoda Goose-necked barnacle	4	6	10	2.4
Ulcerations	16	20	36	8.9
Remora osteochir (?)	22	16	38	9.4

[a] All body measurements from tip of mandible to tail fork. (Summarized from data and specimens collected by R. C. Richardson.)

	Inf/Ex		%	Length (m)			Weight (lb)		
				Minimum	Maximum	Average	Minimum	Maximum	Average
Male	183	183	100	0.855	1.855	1.494	4	57	33.4
Female	221	221	100	1.015	1.860	1.582	7	82.5	42.9
Total	404	404	100						

greatest during that time. Few species (*Bothriocephalus, Gloiopotes, T. pricei*) failed to show an increase corresponding to the warmer months of the year. Irregular fluctuations, apparently unrelated to season, were noted for *T. pricei* (provis), *H. ventricosa,* and *Makairatrema* sp. Infections by *Pseudomonas* were observed most often during summer months with increasing water temperature. Impaired clotting was noted in cases of septicemia. The diversity of the parasite fauna increased with size of the fish. Prey of these sailfish in order of decreasing occurrence in stomach contents were bonito (Scombridae), ballyho or needlefish

TABLE VII

Monthly Prevalence, Expressed as Percentage, of Some Parasites of the Atlantic Sailfish, *Istiophorus platypterus*

	Month											
	1	2	3	4	5	6	7	8	9	10	11	12
Monogenea												
T. laevis	23	16	29	37	52	64	50	70	54	58	35	18
T. pricei	45	39	48	52	50	48	56	59	54	58	44	36
Digenea												
Makairatrema	23	24	25	42	52	24	41	33	54	42	18	28
Metadidymozoon	41	42	50	82	83	64	68	89	77	58	71	28
Neonematobothrioides	18	24	23	70	59	40	53	37	15	37	22	18
Didymozoid spp.	54	76	77	90	72	52	53	67	69	47	49	41
Hirudinella	9	18	6	22	22	4	26	11	15	0	13	20
Cestoda												
Fistulicola / *Bothriocephalus*	82	74	67	82	87	64	82	89	77	89	73	90
Nematoda												
T. histiophori	4	10	17	30	39	28	23	11	15	26	13	8
Copepoda												
Caligus	0	5	10	10	22	32	32	26	46	0	0	0
Gloiopotes	95	79	81	85	76	80	73	74	77	79	93	90
Total fish	22	38	48	40	54	25	34	27	13	19	45	39
Males	13	19	22	19	25	12	16	10	5	10	19	13
Females	9	19	26	21	29	13	18	17	8	9	26	26
Parasite species/fish												
X	4.2	4.6	4.8	6.7	6.9	5.4	6.1	6.3	5.9	5.3	4.6	4.5
Range	1 to 7	1 to 8	1 to 9	2 to 11	2 to 11	1 to 9	2 to 11	4 to 9	4 to 8	2 to 9	2 to 8	2 to 8

(Exocoetidae or Belonidae), mullet (Mugilidae), squid (*Ilex*), jack (Carangidae), and shrimp or other pelagic crustacea.

Similar lists of 15 species of parasites, including blood protozoa, hemiurid and echinostome digenea, and the postlarvae of the homeoacanthous trypanorhynchs *Nybelinia* and *Tentacularia* (Dollfus, 1942) are reported from dolphinfish, *Coryphaena hippuris,* and swordfish, *Xiphias gladius* from the Atlantic.

E. Deep-Ocean Fishes

The parasite faunas of mesopelagic and bathypelagic fishes can be described quite simply as poor but with subtle differences. Of particular interest is the abundant evidence that their helminth parasites are rarely adults. Adult digenetic trematodes are particularly uncommon, and the usual larvae encountered are juvenile nematodes and cestodes—two groups noted for their use of many paratenic hosts. Some of their same species of helminths are widespread among paratenic hosts of the waters above and below them, where the usual combination of hosts reside for completion of the life cycles. Paxton (1967) compiled a list of marine mammals and fishes that prey on midwater fishes. Intermittent visits into upper or lower levels of the water column provide brief opportunity for infection, but the temporary nature of their visits into the transient planktonic community of the surface layers and lack of their own adult helminths, particularly those having complex life histories, indicates that they are not significant intermediates in cycles to definitive hosts in either epipelagic or benthic communities. Although some cycles will certainly be found to be completed by means of midwater and bathypelagic hosts, our data and those of others discussed below indicate that the number will be very small in contrast to those completed in the epipelagic and deep benthic communities. It is also possible that certain populations of digenea are maintained by completion of development in invertebrates. This alternative cycle is known for some hemiurids and opecoelids.

The distribution of biomass in pelagial waters and the diets and longevity of the fish hosts are predictable causes of variable parasite success. It is well established that biomass decreases with depth in pelagial waters (Marshall, 1954, 1971); as a result, the abundance of prey (intermediate hosts) decreases with the descent from epipelagial waters through the middepths into the bathypelagic environment. Food abundance is especially reduced beneath waters having low productivity such as the central oceanic gyres. Because parasite populations are not randomly distributed throughout the host population (Kennedy, 1975), the probability of the host (fish) encountering an infective intermediate host (prey) is dependent on the population density of all necessary hosts. Indeed, a comparison of published and unpublished data that I have available indicates highest diversity and infection rates among fishes in epipelagic and benthic zones with a two- to fourfold stepwise decrease in infection rate in vertically migrating mesopelagic fishes versus deeper-dwelling nonmigratory mesopelagic fishes to bathypelagic fishes. Also, a horizontal comparison at depth between pelagic ocean communities and the benthic fauna on the continental shelf, slope, and rise (northwestern Atlantic) shows a corresponding decrease in diversity and parasite success with depth that reflects the presence of faunal zones (see Fig. 13) (Haedrich et al. 1975, 1980; Campbell et al. 1980) (see also data and observations by Gartner mentioned below). The probability of parasite success is obviously decreased when one considers the increased water volume in which fewer organisms are dispersed with the descent from the upper mesopelagic into the bathypelagic zone. Biomass of the mesopelagic zone is intermediate between that of the epipelagic and bathypelagic zones, and fishes living there have variously adapted to energy sources not available to bathypelagic fishes. These include diel vertical migrations into overlying waters where food is more abundant; schooling behavior of midwater fishes themselves provides concentrations of prey for piscivorous predators; the organically rich layer created by water density at the permanent thermocline is believed to attract populations of potential hosts; and the nutrient-rich benthic boundary layer over the continental slope of the northwestern Atlantic probably involves midwater fishes at some depths. In contrast, the bathypelagic zone is food poor, with no density barriers to concentrate nutrients or attract organisms and is inhabited by rather lethargic and mostly nonmigrating fishes that apparently are nonschooling. The stomach contents of some bathypelagic fishes indicate that they encounter the rich benthic layer during feeding forays, whether by vertical or horizontal movement into the continental margin. The short life span (1 or 2 yr) of the majority of these deep-living pelagic fishes provides an important regulatory mechanism over their parasite populations in contrast to the much longer life spans of fishes living in epipelagic and benthic zones.

Gusev (1957) examined five species of bathypelagic fishes (41 specimens) taken from 800 to 5720 m in the northwestern Pacific and found only a single copepod. Specimens of the benthic macrourid *Coryphaenoides acrolepis* taken between 2500 and 7000 m in the same area harbored two species of copepods, a monogenetic and digenetic trematode, two species of cestodes, and nematodes. From these limited data he proposed that parasitism of deep-sea fishes from middepths differed in character from that of benthic fishes. Recognizing that bathypelagic species are dispersed in the largest living space on Earth, he concluded that von Linstow's (1888) hypothesis that parasite reproductive potential is insufficient to overcome the water volume of the deep sea was correct and that few parasites were present because of the population density and variety of available food (intermediate hosts).

Subsequent studies of mesopelagic and bathypelagic species have corroborated the fact, in greater detail, that metazoan parasites are rarely found on or in these fishes. Collard (1968) gave a detailed analysis of data obtained from 1122 "midwater" fishes (35 epipelagic, 953 mesopelagic, 134 bathypelagic) from 19 different localities in the eastern Pacific as far south as Antarctica. Most of the fish were caught off California and Mexico. Of the mesopelagic sample, 486 specimens belonged to the same species, *Stenobrachius leucopsarus,* and 760 of the total fish sampled had no parasites. Nine midwater specimens were hosts for mature digenea. Only 10 fishes were simultaneously infected by four or more kinds of parasites. A maximum of eight different parasites was found in one fish. Protozoans were not collected. The hosts, both meso- and bathypelagic, and numbers of parasites recovered on which conclusions were later based (Collard, 1970; Noble and Collard, 1970) is presented in Table VIII. Collard (1968) stated that the parasites were generally in very poor condition and mostly unidentifiable.

Minor discrepancies exist between Collard (1968) and Collard (1970) (see Tables VIII and IX), but the basic characterization remains the same. General

Table VIII
Summary of Hosts and Numbers of Parasites Recovered from "Midwater" Fishes [Derived from Collard (1968)] (All Helminths Are Juveniles Unless Otherwise Indicated; A = Adult; + Indicates Presence)

	Infected/ Examined	Cy	M	D	C	N	A	Co	Fu
	Mesopelagic Species								
Parmaturus xaniurus	0/10	—	—	—	—	—	—	—	—
Cyclothone signata	0/15	—	—	—	—	—	—	—	—
Danaphos oculatus	0/1	—	—	—	—	—	—	—	—
Argyropelecus lynchus	0/5	—	—	—	—	—	—	—	—
A. pacificus	2/12	—	—	1	—	—	1	—	—
A. olfersi	0/1	—	—	—	—	—	—	—	—
Idiacanthis antrostomus	0/1	—	—	—	—	—	—	—	—
Stomias atriventer	9/15	10	—	—	12	—	—	—	—
Stomiatidae sp.	0/1	—	—	—	—	—	—	—	—
Macroparalepis sp.[a]	3/3	1	—	—	1	4	—	—	—
Microstomus pacificus[b]	0/1	—	—	—	—	—	—	—	—
Stenobrachius leucopsarus	153/486	121[e]	—	3	30	184	—	19	—
Triphoturus mexicanus	20/120	7	—	—	7	10	—	—	—
Ceratosopelus townsendi	13/39	2	—	10	2	11	—	1	—
Diaphus theta[c]	27/61	16	—	—	12	28	—	13	—
D. elucens[d]	5/7								
D. coeruleus	2/2	5	—	1	—	12	—	—	—
Diaphus sp.	2/3								
Myctophum spinosum	4/6	—	—	—	1	1	—	2	—
Diogenichthys laternatus	0/1	—	—	—	—	—	—	—	—
P. crockeri	0/1	—	—	—	—	—	—	—	—
Tarletonbeania crenularis	4/35	—	—	—	3	2	—	—	—
Symbolophorus californiensis	6/35	—	—	—	6	2	—	—	—
Lampanyctus ritteri	34/53	123	3	—	15	41	—	—	+
L. macropterus	1/3	—	—	—	—	1	—	—	—
L. australis	6/35	3	—	—	—	6	—	1	—
Myctophum autolaternatum (?)	1/1	—	—	—	—	1	—	—	—
M. brachygnathos (?)	0/2	—	—	—	—	—	—	—	—
Subtotals fish species 28	291/955	288	3	15	89	303	1	36	
	Bathypelagic Species								
Cyclothone acclinidens	1/21	2	—	—	—	1	—	—	—
C. pallida	1/15	2	—	—	—	—	—	—	—
Sternptyx diaphana	2/17	—	—	—	105+	—	—	—	—
Evermanella sp.	1/1	—	—	—	—	1	—	—	—
Nemichthys scolopaceus	0/1	—	—	—	—	—	—	—	—
Avocettina bowersi	0/1	—	—	—	—	—	—	—	—
Serrivomer sector	5/7	—	—	—	—	6	—	—	—
Melamphaes lugubris	0/1	—	—	—	—	—	—	—	—
M. acanthomus	1/1	1	—	—	—	1	—	—	—

TABLE VIII (*Continued*)

	Infected/ Examined	Cy	M	D	C	N	A	Co	Fu
Poromitra crassiceps	0/2	—	—	—	—	—	—	—	—
Scopelogadus mizolepis	3/10	100[e]	—	7	—	—	—	—	—
Melanostigma pammelas	2/2	—	—	4	—	—	—	—	—
Protomyctophum anderssoni	13/29	6	—	—	12	—	—	—	—
Parvilux ingens	3/21	14	—	—	—	1	—	—	+
Lampanyctus regalis	1/5	—	—	—	—	1	—	—	+
Subtotals fish species 15	33/134	125	0	11	117	10	0	0	
Totals fish species 43	323/1089	413	3	26	206	313	1	36	

[a] Habitat not determined.
[b] Pelagic juvenile.
[c] Discrepancies were found in the number of *D. theta* examined, the total number of mesopelagic hosts, and the total number of fish examined.
[d] Data for *Diaphus elucens* to *Diaphus* sp. combined.
[e] Approximate number.

conclusions made by Collard (1970) were: (1) a general scarcity of adult metazoan parasites (44.4%), especially those with direct development; (2) the helminth fauna is comprised principally of larval nematodes and cestodes; (3) adult fishes have more diverse faunas and heavier infections than do preadults; (4) female fishes are more heavily parasitized than males; (5) seasons exert less effect on the parasite fauna with increasing depth; and (6) mesopelagic fishes serve mainly as transport hosts for larval helminths. It was suggested that invertebrate intermediate host populations are sufficiently abundant to support the parasite fauna of midwater fishes throughout the year, and the only evidence of seasonal change is associated with a seasonal fluctuation in the abundance of an obligatory intermediate host. He also envisioned mesopelagic fishes as serving to disperse parasites to predatory fishes in the epipelagic and bathypelagic zones. Scarcity of parasites was noted despite nonpreferential feeding habits for the fishes. Prey-size preferences were indicated for *Stenobrachius leucopsarus,* but not prey-species selection preferences. Paxton (1967) concluded that this fish showed specific prey selectivity. Larger bathypelagic hosts supposedly acquired cestode larvae by eating smaller midwater fishes, but nematodes supposedly could not survive the fish-to-fish transfer. The latter idea is contrary to the literature, his idea of vertical transport and some of our deep benthic data (see below). In fact, the data show little, if anything, of specific value because more than half of the midwater fishes examined belonged to the same species (486 of 955), most were vertical migrators, and small samples of preserved fishes were pooled from 19 widespread geographic localities. On the other hand, the study is of value in pointing out the relative impermanence of the midwater communities in terms of maintaining parasite populations with complex life histories.

TABLE IX
Parasites of Fishes from Mesopelagic and Bathypelagic Zones
[After Collard (1970)]

Parasites	Mesopelagic		Bathypelagic		Total
	Number	Percent	Number	Percent	
Nematoda (total)	211	22.1	10	7.4	221
Anisakis sp.	82	8.7	6	4.4	88
Contracaecum sp.	48	5.0	1	0.7	49
Paranisakis sp.	2	0.2	1	0.7	3
Terranova sp.	10	1.1	1	0.7	11
Ascarophis sp.	2	0.2	1	0.7	3
Anisakinae	18	1.9	—	—	18
Unidentified	49	5.1	—	—	49
Cestoda (total)	67	7.0	12	8.9	78
Tetraphyllidea	27	2.8	6	4.4	32
Pseudophyllidea	20	2.1	3	2.2	23
Trypanorhyncha	1	0.1	—	—	1
Unidentified	18	1.8	3	2.1	21
Trematoda (total)	13	1.3	4	2.9	17
Monilicaecum sp.	2	0.2	—	—	2
Hemiuridae	5	0.5	4	2.9	9
Metacercaria sp.	3	0.4	—	—	3
Macrovalvitrematidae	3	0.4	—	—	3
Copepoda (total)	33	3.5	0	0	33
Cardiodectes medusaeus	27	2.8	—	—	27
Bomolochinae	3	0.4	—	—	3
Caligidae (*Chalimus*)	2	0.2	—	—	2
Acanthocephala	1	0.1	0	0	1
Fungi	20	2.1	1	0.7	21
Gill cysts	33	3.5	2	1.4	35
Unidentified cysts	45	4.9	10	7.4	55
Number of fishes examined	953[a]		134		1087[a]
Number of fishes parasitized	560[a]	58.7	33	24.6	593

[a]Numbers from text and tables of Collard (1968) indicate totals of 955 examined and 291 parasitized for fishes he considered mesopelagic of 1089 meso- and bathypelagic fishes examined.

In another version of their data, Noble and Collard (1970) assessed the collective metazoan parasite fauna of 1087 (42 species) midwater fishes collected from eastern Pacific waters off southern California to Mexico and compared it with material from the Peru–Chile Trench, the central Pacific, and the Antarctic. Tabulation was made of protozoan genera found in 594 fishes, including some macrourids. Twenty-eight percent were infected. Among metazoan parasites, three monogenea were found on three *Lampanyctus ritteri*, 14 of 1087 harbored digenea (hemiurids and one fellodistomid), one mesopelagic and one bathypelagic species had adult digenea, 221 (20.3%) contained nematode larvae, 77 (7%) harbored cestodes of which only three were adults (pseudophyllideans), and 33 had copepods (5 species), of which *Cardiodectes medusae* occurred 27 times. None of the nematodes or cestodes could be identified to species, but the nematode genera (*Anisakis, Contracaecum, Terranova, Paranisakis*) and cestode orders (Pseudophyllidea, Tetraphyllidea, Trypanorhyncha) reported are commonly found as larvae or adults among many teleosts, elasmobranchs, and marine mammals of the epipelagic zone. Mamaev and Parukhin (1975) and Mamaev (1976) described new genera of diclidophorid monogenea from midwater fishes. Gartner (personal communication) found larval *Botulus* (D) and *Ceratobothrium* (C) in midwater fishes. *Ascarophis, Anisakis, Thynnascaris* and *Contracaecum* are among the nematodes common in deep benthic fishes (Campbell et al., 1979). Noble and Collard stated that "The presence of only three adult tapeworms out of about 1100 fish makes it clear that mesopelagic fishes do not normally serve as definitive hosts for cestodes." In a communication to them Berland suggested that the larval nematodes were using the midwater fishes as paratenic hosts. This observation may be extended to other larval helminths as well, as suggested by the abundance and ubiquitous occurrence of phyllobothriid tetraphyllidean plerocercoids in their collection. Phyllobothriids are the most commonly encountered tetraphyllideans among elasmobranchs in general. Furthermore, the hemiurids and fellodistomids are among the most common digenea in marine teleosts. Collard's (1970) speculation that midwater fishes are suitable definitive hosts for tetraphyllideans but are too small and short-lived (1 year) for adult cestode populations to be realized is doubtful because most cestode populations show annual fluctuations, usually in response to temperature extremes. The meso- and bathypelagic fishes I have examined (Table X) were capable of easily swallowing the prey (cumaceans and gammarid amphipods) that I found in the stomachs of young *Raja erinacea* (14 cm total length). The little skates were infected with mature *Echeneibothrium* at this early age. Small midwater fishes, even *Myctophum punctatum* of 5 cm total length, have mouths as large as these small skates. In addition, the infection rate noted by Noble and Collard (1970) was not unusually high (32 of 1087 fishes) for fishes supposedly of nonpreferential diet, or different in terms of location of plerocercoids (intestinal lumen) when compared with the overwhelming numbers encountered seasonally in inshore teleosts or their occurrence in long-lived deep-sea benthic fishes (see below). Growth of the larvae without further development indicates that midwater fishes are paratenic rather than definitive hosts. The post-larvae of *Pelichnibothrium speciosum* from *Alepisaurus ferox,* a large midwater fish, undergo strobilization and maturation, but the tail-like bladder of a larva is still retained and the only report of an adult was from the great blue shark, *Prionace glauca* (Yamaguti, 1934). I have collected several *P. speciosum* with up to 124 proglottids but none contained ova. Yamaguti (1959) lists this cestode larva from cephalopods and the teleosts *Lampris regia* and *Thunnus thynnus* (L.). Extensive growth of nematode larvae is also evident in deep-sea fishes.

Mary H. Pritchard (University of Nebraska, Lincoln; personal communication) and Juan Carvajal G. (Universidad Catolica de Chile, Santiago) examined 152 fishes (representing 30 families, 40 genera, and 44 species) for metazoan parasites from the Peru–Chile trench. Of a total of 104 (68%) uninfected fishes, 90 were midwater specimens. Pritchard also noted that this and data from three previous cruises from which midwater fishes were examined demonstrated that the major-

TABLE X
Parasitism in Some Meso- and Bathypelagic Fishes from North Atlantic[a]

Species	Infected/Examined		Protozoa	Nematoda	Cestoda	Trematoda
Diaphnus splendidus	20	26	16	15(22)	—	—
Argyropelecus aculeatus	0	5	—	—	—	—
Symbolophorus veranyi	1	4	1	—	—	—
Gonostomum elongatum	0	6	—	—	—	—
Myctophum punctatum	0	20	—	—	—	—
Benthosema glaciale	0	20	—	—	—	—
Diaphnus holti	3	20	—	—	2(2)	1(1)
Lampanyctus macdonaldi	1	1	—	1(1)	—	—
Melanostigmum sp.	1	1	—	1(2)	1(7)	—
Penopus macdonaldi	1	1	1	—	—	—
Alepisaurus ferox	1	1	—	—	1(2)	—
Chauliodus danae	0	11	—	—	—	—
Eurypharynx pelecanoides	11[b]	60	—	1(1)	11(15)	—
Bathylagus euryops (?)	0	3	—	—	—	—
Malacosteus niger	0	2	—	—	—	—
Anoplogaster cornutus	1	2	—	1(1)	—	—
Stomias boa ferox	0	50	—	—	—	—

[a]Numbers indicate the number of infected fish followed by actual numbers of parasites in parentheses. All helminths are larvae except the single hemiurid (D) from *Diaphnus* and the pseudophyllidean (C) from *Eurypharynx*.
[b]All hosts occurred between 23° and 40°N (Campbell and Gartner, Jr., 1982).

ity of midwater fishes are not infected by metazoan parasites and that when helminths were present, they were larval. Adult digenea were found only in *Binghamichthys aphos* (twice), indicating that it is a normal definitive host. Pritchard surmised that their collections indicated a generally sparse fauna in waters underlying the Humboldt Current in the region studied (23°30′S, 70°30′W to 12°5′S, 75°10′W) and stated that "Hydrographic data indicated an exceedingly low oxygen content which accounts for virtual sterility in certain waters." Results of my examinations of meso- and bathypelagic species (Table X) echo the findings of others. Anisakid nematodes are by far the most common helminths encountered among epipelagial migrants along the Chilean coast between Puerto Montt and Antofagasta, Chile (Campbell and Carvajal, unpublished data). Digenea, acanthocephalans, and cestodes were rarely encountered, indicating the importance of a continental shelf benthos and fish populations with more stationary behavior to the proliferation of these helminth populations.

Collard (1970) stated that bathypelagic fishes were more heavily parasitized than mesopelagic fishes (Table IX). The 134 bathypelagic types he examined (Table VIII) actually were insufficient as a basis for such a conclusion for digenetic trematodes (4 fish) or cestodes (12 fish) when compared with samples of midwater species. Noble and Orias (1975) and Orias et al. (1978) found bathypelagic fishes rather parasite poor and ecologically quite different from benthic species. Gusev (1957) made similar comments. From studies of *Melanostigma pammelas* off southern California and 84 fishes (four species) from the northern Atlantic, Noble and Orias (1975) and Orias et al. (1978) claim evidence of a relatively high percentage of infection by adult digenetic trematodes and low incidences of other helminths and protozoa than in midwater fish. Eighty of the 84 fishes examined in the latter study were not infected, and the six digenea, *Lecithophyllum irelandeum,* were obtained from one fish. More evidence and larger samples are needed from future studies to accurately indicate any true differences in community interaction for fishes in this zone. The apparent lack of most parasites from bathypelagic fishes provides a sharp contrast to the richness of the parasite fauna found in fishes of the benthic boundary layer below.

Markle and Wenner (1979) used digenea in *Melanostigma atlanticum* and *Xenodermichthys copei* as partial evidence of demersal spawning. On examining their specimens from *X. copei,* I found them to be *Steringophorus blackeri* (= *Abyssotrema* sp.) and the hemiurid *Dinosoma* sp. Bray (personal communication) has found the same digenea in *X. copei* off Scotland. Their assumption of association of these fish with ophiuroids (intermediate host for some fellodistome digenea) because of the presence of *S. blackeri* was not verified by stomach content data and did not take into consideration the fact that jelly animals also serve as intermediate hosts for fellodistomes.

The findings on parasitism and feeding habits of meso- and bathypelagic fishes by J. Gartner (personal communication) are not in complete accord with those of Collard or Noble and co-workers (see references cited). Gartner found 303 of 668 (45.4%) fishes of 18 species (11 families; 16 genera), taken in the area of Norfolk Submarine Canyon in the northwestern Atlantic, parasitized by metazoans or fungi. Only a single species, the bathypelagic eel *Serrivomer brevidentatus,* was not parasitized. Larval helminths, mostly tetraphyllidean and trypanorhynch cestodes and anisakid nematodes (*Anisakis, Thynnascaris, Contracaecum*), were most commonly encountered. All the trypanorhynch larvae belonged to a single family, Tentaculariidae. Unlike the conclusions of the aforementioned workers, he found that trypanorhynch cestode larvae were very common and that bathypelagic fishes (149 fish of 6 species) had fewer parasites and a less diverse fauna than did the 504 (12 species) mesopelagic fishes he examined (excluding 15 fish of 2 species he considered transients). By subdividing the fish species according to depth and taxon, he found infection rates at all levels in the mesopelagic zone to be about two- to fourfold greater than for hosts in the bathypelagic zone (average 63.7 vs. 15%). Gartner obtained some adult digenea (31 of 33 infections), nematodes (8 of 41 infections), and cestodes (2 of 151 infections) among all hosts. Forty-three fish (6.4% in 5 fish species) had internal fungal infections, but only two of the 668 examined carried parasitic copepods. No monogenea were found and only a single acanthocephalan (adult), *Echinorhynchus* sp., was discovered. Overall helminth incidence among mesopelagic hosts was 41.6% (210 of 504). Only 6.1% of Gartner's mesopelagic hosts were infected with nematodes, in contrast to the 19.6% reported by Collard (1968); however, only 18 specimens of Gartner's hosts were surface-migrating myctophids, whereas 944 of Collard's fish were vertical migrators. Gartner's results are probably a more realistic characterization of parasitism in midwater fishes because of the deeper dwelling, nonmigratory nature of the fishes he sampled. A maximum of seven parasite species was recovered from any of the 18 host species he examined. Of particular interest are his observations of reduced helminth infections in bathypelagic hosts (10%; 15 of 149) and their diets, which included mostly crustaceans and some bottom-living organisms (clam, polychaetes, holothurian, and echinoderm fragments) instead of fish. Thus Collard's (1970) belief that bathypelagic fishes are piscivorous is contraindicated by Gartner's data. In fact, Collard's (1968) stomach content data from 38 bathypelagic hosts also indicated that about 20% contained crustacean fragments and less than 10% contained fish remains. Evidence of benthic feeding by fishes usually found well above the bottom was also noted by Marshall (1954; sea urchin) and may be the result of a horizontal encounter with the continental slope instead of a vertical migration.

Wolff (1977) has reviewed the ideas proposed to explain the diversity in the deep-sea benthos. In all areas studied the most varied and abundant life is found on the continental slope and decreases with depth and distance from land. An abundance of invertebrates live below 200 m, with an abundant and diverse assemblage of benthopelagic fishes concentrated along the continental slopes to a depth of about 2000 m (Haedrich and Rowe, 1977). Some 300 species of macrourids, 65 species in the Atlantic, and 150 species of brotulids comprise more than half of the benthopelagic piscine fauna. All but a few macrourids live just above the bottom, where they dominate as some of the larger predators on the sea floor. Of the four subfamilies of macrourids more information has been obtained for the more common Macrourinae (270 species) than for bathygadine macrourids. Possible roles in trophic ecology center around food webs in the deep-sea benthic boundary layer. The roles include (1) a few species that transport energy from midwaters to bottom by vertical migrations, (2) others scavenging on detritus that falls from waters above them, and (3) those involved directly in predation on the benthic fauna as benthic pelagic species found just off the bottom. Feeding behavior may be directly related to the amount of available energy with increasing depth and distance from land. The parasite fauna of macrourids and other deep-sea fishes in general has not received thorough taxonomic or ecological study and only recently has begun to receive attention. The majority of parasites described to date from deep-sea hosts are from macrourids, thereby providing some information about the involvement of this group and serving as a base for zoogeographic study.

Campbell et al. (1980) have shown conclusively that deep-living benthic fishes to 5000 m in the New York Bight (39 to 40°N, 70 to 72°W) are heavily parasitized. A total of 1763 fishes of 52 species (22 families) were taken from 97 stations in the environs of Hudson Submarine Canyon (Fig. 8) as a part of benthic faunal studies, summarized in Table XI. Eighty percent of the fishes examined were infected with 1 to 12 ($\bar{x}$ = 2.6) species of helminths. About 17,300 helminths (80 species) and 137 copepods (8 species) were recovered. The helminth fauna (Table XII) is well represented in nematodes (53.9%), digenetic trematodes (47.7%), cestodes (21.5%), monogenetic trematodes (12.7%), and acanthocephalans (3.7%). Our study concentrated on comparison of the parasite faunas of the 10 dominant

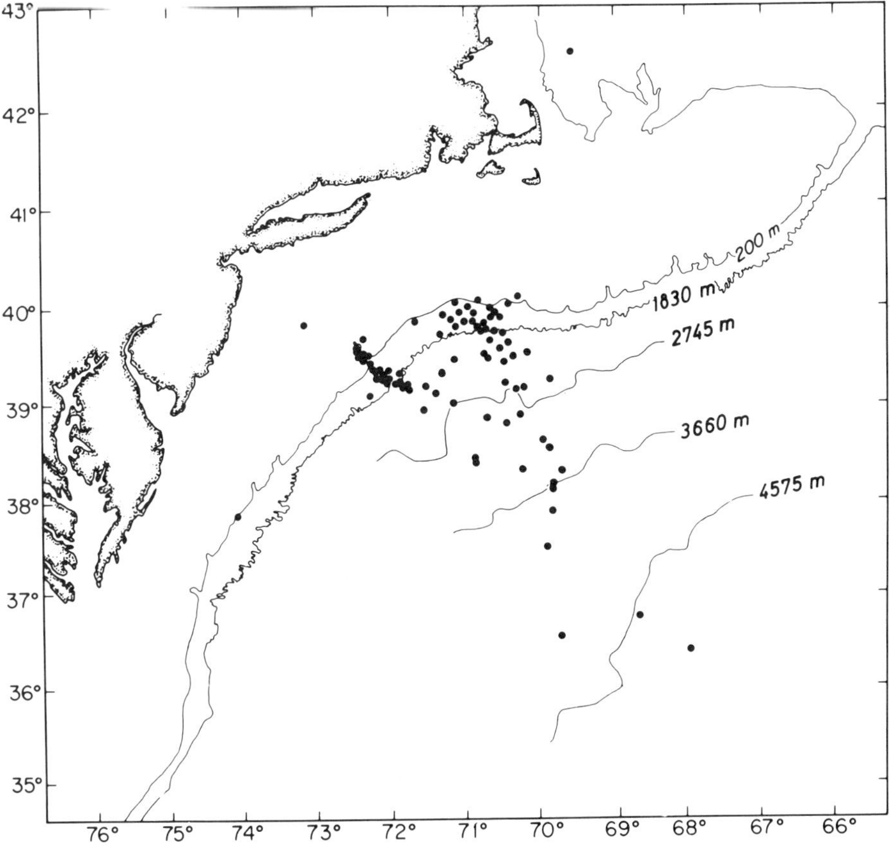

Fig. 8. Chart of positions of deep-ocean trawl stations that took fishes examined for parasites in the New York Bight. [Reprinted with permission of Springer-Verlag New York, Inc. (*Marine Biology*) from Campbell, R. A., R. L. Haedrich, and T. A. Monroe, 1980. Parasitism and ecological relationships among deep-sea benthic fishes. *Mar. Biol.*, **57**, 302.]

TABLE XI
Summary of Parasite Data for Fishes Examined in This Study, New York Bight Area[a]

Fish Species	Infected/Examined	M	D	C	N	A	Co
Centroscyllium fabricii	0/1	—	—	—	—	—	—
Raja bathyphila	3/3	—	—	3(15)	2(4)	—	—
R. jenseni	1/1	—	—	—	1(9)	—	—
R. radiata	1/1	—	—	1(12)	—	—	—
R. spinicauda	1/1	—	—	1(72)	—	—	—
Bathyraja richardsoni	1/1	—	—	1(10)	1(10)	—	—
Veneficia procera	0/4	—	—	—	—	—	—
Omochelys cruentifer	1/10	—	—	—	1(1)	—	—
Nemichthys scolopaceus	1/1	—	—	—	1(1)	—	—
Histiobranchus bathybius	0/2	—	—	—	—	—	—
Synaphobranchus kaupi	20/32	4(20)	1(1)	4(6)	18(63)	—	5(5)
Aldrovandia affinis	2/7	—	1(1)	1(1)	—	—	—
A. phalacra	5/33	—	—	4(6)	1(1)	—	—
Halosauropsis macrochir	41/47	—	39(446)	4(7)	24(100)	—	—
Notacanthus chemnitzi	0/1	—	—	—	—	—	—
Polyacanthonotus africanus	1/2	—	1(33)	—	1(2)	—	—
P. rissoanus	2/2	—	2(25)	—	2(2)	—	—
Alepocephalus agassizi	62/62	30(54)	62(2671)	37(180)	24(46)	—	10(16)
A. productus	1/2	—	1(1)	—	—	—	—
Bathytroctes koefoedi	1/1	—	—	1(1)	—	—	—
Bathypterois phenax	1/3	—	—	—	—	1(2)	—
Bathysaurus agassizi	7/9	—	1(6)	2(2)	7(54)	—	—
Benthosaurus grallator	0/1	—	—	—	—	—	—
Lophius piscatorius	1/3	—	—	—	1(2)	—	—
Antimora rostrata	112/124	4(6)	83(749)	13(20)	89(393)	—	4(4)
Phycis chesteri	47/52	7(11)	34(262)	9(87)	27(178)	—	—

Urophycis chuss	89/98	45(138)	7(9)	2(2)	76(699)	—	23(38)
U. regius	2/3	1(6)	1(2)	—	2(6)	—	—
Merluccius albidus	1/1	1(1)	—	—	1(1)	—	—
Dicrolene intronigra	72/81	—	54(252)	12(21)	17(39)	50(1551)	—
Lepophidium cervinum	39/68	—	—	1(1)	38(128)	—	—
Penopus macdonaldi	0/1	—	—	—	—	—	—
Porogadus milesi	2/5	—	2(6)	—	—	—	3(8)
Lycenchelys paxillus	19/25	—	10(19)	—	15(26)	—	—
L. verrilli	38/104	—	31(213)	—	15(47)	—	—
Lycodes atlanticus	11/11	—	6(16)	—	5(10)	6(91)	—
L. esmarki	1/1	—	—	1(1)	—	—	—
Melanostigma atlanticus	1/1	—	—	1(7)	1(2)	—	—
Coryphaenoides armatus	181/213	5(6)	112(543)	106(557)	115(647)	—	6(14)
C. carapinus	209/222	7(10)	182(917)	104(305)	140(831)	3(11)	13(37)
C. leptolepis	17/20	—	16(105)	3(9)	11(144)	—	1(2)
C. brevibarbis	4/5	—	1(3)	—	3(10)	—	1(2)
C. rupestris	36/36	14(47)	35(288)	5(6)	12(36)	—	—
C. guentheri	2/2	—	2(3)	—	2(7)	—	—
Coelorhinchus carminatus	3/4	—	3(6)	2(16)	1(1)	—	—
Nezumia aequalis	19/25	1(1)	15(40)	3(7)	9(38)	—	—
N. bairdi	323/378	105(336)	128(533)	52(152)	271(2722)	6(6)	11(11)
Ventrifossa occidentalis	1/2	—	—	1(3)	1(3)	—	—
Acanthochaenus lutkenii	3/3	—	2(4)	3(22)	—	—	—
Helicolenus dactylopterus	3/20	—	2(2)	1(1)	—	—	—
Paralichthys oblongus	2/2	—	—	—	2(10)	—	—
Glyptocephalus cynoglossus	18/26	—	7(13)	1(2)	14(36)	—	—
Total fish species 52	1408/1763	224(636)	841(7169)	379(1531)	951(6309)	66(1661)	77(137)

[a] Entries represent the number of fish infected in each taxon and the number of parasites recovered in parentheses (M, Monogenea; D, Digenea; C, Cestoda; N, Nematoda; A, Acanthocephala; Co, Copepoda).

TABLE XII

Comparison of Parasitism of Fishes from Inshore and Deep-Water Habitats (Entries = Number Infected; Percent Infected Given in Parentheses)

	Tortugas[a]		New York Bight					
	Inshore	Deep	Nonmacrourid Teleosts	Macrourids	All spp.	Gulf and Caribbean Macrourids[b]	Mesopelagic[c]	Bathypelagic[c]
Number fish spp.	237	90	37[d]	9	52	22	28	15
Number examined	2039	721	850	905	1763	275	953	134
Percent infected	77	80	71.4	87.7	79.9	90	59	25
Monogenea	—	—	92	132	224	14	3	0
	—	—	(10.8)	(14.6)	(12.7)	(5)	(0.4)	0
Digenea	1835	264	347	494	842	165	10	4
	(90)	(37)	(40.8)	(54.6)	(47.7)	(60.0)	(1.1)	(2.9)
Cestoda	—	183	98	275	379	220	67	12
	—	(25)	(11.5)	(30.3)	(21.5)	(80.0)	(7.0)	(8.9)
Nematoda	—	127	383	564	951	206	211	10
	—	(18)	(45.0)	(62.3)	(53.9)	(80.0)	(22.1)	(7.4)
Acanthocephala	—	13	57	9	66	7	1	0
	—	(2)	(6.7)	(0.9)	(3.7)	(2)	(0.1)	0
Copepoda	—	—	45	32	77	—	33	0
	—	—	(5.3)	(3.5)	(4.4)	—	(3.5)	0

[a] Manter (1934, 1947).
[b] Armstrong (1974).
[c] Collard (1970).
[d] Deep demersal species.

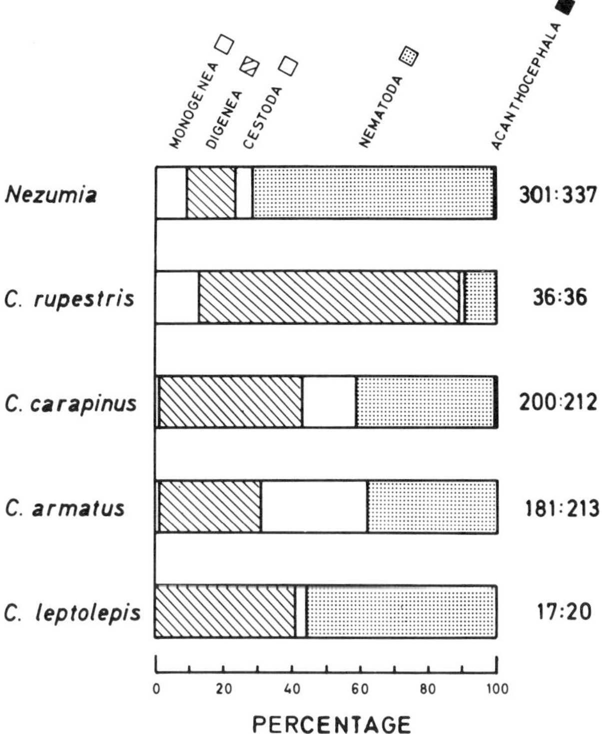

Fig. 9. Percentage occurrence of the major groups of helminth parasites in macrourid fishes, *Nezumia bairdi* and four species of *Coryphaenoides*. Fishes arranged according to depth of occurrence. The ratio to the right is number infected:number examined. [Reprinted with permission of Springer-Verlag New York, Inc. (*Marine Biology*) from Campbell, R. A., R. L. Haedrich, and T. A. Monroe, 1980. Parasitism and ecological relationships among deep-sea benthic fishes. *Mar. Biol.*, **57**, 304.]

species of deep-water bottom fishes of the New York Bight (Figs. 9 and 10). These were the dominant species from the full range of depths sampled and represent a diversity of feeding strategies among fishes living in association with the bottom. The macrourids (rattails), the dominant family, provided the greatest number of related species and specimens for comparison. Predominance of species in order of increasing depth was *Nezumia bairdi, Coryphaenoides rupestris, C. (Lionurus) carapinus, C. (Nematonurus) armatus,* and *C. (Chalinura) leptolepis.* Samples included many juvenile fish less than 10 cm in total length for comparison with adults taken both within submarine canyons and on the adjacent continental slope and rise.

Individual species of fish showed marked differences in the composition of their parasite faunas (Figs. 9 and 10) that appear directly related to diet. Generalized predators, such as *Halosauropsis macrochir,* which feeds on benthos (McDowell, 1973), and *Antimora rostrata,* which appears to feed mostly on pelagic food (Sedberry and Musick, 1978), have more than 60% of their total faunas comprised of digenetic trematodes. Digenea were more abundant and diverse in *Antimora, Alepocephalus,* and *Halosauropsis. Antimora rostrata* and

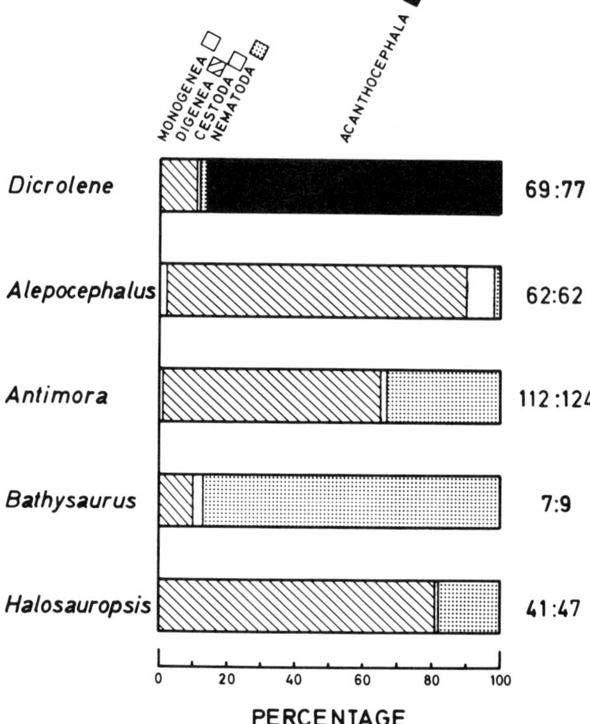

Fig. 10. Percentage occurrence of the major groups of helminth parasites in five unrelated species of deep-ocean fishes, *Dicrolene intronigra, Alepocephalus bairdii, Antimora rostrata, Bathysaurus agassizi,* and *Halosauropsis macrochir*. Species arranged according to depth of occurrence. Ratio to the right is number infected : number examined. [Reprinted with permission of Springer-Verlag New York, Inc. (*Marine Biology*) from Campbell, R. A., R. L. Haedrich, and T. A. Monroe, 1980. Parasitism and ecological relationships among deep-sea benthic fishes. *Mar. Biol.*, **57**, 304.]

Halosauropsis macrochir shared congeneric helminths with up to seven other fishes, indicating that these two species shared more types of prey with all fishes than any others in the study. Macrourids have generalized feeding habits, and their helminth fauna consists of many nematodes, cestodes, and digenetic trematodes. The higher incidences and worm burdens of macrourids indicated that this group of teleosts was more frequently infected, but nonmacrourid teleosts carried the bulk of the parasites recovered (Table XII). The overlaps in diet between the species of *Coryphaenoides* indicated by Sorenson's similarity index is in agreement with the findings of Pearcy and Ambler (1974) that small *C. armatus* and *C. leptolepis* have broad overlaps in diet, not unlike overlaps in the diets of shallow-water demersal fishes. It is interesting to note that the degrees of overlap in the parasite faunas by parasite groups (Fig. 9) and helminth species shared of co-occurring rattails are 10 to 57%. This fact, compared with stomach content analyses (Haedrich and Henderson, 1974; Haedrich and Polloni, 1976; Sedberry and Musick, 1978), indicates the reduction of competition by varying degrees of food selection among these generalized feeders. *Nezumia bairdi, C. carapinus,* and *C. leptolepis* feed to a greater extent on benthos than do *C. armatus* or *C. rupestris*. Haedrich and Polloni (1976) showed *C. carapinus* to be

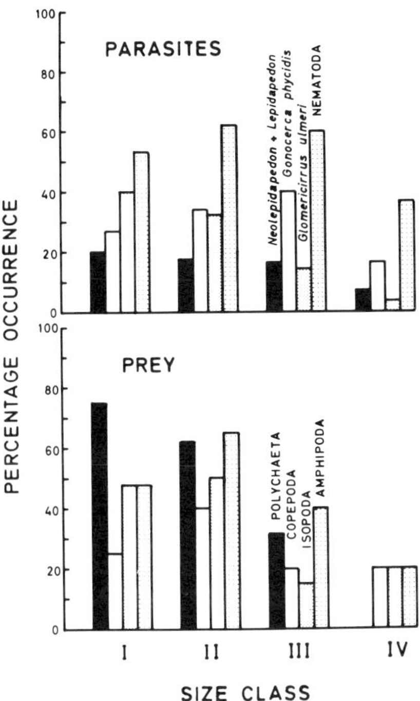

Fig. 11. Percentage occurrence of certain helminth parasites and the prey with which these may be associated in *Coryphaenoides armatus*, arranged according to size (snout to anal fin distance) of fish: (I) 69 to 99 mm; (II) 100 to 139 mm; (III) 140 to 179 mm; (IV) 180 to 229 mm. The prey here tend to be benthic [data from Haedrich and Henderson (1974)]. *Glomericirrus ulmeri* is a junior synonym of *G. macrouri* (= *Hemiuris macrouri* Gaevskaya, 1979). [Reprinted with permission of Springer-Verlag New York, Inc. (*Marine Biology*) from Campbell, R. A., R. L. Haedvich, and T. A. Monroe, 1980. Parasitism and ecological relationships among deep-sea benthic fishes. *Mar. Biol.*, **57**, 307.]

selective for amphipods and ophiuroids, particularly *Ophiura ljungmani*, whereas *C. armatus* and *C. rupestris* feed off the bottom. Thus the parasite data (Figs. 9 and 10) and the feeding habits of *Coryphaenoides* spp. together indicate modest radiation within overlapping yet distinctive adaptative zones, with emphasis on different prey species an important consideration.

Coryphaenoides armatus is a generalized feeder that undergoes a change in diet as it grows. Young individuals feed almost entirely on benthic animals, but as the fish grows there is a shift in diet to more pelagic organisms, especially cephalopods and fishes. Large individuals may move considerable distances off the bottom to forage. Declines in occurrence of helminths with host size appeared to result from changes in the diet of *C. armatus* (Figs. 11 and 12) corresponding to the decreased ingestion of suitable intermediate hosts. Decreased ingestion of polychaetes, isopods, amphipods, and small crustacea paralleled decreased incidence and relative abundance of trematodes and nematodes (Figs. 11 and 12). Cestode larvae, *Grillotia rowei*, gradually accumulated until they dominated the fauna of larger hosts.

Parasite recruitment through the life history of some fishes indicates more consistent feeding habits. *Gonocerca phycidis*, *G. haedrichi*, *Glomericirrus macrouri*, *Grillotia rowei*, and *Thynnascaris* sp. parasitized both *C. armatus* and *C.*

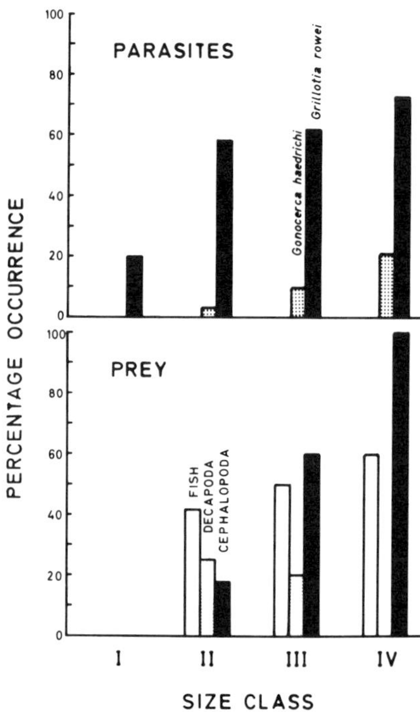

Fig. 12. Percentage occurrence of certain helminth parasites and the prey with which these may be associated in *Coryphaenoides armatus*, arranged according to size (snout to anal fin distance) of fish: (I) 69 to 99 mm; (II) 100 to 139 mm; (III) 140 to 179 mm; (IV) 180 to 229 mm. The prey here tend to be pelagic; data from Haedrich and Henderson (1974) [Reprinted with permission of Springer-Verlag New York, Inc. (*Marine Biology*) from Campbell, R. A., R. L. Haedrich, and T. A. Monroe, 1980. Parasitism and ecological relationships among deep-sea benthic fishes. *Mar. Biol.*, **57**, 307.]

carapinus. However, the trend for *C. carapinus* involved increasing parasite diversity and abundance with host size, which remained relatively constant through maturity. This is consistent with the findings of Haedrich and Polloni (1976) indicating the same benthic food preferences throughout the life of *C. carapinus*. *Nezumia bairdi*, and *Alepocephalus agassizi* show a similar consistency in diet and recruitment of their parasite faunas with increased size. The helminth fauna of *C. rupestris* was less diverse (7 species) than that of other rattails studied but suggests a pattern of increasing recruitment and consistent diet as noted for *C. carapinus* and *N. bairdi*. *Coryphaenoides leptolepis*, a very deep living rattail, undergoes a diet change from benthic to pelagic invertebrates as it grows, but the scarcity of specimens prevented indication of any trend for the parasite fauna.

Two of the more specialized feeders were *Dicrolene intronigra* and *Bathysaurus agassizi* (Fig. 10). *Dicrolene* had a fauna comprised almost entirely of acanthocephalans (80%) and one numerically dominated by a single helminth species (*Acanthocephalus* sp.). The acanthocephalan infestations in *Dicrolene* suggest that this fish fed extensively on amphipods, the known intermediate hosts for these helminths. *Bathysaurus*, a large piscivorous predator that feeds on fishes and decapod crustaceans, had a helminth fauna dominated by nematodes (87%). The poor helminth fauna (5 species) of *Bathysaurus* contrasts markedly with the rich and varied faunas typical of inshore, piscivorous fishes (Polyanski, 1958).

This might indicate that the animals it preys on are not important intermediate hosts for deep-sea helminths such as digenea. On the other hand, we do know that *Bathysaurus* bites the tails off macrourids. By not eating the body of its prey, this fish would avoid infection by most helminths.

Alepocephalus agassizi may feed on a wide variety of macroplankton, especially coelenterates, as does *A. bairdii* (Golovan and Pakhorukov, 1975), but we have recovered a variety of benthic organisms (polychaetes, decapod and crustacean fragments, gastropod shells, a benthic foraminiferan) and sediment from this fish. *Alepocephalus* overlapped little in helminth fauna with other fishes except *Antimora,* suggesting that it exhibits some prey selectivity. Some benthic feeding for *A. agassizi* is suggested by the numerous zoogonids, *Hudsonia agassizi,* commonly encountered in this host. Known zoogonid cycles result in the production of a cercariaeum (tailless cercaria) that must crawl on the bottom to reach its polychaete, urchin, or benthic crustacean second intermediate hosts (Yamaguti, 1975). The fellodistome *Steringophorus pritchardae,* like the three species of hemiurids and two pseudophyllideans harbored by *Alepocephalus,* all suggest the taking of pelagic prey.

Comparison of this data with the studies by Manter (1934, 1947, 1954), Yamaguti (1934, 1938), and others indicates that helminth parasitism of deep-sea benthic fishes is not peculiar in terms of abundance or specificity compared to fishes living in shallow water. However, we found evidence of a lesser parasite diversity. Host size (age) and the environmental gradient provided by the canyon and transition to the depths of the abyssal plain, is apparent in both parasite and free-living populations. Resolution of questions of alpha (within-habitat) and gamma (between-habitat) diversity remain intriguing and elusive at present.

Infection rate indicates that parasite success in the deep ocean is directly related to the abundance and diversity of hosts in the benthic community. Recent data (Wolff, 1977) indicate that biomass and population density undergo greatest changes with depth. Fish, like *C. armatus,* captured on the lower rise or abyssal plain typically do not harbor as many or diverse infections as do individuals from the lower slope and upper rise regions (Fig. 13). Larger *C. armatus* tend to be found with increasing depth [''bigger–deeper'' relationships; see Polloni et al. (1979)], and the incidence of infection generally decreases in this species with increasing host size ($p < 0.01$). Infections decreased as the fish grew and turned more to pelagic animals as food (Haedrich and Henderson, 1974).

Catch rates of epibenthic macrofaunal animals were slightly greater in Alvin and Hudson Submarine canyons than at comparable depths on the slope (Haedrich et al., 1975). The contrast was also suggested by differences in the faunas of hosts taken from comparable depths within and outside the canyon. For example, *C. armatus* 41 to 50 cm long were more frequently infected (100 vs. 78%; $p < 0.05$) and carried more parasites (24 vs. 7 helminths per host; $p < 0.001$) than did those from outside the canyon.

Infection rates of all size groups of *Nezumia bairdi* from the canyon were equal to or 5 to 18% greater than those of comparable-sized fish from the slope. Analysis of three size groups of *Antimora rostrata* from the canyon and slope showed that 41 to 88-cm fish from the canyon harbored more species per host than did comparable-sized hosts collected on the slope. The abundance and incidence of parasites with complex life histories are directly related to the abundance of the benthic

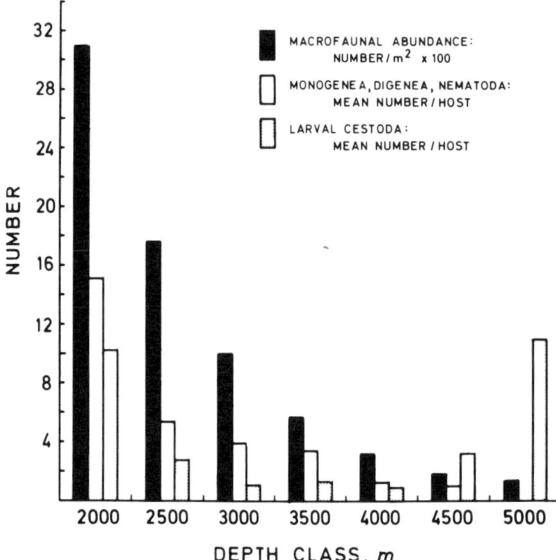

Fig. 13. Rates of infection by helminth parasites in *Coryphaenoides armatus* and overall abundance of macrofaunal invertebrates, arranged by depth. The macrofaunal values are calculated from a regression in Rowe et al. (1982) for Hudson Submarine Canyon. The numbers of *C. armatus* examined in each depth class, from shallow to deep, are 17, 86, 85, 11, 10, 3, and 1. [Reprinted with permission of Springer-Verlag New York, Inc. (*Marine Biology*) from Campbell, R. A., R. L. Haedrich, and T. A. Monroe, 1980. Parasitism and ecological relationships among deep-sea benthic fishes. *Mar. Biol.*, **57**, 311.]

fauna as a whole and decline dramatically with depth (Fig. 13). Parasites without intermediate hosts (monogenetic trematodes) do not vary in abundance until the lower slope and abyssal plain are reached (4000 to 5000 m), where they disappear. Similarities of the helminth faunas indicate generalized predation for most hosts, with some varying degrees of trophic selectivity. This observation is in keeping with the prediction by Dayton and Hessler (1972) that deep-ocean predators should tend to be generalists.

Proliferation of digenea in shallow- versus deep-water environments is demonstrated by comparison of the families : genera : species ratios recovered from several large surveys and compiled lists based on records from Maine to Florida. Manter (1947) found 18 : 106 : 189 digenea in 237 fish species (2039 examined) from shallow water at Tortugas, Florida. Lists compiled from the literature for Beaufort, North Carolina yield 13 : 40 : 50; Bermuda, 18 : 42 : 52; and Woods Hole, 14 : 52 : 77. For deep water, we find Manter (1934), 8 : 32 : 49 at Tortugas, Florida; Armstrong (1974), 4 : 17 : 32 in the Gulf and the Caribbean; and Hudson Canyon, 8 : 24 : 41. Prudhoe and Bray (1973) found a distribution of 5 : 14 : 24 among 37 species of benthic fishes in the Antarctic. As extensive as these lists are, they cover only major works and my own data but clearly show that the contrast with the deep sea would only be increased were every record included. The implication that the deep-water parasite fauna in absolute numbers of taxa is less diverse than that of shallow-water areas is obvious. Only four families of digenea predominate in deep-water demersal fishes of the New York Bight—the hemiurids, fellodistomes, lepocreadiids, and opecoelids. These families are also well represented in shallow water, but not in the same relative proportions. Members of other families (bucephalids, acanthocolpids, cryptogonomids, haplosplanchnids, didymozoids, azygiids, zoogonids, monorchiids, gorgoderids, echinostomes, etc.) are also present and, in some cases, well represented. Among the families of digenea commonly encountered in deep-sea hosts, relatively few of the known genera seem to be particularly common. Most of these same genera and some of the species can be found in shallower living hosts as well (*Gonocerca, Dinosoma, Genolinea, Dissosaccus, Hemiuris, Derogenes, Lepidapedon, Steringophorus, Podocotyle, Plagioporus*). Relatively few genera described are unique to deep-sea hosts thus far. However, despite their more select group of species, relatively few are found in both shallow- and deep-sea habitats. Notable exceptions are hemiurids (*Gonocerca phycidis, Derogenes varicus, Hemiuris communis, Genolinea laticauda, Lecithophyllum botryophoron*), a lepocreadiid (*Lepidapedon elongatum*), and fellodistome (*Steringophorus furciger*). Nematodes of the genera *Thynnascaris, Ascarophis,* and *Paranisakiopsis* are common. *Thynnascaris aduncum* is widespread in shallow waters, as are acanthocephalans of the genera *Echinorhynchus* and *Acanthocephalus*. The most widely distributed are the hemiurids and in all probability are dispersed by pelagic intermediate hosts. Four trematode species were found in macrourids from the Gulf and Caribbean and at Tortugas, Florida. Hudson Canyon hosts share four species with each of these studies, and it appears that several species unnamed by Armstrong are conspecific with Hudson Canyon species. Bray (personal communication) has found 11 species that are conspecific or very close to those found in deep-sea hosts off the Atlantic coast of the United States. His fishes were caught in subarctic waters and the parasites are from either the same or related hosts as those from Hudson Canyon.

Manter (1934) found 61% of the trematodes from deep water at Tortugas restricted to a single host (Table XIII) and about 56% in shallow water. Linton's (1940) studies in continental waters at Woods Hole showed about 54% specific for a single host species. Deep-sea hosts of Hudson Canyon were much more specific for a single host (73%), a result that is very similar to the degree of specificity derived from Armstrong's study of macrourids (71.8%). Rohde (1978) found data from 15 shallow-water surveys of marine digenea [average of 91 trematode species and more than 80 (25 to 237) fish species per survey] show a decrease in host-specificity with increasing latitude. Sea temperature was used as an approximate indicator of changing specificity. In the physical stability of the deep-sea where annual changes in temperature are fractions of a degree it would appear that such a gradient may not exist or, if present, be due to biological factors unrelated to temperature at depth (Table XIII). For marine monogenea (12 surveys averaging 52 species of monogenea and more than 49 fish species per survey) Rohde found no gradient of reduced host specificity at higher latitudes. Nine of the 10 species of monogenea from Hudson Canyon hosts were host-specific, as were 3 of 4 species found by Armstrong (1974). However, some species such as *Diclidophoropsis tissieri, Cyclocotyloides pinguis,* and *Paracyclocotyla cherbonnieri* are found on closely related fishes, indicating specificity for the host families, if not the genera.

An interesting result of Manter's (1934) study of deep water (75 to 1064 m) trematodes from Tortugas, Florida was the similarity of the fauna to that of fishes

TABLE XIII
Host Specificity (Preference for a Certain Number of Host Species) of Adult Digenetic Trematodes of Marine Fishes from Shallow and Deep Water

	Number of Fish Species/Specimens	Number of Trematode Species in Number of Host Species						Total	Reference
		1 Fish sp.	2	3	4	5	>5		
Woods Hole	≥94/many	37 (53.6)	7 (10.1)	5 (7.2)	4 (5.8)	1 (1.4)	15 (21.7)	69	Linton, 1940
Tortugas, Fla.	237/2039	105 (55.8)	43 (22.8)	14 (7.4)	8 (4.2)	5 (2.6)	13 (6.9)	188	Manter, 1947
Tortugas (deep)	≤90/721	30 (61.2)	5 (10.2)	6 (12.2)	2 (4.1)	1 (2)	5 (10.2)	49	Manter, 1934
Gulf and Caribbean	21/275	23 (71.8)	5 (15.6)	2 (6.2)	0 (0)	0 (0)	2 (6.2)	32	Armstrong, 1974
Hudson Canyon	52/1763	30 (73.1)	7 (17)	1 (2.4)	1 (2.4)	1 (2.4)	1 (2.4)	41	Campbell, present study

around the British Isles and Japan. Temperature was considered important in determining the distribution, and he suggested that some species may have a continuous distribution from the Arctic to the Antarctic through deep-sea hosts. Of interest in this regard is current evidence indicating the predominance and continuity of certain genera in the depths (*Lepidapedon, Neolepidapedon,* etc.) and that bottom-living fishes are the most important hosts for digeneans. Bathypelagic fishes are apparently only occasionally involved as final hosts, but evidently more often than mesopelagic species. *Gonocerca phycidis* and *Derogenes varicus,* both hemiurids, are distributed from pole to pole, but congeneric and very similar species of other trematodes are commonly found in distant regions among deep-sea hosts (*Glomericirrus, Dinosoma, Steringophorus, Plagioporus, Lepidapedon, Neolepidapedon*). Armstrong (1974), like Manter, found the trematode fauna of his deep-sea hosts more like that of distant colder waters than nearby inshore and warmer areas. He found the fauna of 21 species of macrourids more like that of the northern Pacific and northwestern Atlantic than that around the British Isles or Australia. Such evidence is supportive of the wide distribution of certain helminth species in the depths but is inconclusive because of the lack of extensive sampling from other geographic regions. Besides noting species common to widely separated geographic areas, he also noted trematode genera containing very similar species from these same areas as I have: *Prosomicrocotyla gotoi* (M), *Gonocerca crassa* (D), *Glomericirrus amadai* (D), *Paranisakiopsis coelorhynchi* (N), and the *Lepidapedon elongatum* complex (D) from the northwestern Pacific; *Paranisakiopsis lintoni* and *L. elongatum* complex off New England; *Paranisakiopsis australis* off Australia; and *Podocotyle levinseni* (D) and *L. elongatum* complex of the northeastern Atlantic. To this list can now be added species of *Lepidapedon, Neolepidapedon, Gonocerca, Lecithophyllum* (D), *Steringophorus* (D), *Glomericirrus* (D), *Hudsonia* (D), *Paraccacladium* (D), *Degeneria* (D), *Diclidophora* (M), *Diclidophoropsis* (M), *Ascarophis* (N), *Thynnascaris* (N), and *Heterotyphlum* (N) from the New York Bight.

Arctic fishes caught by commercial trawlers occur in greatest abundance in shallower depths of higher latitudes, but in subarctic waters they are found deeper and are caught with some of the shallower-living deep-sea types found on the upper continental slope (1400 m). Comparison of published and unpublished data (Bray and Gibson, personal communication) from Arctic and sub-Arctic collections is consistent in showing one group with a distribution among shelf fish to deep-sea types [*Gonocerca phycidis, Derogenes varicus, Hemiuris communis, Genolinea laticauda, Dissosaccus laevis, Lecithophyllum botryophoron* (= *anteroporum*), *Lepidapedon elongatum, Steringophorus furciger*], and a second group among fish having deep-sea affinities [*Glomericirrus* sp., *Steringophorus* (= *Abyssotrema*) *pritchardae, Hudsonia* sp., *Dinosoma* sp., *Steringophorus blackeri, Bathycreadium flexicollis*]. A third larger and more diverse group is trans-Atlantic in distribution among hosts at less than 400 m. These records—in addition to Armstrong (1974), those from the literature, and our own unpublished data—indicate zoogeographic continuity among deep-water hosts in the Northern Hemisphere (*Diclidophorpsis tissieri, Cyclocotyloides pinguis, Paracyclocotyla cherbonnieri, Steringophorus pritchardae, Hudsonia* sp., *Lepidapedon abyssorum, Lepidapedon elongatum, Steringotrema crassum, Gonocerca crassa, Lomasoma wardi, Paraccacladium jamiesoni,* and the closely related *Gonocerca oshoro*/*G. haedrichi* and *Plagioporus lobatus*/*Plagioporus* sp.). Prudhoe and Bray (1973) pointed out that the presence of closely related species in the same host suggests speciation through sibling species instead of allopatric speciation [between two species of *Neolepidapedon* (D) in the macrourid *Coryphaenoides whitsoni*]. This phenomenon is likely widespread because we have observed similar associations among *Lepidapedon* and *Steringophorus* in macrourids and *Antimora rostrata*. The true distribution among similar species of *Steringophorus, Lepidapedon, Neolepidapedon,* and *Plagioporus* may ultimately reveal that differentiation of some species has been made with features subject to intraspecific variation. It remains to be determined whether the distribution of parasites with complex life cycles will be concentrated along continental margins or have a more general, bipolar, and ocean-wide distribution. At present it appears that the most widespread species use pelagic intermediate and definitive hosts. The decrease in total abundance of free-living organisms with depth (see Chapter 3) appears to cause decreased parasite success, even among deep-living vertical migrators, as one leaves the continental margins (Gaevskaya, 1979; Campbell et al., 1980).

Manter (1934) emphasized that environmental gradients affect the distribution of parasites as well as free-living organisms. Cape Cod has traditionally been considered to mark the boundary between warm–temperate and cold–temperate (boreal) faunas of the northwestern Atlantic (Ekman, 1953). It is not a major zoogeographic boundary but provides a striking demarcation that extends to both sides of the Atlantic for some groups and is especially striking for the most motile, eurythermic species. Stephenson and Stephenson (1954) noted that the fauna of the mid-Atlantic seacoast of the United States in general did not appear to represent a province with a distinct population of its own. Most fishes occur there seasonally, about equally represented by inhabitants of boreal and warm-water areas. Taylor (1962) found the flora from New Jersey to New England to be boreal rather than tropical. Therefore, the mid-Atlantic fauna is the southern extent of the cold-temperate faunal area that extends north of Cape Cod. Evidence from mollusks, asteroids, polychaetes, fishes, and mammals indicates that the fauna of Labrador is Arctic and the boundary between the arctic and boreal faunas is around the Strait of Belle Isle, the northern entrance to the Gulf of St. Lawrence.

Hayden (unpublished data) has found the pelagic stingray, *Dasyatis violacea,* taken from George's Bank off Cape Cod, to have the same cestodes as reported for that ray off the coast of France. The cestode genus *Echeneibothrium* is more common from Cape Cod northward. South of Cape Cod we have found it limited to skates in deeper (colder) water as far south as Virginia. *Echeneibothrium* and *Pseudanthobothrium* have a trans-Atlantic distribution in benthic skates (i.e., *Raja radiata*), being most apparent at George's Bank north to the Labrador–Greenland area across the northern Atlantic basin to northern Europe. Interestingly, current records show that the genus *Echeneibothrium* is most prevalent above 40°N in the Atlantic. All the Mediterranean Sea is probably included in this distribution, but little is known about the southeastern portion of the Atlantic gyre in shallow or deep water. In the western Atlantic the genus *Acanthobothrium* is more common in skates and rays south of Cape Cod than it is to the north, and the elasmobranch cestode fauna generally increases in richness as collections are made southward to the equator. The seasonal occurrence of fish such as cownose rays, *Rhinoptera bonasus,* demonstrate well the decreasing prevalence and diversity of the cestode fauna that parallels the distribution of migrant batoids reaching Cape Cod during the summer months (Table IV).

The Gulf Stream forms an effective barrier between the cold-water area to the northwest and the warm water of the Sargasso Sea to the southeast. Between the Gulf Stream and the shore of North America a coastal current flows southwest, forming an elongated, counterclockwise circulating eddy. This coastal current is cold low-salinity water from the Labrador Current that flows south from the Labrador Sea around the eastern tip of Newfoundland and plays an important role in determining the fauna from Newfoundland to Cape Hatteras. Parr (1933) pointed out that the extreme temperature fluctuations in shallow waters between Cape Cod and Cape Hatteras make it difficult for either the northern or southern faunas to contribute any permanent residents to this area. There is little evidence of dramatic changes in the parasite fauna of deep-sea fishes throughout this area to date except between the Grand Bank of Newfoundland (44 to 45°N, 48 to 49°W) and the Flemish Cap (45 to 47°N, 44 to 46°W). The Flemish Cap is physically isolated from the Grand Bank and the extremely cold waters of the Labrador Current form an effective barrier to the west. Warm water of the Gulf Stream influences water temperatures and the fauna to the south of the Cap. Campbell and Haedrich (unpublished data) found an even greater contrast in the faunas of deep-sea fishes than did Zubchenko (1981), who compared small samples of *Nezumia bairdi* (23 fish) and *Macrourus berglax* (30 fish) taken off southern Labrador and the Flemish Cap. Both our data (34 *N. bairdi,* 46 *M. berglax*) and his indicate that the fish represent distinct host populations despite the fact that his samples were taken over a greater depth range (400 vs. 1100 m on Flemish Cap) than ours (400 to 700 m vs. 400 to 600 m on Flemish Cap). He noted that the parasite fauna of *N. bairdi* (15 fish from each area) was richer by three species (two of which have simple life cycles) on Flemish Cap. However, he found the fauna of *M. berglax* (7 fish from Labrador at 400 m; 16 fish from Flemish Cap at 1100 m) from Flemish Cap to be depauperate and speculated that the differences were due to isolation and a more diverse diet, incorporating more amphipods, polychaetes, and fish for hosts from southern Labrador. The fauna of *N. bairdi* was considered influenced by its preference for plankton. [Host size (age) should be correlated with trophic behavior and parasite fauna; see Campbell et al. (1980) and Munroe et al. (1981).]

Campbell and Haedrich (unpublished data) also found marked differences in the parasite faunas of *N. bairdi* and *M. berglax* from Flemish Cap and the nearby eastern portion of the Grand Bank (44 to 45°N, 48 to 49°W). However, we found both qualitative and quantitative differences in the parasite faunas of these fishes from the two areas. Our samples, taken over comparable depths in both locations, included 22 *N. bairdi* and 28 *M. berglax* from the Grand Bank and 12 *N. bairdi* and 18 *M. berglax* from Flemish Cap. We found the parasite faunas of both fish species to be richer on Flemish Cap. Remarkably, the metazoan faunas of these fishes differ from Zubchenko's (1981) from either area. *Nezumia bairdi* from Flemish Cap have a fauna dominated by nematodes (52%) and the larval cestode *Grillotia erinaceus* (39%), but different species of digenea were found in fish on the Cap and none in 22 *N. bairdi* from the Grand Bank. We also found intramuscular cysts of the microsporidan, *Pleistophora* sp., in *N. bairdi* from the Grand Bank (this parasite is common in *N. bairdi* from Hudson Canyon). Because the fish samples are small, this probably accounts for the different species of digenea encountered, but our data and Zubchenko's indicate a more diverse diet for these fish on Flemish Cap. It is interesting to note that samples of 25 hosts were typically sufficient to demonstrate almost the entire metazoan parasite fauna of fishes from Hudson Canyon (Campbell et al., 1980). Appy (unpublished data) found the same to be true in his studies of cod stocks (personal communication). Zubchenko (1981) found the parasite fauna of *M. berglax* on Flemish Cap to be depauperate in comparison to his sample from southern Labrador. He listed the following numbers of parasite species for 15 *M. berglax* from Labrador: Myxosporida, 3; Monogenea, 0; Digenea, 6; Cestoda, 3; Nematoda, 4; Acanthocephala, 1; and Copepoda, 3. The 15 *M. berglax* he sampled from Flemish Cap had the following numbers of species of parasites: Myxosporida, 1; Monogenea, 1; Digenea, 1; Cestoda, 0; Nematoda, 1; Acanthocephala, 1; and Copepoda, 3. Our samples of *M. berglax* from the Grand Bank (22 fish) and Flemish Cap (18 fish) show a rich fauna for this fish in both areas and includes different species of Monogenea (1), Digenea (2), Cestoda (2), and Copepoda (3) than observed by Zubchenko. Not only were the parasites distinctly different (Campbell et al., unpublished), but they were more common, or, in some instances new (Campbell et al., 1982). Prevalence of these parasites ranged from 16 to 38% of the *M. berglax* sampled on the Cap. The fauna of *M. berglax* emphasizes benthic feeding, but the fauna is considerably more diverse on Flemish Cap than predicted by Zubchenko (1981). Differences in parasite faunas of the fish population should be correlated with fish size (age) and their trophic relationships at different depths [see *Coryphaenoides armatus* in Campbell et al. (1980)]. This is especially important in this locality because the shallower portions of Flemish Cap are influenced by its own unique circulation. We also found that plerocercoids of *Grillotia erinaceus* were common in both *N. bairdi* and *M. berglax* from Flemish Cap but were never found in samples from the Grand Bank. Frequencies of 22% (*M. berglax*) and 92% (*N. bairdi*) indicate that this cestode larva would make a useful biological tag for studies in that area. A more thorough comparison of these fishes from the entire perimeter of Flemish Cap may provide further contrasts in the parasite faunas, indicating the importance of water masses on the community structure found there.

Antarctica has its own exceedingly distinct cold-water fish fauna with no geographic barrier. The isolation of these fishes is indicated by their systematic distinction at the subordinal level, Notothenioidei. However, this distinction does not apply to their parasites. Only species differences appear among digenea, nematodes, and cestodes of these benthic fishes and the parasite genera common among deep-sea benthic and bottom fishes from other cold-water regions of the world. The intrusion of some of these genera into shallower water (*Lepidapedon, Gonocerca, Derogenes*), or vice versa, is apparently through predominant benthic invertebrate groups or through vertical movements of pelagic invertebrates, especially with certain hemiurids and opecoeliids whose precocious development may allow completion of the life cycle without the fish host.

The wide distribution of genera like *Lepidapedon* and *Neolepidapedon* among macrourids may have resulted in their introduction into the Antarctic fish fauna by fish migrations. Among Antarctic and sub-Antarctic fishes digenea have only exceptionally been reported from other than bottom-living fishes. Prudhoe and Bray

(1973) found a single hemiurid in a bathypelagic fish, *Bathylagus antarcticus*. Comparison of parasite faunas of related shallow-dwelling teleosts from widely separated regions often provides contrasts involving diversity at the generic or even family level. Such variability in shallow water markedly contrasts with the decreased variability and predominance of certain groups in the deep ocean. The parasite success appears to be in response to fish size and age, and their degree of trophic specialization, for benthic or pelagic food, in communities comprised of fewer suitable intermediate hosts. As more information is obtained, it appears that decreased host specificity will be the rule for the most common helminth parasites with complex life cycles. In deep-living fishes, their widespread success is directly related to the generalized feeding habits of their hosts.

5. Final Questions

The relative size of the ecosystem in which deep-sea fishes live and the permanence of their association with invertebrate communities is a major factor in determining parasite success. The more diverse communities provide opportunities for more diverse parasite faunas. Most deep-sea fishes are widely distributed, and many may be far-ranging. Midwater and bathypelagic fishes are dispersed in a tremendous volume of water many thousands of kilometers in range and are in a continuous state of transition. Benthic fishes live in a more restricted zone that, in slope areas, offers a concentrated food supply. When the dominant fishes are studied, the community appears to be much the same from one region to another but in actuality differs when the entire fish and invertebrate faunas are itemized. Deep-sea fish faunas generally seem limited to depth zones, a fact supported by parasite data as well, but are far ranging horizontally. The relative area encompassed by the benthic fish–invertebrate associations may be much more limited geographically than previously expected in that certain parasites are found only at certain depths and are geographically limited within the range of the fish host. Therefore, the same fish species may have a decidedly different incidence, intensity, and diversity of their parasite faunas over relatively short (geographically) distances. The lower diversity and success of parasites in northern waters is not indicated by a decrease in invertebrate diversity or water temperature because the invertebrate fauna is rich and the permanent thermocline is less than 500 m. A comparison of the parasite faunas of a deep benthic fish species from communities with very different fish and invertebrate faunas would be of interest in determining whether the parasite fauna changes because of differences in invertebrate community structure or a change in feeding strategy on the part of the fish. Hierarchy in the food webs may also differ. Stomach contents provide no evidence of higher relationships in the food chain as larval parasites may. Despite the apparent physical stability of the deep ocean, will the metazoan parasite fauna not show the influence of events occurring in overlying waters or in nearby shelf areas? Seasonal fluxes of phyto- and zooplankton above the permanent thermocline attract pelagic fishes and marine mammals into specific areas, and those seasonal events very likely affect the success of populations that live in the depths below. Parasite populations increase in shallow-living hosts during that time, that is, when hosts are behaviorally and numerically available. Does this affect show in parasite populations of deep-sea fishes in latitudes where productivity is less continuous and more seasonal? How consistent are faunal zones? There are unexplained changes in faunal distribution with unknown impact on parasite distribution. Does the instability of the community low population densities of hosts and irregular distribution explain the comparatively low supraspecific diversity of the parasite fauna? What effect does upwelling have on the parasite faunas of slope dwelling or pelagic communities? Are horizontal encounters with the continental slopes significant to the presence of parasites in bathypelagic or certain mesopelagic fishes? Have the parasites retained adaptability instead of specializing and diverging in the face of low host population densities and irregular distributions of intermediate hosts? Is evolution of parasite species proceeding at a much reduced rate because of the uncertainties of invertebrate community structure compared to the distribution of the fishes? Is the lack of exposure to sunlight and its ultraviolet component of significance to DNA variability in helminths? Is enzymatic photoreactivation of pyrimidine dimers really "universal"? Rohde (1978) pointed out the phenomenon of decreased host specificity and diversity of trematodes (Digenea) among shallow-living fishes with increasing latitude. This does not appear to be true for deep-sea fishes. Are deep-sea helminth parasites generally less specific for their intermediate hosts than those common to shelf communities? What species and what proportion of the parasite faunas of benthic fishes are found in the various pelagic zones? What is the reason for the scarcity of cestodes in deep-living elasmobranchs? Are scavenging invertebrates involved as definitive or paratenic hosts in parasite life cycles at the greatest depths? Are helminths of deep benthic fishes randomly distributed around the hemispheres or along the continental margins? Is the parasite fauna of the megafauna depauperate far from the continental margins? Does the phylogeny of the metazoan parasites reflect the phylogeny of their hosts? Are mesopelagic fishes significant in cycling helminths into the depths, or are pelagic invertebrates the primary mode of transmission to the deeper-living fishes? Is such transmission of major importance to the benthic community? The degree of parasitism in the bathypelagic community relative to other fish communities has not been resolved. Incidence data from invertebrate populations and life-cycle studies are badly needed. What, if any, are the parasite population regulatory mechanisms? Our knowledge about these and other basic questions about the nature of parasitism and its role in deep-sea ecology will remain superficial until sufficiently large samples are taken and studies continued over prolonged periods of time.

Acknowledgments

My appreciation is expressed to the following who have provided information, specimens, and facilities; have confirmed identifications; or given technical assistance in these studies: R. Haedrich and J. Lien, Memorial University of Newfoundland; G. Rowe, Brookhaven National Laboratory; B. Robison, Marine Science Institute, University of California at Santa Barbara; R. Backus, P. Polloni, P. Sorenson, M. Lakich, and C. Clifford, Woods Hole Oceanographic Institution; N. Kingston, University of Wyoming; D. Zwerner, W. Hargis, Jr., J. Gartner, J. Smith, D. Markle, C. Wenner, and L. Suydam, Virginia Institute of Marine Science; A. Lawler and T. Mattis, Gulf Coast Research Laboratory; J. Scott and the *Journal Fisheries Research Board of Canada*; Springer-Verlag, New York for permission to use illustrations from *Marine Biology*; G. McClelland, Dalhousie University; T. McLellan and K. Hartel, Harvard University; M. H. Pritchard and the H. W. Manter Laboratory of Parasitology, University of Nebraska State Museum; J. Bier, U.S. Food and Drug Administration, Division of Microbiology; R. Lichtenfels and A. Fusco, Animal Parasitology Institute, Beltsville; R. Bray and D. Gibson, British Museum (Natural History); P. Gooch and the Commonwealth Bureau of Helminthology, England; H.-P. Fagerholm, Åbo Akademi, Finland; R. Richardson, Florida Department of Natural Resources; D. Huffman, Southwest Texas State University; M. Ulmer, Iowa State University; M. Moser, University of California, Santa Cruz; R. Appy, University of Guelph; J. Carvajal G., Universidad Catolica de Chile, Santiago; J. Amato, Universidad Federal Rural do Rio de Janiero, Brasil; Yu. V. Kurochkin, Pacific Research Institute of Fisheries and Oceanography, U.S.S.R.; R. Griffith, Southeastern Massachusetts University; and students T. Munroe, B. Hayden, A. Williams, and S. Correia. R. Sasseville and M. A. Campbell assisted with typing of the manuscript. Portions of this work were supported by the Research Foundation of Southeastern Massachusetts University and National Science Foundation Grants DEB 76-20103 and OCE 76-21878; the latter was granted to R. Haedrich and G. Rowe, principal investigators.

References

Anderson, R. M., 1974. Mathematical models of host–helminth interactions. In *Ecological Stability*, M. B. Usher and M. H. Williamson, eds. Chapman and Hall, London.

Armstrong, H. W., 1974. A study of the helminth parasites of the family Macrouridae from the Gulf of Mexico and Caribbean Sea: Their systematics, ecology and zoogeographical implications. Doctoral dissertation, Texas A & M University.

Bakke, T. A., 1975. Studies of the helminth fauna of Norway XXIX: The common gull, *Larus canus* L., as final host for *Syngamus (Cyathostoma) lari* (Blanchard, 1849) (Nematoda, Strongyloidea). *Norwegian J. Zool.*, **23,** 37–44.

Ball, I. R., and R. A. Khan, 1976. On *Micropharynx parasitica* Jägerskiöld, a marine planarian ectoparasitic on the thorny skate, *Raja radiata* Donovan, from the North Atlantic Ocean. *J. Fish. Biol.*, **8,** 419–426.

Bell, F. J., 1887. Description of a new species of *Distomum*. *Ann. Mag. Nat. Hist.* (5th ser.), **19,** 116–117.

Berland, B., 1961. Nematodes from some Norwegian marine fishes. *Sarsia*, **2,** 1–50.

Beverley-Burton, M., O. L. Nyman, and J. H. C. Pippy, 1977. The morphology, and some observations on the population genetics of *Anisakis simplex* larvae (Nematoda: Ascaridata) from fishes of the North Atlantic. *J. Fish. Res. Board Can.*, **34,** 105–112.

Bier, J. W., 1976. Experimental anisakiasis: Cultivation and temperature tolerance determinations. *J. Milk Food Technol.*, **39,** 132–137.

Bigelow, H. B., and W. C. Schroeder, 1953. *Memoir Sears Foundation for Marine Research*, No. I, *Fishes of the Western North Atlantic*, Part II. Sears Foundation for Marine Research, Yale University, New Haven, CT.

Blankespoor, H. D., 1971. Host parasite relationships of an avian trematode, *Plagiorchis noblei* Park, 1936. Doctoral dissertation, Iowa State University.

Blaxter, J. H. S., C. S. Wardle, and B. L. Roberts, 1971. Aspects of the circulatory physiology and muscle systems of deep-sea fish. *J. Mar. Biol. Assoc., U. K.*, **51,** 991–1006.

Bowman, R. E., 1977. *Seasonal Food Habits of Demersal Fish in the Northwest Atlantic—1972*. Food Chain Investigation No. 77-01. Northeast Fisheries Center, Woods Hole, Massachusetts, 31 pp.

Bowman, R. E., R. O. Maurer, Jr., and J. A. Murphy, 1976. *Stomach Contents of Twenty-Nine Fish*

Species from Five Regions in the Northwest Atlantic—Data Report. Food Chain Investigation No. 76-10. Northeast Fisheries Center, Woods Hole, Massachusetts, 37 pp.

Bray, R. A., 1973. Some digenetic trematodes in fishes from the Bay of Biscay and nearby waters. *Bull. Br. Mus. (Nat. Hist.) Zool.*, **26**, 151–183.

Bray, R. A. and D. I. Gibson, 1977. The Accacoeliidae (Digenea) of fishes from the north-east Atlantic. *Bull. Br. Mus. (Nat. Hist.) Zool.*, **31**, 53–99.

Brinkmann, A., Jr., 1942. On some new and little known *Dactylocotyle* species, with a discussion on the relations between the genus *Dactylocotyle* and the "Family" Diclidophoridae. *Goteborgs Mus. Zool. Avdelnig.* (92). *Goteborgs K, Vetensk-O. Vitterhets-Samh Handl.*, **6** [Ser. B., Bd. I (No. 1)], 32 pp.

Brinkmann, A., Jr., 1952. Fish trematodes from Norwegian waters I. The history of fish trematode investigations in Norway and the Norwegian species of the order Monogenea. *Arbok. Bergen (Naturvit. Rekke)*, **1**, 1–134.

Burreson, E. M., 1975. Biological studies on the hemoflagellates of Oregon marine fishes and their potential leech vectors. *Diss. Abstr.*, **36B**, 1609.

Bychowsky, B. E., 1957. *Monogenetic Trematodes (Engl. Transl.)*. American Institute of Biological Sciences, Washington, DC (1961), 627 pp.

Cable, R. M., 1954. Studies on marine digenetic trematodes of Puerto Rico. The life cycle in the family Haplosplanchnidae. *J. Parasitol.*, **40**, 71–75.

Cake, E. W., Jr., 1976. A key to larval cestodes of shallow-water benthic mollusks of the northern Gulf of Mexico. *Proc. Helminthol. Soc. Wash.*, **43**, 160–171.

Cake, E. W., Jr., 1977a. Experimental infection studies with bothridioplerocercoids of *Rhinebothrium* sp. (Cestoda: Tetraphyllidea) and two intermediate molluscan hosts. *Northeast Gulf Sci.*, **1**, 55–59.

Cake, E. W., Jr., 1977b. Larval cestode parasites of edible mollusks of the northeastern Gulf of Mexico. *Gulf Res. Rep.*, **6**, 1–8.

Cake, E. W., Jr., and R. W. Menzel, 1980. Infections of *Tylocephalum* metacestodes in commercial oysters and three predaceous gastropods of the eastern Gulf of Mexico. *Proc. Natn. Shellfish Assoc.*, **70**, 94–104.

Cameron, T. W. M., 1950. Parasitology and Evolution. *Trans. Roy. Soc. Can.* (3rd series, XLIV, Sect. 5), 1–20.

Campbell, R. A., 1972. New experimental hosts of *Posthodiplostomum minimum* (Trematoda: Diplostomatidae). *J. Parasitol.*, **58**, 1051.

Campbell, R. A., 1973. Studies on the host-specificity and development of the adult strigeoid trematode *Cotylurus flabelliformis*. *Trans. Am. Microsc. Soc.*, **92**, 256–265.

Campbell, R. A., 1975a. Two new species of *Echeneibothrium* (Cestoda: Tetraphyllidea) from skates in the western North Atlantic. *J. Parasitol.*, **61**, 95–99.

Campbell, R. A., 1975b. *Hudsonia agassizi* gen. et sp. n. (Zoogonidae: Hudsoniinae subf. n.) from a deep-sea fish in the western North Atlantic. *J. Parasitol.*, **61**, 409–412.

Campbell, R. A., 1975c. *Abyssotrema pritchardae* gen. et sp. n. (Digenea: Fellodistomidae) from the deep-sea fish *Alepocephalus agassizi* Goode and Bean, 1883. *J. Parasitol.*, **61**, 661–664.

Campbell, R. A., 1977a. *Degeneria halosauri* (Bell 1887) gen. et comb. n. (Digenea: Gorgoderidae) from *Halosauropsis macrochir*. *J. Parasitol.*, **63**, 76–79.

Campbell, R. A., 1977b. A new family of pseudophyllidean cestodes from the deep-sea teleost *Acanthochaenus lutkenii*. *J. Parasitol.*, **63**, 301–305.

Campbell, R. A., 1977c. New Tetraphyllidean and trypanorhynch cestodes from deep-sea skates in the western North Atlantic. *Proc. Helminthol. Soc. Wash.*, **44**, 191–197.

Campbell, R. A., 1979. Two new genera of pseudophyllidean cestodes from deep-sea fishes. *Proc. Helminthol. Soc. Wash.*, **46**, 74–78.

Campbell, R. A., S. J. Correia, and R. L. Haedrich, 1982. A new monogenean and cestode from the deep-sea fish, *Macrourus berglax* Lacépède, 1802, from the Flemish Cap off Newfoundland. *Proc. Helminthol. Soc. Wash.*, **49**, 169–175.

Campbell, R. A. and J. V. Gartner, Jr., 1982. *Pistana eurypharyngis* gen. et sp. n. (Cestoda:

Pseudophyllidea) from the bathypelagic gulper eel, *Eurypharynx pelecanoides* Vaillant, 1882, with comments on host and parasite ecology. *Proc. Helminthol. Soc. Wash.*, **49,** 218–225.

Campbell, R. A. and T. A. Munroe, 1977. New hemiurid trematodes (Digenea: Hemiuridae) from deep-sea fishes of the western North Atlantic. *J. Parasitol.*, **62,** 285–294.

Campbell, R. A., R. L. Haedrich, and T. A. Munroe, 1980. Parasitism and ecological relationships among deep-sea benthic fishes. *Mar. Biol.*, **57,** 301–313.

Cheng, T. C., 1967. Marine molluscs as hosts for symbioses with a review of known parasites of commercially important species. *Adv. Mar. Biol.*, **5,** 1–424.

Cheng, T. C., 1976. The natural history of anisakiasis in animals. *J. Milk Food Technol.*, **39,** 32–46.

Chubb, J. C., 1982. Seasonal occurrence of helminths in freshwater fishes. Part IV. Adult Cestoda, Nematoda, and Acanthocephala. In *Advances in Parasitology,* Vol. 20, W. H. R. Lunsden, R. Muller, J. R. Baker, eds. Academic Press, New York, pp. 1–292.

Chubrik, G. K., 1952. Larval stages of the trematode *Fellodistomum fellis* from invertebrates of the Barents Sea. *Zool. Zh.*, **31,** 653–658.

Chubrik, G. K., 1966. Fauna and ecology of larval trematodes in molluscs of Barents and White Seas. Life cycle of parasitic worms of North Seas. *Trudy Akad. Nauk SSSR,* **10,** 78–158.

Collard, S., 1968. A study of parasitism in mesopelagic fishes. Doctoral dissertation, University of California at Los Angeles.

Collard, S., 1970. Some aspects of host–parasite relationships in mesopelagic fishes. In *A Symposium on Diseases of Fishes and Shellfishes,* S. F. Snieszko, ed. Am. Fish. Soc. Spec. Publ. 5, pp. 41–56.

Dailey, M. D., 1975. The distribution and intraspecific variation of helminth parasites in pinnipeds. *Rapp. P.-v. Reun. Cons. Int. Explor. Mer.*, **169,** 338–352.

Dailey, M. D. and R. L. Brownell, Jr., 1972. A checklist of marine mammals and parasites. In *Mammals of the Sea, Biology and Medicine,* S. H. Ridgeway, ed. Thomas, Springfield, IL, pp. 528–589.

Dailey, M. D. and W. Vogelbein, 1982. Mixodigmatidae, a new family of cestode (Trypanorhyncha) from a deep sea, planktivorous shark. *J. Parasitol.*, **68,** 145–149.

Davey, J. T., 1969. The early development of *Contracaecum osculatum. J. Helminthol.*, **43,** 293–298.

Dawes, B., 1958. *Sagitta* as a host of larval trematodes, including a new and unique type of cercaria. *Nature,* **182** (4640), 960–961.

Dayton, P. K. and R. R. Hessler, 1972. Role of biological disturbance in maintaining diversity in the deep-sea. *Deep-Sea Res.*, **19,** 199–209.

Deardorff, T. L. and R. M. Overstreet, 1981. Review of *Hysterothylacium* and *Iheringascaris* (both previously = *Thynnascaris*) (Nematoda: Anisakidae) from the northern Gulf of Mexico. *Proc. Biol. Soc. Wash.* (1980), **93,** 1035–1079.

Delyamure, S. L., 1968. *Helminthofauna of Marine Mammals: Ecology and Phylogeny*. Academy of Science, USSR, Moscow, 1955 (transl. Israel Program Scientific Translations).

Dienske, H., 1968. A survey of the metazoan parasites of the rabbit-fish, *Chimaera montrosa* L. (Holocephali). *Netherl. J. Sea Res.*, **4,** 32–58.

Dillon, W. A. and W. J. Hargis, Jr., 1968. Monogenetic trematodes from the southern Pacific Ocean. Part IV. Polyopisthocotyleids from New Zealand fishes: The families Mazocraediae, Diclidophoridae, and Hexabothriidae. *Proc. Biol. Soc. Wash.*, **81,** 351–366.

Dogiel, V. A., 1962. *General Parasitology* (transl. Z. Kabata, 1964). Oliver and Boyd, London.

Dogiel, V. A. and G. K. Petrushevski, 1933. Parasite fauna of fishes of the Gulf of Neva. *Trudy Leningr. Obshch. Estest.*, **42,** 366–434.

Dogiel, V. A. and G. K. Petrushevski, 1935. An attempt at the ecological study of the parasite fauna of salmon in the White Sea. *Vop. Ekol. Biotsenol.*, **2,** 137–169.

Dollfus, R. Ph., 1923. Énumération des cestodes du Plancton et des invertébrés marins. *Ann. Parasitol. Hum. Comp.*, **1,** 276–300.

Dollfus, R. Ph., 1931. Acanthocephalidae d'un poisson capture par 4,785 m de profondeur. *Ann. Parasitol. Hum. Comp.*, **9,** 185–187.

Dollfus, R. Ph., 1942. Études critiques sur les tetrarhynques du Muséum de Paris. *Arch. Mus. Natl. Hist. Nat. (Paris)*, **19,** 1–466.

Dollfus, R. Ph., 1953. Apercu général sur l'histoire naturelle des parasites animaux de la morue atlanto-arctique, *Gadus callarias* L. (= *morhua* L.) et leur distribution géographique. *Encyclopedie Biol. (Paris)*, **43,** 428 pp.

Dollfus, R. Ph., 1963. List des coelentérés marins, Paléarctiques et Indiens, on ont été trouvés des trématodes digénétiques. *Bull. Inst. Pech. Marit. Maroc,* **9–10,** 33–57.

Dollfus, R. Ph., 1970. Campagne d'essais du "Jean Charcot" (3–8 Decembre 1968). 4. D'un trematode monogénétique trouvé libre en dix exemplaires parmi les matériaux récoltes par un chalutage au large de la Bretagne. *Bull. Mus. Hist. Nat. Paris (Ser. 2),* **41,** 1522–1530.

Dollfus, R. Ph., 1971. De quelques trématodes digénétiques de sélaciens du Sénégal et considérations anatomiques sur les Anaporrhutinae A. Looss, 1901. *Bull. Inst. Fr. Afr. Noire, Ser. A,* **33,** 347–370.

Donets, Z. S., 1969. Distribution of Myxosporida in Ukrainian water basins in relation to ecology of the host fishes and to spore structure (abstract): 228. Third International Congress of Parasitology, Leningrad, U.S.S.R.

Eagle, R. J. and J. R. McCauley, 1965. Collecting and preparing deep-sea trematodes. *Turtox News,* **43,** 220–221.

Edgar, R. K. and J. G. Hoff, 1977. Grazing of freshwater and estuarine, benthic diatoms by adult Atlantic Menhaden, *Brevoortia tyrannus. Fish Bull.,* **74,** 689–693.

Ekman, S., 1953. *Zoogeography of the Sea.* Sidgwick and Jackson, London, 417 pp.

Emery, K. O., 1969. The continental shelves. In *Readings from Scientific American, Oceanography.* Freeman, San Francisco, pp. 143–154.

Esch, G. W., T. C. Hazen, and J. M. Aho, 1977. Parasitism and *r*- and *K*-selection. In *Regulation of Parasite Populations,* G. W. Esch ed. Academic Press, New York, pp. 9–62.

Fagerholm, H.-P., 1982. Parasites of fish in Finland. VI. Nematodes. *Acta Acad. Aboensis, Ser. B,* **40,** 1–126.

Fange, R., A. Larsson, and U. Lidman, 1972. Fluids and jellies of the acousticolateralis system in relation to body fluids in *Coryphaenoides rupestris* and other fishes. *Mar. Biol.,* **17,** 180–185.

Fischthal, J. H., 1977. Some digenetic trematodes of marine fishes from the Barrier Reef and Reef Lagoon of Belize. *Zoologica Scr.,* **6,** 81–88.

Gaevskaya, A. V., 1979. The systematic position of *Hemiuris macrouri* Gaevskaya, 1975 (Trematoda: Hemiuridae) (in Russian). *Parazitologiya* **13,** 269–270.

Gallien, L., 1937. Recherches sur quelques trematodes monogeneses nouveaux ou peu connus. *Ann. Parasitol. Hum. Comp.,* **15,** 9–28.

Geraci, J. R., 1978. The enigma of marine mammal strandings. *Oceanus,* **21,** 38–47.

Gibson, D. I., 1972. Flounder parasites as biological tags. *J. Fish Biol.,* **4,** 1–9.

Golovan, G. A. and N. B. Pakhorukov, 1975. Some data on the morphology and ecology of *Alepocephalus bairdi* (Alepocephalidae) of the central and eastern Atlantic. *J. Ichthyol.,* **15,** 44–50.

Griffith, R. W., 1981. Composition of the blood serum of deep-sea fishes. *Biol. Bull. Mar. Biol. Lab., Woods Hole,* **160,** 250–264.

Guiart, J., 1935. Cestode parasites provenant des campagnes scientifiques de S. A. S. le Prince Albert Ier de Monaco (1886–1913). *Res. Camp. Scient. Albert Ier Prince Monaco,* **91,** 1–100.

Guiart, J., 1938. Trematodes parasites provenant des campagnes scientifiques du Prince Albert Ier de Monaco (1886–1912). *Res. Camp. Scient. Albert Ier Prince Monaco, Ser. C,* **75,** 1–84.

Guildal, J. A., 1964. Some qualitative and quantitative investigations on the endoparasitic fauna of the Scandinavian-Baltic population of the black headed gull (*Larus ridibundus*). *Royal Vet. and Agric. Coll., Copenhagen Yearbook.*

Gusev, A. V., 1957. Parasitological investigation of some deep-sea fishes in the Pacific Ocean (in Russian). *Trudy Inst. Okeanol.,* **27,** 362–366.

Haedrich, R. L. and N. R. Henderson, 1974. Pelagic food of *Coryphaenoides armatus,* a deep benthic rattail. *Deep-Sea Res.,* **21,** 739–744.

Haedrich, R. L. and P. T. Polloni, 1976. A contribution to the life history of a small rattail fish, *Coryphaenoides carapinus*. *Bull. S. Calif. Acad. Sci.*, **75**, 203–211.

Haedrich, R. L. and G. T. Rowe, 1977. Megafaunal biomass in the deep sea. *Nature*, **269** (5624), 141–142.

Haedrich, R. L., G. T. Rowe, and P. T. Polloni, 1975. Zonation and faunal composition of epibenthic populations on the continental slope south of New England. *Mar. Research*, **33**, 191–212.

Haedrich, R. L., G. T. Rowe, and P. T. Polloni, 1980. The megabenthic fauna in the deep sea south of New England. *Mar. Biol.* **57**, 165–179.

Halvorsen, O. and H. H. Williams, 1968. Studies on the helminth fauna of Norway. IX. *Gyrocotyle* (Platyhelminthes) in *Chimaera montrosa* from Oslo Fjord, with emphasis on its mode of attachment and a regulation in the degree of infection. *Nytt. Mag. Zool.*, **15**, 130–142.

Hare, G. M., and M. D. B. Burt, 1976. Parasites as potential biological tags of Atlantic salmon (*Salmo salar*) smolts in the Miramichi River system, New Brunswick. *J. Fish. Res. Board Can.*, **33**, 1139–1143.

Hargis, W. J., 1957. The host specificity of monogenetic trematodes. *Expl. Parasitol.*, **6**, 610–625.

Harris, J. E., 1973. The immune responses of dace, *Leuciscus leuciscus* (L.), to injected antigenic materials. *J. Fish. Biol.*, **5**, 261–276.

Harrison, L., 1928. Host and parasite. *Proc. Linn. Soc. N. S. W.*, **53**, 9–29.

Ho, J.-S., 1970. Revision of the genera of the Chondracanthidae, a copepod family parasitic on marine fishes. *Beaufortia*, **17**, 105–218.

Ho, J.-S., 1971. Parasitic copepods of the family Chondracanthidae from fishes of eastern North America. *Smithsonian Contrib. Zool.*, No. 87, 39 pp.

Ho, J.-S., 1975. Copepod parasites of deep-sea fish off the Galapagos Islands. *Parasitology*, **70**, 359–375.

Ho, J.-S., 1977. *Marine Flora and Fauna of the Northeastern United States. Copepoda: Lernaeopodidae and Sphyriidae.* NOAA Technical Report NMFS, Circular 406, 14 pp.

Ho, J.-S., 1978. *Marine Flora and Fauna of the Northeastern United States. Copepoda: Cyclopoids Parasitic on Fishes.* NOAA Technical Report NMFS, Circular 409, 12 pp.

Holmes, W. N. and E. M. Donaldson, 1969. The body compartments and the distribution of electrolytes. In *Fish Physiology*, Vol. 1, W. S. Hoar and D. J. Randall, eds. Academic Press, New York, pp. 1–89.

Hyman, L. H., 1951. *The Invertebrates. Platyhelminthes and Rhynchocoela. The Acoelomate Bilateria*, Vol. II, McGraw-Hill, New York, 550 pp.

Jackson, G. J., 1975. The "new disease" status of human anisakiasis and North American cases: A review. *J. Milk Food Technol.*, **38**, 769–773.

Johnston, T. H. and P. M. Mawson, 1945. Parasitic nematodes. B. A. N. Z. Antarctic Res. Exped. (1929–1931). *Rep. Br. Mus. (Nat. Hist.), Ser. B*, **5**, 73–160.

Jorotaeva, V. D. and V. G. Leont'eva, 1972. Helminth fauna of *Marcruronus novaezealandiae*. Problemy parasitologii. *Trudy VII Nauchnoi Konferentsii Parazitologov USSR*, Part 1. *Izdatel'stov "Naukovo Pumka,"* Kiev, USSR, pp. 403–405.

Kabata, Z., 1959. Ecology of the genus *Acanthochondria* Oakley, 1927 (*Copepoda parasitica*). *J. Mar. Biol. Assoc. U. K.*, **38**, 249–262.

Kabata, Z., 1963. Parasites as biological tags. In *North Atlantic Fish Marking Symposium*. Int. Comm. Northwest Atlantic Fisheries, Special Publication No. 4, pp. 31–37.

Kabata, Z., 1965. Parasitic copepoda of fishes. *Rep. B. A. N. Z. Antarct. Res. Exped., Ser. B.*, **8**, 16 pp.

Kabata, Z., 1969. Four Lernaeopodidae (Copepoda) parasitic on fishes from Newfoundland and West Greenland. *J. Fish. Res. Board Can.*, **26**, 311–324.

Kabata, Z., 1970. Discovery of *Branchiella logeniformis* (Copepoda: Lernaeopodidae) in the Canadian Pacific and its significance to the zoogeography of the genus *Merluccius* (Pisces: Teleostei). *J. Fish. Res. Board Can.*, **27**, 12–20.

Kabata, Z., and B. Cousens, 1973. Life cycle of *Salmincola californiensis* (Dana 1852) (Copepoda: Lernaeopodidae). *J. Fish. Res. Board Can.*, **30**, 881–903.

Kennedy, C. R., 1974. Regulation of fish parasite populations. In *Regulation of Parasite Populations,* G. W. Esch, ed. Academic Press, New York, pp. 63–109.

Kennedy, C. R., 1975. *Ecological Animal Parasitology.* Blackwell, Oxford.

Khan, R. A., 1974. Transmission and development of the trypanosome of the Atlantic cod by a marine leech. In *Proceedings of the Third International Congress of Parasitology, Munich,* Vol. 3. FACTA Publ., Vienna, Austria, 1608.

Laird, M., 1953. The protozoa of New Zealand intertidal zone fishes. *Trans. Roy. Soc. N. Z.,* **81,** 79–143.

Lapage, G., 1961. A list of the parasitic protozoa, helminths and Arthropoda recorded from species of the family Anatidae (ducks, geese, and swans). *Parasitology,* **51,** 1–109.

Leonov, V. A., 1960. Dynamics of the helminth fauna of the herring gull nesting in the territory of the Black Sea reserve. *Uchen. Zap. Gorkov. Gos. Pedagog. Inst.,* **27,** 38–57.

Linton, E., 1898. Notes on trematode parasites of fishes. *Proc. U.S. Nat. Mus.,* **20,** 507–548.

Linton, E., 1900. Fish parasites collected at Woods Hole in 1898. *Bull. U.S. Comm. Fish Fisheries* (for 1899), **19,** 267–304.

Linton, E., 1901. Parasites of fishes of the Woods Hole region. *Bull. U.S. Comm. Fish Fisheries* (for 1899), **19,** 405–497.

Linton, E., 1905. Parasites of fishes of Beaufort, North Carolina. *Bull. U.S. Bur. Fish.* (for 1904), **24,** 321–428.

Linton, E., 1909. Helminth fauna of the Dry Tortugas. I. Cestoda. *Carnegie Inst. Wash. Publ.,* **102,** 157–190.

Linton, E., 1910. Helminth fauna of the Dry Tortugas. II. Trematodes. *Carnegie Inst. Wash. Publ.,* **133;** *Pap. Tortugas Lab.,* **5,** 1–98.

Linton, E., 1924. Notes on cestode parasites of sharks and rays. *Proc. U.S. Nat. Mus.,* **64,** 1–114.

Linton, E., 1940. Trematodes of fishes mainly from the Woods Hole region. *Proc. U.S. Nat. Mus.,* **88,** 1–172.

Llewellyn, J., 1965. The evolution of parasitic platyhelminths. In *Evolution of Parasites, Proceedings of 3rd Symposium of the British Society for Parasitology.* Blackwell, Oxford, pp. 47–78.

Lom, J., 1970. Protozoa causing diseases in marine fishes. In *A Symposium on Diseases of Fishes and Shellfishes,* S. F. Snieszko, ed. Am. Fish. Soc. Spec. Publ. 5, pp. 101–123.

Lom, J., A. V. Gayevskaya, and I. Dykova, 1980. Two microsporidian parasites found in marine fishes in the Atlantic Ocean. *Folia Parasitol.,* **27,** 197–202.

Lubieniecki, B., 1976. Aspects of the biology of the plerocercoid of *Grillotia erinaceus* (van Beneden, 1858) (Cestoda: Trypanorhyncha) in haddock *Melanogrammus aeglefinus* (L.). *J. Fish. Biol.,* **8,** 431–439.

MacKenzie, K. and D. I. Gibson, 1970. Ecological studies of some parasites of plaice, *Pleuronectes platessa* (L.) and flounder, *Platichthys flesus* (L.). In *Aspects of Fish Parasitology,* A. E. R. Taylor, and R. Muller, eds. Blackwell, Oxford, pp. 1–42.

Mamaev, Yu. L., 1976. Some new monogeneans from the family Diclidophoridae. *Proc. Inst. Biol. Pedol., Vladivostok, New Ser.,* **34,** 92–103.

Mamaev, Yu. L., and A. M. Parukhin, 1975. New monogeneans of the subfamily Diclidophoridae (Monogenoidea, Diclidophoridae). Helminth. Res. Animals and Plants (25th Anniv. Lab. Gen. Helminthol.). *Proc. Inst. Biol. Pedol., Vladivostok, New Ser.,* **26,** 126–142.

Manter, H. W., 1925. Some fish trematodes of Maine. *J. Parasitol.,* **12,** 11–18.

Manter, H. W., 1931. Some digenetic trematodes of marine fishes of Beaufort, North Carolina. *Parasitology,* **23,** 396–411.

Manter, H. W., 1934. *Some Digenetic Trematodes from Deep-Water Fish of Tortugas, Florida.* Pap. Tortugas Lab., Carnegie Inst. Wash. Publ. 28, pp. 257–345.

Manter, H. W., 1940. Digenetic trematodes of fishes from the Galapagos Islands and the neighboring Pacific. *Allan Hancock Pacif. Exped.,* **2,** 329–497.

Manter, H. W., 1947. The digenetic trematodes of marine fishes of Tortugas, Florida. *Am. Midl. Nat.,* **38,** 257–416.

Manter, H. W., 1954. Some digenetic trematodes from fishes of New Zealand. *Trans. Roy. Soc. N. Z.*, **82,** 475–568.

Manter, H. W., 1955. The zoogeography of trematodes of marine fishes. *Expl. Parasitol.*, **4,** 62–86.

Manter, H. W., 1960. Some additional digenea (Trematodes) from New Zealand fishes. *Libr. Hom. al Dr. Caballero,* 197–201.

Manter, H. W., 1963. The zoogeographical affinities of trematodes of South American freshwater fishes. *Syst. Zool.,* **12,** 45–70.

Manter, H. W., 1967. Some aspects of the geographical distribution of parasites. *J. Parasitol.*, **52,** 1–9.

Manter, H. W., 1969. Problems in systematics of trematode parasites. In *Problems in Systematics of Parasites,* G. D. Schmidt, ed. University Park Press, Baltimore, Maryland, pp. 93–105.

Marcial-Rojas, P. A., 1975. *Pathology of Protozoal and Helminthic Diseases with Clinical Correlation.* Kreiger, New York.

Margolis, L., 1971. Polychaetes as intermediate hosts of helminth parasites of vertebrates: A review. *J. Fish. Res. Board Can.*, **28,** 1385–1392.

Margolis, L., 1977. Public health aspects of "Codworm" infection: A review. *J. Fish. Res. Board Can.,* **34,** 887–898.

Margolis, L., 1982. Parasites as biological tags for salmonid populations. Fifth International Congress of Parasitology, Toronto, Canada.

Markle, D. F. and C. A. Wenner, 1979. Evidence of demersal spawning in the mesopelagic zoarcid fish *Melanostigma atlanticum* with comments on demersal spawning in the alepocephalid fish *Xenodermichthys copei. Copeia No. 2,* pp. 363–366.

Marshall, N. B., 1954. *Aspects of Deep Sea Biology.* Hutchinsons, London.

Marshall, N. B., 1971. *Explorations in the Life of Fishes.* Harvard University Press, Cambridge, MA.

Maurer, R. O., Jr. and R. E. Bowman, 1975. *Food Habits of Marine Fishes of the Northwest Atlantic—Data Report.* Food Chain Investigation No. 75-3. Northeast Fisheries Center, Woods Hole, MA, 90 pp.

May, R. M., 1973. *Stability and Complexity in Model Ecosystems.* Princeton University Press, Princeton, NJ.

May, R. M., G. R. Conway, M. P. Hassell, and T. R. E. Southwood, 1974. Time delays, density-dependent and single species oscillations. *J. Anim. Ecol.*, **43,** 747–770.

McClelland, G. and K. Ronald, 1974. *In vitro* development of the nematode *Contracaecum osculatum,* Rudolphi 1802 (Nematoda: Anisakinae). *Can. J. Zool.*, **52,** 847–855.

McDowell, S. B., 1973. Family Halosauridae. In *Fishes of the Western North Atlantic,* D. M. Cohen, ed., Memoir No. 1, Part 3. Sears Foundation for Marine Research, New Haven, CT, pp. 32–123.

McVicar, A. H., 1977. Intestinal helminth parasites of the ray *Raja naevus* in British waters. *J. Helminthol.*, **51,** 11–21.

Mead, G. W., E. Bertelson, and D. M. Cohen, 1964. Reproduction among deep-sea fishes. *Deep-Sea Res.*, **1,** 569–596.

Meglitsch, P. A., 1960. Some coelozoic myxosporidia from New Zealand fishes. I. General, and family Ceratomyxidae. *Trans. Roy. Soc. N. Z.*, **88,** 265–356.

Metcalf, M., 1923. The opalinid ciliate infusorians. *Bull. U.S. Nat. Mus.*, **120,** 1–484.

Metcalf, M., 1929. Parasites and the aid they give in problems of taxonomy, geographical distribution and paleogeography. *Smithsonian Misc. Collect.*, **81,** 1–36.

Metcalf, M., 1940. Further studies on the opalinid ciliate infusorians and their hosts. *Proc. U.S. Nat. Mus.*, **87,** 465–634.

Moller, H., 1975. Parasites of the cod (*Gadus morhua* L.) from the Kiel Fjord. *Ber. Deutsch. Wissenschaft. Komm. Meeresforsch.*, **24,** 71–78.

Moser, M., 1977a. Meglitsch's hypothesis: A critical evaluation. Folia Parasit., Praha, **24,** 177–178.

Moser, M., 1977b. Myxosporida (Protozoa): The determination and maintenance of spore size and shape. *Int. J. Parasit.*, **7,** 389–391.

Moser, M. and E. R. Noble, 1976. The genus *Ceratomyxa* (Protozoa: Myxosporida) in macrourid fishes. *Can. J. Zool.*, **54,** 1535–1537.

Moser, M., E. R. Noble, and R. S. Lee, 1976. The genus *Myxidium* (Protozoa: Myxosporida) in macrourid fishes. *J. Parasitol.*, **62,** 685–689.

Mudrey, D. W. and M. D. Dailey, 1971. Postembryonic development of certain tetraphyllidean and trypanorhynchan cestodes with a possible alternative life cycle for the order Trypanorhyncha. *Can. J. Zool.*, **49,** 1249–1253.

Munroe, T., 1976. Helminth parasites of deep-sea benthic fishes of Hudson Submarine Canyon: Taxonomy and host–parasite relationships. Master's thesis, Southeastern Massachusetts University.

Munroe, T., R. A. Campbell, and D. E. Zwerner, 1981. *Diclidophora nezumiae* sp. n. (Monogenea: Diclidophoridae) and its ecological relationships with the macrourid fish *Nezumia bairdii* (Goode and Bean, 1877). *Biol. Bull. Mar. Biol. Lab., Woods Hole,* **161,** 261–290.

Myers, B. J., 1975. The nematodes that cause anisakiasis. *J. Milk Food Technol.*, **38,** 774–782.

Nikolaeva, V. M., 1968. On finding Accacoeliidae larvae in fishes and invertebrates. *Biol. Morya, Kiev,* **14,** 83–89.

Noble, E. R., 1973. Parasites and fishes in a deep-sea environment. In *Advances in Marine Biology,* Vol. II. F. S. Russell and M. Yonge, eds. Academic Press, London, pp. 121–195.

Noble, E. R., and S. B. Collard, 1970. The parasites of midwater fishes. In *A Symposium of the American Fisheries Society on Diseases of Fishes and Shellfishes,* S. F. Snieszko, ed. Special Publication No. 5, Am. Fish. Soc., Washington, DC, pp. 57–68.

Noble, R. R. and J. Orias, 1975. Parasitism in the bathypelagic fish, *Melanostigma pammelas. Int. J. Parasitol.,* **5,** 89–93.

Noble, R. R., J. D. Orias, and T. D. Rodella, 1972. Parasitic fauna of the deep-sea fish, *Macrourus rupestris* (Gunnerus) from Korsfjorden, Norway. *Sarsia,* **50,** 47–50.

Nyman, O. L. and J. H. Pippy, 1972. Differences in Atlantic salmon (*Salmo salar*) from North America and Europe. *J. Fish. Res. Board Can.*, **29,** 179–185.

Odum, W. S., 1966. The food and feeding of the striped mullet, *Mugil cephalus* Linnaeus, in relation to the environment. Master's thesis, University of Miami.

Orias, J. D., E. R. Noble, and G. D. Alderson, 1978. Parasitism in some east Atlantic bathypelagic fishes with a description of *Lecithophyllum irelandeum* sp. n. (Trematoda). *J. Parasitol.*, **64,** 49–51.

Oshima, T., 1972. *Anisakis* and anisakiasis in Japan and adjacent area. In *Progress of Medical Parasitology in Japan,* Vol. 4, K. Morichia and Y. Komiya, eds. Meguro Parasitological Museum, Tokyo.

Osmanov, S. V., 1940. Materials on the parasite fauna of fish in the Black Sea. *Uchen. Zap. Leningr. Gos. Inst.*, **30,** 187–265.

Overstreet, R. M., 1977. *Poecilancistrum caryophyllum* and other trypanorhynch cestode plerocercoids from the musculature of *Cynoscion nebulosus* and other sciaenid fishes in the Gulf of Mexico. *J. Parasitol.*, **63,** 780–789.

Overstreet, R. M. and F. G. Hochberg, 1975. Digenetic trematodes in cephalopods. *J. Mar. Biol. Assoc. U. K.*, **55,** 893–910.

Overstreet, R. M. and D. W. Martin, 1974. Some digenetic trematodes from synaphobranchid eels. *J. Parasitol.*, **60,** 80–84.

Overstreet, R. M. and M. H. Pritchard, 1977. Two new zoogonid digenea from deep-sea fishes in the Gulf of Panama. *J. Parasitol.*, **63,** 840–844.

Palmieri, J. R., 1976. Host–parasite relationships and intraspecific variation in *Posthodiplostomum minimum* (Trematoda: Diplostomatidae). *Great Basin Nat.*, **36,** 334–346.

Palmieri, J. R., 1977a. Host-induced morphological variations in the strigeoid trematode *Posthodiplostomum minimum* (Trematoda: Diplostomatidae). II. Body measurements and tegument modifications. *Great Basin Nat.*, **37,** 129–137.

Palmieri, J. R., 1977b. Host-induced morphological variations in the strigeoid trematode *Posthodiplostomum minimum* (Trematoda: Diplostomatidae). III. Organs of attachment. *Great Basin Nat.*, **37,** 375–382.

Palmieri, J. R., 1977c. Host-induced morphological variations in the strigeoid trematode *Posthodiplo-*

stomum minimum (Trematoda: Diplostomatidae). IV. Organs of reproduction (ovary and testes), vitelline gland, and egg. *Great Basin Nat.*, **34**, 481–488.

Palmieri, J. R. and J. T. Sullivan, 1977. A technique for the location of microsporida in host tissues. *J. Invert. Pathol.*, **30**, 276.

Palmieri, J. R., W. Kelderman, and J. T. Sullivan, 1978. A straining technique for smears containing inclusing body viruses. *J. Invert. Pathol.*, **31**, 264.

Parr, A. E., 1933. A geographic-ecological analysis of the seasonal changes in temperature conditions in shallow water along the Atlantic coast of the United States. *Bull. Bingham Oceanogr. Lab.*, **4**, 1–90.

Paxton, J. R., 1967. Biological notes on southern California lanternfishes (Family Myctophidae). *Calif. Fish. Game*, **53**, 214–217.

Pearcy, W. G. and J. W. Ambler, 1974. Food habits of deep-sea macrourid fishes off the Oregon coast. *Deep-Sea Res.*, **21**, 745–759.

Pemberton, R. T., 1963. Helminth parasites of three species of British gulls, *Larus argentatus* Pont., *L. fuscus* L. and *L. ridibundus* L. *J. Helminthol.*, **37**, 57–88.

Petrushevski, G. K., 1957. Parasite fauna of the clupeids of the Black Sea and its changes with migrations of the fish. *Bull. Inst. Freshwater Fish., Leningr.*, XLII.

Pianka, E. R., 1970. On *r*- and *K*-selection. *Am. Natur.*, **104**, 592–597.

Pippy, J. H. C., 1969a. *Pomphorhynchus laevis* (Zoega) Muller 1776 (Acanthocephala) in Atlantic salmon (*Salmo salar*) and its use as a biological tag. *J. Fish. Res. Board Can.*, **26**, 909–918.

Pippy, J. H. C., 1969b. Preliminary report on parasites as biological tags in Atlantic salmon (*Salmo salar*). I. Investigations 1966–1968. *J. Fish. Res. Board Can.*, Technical Report 134.

Platt, N. E., 1975. Infestation of cod (*Gadus morhua* L.) with larvae of cod worm (*Terranova decipiens* Krabbe) and herringworm, *Anisakis* sp. (Nematoda, Ascaridata), in North Atlantic and Arctic waters. *J. Appl. Ecol.*, **12**, 437–450.

Platt, N. E., 1976. Cod worm—a possible indicator of the degree of mixing of Greenland and Iceland cod stocks. *J. Con. Int. Explor. Mer*, **37**, 41–45.

Polloni, P. T., R. L. Haedrich, G. T. Rowe, and C. H. Clifford, 1979. The size–depth relationship in deep-ocean animals. *Int. Rev. Ges. Hydrobiol.*, **64**, 39–46.

Polyanski, Y. I., 1955. Parasites of fish of the Barents Sea. *Transl. Israel Progr. Sci. Transl.*, **1966**, 158 pp.

Polyanski, Y. I., 1958. Ecology of parasites of marine fishes. In *Parasitology of Fishes*, V. A. Dogiel, G. K. Petrushevski, and Y. I. Polyanski, eds. Leningrad University Press, 384 pp.

Polyanski, Y. I. and I. V. Kulemina, 1963. Parasite fauna of young *Gadus morhua* in the Barents Sea. *Vest. Leningr. Gos. Univ. Seriya Biologii*, **18**, 12–21.

Price, E. W., 1934. Reports on the collections obtained by the first Johnson–Smithsonian deep-sea expedition to the Puerto Rican deep: New digenetic trematodes from marine fishes. *Smithsonian Misc. Coll.*, **91**, 1–8, 1 pl.

Prudhoe, S. and R. A. Bray, 1973. Digenetic trematodes from fishes. *B. A. N. Z. Antarctic Res. Exp. 1929–1931, Reports, Ser. B.*, Vol. 8, Part 10, pp. 199–225.

Rae, B. B., 1960. Seals and Scottish fisheries. *Mar. Res. Dept. Agr. Fish. Scotland*, No. 2, 39 pp.

Rau, M. E. and D. M. Gordon, 1977. A technique for the demonstration of the metacercariae of *Apatemon gracilis pellucidus* (Yamaguti, 1933) in the deep tissues of the brook stickleback (*Culaea inconstrans*). *Can. J. Zool.*, **55**, 1200–1201.

Rawson, M. V., 1973. The development and seasonal abundance of the parasites of striped mullet, *Mugil cephalus* L., and mummichogs, *Fundulus heteroclitis* (L.). Doctoral dissertation, University of Georgia.

Rebecq, J., 1965. Considerations sur la place des trematodes dans le zooplankton marin. *Annals Fac. Sci. Marseille*, **38**, 61–84.

Rex, M. A., 1976. Biological accommodation in the deep-sea benthos: Comparative evidence on the importance of predation and productivity. *Deep-Sea Res.*, **23**, 975–987.

Ridgeway, S. H. and M. D. Dailey, 1972. Cerebral and cerebellar involvement of trematode parasites in dolphins and their possible role in stranding. *J. Wildlife Dis.*, **8**, 33–43.

Robins, C. R., R. M. Bailey, C. E. Bond, J. R. Brooker, E. A. Lachner, R. N. Lea, and W. B. Scott, 1980. A List of Common and Scientific Names of Fishes from the United States and Canada. Am. Fish. Soc. Spec. Publ. No. 12, Bethesda, Maryland, 174 pp.

Rohde, K., 1977. Species diversity of monogenean gill parasites of fish on the Great Barrier Reef. *Proc. Int. Symp. Coral Reefs*, **3**, 585–591.

Rohde, K., 1978. Latitudinal differences in host-specificity of marine Monogenea and Digenea. *Mar. Biol.*, **47**, 125–134.

Ronald, K., 1959. *A Checklist of the Metazoan Parasites of the Heterosomata*. Contrib. Dep. Pech., Queb., No. 167, 152 pp.

Ruszkowski, J. S., 1932. Études sur le cycle évolutif et la structure des cestodes de mer. III. Le cycle évolutif du tetrarhynque *Grillotia erinaceus* (van Ben., 1858). *Compt. Rend. Mensuels Sci. Math. Nat. Acad. Polon. Classe Sci. Lettres*, **9**, 6.

Ruszkowski, J. S., 1934. Études sur le cycle évolutif et la structure des cestodes de mer. III. Le cycle évolutif du tetrarhynque *Grillotia erinaceus* (van Ben., 1858). *Mem. Int. Acad. Polon. Sci. Classe Sci. Math. Nat. B*, **6**, 1–9.

Rysavy, B., 1966. The occurrence of cestodes in the individual orders of birds and the influence of food on the composition of the fauna of bird cestodes. *Folia Parasitol. (Praha)*, **13**, 158–169.

Schell, S. C., 1973. *Rugogaster hydrolagi* gen. et sp. n. (Trematoda: Aspidobothria: Rugogastridae fam. n.) from the Ratfish, *Hydrolagus colliei* (Lay and Bennett, 1839). *J. Parasitol.*, **59**, 803–805.

Scott, D. M., 1954. Experimental infection of Atlantic cod with a larval nematode from smelt. *J. Fish. Res. Board Can.*, **11**, 894–900.

Scott, J. S., 1969. Trematode populations in the Atlantic argentine, *Argentina silus*, and their use as biological indicators. *J. Fish. Res. Board Can.*, **26**, 879–891.

Scott, J. S., 1975. Incidence of trematode parasites of American plaice (*Hippoglossoides platessoides*) of the Scotian Shelf and Gulf of St. Lawrence in relation to fish length and food. *J. Fish. Res. Board Can.*, **32**, 479–483.

Scott, J. S. and W. F. Black, 1960. Studies on the life-history of the ascarid *Porrocaecum decipiens* in the Bass d'Or Lakes, Nova Scotia, Canada. *J. Fish. Res. Board Can.*, **17**, 763–774.

Scott, J. S. and W. R. Martin, 1957. Variation in the incidence of larval nematodes in Atlantic cod fillets along the southern Canadian mainland. *J. Fish. Res. Board Can.*, **14**, 966–975.

Scott, J. S., and W. R. Martin, 1959. The incidence of nematodes in the fillets of small cod from Lockeport, Nova Scotia, and the southwestern Gulf of St. Lawrence. *J. Fish. Res. Board Can.*, **16**, 213–221.

Sedberry, G. R. and J. A. Musick, 1978. Feeding strategies of some demersal fishes of the continental slope and rise off the mid-Atlantic Coast of the U.S.A. *Mar. Biol.*, **44**, 357–375.

Shotter, R. A., 1973. Changes in the parasite fauna of whiting *Odontogadus merlangus* L. with age and sex of host, season, and from different areas in the vicinity of the Isle of Man. *J. Fish. Biol.*, **5**, 559–573.

Shulman, S. S. and R. E. Shulman-Albova, 1953. *Parasites of Fishes of the White Sea*. USSR Academy of Science, Moscow, 199 pp.

Sinclair, N. R. and D. T. John, 1973. Hot water as a tool in mass field preparation of platyhelminthic parasites. *J. Parasitol.*, **59**, 935–936.

Sindermann, C. J., 1970. *Principal Diseases of Marine Fish and Shellfish*. Academic Press, New York.

Skinner, R., 1975. Parasites of the striped mullet, *Mugil cephalus*, from Biscayne Bay, Florida, with descriptions of a new genus and three new species of trematodes. *Bull. Mar. Sci.*, **25**, 318–345.

Sogandares-Bernal, F., 1959. Digenetic trematodes of marine fishes from the Gulf of Panama and Bimini, British West Indies. *Tulane Stud. Zool.*, **7**, 70–117.

Stafford, J., 1904. Trematodes from Canadian fishes. *Zool. Anz.*, **27**, 481–495.

Stephenson, T. A. and A. Stephenson, 1954. Life between tide-marks in North America. III B. Nova Scotia and Prince Edward Island: The geographical features of the region. *J. Ecol.*, **42**, 46–70.

Strelkov, Y. A., 1960. Endoparasitic worms of marine fish of eastern Kamchatka. *Trudy Zool. Inst., Leningr.*, **28,** 147–196.

Stunkard, H. W., 1957. Intraspecific variation in parasitic flatworms. *Syst. Zool.*, **6,** 7–18.

Stunkard, H. W., 1977. Studies on tetraphyllidean and tetrarhynchidean metacestodes from squids taken on the New England coast. *Biol. Bull.*, **153,** 387–412.

Taylor, W. R., 1962. *Marine Algae of the Northeastern Coast of North America.* University of Michigan Press, Ann Arbor, 509 pp.

Templeman, W. and A. M. Fleming, 1963. *Distribution of Lernaeocera branchialis in Cod as an Indicator of Cod Movements in the Newfoundland Area.* Int. Comm. Northwest Atlantic Fish., Spec. Publ. No. 4, pp. 318–322.

Threlfall, W., 1964. Factors concerned in the mortality of some birds which perished in Anglesey and northern Caernarvonshire during the winter of 1963, with special reference to parasitism by helminths. *Ann. Mag. Nat. Hist. Ser.*, **13,** 721–737.

Threlfall, W., 1965. Studies on the helminth parasites of herring gulls (*Larus argentatus* Pontopp.) on the Newborough Warren Nature Reserve, Anglesey. Doctoral thesis, University of Wales.

Threlfall, W., 1966. *The Helminth Parasites of the Herring Gul (Larus argentatus Pontopp.).* Commonwealth Bureau of Helminthology Technical Communication 37, 23 pp.

Threlfall, W., 1969. Some parasites from elasmobranchs in Newfoundland. *J. Fish. Res. Board Can.*, **26,** 805–811.

Tripathi, Y. R., 1957. Monogenetic trematodes from fishes of India. *Indian J. Helminthol.*, **9,** 1–49.

Unnithan, R. V., 1957. On the functional morphology of a new fauna of Monogea on fishes from Trivandarum and environs. Part I. *Axinidae Fam. Nov. Bull. Cent. Res. Inst. Univ. Travancore, Ser. C,* **5,** 27–122.

Unnithan, R. V., 1971. On the functional morphology of a new fauna of Monogenoidea on fishes from Trivandarum and environs. Part IV. Microcotylidae *sensu stricto* and its repartition into subsidiary taxa. *Am. Midl. Nat.*, **85,** 366–398.

Vinogradov, M. E., 1968. Vertical distribution of the oceanic zooplankton. *Akad. Nauk SSSR Inst. Okeanol.*, 1970 (transl. Israel Program Scientific Translations).

Von Ihering, H., 1891. On the ancient relations between New Zealand and South America. *Trans. Proc. N. Z. Inst.*, **24,** 431–455.

Von Ihering, H., 1902. Die helminthen als hilfsmittel der zoogeographischen forschung. *Zool. Anz.*, **26,** 42–51.

Von Linstow, O., 1888. Report on the entozoa collected by H.M.S. *Challenger* during the years 1873–76. *Rep. Voyage H.M.S. Challenger* (1873–76), **23,** 1–18.

Watertor, J. L., 1967. Intraspecific variation of adult *Telorchis bonnerensis* (Trematoda: Telorchiidae) in amphibian and reptilian hosts. *J. Parasitol.*, **53,** 962–968.

Wiles, M., 1968. Possible effects of the harbor seal bounty on codworm infestations of Atlantic cod in the Gulf of St. Lawrence, the Strait of Belle Isle, and Labrador Sea. *J. Fish. Res. Board Can.*, **25,** 2749–2753.

Williams, H. H., 1960. The intestine in members of the genus *Raja* and host-specificity in the Tetraphyllidea. *Nature,* **188** (4749), 514–516.

Williams, H. H., 1963. Observations on *Echeneibothrium* (Cestoda: Tetraphyllidea) from various species of *Raja*. *Parasitology,* **53,** abstract.

Williams, H. H., 1964. Observations on the helminths of *Raja*. *Parasitology,* **54,** abstract.

Williams, H. H., 1966. The ecology, functional morphology and taxonomy of *Echeneibothrium* Beneden, 1849 (Cestoda: Tetraphyllidea), a revision of the genus and comments on *Discobothrium* Beneden, 1870, *Pseudanthobothrium* Baer, 1956, and *Phormobothrium* Alexander, 1963. *Parasitology,* **56,** 227–285.

Williams, H. H., 1968a. The taxonomy, ecology and host-specificity of some Phyllobothriidae (Cestoda: Tetraphyllidea), a critical revision of *Phyllobothrium* Beneden, 1849 and comments on some allied genera. *Phil. Trans. Roy. Soc., Ser. B,* **253,** 231–307.

Williams, H. H., 1968b. *Acanthobothrium quadripartitum* sp. nov. (Cestoda: Tetraphyllidea) from *Raja naevus* in the North Sea and English Channel. *Parasitology,* **58,** 105–110.

Williams, H. H., 1969. The genus *Acanthobothrium* Beneden 1849 (Cestoda: Tetraphyllidea). *Nytt. Mag. Zool.*, **17**, 1–56.

Williams, I. C., 1961. A list of parasitic worms, including twenty-two new records, from British birds. *Ann. Mag. Nat. Hist., Ser. 13*, **44**, 467–480.

Wolff, T., 1977. Diversity and faunal composition of the deep-sea benthos. *Nature*, **267** (5614), 780–785.

Yamaguti, S., 1934. Studies on the helminth fauna of Japan. Part 4. Cestodes of fishes. *Jap. J. Zool.*, **6**, 1–112.

Yamaguti, S., 1938. *Studies on the Helminth Fauna of Japan*, Part 21, *Trematodes of Fishes*, Vol. IV. Naruzen Company, Tokyo, 139 pp.

Yamaguti, S., 1941. Studies on the helminth fauna of Japan. Part 33. Nematodes of fishes, II. *Jap. J. Zool.*, **9**, 343–396.

Yamaguti, S., 1951. Studies on the helminth fauna of Japan. Part 44. Trematodes of fishes, IX. *Arb. Med. Fak. Okayama*, **7**, 247–282.

Yamaguti, S., 1959. *Systema Helminthum*. Vols. I–V. Interscience. New York.

Yamaguti, S., 1963. *Parasitic Copepoda and Branchiura of Fishes*. Interscience, New York.

Yamaguti, S., 1968. *Monogenetic Trematodes of Hawaiian Fishes*. University of Hawaii Press, Honolulu.

Yamaguti, S., 1970. *Digenetic Trematodes of Hawaiian Fishes*. Keigaku Publishing Company, Tokyo.

Yamaguti, S., 1975. *A Synoptical Review of Life Histories of Digenetic Trematodes of Vertebrates*. Keigaku Publishing Company, Tokyo.

Young, P. C., 1972. The relationship between the presence of larval anisakine nematodes in cod and marine mammals in British home waters. *J. Applied Ecology*, **9**, 459–485.

Young, P. C., 1967. New Monogenoidea from Australian brackish water and reef fishes. *J. Parasitol.*, **53**, 1008–1015.

Young, P. C., 1970. The species of Monogenoidea recorded from Australian fishes and notes on their zoogeography. *An. Inst. Biol. Univ. Mex., Ser. Zool.*, **41**, 163–176.

Zhukov, E. V., 1953. Endoparasitic helminths of fishes of the Sea of Japan and the Kuril shallows. Report, Zool. Inst. Acad. Sci., U.S.S.R.

Zhukov, E. V., 1960. On the fauna of parasites of fishes of the Chukotsk Peninsula and the adjoining seas. I. Monogenetic trematodes of marine and freshwater fishes. *Parazitol. Sborn. Zool. Inst., Akad. Nauk SSSR*, **19**, 308–332.

Zubchenko, A. V., 1981. Parasitic fauna of some Macrouridae in Northwest Atlantic. *J. Northw. Atl. Fish. Sci.*, **2**, 67–72.

Zubchenko, A. V. and V. K. Krasin, 1980. Myxosporidia of the genus *Myxidium* in some macrourids from Northern Atlantic and Pacific Oceans. *Parazitologiya*, **14**, 168–176.

INDEX